MATERIALS SCIENCE OF THE EARTH'S INTERIOR

MSMR

Materials Science of Minerals and Rocks

MATERIALS SCIENCE OF THE EARTH'S INTERIOR

Edited by Ichiro Sunagawa

Terra Scientific Publishing Company
Tokyo, Japan

D. Reidel Publishing Company
A MEMBER OF THE
KLUWER ACADEMIC PUBLISHERS GROUP
Dordrecht / Boston / Lancaster

Library of Congress Cataloging in Publication Data

CIP

Main entry under title:

Materials science of the Earth's interior.

(Materials science of minerals and rocks)
Includes index.
1. Earth--Internal structure. I. Sunagawa, Ichiro.
II. Series.
QE509.M34 1983 551.1'1 83-13881
ISBN 90-277-1649-8 (D. Reidel Pub. Co.)

Published by Terra Scientific Publishing Company (TERRAPUB),
307 Shibuyadai-haim, 4-17 Sakuragaoka-cho, Shibuya-ku, Tokyo 150, Japan,
in co-publication with D. Reidel Publishing Company, Dordrecht, Holland

Sold and distributed in the U.S.A. and Canada
by Kluwer Boston Inc.,
190 Old Derby Street, Hingham, MA 02043, U.S.A.,
in Japan by Terra Scientific Publishing Company (TERRAPUB),
307 Shibuyadai-haim, 4-17 Sakuragaoka-cho, Shibuya-ku, Tokyo 150, Japan

In all other countries, sold and distributed
by Kluwer Academic Publishers Group,
P. O. Box 322, 3300 AH Dordrecht, Holland

D. Reidel Publishing Company is a member of the Kluwer Academic Publishers Group

This is the report of Special Research Project No. 56122002 on the Material Science of the Earth's Interior which was carried out during 1978 to 1981 and defrayed by the Ministry of Education, Science and Culture of Japan.

Printed in Japan

PREFACE

In these days of numerous satellites moving around the Earth, we have full of information from the space, far more than we imagined in the past. We now can have even rock samples from the moon.

As compared to a tremendous amount of information from the space, it is almost unbelievable to realize how little we know about the depths of the Earth on which we stand and from where we obtain all natural resources necessary for our daily life. The deepest point that the human being could reach so far is perhaps less than a few kilometers. Even the deepest bore hole reaches no more than ten kilometers. Just pores of the Earth's skin, or even less than that. It is no exaggeration to say that the Earth's interior is the most unknown and unrevealed place for us. This is the place worthwhile to pay our effort.

There have been three different ways of approach to understand our Earth. The oldest science about the Earth is perhaps geology. By surveying mutual relations of rocks exposed on the Earth's surface and investigating fossils therein, geologists have tried to construct the history of the Earth's crust. This is an approach of historical science. In geophysics, dynamical approach has been taken, by observing and analysing seismic waves and other data, and effort has been paid to know the layer structure of the Earth. The concept of plate tectonics, the most exciting idea about the Earth in the past few decades, has emerged from and witnessed by these studies. However, in understanding the Earth, we should not forget the third way of approach, that is materials science of the Earth. Unless we understand the constituents of the Earth, and how and under what conditions are they formed, we do not understand the Earth, neither do we understand the history and the dynamics of the Earth. If dynamical approach could be regarded as diagnostics of the Earth by means of stethoscope, material scientific approach can be regarded as a combination of pathologic investigation of pieces taken from the affected part and laboratorical generation of cancer using guinea pig. Cooperation from both sides is absolutely necessary to understand the Earth's interior.

Classical mineralogy and petrology are the roots of modern materials science of the Earth. These were basically the science of description and classification. However, drastic and even revolutional changes have been made in these sciences since the beginning of this century, particularly after the 2nd World War. Equilibrium phase diagrams have been prepared for not only oxides, silicates but also sulfides, and mineral assemblages in igneous and metamorphic rocks and in metallic ore deposits have been analysed based on these phase diagrams. Experimental techniques to realize high pressure and high temperature conditions in laboratories have made an impressive progress in the last few decades, and we are now able to reproduce in laboratories the conditions corresponding to even the lower mantle. Not only static equilibrium thermodynamics studies but also studies on kinetic problems on the generation and properties of magmas, or on the nucleation and growth of crystals in such systems have made a significant progress. Theoretical and experimental studies on the mechanism of crystal growth have been advanced impressively particularly after the war, and the knowledge thereof have been applied to understand how mineral crystals grow in nature. Atomic structures of most mineral crystals have been analysed during the past seventy years after the discovery of X-ray diffraction by Max von Laue, and at present effort is being paid to analyse the structures of crystals under high temperature, high pressure conditions, and even of melt phase. Imperfections in natural crystals, such as various lattice defects, physical and chemical inhomogeneities, have been studied intensively in the scale of nanometers using a wide range of sophisticated apparatus. Through these studies, i.e. through detailed characterization of mineral crystals, growth and post growth histories that mineral crystals experienced are being analysed. In short, materials science of the Earth has been changed almost revolutionally in the past few decades.

Under these circumstances, it was felt very important among mineralogists, petrologists and economic geologists in Japan that the efforts from diverged disciplines be united and concentrated to solve the common problem. Thus, we planned a co-operative research project entitled "Materials Science of the Earth's Interior," and applied to the Ministry of Education in Japan for a research grant. The application was accepted and a big research grant, amounting 8×10^8 yen, was provided, with which expensive and sophisticated apparatuses could be installed at several major universities. The three years' special research project was thus realized and carried out from 1978 to 1980. The year of 1981 was devoted to prepare the report of the project, which was published in Japanese as a special issue of the Journal of the Japanese Association of Mineralogists, Petrologists and Economic Geologists, No. 3, in which 40 scientific papers were included.

In planning the project, we set up nine pre-planned research groups along three principal lines, about which a brief description will be given below.

[A] *Experimental studies of the materials of the Earth's interior*

The main purposes of four groups belonging to this category are 1) laboratorical reproduction of the conditions corresponding to the Earth's interior, and 2) attempts to synthesize large single crystals of the materials of the Earth's interior, which can be used for precise measurements of their physical properties. Both phase relation studies and studies on kinetic problems were carried out in the former studies. In the latter, it was aimed to synthesize large single crystals of olivine, pyroxene and other minerals, using various techniques of melt growth and high temperature, high pressure flux growth. The following four groups were set up along this line, and each group was represented by two scientists.

[A1] *Syntheses of large single crystals of materials of the Earth's interior and measurements of their physical properties*
S. Akimoto, University of Tokyo
F. Takei, Tohoku University

[A2] *Generation of ultra high pressure and synthesis and measurements of properties of materials corresponding to the lower mantle and the core*
N. Kawai† and later S. Endo, Osaka University
M. Kumazawa, Nagoya University

[A3] *Phase equilibrium and kinetic studies of the systems corresponding to lower crust and upper mantle*
I. Kushiro, University of Tokyo
Y. Hariya, Hokkaido University

[A4] *Mineral synthesis and phase equilibrium studies under hydrothermal conditions*
A. Sugaki, Tohoku University
T. Iiyama, University of Tokyo

[B] *Characterization of materials derived from the Earth's interior*

Through detailed characterization of mineral crystals and rocks derived from the Earth's interior, it is aimed to understand how and under what conditions mineral crystals are formed and what kind of thermal or stress histories have they experienced. Along this line, the following four groups were set up.

[B1] *Crystal structure analyses under high temperature and/or high pressure*
R. Sadanaga and Y. Takéuchi, University of Tokyo

[B2] *Analyses of crystal growth mechanisms, thermal and stress histories of mineral crystals*
I. Sunagawa, Tohoku University
N. Morimoto, Kyoto University

[B3] *Analyses of mineral assemblages and the conditions of formation of minerals and rocks in the Earth's interior*
K. Aoki, Tohoku University
S. Banno, Kyoto University

[B4] *Minerals of hydrothermal origin and the behavior of water in the Earth's interior*
S. Takenouchi, University of Tokyo
Y. Kuroda, Shinshu University

[C] *Theoretical group*

Since theoretical studies have been rather weak in geological sciences, this group was purposely set up in order to encourage such studies, with a title of simulation of the Earth's interior. Main effort was paid to computer simulation of the structures of hypothetical compounds and silicate melts, and to understanding of electron theory under high pressure.

Y. Matsui, Okayama University

In addition to the above nine pre-planned groups, twenty six research groups were selected from applications and about 30% of the budget was spent on these groups.

The whole project was represented and organized by I. Sunagawa, Tohoku University.

Scientists involved in this special research project came from a very wide spectrum of desciplines, ranging from crystallography, mineralogy, petrology, economic geology, geochemistry, solid state geophysics to solid state physics, solid state chemistry, physical metallurgy, ceramic science, inorganic material science and even theoretical physics. Every year, three days symposium was organized and held in Sendai. The symposia gave an impression that a new well-cooperated descipline was emerging through this research project in earth sciences.

In preparing the reports of the special research project "Materials Science of the Earth's Interior," it was strongly felt that the essence of the results obtained through the project, together with critical reviews on respective problems, should be published in English in a form of advanced text book. Thus the publication of this book was realized as Volume 1 of the new series of books on "Materials Science of Minerals and Rocks." It will be more than delighted if this book would provide the readers with useful idea and information as to the recent advancements particularly those achieved in Japan, and the future prospect in materials science of the Earth's interior.

This book consists of seven chapters, starting from the structure and physical properties of silicate melts, to discussions on crystal growth mechanism and syntheses of large single crystals, electron and crystal structures under high temperature and/or high pressure, analysis of thermal and stress histories, discussions on particular materials of the Earth's interior, interaction between soild and fluid components in the Earth's interior, and ending by a chapter on technical development achieved in the present study.

Chapter 1 covers the theoretical and experimental studies on silicate melt, and includes sections on computer simulation of structures of silicate melts and glasses by Matsui and Kawamura, on X-ray structural studies of molten anorthite and albite by Marumo and Okuno, on the structures and physical properties of silicate melts of geological interest by Kushiro. In this chapter, it is intended to present the reults of the present special research project with the hope of deeper understanding of the structure and physical properties of silicate melts, since such understandings are particularly important in the studies of future materials science of the Earth's interior. Without these, dynamic movement and kinetic process of mineral formation would be unable to understand properly. However, studies along this line are just at their initial stage, and everything is still in vague. It is hoped that the achievement obtained and described in this chapter will form a good mile stone for the future development along this line.

Chapter 2 deals with the mechanism of crystal growth and the synthesis of large single crystals. In this chapter, characteristics of crystal growth mechanism in natural crystallization is discussed by Sunagawa, and syntheses of large single crystals of silicates and titanates by melt growth method are reported by Takei *et al.*, those of single crystals under high pressure by Akimoto *et al.* Perfections of single crystals synthesized by these techniques are critically studied and reported by Komatsu and Inoue. It is to be prouded that large single crystals of high perfection, attaining 1 cm across and 10 cm long, are now available for measurements of physical properties.

Chapter 3 covers electron structure and crystal structures under high pressure and/or high temperature conditions. Sugano and Ohnishi discuss in the first section of this chapter on the electron theory of transition-metal compound under high pressure. Takéuchi *et al.* discuss high temperature crystallography of olivine and spinel, and Yagi and Akimoto report in-situ observation of crystal structure and phase transformation under pressure, using diamond anvils.

Chapter 4 deals with the analyses of thermal and stress histories of materials derived from the Earth's interior, and consists of three sections. In the first section, development and application of high resolution analytical electron microscopy to the decomposition and exsolution textures in minerals are discussed by Morimoto and Kitamura. We are now able to analyse chemical compositions of lamellae with a few hundred Angstrom thickness. In the second section, Oishi and Ando report and discuss the results of measurements of oxygen self-diffusion coefficient in single crystal forsterite, which should be important for the study of plastic deformation of olivines and was made possible because large single crystals of forsterite could be synthesized in the present study. In the final section, transition and steady state creep of olivine are discussed by Toriumi *et al.*

Chapter 5 is for the discussions on particular materials of the Earth's interior. Diamond, a representative mineral derived from the upper mantle, is discussed in a great detail in four sections. Morphologies of natural and synthetic diamond crystals are compared and critically analysed in relation to growth parameters by Sunagawa; Surface microtopographic and X-ray topographic evidence to support spiral growth of natural octahedral diamond crystals are presented by Sunagawa *et al.*; diamond syntheses and the perfection of, particularly inclusions in, synthetic crystals are discussed by Wakatsuki; and Ar isotopic ratio and K, Na and other trace element contents in natural South African diamonds are reported by Ozima *et al.* Ultra-high pressure phase relations of $MgO-FeO-SiO_2$ system are discussed by Ito, shock induced phase relation in oxide and silicates by Syono, petrology of materials derived from the upper mantle by Aoki, and igneous mineral equilibria in some Japanese Alpine-type peridotite by Arai.

Chapter 6 is for the problem relating to the interaction between solid and fluid components, particularly water, in the Earth's interior. Hariya, and Akimoto and Akaogi discuss hydrous silicates in the Earth's interior, Kuroda *et al.* report the results of measurements on D/H ratio in granite intrusions. Iiyama discusses hydrothermal solution-mineral equilibria in order to understand geochemical transport phenomena in the Earth's crust. Takenouchi discusses physical and chemical properties of hydrothermal fluid based on the studies of fluid inclusions, and Sugaki *et al.* report hydrothermal synthesis and phase relations of polymetallic sulfide systems.

The final Chapter 7 is devoted to report the technical development achieved in the present special research project. This chapter consists of 4 sections. Kumazawa and Endo discuss the recent progresses in the generation of static high pressure by means of the multiple anvil, whereas Goto and Syono report shock compression experiments using the gun method. The final two sections by Yamanaka, and by Ohsumi *et al.* are on the techniques of high temperature single crystal X-ray diffraction studies, using gas-flame heater and laser-heating device, respectively.

It is hoped that the present volume will give the reader a clear insight into various problems connected with materials science of the Earth's interior, and that this volume will form a good starting point of future development of the science and the new series of books on "Materials Science of Minerals and Rocks."

In this book, where the contributions contain critical reviews of respective problems, many diagrams and illustrations have been copied from various sources. The permission for reproduction, obtained from the authors, editors and publishers concerned, is greatly acknowledged herewith. In all cases, reference is made to the original publication.

Finally, the editor express his sympathy over the death of Prof. Naoto Kawai, Osaka University, who passed away during the course of this study.

March 10, 1982

I. Sunagawa
Editor in Chief

MATERIALS SCIENCE OF THE EARTH'S INTERIOR

CONTENTS

Chapter 1: Silicate Melt

Materials Science of the Earth's Interior, edited by I. Sunagawa, pp. 3–23.

COMPUTER SIMULATION OF STRUCTURES OF SILICATE MELTS AND GLASSES

Yoshito MATSUI[1] and Katsuyuki KAWAMURA[2,*]

[1]*Institute for Thermal Spring Research, Okayama University, Misasa, Tottori-ken, Japan*
[2]*Department of Geology, The University of Manchester, Manchester, England*

1. *Introduction*

The X-ray and neutron diffraction techniques have been extensively applied to the molten and vitreous materials to reveal their structures (see MARUMO and OKUNO, 1982). In principle, however, it is almost impossible to obtain three-dimensional (3-D) structures of non-crystalline materials by X-ray or neutron studies alone, because the diffraction measurements provide us with only one-dimensional information of the relevant material.

Thanks to developments in high-speed computer technology, we can now calculate the 3-D configuration of atoms or molecules in amorphous substances on the basis of classical mechanics. Since 1976, calculations have been carried out, with considerable success, on molten and vitreous silica and silicates (WOODCOCK *et al.*, 1976; SOULES, 1979; MATSUI and KAWAMURA, 1980; MATSUI *et al.*, 1982). The calculated configurations are generally consistent with X-ray and neutron observations. In the present article the method of computer 'experiments' to construct 3-D configurations will be outlined and this will be followed by a discussion on several 'computer products' to demonstrate the potential usefulness of these calculations.

2. *Monte Carlo and Molecular Dynamics Calculations*

The following two methods are known to be effective in generating 3-D configurations of atoms or molecules in liquids and glasses: Monte Carlo (MC) and molecular dynamics (MD) calculations.** Both methods assume

*On leave of absence from Geological Institute, Faculty of Science, The University of Tokyo, Tokyo, Japan.

**The readers are referred to a general review by WOODCOCK (1975), although great care must be paid to the numerous errors, even in the fundamental equations.

FIG. 1. A two dimensional illustration of the basic (centre) and adjacent ghost cells with an edge length L and five particles within a cell ($N = 5$). For an ion i (shaded) in the basic cell, the summation over the surrounding ions j is necessary when the cut-off distance is equal to L in Eqs. (1) and (2).

hypothetical periodic boundary conditions as illustrated in Fig. 1. The space is divided by a 3-D grid into cubic cells (edge length, L) which are strictly equivalent to each other. Thus, as far as the computation is concerned, the "model space" possesses the $P1$ symmetry. Among these cells, one is the *basic cell*, and others are the "ghost cells" or "replica cells." Into the basic cell we put particles (model ions) with an arbitrary configuration. The number of particles in a cell (N) ranges from 216 ($= 6^3$; in most runs for the alkali halides) up to 1,000. The local $P1$ symmetry automatically "generates" the same atomic configuration in ghost cells.

As the model system can be treated as a "crystal" with the space group $P1$, we can readily calculate the Madelung (electrostatic) energy of the system. If we assume appropriate repulsive and dispersive potentials between constituent model ions, the cohesive energy (U, Madelung plus repulsive, van der Waals energies, etc.) can be calculated. In this calculation, no distinction is made whether the ion pair is across the cell boundary or not:

$$U_{\text{cell}} = \frac{1}{2}\sum_{j \neq i}\sum \phi_{ij}(r_{ij}) \tag{1}$$

where $\phi_{ij}(r_{ij})$ is the assumed interionic (pair) potential between ion i (within the basic cell) and ion j (all ions inside and outside the basic cell) with a distance r_{ij}, and U_{cell} is the cohesive energy per cell. In carrying out the summation it is a common practise to set a cut-off distance beyond which the assumed potential is safely neglected when a proper (small) correction is made. The cut-off distance is most conveniently chosen to be equal to $L/2$. The instantaneous pressure of the system, p, can be given by the virial theorem as follows (cf. SCHÄFER and KLEMM, 1976):

$$p = \frac{NkT}{V} - \frac{1}{3V}\sum_{ij} r_{ij}\frac{\partial \phi_{ij}}{\partial r_{ij}} = \frac{NkT}{V} + \frac{1}{3V}\sum_{ij} r_{ij}F_{ij} \tag{2a}$$

where k is the Boltzmann's constant, V is the volume of a cell, and F_{ij} is the force between atoms i and j. Pressure (P) is given by taking an average of p over a sufficiently long time interval:

$$P = \langle p \rangle \tag{2b}$$

in which the brackets denote the time-average.

The "stable" (and, therefore, most likely) configuration should have a minimum value of U at a given temperature (T) and density (and lowest p within a range of fluctuation). The MC calculation, which has been successfully applied to alkali halide melts and even to aqueous solutions, is carried out in the following manner. The configuration of particles within the basic cell (and consequently in the ghost cells) is modified by a random displacement of one randomly chosen particle. The modified configuration may have a lower or higher U than the preceding U. When the change in $U (\Delta U)$ is positive, the value of $\exp(-\Delta U/NkT)$ is compared with rand (1), where rand (1) is a uniform random (by definition, unpredictable) number within a range from 0 to 1. If the random number happens to be greater than $\exp(-\Delta U/NkT)$, the change is abandoned as an unsuccessful trial. The calculation proceeds from a successful trial in search of configurations with a lower U.

The advantage of the MC calculation, namely its relatively simple computation scheme, is largely offset by its shortcomings. The approach to the "equilibrium" state is slow, and it is impossible to obtain time-dependent (in particular, transport) properties, such as diffusion coefficients, viscosity coefficients, etc. The MC calculation has not been applied to the silicate systems and will not be discussed further.

The molecular dynamics (MD) calculation, on the other hand, can give us not only a static 3-D configuration, but also its variation as a function of time, so that transport coefficients can also be calculated. The MD calculation is nothing but the numerical solution of the classical Newtonian equation of motion for a many-body system under the constraint of periodic boundry conditions (see Fig. 1). For the calculation by computers, the Newtonian differential equation is transformed into a finite difference equation (VERLET, 1967), such that

$$x_i(t + \Delta t) = 2x_i(t) - x_i(t - \Delta t) + \Delta t^2 F_i(t)/m_i \tag{3}$$

where Δt is a time increment of the order of 10^{-15} second (= 1 femtosecond, fs), $x_i(t)$ is the position of ion i at $t = t$, $F_i(t)$ is the overall force acting on ion i at t (derived from the assumed potential function), and m_i is the mass of ion i. When the calculation yields x_i beyond the region of the basic cell ($x < 0$ or $x > L$), the particle is regarded to have migrated out of the basic cell, and at the same time, automatically due to the $P1$ symmetry, a "ghost" ion enters the basic cell to become a member of the basic configuration (the coordinate being $x + L$ or $x - L$). This assures the constancy of N and, hence, the conservation of energy.

In sharp contrast to the MC calculation, any movement of ions is always physically meaningful and therefore the approach to the equilibrium state is generally much more rapid. One may wonder, however, why the time-reversible classical mechanics (see Eq. 3) can lead a system to a state close to thermal equilibrium, even though we start the calculation with an unrealistically chaotic configuration. In order to answer this question, an example of the progress of an MD calculation of an $MgSiO_3$ melt, starting from a chaotic configuration, is shown in Fig. 2. In this run, the total energy was initially positive, i.e., the system was absolutely unstable. During the first ten steps (1 step corresponds to 0.1 fs) the "lattice" (potential) energy decreased rapidly and the thermal (kinetic) energy increased satisfying the law of the conservation of energy (the slight drift in total energy is ascribed to computational error). The temperature was compulsorily regulated after every ten steps to a prescribed value of 10,000 K. This regulation means that the system was immersed into a heat source of an infinite capacity at every 1 fs. At this moment the kinetic energy (and hence total energy) changes irreversibly. This artificial "dissipation" of thermal energy, which is obviously an irreversible process, is considered to be the chief reason for the approach towards thermal equilibrium despite the calculation being based on time-reversible mechanics.

3. *Potential Functions and Their Parameters*

Any realistic MD (and MC as well) calculations require realistic potential functions among particles as is evident in Eq. 1 to 3. From a practical point of view, however, only the two-body (pair) potentials are acceptable because more accurate many-body potentials require a much longer time to calculate the forces acting on the ions, since in cases of many-body potentials the force acting on every ion can be expressed only as a complicated function of the configuration of the whole system. It is customary to assume the pair potential function of the Born-Mayer-Huggins type for alkali halide systems

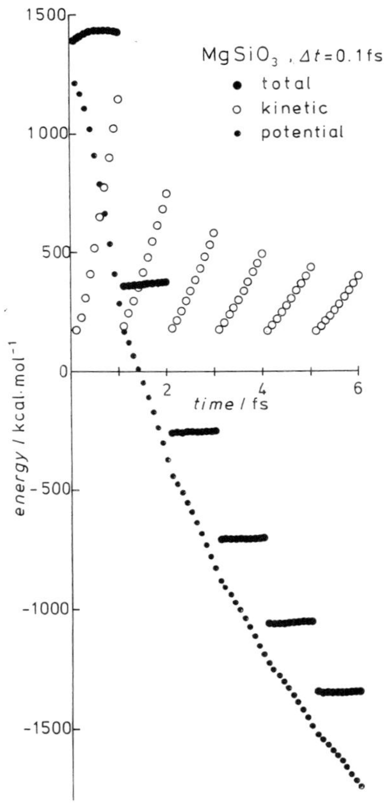

FIG. 2. The evolution of an MD calculation started from a random, chaotic configuration (MATSUI *et al.*, 1982, Fig. 1). The system approaches asymptotically a total energy of around −3,800 kcal mol^{-1} at 10,000 K within 50 fs.

$$\phi_{ij}(r_{ij}) = \frac{z_i z_j e^2}{r_{ij}} + b_{ij} A_{ij} \exp\left(\frac{\sigma_i + \sigma_j - r_{ij}}{\rho_{ij}}\right) - \frac{c_{ij}}{r_{ij}^6} - \frac{d_{ij}}{r_{ij}^8} \tag{4}$$

in which the first term in the right hand side represents the Coulomb energy (e is the electronic charge and z_i is the electrostatic valence of ion i, A_{ij} denotes the Pauling factor, b_{ij} is a constant, σ_i and ρ_{ij} are the "size" and "softness" of ion i or ion pair ij, and c_{ij} and d_{ij} are coefficients for the van der Waals (dipole-dipole) and the dipole-quadrupole dispersive energies, respectively. For the alkali metal and the halide ions, numerous empirical and theoretical sets of parameters for Eq. (4) have been proposed (see, for example, TOSI, 1964) on the basis of lattice dynamics. For ions which constitute silicates, which are obviously not-so-perfectly ionic, no systematic studies have been made of their potential functions and parameters. Under the circumstance it is best to perform MD calculations using potential functions in their simplest form. As is shown by IDA (1977) for alkali halide crystals, the Born-Mayer-Huggins type potential function, omitting the dispersion terms, can be expressed in terms of pairwise additive "ionic radius" a_i and "ionic compressibility" b_i as follows:

$$\phi_{ij}(r_{ij}) = \frac{z_i z_j e^2}{r_{ij}} + f_0 A_{ij}(b_i + b_j) \exp\left(\frac{a_i + a_j - r_{ij}}{b_i + b_j}\right) \tag{5}$$

where f_0 is an arbitrary constant.

MIYAMOTO and TAKEDA (1980) tried to estimate the parameters for Mg^{2+}, Si^{4+} and O^{2-} in a more simplified version of a pair potential function (putting A_{ij} to be unity and omitting the dispersion terms)

$$\phi_{ij}(r_{ij}) = \frac{z_i z_j e^2}{r_{ij}} + f_0(b_i + b_j) \exp\left(\frac{a_i + a_j - r_{ij}}{b_i + b_j}\right) \tag{6}$$

by means of the lattice energy minimization (WMIN) technique after the method of BUSING (1970), using a computer program written by Busing (unpublished). The principle of the WMIN is simple. Whenever the crystalline system exists as a stable or metastable phase, it must have a minimum value in U with respect to minute changes in the position of ions or any parameters determining the structure of the relevant phase, so that

$$\sum_k \left(\frac{\partial U}{\partial x_k}\right)^2 = \min \tag{7}$$

where the x_k are the structural parameters. MIYAMOTO and TAKEDA (1980, and private communication) applied the WMIN technique to α-Mg_2SiO_4 (forsterite), β-Mg_2SiO_4 (modified spinel), and SiO_2 (stishovite) assuming full charges for ions (e.g., +4 for Si, −2 for O). The resultant values showed an impressive agreement for two or three structures within 3σ (Table 1A). The potential function (Eq. 6) with the resultant parameters was immediately adopted for a reconnaissance MD calculation of an $MgSiO_3$ melt/glass in order to assess whether these parameters might reproduce interionic distances within an acceptable range. The proper interionic separations were obtained by MD calculation after a slight modification of Miyamoto and Takeda's original values (MATSUI and KAWAMURA, 1980).

To date, two sets of parameters have been shown to be successful for use in MD calculations (Table 1B). The C-2a set also contains provisional values for Na, Ca and Al, estimated after consideration of metal-oxygen distances.

4. *Procedure of MD Calculations of Silicate Melts/Glasses*

Woodcock, the pioneer in MD calculations of molten ionic salts, was the first to extend the calculation to the not-so-perfectly ionic systems such as

TABLE 1. Potential Parameters for Eq. (6).

A. The WMIN-derived values

		Mg^{2+}	Si^{4+}	O^{2-}	Reference and crystal data
1	a/Å	0.97*	0.608(0.211)**	1.770(0.084)	MIYAMOTO and TAKEDA (1980):
	b/Å	0.065*	0.017(0.004)	0.105(0.020)	α-Mg_2SiO_4 (SMYTH and HAZEN, 1973)
2	a/Å	0.948(0.034)	0.595(0.020)	1.798(0.025)	Miyamoto (unpublished): β-Mg_2SiO_4
	b/Å	0.063(0.011)	0.021(0.005)	0.104(0.010)	(HORIUCHI and SAWAMOTO, 1981)
3	a/Å		0.60 (0.01)	1.84 (0.03)	Takeda and Miyamoto (unpublished):
	b/Å		0.015(0.002)	0.130(0.006)	SiO_2 (stishovite, SINCLAIR and RINGWOOD, 1978)

$f_0 = 1\ \text{kcal} \cdot \text{mol}^{-1} \cdot \text{Å}^{-1} = 6.948 \times 10^{-6}$ dyn.
*Fixed values taken from BUSING (1970).
**Standard deviation are given in parentheses.

B. The parameter sets adopted in the MD calculations

Set		Mg	Ca	Al	Na	Si	O
	z	+2	+2	+3	+1	+4	−2
C-lc	a/Å	0.965				0.605	1.792
	b/Å	0.0428				0.0130	0.114
C-2a	a/Å	0.89	1.25	0.65	1.15	0.64	1.84
	b/Å	0.03	0.05	0.013	0.065	0.0125	0.130

$f_0 = 6.948 \times 10^{-6}$ dyn.

SiO_2 (WOODCOCK *et al.*, 1976). Although their "SiO_2 glass" might be easily distinguished from actual SiO_2 by comparison of calculated and observed X-ray diffraction intensities (see Fig. 3), it should be emphasized that their (partial) success was so stimulating that we decided to start our own calculations with the improved potential parameters. Our procedure is essentially the same as that described by WOODCOCK *et al.* (1976). The calculation is performed in several successive stages. The first stage is to obtain a "realistic" configuration at a very high temperature (taken to be 10,000 K) at which ions can migrate freely. During the process the chaotic initial configuration is brought to a near-equilibrium state (cf. Fig. 2). The time increment, Δt in Eq. 3, must be extremely small (0.1 to 0.25 fs) to prevent an "explosive" displacement of ions. The next stage is the high-temperature equilibration run, where Δt is safely taken to be 0.5–1.25 fs. Since the silicate framework is largely determined at this stage, the duration of the run has a decisive effect on the final results. Next comes the "cooling" stage, where the system is "cooled" down to temperatures of a practical interest ($\leq$2,000 K). This is the most problematic process because, even in a "very mild cooling" of the system by only 1 K per step (1 fs), the cooling rate is still unrealistic (10^{15} Ks^{-1}), and slower cooling is obviously impractical because of economic limitations as will be discussed later. Our "ordinary" cooling has, therefore, been much more severe: 100 K per 10 steps (10^{16} Ks^{-1}). It is thus doubtful whether the final, "annealing" stage can actually anneal the system, or, remove thermal stress effectively; and whether the "annealed" product does represent the near-equilibrium state at the nominal temperature. In cases of systems with high diffusivity and low viscosity such as alkali halide melts, there are no such difficulties. It is almost certain that the silicate systems require a much greater number of steps in the MD calculation than those in alkali halides presumably by a factor of 10^3 to 10^5. Details of our calculation scheme are described elsewhere (MATSUI *et al.*, 1982).

5. *Presentation of Results: PCF, $s \cdot i(s)$ and 3-D Views*

In ordinary X-ray and neutron diffraction measurements, the intensity of the diffracted X-ray (or neutron) beam is the primary information (see MARUMO and OKUNO, 1982). In contrast, the primary information from MD calculations is the positions of ions as a function of time. Therefore, the instantaneous ("snap-shot") structure is the most straightforward way of presentation of MD-derived results. The pair correlation function (PCF), or pairwise discriminate radial distribution function for each pair of ionic species i and j, $g_{ij}(r)$, is obtained from the positional data accumulated during the final equilibration run (in the actual computation, the distance interval, Δr, is usually taken to be 0.1 Å)

$$g_{ij}(r) = \frac{V}{N_i N_j} \frac{n_{ij}(r)}{4\pi r^2} \tag{8}$$

in which $n_{ij}(r)$ is time-averaged number of ion pairs between i and j within a distance range from $r - \Delta r/2$ to $r + \Delta r/2$, N_i is the number of ion species i in the basic cell and V is the volume of the basic cell. The running coordination number (RCN), $N_i(r)$, which is important in characterizing the coordination state of cations, is give by

$$N_i(r) = \sum_{r=0}^{r} n_{ij}(r)/N_i \tag{9}$$

where the subscript i refers to the cation species i, and j to the surrounding anion species j.

The intensity of coherent, scattered X-ray (or neutron) beam at a diffraction angle θ, $i(s)$, of the model melt/glass, where $s = 4\pi \sin(\theta/\lambda)$ and λ is the wavelength, can be calculated by a Fourier integral of $g_{ij}(r)$. It is important to note that the transformation of $g_{ij}(r)$ to $i(s)$ can be made, in effect, free from the troublesome Fourier noise (see Fig. 6 in MARUMO and OKUNO, 1982) because $g_{ij}(r)$ approaches unity rapidly with an increasing distance ($r_{ij} \geq 8$ Å), whereas this is not the case for the inverse Fourier transformation from $i(s)$ to $g(r)$, which is routinely performed in the ordinary X-ray study of amorphous materials. It follows that the comparison between actual and calculated configurations should be done on $i(s)$ or $s \cdot i(s)$. The $s \cdot i(s)$ can be obtained in the following manner:

$$s \cdot i(s) = \frac{4\pi N}{V} \frac{\sum_i \sum_j N_i N_j f_i(s) f_j(s)}{\left[\sum_{\text{all}} N_k f_k(s)\right]^2} \sum_{r=0}^{r_{\max}} \sum_{ij} [r(g_{ij}(r) - 1)\sin(s \cdot r)] \tag{10}$$

where $f_i(s)$ denotes the atomic scattering factor for ion species i at s.

6. *Examples of MD Results*

6.1 *Silica, SiO_2, melt and glass*

As already mentioned, the first MD calculation on the "model SiO_2" was carried out by WOODCOCK *et al.* (1976) in a density range of 2.2 to 5.5 $g \cdot cm^{-3}$. A part of our calculation has been reported elsewhere (MATSUI *et al.*, 1982) and a more comprehensive computation are described in a separate paper (OKADA *et al.*, 1982). The following is a summary of our latest results.

Unlike the model SiO_2 by Woodcock *et al.*, our model SiO_2, using the C-2a potential parameters, gave a configuration which is in satisfactory agreement with the X-ray (MOZZI and WARREN, 1969; DA SILVA *et al.*, 1975;

WASEDA and TOGURI, 1977) and neutron (WASEDA and SUITO, 1977; WRIGHT and SINCLAIR, 1978) diffraction measurements. This is shown by comparison of the $s \cdot i(s)$ functions in Fig. 3. For the model system with 324 ions (108 SiO_2 units) in the basic cell of $L = 16.98$ Å (density, d, being 2.20 $g \cdot cm^{-3}$), the calculated PCF's for Si–O, Si–Si and O–O, and RCN for Si–O pairs are illustrated in Fig. 4. Although all the Si^{4+} ions are surrounded by four O^{2-} ions, the shape of the SiO_4 tetrahedra is more irregular than that in $MgSiO_3$ and Mg_2SiO_4 systems. The O(bridge)–Si–O(bridge) angles (ideal value: 109.5°) range from 85° to 140° with a peak at 106.5°. The Si–O(bridge)–Si angles range from 120° to 180° with a broad but clear peak at 153°. This value compares favourably with that of a re-evaluation of the X-ray data reported by Mozzi and WARREN (1969) by DA SILVA *et al.* (1975), who concluded that the most probable Si–O–Si angle is 152° in SiO_2 glass.

A part of the instantaneous configuration at 2,000 K is displayed as a stereoscopic pair drawing in Fig. 5. Three- to nine-(or more) membered rings can be seen. Among the rings, five- to eight-membered rings are predominant. All the above characteristics of the MD-derived model SiO_2 strongly indicate that the calculation is successful, even with all the inherent difficulties in this type of calculation on silicates, as stated above. The "compression experiment" (by decreasing L with N kept constant) resulted in a partial change in the coordination state of Si^{4+} ion from usual fourfold to sixfold coordination. The change is similar to that observed in the model $MgSiO_3$ system which will be discussed later.

Unlike the cases of $MgSiO_3$ or Mg_2SiO_4 melts, the model SiO_2 melt (see Fig. 5) with a "realistic" configuration, is rather difficult to obtain by MD

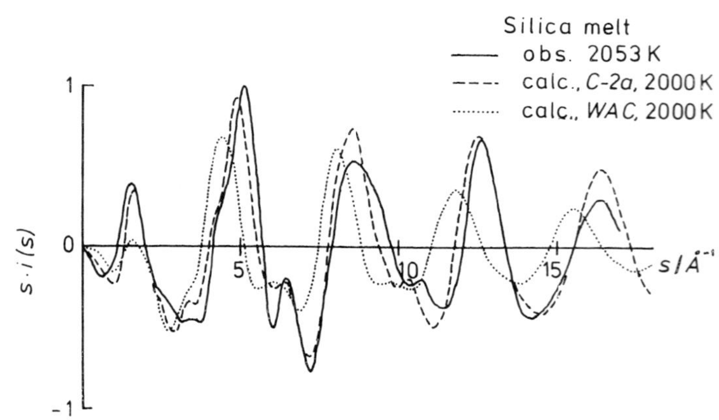

FIG. 3. Comparison of the $s \cdot i(s)$ function of molten SiO_2 for X-ray "observation." Observed curve: WASEDA and TOGURI (1977). The MD-derived curves: OKADA *et al.* (1982), assuming the C-2a potential parameters (Table 1B) and assuming parameters (WAC) adopted by Woodcock *et al.* (1976), recalculated by Okada *et al.* (1982). Compare also parameters the $s \cdot i(s)$ curves shown in Matsui *et al.* (1982), Fig. 6.

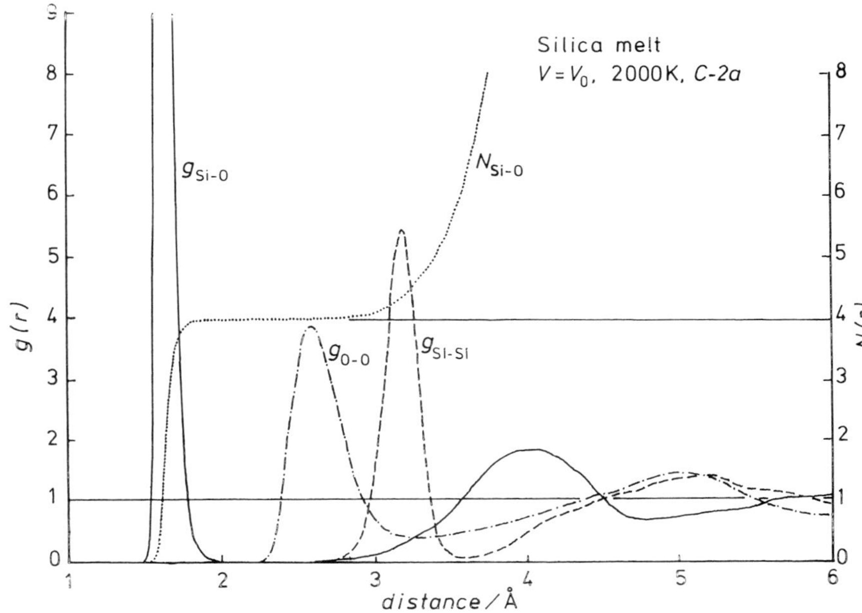

FIG. 4. Calculated PCF's and RCN (for Si–O) for a model SiO_2 melt assuming C-2a potential parameters at V_0 (2.20 g · cm^{-3}). The C-2a based $s \cdot i(s)$ curve in Fig. 3 corresponds to the present PCF's.

calculation. When we start from the "random" initial configuration, the "free" O^{2-} ions, which cannot find an Si^{4+} ion to complete an SiO_4 tetrahedron, often persist for a very long time at the high-temperature equilibration stage. On the other hand, if we start with an ordered configuration such as the β-cristobalite arrangement*, no essential change, such as dissociation of Si–O bonds, occurred during an extremely long high-temperature run within economic limit. This implies that the first-order transition (in this case, fusion) of the "perfect crystal" (i.e., a crystal without any defect) cannot be realized during MD runs for systems with high potential barriers, such as the framework silicates. Starting from a random configuration, the "perfect amorphous" configuration containing no "stray" O^{2-} ions and consequently no SiO_3 groups is realized as a rare event. The "perfect product" behaves as if it were a perfect (pseudocubic) crystal so that any migration of ions is no longer possible. Since the self-diffusion coefficient, D_i, for every ionic species i is derived through the Einstein equation for Brownian motion

*The structure has a cubic unit cell containing 8 SiO_2 units, and consists of six-membered rings only.

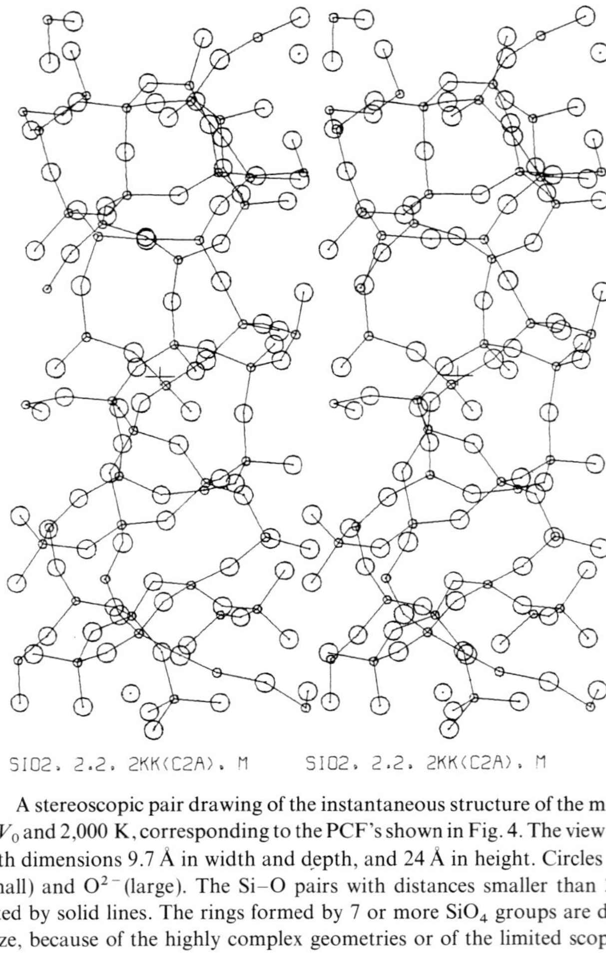

FIG. 5. A stereoscopic pair drawing of the instantaneous structure of the model SiO_2 melt at V_0 and 2,000 K, corresponding to the PCF's shown in Fig. 4. The view covers an area with dimensions 9.7 Å in width and depth, and 24 Å in height. Circles represent Si^{4+}(small) and O^{2-}(large). The Si–O pairs with distances smaller than 2.0 Å are connected by solid lines. The rings formed by 7 or more SiO_4 groups are difficult to recognize, because of the highly complex geometries or of the limited scope.

$$D_i = \frac{1}{6\tau N_i} \sum_i \langle [\boldsymbol{x}_i(t) - \boldsymbol{x}_i(t+\tau)]^2 \rangle \tag{11}$$

where τ is an arbitrary time interval, it follows that the model SiO_2 may give rise to zero diffusion coefficients at temperatures well above the melting point, while the $s \cdot i(s)$ function may be close enough to that of the actual melt.

6.2 *Enstatite, $MgSiO_3$, melt and glass*

A 3-D view of a model $MgSiO_3$ at 2,000 K is shown in Fig. 6. The structure represents a part of the result of a calculation of the system with a basic cell of edge length 16.24 Å, containing 320 ions (65 $MgSiO_3$ units, $d = 2.49\ \mathrm{g \cdot cm^{-3}}$). Every Si^{4+} ion is surrounded by four O^{2-} ions to form SiO_4

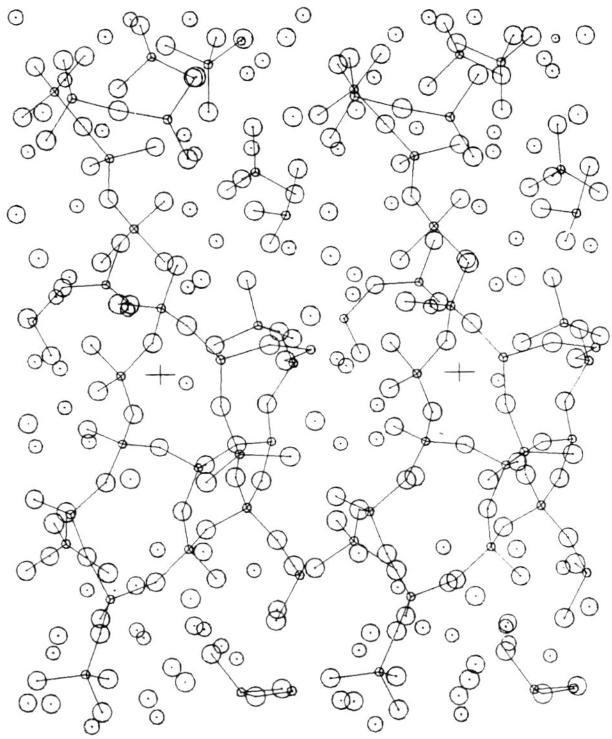

FIG. 6. A stereoscopic pair drawing of the instantaneous structure of a model $MgSiO_3$ at V_0 (2.20 g · cm^{-3}) and 2,000 K, assuming the C2-a potential parameters. Circles represent Si^{4+}, Mg^{2+} and O^{2-} in the increasing order in sizes. The Si–O pairs with distances smaller than 1.85 Å are connected by solid lines. The dimensions of the view are the same as those for Figs. 5, 7 and 10.

tetrahedra, as is the case with SiO_2. The Si–O(bridge)–Si angles show a distribution similar to that of SiO_2 with a peak around 150°. On the other hand, due to the increased O/Si ratio, the SiO_4 tetrahedra can exist in five different modes, according to the number of shared (bridging) O^{2-}. These are discrete SiO_4, SiO_3O– (O– denotes the bridging O), $SiO_2(O–)_2$, $SiO(O–)_3$, and $Si(O–)_4$. The relative frequency of occurrence of these modes is subject to a great degree of fluctuation and the following values are rough estimates of the typical values at 2,000 K: SiO_4, 2%; SiO_3O–, 20%, $SiO_2(O–)_2$, 44%; $SiO(O–)_3$, 28%; and $Si(O–)_4$, 6%. The number of bridging oxygens is therefore 2.2 per SiO_4 unit on average. The comparison of the above frequency distribution with the Raman spectroscopy-derived "structon distribution" (see KUSHIRO, 1982) appears to be intriguing.* However, it should be remembered that the SiO_3O- mode is not equivalent to a "dimer" in the "structon analysis," because the SiO_3O- mode fulfils only a necessary condition of being a constituent of a "dimer." A similar caution applies to the correspondence between the $SiO_2(O-)_2$ mode and "chain," the $SiO(O-)_3$ and "sheet," and the $Si(O-)_4$ and "3-D network", so that direct comparison of the present results with that of the "structon analysis" cannot be justified. Nevertheless, it is interesting to note a definite parallel between the Raman spectroscopy data and the frequency of the modes of linkage among SiO_4 groups calculated for model silicates covering a range from Mg_2SiO_4 (MATSUI *et al.*, 1982), where the dominant modes are SiO_4 and $SiO_3(O-)$, to $MgSiO_3$, where $SiO_2(O-)_2$ and $SiO(O-)_3$ are predominant.

The distribution of modes of occurrence of O^{2-} ions, which are derived from the population of modes of the SiO_4 groups, is subject to less fluctuation. The "free" or fully-active oxygens "O^{2-}" (i.e., not coordinated to Si^{4+}), "O^-" (coordinated to one Si^{4+}), and "O^0" (the bridging oxygen) exist in the following proportions at 2,000 K : O^{2-}, 3%; O^-, 60%; and O^0, 37%. It is again interesting to note that the above proportions yield the "dissociation constant," K, for the following hypothetical "reaction" first proposed by TOOP and SAMIS (1962):

$$2O^- = O^0 + O^{2-} \tag{12}$$

of around 0.03 ($K = [O^0][O^{2-}]/[O^-]^2$, where [ ] denotes the hypothetical "activity" of the relevant oxygen ion) by putting the activity coefficients to be unity, which is in reasonable agreement with 0.14–0.003 estimated by YOKOKAWA and NIWA (1969) for the $M^{2+}O–SiO_2$ binary systems (M : Mn, Pb and Ca).

The "compression experiments" have been carried out by MATSUI and KAWAMURA (1980) and MATSUI *et al.* (1982) on a model $MgSiO_3$ system. The results can be summarized as follows. At 2,000 K, SiO_4 groups persist up to 0.8 V_0 with an increasing degree of distortion. Compaction beyond this stage results in a partial collapse of SiO_4 tetrahedra and at 0.6 V_0 ($d =$ 4.15 $g \cdot cm^{-3}$) about two-thirds of the Si^{4+} ions assume more or less regular sixfold coordination. In Fig. 7 the 3-D view of a "quenched" product at 0.6 V_0 and 500 K is shown. During the course of compression, the average Si–O(nearest)–Si angles show a remarkable decrease down to 130°, but the distribution of angles is quite complex. As the volume is reduced from V_0 to 0.8 V_0 the average Si–O–Si angles show a steady decrease. However, the occurrence of Si^{4+} ions in sixfold coordination (SiO_6 octahedra) gives rise to

*Our provisional calculations on $CaSiO_3$ and Ca_2SiO_4 systems gave images not drastically different from those of Mg-analogues. The degree of polymerization of SiO_4 groups, however, is systematically less pronounced in Ca-analogues.

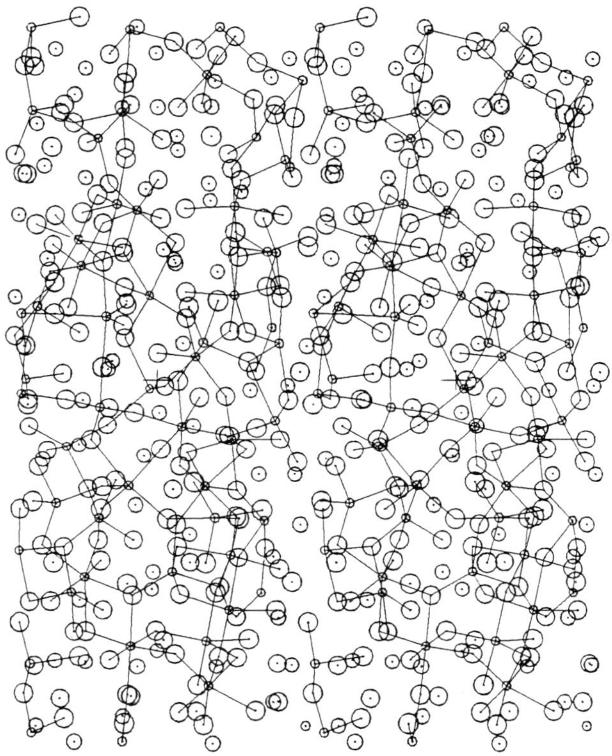

FIG. 7. A stereoscopic pair drawing of the instantaneous structure of a model $MgSiO_3$ glass at 0.6 V_0 (4.15 g · cm^{-3}) and 500 K. Circles represent Si^{4+}, Mg^{2+} and O^{2-} in an increasing order in sizes. The Si–O pairs with distances smaller than 2.2 Å are connected by solid lines. Note the present of more or less regularly shaped SiO_6 groups, linked together by sharing edges and faces.

peaks in the distribution at about 100° which correspond to the rutile- and perovskite-like structures. At 0.6 V_0, therefore, the angles have a bimodal distribution with peaks located at around 100° and 150°. The latter peak may be a composite one composed of at least two components; one with a peak at around 140° (largely due to the polymerized SiO_4 groups) and another with a peak at > 160° (partly due to the straight alignment characteristic of the rutile and perovskite structures). In the PCF, a significant increase in peak position for the Si–O pair is observed from 1.62 Å at V_0 to 1.68 Å at 0.6 V_0. On the other hand, the peak position for the Mg–O pair shifts steadily to a smaller value: 1.98 Å at V_0 at V_0 to 1.94 Å at 0.6 V_0. The coordination around Mg^{2+} is ill-defined throughout, but the RCN (see MATSUI *et al.*, 1982, Figs. 2 and 4) indicates that the number of surrounding O^{2-} ions increases on compaction.

6.3 Anorthite, $CaAl_2Si_2O_8$, melt

Several reconnaissance calculations, assuming the C-2a parameters, were performed by one of the present authors (K.K.) on the model $CaAl_2Si_2O_8$ melt with an unusually large number of ions within the basic cell ($N = 780$, or 60 $CaAl_2Si_2O_8$ units). Two sizes of basie cell were used; one with $L = 22.14$ Å (V_0; $d = 2.55$ g · cm^{-3}) and another with $L = 20.55$ Å (0.8 V_0; $d = 3.19$ g · cm^{-3}). Resultant PCF's at V_0 and 2,000 K are shown in Fig. 8, and the calculated $s \cdot i(s)$ function is compared with that observed by MARUMO and OKUNO (1982) in Fig. 9. Agreement between the calculated and the observed curves is reasonable. It is noted that, for a melt of such complex chemistry, the derivation of PCF's (in the present case, for 10 pairs) from diffraction measurements alone is practically impossible. In this context, the MD results presented here will be of great help in the analysis of the observed data in terms of PCF's.

There are three main points of interest in such a calculation; firstly to determine coordination states of Al^{3+} at low and high densities, secondly to

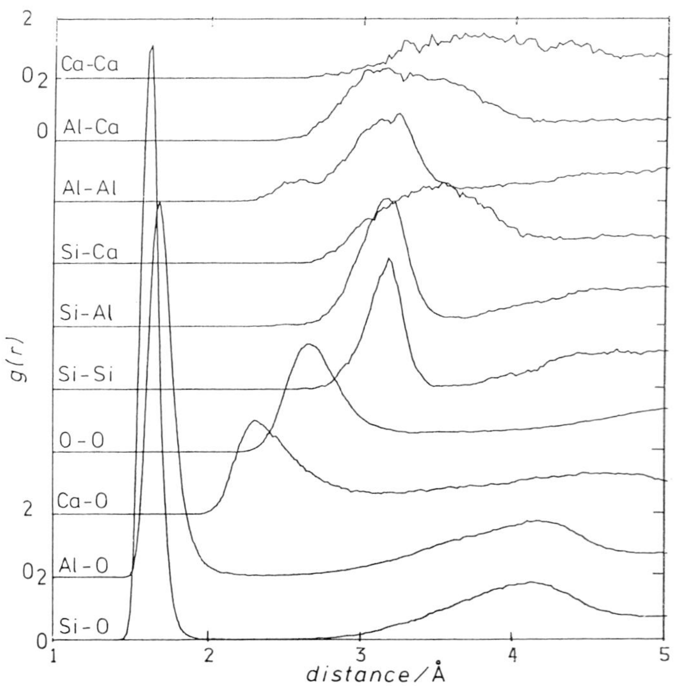

FIG. 8. Calculated PCF's for a model $CaAl_2Si_2O_8$ (anorthite) melt, assuming the C-2a potential parameters. No smoothing is applied in drawing curves so that the fluctuation is apparent, especially for pairs with low probability of occurrence. The base line ($g = 0$) is different for each PCF: The interval between two (adjacent) base lines corresponds to difference of 2 in $g(r)$.

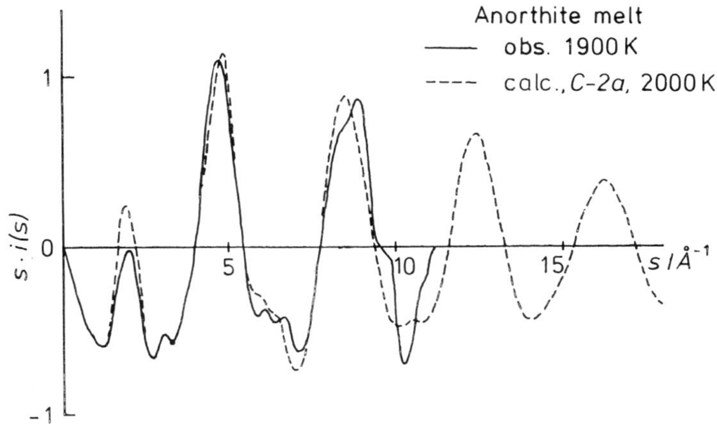

FIG. 9. Comparison between observed $s \cdot i(s)$ by MARUMO and OKUNO (1982) and MD-calculated $s \cdot i(s)$ derived from the present PCF's shown in Fig. 8 for a $CaAl_2Si_2O_8$ melt. The corresponding 3-D structure is shown in Fig. 10.

observe variations, if any, in the ring size of the aluminosilicate framework under compaction, and thirdly to check whether the aluminium avoidance rule, which is generally accepted to hold in the structures of aluminosilicate crystals, also hold in the melt. A part of the calculated model melt at V_0 and 2,000 K is given as Fig. 10.

The SiO_4 tetrahedra are again observed with little deformation, whereas the AlO_4 tetrahedra are much more irregular in their shape. Out of 120 Al^{3+} ions, 8 to 10 are found in threefold coordination, indicating a relative instability of the regular AlO_4 group (cf. the ill-defined MgO_n polyhedra in $MgSiO_3$), presumably as a result of the low valence (+3) and long Al–O distance compared with the SiO_4 group. The irregularity in coordination of O^{2-} around Al^{3+} becomes more pronounced upon compaction to 0.8 V_0, where five-coordinated Al^{3+} ions are observed. The most remarkable feature observed under compaction is the change in size of the aluminosilicate rings. At V_0, 4- or less-membered rings are rather rare and 5- to 8-membered rings are predominant. At 0.8 V_0, a significant increase in the number of 4- to 6-membered rings is observed, suggesting a marked decrease in the size of the effective framework polyanions.

The coordination change of Al^{3+} appears to be gradual under compaction, as is the case for Si^{4+} and Mg^{2+} in $MgSiO_3$. Therefore we can suppose that the change in coordination states of cations in molten and vitreous states does not occur in a sudden manner, in contrast to that accompanying the high-pressure induced phase transitions in crystals.

The aluminium avoidance rule requires that every SiO_4 tetrahedron is linked by four AlO_4 tetrahedra in the case of $CaAl_2Si_2O_8$, where the Al:Si

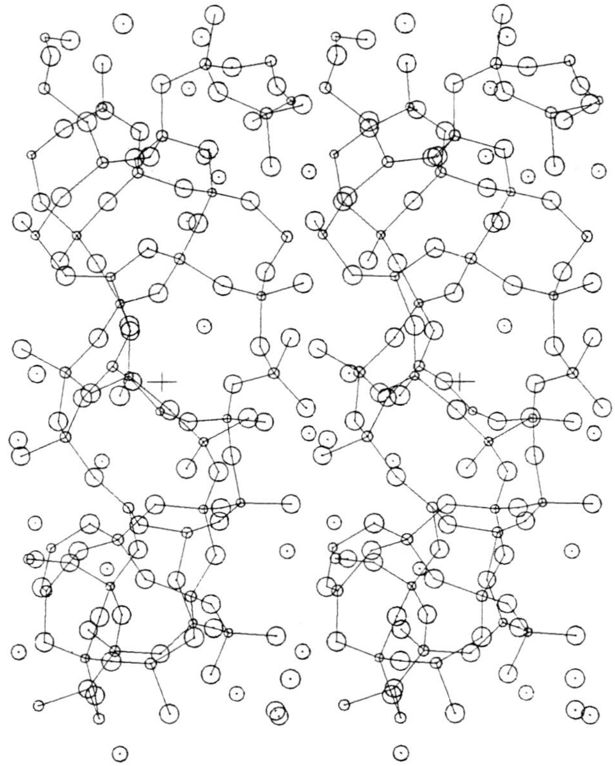

FIG. 10. A stereoscopic pair drawing of the instantaneous structure of the model $CaAl_2Si_2O_8$ melt at V_0 (2.55 g · cm^{-3}) and 2,000 K. Circles represent Si^{4+}, Al^{3+}, Ca^{2+} and O^{2-} in the increasing order in sizes. The Si–O and Al–O pairs are connected by solid lines if the distances are smaller than 1.9 Å for Si–O and 2.2 Å for Al–O.

ratio is equal to one. As is clear in the 3-D view (Fig. 10), this requirement is not satisfied for most SiO_4 groups in the model melt. The distribution of bridging oxygens (forming T–O–T, where T denotes Si and/or Al), however, indicates a slight, but energetically meaningful, preference for the Al–O–Si bridge (55%) over Si–O–Si (23%) and Al–O–Al (22%) bridges. (A statistically random configuration should have the following distribution: Al–O–Si: 50%, Si–O–Si and Al–O–Al: 25% each.) The average bridging angles are 151° for Si–O–Si, 143° for Al–O–Si, and 133° for Al–O–Al at V_0, are all found to decrease under compaction. Most notable is the decrease in the average Si–O–Si angle; the mean value being 145° at 0.8 V_0 and 2,000 K.

7. *Concluding Statements*

In 1976, Woodcock and his co-workers tried their MD calculation on the model SiO_2 "with little expectation of success" (WOODCOCK *et al.*, 1976, p. 1,572). Now, bearing in mind what has been stated above, we can say instead that the MD calculation provides virtually the only way of obtaining realistic 3-D images of silicate melts, automatically satisfying numerous limiting conditions such as interatomic distances, Pauling's electrostatic valence rule, and so on, with little fear of complete failure. However, the periodic boundary conditions, which is essential for both the MD and MC calculations, make it very difficult to obtain some important information about silicate chemistry, such as the mean molecular weight of silicate (and aluminosilicate) polyanions. To calculate this, a basic cell with an L of only 15 to 22 Å is obviously too small, because the size of a discrete SiO_4 tetrahedron is about 5 Å in diameter, and the sizes of polyanions may be of the order of 20 Å or more in diameter (cf. FRAY, 1970). Here the most serious difficulty appears. The calculation is extremely time-consuming and, hence, is tremendously expensive. In the case of a system with the basic cell of $N = 320$, it takes about 1 second to complete one step of a calculation using a HITAC M200-H at the Computer Centre of the University of Tokyo, which is one of the fastest computers available. As has been noted already, even an "absurd" cooling stage requires 8,000 or more steps, and the further "annealing" stage would require a far greater number of steps. It follows that, even on a system whose L and N are "too small," calculation which covers a sufficient time interval is still economically impractical. This explains why we have had to restrict our MD trials to within $N = 780$ ($L \sim 22$ Å). Since the computation time is roughly proportional to the square of N, a doubled L immediately implies that the computation time should be increased by a prohibitive factor of ~ 64.

The present results on model systems, though satisfactory in general, appear to indicate discrepancies with actual systems in several respects, not only in the $s \cdot i(s)$ function but also in pressure (usually too high, by a factor of 3 or more), in the self-diffusion coefficients, as has been already noted for SiO_2, and in the degree of polymerization of SiO_4 groups. It is not yet clear as to what extent the "unrealistic" assumptions and procedures are responsible for these discrepancies. It is expected that the prolonged equilibration would serve to reduce local strain which would give rise to erroneously high pressures, and that a larger cell may be effective in reducing the degree of polymerization (which is thought to have a similar effect on pressure). Nevertheless we still have no decisive evidence to support these expectations, again due to the problem of economic limitations. Thus, we are forced to conclude that the most serious problem in MD calculations of silicate melts is not a purely scientific one but to discover an "acceptable" condition of compromise between science and economy.

From a purely scientific point of view, the most urgent problem is, without doubt, to find a more realistic, but at the same time practical, potential function which is readily applicable for silicate systems. To this aim, closer collaboration among mineralogists, crystallographers, and inorganic and physical chemists is thought to be essential.

REFERENCES

BUSING, W. R. (1970): An interpretaion of the structures of alkaline earth chlorides in terms of interionic forces, *Trans. Am. Crystallogr. Assoc.*, **6**, 57–72.

DA SILVA, J. R. G., PINATTI, D. G., ANDERSON, C. E. and RUDEE, M. L. (1975): A refinement of the structure of vitreous silica, *Phil. Mag.*, **31**, 713–717.

FRAY, D. J. (1970): The structure of alkali silicate melts, *Phys. Chem. Glasses*, **11**, 219–222.

HORIUCHI, H. and SAWAMOTO, H. (1981): β–Mg_2SiO_4: Single crystal X-ray diffraction study, *Am. Mineral.*, **66**, 568–575.

IDA, Y. (1976): Interionic repulsive force and compressibility of ions, *Phys. Earth Planet. Inter.*, **13**, 97–104.

KUSHIRO, I. (1982): Structures and some physical properties of silicate melts of geological interest, this volume, pp. 39–60.

MARUMO, F. and OKUNO, M. (1982): X-ray structural studies of molten silicates: Anorthite and albite melts, this volume, pp. 25–38.

MATSUI, Y. and KAWAMURA, K. (1980): Instantaneous structure of an $MgSiO_3$ melt simulated by molecular dynamics, *Nature*, **285**, 648–649.

MATSUI, Y., KAWAMURA, K. and SYONO, Y. (1982): Molecular dynamics calculations applied to silicate systems: Molten and vitreous $MgSiO_3$ and Mg_2SiO_4 under low and high pressures, in *High Pressure Research in Geophysics*, Eds. S. Akimoto and M. H. Manghani, Center for Academic Publications Japan/Reidel, Tokyo/Dordrecht, pp. 511–524.

MIYAMOTO, M. and TAKEDA, H. (1980): An interpretation of the structures of mantle minerals at high pressures in terms of interionic forces, *Geochem. J.*, **14**, 23 248.

MOZZI, R. L. and WARREN, B. E. (1969): The structure of vitreous silica, *J. Appl. Crystallogr.*, **2**, 164–172.

OKADA, I., MATSUI, Y. and KAWAMURA, K. (1982): Structure of molten SiO_2 studied by molecular dynamics simulation, *Nippon Kagaku Kaishi*, **1982**, 910–919 (in Japanese with English abstract).

SCHÄFER, L. and KLEMM, A. (1976): Der Druck bei molekular-dynamischen Simulationen von Ionenflüssigkeiten, *Z. Naturforsch.*, **31a**, 1068–1072.

SINCLAIR, W. and RINGWOOD, A. E. (1978): Single crystal analysis of the structure of stishovite, *Nature*, **272**, 714–715.

SOULES, T. F. (1979): A molecular dynamic calculation of the structure of sodium silicate glass, *J. Chem. Phys.*, **71**: 4570–4578.

SMYTH, J. R. and HAZEN, R. M. (1973): The crystal structures of forsterite and hortonolite at several temperatures up to 900°C, *Am. Mineral.*, **58**, 588–593.

TOSI, M. P. (1964): Cohesion of ionic solids in the Born model, in *Solid State Physics*, Eds. F. Seitz and D. Turnbull, vol. 16, Academic Press, London, pp. 1–120.

TOOP, C. W. and SAMIS, C. S. (1962): Activities of ions in silicate melts, *Trans. Metallurg. Soc. AIME*, **224**, 878–887.

VERLET, L. (1967): Computer 'experiments' on classical fluids, I. Thermodynamic properties of Lennard-Jones molecules, *Phys. Rev.*, **159**, 98–103.

WASEDA, Y. and SUITO, H. (1977): The structure of molten alkali metal silicates, *Trans. ISIJ*, **17**, 82–91.

WASEDA, Y. and TOGURI, J. M. (1977): The structure of molten binary systems $CaO-SiO_2$ and $MgO-SiO_2$, *Metalurg. Trans.*, **3B**, 563–568.

WOODCOCK, L. V. (1975): Molecular dynamics calculations on molten ionic salts, in *Advances in Molten Salt Chemistry*, Eds. J. Braunstein, G. Mamantov, and G. P. Smith, Plenum Press, New York, pp. 1–74.

WOODCOCK, L. V., ANGELL, C. A. and CHEESEMAN, P. (1976): Molecular dynamics studies of the vitreous states: Simple ionic systems and silica, *J. Chem. Phys.*, **65**, 1565–1577.

WRIGHT, A. C. and SINCLAIR, R. N. (1978): Neutron diffraction by vitreous silica, in *The Physics of Silicon Oxide and Its Interfaces*, Ed. S. T. Pantelides, Pergamon Press, London, pp. 133–138.

YOKOKAWA, T. and NIWA, K. (1969): Free energy of solution in binary silicate melts, *Trans, JIM*, **10**, 3–7.

Materials Science of the Earth's Interior, edited by I. Sunagawa, pp. 25–38.

X-RAY STRUCTURAL STUDIES OF MOLTEN SILICATES: ANORTHITE AND ALBITE MELTS

Fumiyuki MARUMO and Masayuki OKUNO

Tokyo Institute of Technology, Midori-ku, Yokohama, Japan

1. *Introduction*

The structures of molten substances have stimulated the interest of material scientists, since they are expected to afford not only the basis of understanding the properties of high-temperature liquids, but also to provide a clue to the elucidation of the detailed mechanism of crystal growth from melts. Silicate melts are especially important in petrological as well as in metallurgical fields and efforts have been made to reveal the short-range orderings of their structures.

There are various experimental methods which can be used for the structural investigation of amorphous materials. However, X-ray and neutron diffraction techniques afford the most direct information of the spacial arrangements of atoms, and they play a leading part in structure determination of amorphous materials. The theoretical basis for structure determination of amorphous materials from X-ray diffraction data was given by ZERNIKE and PRINS (1927) . They employed the Fourier transform technique to deduce the radial pair distribution of electrons from X-ray data. The method was first applied successfully on liquid mercury by DEBYE and MENKE (1930, 1931) and on vitreous SiO_2 and B_2O_3 by WARREN *et al.* (1936). Since then this method of calculating the radial distribution function has become the standard technique in the determination of the structure of amorphous materials from diffraction data.

2. *Experimental Techniques*

Both reflection and transmission geometries are employed for intensity measurements of diffracted X-rays from high-temperature melts. The technique with the former geometry, which is more widely adopted than the latter, explores scattered X-rays from a flat free-surface of a melt put into a metal vessel. The transmission geometry is easily realized for room temperature liquids by utilizing a transparent vessel for X-rays. However, it is practically impossible to use the same scheme for high-temperature melts, because no adequate material is avibable for making a vessel which is transparent for X-rays and, at the same time is also durable for high-temperature experiments. Therefore, we make use of a drop of the material to be examined fixed at the top of a rod, which is made of an appropriate material. An X-ray beam with a larger cross-section than that of the drop is irradiated on to the drop and the scattered X-rays are measured in the same way as the usual diffractometry. In the following, outlines of these two techniques are described.

2.1 Diffractometry with a reflection-type equipment

The Bragg-Brentano focusing arrangement is used as in ordinary powder diffractometry. In the case of liquid diffraction, the flat surface of the sample must always be kept horizontal. Therefore, the primary beam and detector are rotated around a horizontal axis on the surface of the specimen in opposite directions to each other to change the diffraction angle in keeping with the angle between the scattered X-rays to be measured and the specimen surface equal to the incidence angle of the primary X-rays. Figure 1 schematically shows the geometry of the so-called θ-θ type diffractometer. Slit 1 and Slit 2 are horizontal slits with vertical Soller slit systems. Advantages of this geometry lie in the following points: (1) a relatively high flux of the scattered beam reaches the detector because of the parafocusing arrangement, and (2) the absorption correction is practically unnecessary when the absorption coefficient is large.

The sample is placed in a platinum vessel and heated in a small electric furnace, as shown in Fig. 2. (For diffraction experiments at temperatures higher than about 1,300°C, samples are put on a molybdenum or tungsten plate, and heated by passing an electric current directly through the plate in a reducing atmosphere.) Platinum and pyromax wires are used for the main and subsidiary heaters, respectively. The main heater is placed in a case made of calcined pyrophillite and the case itself is sealed into a container made of firebrick with a window for X-ray path. The window is covered with a aluminum or nickel foil. With this furnace it is possible to heat the sample up to about 1,300°C.

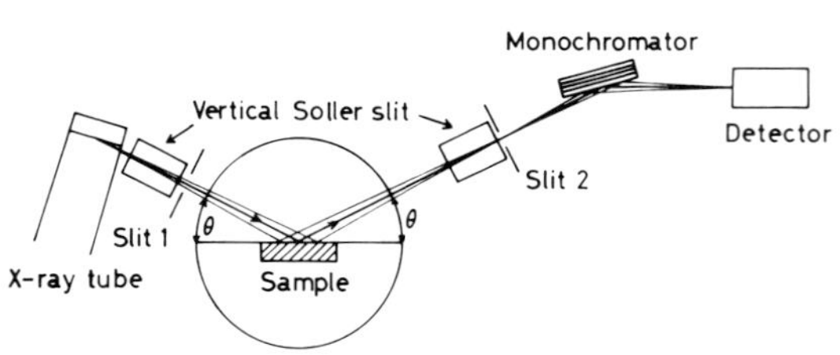

FIG. 1. The geometry of a θ-θ type diffractometer.

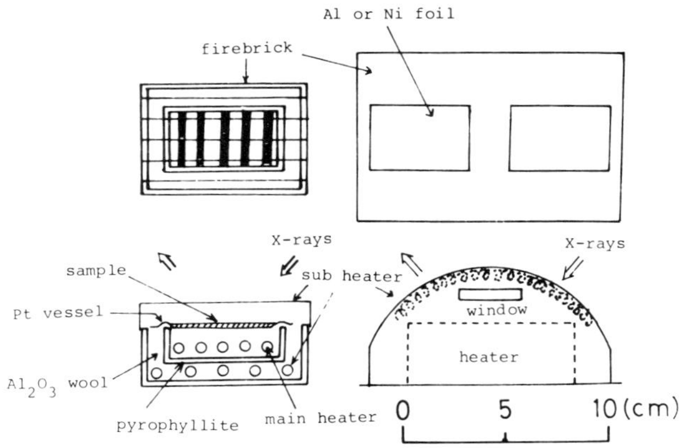

FIG. 2. A sample heating device for the θ-θ type diffractometer.

For the intensity measurements, a divergence slit (Slit 1) and a receiving slit (Slit 2) of 1/2° are used in the region of the diffraction angle 2θ lower than 30°, and those of 1° in the higher diffraction angle region. In actual experiments, measurements are carried out with both sizes of slits in the intermediate region for the scaling of intensity data. Usually the step scan technique is adopted in a fixed time mode.

2.2 *Diffractometry with a small spherical sample*

The geometry of diffractometry with a small spherical sample is shown Fig. 3. The specimen is melted with a micro gas burner and the droplet is attached to the top of a rod by its viscosity. Usually we use a rod made of the same material as the specimen. When an oxygen-propane mixed gas is used the specimen can be heated up to 2,300°C. A detailed description of a gas-flame heater is given by MIYATA *et al.* (1979).

Diffracted X-ray intensities are measured in the 2θ step scan mode. Namely, only the detector is moved to change the scattering angle of the

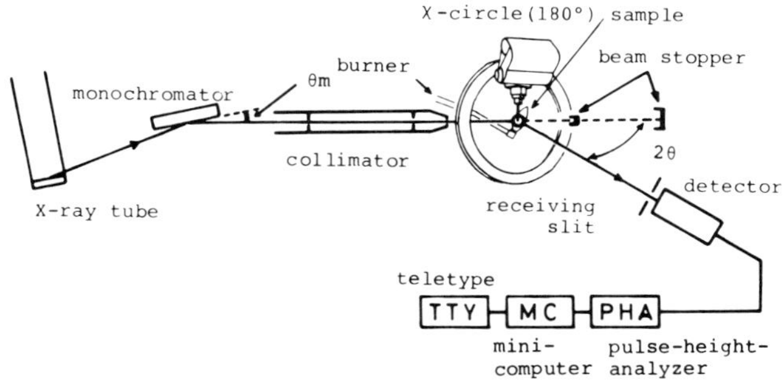

FIG. 3. The geometry in diffractometry with a small spherical sample.

diffracted X-rays to be counted. The following points are the merits of this technique: (1) only a small amount of the sample is required to collect intensity data and (2) the specimen is easily heated to a high temperature of up to 2,300°C. On the other hand, this technique cannot be applied to materials having high vapour pressures because of rapid evaporation and, further, it is not suitable for materials with large absorption coefficients. Intensity data were collected on a melt of Miyake-jima anorthite, using this method, and will be described in the forthcoming section.

3. *Data Processing*

3.1 *Corrections for polarization and absorption factors*

The intensity of X-rays scattered from an amorphous material is given by Debye's formula,

$$I(S) = \sum_i f_i^2(S) + \sum\sum_{i \neq j} f_i(S) f_j(S) \frac{\sin(Sr_{ij})}{Sr_{ij}} + I_{\text{inc}}(S) \tag{1}$$

where r_{ij} is the distance between the *i th and j* th and *j* th atoms, $f_i(S)$ is the atomic scattering factor of *i* th atom, and $S = 4\pi \sin\theta/\lambda$. $I_{\text{inc}}(S)$ is the intensity of the incoherently scattered X-rays. The second term is the structure sensitive part, representing scattering interactions between different atoms. The experimentally-observed intensity $I_{\text{obs}}(S)$ is related to $I(S)$ with the equation

$$\alpha I(S) = \{I_{\text{obs}}(S) - I_{\text{B}}(S)\} \cdot A^{-1} \cdot P^{-1} \tag{2}$$

where α, A, P, and $I_{\text{B}}(S)$ are the scale factor, absorption factor, polarization factor, and the background intensity, respectively. $I_{\text{B}}(S)$ is often assumed to be zero. When an ideally mosaic crystal is used as a monochromator to monochromate an unpolarized incident beam, P has the form

$$P(\theta) = (1 + \cos^2 2\theta_{\text{m}} \cdot \cos^2 2\theta)/2 \tag{3}$$

where θ_{m} is the Bragg angle of the monochromator. If a perfect crystal is used as a monochromator, $\cos^2 2\theta_{\text{m}}$ should be replaced by $\cos 2\theta_{\text{m}}$ in Eq. (3).

The absorption factor for the intensity data, collected with the Bragg-Brentano para-focusing geometry, is given by

$$A = \frac{1}{2\mu}[1 - \{1 - \exp(-Z)\}/Z] \tag{4}$$

where μ is the linear absorption coefficient and $Z = 2\alpha\mu/\sin 2\theta$, with α the beam divergence. For a large μ, A becomes approximately $1/2\mu$.

The absorption factors for spherical samples are a function of μr and θ, r being the radius of the sphere. The absorption correction factors $A^*(= A^{-1})$ for spherical samples are given in INTERNATIONAL TABLES FOR X-RAY CRYSTALLOGRAPHY Vol. 2 (1972).

The multiple scattering is usually ignored because of its low effects. A detailed discussion on this problem is given by MALET *et al.* (1973).

3.2 *Scaling of intensity data*

Several procedures have been proposed for determining the scale factor to normalize the observed intensities. Among them, the Krogh-Moe-Norman method (KROGH-MOE, 1956; NORMAN, 1957) and the high-angle method are the standard procedures. The former method utilizes the constraint that atoms cannot approach each other within a distance shorter than the sum of the relevant atomic radii. In this method, the scale factor α is given with the relation

$$\alpha = \frac{\int_0^{S_{max}} S^2\left[\left\{\sum_j f_j^2(S) + I_{inc}(S)\right\} \Big/ \left\{\sum_j f_j(S)\right\}^2\right] \mathrm{d}S - \left\{2\pi^2\rho_0 \Big/ \left(\sum_j \bar{K}_j\right)^2\right\}}{\int_0^{S_{max}} S^2\left[I_{obs}(S)' \Big/ \left\{\sum_j f_j(S)\right\}^2\right] \mathrm{d}S} \tag{5}$$

where $I_{obs}(S)'$ is the observed intensity corrected for the polarization and absorption factors, S_{max} the maximum value of S in $I_{obs}(S)'$, ρ_0 the average electron-pair density of the specimen, $\bar{K}_j$ the effective number of electrons in j th atom, and the summation for j is taken over the atoms in a stoichiometric unit. The calculation of ρ_0 requires the density of the substance.

The high-angle method makes use of the fact that the coherent scattering from different atoms approaches zero at the large S value region for non-crystalline materials. Accordingly, the intensity of a coherent scattering is approximated with the sum of intensities given by the respective atoms at the large S region. Then the scale factor is given by

$$\alpha = \frac{\int_{S_{ini}}^{S_{max}} \left[\sum_j f_j^2(S) + I_{inc}(S)\right] \mathrm{d}S}{\int_{S_{ini}}^{S_{max}} \left[I_{obs}(S)'\right] \mathrm{d}s} \tag{6}$$

where S_{ini} is the lower cut-off value of S, and usually chosen as $S_{ini} \geqq 10\,\text{Å}^{-1}$.

3.3 *Correction for Compton scattering*

Intensities of Compton scattering due to the elements from He to Cs, are calculated with the formula given by HAJDU (1971),

$$I_{inc}(S/4\pi) = \left[Z - \frac{f^2(S/4\pi)}{2}\right] \times [1 - M\{\exp(-KS/4\pi) - \exp(-LS/4\pi)\}] \tag{7}$$

where Z is the number of electrons in the atom or ion, f the atomic scattering factor. M, K, and L are parameters to be determined for individual atoms or ions, and the values are listed in the literature (HAJDU, 1971). For the elements from Ca to Am, the following formula given by PÁLINKÁS (1972) can be used,

$$I_{\text{inc}}(S/4\pi) = Z\left[1 - a\cdot\left(1 + \frac{bS}{4\pi}\right)^{-c}\right]. \tag{8}$$

The values of a, b, and c are also listed by PÁLINKÁS (1972).

When a post-specimen monochromator is used, the Compton scattering is largely eliminated in the higher scattering angle region, whereas it is scarcely affected in the lower angle region. In such a case, correction for the Compton scattering is carried out using a semi-empirical procedure.

First, the calculated intensities of the Compton scattering are subtracted from the normalized values of observed intensities I_{obs}'',

$$I_{\text{obs}}^{\text{coh}} = I_{\text{obs}}'' - I_{\text{inc}}. \tag{9}$$

Then, the fine correction is carried out as follows. (1) The best fitting curve, which monotonously increases or decreases with S, is determined for $I_{\text{obs}}^{\text{coh}}(S) - \sum_j f_j^2(S)$ in an appropriately high S region. (2) The value of the best fitting curve is subtracted from $I_{\text{obs}}^{\text{coh}}(S)$ at each S. We denote the obtained values with $I_{\text{obs}}^{\text{coh}}(S)'$. (3) $I_{\text{obs}}^{\text{coh}}(S)'$ is further treated with the procedure described by LEVY *et al.* (1966), namely, the values of $I_{\text{obs}}^{\text{coh}}(S)'$ are so changed that the radial distribution function calculated from the new $I_{\text{obs}}^{\text{coh}}(S)''$ gives the same values as those obtained from $I_{\text{obs}}^{\text{coh}}(S)'$ in the range of r longer than the first significant peak, and zero in the shorter r range. This procedure also eliminates systematic errors with long periods other than those due to Compton scattering.

3.4 *Calculation of radial distribution function*

The radial distribution function $D(r) = 4\pi r^2\rho(r)$ is related to the interference function $i(S)$ with the well known formula

$$D(r) = 4\pi r^2\rho_0 + \frac{2r}{\pi}\left(\sum_j \bar{K}_j\right)^2 \int_0^\infty S\cdot i(S)\sin(Sr)\,\mathrm{d}S. \tag{10}$$

The interference function $i(S)$ is the structure-sensitive part of scattering and is directly obtained from the reduced intensity $I_{\text{obs}}^{\text{coh}}(S)''$, the values of $I_{\text{obs}}(S)$ after the final treatment as described above,

$$i(S) = \left[I_{\text{obs}}^{\text{coh}}(S)'' - \sum_j f_j^2(S)\right] \Big/ \left\{\sum_j f_j(S)\right\}^2. \tag{11}$$

The correlation function $G(r)$ is often used for the examination of a structure model, which is related to $D(r)$ with the relation

$$G(r) = \frac{D(r)}{4\pi r^2 \rho_0} = 1 + \frac{\left(\sum_j \bar{K}_j\right)^2}{2\pi^2 r \rho_0} \int_0^\infty S \cdot i(S) \sin(Sr) \mathrm{d}S. \tag{12}$$

3.5 Calculation of intensity from a model

For a detailed structure analysis, it is desirable to compare the calculated intensities of coherent scattering with the observed values. The coherent scattering intensity from an amorphous material is expressed with Eq. (13) by rewriting Eq. (1),

$$I^{\mathrm{coh}}(S) = \sum_i \sum_j N_{ij} f_{0,i}(S) f_{0,j}(S) [\sin(Sr_{ij})/Sr_{ij}] \exp\left(-\frac{\Delta r_{ij}^2}{2} \cdot S^2\right) \tag{13}$$

where $f_{0,i}$ is the atomic scattering factors without the effect of thermal vibration, N_{ij} the multiplicity of the *i-j* atomic pair, and $\Delta r_{ij}{}^2$ is called the temperature factor and gives the mean square deviation of the *i-j* atomic pair. From Eq. (13), the interference function $i(S)$ is given by

$$i(S) = \sum_{i \neq j} \sum N_{ij} f_{0,i}(S) f_{0,j}(S) [\sin(Sr_{ij})/Sr_{ij}] \exp\left(-\frac{\Delta r_{ij}{}^2}{2} \cdot S^2\right). \tag{14}$$

Comparing the calculated $i(S)$ with the observed one, we can change the model so as to give the best fit between them.

4. *Structure Determination of an Anorthite Melt with X-Ray Diffraction*

4.1 Data collection

Anorthite from Miyake-jima, Izu islands, Japan (An_{96}; ISSHIKI, 1959) was used for the structure determination of the melt. A large phenocryst of anorthite in a basaltic lava was picked out and cut into a square prism with dimensions 0.5 × 0.5 × 6.5 mm. The top of the prism was melted at 1,600°C (just above melting point) with a micro gas-flame. Then, a spherical melt with a diameter of about 1.0 mm was obtained at the top of the specimen. X-rays were irradiated on this part to collect the diffraction data.

A four-circle diffractometer with a micro gas-flame heating device was used for data collection (MIYATA *et al.*, 1979). Intensities were measured in the 2θ step scan mode at intervals of 0.5° in the 2θ range from 5° to 50°, and 1.0° in the range from 50° to 100°, utilizing Mo $K\alpha$ radiation monochromated with a pyrolitic graphite (Fig. 3). The measurements were carried out for a fixed time of 30 s, and repeated 5 times in the lower angle region and 10 times in the higher angle region. The minimum total counts of a datum point was about 1,000.

The observed intensities were corrected for background, polarization and absorption factors, and for Compton scattering. Background correction was performed experimentally, by counting without specimen. Absorption corrections were carried out by using A^* values for a spherical sample in INTERNATIONAL TABLES FOR X-RAY CRYSTALLOGRAPHY, Vol. 2 (1972). The data were normalized by the Krogh-Moe-Norman method. The atomic scattering factors were taken from INTERNATIONAL TABLES FOR X-RAY CRYSTALLOGRAPHY, Vol. 4 (1974). The density of the substance used for the calculation of the average electron-pair density was 2.55 g cm^{-3}, the value being calculated from the partial molar volume (BOTTINGA and WEILL, 1970).

4.2 *Radial distribution function*

Figures 4 and 5 show the reduced intensity and the $S \cdot i(S)$ curves calculated from the intensity data of the anorthite melt, respectively. The

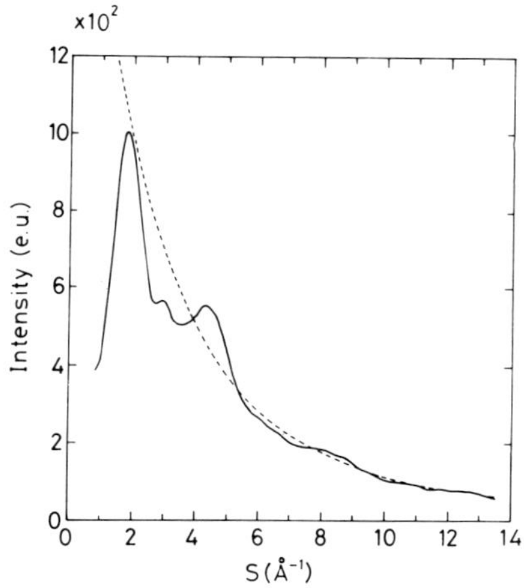

FIG. 4. The reduced intensity of anorthite melt.

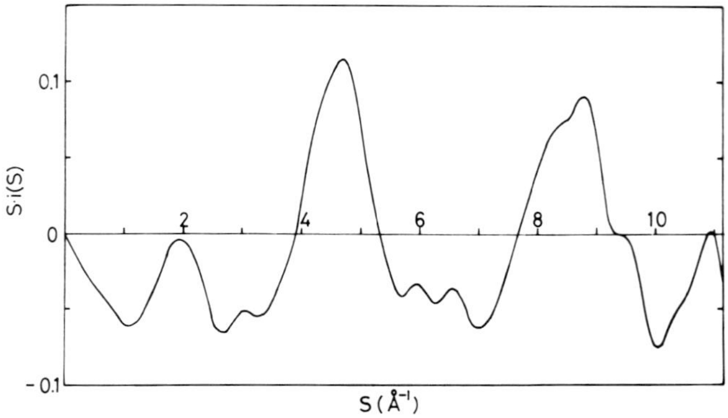

FIG. 5. The observed $S \cdot i(S)$ curve of anorthite melt.

$S \cdot i(S)$ curve has clear maxima at $S = 1.95$, 4.7, and 8.5 $Å^{-1}$. In Fig. 6, the radial distribution function $D(r)$, obtained from $S \cdot i(S)$ in the range $S \leqq 11.15$ $Å^{-1}$, is depicted together with the corresponding correlation function $G(r)$. Peaks are observed in $D(r)$ at $r = 1.68$, 2.40, and 3.12 Å and a shoulder at $r = 4.25$ Å. The first peak at $r = 1.68$ Å agrees well with the average T-O distance (T: Al and Si) in the anorthite structure at 1,430°C (SMITH, 1974), indicating that the T-O distances hardly change on phase transition. The number of oxygen atoms around the T atom was calculated to be 4.3 from the area of the first peak. These facts indicate that the structure of an anorthite melt is based on frameworks of SiO_4 and AlO_4 tetrahedra like in anorthite crystals. The second peak at $r = 2.40$ Å can be explained, if we assume the peak is due to the Ca-O pairs. The third peak at $r = 3.12$ Å can be assigned to the neighboring T-T pairs. However, the mean O–O distance (2.74 Å) in the TO_4 tetrahedra and the mean Ca–T distance (3.3 Å), observed in the anorthite structure at 1,430°C, fall on either side of the peak. Supposedly they contribute to the formation of the peak. The shoulder at $r = 4.25$ Å corresponds to the second neighbor T–O pairs. In the region of r longer than 5 Å, $\rho(r)$ approaches ρ_0, and the $G(r)$ converges rapidly to unity.

These facts show that the structure of an anorthite melt resembles the crystal structure of anorthite at 1,430°C within the range $r \leqq 5$ Å, consisting of networks of SiO_4 and AlO_4 tetrahedra with Ca in the interstices.

4.3 *Refinement of the model with $S \cdot i(S)$ fitting*

We chose the crystal structure of anorthite at 1,430°C (SMITH, 1974) for

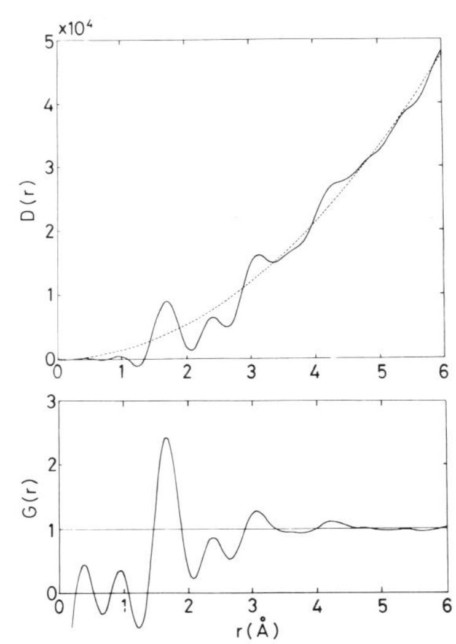

FIG. 6. The radial distribution function $D(r)$ and the correlation function $G(r)$ of an anorthite melt.

the starting model of the melt structure for the following reasons:

1. Peaks in $D(r)$ obtained above correspond well with the distribution of atomic pairs in the anorthite structure at 1,430°C.
2. Anorthite crystalizes relatively easily from the melt among feldspars.
3. The difference between densities of the crystal at 1,430°C and the melt is small.
4. TAYLOR and BROWN (1979) have shown that the anorthite glass gives a radial distribution function which can be explained by four-membered rings of TO_4 tetrahedra observed in the feldspar structure.

Anorthite has a density of 2.64 g cm^{-3} at 1,430°C, which is close to that of the anorthite melt calculated above (2.55 g cm^{-3}). Individual thermal parameters of atoms in the model were refined with the least-squares method by fixing other parameters to give a better agreement between the observed and calculated $S \cdot i(S)$ curves in the range $1.5 \leqq S \leqq 11.15$ Å^{-1}. The calculated $S \cdot i(S)$ curve of the final model is shown in Fig. 7 with the observed $S \cdot i(S)$ curve. The good agreement between two curves also supports the correctness of the structure model, confirming that the four-membered rings of TO_4 tetrahedra are the basic structural unit in an anorthite melt (Fig. 8). FLOOD and KNAPP (1968) reported that the existence of $4AlO_2 \cdot 3SiO_2^{4-}$ clusters can give a good explanation of the shape of the liquidus curve in the $CaAl_2Si_2O_8$–SiO_2 system. Although the $4AlO_2 \cdot 3SiO_2$ group is not the same as the group $Al_4Si_4O_{16}$ of two 4-membered rings joined by sharing two apices, their idea is consistent with the existence of the 4-membered ring. Since no peak is observed on the radial distribution curve beyond 5 Å, polymerization of the rings is supposed to occur in a fairly random way.

5. *Structure Determination of an Albite Melt with X-Ray Diffraction*

5.1 *Data collection*

For comparison of the structure to that of an anorthite melt, the

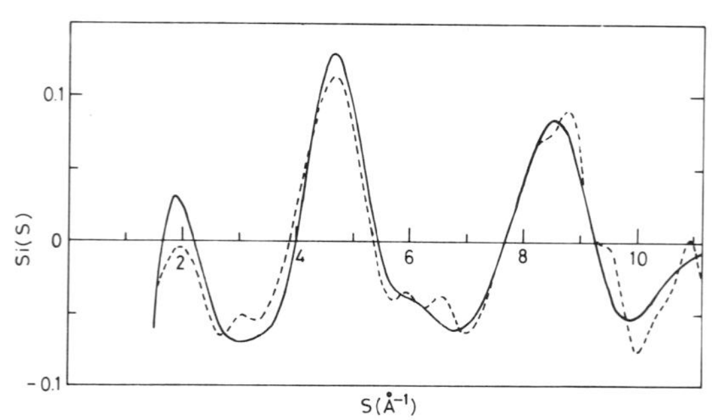

FIG. 7. The calculated and observed $S \cdot i(S)$ curves of an anorthite melt. The dashed line shows the observed one.

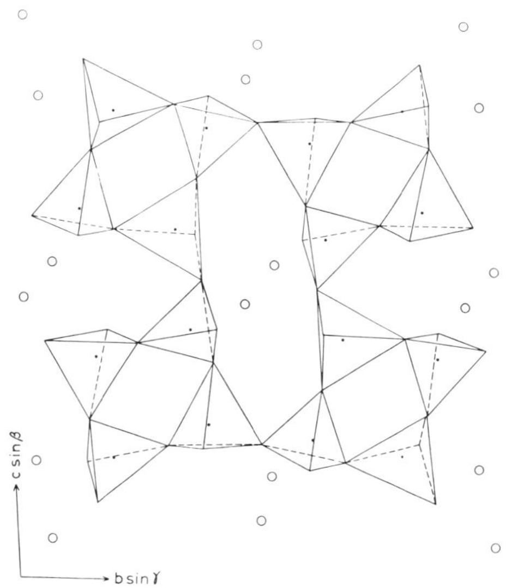

FIG. 8. The 4-membered rings in the anorthite crystal structure.

experiment of an albite melt was performed. $NaAlSi_3O_8$ glass was used for this purpose. The glass was prepared by melting a mixture of SiO_2, Al_2O_3, and Na_2CO_3 in the molar ratios 3:0.5:0.5 in a platinum crucible for 6 h at 1,450°C, and subsequently quenching at room temperature in the air. The weight of the material did not change on melting, except for the loss of CO_2. The glass was crushed into powder and used for the X-ray diffraction experiment. Scatterred intensities were measured on an X-ray diffractometer of the θ-θ type, which has a parafocusing geometry.

The specimen was placed in a platinum vessel and heated in an electric furnace (Pt/13%Rh, 0.7 mm ϕ coil heater) placed in an air-tight container having a window of Ni foil (5 μm in thickness). The temperature was measured with a Pt-Pt/13%Rh thermocouple and controlled to 1,200°C during the experiment.

Measurements were carried out with Mo $K\alpha$ radiation monochromated with a pyrolitic graphite mounted on the path of the diffracted beam (Fig. 1). Slit systems of 1/2°–1/2° and 1°–1° were employed in the lower ($3° \leqq \theta \leqq 15°$) and higher angle ($12° \leqq \theta \leqq 50°$) ranges, respectively. The θ-θ step scan technique was adopted with the fixed counting time of 40 s and the step interval of 0.2°. Scanning was carried out in the range $3° \leqq \theta \leqq 50°$, and repeated 4 times so as to make the minimum of the total counts of a datum point larger than 15,000.

The intensity data were corrected for polarization and absorption factors, and for Compton scattering by the method already described. Normalization of the data was performed with the Krogh-Moe-Norman method. The density was calculated from the partial molar volumes (BOTTINGA and WEILL, 1970) to be 2.32 g cm^{-3}, and used for calculation of the radial distribution function.

5.2 Radial distribution function

Figures 9 and 10 show the reduced intensity and $S \cdot i(S)$ curves of the albite melt, respectively. The $S \cdot i(S)$ curve shows clear maxima at $S = 1.65$, 4.90, 8.50, and 12.50 $Å^{-1}$. Figure 11 shows the radial distribution function $D(r)$ and the correlation function $G(r)$ calculated from $S \cdot i(S)$ in the range $S \leqq 13.50$ $Å^{-1}$. Peaks are observed at $r = 1.64$, 2.25, 3.14, and 4.26 Å. The reduced intensity curve and radial distribution function are in good agreements with those of a super-cooled liquid of $NaAlSi_3O_8$ at 805°C (TAYLOR *et al.*, 1980). The observed peaks in $D(r)$ up to about 4 Å, resemble the atomic pairs in the monalbite structure at 1,060°C (WINTER *et al.*, 1979), and were

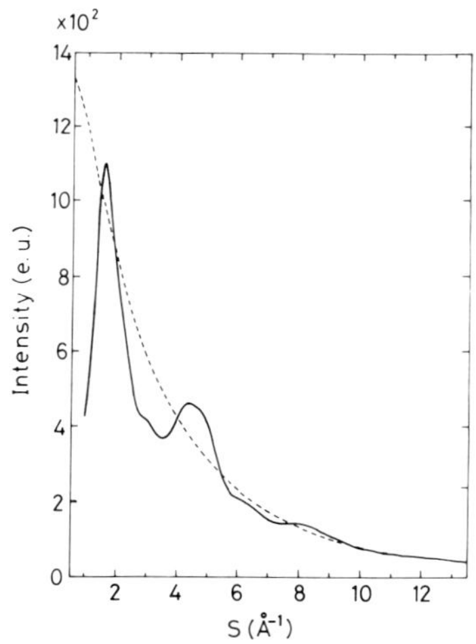

FIG. 9. The reduced intensity of an albite melt.

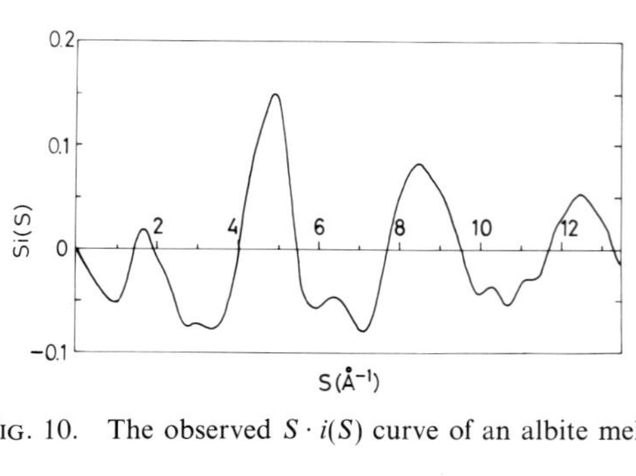

FIG. 10. The observed $S \cdot i(S)$ curve of an albite melt.

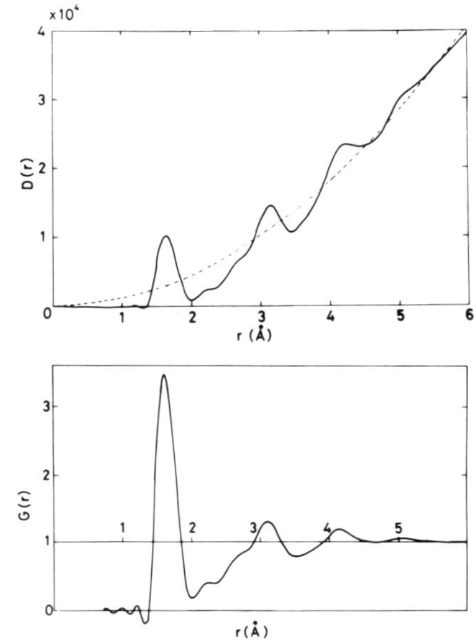

FIG. 11. The radial distribution function $D(r)$ and the correlation function $G(r)$ of an albite melt.

assigned as follows: The first peak at $r = 1.64$ Å can be assigned to T–O (T is the average atom of Si and Al) pairs, indicating that the T–O bond distances do not change on the phase transition from the crystalline to molten state. The number of oxygen atoms around the T atom is found to be 3.93 from the peak area of the first peak. The distance and coordination number confirm the existence of SiO_4 and AlO_4 tetrahedra in the albite melt. The second peak at $r = 2.25$ Å is supposed to be due to Na–O pairs, and a little shorter than those in the albite crystal. The third peak at $r = 3.14$ Å should be mainly due to the contribution of T–T pairs. The shoulder at $r = 2.66$ Å corresponds to O–O pairs in AlO_4 and SiO_4 tetrahedra. These facts suggest the structure of the albite melt is based on a framework of SiO_4 and AlO_4 tetrahedra, having Na^+ ions in its interstices. The fourth peak at $r = 4.28$ Å is supposedly due to the second neighbor of T–O pairs. No distinct correlations are observed in the larger r region beyond 4.3 Å.

5.3 Comparison of the albite melt structure to that of an anorthite melt

The positions of maxima in the $G(r)$ curves are in good agreement with the albite and anorthite melts up to 3.5 Å, suggesting that the two substances have similar structures in the short range below 3.5 Å. However, the $G(r)$ curves show a little difference in the larger r region, indicating a difference in the long range structure between these two melts. Actually, the bridging angle

(α) of neighboring tetrahedra in the albite melt is estimated to be 146.4° from the mean T–O and T–T distances according to the equation $\alpha = 2\text{arc sin}\,[(T-T)/2(T-\text{O})]$, which is much larger than the corresponding angle of 136.4° in the anorthite melt and 141.7° in the structure of monalbite at 1,060°C.

REFERENCES

BOTTINGA, Y. and WEILL, D. F. (1970): Densities of liquid silicate systems calculated from partial molar volumes of oxide components, *Am. J. Sci.*, **269**, 169–182.

DEBYE, P. and MENKE, H. (1930): The determination of the inner structure of liquids by X-ray means, *Physik. Z.*, **31**, 797–798.

DEBYE, P. and MENKE, H. (1931): Untersuchung der Molekularen Ordnung in Flussingkeiten mit Röntgenstrahlung, *Ergeb. Tech. Röntgenk*, **2**, 1–6.

FLOOD, H. and KNAPP, W. J. (1968): Structural characteristics of liquid mixtures of feldspar and silica, *J. Am. Ceram. Soc.*, **51**, 259–263.

HAJDU, F. (1972): Analytic appraximation for incoherent scattered X-ray intensities, *Acta Cryst.*, **A27**, 73–74.

International Tables for X-ray Crystallography, Vol. 2, 1972: Kynoch Press, Birmingham.

International Tables for X-ray Crystallography, Vol. 4, 1974: Kynoch Press, Birmingham.

ISSHIKI, N. (1959): Explanatory text of the geological maps of Japan, Miyake-jima, Geological Survey of Japan.

KROGH-MOE, J. (1956): A method for converting experimental X-ray intensities to an absolute scale, *Acta Cryst.*, **9**, 951–953.

LEVY, H. A., DANFORD, M. D. and NARTEN, A. H. (1966): Data collection and evaluation with an X-ray diffractometer designed for the study of liquid structure, *ORNL*-3960.

MALET, G., CABOS, C., ESCADE, A. and DELORD, P. (1973): Double diffraction des Rayons X par une substance amorphe, *J. Appl. Cryst.*, **6**, 139–144.

MIYATA, T., ISHIZAWA, N., MINATO, I. and IWAI, S. (1979): Gas-flame heating equipment providing temperatures up to 2,600 K for the four-circle diffractometer, *J. Appl. Cryst.*, **12**, 303–305.

NORMAN, N. (1957): A Fourier transform method for normalizing intensities, *Acta Cryst.*, **10**, 370–373.

PÁLINKÁS, G. (1973): Analytic approximation for the incoherent X-ray intensities of the atoms from Ca to Am. *Acta Cryst.*, **A29**, 10–12.

SMITH, J. V. (1974): Crystal structure and physical properties, in *Feldspar Minerals*, Vol. 1, Springer, Berlin, Heidelberg, New York.

TAYLOR, M. and BROWN, G. E. Jr. (1979): Structure of mineral glasses-I. The feldspar glasses $NaAlSi_3O_8$, $KAlSi_3O_8$, $CaAl_2Si_2O_8$, *Geochim. Cosmochim. Acta*, **43**, 61–75.

TAYLOR, M., BROWN, G. E. Jr. and FENN, P. M. (1980): Structure of mineral glasses-III. $NaAlSi_3O_8$ supercooled liquid at 805°C and the effects of thermal history, *Geochim. Cosmochim. Acta*, **44**, 109–117.

WARREN, B. E., KRUTTER, H. and MORNINGSTAR, O. (1936): Fourier analysis of X-ray patterns of vitreous SiO_2 and B_2O_3, *J. Am. Ceram. Soc.*, **19**, 202–206.

WINTER, J. K., OKAMURA, F. P. and GHOSE, S. (1979): A high-temperature structural study of high albite, monalbite, and the analbite→monalbite phase transition, *Am. Mineral.*, **64**, 409–423.

ZERNIKE, F. and PRINS, J. A. (1927): The bending of X-rays in liquids as an effect of molecular arrangement, *Z. Physik*, **41**, 184–194.

Materials Science of the Earth's Interior, edited by I. Sunagawa, pp. 39–60.

STRUCTURES AND SOME PHYSICAL PROPERTIES OF SILICATE MELTS OF GEOLOGICAL INTEREST

Ikuo KUSHIRO

Geological Institute, The University of Tokyo, Tokyo, Japan

1. *Introduction*

Experimental studies on igneous petrology have long been concerned mainly with the phase equilibrium relations of silicate systems under various different physico-chemical conditions. Recently, the importance of the structures and physical properties of silicate melts of geological interest has been emphasized more than ever before for the understanding of magmatic processes such as generation, ascent and eruption of magmas. In consequence, experimental studies related to these problems have been conducted extensively during the last decade. In our three-year project on the 'material sciences of the earth's interior,' we have carried out experimental studies on the physical properties and kinetics of silicate melts of geological interest, mainly at high pressures. Our experimental studies concern changes in viscosity and density with the pressure of some synthetic silicate melts, as well as natural basaltic and andesitic melts, diffusion of alkali earth elements in some silicate melts at high pressures, changes of the liquid immiscibility region in the system Fe_2SiO_4–$KAlSi_2O_6$–SiO_2 with pressures, changes of the liquidus phase or the shift of the liquidus boundaries with pressure in some silicate systems, vesiculation of hydrous andesite melts upon decompression and vapor transport of alkalies, vaporization of Na from chondritic melts, and textural reproduction of chondrules. Some of these studies have been completed but others are still being continued. The studies of the structures of silicate melts have been conducted in collaboration with a group at the Geophysical Laboratory of the Carnegie Institution of Washington. In the present article, the results of the above experimental studies, relating to the structures and physical properties of silicate melts, are summarized.

2. *Structures of Silicate Melts in the System CaO–MgO–SiO_2*

The system CaO–MgO–SiO_2 includes the petrologically important minerals olivine and pyroxene, and the crystallization of these minerals from melts in this system has been considered useful for understanding the crystallization process of natural basaltic magmas (BOWEN, 1914, 1928). It is important, therefore, to determine the structure of the melts in this system and its relationship to the phases crystallizing from them. Recently, we have studied the anionic structures of the quenched melts (glasses) in a portion of this system using the Raman spectroscopic technique (VIRGO *et al.*, 1980). During the three-year project, the structures of melts of compositions Mg_2SiO_4, $MgSiO_3$, and $CaSiO_3$ have also been obtained by molecular dynamics simulations, as shown in the preceding article (MATSUI and KAWAMURA, 1982). It is worthwhile comparing the structures of the melts obtained by the two different methods.

Figure 1 shows the compounds stable at 1 atm in the system $CaO-MgO-SiO_2$. These silicate minerals are classified into four groups based on the degree of polymerization of $(SiO_4)^{4-}$ tetrahedra; that is, they are nesosilicates (Ca_2SiO_4, $Ca_3MgSi_2O_8$, $CaMgSiO_4$, and Mg_2SiO_4), sorosil-

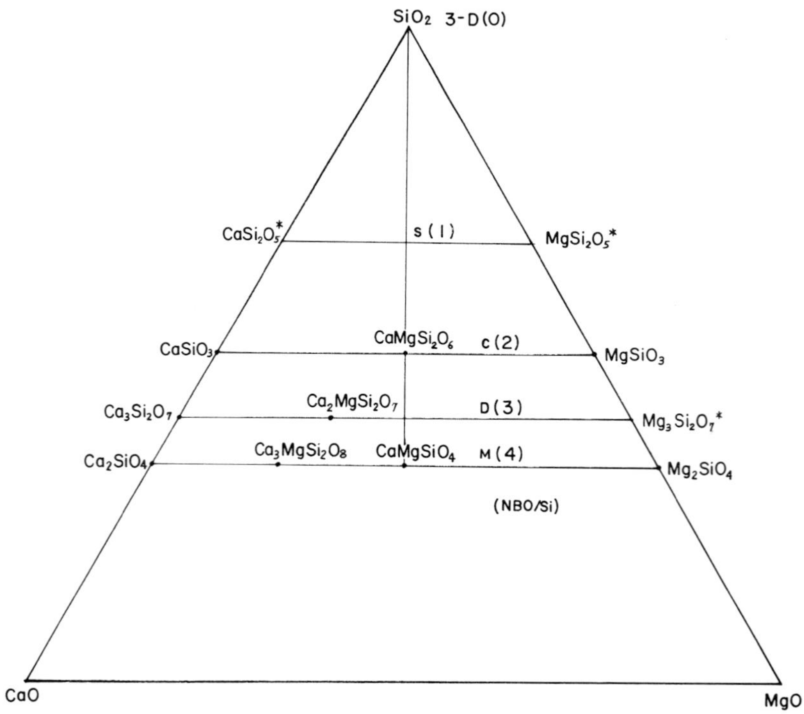

FIG. 1. Compounds in a portion of the system $CaO-MgO-SiO_2$. Those with stars are hypothetical compounds. Abbreviations: M, monomer; D, dimer; C, chain; S, sheet; 3-D, 3 dimensional network; NBO, number of non-bridging oxygens.

icates ($Ca_3Si_2O_7$ and $Ca_2MgSi_2O_7$), inosilicates ($CaSiO_3$, $CaMgSi_2O_6$, and $MgSiO_3$), and tectosilicate (SiO_2). The amount of non-bridging oxygen per Si is 4 for nesosilicate, 3 for sorosilicates, 2 for inosilicates, and 0 for tectosilicates. Hypothetical phyllosilicates or sheet silicates $CaSi_2O_5$ and $MgSi_2O_5$, which have 1 non-bridging oxygen per Si, can be considered as similar to $Na_2Si_2O_5$ in the join Na_2O–SiO_2; the melts in this ternary system may also consist of these structural units. The structural units in the melts, which have non-bridging O/Si values 4, 3, 2, 1, and 0, are called monomer (M), dimer (D), chain (C), sheet (S), and 3-dimensional network (3-D), respectively. The structural units (structons) are useful for describing the structures of silicate melts.

The Raman spectra have been obtained for quenched melts (glasses) of compositions between CaO 56.5 SiO_2 43.5 wt.% (WL50) and CaO 31.9 SiO_2 68.1% ($CaSi_2O_5$) on the join CaO–SiO_2. Figure 2 shows the vibrational spectra of the Si–O stretch mode of the Raman spectra for the above glasses. Assignment of the frequencies of these spectra to the structural units with different degrees of polymerization of SiO_4 tetrahedra is based on the frequencies for silicate minerals of known structures and those predicted theoretically (e.g., BRAWER and WHITE, 1975; VERWEIJ, 1979; LAZAREL, 1964; BRAWER, 1975; SEN and THORPE, 1977). As shown in Fig. 2, the glass of the composition WL50 has vibrational spectra for M, D, and C. Between WL50 and Wo, the spectrum for C becomes more intense with increasing SiO_2 in the glass, while those for M and D become weaker. Between Wo and $CaSi_2O_5$, the spectrum for S becomes stronger, whereas those for C and M become weaker. Figure 3 shows the approximate ranges of five different structons in the melts along the join CaO–SiO_2 estimated from the spectra of Fig. 2. It is interesting that at least three different structons always coexist together in each glass. This is fundamentally different from the structures of common silicate minerals which usually consist of one structural unit. An exception has recently been found in a biopyribole, which has three different structural units.

Viscosity along the join CaO–SiO_2 increases with SiO_2 (BOCKRIS *et al.*, 1955). Such viscosity increase would correspond to the increase of more polymerized units with increasing SiO_2.

Liquid immiscibility exists in the silica-rich portion of this join (GREIG, 1927). In this region, two liquids coexist together; one is enriched in 3-D and the other consists mainly of S and C with small amounts of M and 3-D.

The structure of a $CaSiO_3$ melt obtained by the molecular dynamics calculations (unpublished data; See also MATSUI and KAWAMURA, 1982), comprises M (3), D (16), C (24), S (20), and 3-D (2), where the numbers in parentheses indicate the number of each structons in the cell of 320 atoms. These numbers seem to be consistent with the results obtained by the Raman spectra, except that the proportion of M appears to be larger than that of D in

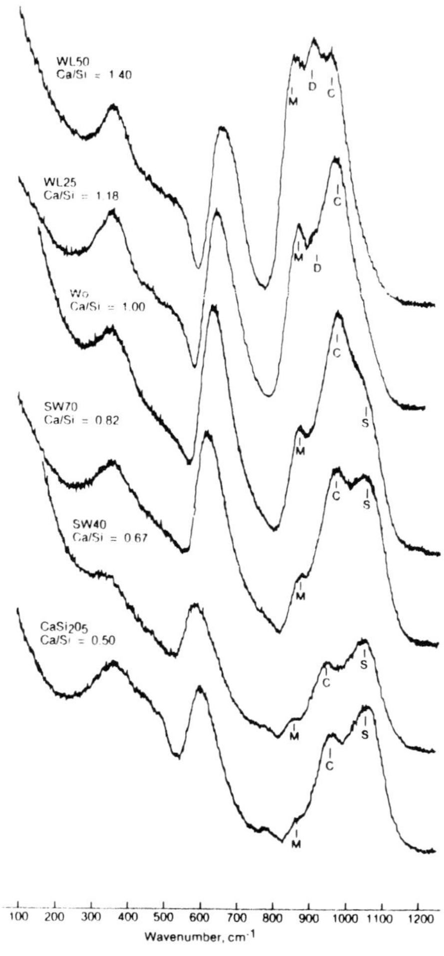

FIG. 2. Raman spectra for quenched melt (glass) along the join $CaO-SiO_2$ (VIRGO *et al.*, 1980). WL50 and 25; $CaSiO_3$ 50, Ca_2SiO_4 50 and 75, 25 respectively; Wo, wollastonite; SW 40, $CaSi_2O_5$ 40, $CaSiO_3$ 60 (wt.%).

the Raman spectra. Better agreement may be expected if the number of atoms is increased in the molecular dynamics calculations.

The Raman spectra for the glasses quenched from melts along the join $CaMgSiO_4-SiO_2$ have also been obtained. The spectra for the glass of composition close to $CaMgSiO_4$ (DM95) shows a strong peak of M, but those of D and C also appear. Presence of D and C suggests that there must be some free oxygens in this glass. The intensity of M decreases with increasing SiO_2, whereas those of D and C increase, similar to the join $CaO-SiO_2$. Near $CaMgSi_2O_6$, the spectrum of S appears and increases with increasing SiO_2.

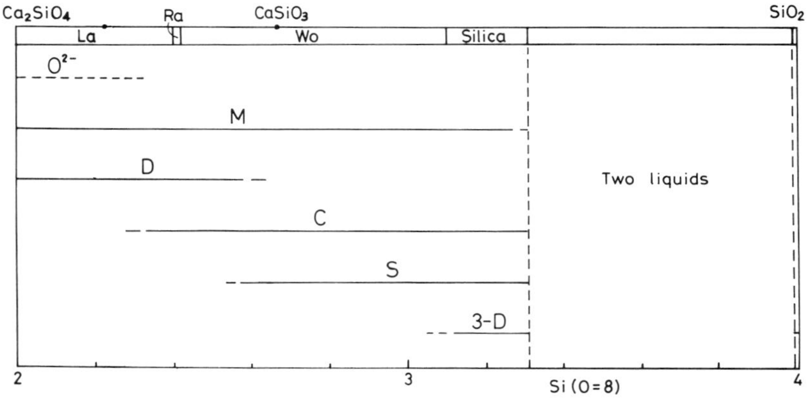

FIG. 3. Approximate ranges of the five different structons M, D, C, S, and 3-D. Abbreviations are the same as those in Fig. 1.

Separation of the spectra is, however, not conspicuous compared with those in the join $CaO-SiO_2$. Liquid immiscibility also appears in the silica rich portion of the join, where 3-D rich melt and a melt rich in S and C coexist together. The probable ranges of the five different structons in melts on the join $CaMgSiO_4-SiO_2$ are similar to those shown in Fig. 3.

YOKOKAWA and NIWA (1969) calculated the activities of SiO_2 and MO in the melts of the join SiO_2-MO, based on the statistical thermodynamics of these binary silicate systems developed by TOOP and SAMIS (1962). In a hypothetical reaction $2O_1 = O_0 + O_2$, where O_0, O_1, and O_2 are bridging, nonbridging, and free oxygens, respectively, equilibrium constant K is given as n_0n_2/n_1^2, where n_0, n_1, and n_2 are numbers of O_0, O_1, and O_2, respectively. They derived a relationship between activities of oxides (SiO_2 and MO) and equilibrium constant K. Using this relation, they calculated the activities of SiO_2 and MO for the joins SiO_2-CaO and SiO_2-PbO and compared them with the activity values actually measured. It was demonstrated that the calculated and measured activities are close together, when $K = e^{-6}$ for the join $CaO-SiO_2$. Based on this K value (e^{-6}), NAKAMURA (1978) calculated the relative proportions of O_0, O_1, and O_2 as a function of mole fraction of SiO_2 or MO (Fig. 4). The figure indicates that the melts between Ca_2SiO_4 and SiO_2 consist mainly of O_0 and O_1, and their proportions change almost linearly. It should be noted that free oxygen exists in melts between Ca_2SiO_4 and $CaSiO_3$, although its amount is small. The O_0/O_1 proportion for $CaSiO_3$ melt is close to 1/2. It is noted that in the structure of $CaSiO_3$ melt obtained by the molecular dynamics calculations, O_0/O_1 ratio is about 1/2, which is in good agreement with that obtained by YOKOKAWA and NIWA (1969) and NAKAMURA (1978).

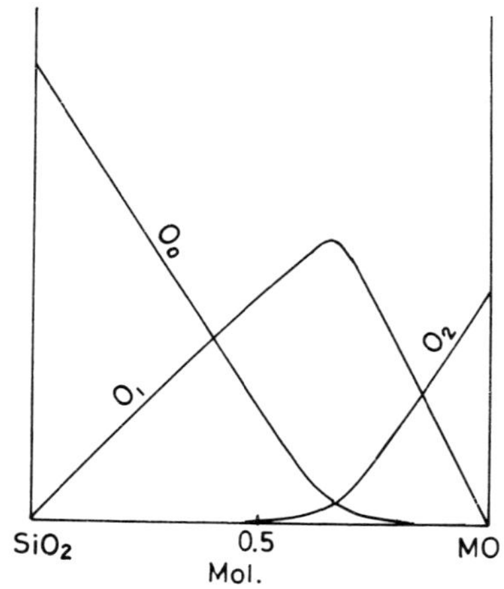

FIG. 4. Calculated ratios of bridging oxygen (O_0), non-bridging oxygen (O_1) and free oxygen (O_2) in the melts along the join $MO-SiO_2$ for partition coefficient $K = e^{-6}$, based on YOKOKAWA and NIWA (1969) and NAKAMURA (1978).

3. Relationships Between Structures of Melts and Liquidus Phases

There may be close relationships between the structures of melts and silicate minerals crystallizing from them. It is expected that silicate minerals whose structures are present in melts in appreciable amounts would crystallize. The studies of the melts in the system $CaO-MgO-SiO_2$, described in the preceding section, give some information on this problem. In the join $CaO-SiO_2$, larnite crystallizes from melts enriched in M and wollastonite crystallizes from melts enriched in C; however, the structure of silicate minerals and the most abundant structon in the melts do not exactly correlate. For example, rankinite crystallizes from melts in which the amount of D is already much reduced, and wollastonite crystallizes from melts in which the relative proportion of C/S is also much reduced. In the join $CaMgSiO_4-SiO_2$, monticellite and forsterite crystallize from melts enriched in M, and diopside crystallizes from melts enriched in C; however, åkermanite does not crystallize in melts enriched in D. In both the joins, silicate minerals with S do not crystallize, although the melts are enriched in S for certain ranges of compositions.

At high pressures, both the crystallizing phases and the compositional ranges of crystallization of minerals change in this ternary system. For example, monticellite breaks down to forsterite and merwinite at pressures above 9 kbar, and åkermanite breaks down to diopside and merwinite at pressures above 15 kbar. Also, the compositional ranges of crystallization of pyroxene (diopside and enstatite) expand relative to that of forsterite. It may be expected, therefore, that the relative proportion of M/C decreases at high pressures. The Raman spectra of the glass of diopside composition, quenched at 20 kbar, which has a higher density than the 1 atm glass, is, however, identical to that quenched at 1 atm (MYSEN *et al.*, 1980). The glass of åkermanite composition, quenched at high pressures, also shows the Raman spectra identical to those at 1 atm. These observations may indicate that the changes of the liquidus relations with pressure are much more sensitive than the structural changes of melts detected by the Raman spectra, or the structures of melts are not preserved during the quenching in high-pressure runs. The latter possibility is not likely, as will be mentioned in the next section. At 1 atm the rate of quenching is much faster and, as mentioned before, the structures of melt and glass are essentially the same (SHARMA *et al.*, 1978).

The most important change in the liquidus phase relations in this system with pressure would be the shift of the forsterite-enstatite liquidus boundaries or the expansion of the liquidus field of enstatite relative to that of forsterite with increasing pressure. Such a shift of the liquidus boundaries might be related to the structural change of the melts with pressure. During the present studies, the shift of the olivine-orthopyroxene liquidus boundary has been examined in the system $CaO-MgO-SiO_2$, as well as several other systems. KUSHIRO (1975) demonstrated that at 1 atm, this boundary shifts toward silica-rich compositions with the addition of oxides of monovalent cations, and toward silica poor compositions with the addition of oxides of polyvalent cations. For petrological problems, silicate components in major rock-forming minerals, rather than oxide components, are more useful. Figure 5 summarizes the shifting of the forsterite-protoenstatite liquidus boundary in the systems $Mg_2SiO_4-SiO_2-X$, where X represents $KAlSiO_4$, (kalsilite), $NaAlSiO_4$ (nepheline), $CaAl_2Si_2O_8$ (anorthite), $MgAl_2Si_2O_8$ and $CaMgSi_2O_6$ (diopside). The liquidus boundary shifts toward silica most strongly with the addition of the kalsilite component among these five different systems. The order of the shift of the liquidus boundary with these four Al-bearing components is the same as that for single oxides (K_2O, Na_2O, CaO, and MgO). Such shift of the olivine-pyroxene liquidus boundary was explained by the depolymerization of the melts by the addition of oxides of monovaliant elements (KUSHIRO, 1975).

The effect of pressure on the shift of the liquidus boundary in these five different systems is shown in Fig. 6. The degree of the shift is compared at 0.2 mole of each component (points a, b, c, d, and e in Fig. 5). It is evident from the figure that the forsterite-enstatite (protoenstatite at low pressures) liquidus boundary shifts away from silica with increasing pressure in all the systems examined. Another interesting observation is that boundaries located in a more silica-rich region at 1 atm, shift away from silica at a greater rate. Thus, the degree of shift is largest in the system $Mg_2SiO_4-KAlSiO_4-SiO_2$ and smallest in the system $Mg_2SiO_4-CaMgSi_2O_6-SiO_2$.

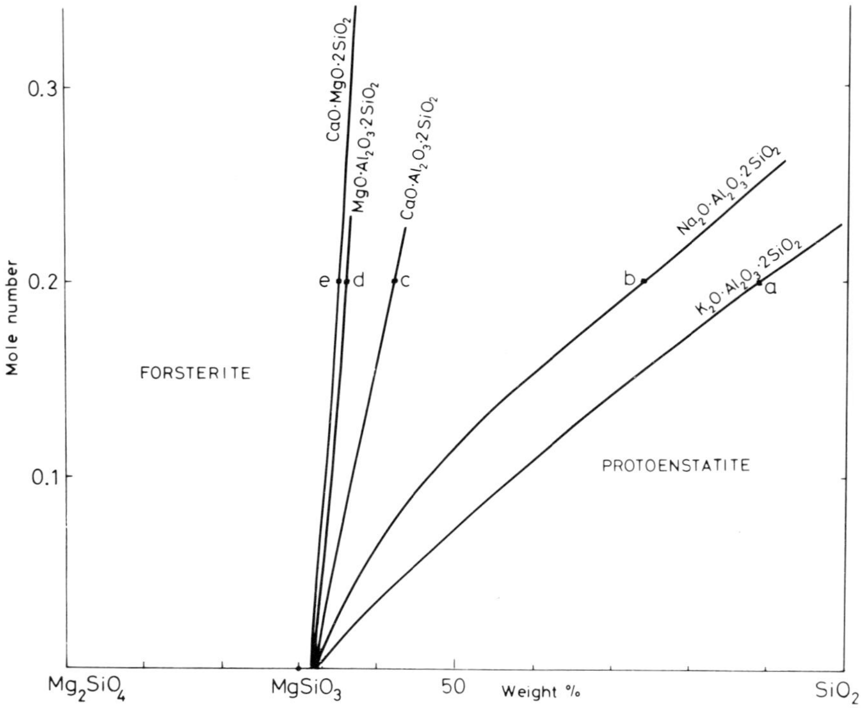

FIG. 5. Shift of the forsterite-protoenstatite liquidus boundary at 1 atm with the addition of $K_2O \cdot Al_2O_3 \cdot 2SiO_2$ (kalsilite), $Na_2O \cdot Al_2O_3 \cdot 2SiO_2$ (nepheline), $CaO \cdot Al_2O_3 \cdot 2SiO_2$ (anorthite), $MgO \cdot Al_2O_3 \cdot 2SiO_2$, and $CaO \cdot MgO \cdot 2SiO_2$ (diopside) components. Based on the data by SCHAIRER (1957), SCHAIRER and YODER (1961), ANDERSON (1915), RANKIN and MERWIN (1918), KEITH and SCHAIRER (1952), BOWEN (1914), and KUSHIRO (1972).

In a paper by KUSHIRO (1975), it was suggested that the silicate melts coexisting with forsterite and enstatite might be more polymerized with increasing pressure, so that the liquidus field of enstatite, which is more polymerized than olivine, might expand relative to that of forsterite. This possibility has been examined using the Raman spectroscopic technique. The glasses quenched from super-liquidus temperatures at high pressures in the systems CaO–MgO–SiO_2 and MgO–Al_2O_3–SiO_2 were examined (MYSEN *et al.*, 1980; Seifert *et al.*, unpublished data). The results indicate that both the frequencies and the intensities of the Raman spectra do not change significantly with increasing pressure of quenching; the polymerized structural units (C, S, and 3-D) do not increase relative to the depolymerized units (M). Although the data are limited at present, it is unlikely that the melts in these systems are more polymerized with increasing pressure, and that the shift of the forsterite-enstatite liquidus boundary is caused by the polymer-

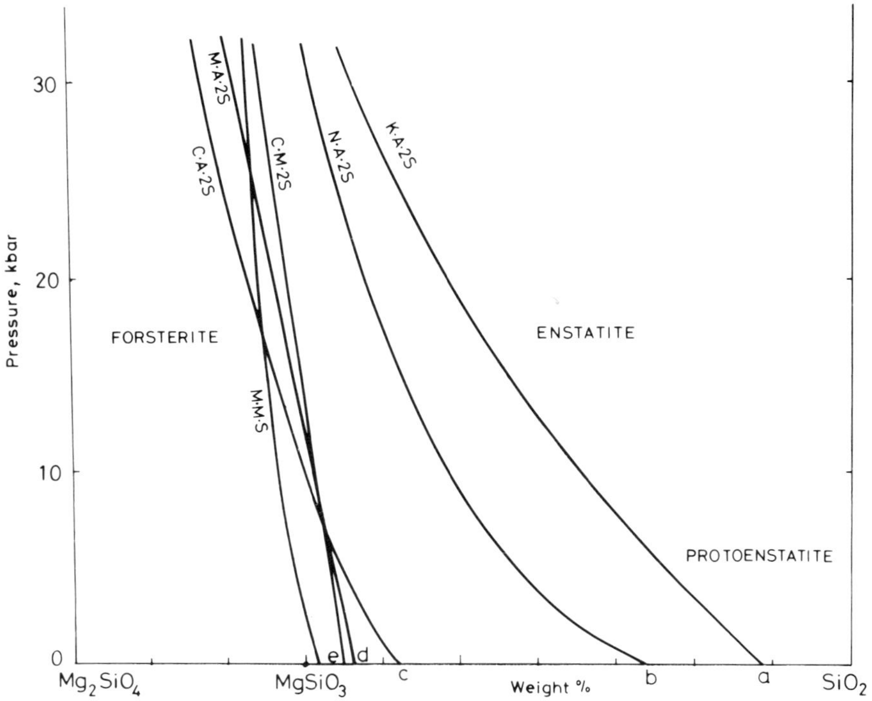

FIG. 6. Shift of the forsterite-enstatite (protoenstatite at low pressures) liquidus boundary (at a, b, c, d, and e of Fig. 5) with pressure in the systems Mg_2SiO_4–SiO_2–X, where X represents the components in Fig. 5. Based on the data by KUSHIRO (1968, 1980), MCGREGOR (1964), KUSHIRO (1969) and CHEN and PRESNALL (1975).

ization of the SiO_4 tetrahedra in the melts.

The shift of this boundary is probably caused by the volume change of the crystal + melt assemblage under pressure. In the system $NaAlSiO_4$–Mg_2SiO_4–SiO_2, the density increases from forsterite + liquid assemblage to enstatite + liquid assemblage (KUSHIRO, 1980). The reactions involving liquid are essentially the same as the solid reactions, and a crystal + liquid assemblage with smaller volume would be stable at higher pressures. If the enstatite + liquid assemblage is stable relative to the forsterite + liquid assemblage, the enstatite field would expand relative to the forsterite field and, consequently, the forsterite-enstatite liquidus boundary shifts away from silica.

NICHOLLS *et al.* (1971) calculated the free energy of the reaction $Mg_2SiO_{4(S)} + SiO_{2(L)} = 2MgSiO_{3(S)}$ as a function of pressure and temperature and demonstrated that silica activity in melts coexisting with forsterite and enstatite decreases with increasing pressure. The decrease of silica activity in melts is due to the fact that the molar volume of enstatite is smaller than that of the forsterite + silica (melt) assemblage. The present discussion on the shift of the forsterite-enstatite liquidus boundary is essentially based on the same calculation. The fact that the forsterite – enstatite liquidus boundary shifts with a greater rate in the $KAlSiO_4$–Mg_2SiO_4–SiO_2 than those in other systems at pressures higher than 10 kbar, may be due to the larger compression of K_2O–containing melts at higher pressures.

It is suggested that the changes of the liquidus phase relations with pressure are controlled more strongly by the total volume of crystal melt assemblage than the structural changes of the melts.

4. *Structural Changes of Some 3-D Silicate Melts under Pressure*

The structural changes of silicate melts under pressure have been suggested from the viscosity measurements of some aluminosilicate melts. The viscosities of melts of several different simple compositions were measured at high pressures (e.g., Kushiro, 1976, 1977, 1978a, b; Scarfe *et al.*, 1979). It is shown that GeO_2, $NaAlSi_2O_6$, $NaAlSi_3O_8$, $K_2MgSi_5O_{12}$ and $Na_2Si_3O_7$ melts show a viscosity decrease with increasing pressure at a constant temperature, whereas $CaMgSi_2O_6$, Na_2SiO_3, and $Na_2Si_2O_5$ melts show a viscosity increase. Among these melts, those containing Al, particularly a $NaAlSi_2O_6$ melt, show a relatively large viscosity decrease. From this evidence and the shift of the wavelength of AlK_α and K_β radiations with pressure, it was suggested that the change of coordination of Al is responsible for the structural change and viscosity decrease (Kushiro, 1976; Velde and Kushiro, 1978). If part of Al shifts from 4-fold to 6-fold coordination, non-bridging oxygen must be formed and, in consequence, the spectra of S and C must appear. However, the Raman spectra of the $NaAlSi_2O_6$ glass quenched at 1 atm and at high pressures (up to 38 kbar) do not show any significant differences; both have 3-D structures, although changes of polarization and a slight shift of some bands were observed (Sharma *et al.*, 1979; Mysen *et al.*, 1980). Jadeite (C) crystallizes from $NaAlSi_2O_6$ melt at pressures higher than about 30 kbar (Bell and Roseboom, 1969). As far as the results of the Raman spectra are concerned, therefore, the structures of the melts and mineral crystallizing from it are not directly related, as mentioned before in the system CaO–MgO–SiO_2 at high pressures.

All the melts that show a viscosity decrease with increasing pressure are highly polymerized ($NBO/T < 1$), whereas those showing a viscosity increase are less polymerized ($NBO/T > 1$) (Scarfe *et al.*, 1979; Mysen *et al.*, 1980). From these facts, it is suggested that the viscosity decrease with increasing pressure may be due to the collapse of the 3-D around the metal cations. The collapse of the network would result in the weakening of the (Si, Al)–O bonds, and, thereby, decreasing viscosity. Seifert (personal comm., 1980) calculated the T–O–T angle (T: tetrahedrally coordinated cation) of the $NaAlSi_2O_6$ glass quenched at different pressures and found that the T–O–T angle decreases with increasing pressure of quenching. This suggests that the ring size of the 3-D structure decreases with increasing pressure. For example, 6-membered rings change to 4-membered rings with increasing pressure, such as the transformation from a cristobalite-type structure to a quartz or coesite type structure. These results also suggest that the structures of glass quenched in the high-pressure runs probably repressent those of melts under high-pressure conditions.

A relatively large viscosity decrease was also observed for the GeO_2 melt (KUSHIRO, 1978b). The Raman spectra of the GeO_2 glass quenched at 1 atm and at high pressure (up to 15 kbar) are again the same (SHARMA *et al.*, 1978), although the T–O–T angles in these glasses have not been measured. Weakening of the GeO_2 bond with increasing pressure may also be expected, as in the alumino-silicate melts.

It is not certain, however, whether the change in the ring size can explain the shift of the wavelength of AlK_α and K_β radiations. Recently, SEIFERT *et al.* (1981) studied the join SiO_2–$CaAl_2O_4$, using the Raman spectroscopic technique. The proportion of 6-membered ring/4-membered ring decreases with the increasing $CaAl_2O_4$ component in the melt. The preliminary measurements of AlK_α of the glass of $CaAl_2Si_2O_8$ and $CaAl_2SiO_6$ compositions, both of which have 3-D structure at 1 atm, show that the wave length of AlK_α is smaller for the latter than the former (Kushiro and Fujii, unpublished data). These results may be consistent with the suggestion that the decrease in the ring size of the 3-D melt with increasing pressure is related to the shift of the wavelength of AlK_α. The viscosity of melts along the join SiO_2–$CaAl_2O_4$ at 1 atm decreases with increasing $CaAl_2O_4$ component, as well as the activation energy for viscous flow (ROSSIN *et al.*, 1964), similar to the changes of these properties of $NaAlSi_2O_6$ melt with increasing pressure.

The structural change of albite melt with pressure is also indicated by the changes of viscosity and density (KUSHIRO, 1978c) and of diffusivity of alkali earth elements with pressure. Fujii (unpublished data) measured the diffusion rates of Ca and Sr in albite melt at high pressures. Albite glass doped with 2 wt.% CaO and that doped with 2% SrO were placed in contact with each other with a flat plane in a graphite capsule. The charges were kept at 1,400°C (above the liquidus temperature) in the pressure range between 5 and 20 kbar under anhydrous conditions. The furnace assemblies are identical to those used for the viscosity measurements of jadeite and albite melts (KUSHIRO, 1976, 1978c). After the runs, the charges were polished and analyzed with the electron microprobe analyzer. Diffusion patterns of Ca and Sr were obtained. Figure 7 shows the patterns of Ca and Sr for 2 and 4 hrs against $1/\sqrt{t}$, where t is the time of the run. Evidently, counter-diffusion of Ca and Sr took place during the runs. As shown in Fig. 8 the patterns obtained by the 2-and 4-hr

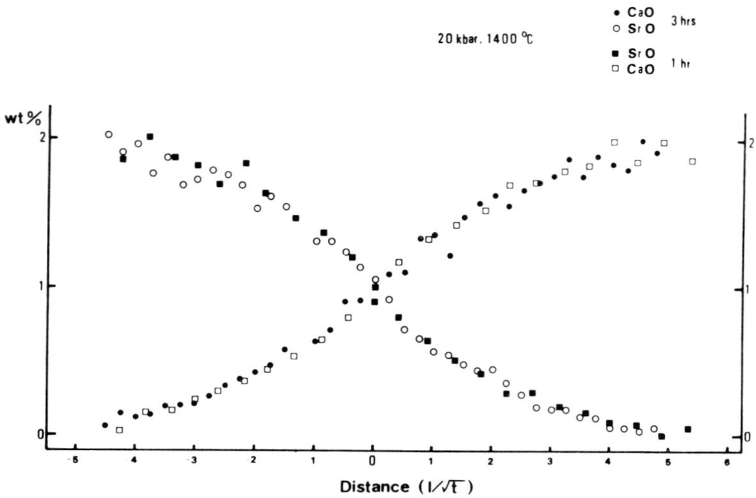

FIG. 7. Diffusion (counter diffusion) pattern of Ca and Sr in albite melt at 1,400°C at 20 kbar under anhydrous conditions (T. Fujii, unpublished data). Each analysis is plotted against $1/\sqrt{t}$, where t is time of the runs.

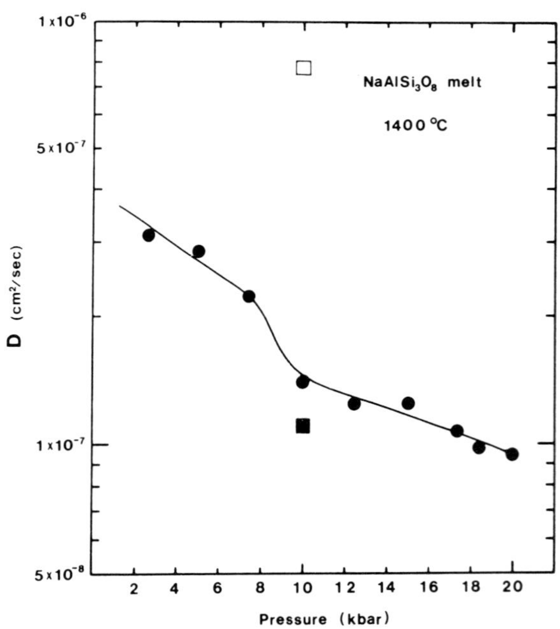

FIG. 8. Diffusion (counter diffusion) coefficient for Ca and Sr′ in albite melt at 1,400°C in the pressure range 2–20 kbar (T. Fujii, unpublished data). Open and solid squares are diffusion coefficients for Ca and Sr in hydrous (4 wt.% H_2O) and anhydrous albite melts, respectively at 1,300°C and 10 kbar.

runs normalized by $1/\sqrt{t}$ are nearly identical, indicating that the diffusion rate was constant during these runs. Analyses at several different parts of the charges indicate that the diffusion patterns are the same along different vertical sections. It means that convection did not occur during the runs. From these diffusion patterns, the diffusion constant was obtained by applying the Fick's law. Figure 8 shows the diffusion coefficient in the pressure range between 5 and 20 kbar. The diffusion coefficient decreases from 3×10^{-7} at 2 kbar to 9.5×10^{-8} cm^2/sec at 20 kbar. The rate of decrease is not constant but is larger between 8 and 10 kbar. This pattern is very similar to those of viscosity and density changes of albite melt and its quenched glass. The pressure range for the rapid change of the diffusion rate is, however, lower than those for the viscosity decrease and density increase of the albite melt. This may be partly due to the change in composition by the addition of 2 wt% of CaO or SrO to albite composition.

The change of the diffusion rate with increasing pressure would be related to the structural change of the melt. As discussed above, the structural change of albite and jadeite melts with increasing pressure may be due to the reduction of the ring size of the 3-D network (e.g., 6-membered ring → 4-membered ring). It is likely that the rate of diffusion of large cations (Ca^{2+} and Sr^{2+}) through the 3-D network, becomes smaller with the reduction of ring size of the network in the melt.

WATSON (1979) measured the diffusivity of Ca in a melt of composition Na_2O 29, CaO 10, Al_2O_3 10, SiO_2 56 wt.% at pressures between 1 atm and 30 kbar and in the temperature range between 1,100 and 1,400°C, using ^{45}Ca as a tracer. The results show that the diffusivity decreases with increasing pressure at constant temperature. For example, the diffusion coefficient decreases from $9.48 \pm 1.61 \times 10^{-7}$ at 1 kbar to $3.55 \pm 0.60 \times 10^{-7}$ cm^2/sec at 20 kbar at 1,200°C. Although the rate of decrease with pressure is smaller at higher temperatures, the results obtained by Fujii in the present project are consistent with those obtained by WATSON (1979).

The counter diffution of Ca and Sr has also been measured in a hydrous albite melt with 4 wt.% H_2O at 10 kbar and at 1,300°C. The diffusion coefficient under this condition is about 8×10^{-6} cm^2/sec, which is an order of magnitude larger than those obtained under anhydrous conditions at same pressure and temperature (Fig. 8). The addition of H_2O would significantly depolymerize the melt, so that the diffusivity becomes much greater.

5. *Possible Coordination Change of Al in Silicate Melts at High Pressures*

A coordination change of Al may take place at higher pressures. If Al in the melt changes from 4-fold to 6-fold coordination, the viscosity would decrease as suggested by WAFF (1975), because of the increase of non-bridging oxygens in the melt, thereby depolymerizing the melt. The density of the melt would also increase because of the reduction of the molar volume, due to the increase of 6-fold coordinated Al, as commonly observed in the Al-bearing silicate minerals (e.g., jadeite, garnet, and kyanite). Such a coordination change of Al in silicate melts may take place at pressures greater than 50 kbar. Recently, OHTANI *et al.* (1981) determined the melting curve of pyrope ($Mg_3Al_2Si_3O_{12}$) up to about 100 kbar. The melting curve shows a distinct inflection (the curve becomes flattened) at about 80 kbar. This may be related to the shift of Al from 4-fold to 6-fold coordination with an increase of density of the melt. In addition, the quench crystals formed in the pyrope melt are pyroxene up to about 80 kbar, but are garnet above this pressure, suggesting the change of structure of the melt due to the coordination change of Al. A hypothetical diagram illustrating the changes of viscosity of Al-bearing silicate melts enriched in 3-D with pressure, is shown in Fig. 9.

Silicon in the silicate melts may also change from 4-fold to 6-fold coordination at very high pressures, as is observed in crystalline silicates (e.g., coesite → stishovite, enstatite → $MgSiO_3$ ilmenite, pyroxene → garnet) (AKIMOTO and SYONO, 1969; KAWAI *et al.*, 1974; RINGWOOD, 1967). The molecular dynamics calculations of $MgSiO_3$ melt demonstrate that part of Si shifts from 4-fold to 6-fold coordination when the volume of the melt is reduced by 33% (MATSUI and KAWAMURA, 1980). Such transformation would take place at pressures greater than 100 kbar.

The Raman spectra for the GeO_2 glass quenched at pressures up to 15 kbar are still those of 3-D structure (SHARMA *et al.*, 1978). The crystalline GeO_2 transforms from quartz structure to rutile structure at about 3 kbar at

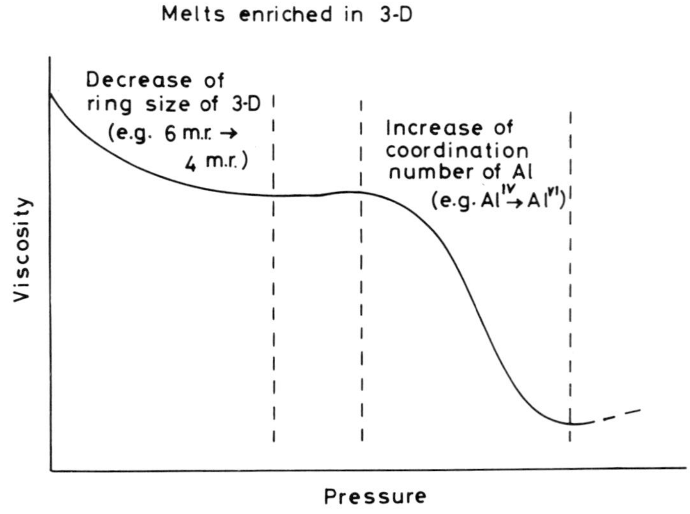

FIG. 9. A hypothetical diagram illustrating the viscosity change of melts enriched in 3-D structural unit with pressure.

the solidus. The transformation of Ge from 4-fold to 6-fold coordination in the melt would take place at pressures much higher than that for the transformation of the crystalline GeO_2. The SiO_2 melt may also be similar to the GeO_2 melt; the transformation of Si from 4-fold to 6-fold coordination may take place at pressures greater than that for the coesite → stishovite transformation.

6. *Viscosity and Density Changes of Natural Basaltic and Andesitic Melts*

A number of measurements have been made of the viscosities of silicate melts of geological interest for the purpose of understanding the behavior of magma and lava. The measurements were made for both lava flows in the field and for natural rock melts in laboratories (e.g., DOELTER, 1902; PALMER, 1927; MINAKAMI, 1951; KANI, 1934a, b; SHAW *et al.*, 1968; MURASE and MCBIRNEY, 1973). These measurements were made at 1 atm.

Measurements of viscosity of silicate melts of natural rock compositions under pressure have been relatively few because of the difficulty of the measurements. Most of them were made on hydrous granitic or rhyolitic melts at pressures of up to 7.5 kbar (SABATIER, 1956; FRIEDMAN *et al.*, 1963, SHAW, 1963; BURNHAM, 1963). SCARFE (1973) measured viscosities of hydrous basaltic melts at 1 kbar. In these measurements, which were made at relatively low pressures, the effect of pressure on the viscosity of silicate melts was not very clear. Since 1976, viscosity of silicate melts of natural rock compositions have been measured at pressures above 10 kbar and the effect of pressure has been discussed. KHITAROV *et al.* (1976) measured the viscosity of a basaltic melt up to 30 kbar and showed that the viscosity increases with increasing pressure at constant temperature. KUSHIRO *et al.* (1976) measured viscosities of melts of Kilauea 1921 olivine tholeiite and Crater Lake andesite at pressures of up to 30 kbar under anhydrous conditions. They showed that the viscosities decrease with increasing pressure at constant temperature (Fig. 10). FUJII and KUSHIRO (1977) also showed that the viscosity of melt of abyssal tholeiite decreases with increasing pressure at constant temperature. KHITAROV *et al.* (1978) remeasured the viscosity of a basaltic melt and showed that the viscosity also decreases with increasing pressure at constant temperature. The cause of the discrepancy between the previous and the latter measurements is uncertain.

The results of these viscosity measurements suggest that the structures of silicate melts of natural rock compositions also change at high pressures. For understanding the nature of the structural change of these melts, the densities of basaltic and andesitic melts have also been measured at high pressures (FUJII and KUSHIRO, 1977; KUSHIRO, 1978a). The results are shown in Fig. 11. The densities of basaltic and hydrous andesitic melts increase with increasing pressure. The rate of density increase of these melts is not constant, but is

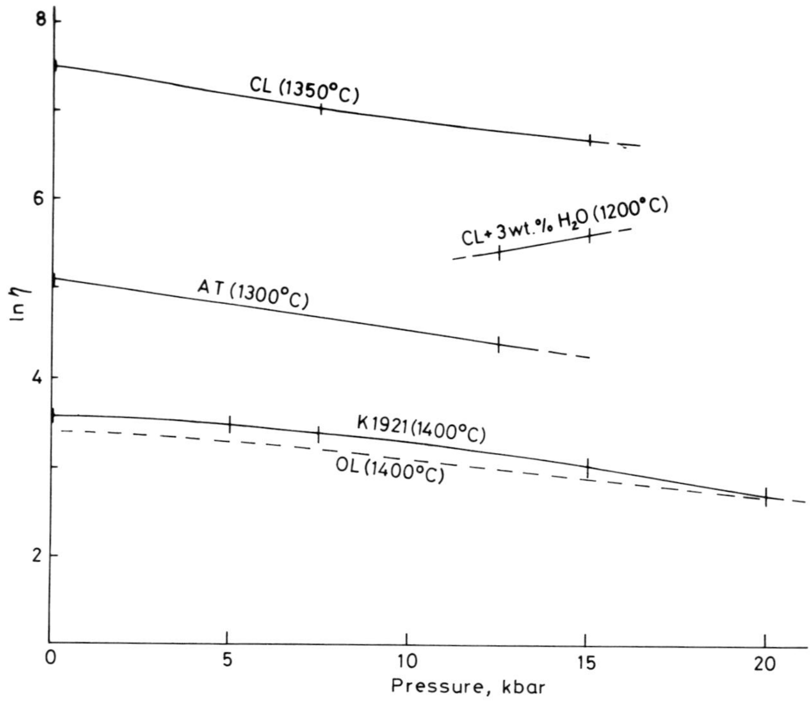

FIG. 10. Viscosity changes of melts of Crater Lake andesite (CL), abyssal tholeiite (AT), Kilauea olivine tholeiite (K1921) under anhydrous conditions (KUSHIRO *et al.*, 1976; FUJII and KUSHIRO, 1977). Dashed line is a viscosity change of an olivine tholeiite melt under anhydrous conditions by KHITAROV *et al.*, (1978). Viscosity of a hydrous Crater Lake andesitic melt (about 3 wt.% H_2O) is from KUSHIRO (1978a).

largest at pressures between 10 and 15 kbar.

The viscosity decrease and the density increase of the melt of Kilauea 1921 olivine tholeiite are similar to those observed for the $NaAlSi_2O_6$ and $NaAlSi_3O_8$ melts, although the rate of the viscosity decrease of the basalt melt is smaller than that of the $NaAlSi_2O_6$ and $NaAlSi_3O_8$ melts. The viscosity and density changes would result from the structural changes of the melts at high pressures. The nature of the structural change is not fully understood at present. The Kilauea olivine tholeiite and Crater Lake andesite have NBO/T ratios 0.76 and 0.27 respectively (both less than 1), suggesting that the melts are relatively enriched in the 3-D structural units. It is probable that the structural change of the 3-D units with pressure suggested for the 3-D network melts (i.e., changes of the ring size) may be the cause of the observed viscosity and density changes.

A hydrous andesite melt has much lower viscosity than an anhydrous andesite melt of the same composition. The anhydrous andesite melt shows a

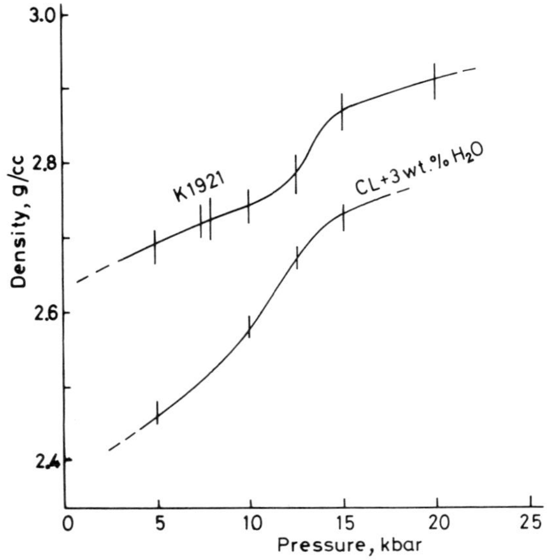

FIG. 11. Density changes of melts of Kilauea 1921 olivine tholeiite (K1921) and Crater Lake andesite (CL) with about 3 wt.% H_2O (FUJII and KUSHIRO, 1977; KUSHIRO, 1978a).

viscosity decrease, whereas the same andesite melt with about 3 wt.% H_2O (~10 mole% H_2O) shows a slight viscosity increase at constant temperature (KUSHIRO, 1978a). Water effectively depolymerizes the silicate melts and the NBO/T becomes much smaller than that under anhydrous conditions. As mentioned before, the depolymerized melts such as diopside and sodium metasilicate melts show viscosity increase with increasing pressure at constant temperature. The hydrous andesite melt with 3 wt.% H_2O has NBO/T ratio which is nearly the same as metasilicate melts. This is probably the reason for the viscosity increase of this hydrous andesite melt. A granitic melt shows a drastic decrease of viscosity with the addition of H_2O. For example, viscosity of a simple granitic melt (near the lowest temperature region in the system $KAlSi_3O_8$–$NaAlSi_3O_8$–SiO_2 decreases from about 10^{10} poise with no H_2O to about 10^5 poise with 3 wt% H_2O (SHAW, 1965). SHAW (1963) observed a slight viscosity increase in a hydrous granitic melt with increasing pressure for a constant H_2O content. Hydrous andesite and granitic melts are significantly depolymerized despite that they are rich in SiO_2 and Al_2O_3.

7. *Vesiculation of a Hydrous Andesitic Melt and Transport of Alkalies*

Calc-alkalic magmas often cause explosive volcanism in orogenic belts and are believed to contain significant amounts of H_2O. Such explosive volcanism would result from the increase of vapor pressure in the magma reservoirs due to vesiculation of magma. As mentioned in the preceding section, H_2O greatly reduces the viscosities of magmas and it is expected that the H_2O-saturated calc-alkalic magmas would have low viscosities and, in consequence, the separated vapor phase might move appreciably in the magma reserv‹ irs. To examine this possibility, experiments have been conducted on the vesiculation of a hydrous andesitic melt as part of the present project (SAKUYAMA and KUSHIRO, 1979). A calc-alkalic andesite from Asama volcano in central Japan was used as a starting material. The andesite was sealed in Pt capsules (5 mm in diameter and 15 mm in length) with 7–12 wt.% H_2O and melted at 1,150°C at pressures between 15 and 18 kbar. The homogenized hydrous andesitic melts were decompressed slowly to lower pressures. Figure 12 shows the summary of the runs. Thin lines indicate the decompression processes during the runs, and a thick solid curve indicates the conditions of vesiculation. The runs quenched in the lower pressure side of the curve have numerous vesicles, whereas those quenched above the curve have no vesicles. The curve should be very close to the solubility curve of H_2O in this andesite melt.

The distribution of vesicles in the vesiculated changes was examined under the microscope. It is found that the vesicles are concentrated in the upper part of the charges; the vesicles become large both in number and size with increasing height from the bottom of the charges. Apparently, the separated vapor phase moved upward and coaggregated together during the ascent after the vesiculation of the melt (about 30 min). No sign of convection was observed in all the charges examined. The viscosity of this hydrous andesitic melt (7–12 wt.% H_2O) must be less than 30 poise, based on the distance of movement of vesicles.

The charges were analyzed with an electron microprobe. The analyses were made with a defocused electron beam ($\sim$40 μm in diameter) so that both glass and vesicles were included. The results of the analyses along the vertical axis of a charge are shown in Fig. 13. It is demonstrated that the composition varies with ʻhe height of the charge. In particular, the content of Na_2O varies from abouι 2 wt.% near the bottom to about 5% near the top. For comparison, the analyses of the unvesiculated charge are shown in the same figure (lines with solid circles). It is evident that the bottom is depleted in Na_2O, whereas the upper part is significantly enriched in Na_2O and slightly enriched in K_2O. The variations along the horizontal axis are almost undetactable, except for the depletion of Fe near the Pt capsule due to the diffusion of Fe into Pt. The fact that the unvesiculated charges do not show such vertical chemical variations indicates that the observed variations, particularly alkalies must be caused by the transport of alkalies by the separated vapor phase. From these experiments, it is suggested that the transport of alkalies, especially Na, by the separated vapor phase in the

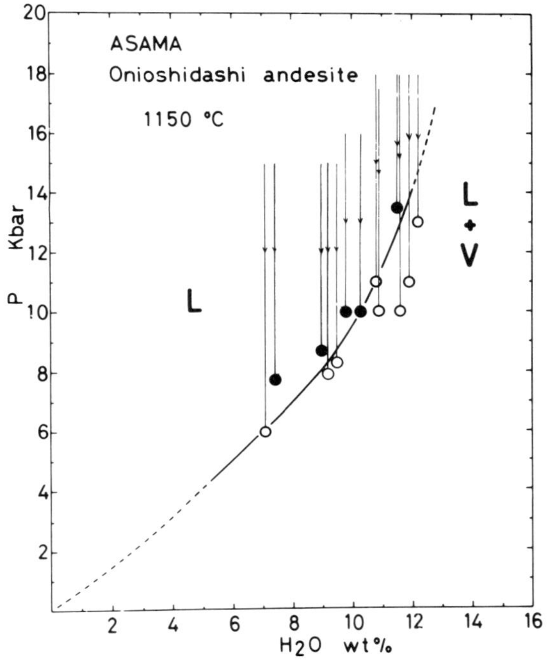

FIG. 12. Summary of the decompression experiments on a hydrous Asama andesite melt at 1,150°C. Symbols: solid circles, runs with no vesicles; open circles, runs with vesicles. Solid lines with arrow indicate the processes of decompression.

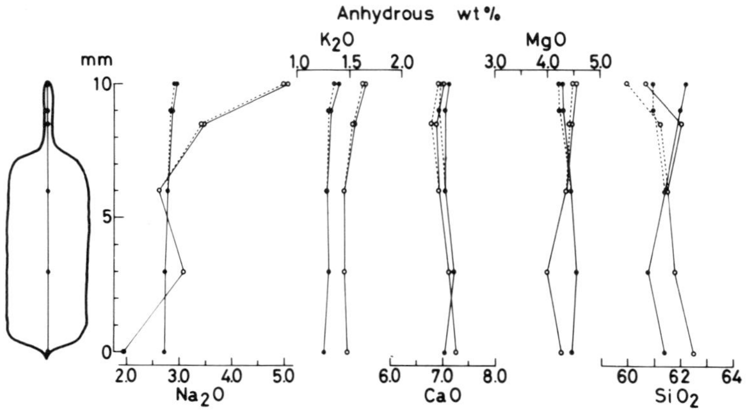

FIG. 13. Compositional variations along the vertical axis of a vesiculated charge, which was decompressed from 18 to 10 kbar at 1,150°C with 11.6 wt.% H_2O (lines with open circles). Lines with solid circles show the compositional variations of an unvesiculated charge.

magma chambers may be significant. The rate of the transport of alkalies in the magma reservoirs cannot be calculated in detail, because it depends on several factors, such as the depth of magma reservoir (pressure) and the solubility of H_2O in magma. If vesicles move for about 1 cm in 30 min as observed in some of the runs, alkalies would be transported for 1 km in about 200 days. This rate is much faster than the diffusion of alkalies in magmas.

REFERENCES

AKIMOTO, S. and SYONO, Y. (1969): Coesite-stishovite transition, *J. Geophys. Res.*, **74**, 1653–1659.

BELL, P. M. and ROSEBOOM, E. H. (1969): Melting relationships of jadeite and albite to 45 kilobars with comments on melting diagrams of binary systems at high pressures, *Mineral Soc. Am. Spec. Paper*, **2**, 151–161

BOCKRIS, J. O'M., MACKENZIE, J. D. and KITCHNER, J. A. (1955): Viscous flow in silica and binary silicates, *Trans. Faraday Soc.*, **51**, 1734–1748.

BOWEN, N. L. (1914): The ternary system diopside-forsterite-silica, *Am. J. Sci., 4th ser.*, **38**, 207–264.

BOWEN, N. L. (1928): *The Evolution of the Igneous Rocks*, Princeton Univ. Press.

BRAWER, S. A. (1975): Theory of the vibrational spectra of some network and molecular glasses, *Phys. Rev.*, **B11**, 3137–3194.

BRAWER, S. A. and WHITE, W. B. (1975): Raman spectroscopic investigation of the structure of silicate glasses I, The binary silicate glasses, *J. Chem. Phys.*, **63**, 2421–2432.

BURNHAM, C. W. (1963): Viscosity of a water rich pegmatite melt at high pressure [abs], *Geol. Soc. Am. Spec. Paper*, **76**, 26.

CHEN, C.-H and PRESNALL, D.C. (1975): The system Mg_2SiO_4–SiO_2 at pressures up to 25 kilobars, *Am. Mineral.*, **60**, 398–406.

DOELTER, C. (1902): Die chemische Zusammensetzung und die Genesis Monzonitgesteine, *Tscher. Min. Pet. Mitt.*, **21**, 191–225.

FRIEDMAN, I., WILLIAM, L. and SMITH, R. L. (1963): Viscosity and water content of rhyolite glass, *J. Geophy. Res.*, **68**, 6523–6535.

FUJII, T. and KUSHIRO, I. (1977): Melting relations and viscosity of an abyssal tholeiite, *Carnegie Inst. Wash. Year Book*, **76**, 461–465.

GREIG, J. W. (1927): Immiscibility in silicat melts, *Am. J. Sci.*, **13**, 1–44, 133–154.

KANI K. (1934): The measurement of the viscosity of basalt glass at high temperatures, I and II, *Proc. Imp. Acad. Japan*, **10**, 29–32, 79–82.

KEITH, M. L. and SCHAIRER, J. F. (1952): Stability field of sapphirine in the system MgO–Al_2O_3–SiO_2, *J. Geol.*, **60**, 181–186.

KAWAI, N., TACHIMORI, M. and ITO, E. (1974): A high pressure hexagonal form of $MgSiO_3$, *Proc. Japan Acad.*, **50**, 378–380.

KHITAROV, N. I., LEBEDEV, E. B., SLUTSKY, A. B., DORFMAN, A. M., SOLDATOV, I. A. and REVIN, N. I. (1976): Pressure influence upon basalt melt viscosity, *Geokhimiya*, **1976**, 1489–1497.

KHITAROV, N. I., LEBEDEV, E. B., DORFMAN, A. M. and SLUTSKY, A. B. (1978): Viscosity of dry and water-bearing basalt melts under the pressure, *Geokhimiya*, **1978**, 900–905.

KUSHIRO, I. (1968): Compositions of magmas formed by partial zone melting of the earth's upper mantle, *J. Geophys. Res.*, **73**, 619–634.

KUSHIRO, I. (1969): The system forserite-diopside-silica with and without water at high pressures, *Am. J. Sci.*, **267A**, 269–294.

KUSHIRO, I. (1972): Determination of liquidus relations in synthetic silicate systems with electron probe analysis: the system forsterite-diopside-silica at 1 atmosphere, *Am. Mineral.*, **57**, 1260–1271.

KUSHIRO, I. (1975): On the nature of silicate melts and its significance in magma genesis: Regularities in the shift of the liquidus boundaries involving olivine, pyroxene, and silica minerals, *Am. J. Sci.*, **275**, 411–431.

KUSHIRO, I. (1976): Changes in viscosity and structure of melt of $NaAlSi_2O_6$ composition at high pressures, *J. Geophys. Res.*, **81**, 6347–6350.

KUSHIRO, I. (1977): Phase transformation in silicate melts under upper mantle conditions, in *High-Pressure Research Applications in Geophysics*, Eds. M. H. Manghnani, and S. Akimoto, Academic Press, New York, pp. 25–37.

KUSHIRO, I. (1978a): Density and viscosity of hydrous calcalkalic andesite magma at high pressures, *Carnegie Inst. Wash. Year Book*, **77**, 675–677.

KUSHIRO, I. (1978b): Viscosity changes of GeO_2 melt with pressure as a model of SiO_2 melt, *Carnegie Inst. Wash. Year Book*, **77**, 672–675.

KUSHIRO, I. (1978c): Viscosity and structural changes of albite ($NaAlSi_3O_8$) melt at high pressures, *Earth Planet. Sci. Lett.*, **41**, 87–90.

KUSHIRO, I. (1980): Changes with pressure of degree of partial melting and K_2O content of liquids in the system Mg_2SiO_4–$KAlSiO_4$–SiO_2, *Carnegie Inst. Wash. Year Book*, **79**, 267–271.

KUSHIRO, I., YODER, H. S. and MYSEN, B. O. (1976): Viscosities of basalt and andesite melts at high pressures, *J. Geophys. Res.*, **81**, 6351–6356.

LAZAREV, A. N. (1972): *Vibrational Specta and Structures of Silicates*, Consultants Bureau, New York.

MATSUI, Y. and KAWAMURA, K. (1980): Instantaneous structure of an $MgSiO_3$ melt simulated by molecular dynamics, *Nature*, **285**, 648–649.

MATSUI, Y. and KAWAMURA, K. (1982): Computer simulation of structure of silicate melt and glasses, this volume, pp. 3–23.

MCGREGOR, I. D. (1964): The reaction 4 enstatite + spinel ⇌ forsterite + pyrope, *Carnegie Inst. Wash. Year Book*, **63**, 156–157.

MINAKAMI, T. (1951): On the temperature and viscosity of the fresh lava extruded in the 1951 Oo-sima eruption, *Bull. Earthq. Res. Inst., Univ. Tokyo*, **29**, 487–498.

MURASE, T. and MCBIRNEY, A. R. (1973): Properties of some common igneous rocks and their melts at high temperatures, *Geol. Soc. Am. Bull.*, **84**, 3563–3592.

MYSEN, B. O., VIRGO, D. and SCAEFE, C. M. (1980): Relations between the anionic structure and viscosity of silicate melts—a Raman spectroscopic study, *Am. Mineral.*, **65**, 690–710.

NAKAMURA, Y. (1978): Thermodynamics of silicate melts, in *Material Sciences of the Earth II*, Eds. I. Kushiro and S. Aramaki, Iwanami-Shoten, pp. 206–225 (in Japanese).

NICHOLLS, J., CARMICHAEL, I. S. E. and STOMER, J. C. (1971): Silica activity and P_{total} in igneous rocks, *Contr. Min. Petrol.*, **33**, 1–20.

OHTANI, F., and IRIFUNE, T. and FUJINO, K. (1981): Fusion of pyrope at high pressures and rapid crystal growth from the pyrope melt, *Nature*, **294**, 62–64.

PALMER, H. S. (1927): A study of the viscosity of lava, *Hawaiian Volc. Obs. Bull.*, **15**, 1–4.

RANKIN, G. A. and MERWIN, H. E. (1918): The ternary system MgO–Al_2O_3–SiO_2, *Am. J. Sci., 4th ser.*, **45**, 301–325.

RINGWOOD, A. E. (1967): The pyroxene-garnet transformation in the earth's mantle, *Earth Planet. Sci. Lett.*, **2**, 255–263.

ROSSIN, R., BERSAN, J. and URBAIN, G. (1964): Étude de la viscosité de laitiers liquides appartenant en système terunaire SiO_2–Al_2O_3–CaO, *Rev. Hautes Temp. Refract.*, **1**, 159–170.

SABATIER, G. (1956): Influence de la teneur en eau sur la viscosité d'une rétinite, verre ayant la composition d'un granite, *Comptes Rendus, Acad. Sci. Paris*, **242**, 1340–1342.

SAKUYAMA, M. and KUSHIRO, I. (1979): Vesiculation of hydrous andesitic melt and transport of alkalies by separated vapor phase, *Contr. Min. Pet.*, **71**, 61–66.

SCARFE, C. M. (1973): Viscosity of basic magmas at varying pressure, *Nature*, **241**, 101–102.

SCARFE, C. M., MYSEN, B. O. and VIRGO, D. (1979): Changes in viscosity and density of melts of sodium disilicate, sodium metasilicate and diopside composition with pressure, *Carnegie Inst. Wash. Year Book*, **78**, 547–551.

SCHAIRER, J. F. (1954): The system K_2O–MgO–Al_2O_3–SiO_2: I. Results of quenching experiments on four joins in the tetrahedron cordierite-forsterite-leucite-silica and on the join cordierite-mullite-potash feldspar, *J. Am. Ceram. Soc.*, **37**, 501–533.

SCHAIRER, J. F. and YODER, H. S. (1961): Crystallization in the system nepheline-forterite-silica at one atmosphere pressure, *Carnegie Inst. Wash. Year Book*, **61**, 75–82.

SEIFERT, F. D., MYSEN, B. O. and VIRGO, D. (1981): Structure and properties of aluminosilicate melts with three-dimentional network structure, *Carnegie Inst. Wash. Year Book*, **80**, 305–308.

SEN, P. N. and THORPE, M. F. (1977): Phonons in X_2 glasses: from molecular to band-like modes, *Phys. Rev. B*, **15**, 4030–4038.

SHARMA, S. K., VIRGO, D. and MYSEN, B. O. (1978): Structure of glasses and melts of $Na_2O \cdot xSiO_2$ ($x = 1, 2, 3$) composition from Raman spectroscopy, *Carnegie Inst. Wash, Year Book*, **77**, 649–652.

SHARMA, S. K., VIRGO, D. and MYSEN, B. O. (1979): Raman study of the coordination of aluminum in jadeite melts as a function of pressure, *Am. Mineral*, **64**, 779–787.

SHAW, H. R. (1963): Obsidian-H_2O viscosities at 1,000 and 2,000 bars in the temperature range 700° to 900°C, *J. Geophys. Res.*, **68**, 6337–6343.

SHAW, H. R. (1965): Comments on viscosity, crystal settling, and convection in granitic magmas, *Am. J. Sci.*, **263**, 120–152.

SHAW, H. R., WRIGHT, T. L., PECK, D. L. and OKAMURA, R. (1968): The viscosity of basaltic magma: An analysis of field measurements in Makaopuhi lava lake, Hawaii, *Am. J. Sci.*, **266**, 225–264.

TOOPS, G. W. and SAMIS, C. S. (1962): Activities of ions in silica melts, *Trans. Metallurg. Soc. AIME*, **224**, 87–887.

VELDE, B. and KUSHIRO, I. (1978): Structure of sodium aluminosilicate melts quenched at high pressure; Infrared and aluminum *K*-radiation data, *Earth Planet. Sci. Lett.*, **40**, 137–140.

VERWEIJ, H. (1979): Raman study of the structure of alkali germanosilicate glasses. I. Sodium and potassium metagermanosilicate glasses, *J. Non-Cryst. Solids*, **33**, 41–53.

VIRGO, D., MYSEN, B. O. and KUSHIRO, I. (1980): Anionic constitution of 1-atmosphere silicate melts: Implications for the structure of igneous melts, *Science*, **208**, 1371–1373.

WAFF, H. S. (1975): Pressure-induced coordination changes in magmatic liquids, *Geophys. Res. Lett.*, **2**, 193–196.

WATSON, E. B. (1979): Calcium diffusion in a simple silicate melt to 30 kilobars, *Geochim. Cosmochim. Acta*, **43**, 313–322.

YOKOKAWA, T. and NIWA, K. (1969): Free energy of solution in binary silicate melts, *Trans. J. I. M.*, **10**, 3–7.

Chapter 2: Crystal Growth and Synthesis of Large Single Crystals

Materials Science of the Earth's Interior, edited by I. Sunagawa, pp. 63–105.

GROWTH OF CRYSTALS IN NATURE

Ichiro SUNAGAWA

Institute of Mineralogy, Petrology and Economic Geology, Tohoku University, Aoba, Sendai, Japan

1. *Introduction*

Mineral crystals grow in nature under a very broad spectrum of conditions and in widely different environments. They crystallize in magmas of various compositions at different depths, in vapour with complex compositions, in hydrothermal solutions, in low temperature aqueous solutions, and in metamorphic, metasomatic environments. Growth of crystals occurs in every geological process taking place both on the surface and in the interior of the Earth.

How beautiful crystals of natural minerals, like quartz or diamond, nucleate and grow in the depths of the Earth has attracted the attention of many people from the time of pre-science, and all sorts of speculation have been put forward. The following charming quotation of the 14th Century vividly demonstrates how deeply medieval people were fascinated by the beauty of natural crystals and how they tried to understand the formation of crystals in Nature.

"Diamonds grow in India,
Some as big as a bean,
Some like a hazel nut.
They are male and female,
And from the falling dew, they multiply
And bring forth small children."

Sir John Mandeville, 1360

The interest continued further after the Renaissance, which led to the establishment of classical crystal morphology, mineralogy and petrology. It seems, however, that studies on these problems have been neglected during the past several decades, due to a shift of interest towards crystal structure analyses, thermodynamics and phase relation studies.

When thermodynamics and phase studies of mineral assemblages advanced to some extent, people gradually started to realize that thermodynamics or phase studies represent only half of the natural phenomena and that a kinetic approach is needed in understanding the remainder. This is partly due to the rapid advance in science and technology of crystal growth in recent years, and partly due to the fact that earth scientists have become increasingly aware of the implications of imperfections and inhomogeneities that mineral crystals exhibit. It seems that interest among earth and planetary scientists in kinetic processes of crystallization taking place in nature has come to the forefront again.

Since, unlike laboratory crystallization, *in-situ* observations or rate measurements are practically impossible in the studies of kinetic processes of natural crystallization, indirect ways of approach are unavoidable. There are two general approaches; (1) experimentally simulating natural crystallization, and (2) interpreting the paragenic history of natural crystals.

Experimental studies to reproduce textures of igneous rocks and morphological variations of rock-forming minerals under controlled growth conditions —what are generally called dynamical crystallization experiments— have recently been extensively carried out. Studies of this type were initiated by the Appolo Project with a hope of understanding the textures seen in lunar minerals and rocks, and have been extended to terrestrial rocks and minerals. Various types of igneous rock textures, including pegmatite textures, have now been reproduced in laboratories, and the results are utilized to interpret natural igneous rock textures. DOWTY (1980) and LOFGREN (1980) are important review papers on this topic.

In interpreting the paragenic history of natural crystals, morphological aspects of crystals provide the most important key codes, since the morphology of crystals is the direct result of growth or dissolution and thus reflects the differences in growth conditions and environments. Overall morphology of crystals (Why can crystals take morphologies of dendritic, hopper or polyhedral bounded by flat faces? Why does variation in crystal habit occur?, etc.) is the result of the total on the growth history of crystals. Surface microtopographs of crystal faces show up the growth mechanism on the molecular scale. Internal morphologies of crystals (growth bandings, sector structures, dislocation configurations, etc.) reflect the whole history of growth. All these are aspects directly relating to the morphology of crystals. If these are combined, one may decipher the letter sent from the depths of the Earth, conveying how and under what conditions mineral crystals grew, provided that the relation between the morphology and growth conditions are well understood. Thanks to the development of various techniques to characterize crystal perfections, these properties can be analysed precisely, in some cases with an accuracy of the order of 1 Å. Recent theoretical and experimental development on crystal growth mechanisms have also provided a good understanding of the relation between the morphology of crystals and growth conditions. Thus we are now in a position to apply the present knowledge of the growth mechanisms and morphology of crystals of simple systems to the very complicated systems of crystal growth taking place in Nature.

It is the purpose of this chapter to analyse the characteristics of natural crystallization based on these understandings. Before doing so, it is perhaps useful to summarize our understandings of the conditions of various geological processes. Figure 1 schematically illustrates the temperature and pressure conditions under which various geological processes take place (WYLLIE, 1971). We have to understand how mineral crystals nucleate and grow in these different geological processes, and what are the essential features of natural crystallization.

We shall not discuss the nucleation problem in this chapter, since the allotted space is too small. However, we note that in natural crystallization, any nucleation phenomena can be regarded as heterogeneous nucleation, and thus nucleation can occur at much smaller chemical potential differences between solid and liquid phases than that expected for homogeneous nucleation. Any natural crystal growth systems contain inhibitors for heterogeneous nucleation. We also note that grain-size distributions and textures of mineral assemblages are directly related to the rate of nucleation, and that chemical potential differences between solid and liquid phases, densities of effective inhibitors and the viscosity of the system are the major factors in determining the nucleation rates.

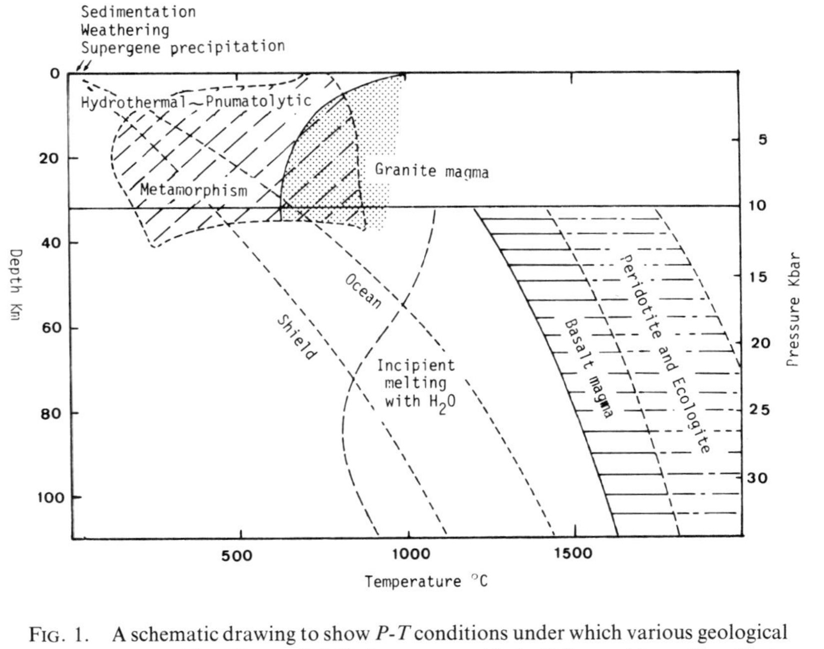

FIG. 1. A schematic drawing to show *P-T* conditions under which various geological processes occur. After WYLLIE (1971). By courtesy of John Wiley and Sons, New York.

2. *Growth of Crystals from Different Phases*

Growth of crystals may be classified into (1) solid-solid crystallization and (2) liquid-solid crystallization. The latter is further classified into (2a) melt growth, (2b) solution growth, and (2c) vapour growth.

In solid-solid crystallization, which is often called re-crystallization in the field of physical sciences, the initial and the final solids have the same crystal structure and chemistry. Only grain sizes increase through the movement of grain boundaries, which occurs when the strain energies in the grain boundaries are released by the application of thermal energies. No liquid phase is assumed to be present in the grain boundaries, which consist of dislocation arrays or nets. Grains grow larger and become equi-dimensional, having equilibrium angles of 120° at grain boundary junctions when primary re-crystallization is completed. If further thermal energies are applied, a few grains start to grow much larger. This is called secondary recrystallization and is due to impurities in the grain boundaries becoming mobile through thermal energies.

In the case of liquid-solid crystallization, the growth of crystals occurs because of the re-arrangement of structures, i.e. phase transition, either in or from the liquid, which has only a short-range ordering. The processes of and the driving forces for such re-arrangements are different, depending on the state of the liquid phase. If the liquid phase is in a very condensed state, e.g. pure melt, the re-arrangement cannot occur through the processes of mass transport and condensation, since both solid and liquid phases have nearly the same densities and similarly close interatomic distances. Thus heat transfer, rather than mass transfer, plays the major role in crystallization. If, on the other hand, the fluid phase is dilute and the crystallizing species exist in a widely dispersed state in the liquid, e.g. in solution or vapour phase, it is necessary to transport and condense them to nucleate and grow crystals.

In these cases, mass transfer plays an essential role. Although both mass and heat transfer are combined in any types of crystal growth, we may simplify the situation by assuming that heat transfer plays an essential role in melt growth, whereas mass transfer is essential in solution and vapour growth.

Other important differences among melt, solution and vapour growths are as follows. Firstly, as can be understood from the phase diagram of an eutectic system, the growth of crystals occurs at much lower temperatures in solution growth than in melt growth. A lower temperature is also expected for vapour growth. Secondly, the interactions between solid and liquid phases are markdly different among the three types. No solid-liquid interactions are assumed for pure melt and vapour growths. In a solution growth, a strong solid-liquid interaction, i.e. interaction between solute and solvent components, is always involved. The solute component is bounded by the solvent component and together they move as one entity in the solution until the former is incorporated into the crystal. Therefore, we have to assume several desolvation steps before the crystallizing particles are incorporated into the solid phase (BENNEMA and GILMER, 1973). The strength of the solid-liquid interaction is different depending on the type of solvent. In general, this is stronger in an aqueous solution than in a high temperature solution (flux growth system; ref. to ELWELL and SCHEEL, 1975). In a similar manner, solid-liquid interaction is assumed for crystal growth from vapour phases, in which chemical reaction is involved, such as in chemical vapour transport processes. Vapour growth in which chemical reaction is involved, is regarded as representing an intermediate state between solution and pure vapour growth.

The above differences also result in the difference in edge free energies per growth unit, γ. It is generally believed that γ is much lower in solution growth than in vapour growth.

As will be discussed in some details in Section 5, the α-factor, which is a criterion for predicting the roughness of solid-liquid interface, is assumed to be low for melt growth and high for vapour growth. Solution growth is situated in between the two. It follows from this that, in melt growth, the solid-liquid interface will, in general, be rough when crystallization occurs and smooth in vapour and solution growths.

The essential differences among crystal growth from melt, solution and vapour phases may be summarized as in Table 1. Based on this, we shall next discuss how we may characterize crystal growths taking place in Nature.

It is evident from the preceding discussions that pure melt growth is absent in any geological processes, since complicated components are always involved in natural crystallization. The growth of rock-forming minerals in magma, which has been customary called silicate melt, is not a crystal growth from the melt phase, since magmas are multi-component systems and mass transfer and condensation are required to grow crystals of rock-forming minerals. Crystallization in magma is regarded as a crystal growth from high temperature solutions.

The growth of crystals in hydrothermal solutions and in low temperature aqueous solutions in a surficial or sedimentary environment, are, without doubt, crystal growths from the solution phase, in which H_2O is the major solvent component. In a growth of crystals from the vapour phase taking place in Nature, such as the formation of pegmatite minerals, fumarole minerals, or minerals in druses of lava, when they are crystallized at supercritical temperatures, solid-liquid interaction is always involved, although this is, in general, much weaker than in aqueous and high temperature solutions. Therefore, natural vapour growth is regarded as representing an intermediate situation between pure vapour and aqueous solution growth.

Geological processes, such as the formation of marble from limestone or of quartzite from arkose sandstone, may be regarded as solid-solid crystallization. In these processes, if impurity components are neglected, the initial

TABLE 1. Qualitative comparison of characteristics of crystal growth in different phases.

	Melt	Solution		Vapour	
		High temperature solution (flux)	Low temperature aqueous solution	Involving chemical reaction	Pure vapour phase
State	Very condensed	———————————————			→ Very dispersed
Driving force	Heat transfer Supercooling ΔT	Mass transfer (transportation and condensation) Supersaturation σ			
Growth temperature	High	Medium	Low	Medium ~	Low
Interaction with solvent	None	Strong	Strongest	Strong	None
Edge free energy γ	High	Low	Low	Medium	High
α	Small	Medium	Medium	Large	Largest
Interface roughness	Rough	Smooth	Smooth	Smooth	Smooth
Principal growth mechanism	Continuous	Two-dimensional nucleation growth to spiral growth			

and the final materials have the same chemical compositions and crystal structures, and only grain sizes increase. It is, however, evident that crystal growths in metamorphic and metasomatic environments cannot be treated as a pure solid-solid crystallization, since the initial and final materials are different. In these processes, solvent components, like volatile elements, will play an essential role in the growth of crystals, and thus actual processes may be similar to solution growth. This will be discussed in more details in a later section.

Summarizing the above, we can assume that, in natural crystallization, the most important and universally encountered growth of crystals is solution growth. Pure melt growth is absent and pure solid-solid crystallization is also restricted in natural crystallization. Natural crystal growth from vapour phases is close to solution growth in fundamental character. We shall therefore pay attention to the growth of crystals from solution phases in the following discussions.

3. *Driving Force in Crystal Growth*

Neither nucleation, nor growth of crystals occurs when solid and liquid phases are in equilibrium. When crystallizing particles, arriving onto the surface of a crystal, exceed the amount of those leaving the surface, the growth of crystals occurs, and vice versa in dissolution. Therefore, we need a situation which deviates from the equilibrium for growth or dissolution to take place, and this deviation constitutes the driving force for nucleation and growth. If there is no deviation, neither nucleation nor growth will occur. When a system is placed apart from the equilibrium line, either by taking off the heat or by condensing the system, nucleation and growth can occur. There is a gap between the equilibrium line and the positions where nucleation can occur, which is called Mier's region. This is schematically shown in Fig. 2.

In melt growth, where heat transfer plays the major role, supercooling ΔT represents the driving force. In solution and vapour growths, where mass transfer is important, supersaturation σ represents the driving force. We may express the driving force by the difference of chemical potentials between solid and liquid phases $\Delta \mu$, divided by Boltzmann constant k and absolute temperature T; $\Delta \mu / kT$.

4. *Mass Transport*

In solution and vapour growths, there is a difference in concentrations between the bulk phase and the surface of a growing crystal, with which crystallizing particles can be transported by diffusion from the bulk onto the surface of a growing crystal. This difference is called bulk supersaturation σ_B. There is also a difference in supersaturation between the adsorption layer

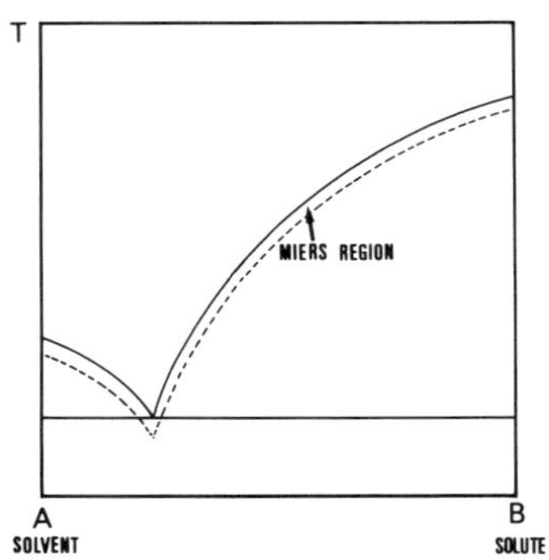

FIG. 2. A hypothetical phase diagram of an eutectic system to show Miers' region.

around the crystal and the surface of the crystal. This is called surface supersaturation σ_S, $\sigma_B > \sigma_S$. The supersaturation which controls the growth rate of polyhedral crystals, is σ_S, and is difficult to measure experimentally. In dendritic growth, however, the growth rate is controlled by σ_B. Thus, such a growth is called volume diffusion controlled growth.

In case of a smooth solid-liquid interface (low index face), there is also a variation in supersaturation over a surface. BERG (1938) showed that supersaturation over a flat surface is not uniform, and is highest at the edges or corners of the face, the lowest at the centre of a face. This is later called Berg effect or phenomenon. FOLLENIUS (1959) found that equi-concentration curves truncate the corners in the case of a polyhedral crystal, whereas they follow the outline of the crystal once the corner protracts, i.e. in hopper or dendritic interfaces. These are schematically shown in Fig. 3.

The adsorption layer has been assumed to exist around a growing crystal. The thickness of adsorption layer changes, depending on the state of movement in the liquid phase. Convection and agitation have an effect of diminishing the thickness.

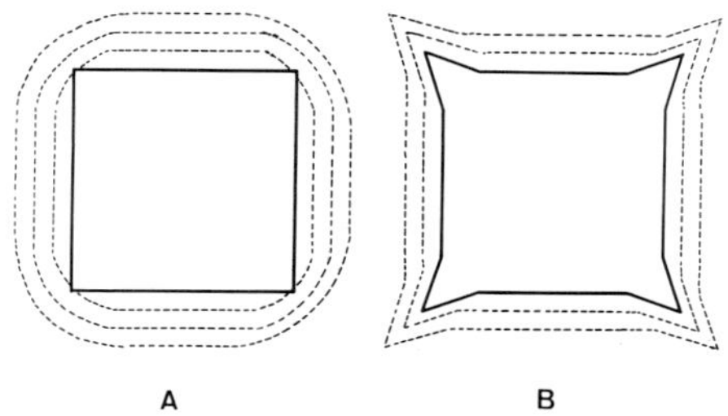

FIG. 3. Equi-concentration curves (Berg effect) around a growing polyhedral (A) and dendritic (B) crystal.

5. *Roughness of Solid-Liquid Interface*

Since the growth of crystals occurs exclusively on the solid-liquid interface, understanding of the state of solid-liquid interfacing has primal importance in the study of crystal growth.

The state of solid-liquid interface may be classified into two extreme types: rough interface and smooth interface. The two types are also called non-singular and singular (BURTON *et al.* 1951), or incomplete and complete (KOSSEL, 1930) interfaces, respectively.

When a solid-liquid interface is atomistically rough, consisting only of kinks, the crystallizing particles arriving onto the interface can immediately be incorporated into the crystal, since they can find their stable positions for settlement. In this case, the interface advances continuously and, the growth rate of the interface R has a linear relation with the driving force, $R \propto A$ $(\Delta \mu/kT)$. Such growth mechanism is called continuous growth or adhesive growth.

If a solid-liquid interface is atomistically smooth, with a small number of steps and kinks on the flat surface, the crystallizing particles meet with difficulty in finding their stable settling positions. They migrate on the flat surface for some distances, during which some will again leave from the surface to the environment, but others will have the opportunity to meet kinks or steps, where they will be incorporated into the crystal. The growth process will then be of a two-dimensional manner, through lateral spreading of the steps, parallel to the interface. In such a case, there are two possibilities of providing sources for new growth layers to commence; two-dimensional nucleation growth mechanism (2DNG mechanism, or KOSSEL (1930)–STRANSKI (1928)–VOLMER (1922, 1926) mechanism) and spiral growth mechanism (BURTON–CABRERA–FRANK (1951) or BCF mechanism). The 2DNG mechanism may be further divided into three different models: mono-nuclear, poly-nuclear, and birth-and-spread (OHARA and REID, 1973). In the case of a 2DNG mechanism, it is necessary to overcome an energy barrier so that growth may continue, since two-dimensional nucleation should occur before commencing a new step. The growth rate is determined by the rate of two-dimensional nucleation. In the case of a 2DNG mechanism, it follows that growth will occur only above a certain critical supersaturation. In spiral growth mechanism, this is not necessary since screw dislocation provides a selfperpetuating step and, thus, growth occurs even with an extremely small driving force. In both 2DNG and spiral growth mechanisms, growth is governed by the interface kinetics. The relation between the growth rate R and the driving force $\Delta \mu/kT$ is not linear, but exponential; $R \propto (\Delta \mu/kT)^n$, where n is different for spiral growth and 2DNG.

Since solid-liquid interface roughness determines the growth mechanisms, several theories have been put forward to predict interface roughness. WOODRUFF (1973) summarized the earlier works on solid-liquid interfaces, and in recent years, extensive computer simulation works have been carried out to simulate the interface roughness. Recent results obtained by computer simulation have been reviewed by GILMER and JACKSON (1977).

The earliest theoretical treatment of this problem was made by Burton *et al.* (1951), who found that the transition temperatures from singular (smooth) to non-singular (rough) interfaces in any directions of representative metals, are always much lower than the growth temperatures from the melt. They predicted rough interfaces for these metals when grown from the melt.

JACKSON (1958) proposed α-factor as criterion for predicting interface roughness, and concluded that the interface will be rough for materials with α smaller than 2 and smooth for $\alpha > 3$. This is shown in Fig. 4. As is seen in this figure, materials with α larger than 3 have two minima at 0 and 1 on the

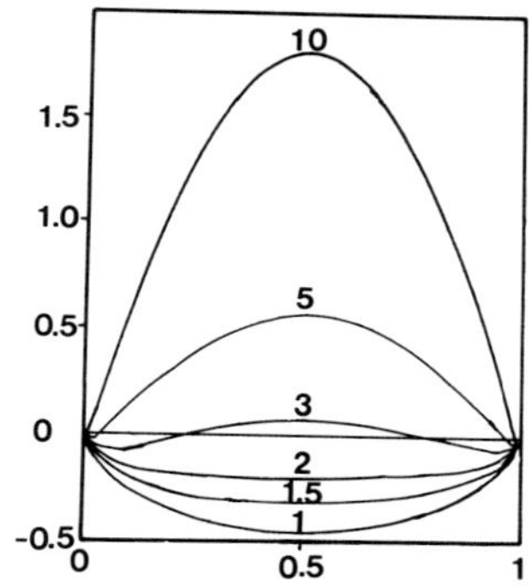

FIG. 4. Jackson's diagram. The vertical axis represents relative free energy, and the horizontal axis, the occupied fraction of surface sites. JACKSON (1958).

horizontal axis, and those with $\alpha < 2$ only one minimum at 0.5. The former corresponds to smooth interface, the latter to rough interface. The Jackson's theory has been a very successful one in that it can classify materials into two groups, depending on α values. However, this theory assumes equilibrium temperature, and thus cannot directly apply to real growth systems, nor to solution growth.

The α-factor was then generalized by BENNEMA and GILMER (1973), and further by BENNEMA and VAN DER EERDEN (1977). The expressions of α-factor are summarized as follows:

$$\alpha = \xi \frac{L}{kT_{\mathrm{M}}} \tag{1}$$

$$= -\frac{2\phi_{\mathrm{ss}}}{kT} \tag{2}$$

$$= \frac{4\{\phi_{\mathrm{sf}} - 1/2(\phi_{\mathrm{ss}} + \phi_{\mathrm{ff}})\}}{kT} \tag{3}$$

$$= \xi(1 - x_{\mathrm{s}})^2(-\ln x_{\mathrm{s}} + C). \tag{4}$$

Equation (1) is Jackson's expression, Eqs. (2) and (3) are generalized expressions by BENNEMA and GILMER (1973), and assume (100) interface of a Kossel crystal, and Eq. (4) is an expression by BENNEMA and VAN DER EERDEN (1977), applicable for solution growth.

In these expressions, ϕ_{ss} is the potential energy of the nearest pair in solid and corresponds to the average bond energy per pair, ϕ_{sf} the average potential energy of solid-liquid, ϕ_{ff} the interaction energy of two nearest blocks in the liquid phase, L latent heat, k Boltzmann constant, T_{M} melting (equilibrium) temperature, T growth temperature, ξ anisotropic factor (which is smaller than unity and can be expressed by $\xi = E^{sl}/E^{cr}$, where E^{sl} is an attachment energy of a slice defined by HARTMAN and PERDOK (1955)), E^{cr} crystallization energy ($\simeq$ bond energy), x_s the solubility expressed by molar fraction, and C a constant which ranges between 1 and 2.5.

In the case of Jackson's equation, α is a kind of material constant, since it assumes equilibrium temperature. Therefore, with this value, one can predict whether the interface is rough or smooth at equilibrium temperature for a particular material. In other generalized expressions the α-values are applicable to actual crystal growth, not only for pure melt growth or pure vapour growth, but also for solution growth, since in these expressions T is growth temperature and ϕ_{sf} and ϕ_{ff} are involved. In fact, an estimation of the α values for different solvents and their correlations with observed interface roughness of biphenyl, have been made recently by HUMAN *et al.* (1981). Such estimations should be useful for natural crystal growth systems, but have not been made so far.

In both Jackson's and generalized α-factors, an interface will be smooth for a large α values, and rough for a small α value. Jackson predicted $2 < \alpha < 3$ as the transitional value from smooth to rough. Recent computer simulations indicate $\alpha \approx 4.9$ as the transitional value (GILMER and JACKSON, 1977).

There is a transition from smooth to rough as α decreases and thus the temperature increases. Such a transition is called the roughening transition. Extensive computer simulations have been carried out recently, using the Monte Carlo technique, and the roughening transition has been visualized (e.g. GILMAR and JACKSON, 1977). An example of Monte Carlo simulated interfaces depending on α values, are shown in Fig. 5.

The solid-liquid interface will also become rough when the probability of crystallizing particles arriving onto the interface increases (namely chemical potential difference increases). This was treated by TEMKIN (1969, 1971), who showed that on a diagram of $\beta(\Delta\mu/kT)$ and γ(related to α), two regions A and B can be divided, corresponding to smooth and rough interface regions, respectively. This is shown in Fig. 6, which indicates that as $\beta(\Delta\mu/kT)$ increases, so region B expands. Such a roughening transition from smooth to rough interface may be called a kinetic roughening.

From the preceding discussions, we may assume the following relations as to the interface roughnesses:

(1) As α increases, the interface becomes smoother.

(2) Materials with Jackson's α smaller than 2 have rough interfaces in all directions.

(3) Crystals with weaker bonds, and also simpler structures, have smaller Jackson's α values than those with stronger bonds and complicated structures.

(4) For the same crystal, the interface becomes rougher as growth

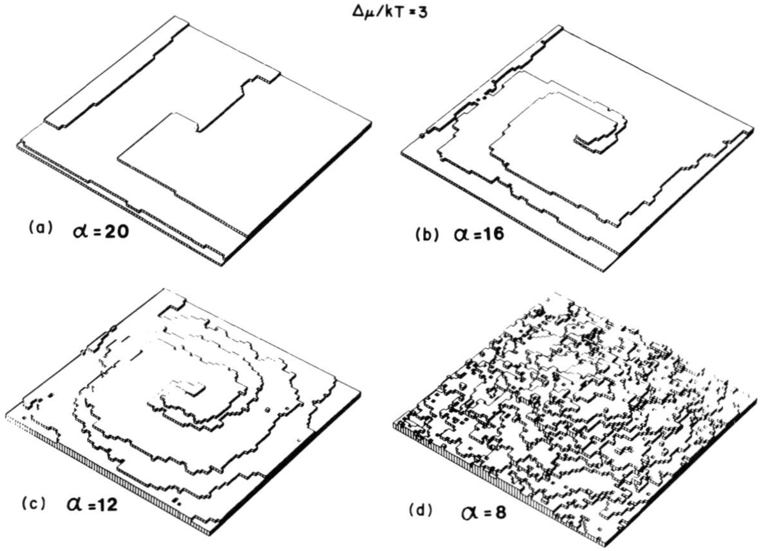

FIG. 5. Computer simulations showing the roughening transition as α decreases. After GILMER and JACKSON (1977). By courtesy of North-Holland Pub. Co., Amsterdam.

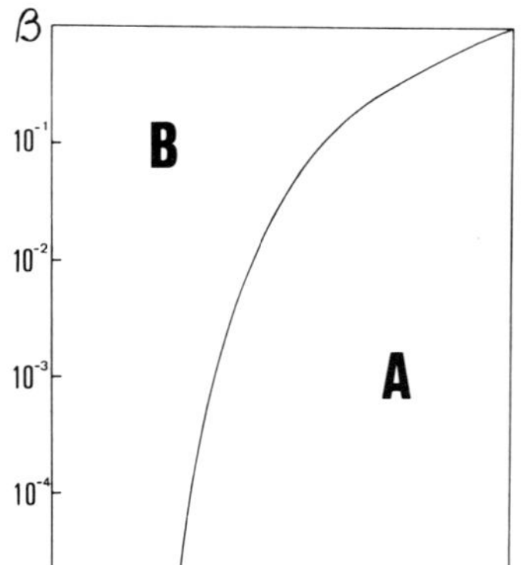

FIG. 6. Temkin's diagram of kinetic roughening transition. Region A corresponds to smooth interface, B to rough interface. After BENNEMA and GILMER (1973). By courtesy of North-Holland Publ. Co., Amsterdam.

temperature increases.

(5) Under the same temperature conditions, closer packed interfaces have higher α values and thus show a smoother interface than less closely packed interfaces. In other words, the degree of roughness follows the order of morphological importance.

(6) Materials having larger solubility have smaller α values and thus show rougher interfaces than those having smaller solubility.

(7) If vapour, solution and melt growths are compared at the same temperature, α values are smaller and the interfaces rougher in solution growth than in vapour and melt growths, because ϕ_{sf} is involved in the former which diminishes the α value. However, the decrease of the α value due to ϕ_{sf} will be greatly compensated by much lower growth temperatures in a solution growth as compared to a melt growth. Thus, higher α values are expected for vapour and solution growths than for melt growth.

(8) Higher $\Delta\mu/kT$ results in a rougher interface. If this is taken into consideration, the solid-liquid interface will be smooth in case of vapour growth, followed by solution growth, and it will, in general, be rough in melt growth, since the vapour phase is the most dilute phase, followed by the solution phase, and the melt phase is the most condensed phase.

(9) Among solution growths, the interface roughnesses will vary depending on the strength of ϕ_{sf} and on the solubilities, i.e. on types of solvent. Systems with stronger ϕ_{sf} will show a rougher interface than the system of weaker ϕ_{sf}, because α values are smaller.

Summarizing these analyses, we may expect the following relations:

$$\alpha_{\text{vapour}} > \alpha_{\text{solution}} > \alpha_{\text{melt}}$$

and thus rougher interfaces for a melt growth as compared to a smoother interface for solution and vapour growths. Among the three types of growths, interfaces will be the smoothest in vapour growth.

6. *R(σ) Relation*

The growth rate of a crystal, which determines its morphology, is controlled by the roughness of the solid-liquid interface and the driving force $\Delta\mu/kT$. For a rough interface, the continuous growth mechanism is assumed, whereas for a smooth interface, either a 2DNG mechanism or a spiral growth mechanism, or a combined mechanism, is assumed. Thus, the relation between growth rate R and chemical potential difference $\Delta\mu/kT$ (or supersaturation σ) for the three mechanisms and for rough and smooth interfaces, may be schematically illustrated as in Fig. 7 (SUNAGAWA, 1977, 1981; SUNAGAWA and BENNEMA, 1982). Since in this figure, the BCF curve does not take into account the nucleation between the steps, and whereas the 2DNG curve does not take into account the interaction between steps via their

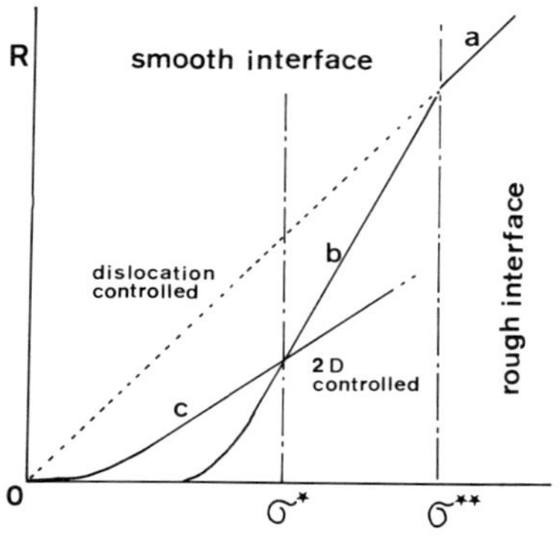

FIG. 7. Schematic illustration of growth rate R vs. chemical potential difference ($\Delta\mu/kT$) or supersaturation (σ) relation. Curve a, continuous growth; b, 2DNG mechanism; c, spiral growth mechanism. After SUNAGAWA (1981).

diffusion fields, the relations are qualitative ones. However, if these arguments are still valid, we see in Fig. 7 the existence of two transitional supersaturations σ^* and σ^{**}. In the supersaturation region over σ^{**}, the solid-liquid interface is rough and thus continuous growth occurs (curve a), whereas below σ^{**} the interface is smooth, where either 2DNG or spiral growth or a combined mechanism operates. In the supersaturation region below σ^*, the BCF curve (curve c) overwhelms the 2DNG curve (curve b), thus a dislocation-controlled spiral growth mechanism operates predominantly. In the supersaturation region between σ^* and σ^{**}, the 2DNG curve (curve b) predominates a BCF curve (curve c), thus a 2DNG mechanism is assumed for this region.

From the discussions given in Section 5, we can expect the following relations as to the values of σ^* and σ^{**}, when melt, solution and vapour growths are compared. It should also be mentioned that the increase of the α value implies the increase of σ^{**}, and thus σ^* (SUNAGAWA and BENNEMA, 1982). Thus, we can relate σ^{**} and σ^* to α.

(1) σ^* and σ^{**} should be very high in vapour growth, as compared to solution and melt growth. According to the calculation by BCF (1951), critical supersaturation at which 2DNG can occur, and thus roughly corresponding to σ^*, was around 25 ~ 50% for growth from pure vapour phase.

(2) σ^* and σ^{**} are much lower for solution growth than for vapour growth. σ^* was assumed to be less than 1% (KURODA *et al.*, 1977) for solution growth in general, and measured to be 2.8% for $NaClO_3$ from an aqueous solution growth (HOSOYA *et al.*, 1978). For high temperature solution growth, σ^* is estimated to be bit higher than in aqueous solution growth.

(3) σ^* and σ^{**} for vapour growth involving chemical reactions will be in between the vapour and solution growths.

(4) Melt growth usually occurs above σ^{**}.

These relations are schematically illustrated in Fig. 8.

In addition to the above, the following may be expected, since ϕ_{ss}, ϕ_{sf}, ϕ_{ff}, mass transfer rate, etc. vary depending on materials and environmental conditions.

(5) σ^* and σ^{**} are different depending on different materials.

(6) σ^* and σ^{**} vary depending on types of solvent.

(7) σ^* and σ^{**} vary depending on viscosity.

(8) σ^* and σ^{**} are different for different faces on the same crystal growing under the same condition.

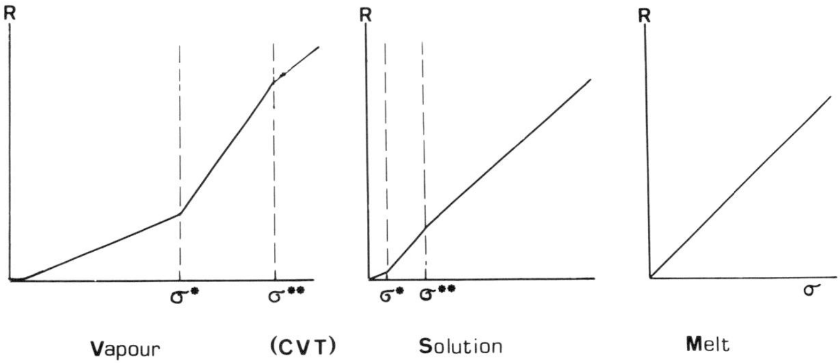

FIG. 8. Schematic illustration of σ^* and σ^{**} for growth from different phases. After SUNAGAWA (1981).

7. *Relations among Dendritic, Hopper and Polyhedral Morphologies*

If a crystal grows by a BCF mechanism, spiral layers spread outwards from the dislocations cropping the surface, which, in general, concentrate in the centre of a face. This will result in a crystal with polyhedral morphology bounded by macroscopically-flat surfaces (microscopically with a hill profile with its summit at the centre of the face). If a crystal grows by a 2DNG mechanism, there is a high possibility that it takes hopper (or skeletal) morphology, since the edges and corners of a face are exposed at a higher supersaturation region than at the centre of the face (Berg effect; refer to Section 4). Therefore, a larger probability of two dimensional nucleation is expected at the corners and edges than at the centre of the face. Growth layers will preferentially spread inward from the edges and corners, resulting in the appearance of hopper faces. If a crystal grows by a continuous growth mechanism, the morphology will be dendritic with curved surfaces and branches, since, in this case, growth is controlled by volume diffusion, and interface perturbation (morphological instability) will occur, principally due to impurity components. As to the origin of branching in dendritic growth, the reader may refer to WOODRUFF's book (1973) and the work by Langer and

MÜLLER-KRUMBHAAR (1977). In entering into a further higher supersaturation region, spherulites of various morphologies will appear, which is a morphology that crystals may take on under very high supersaturation conditions or under the influence of high viscosity.

Summarizing the above analyses and the discussions given in Sections 5 and 6, we may schematically illustrate the changes of morphologies of crystals depending on supersaturation as expressed in Fig. 9. It should be noted here that such morphological changes are also influenced by the sizes of crystals (KURODA *et al.*, 1977). In general, the region for polyhedral crystals expands wider when crystals are small in sizes.

From Figs. 8 and 9, and the summaries given in Sections 5 and 6, we can expect the following relating to the morphologies of crystals:

(1) Crystals grown from a melt phase will have a stronger tendency to show dendritic morphology than those grown from solution and vapour phases when growth occurs in an unconstrained environment.

(2) Crystals grown from a vapour phase will have a stronger tendency to take polyhedral morphology. Those grown from solution phases will have an intermediate tendency.

(3) Since α values and thus σ^* and σ^{**} are different in different materials, different crystals will show different morphologies under the same $\Delta\mu/kT$ condition, even if they all grow from the same kind of solution. This explains why NH_4Cl exclusively takes dendritic morphology, NaCl hopper morphology, whereas $NaClO_3$ takes polyhedral morphology, although they all grow from aqueous solution. The σ^* values of $NaClO_3$, NaCl and NH_4Cl in aqueous solution are estimated to be about 2.8%, 0.5% and 0.1%, respectively.

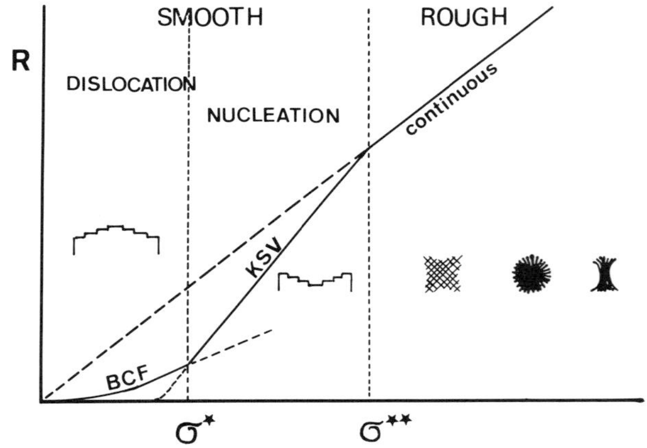

FIG. 9. Schematic drawing to show different morphologies of crystals depending on R (σ) relation.

(4) Due to the difference of ϕ_{sf} and thus α values, the same crystal species will exhibit different morphologies in different solution systems. High temperature solution systems will have a higher tendency to exhibit more polyhedral morphology than in aqueous solution growth.

(5) Since α values and, thus the roughness of interfaces, are different in different crystallographic directions through $\xi = E^{sl}/E^{cr}$, it is possible that on same crystal, one type of face grows by the BCF mechanism and shows a flat surface, whereas the other type grows by the 2DNG mechanism, thus showing a hopper face.

8. *Internal Morphologies*

From Fig. 9, we can expect that the morphology of crystals will change as the crystals grow in a closed system of solution or vapour phase, because supersaturation diminishes in such systems as the crystals grow. Crystals will take on dendritic morphology at the initial stage, followed by hopper morphology, and eventually polyhedral morphology at the later stage. Therefore, when a crystal is bisected at the centre and adequate techniques are applied to reveal physical and chemical heterogeneities in the crystal, it will exhibit internal heterogeneities (internal morphology) like those shown in Fig. 10, depending on the duration of respective stages (SUNAGAWA, 1977). At the centre is a dendritic core, surrounded by a hopper outline, and in the margin are polygonal growth bandings. By dendritic growth, a pattern will appear like the one seen at the centre of Fig. 10a, which corresponds to the so-called lineage structure (BUERGER, 1932). Spiral growth and 2DNG give polygonal growth bandings, since growth occurs on smooth interfaces and the growth rates vary alternatively and, thus, defect or impurity concentrations will also vary alternatively, due to the unbalanced growth and diffusion rates during the growth process or to a change in growth conditions.

It is expected that various internal morphologies will appear, depending on the conditions of growth, as illustrated in Fig. 10. If the major part of a crystal is formed in the region above σ^{**}, internal morphology like Fig. 10a will appear. If growth occurs from the initial stage in the region below σ^{*}, we can expect the pattern to be like Fig. 10d. In this case, we may expect dislocation configurations and surface microtopographs, such as shown in Fig. 11, provided that the crystal grows from either the vapour or solution phase in an open space (TANNER, 1976). For the crystals grown from the melt phase, dislocation configurations will be very different from this figure. When a crystal grows in a confined environment, in association with other solid bodies, dislocation configurations are also different, since there are many sites for dislocation generation, and growth has to take place through the interstitials of the co-existing solid bodies.

If environmental conditions change, further complexities of internal

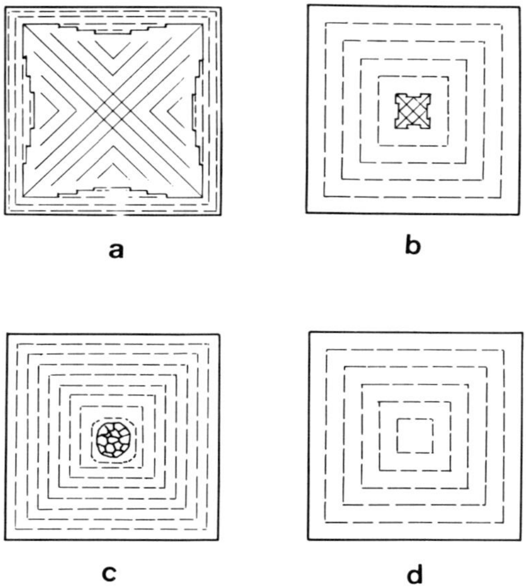

FIG. 10. Various internal morphologies expected for crystals grown in open space under unconstrained conditions. After SUNAGAWA (1977).

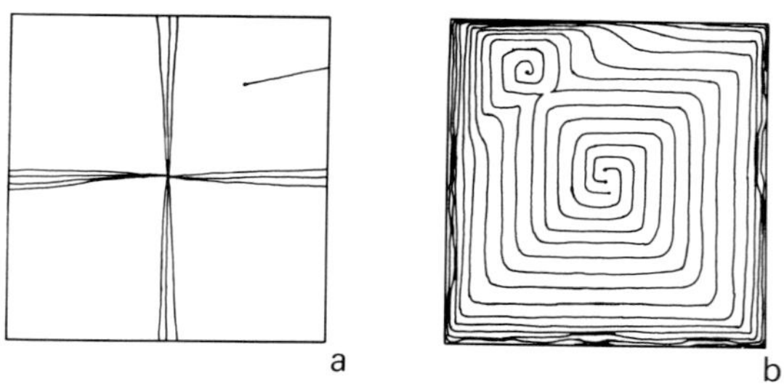

FIG. 11. Dislocation configuration (a) and surface microtopograph (b) expected for a solution grown crystal.

morphologies will appear, which are schematically shown in Fig. 12 (SUNAGAWA, 1981). Due to the change of environmental conditions, the system will again or repeatedly enter into the region over σ^{**}. Such changes in environmental conditions are expected to occur frequently in a natural crystal growth system, due to uplifts or intrusions of magmas, etc.

Since α values of the solid-liquid interfaces are different, depending on crystallographic directions, growth rates, and, thus impurity incorporation rates can vary in different crystallographic directions. This results in the formation of sector structures. Depending on relative growth rates in different

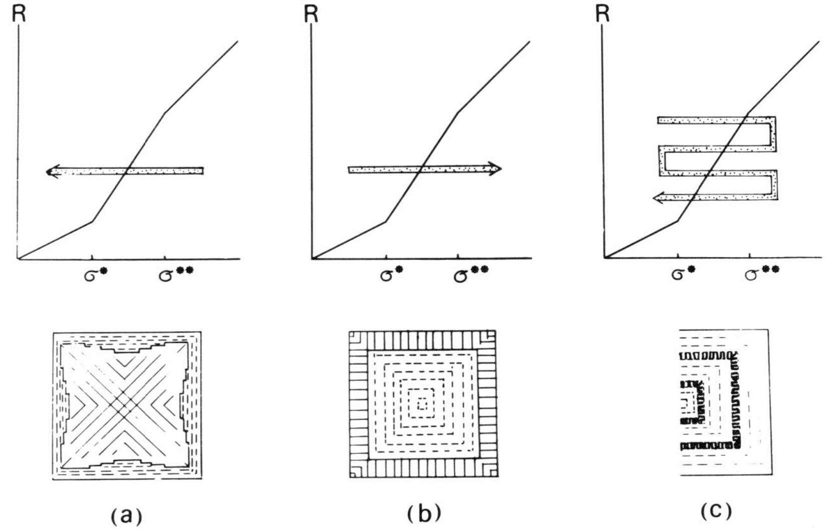

FIG. 12. Changes of internal morphologies of crystals due to the change in environmental conditions. After SUNAGAWA (1981).

directions, various morphologies of sector structures appear as schematically shown in Fig. 13.

Since growth rates are different, the rates of impurity incorporation, i.e. the chemistry, are different in different sectors. This problem has been treated extensively in the field of geological sciences, principally based on equilibrium considerations. However, it has been increasingly realized that kinetic consideration is needed in solving the problem, since effective distribution coefficients K_{eff} are found to be dependent upon growth kinetics and

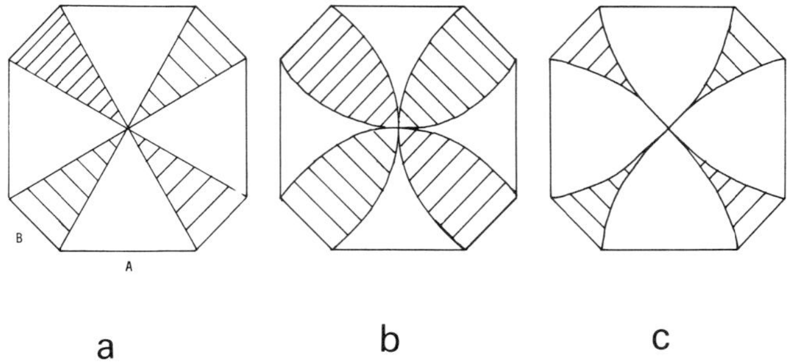

FIG. 13. Three types of sector structures depending on the relative growth rates of the faces concerned, (a) R_A, R_B constant, (b) R_A decreases, (c) R_B decreases as growth proceeds.

crystallographic directions. KITAMURA and SUNAGAWA (1977) have treated the problem, taking interface kinetics and morphological importance into account (this is related to α-factor). They arrived at a conclusion that elements with K_{eff} smaller than 1 concentrate in the sectors which have morphologically more important faces than in less important sectors. This is schematically shown in Fig. 14. Element distributions in solution growth and in olivine crystals have been investigated in relation to interface kinetics (growth rate, chemical potential difference and crystallographic direction) (TSUCHIYAMA *et al.*, 1981; KOUCHI, 1981). These results demonstrate the dependence of K_{eff} upon the interface kinetics.

In the above discussions, sector structures were analysed of the crystals bounded by more than two types of faces. Similar sector structures can be expected for a crystal bounded by only one type of equivalent face, since growth rates are not necessarily the same for crystallographically equivalent faces, due to the difference in dislocation concentrations, and thus strain concentrates along the boundaries of respective growth sectors.

Sector structures may be seen easily with the naked eye or under a polarizing microscope, or can become visible only when special techniques or treatments are applied. Even if they are not seen, this does not prove the absence of sector structures in the crystal.

(a) (b)

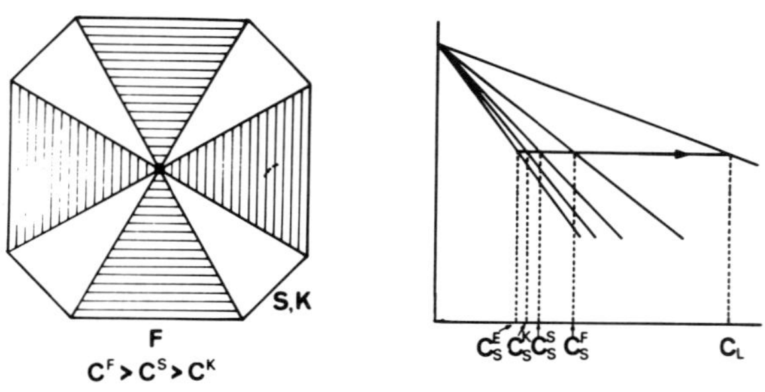

FIG. 14. A schematic illustration of the distribution of elements in different growth sectors depending on the morphological importance of crystal faces. (a) element distribution in different sectors, (b) phase diagram. F, flat face, S, stepped face, K, kinked face, as decreasing morphological importance. F, S, and K faces are according to Hartman and Perdok's definition. C_S concentration in solid, C_L in liquid, C_E in equilibrium.

9. *Morphology of Polyhedral Crystals*

Polyhedral crystals of one species can exhibit a wide variety of forms. This has been known as habit variations for many years. Calcite, pyrite, etc. are representative examples of natural crystals which show a wide variety of forms. It is evident that this is a problem relating to crystals grown in the supersaturation region below σ^*, i.e. dislocation-controlled growth region.

The so-called habit variations imply two different variations; (1) variations due to anisotropic development of crystallographically equivalent faces (Fig. 15), and (2) variations due to the difference in combination and development of different faces (Fig. 16). The former variation is derived from anisotropy involved in either environmental factors or structural factors. A directional flow of solution may cause elongated morphology. The direction of elongation will be reversed, depending on the rates and types of flow. When crystals grow at the bottom of a crucible or on the surface of an other crystal (epitaxial growth), they exhibit flattened morphology. All these modifications are due to the anisotropy involved in growth environment. If screw dislocations are active only in one or two directions among three equivalent directions, the crystal will become acicular or thin platy, respectively. This was exemplified in the case of NaCl and KCl crystals grown from an aqueous solution. This variation is due to structural anisotropy. A VLS mechanism of whisker growth (WAGNER and ELLIS, 1965) is another example of the effect of structural anisotropy.

The latter variations are due to the dependency of growth rates of different faces upon various growth conditions. This is schematically shown in Fig. 17. If the dependency of the growth rate of one face upon a certain growth condition is stronger than that of the other face, a different morphology appears under different growth conditions, as seen in this figure.

The growth rate of a face is determined by the roughness of the face under the same growth condition. Therefore, the problem is to find what are the essential factors to control the roughness of different faces and how the roughnesses change depending on these factors. For overall roughness of an interface, ϕ_{ss}, ϕ_{sf}, ϕ_{ff}, T, $\Delta\mu/kT$ are such factors. GILMER and JACKSON (1977)

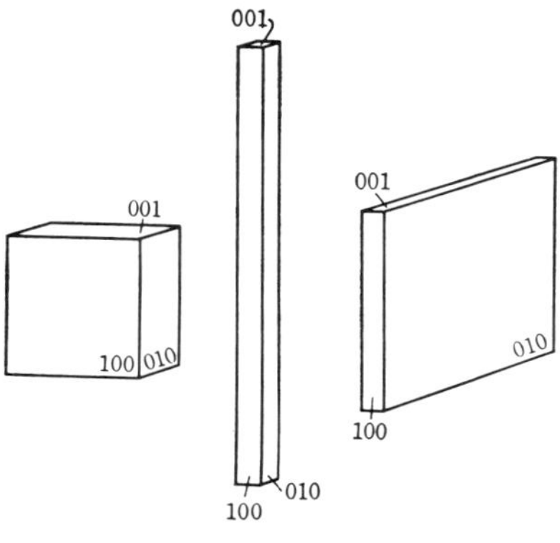

FIG. 15. Changes of morphology of polyhedral crystals depending on the anisotropic development of equivalent faces. Cubic crystals may take cubic, prismatic and thin platy forms depending on the anisotropic development of {100} faces.

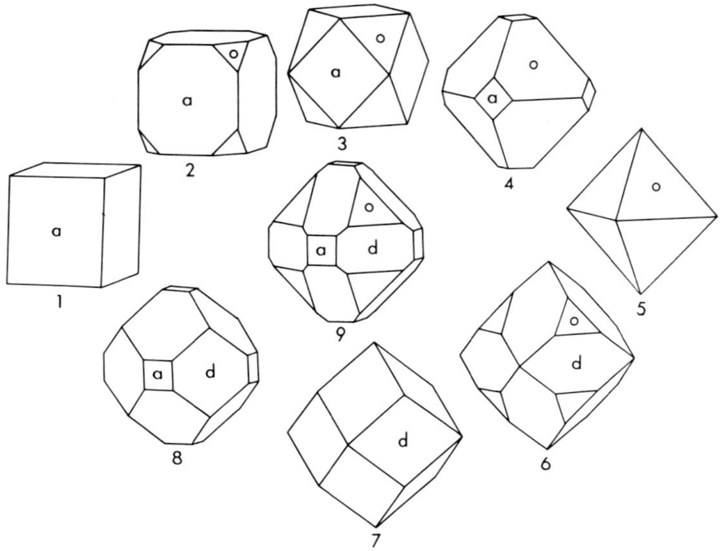

FIG. 16. The change of morphology of polyhedral crystal depending on the combination and development of different faces, a {100}; o {111}, d {110}.

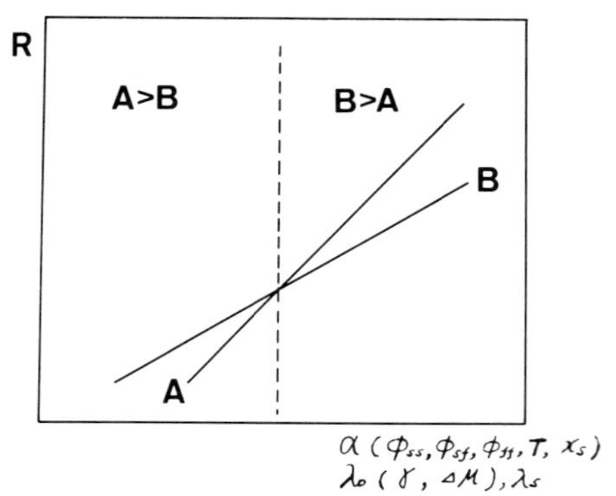

FIG. 17. A schematic drawing to explain habit changes depending on growth conditions. Vertical axis, growth rate R; horizontal axis, growth conditions.

carried out a series of computer simulation experiments on the relations between normalized growth rates R/k_a^+ and the chemical potential difference $\Delta\mu/kT$ for three faces of f.c.c. crystal, (111), (100) and (110), at different α values. Their results are summarized in Fig. 18, which shows that the three faces have different dependences on $\Delta\mu/kT$. Habit variations depending on temperature (T), the type of solvents (ϕ_{sf}, ϕ_{ff}), supersaturation ($\Delta\mu/kT$),

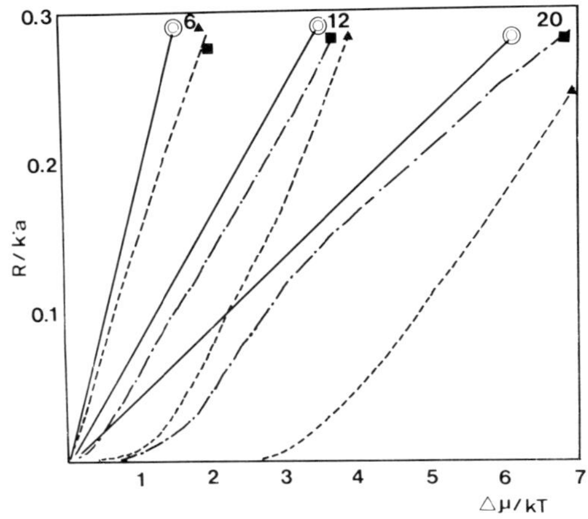

FIG. 18. Normalized growth rate R/k_a^+ vs. chemical potential difference relation. f.c.c. crystal, ▲ (111), ■ (100) and ○ (110) faces. Compiled diagram based on the data after GILMER and JACKSON (1977).

which were discussed in WELL's classical papers (1946), represent such cases.

In addition to the overall interface roughness, we have to pay attention to the roughness of steps of growth layers and to the densities of active dislocations, in the case of habit change, since habit variation is a problem relating to the dislocation-controlled growth. When a crystal grows by spiral growth mechanism under such a condition, the normal growth rate R of a face is determined by the height h, step separation λ_0, and the lateral advancing velocity v of the spiral layer; $R = hv/\lambda_0$. Under very low supersaturation, the number of active screw dislocations have some effect upon the normal growth rate, but this is not so severe under a relatively higher supersaturation region (BURTON *et al.*, 1951). We shall, therefore, analyse the morphology of growth spirals in the following section.

10. Morphology of Growth Spirals

SUNAGAWA and BENNEMA (1982) have recently summarized the present understandings of the morphology of growth spirals, both theoretical and experimental. We shall follow their analyses in this section.

The morphology of a growth spiral is defined by (1) whether it is polygonal or circular, and (2) the width of the step separation of the spiral. These are related to the roughness of the spiral step and the chemical potential difference. Therefore, we can apply the same concept of the α-factor to the spiral morphology.

If a spiral step contains many kinks, i.e. if it is rough, crystallizing particles can find points where they can settle without difficulty. The spiral step will then advance isotropically in any directions on the surface and become circular. If, on the other hand, the step contains less kink sites, i.e. smooth, the advancement will be strongly controlled by crystallographic directions and polygonal spiral will appear. It follows from this and from the discussions in Section 5 that the larger the α-factor, the more polygonized will be the spiral. Computer simulations by MÜLLER-KRUMBHAAR *et al.* (1977), and GILMER and JACKSON (1977) clearly demonstrate this change. Figure 19 shows the change of spiral morphology depending of α-factors for (100) face of a Kossel crystal. The spiral is circular for $\alpha = 6$, and becomes somewhat polygonal for $\alpha = 8$. For $\alpha = 12$, the step apart from the centre is almost polygonal, but the centre is still circular. Only when α is larger than 25, the spiral becomes completely polygonal, even at the centre. This implies that on the same face of the same crystal, spirals will be more polygonal as growth

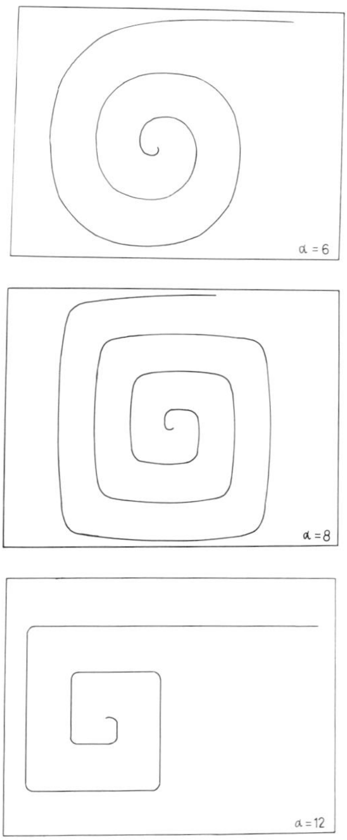

FIG. 19. Polygonization of a spiral as α increases. Computer simulation. After MÜLLER-KRUMBHAAR *et al.* (1977). By courtesy of North-Holland Pub. Co., Amsterdam.

temperatures decrease and as solid-liquid interaction becomes weaker. Similar change is expected when $\Delta\mu/kT$ decreases. For different faces of the same crystal grown under the same condition and in the same environment, morphologically more important faces will exhibit more polygonal spirals than morphologically less important faces. Figure 20 serves as a good example of demonstrating different degrees of polygonization of growth spirals on different faces, observed on an emerald crystal grown from a hydrothermal solution.

When a spiral is polygonized, the symmetry follows that of the structure of a spiral layer (SUNAGAWA, 1975).

According to BCF (1951), OHARA and REID (1973), and MÜLLER-KRUMBHAAR *et al.* (1977), step separation of a growth spiral λ_0 is related to the radius of the critical two-dimensional nuclei; $\lambda_0 = 19r^*$ for a circular spiral, and $\lambda_0 = 5.4r^*$ for a polygonal spiral. The r^* is related to edge-free energy γ chemical potential difference $\Delta\mu$, and molar volume Ω; $r^* = (\gamma/\Delta\mu)\Omega$. Thus, λ_0 will become smaller as:

(1) $\Delta\mu$ increases

(2) γ diminishes.

Since γ is smaller in solution growth than in vapour growth, λ_0 will be smaller

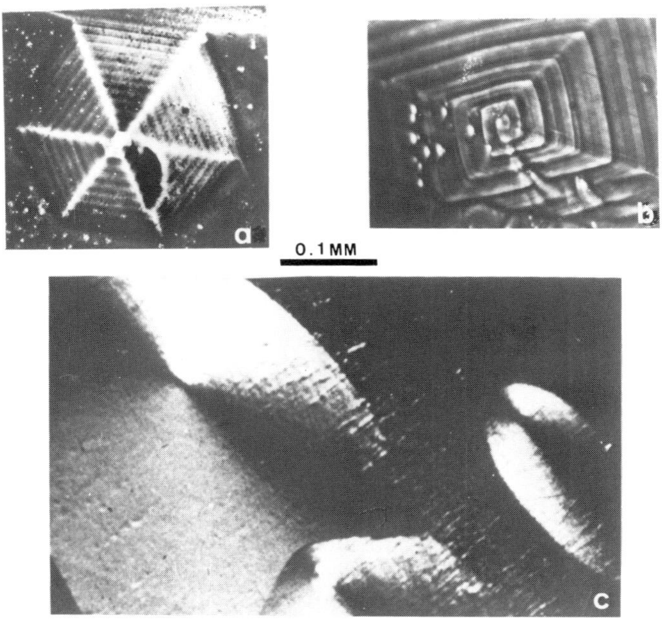

FIG. 20. Growth spirals observed on (0001) (a), $(10\bar{1}0)$ (b) and $(21\bar{3}1)$ (c) faces of an emerald crystal, grown from a hydrothermal solution, to show changes of the degree of polygonization depending on morphological importance.

on a solution grown crystal than on one that is vapour grown. Since impurity will have a strong effect upon γ, λ_0 will change drastically because of impurities. Since γ is different, depending on crystallographic directions, λ_0 will vary on different faces of the same crystal grown under the same conditions. This explains why growth spirals are easily observed on the basal plane of crystals with a sheet structure, but not easy for crystals with more an isotropic structure.

Through the differences in ϕ_{sf}, ϕ_{ff}, T, $\Delta\mu$, and γ, growth spirals on the same kind of faces of crystals grown from different phases, will exhibit different morphologies and step separations. SUNAGAWA (1978a) and SUNAGAWA and BENNEMA (1982) compared morphologies of growth spirals on crystals grown from different phases (pure vapour (PV), chemical transport (CVT), high temperature solution (HTS), and hydrothermal and aqueous solution (HS), as well as between natural and synthetic crystals). They compared the degree of polygonization and normalized step separations of growth spirals ($S = \lambda_0/h$) observed on the equivalent faces among these crystals, and concluded the following relations:

$$\alpha_{\mathrm{PV}} > \alpha_{\mathrm{CVT}} > \alpha_{\mathrm{HTS}} > \alpha_{\mathrm{HS}}$$
$$\gamma_{\mathrm{PV}} > \gamma_{\mathrm{CVT}} > \gamma_{\mathrm{HTS}} \simeq \gamma_{\mathrm{HS}}$$

$$\alpha_{\mathrm{natural}} > \alpha_{\mathrm{laboratory}}$$

$$\Delta\mu_{\mathrm{natural}} < \Delta\mu_{\mathrm{laboratory}}$$

$$T_{\mathrm{natural}} < T_{\mathrm{laboratory}}$$

$$\lambda_{S_{\mathrm{vapour}}} > \lambda_{S_{\mathrm{solution}}}$$
$$D_{S_{\mathrm{vapour}}} > D_{S_{\mathrm{solution}}}$$

$$\tau_{\mathrm{deads}_{\mathrm{vapour}}} > \tau_{\mathrm{deads}_{\mathrm{solution}}}$$

where λ_S is an average surface diffusion distance of admolecules, D_S the surface diffusion coefficient, τ_{deads} the average life time of stay on the surface of a growth unit. Table 2 is cited from SUNAGAWA's summary (1978) and Fig. 21 is an actual example.

It was a general tendency noted in Sunagawa's observations that:

(1) Spirals are more commonly observed in vapour grown crystals than in solution grown crystals. This is due to the lower σ^* in solution growth.

(2) Vapour grown crystals exhibit one- or two-order wider step separations as compared to solution grown crystals.

(3) Natural crystals exhibit, in general, more polygonal spirals with much wider step separations than laboratory-grown crystals, when crystals grown from similar environments are compared. This is due to lower T, $\Delta\mu$, and impure growth environment in natural crystallization.

Since the normal growth rate of a face, when it grows by a spiral growth

TABLE 2. Comparison of morphologies of growth spirals on crystals grown from different conditions After SUNAGAWA (1978a).

			SiC 0001	Fe_2O_3 0001	Al_2O_3 0001	Mica minerals 001		Beryl 0001, $10\bar{1}0$
						Phlogopite	Muscovite	
Vapour	PV		10^4 Circular			10^3 Circular		
	CVT	Natural		10^4 Polygonal		10^5 Polygonal		10^4 Polygonal
	CVT	Synthetic	10^3 Polygonal	10^3 Polygonal to circular	10^4 Circular			
Solution	HTS			10^2 Polygonal to circular	10^2 Polygonal			10^2 Circular
	HS	Natural					10^2 Polygonal	
	HS	Synthetic					10^2 Circular	10^2–10^3 Polygonal

PV pure vapour phase; CVT chemical vapour transport; HTS high temperature solution; HS hydrothermal solution; 10^2–10^4 represent the ratio $S = \lambda_0/h$, where λ_0 is the step separation and h the step height.

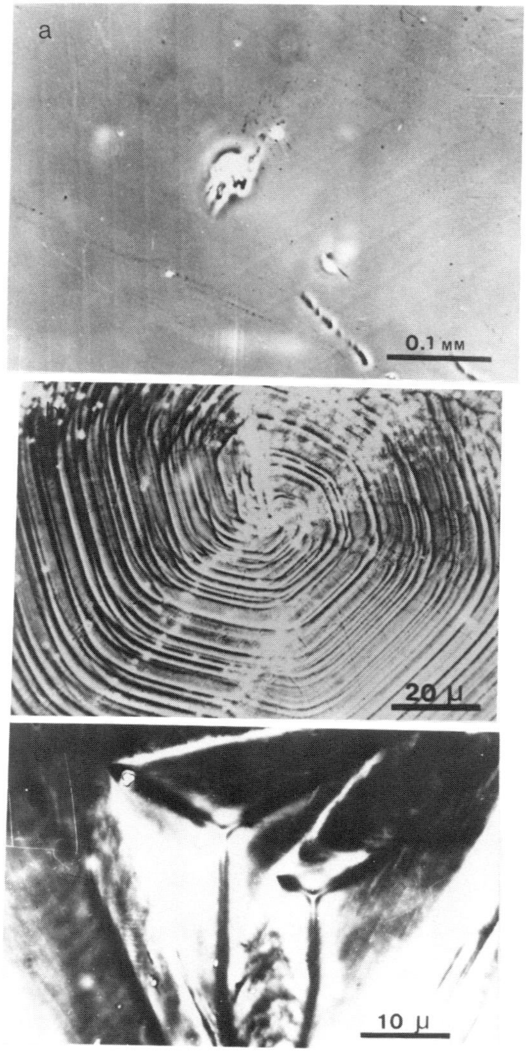

FIG. 21. An example of surface microtopographs to show the changes in the degree of polygonization and step separations depending on growth environments. Hematite. (a) natural, (b) artificial vapour growth (chemical transport), (c) artificial high-temperature solution growth. After SUNAGAWA (1978a).

mechanism, is defined by $R = hv/\lambda_0$, γ and $\Delta\mu$ have an effect upon habit variations, in addition to the factors discussed in Section 9.

From the discussions in Sections 9 and 10, we may assume that the major factors affecting the morphology of polyhedral crystals are supersaturation, temperature, types of solvent, and impurity. We may also expect that vapour-grown crystals will have a more flattened morphology than solution-grown crystals, since λ_0 is much wider in the former. This is exemplified by the characteristic differences in morphologies between hematite and corundum crystals grown from vapour and solution phases. Habit variations depending on supersaturation (NaCl, KCl etc.), temperature (diamond, garnet, etc.), solvent (many examples), impurities (numerous examples), have been known among natural and laboratorically grown crystals.

11. Size of Growth Unit

The preceding discussions in Sections 5–10 are based on an assumption that the growth unit is either an atomic or ionic entity. This is the basic assumption of any existing theories of crystal growth mechanism, including morphology of crystals. In fact, this assumption is justified by numerous observations on surface microtopographs of crystal faces, or by the analysis of rate measurements. However, the assumption is not necessarily true under any growth conditions. There are possibilities that growth units are much larger than atomic or ionic entities and growth occurs through the accumulation process of larger growth units. Such a situation may occur, particularly when the system contains a large amount of impurities that can act as heterogeneous nucleation sites. Recently, GLASNER *et al.* (1974), by analysing thermal bahavior in nucleation and growth of NaCl crystals, put forward a bloc nuclei model in which they assumed the existence of larger growth units (they called these bloc nuclei) and the growth by accumulation of bloc nuclei. Step patterns, including spiral steps, can appear on crystal surfaces, even by such growth mechanism.

If the size of growth units is different, morphology of crystals or growth spirals will be modified to some extent from those expected on the basis of atomic or ionic growth units. Growth spirals will have high step heights, jagged steps and wavy surfaces, as contrast to mono-molecular heights, smooth steps and flat surfaces of growth spirals formed by ionic entities. Morphology of crystals will be modified accordingly. The latter problem has been analysed, taking striated faces and anisotropic development of equivalent faces in polar crystals as examples, by expanding the PBC concept so as to include larger growth units (MIYATA *et al.*, 1980).

In natural crystallization, the existence of larger growth units is highly probable. So we have to check this problem as well.

In the following sections, we shall try to find the characteristics of crystal growth taking place in Nature, principally based on the preceding discussions.

12. Crystal Growth in Magmas

Crystals of most rock-forming minerals in not only plutonic rocks but also those occurring as phenocrysts of volcanic rocks, usually do not show dendritic cores. This is also the case for the internal morphologies of crystals formed in the groundmass of volcanic rocks. Even in the chilled margin of lavas or volcanic rocks intruded into the surrounding strata, dendritic or hopper crystals are seldom observed. Dendritic or hopper crystals are formed only when magmas are solidified under extreme conditions. Feather-like olivine dendrites found in pillow lavas of basalt erupted on the ocean floor, hopper crystals of magnetite in some basaltic lavas, and so-called crystallites observed in glassy volcanic rocks, are the exceptional cases in which dendritic or hopper morphology of crystals is universally noticed. Internal morphology, like that shown in Fig. 10d is a universally encountered morphology of rock-forming minerals. In contrast, morphologies like those shown in Fig. 10a or b are those commonly encountered morphologies in laboratory crystallization. Representative examples are silicate crystals formed in slags. Slow cooling rates in magma, low thermal conductivity, and large volumes of magmas are perhaps responsible for the lack of dendritic or hopper morphologies in most rock-forming minerals. In addition to these factors, we have to take into account the fact that crystallization in magma is growth from solution phase.

The above characteristics seen in the morphologies of rock-forming minerals crystallized in magmas, suggest that a growth of crystals of rock-forming minerals in magma takes place, in most cases, under the supersaturation region below σ^*, i.e. under the condition of dislocation control growth. Since crystals of rock-forming minerals usually contain too many lattice defects to apply X-ray topography, and are difficult to separate from the host while keeping the as-grown surfaces undamaged so that techniques of surface microtopograph can be applied, it is not easy to prove that most rock-forming minerals grow by a spiral growth mechanism. The only exception in which this was definitely proved is diamond crystals in kimberlite. LANG (e.g. 1965) carried out an extensive X-ray topographic study of single crystalline diamond crystals, and demonstrated that they exhibit bundles of dislocations radiating from the central point of a crystal, together with growth bandings parallel to the octahedral faces. As already discussed, such dislocation configurations can be expected when crystals grow from a solution phase in an open space.

SUNAGAWA *et al.* (1982) have made a combined study of X-ray topography and surface microtopography of a few diamond octahedral crystals from Siberia, which received a very little dissolution and thus preserved fresh as-grown octahedral surfaces. They demonstrated conclusively that the summits of growth pyramids, consisting of growth layers thinner than 5 Å correspond to the outcrop of screw dislocations radiating from the centre of the crystal. This observation definitely proved that the natural octahedral crystals of diamond grew from a solution phase of low supersaturation by a spiral growth mechanism through incorporation of atomic entities, from the very beginning of their growth history. Natural diamond crystals of the single crystalline type are not formed by metamorphic process, nor from the melt phase, but from the solution phase of low supersaturation in which enough space for free mass transport is assumed. We note here that some natural diamonds grew under the supersaturation region over σ^{**}, which became polycrystalline diamonds like bort, cuboid or carbonado. However, we believe that the majority of single crystalline diamonds are formed by a spiral growth mechanism.

Ideal growth spirals with a mono-molecular step height have only been observed in diamonds, among the many rock-forming minerals. This is due to the difficulty of securing fresh as-grown crystal faces of these minerals. The only other rock-forming minerals on which growth spirals have been observed are anorthite megacrystals occurring as crystal lapilli ejected by volcanic eruption. TSUCHIYAMA (1976) observed spirals with step heights of a few thousands Å, originating from the terminations of twin lamellae, on the surface of anorthite crystals from several localities in Japan. The spirals (Fig. 22) are not monomolecular in height, suggesting larger growth units in the growth of these crystals.

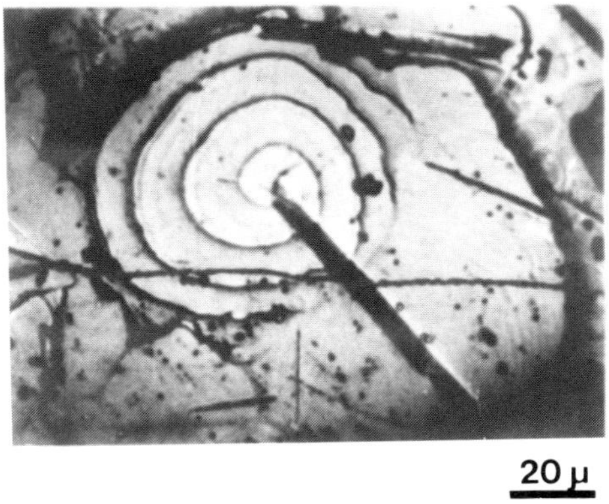

FIG. 22. Growth spirals with high step heights observed on anorthite megacryst. Photo. by A. Tsuchiyama.

13. *Crystal Growth in Pegmatite*

At the final stage of differentiation of magma, volatile components such as H_2O, CO_2, F_2, Cl_2, etc. are enriched to form a high-temperature super-critical vapour phase. Some pegmatite minerals and minerals of pneumatolytic ore deposits are formed at this stage. Crystals formed by post-volcanic action around volcanic fumaroles, such as hematite, or some druse minerals in lava flows, such as phlogopite, are also formed from such vapour phases. It is evident that in these crystallizations, chemical reactions are always involved and, thus, the ϕ_{sf} factor. Therefore, α and σ^* values are lower than those in pure vapour growth, but higher than those in solution growth. Furthermore, T and $\Delta\mu/kT$ are expected to be much lower than those in laboratorical crystallization. Thus, we can expect that growth spirals will be more commonly found on these crystals than in crystals grown from a solution phase and those synthesized in a laboratory. In fact, the first observation to support the spiral growth theory was made on prism faces of beryl crystals occurring from pegmatite (GRIFFIN, 1956). The spiral was mono-molecular in step height and had a very wide step separation. From the λ_0 values of these spirals, Griffin evaluated supersaturation to be as low as 0.2% when the beryl crystal was formed. Growth spirals with a mono-molecular step height and wide-step separations have been observed in topaz crystals from pegmatite and pnumatolytic deposits.

Similar typical growth spirals have been observed on many crystals grown from vapour phases. Hematite crystals occurring in volcanoes are formed by a chemical vapor transport process, Cl_2 being the transport agent, and are the representative example whose surface microtopographs have been investigated extensively. According to Sunagawa's observations on hematite crystals occurring from all over the world, they universally exhibit mono-molecular growth spirals with the thinnest step height being only 2.3Å, which corresponds to the height of a single layer of Fe_2O_3 (SUNAGAWA, 1960, 1961, 1962). Phlogopite occurring in druses of basaltic lava and biotite of similar occurrence are other examples in which typical growth spirals with mono-molecular step heights and wide step separations are universally observed (SUNAGAWA, 1964). Growth spirals with mono-molecular step heights and wide step separations are assumed to be a common feature for crystals grown from the vapour phase in Nature. This indicates that the growth units in these crystallizations are ionic entities and $\Delta\mu/kT$ was very low, compared to laboratorical vapour growth. When the same crystal species and the same face are compared between natural and laboratorically synthesized crystals (e.g., hematite and phlogopite), the step separations are, in general, two orders of magnitude wider in natural crystals than in those laboratorically systhesized crystals (SUNAGAWA, 1978a, b). The growth spirals on natural crystals grown from vapour phase are mostly polygonal, even at the centre of the spirals, and the morphology follows the symmetry of the face on which they develop. Ideally, circular growth spirals are not observed on natural crystals, which makes a good contrast with the spirals observed on artificial crystals grown from pure vapour phase. This difference is due to much lower growth temperatures and thus lower α values in natural crystallization.

Since pegmatite minerals often provide high quality crystals on which informative X-ray topographs may be obtained, some X-ray topographic works have been made on topaz, beryl, spodumene, quartz etc. Growth bandings and dislocation bundles, which demonstrate growth in an open space and on a smooth interface, have been commonly observed. A characteristic feature commonly observed in the X-ray topographs of pegmatite minerals is that two different stages are universally found in single crystals. There is a dissolution period between the two stages, and the morphologies of the crystal change drastically between the two stages. This variation was well demonstrated in topaz crystals by ISOGAMI and SUNAGAWA (1975). It is not yet certain whether this change corresponds to the change from the vapour phase to the hydrothermal solution phase as the temperature decreased or to the change in the chemistry of the vapour phase.

14. *Crystal Growth in Hydrothermal Solution*

In contrast to a common occurrence of mono-molecular growth spirals on vapour grown crystals, such spirals are not commonly observed in the crystals formed in hydrothermal solutions. On natural quartz crystals, typical growth spirals have not been reported so far, in spite of a vast number of observations. The rhombohedral faces exhibit triangular growth hillocks, and the prism faces striations parallel to the edges with the rhombohedral faces, but neither faces exhibit mono-molecular growth spirals with wide step separations. Surface microtopographs of natural quartz crystals change systematically from higher temperature origins in pegmatite and pnumatolytic veins to lower temperature origins in epithermal veins and further in geodes. Crystals of lower temperature origins are characterized by more rounded growth layers of higher step heights on rhombohedral faces and a more pronounced development of striations on prism faces than those observed on the crystals of higher temperature origins. This indicates that the probability of polymerized SiO_2 in a hydrothermal solution increases as the temperature decreases. The formation of chalcedony in geodes is considered to take place by the coagulation and accumulation of such polymerized entities (SUNAGAWA and OHTA, 1976).

Another example of minerals of hydrothermal origin whose surface microtopographs have been studied extensively is pyrite, on which ideal mono-molecular growth spirals have only exceptionally been observed (SUNAGAWA and ENDO, 1968; ENDO, 1978). Growth layers developing on (100), (111) and (210) faces usually have macro-step heights, which are formed not by bunching of mono-molecular layers but from the beginning, and have jagged steps and more or less wavy surfaces. Sphalerite, galena, calcite, barite and other minerals of hydrothermal origin also exhibit similar surface microtopographs. Crystals grown from aqueous solution, such as KCl, NaCl, etc. also have similar characteristics of surface microtopography, unless they are grown under very low supersaturation. Mono-molecular growth spirals are observed only when crystals grow under a very low supersaturation (HUMAN, 1981) in aqueous solutions, and are not a common feature for crystals grown from such phases. This makes a good contrast with the crystals grown from the vapour phase, on which mono-molecular growth spirals are universally observed. The difference between vapour and aqueous solution growth is partly due to much a lower σ^* for a solution growth than for a vapour growth. But we should also take into account that in an aqueous solution growth, larger growth units more probably exist than in a vapour growth. Higher step height, rugged steps and wavy surfaces commonly encountered on the surface microtopographs of crystals grown from aqueous and hydrothermal solutions, are considered as due to such larger growth units.

15. Crystal Growth in Surficial and Sedimentary Environments

Minerals of the following occurrences are formed in a surficial or sedimentary environment, under low temperatures and atmospheric pressure conditions:

Evaporite minerals formed by the evaporation of saline water;

Minerals of secondary enrichment or from the oxidation zone of metallic ore deposits;

Minerals precipitated from ground water, like opal in sedimentary strata and stalactite in limestone caves;

Minerals formed by weathering processes;

Authigenic minerals in sedimentary rocks formed by early diagenesis.

The growth of these minerals occurs exclusively in aqueous solutions at a low temperature and atmospheric pressure. Thus, the understanding on the mechanism of crystal growth from an aqueous solution phase (e.g. refer to the discussions in Sections 2–6; Bennema, 1967; Ohara and Reid, 1973) are applicable in principle to the growth of crystals in these environments. In aqueous solution growth, σ^* is much lower than in vapour growth. Thus the supersaturation region in which a spiral growth mechanism is operative is low and narrow, leading to less probability of encountering typical spirals in these crystals as compared to vapour-grown crystals. Attention should also be given to the following in analysing the characteristics of crystal growth in these environments: Natural aqueous solutions always contain impurity elements or foreign particles which can act as inhibitors for nucleation. It is thus highly possible that natural aqueous solutions contain larger particles than ionic entities, which act as growth units. Due to low temperature conditions and the existence of such growth units, crystals will be small in size and take irregular or fibrous aggregation rather than form polyhedral crystals. The growth of crystals taking place in early diagenesis has theoretically been dealt with by Berner (1980).

16. Crystal Growth in Metasomatic Environment

The term metasomatism is used when the original solid rocks react with permeated high temperature vapours or hydrothermal solutions and change into different mineral assemblages. The original rocks and the products are different both chemically and mineralogically.

Metasomatism can occur under a broad spectrum of *P-T* conditions, as well as chemistries of both original rocks and permeated vapours or solutions. The so-called skarn or greisen minerals and contact metamorphic ore deposits are considered to have been formed by the reaction of high-temperature vapours originating from magmas with pre-existing solid rocks; minerals in alteration zones around hydrothermal veins, in hydrothermal metasomatic deposits such as sericite or kaolin group minerals, and also those in alteration zones around volcanic fumaroles or in geothermal areas by the action of hydrothermal solutions of varying temperatures.

It has not been well understood whether the growth of crystals in metasomatic environments takes place in the solid state or via a solution phase. Not much work has been done in this connection so far, although the fact that very fine crystallographic growth bandings often develop in garnet or other skarn minerals may suggest growth via a solution phase.

To understand how crystals grow in metasomatic environments, the decoration technique of electron microscopy was applied to the surface microtopographs of kaolin group minerals (SUNAGAWA and KOSHINO, 1975) and sericite (SUNAGAWA *et al.*, 1975) formed in hydrothermal metasomatic deposits. It was found that these crystals invariably show polygonal growth spirals with mono-molecular step heights on the (001) faces, and they usually occur as isolated crystals (Fig. 23). The coalescence of several crystals, which is universally observed for the same minerals occurring in hydrothermal veins (Fig. 24), is seldom observed on these crystals. These observations suggest that the crystals must have been precipitated directly from a solution with a low supersaturation which was not moving violently. They are not formed directly from solid phases, and neither are they formed in a violently-moving solution. When crystals grow in a flowing solution a coalescence of crystals can occur frequently. It is therefore suggested that in the metasomatic environment, permeated high temperature vapours or hydrothermal solutions react with the original solid rocks and dissolve the latter forming vapours or solutions with different chemistries, from which new crystals are precipitated. They are not formed by the direct chemical reaction between the original rocks and the vapours or solutions. It is also assumed that such vapours or solutions appear only in a form of thin film, and not in quantity.

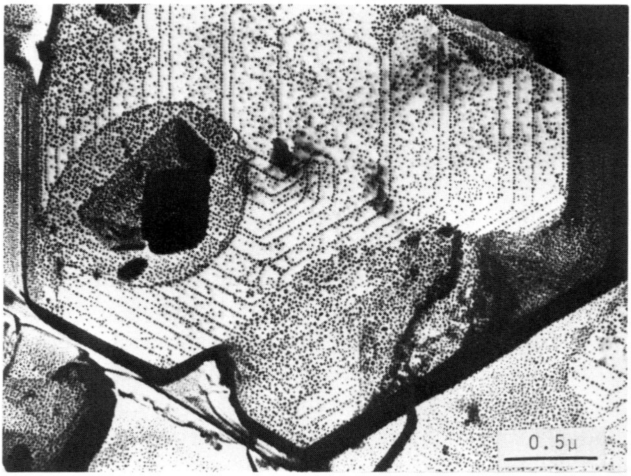

FIG. 23. Growth spirals observed on dickite crystals formed by hydrothermal metasomatism, taken by the decoration technique. Step height of the spiral is 10Å. Photo. by Y. Koshino.

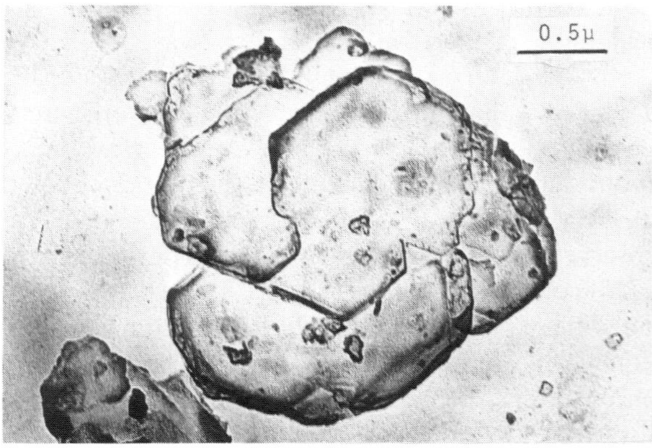

FIG. 24. Electron photomicrograph of chlorite crystals showing coalescence of crystals. The crystals are from a vein type deposit. Decoration technique.

17. Crystal Growth in a Metamorphic Environment

Although thermodynamic studies have been extensively carried out as to the *P-T* conditions of regional metamorphism, studies on how crystals are formed in metamorphic environments have not been done systematically, except for special cases such as rotated garnet crystals (ROSENFELD, 1970).

In pure solid-solid crystallization (recrystallization), grain growth occurs principally due to the release of strain energy by the application of thermal energy, and volatile elements or solvent components do not play an essential role. The formation of crystalline marble from limestone or of quartzite from arkose sandstone belongs to a crystallization of this type. However, in metamorphic processes, volatile components such as H_2O, CO_2, F_2, Cl_2 should have played an important role.

Although crystals of many metamorphic minerals have irregular outlines (xenoblasts), unlike the case of igneous rocks, there are some metamorphic minerals with habit faces (idioblasts), and some developing as large as 30 cm across (porphyroblasts). Characteristics of well-developed metamorphic crystals are that they are always full of foreign mineral grains, showing that crystals grew in the presence of a large number of solid particles, namely in a constrained environment, which is one of the principal differences between crystal growths in metamorphic and igneous environments.

In order to understand the growth mechanism in metamorphic environments, TOMURA *et al.* (1979) applied the decoration technique of electron microscopy to porphyroblastic white mica and sericite occurring in different metamorphic zones in Shiragayama district, Shikoku. The ratio of sericite to porphyroblastic white mica decreases from the chlorite zone, through garnet zone to the biotite zone as the metamorphic grade increases. Abrupt changes in this ratio are noted at the boundaries of neighbouring zones.

In contrast to the common occurrence of typical growth spirals on sericite crystals occurring in hydrothermal veins and metasomatic deposits, neither sericite nor porphyroblastic white mica investigated show spiral patterns. Two different types of step patterns have been observed on these crystals: a parallel step system consisting of closed loops of smooth layers with narrow spacings, and a saw-tooth step system. Examples are shown in Fig. 25. TOMURA *et al.* (1979) interpreted that the parallel step system represents the surface microtopograph formed either by growth or under near equilibrium conditions, whereas the saw-tooth step system was formed either by dissolution or under rigorous kinetic conditions. The parallel step system has been observed on porphyroblastic white micas occurring in the middle portion of each metamorphic zone and on all sericite crystals, whereas the saw-tooth step system has been observed on most sericites and on porphyroblastic white micas occurring along the boundaries of the neighbouring metamorphic zones. Step systems representing growth are observed on both smaller (sericite) and larger muscovite (porphyroblastic white mica) in the middle of each metamorphic zone, whereas those representing dissolution are observed only on smaller crystals in each zone and also on larger crystals occurring at the boundaries of metamorphic zones. TOMURA *et al.* (1979) suggested that the observed variations in the surface microtopographic characteristics are due to Ostwald ripening during a kind of sintering process in which interstitial water

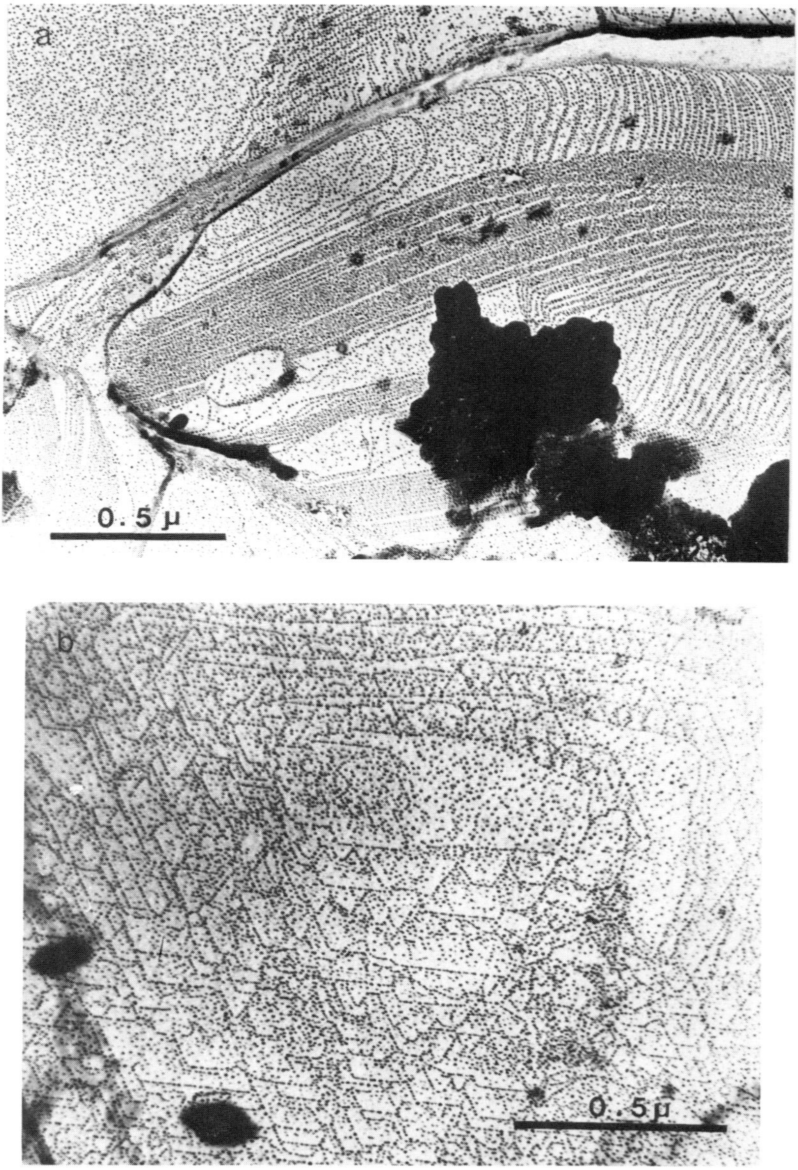

FIG. 25. Parallel step system (a) and saw-tooth step system (b) observed on muscovite crystals in regional metamorphic rocks. Electron photomicrographs taken by the decoration technique. After TOMURA *et al.* (1979).

plays an essential role, and that the drastic changes along the metamorphic zone boundaries are due to dehydration reactions.

Judged from these observations, it is considered that growth of crystals in metamorphic environments is similar to a sintering process with interstitial solvent components, and the Ostwald ripening plays an important role, in which actual growth proceeds through a thin film of interstitial solution. These observations again suggest the importance of solution growth in natural crystallization.

Crystals in regional metamorphic rocks are, in general, too imperfect to offer meaningful X-ray topographs, since they are full of inclusions. However, recent X-ray topographic observations on emerald crystals in metamorphic rocks have shown marked differences in the dislocation configurations from those observed in beryl crystals in pegmatite (GRAZIANI *et al.*, 1980). The difference is due to the former crystals containing many sites for dislocation generation, since the crystals grew in co-existence of other solid particles.

18. Concluding Remarks

Judging from the characteristics of morphologies of natural crystals, the following conclusions may be mentioned:

(1) Crystal growth in magma should be treated as growth from high-temperature solutions, not as growth from the melt phase.

(2) Crystals grow from high-temperature vapours by an ideal spiral growth mechanism via the addition of ionic entities, whereas in hydrothermal solution most crystals grow by the addition of particles of larger sizes which are moving around in solutions, except in solutions of very low supersaturation.

(3) The growth of crystals in surficial and sedimentary environments is treated as the growth from low temperature aqueous solutions, which contain many inhibitors.

(4) In the metasomatic environment, crystals precipitate from the vapour or solutions of low supersaturation, which are formed by the reaction between the original rocks and permeated vapour or solution.

(5) The growth of crystals in a metamorphic environment is regarded as a sintering process, in which interstitial solvent components play an essential role, and the actual growth process takes place via solution phase. Ostwald ripening plays an important role in this process.

It should be stressed that in all natural crystallization, growth of crystals from solution phases, either from high temperature or aqueous solutions, plays the major part. We may summarize the present analysis as schematically illustrated in Fig. 26.

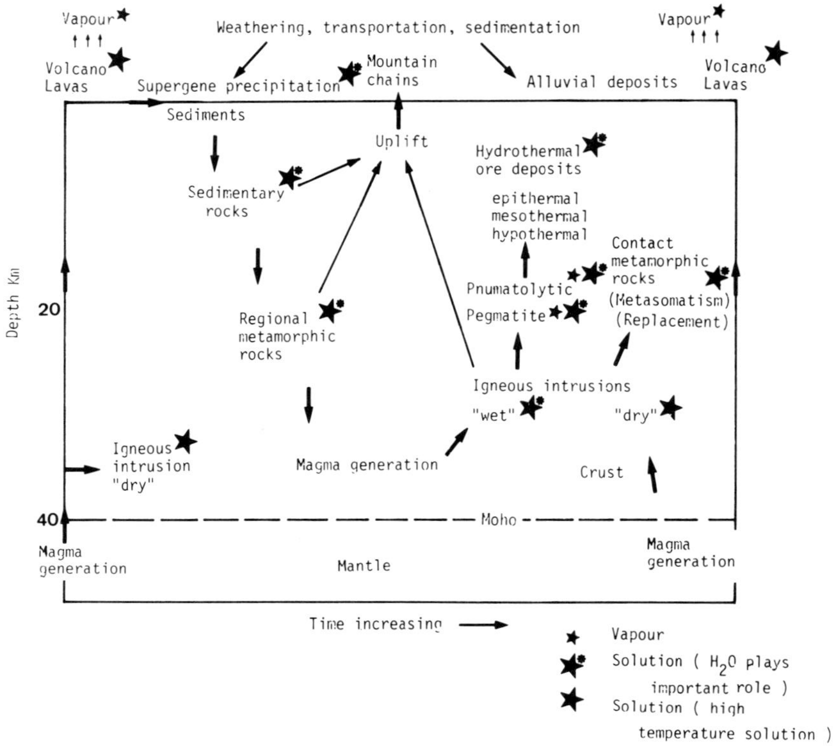

FIG. 26. A schematic diagram to show characteristics of crystal growth in various geological processes. Based on the diagram of WYLLIE (1971). By coutesy of John Wiley and Sons, New York.

REFERENCES

BENNEMA P. (1967): Analysis of crystal growth models for slightly supersaturated solutions, *J. Cryst. Growth*, **1**, 278–286.

BENNEMA, P. and GILMER, G. H. (1973): Kinetics of crystal growth, in *Crystal Growth, An Introduction*, Ed. P. Hartman, North Holland-Pub. Co., Amsterdam, pp. 263–327.

BENNEMA, P. and VAN DER EERDEN, J P. (1977): Crystal growth from solution: development in computer simulation, *J. Cryst. Growth*, **42**, 201—213.

BERG, W. F. (1938): Crystal growth from solutions, *Proc. R. Soc. Lond., A*, **164**, 79–95.

BERNER. R. A. (1980): *Early Diagenesis, A Theoretical Approach*, Princeton Univ. Press, N. J., 241 pp.

BUERGER, M. J. (1932): The significance of "block structure" in crystals, *Am. Mineral.*, **17**, 177–191.

BURTON, W. K., CABRERA, N. and FRANK, F. C. (1951): The growth of crystals and equilibrium structure of their surfaces, *Philos. Trans., A*, **243**, 299–358.

DOWTY, E. (1980): Crystal growth and nucleation theory and the numerical simulation of igneous crystallization, in *Physics of Magmatic Processes*, Ed. R. B. Hargraves, Princeton Univ. Press, N. J., pp. 487–551.

ELWELL, D. and SCHEEL, H. J. (1975): *Crystal Growth from High Temperature Solutions*, Academic Press, London, 634 pp.

ENDO, Y. (1978): Surface microtopographic study of pyrite crystals, *Bull. Geol. Surv. Japan*, **29**, 701–745.

FORRENIUS, M. (1959): Application d'une méthode interférométrique en lumière blanche polarisée à l'etude de la croissance cristalline, *Bull. Soc. Franc. Miner. Cryst.*, **82**, 343–360.

GILMER, G. H. and JACKSON, K. A. (1977): Computer simulation of crystal growth, in *Current Topics in Materials Science*, 2, Eds. E. Kaldis and H. J. Scheel, North-Holland Pub. Co., Amsterdam, pp. 80–114.

GLASNER, A. and TASSA, M. (1974): The thermal effect of nucleation and crystallization of KBr and NaCl solutions, I. Experimental and mathematics, illustrated, *J. Cryst. Growth*, **13**/**14**, 441–444; II. The heat of nucleation and the supersaturated solution, III. The heat of crystallization and the coprecipitation of lead ions, *Israel J. Chem.*, **12**, 799–816, 817–826.

GRAZIANI, G., SCANDALE, E. and ZARKA, A. (1980): Development and genetic medium of a beryl single crystal, *Collected Abstract, IMA–12, Orleans*, p. 12.

GRIFFIN, L. J. (1950): Observation of unimolecular growth steps on crystal surfaces, *Phil. Mag.*, **41**, 196–199.

HARTMAN, P. and PERDOK, W. G. (1955): On the relation between structure and morphology of crystals, *Acta Cryst.* **8**, 49–52, 521–524, 525–529.

HOSOYA, S., KITAMURA, M. and MIYATA, T. (1978): Growth mechanisms of $NaClO_3$ and KBr from aqueous solutions under relatively high supersaturation ranges, *Mineral. J. (Japan)*, **9**, 147–168.

HUMAN, H. J., van der EERDEN, J. P., JETTEN, L. A. M. J. and ODEKERKEN, J. G. M. (1981): On the roughening transition of biphenyl: Transition of faceted to non-faceted growth of biphenyl for growth from different organic solvents and the melt, *J. Cryst. Growth*, **51**, 589–600.

HUMAN, H. J. (1981): Crystal growth from the liquid state, experimental studies on crystal growth mechanisms and kinetics, Ph. D. Thesis, Catholic Univ., Nijmegen, The Netherlands, 212 pp.

ISOGAMI, M. and SUNAGAWA, I. (1975): X-ray topographic study of a topaz crystal, *Am. Mineral.*, **60**, 889–897.

JACKSON, K. A. (1958): Liquid Metals and Solidification, *Am. Soc. Metals.*, p. 174, see also Interface kinetics, in *Growth and Perfection of Crystals*, Eds. R. H. Doremus, B. W. Roberts and D. Turnbull., John Wiley, New York, pp. 319–324.

KITAMURA, M. and SUNAGAWA, I. (1977): The effective distribution of solute and the anisotropy of a crystal, *Collected Abstract, ICCG–5, Boston*, p. 91.

KOSSEL, W. (1930): Uber Kristallwachstrum, *Naturwiss.*, **18**, 901–910.

KOUCHI, A. (1981): Effective distribution coefficient of Ni between olivine and its liquid as a function of growth kinetics, M. Sc. Thesis, Tohoku Univ.

KURODA, T., IRISAWA, T. and OOKAWA, A. (1977): Growth of polyhedral crystal from solution and its morphological stability, *J. Cryst. Growth*, **42**, 41–46.

LANG, A. R. (1965): X-ray topography of diamond, in *Physical Properties of Diamond*, Ed. R. Berman, Clarendon Press, Oxford, pp. 67–115.

LOFGREN, G. (1980): Experimental studies on the dynamic crystallization of silicate melts, in *Physics of Magmatic Processes*, Ed. R. B. Hargraves, Princeton Univ. Press, N. J., pp. 487–551.

MIYATA, T., KITAMURA, M. and SUNAGAWA, I. (1980): An extended PBC method and its application to striations and polar crystals, *Sci. Rept., Tohoku Univ., Ser. III*, **14**, 191–260.

MÜLLER-KRUMBHAAR, H., BURKHARDT, T. and KROLL, D. (1977): A generalized kinetic equation for crystal growth, *J. Cryst. Growth*, **38**, 13–22.

LANGER, J. S. and MÜLLER-KRUMBHAAR, H. (1977): Stability effects in dendritic crystal growth, *J. Cryst. Growth*, **42**, 11–14.

OHARA, M. and REID, P. C. (1973): *Modelling Crystal Growth Rates from Solution*, Prentice Hall, Englewood Cliffs, 272 pp.
ROSENFELD, J. L. (1970): Rotated Garnets in Metamorphic Rocks, *Geol. Soc. Am*, Special Paper **129**, 105 pp.
SUNAGAWA, I. (1957): Variation in crystal habit of pyrite, *Rep. Geol. Surv. Japan*, No. 175. 41 pp.
SUNAGAWA, I. (1960): Mechanism of crystal growth, etching and twin formation of hematite, *Mineral. J. (Japan)*, **3**, 59–89.
SUNAGAWA, I. (1961): Step height of spirals on natural hematite crystals, *Am. Mineral.*, **46**, 1216–1226.
SUNAGAWA, I. (1962): Mechanism of growth of hematite, *Am. Mineral.*, **47**, 1139–1155.
SUNAGAWA, I. (1964): Growth spirals on phlogopite crystals, *Am. Mineral.*, **49**, 1427–1434.
SUNAGAWA, I. (1975): Characterization of crystal surfaces by optical microscopy, II, in *Crystal Growth and Characterization*, Eds. R. Ueda and J. B. Mullin, North-Holland Pub. Co., Amsterdam, pp. 347–359.
SUNAGAWA, I. (1977): Natural crystallization, *J. Cryst. Growth*, **42**, 214–223.
SUNAGAWA, I. (1978a): Vapour growth and epitaxy of minerals and synthetic crystals, *J. Cryst. Growth*, **45**, 3–12.
SUNAGAWA, I. (1978b): Morphology of growth spirals observed on vapour grown crystals, *J. Japan. Assoc. Cryst. Growth*, **5**, 103–118.
SUNAGAWA, I. (1981): Characteristics of crystal growth in nature as seen from the morphology of mineral crystals, *Bull. Mineral. (France)*, **104**, 81–87.
SUNAGAWA, I. and BENNEMA, P. (1982): Morphology of growth spirals, theoretical and experimental, in *Preparation and Properties of Solid State Materials*, Vol. 7, Ed. W. R. Wilcox, Marcel Dekker Inc., New York, pp. 1–129.
SUNAGAWA, I. and ENDO, Y. (1968): Macro-and micro-morphology of quartz and pyrite, Part II, *IMA symposia, Mineral. Soc. London*, pp. 63–84.
SUNAGAWA, I. and KOSHINO, Y. (1975): Growth spirals on kaolin group minerals, *Am. Mineral.*, **60**, 407–412.
SUNAGAWA, I., KOSHINO, Y., ASAKURA, M. and YAMAMOTO, T. (1975): Growth mechanism of some clay minerals, *Fortschr, Miner.*, **52**, 217–224.
SUNAGAWA, I. and OHTA, E. (1976): Mechanism of formation of chalcedony, *Sci. Rept., Tohoku Univ., Ser. III*, **12**, 131–146.
SUNAGAWA, I., TSUKAMOTO, K. and YASUDA, T. (1982): Surface microtopographic and X-ray topographic study of diamonds, this volume, pp. 331–349.
TAMMAN, G. (1903): *Kristallisieren und Schmeltzen*, Leopold Voss, Leipzig.
TANNER, B. K. (1976): *X-Ray Diffraction Topography*, Pergamon Press, London, 174 pp.
TEMKIN, O. E. (1969): The structure and kinetics of advance of a monomolecular step on a crystalline surface, *Soviet Phys. Cryst.*, **14**, 179–185; A Kossel model of the kinetics of growth and evaporation normal to a (100) face, **Ibid.**, **14**, 344–348; Toward a theory of diffusionless crystal growth, **Ibid**, **14**, 349–354.
TEMKIN, O. E. (1971): Phenomenological kinetics of the motion of a phase boundary, *Soviet Phys. Cryst.*, **15**, 767–772; Kinetic phase transition during a phase conversion in a binary alloy, **Ibid.**, **15**, 773–780.
TOMURA, S., KITAMURA, M. and SUNAGAWA, I. (1979): Surface microtopography of metamorphic white micas, *Phys. Chem. Minerals*, **4**, 1–17.
TSUCHIYAMA, A. (1976): A study on anorthite megacrysts, Graduation Thesis, Tohoku Univ. (In Japanese).
TSUCHIYAMA, A., KITAMURA, M. and SUNAGAWA, I. (1981): Distribution of elements in growth of (Ba, Pb) $(NO_3)_2$ crystals from the aqueous solution, *J. Cryst. Growth*, **55**, 510–516.
VOLMER, M. (1922): Crystal growth, *Z. Physik.*, **9**, 193.

VOLMER, M. and WEBER, A. (1926): Keimbildung in übersattigten Gebilden, *Z. phys. Chem.*, **119**, 277–301.

WAGNER, R. S. and ELLIS, W. C. (1965): The vapour-liquid-solid mechanism of crystal growth and its application to silicon, *Trans. Metall. Soc. AIME*, **233**, 1053–1064.

WELLS, A. (1946): Crystal habit and internal structure, I, II., *Phil. Mag. Ser. 7*, **37**, 184–236.

WOODRUFF, D. P. (1973): *The Solid-Liquid Interface*, Cambridge Univ. Press, Cambridge, 182 pp.

WYLLIE, P. J. (1972): *The Dynamic Earth*, John Wiley, New York, 416 pp.

Materials Science of the Earth's Interior, edited by I. Sunagawa, pp. 107–130.

SYNTHESIS OF LARGE SINGLE CRYSTALS OF SILICATES AND TITANATES

Humihiko TAKEI,[1] Syoichi HOSOYA,[1] and Mituko OZIMA[2]

[1] *The Research Institute for Iron, Steel and Other Metals, Tohoku University, Katahira, Sendai, Japan*
[2] *The Institute for Solid State Physics, University of Tokyo, Roppongi, Minato-ku, Tokyo, Japan*

1. *Introduction*

It is widely known that large, high-quality single crystals are essential for material science. Even in the geosciences, good, large single crystals are also needed in order to obtain more information about physical or chemical properties of the earth's materials. Single crystals of silicates and titanates, namely, olivines, pyroxenes, pyroxenoids and ilmenites, are undoubtly most important; the former three are believed to be main constituents of the earth's mantle, and ilmenites are key materials for rock magnetism. Properties of these materials have been measured using natural crystals or synthetic polycrystalline samples. However, some limitations exist; natural crystals contain many kinds of imperfections such as inclusions, cracks, voids, twins, dislocations, undesired impurities, small angle grain boundaries and strains, and polycrystals have many grain boundaries which often hinder the intrinsic property of minerals. Many investigators have tried to prepare single crystals of silicates and titanates of appreciable size. SHANKLAND (1963) first prepared Mg_2SiO_4 (forsterite) crystals by the Verneuil flame-fusion method. The crystals were large enough so that their physical properties could be measured. However, the as-grown boules sometimes contained inclusions of unreacted MgO, and tended to crack during cooling after growth (SHANKLAND, 1967). GRODKIEWICZ and VAN UITERT (1963), VU TIEN *et al.* (1972), and WANKLYN (1977) grew single crystals of Mg_2SiO_4 using flux methods. Crystals obtained were very small and often contaminated with solvent materials. JORDAN and NAUGHTON (1964) and WILSON (1965) tried to prepare Mg_2SiO_4 crystals by the Bridgman technique. It was observed that chemical reactions occurred between the melt and crucible materials, Mo or graphite, and the crystals were contaminated with these materials.

The most successful growth of Mg_2SiO_4 crystals was performed by using the Czochralski-pulling technique. FINCH and CLARK (1971), and TAKEI and KOBAYASHI (1974) synthesized large single crystals using iridium crucibles and RF heating systems. Large, transparent crystals of Mn_2SiO_4 (tephroite) were also synthesized by use of this technique (FINCH *et al.*, 1975 and TAKEI, 1976). It was found that the crystal quality was greatly influenced by the melt stoichiometry and the atmospheric condition in the furnace (TAKEI, 1976). The edge-defined film-fed growth (EFG) method was applied to Mn_2SiO_4 crystallization to understand the effect of the melt stoichiometry (FINCH *et al.*, 1975). Fayalite crystals were also grown by the floating-zone (TAKEI, 1978) and the Czochralski (FINCH, 1980) methods.

Crystals of Co_2SiO_4 with several grains were prepared by NAKA *et al.* (1968) using the Bridgeman method. Small single crystals were obtained from a flux of $PbO-SiO_2$ (WANKLYN, 1977) and by a vapor transport method (STROBEL *et al.*, 1981). Sizable Co_2SiO_4 crystals were first synthesized by HOSOYA and TAKEI (1978) using the floating-zone method under a controlled atmosphere of partial pressure of oxygen, which had a sharp effect on the quality of the growing crystals. Small single crystals of Ni_2SiO_4, which has the incongruent melting point, were grown by OZIMA (1976) using the flux of MoO_3-Li_2O, and by HOFMANN and GRUEHN (1977) using a chemical vapor deposition technique.

Solid-solution crystals of the olivine group minerals have also been prepared by many investigators. SHANKLAND (1966) synthesized single crystals of $[Mg_{1-x}Fe_x]_2SiO_4$ where x was less than 0.01 by using the Verneuil flame-fusion method. TAKEI (1974) also produced the same kind of crystals with $x \leq 0.05$ by using the Czochralski method. Recently, HOSOYA and TAKEI (1980, 1981), and TAKEI and HOSOYA (1981) reported growth of sizable and transparent crystals of $[Mg_{1-x}Fe_x]_2SiO_4$ with a whole range of x using a lamp-image type floating-zone furnace. Other crystal syntheses of olivine solid-solutions were performed using the same type of growth furnace; Kozima and Osada made $[Mg_{1-x}Co_x]_2SiO_4$ and $[Mg_{1-x}Ni_x]_2SiO_4$ crystals (1978), and HOSOYA and TAKEI prepared $[Mn_{1-x}Co_x]_2SiO_4$ and $[Mg_{1-x}Mn_x]_2SiO_4$ (1979a, b).

Fewer studies on crystal syntheses of pyroxenes, or pyroxenoids have been reported. Very small single crystals of $MnSiO_3$ (rhodonite) were made by solid-state reaction of MnO and SiO_2 (AKIMOTO and SYONO, 1972). ITO (1975) synthesized orthopyroxene single crystals with a flux method using $MoO_3-Li_2O-V_2O_5$ flux. OZIMA (1979, 1980) has developed Ito's procedure and obtained clear, needle-shaped crystals a few cm long. The procedure and results will be described in a later section.

Crystals of $NiTiO_3$, belonging to the ilmenite family, were grown by WANKLYN (1970) and GARTON *et al.* (1972) from fluxes of $PbO-V_2O_5$ mixture. BARSUKOVA *et al.* (1972) and SOMIYA *et al.* (1978) synthesized tiny single crystals of $FeTiO_3$ (ilmenite) under hydrothermal conditions. GILEY and BAUGHMAN (1976) produced large crystalline boules of $FeTiO_3$ from the stoichiometric melt using high-pressure type Czochralski pulling equipment. However, the crystals thus obtained do not seem to be well characterized, that is, the composition or perfection of the boules was not clearly described. TAKEI and KITAMURA (1978) and TAKEI (1981) reported the preparation of single crystalline $FeTiO_3$ and $MnTiO_3$, respectively, using a lamp-image type floating-zone furnace, which has already been described in respect to its usefulness for the growth of olivine crystals. TAKEI (1979) also obtained $CoTiO_3$ and $NiTiO_3$ crystal rods, as well as solid solutions of $[Fe_{1-x}Co_x]TiO_3$ and $FeTiO_3$–Fe_2O_3. It became clear that the atmospheric condition is most important in obtaining high quality ilmenite crystals.

This article will focus on the synthesis of large single crystals of olivines, pyroxenes, pyroxenoids and ilmenites with special emphasis on growth with the Czochralski-pulling, floating-zone and flux techniques.

2. *Choice of Growth Methods*

In order to obtain large and high quality single crystals, the most important factor is to choose the best method of crystallizing a target material. In most cases, melt growth, in which a crystal is obtained by solidifying the melt, is a very popular and effective method for growing large crystals, and many kinds of techniques have been developed, such as the Bridgman (BM), the Czochralski-pulling (CZ) and the floating-zone (FZ) techniques. These techniques are also useful for producing crystals of olivines, pyroxenes, pyroxenes, pyroxenoids and ilmenites. Melt growth usually requires that (1) the materials to be crystallized melt congruently (melt without decomposition), and that (2) the melts have low vapor pressures near their melting points. In addition to these conditions, BM and CZ require a crucible which can contain the melt without any chemical reactions. In the case of FZ, the travelling-solvent method (TSM) is applicable when a strong temperature gradient is to be realized around the molten zone (WOLFF and MLAVSKY, 1974). This technique often enables the elimination of the above-described two requirements: congruent melting and low vapor pressure.

It is commonly recognized that the CZ growth is superior to the other two because large single crystals of high quality can be prepared relatively easily and rapidly. However, in principle, solid-solution crystals of homogeneous composition cannot be obtained by this method.

The high temperature solution (flux) growth method is undoubtly another choice for obtaining single crystals of high quality. More detailed descriptions of this method will be presented in a later chapter.

Table 1 summarizes some melting behaviours of olivines, pyroxenes, pyroxenoids and ilmenites. There are six compounds which melt congruently,

TABLE 1. Some melting behaviours of olivines, pyroxenes, pyroxenoids and ilmenites.

Olivines	m.p. (°C)	Pyroxenes & Pyroxenoids	m.p. (°C)	Ilmenites	m.p. (°C)
Mg_2SiO_4	1,890	$MgSiO_3$	(1,590)	$MnTiO_3$	(1,400)
Mn_2SiO_4	1,345	$MnSiO_3$	(1,450)	$FeTiO_3$	1,400
Fe_2SiO_4	1,205	$CaMgSi_2O_6$	(1,392)	$CoTiO_3$	1,463
Co_2SiO_4	1,415			$NiTiO_3$	(1,400)?
Ni_2SiO_4	(1,730)				

( ) means incongruent melting.

and the vapor pressures at their melting points are small enough to allow a melt growth process. However, Fe- and Co-compounds have a tendency to react with noble metals such as Pt or Ir, which are the most useful crucible materials. Consequently, the CZ method is applicable only for two compounds, Mg_2SiO_4 and Mn_2SiO_4.

The floating-zone method is useful for Co- and Fe-compounds as well as Mn-compounds. This method has some advantages over the BM or CZ methods: (a) no use of a crucible, (b) applicable to the incongruently melting compounds, $MnTiO_3$ and $MnSiO_3$, and (c) a uniform distribution of solute in solid-solution systems. In addition to these factors, careful control of oxygen fugacity can be achieved during growth. This factor will be discussed in the next chapter. Table 2 is a list of solid-solution crystals of silicates and titanates referred to in this article.

3. *Crystal Growth*

3.1 *Atmosphere control*

Most of the silicates and titanates described in this article contain

TABLE 2. Solid-solution of olivines, pyroxenoid and ilmenites.

Crystals	Melting range (°C)	Composition range	References
$[MgFe]_2SiO_4$	1,890 ~ 1,205	whole range	BOWEN (1935)
$[MgCo]_2SiO_4$	1,890 ~ 1,415	〃	—
$[MgMn]_2SiO_4$	1,890 ~ 1,345	〃	GLASSER (1960)
$[MgNi]_2SiO_4$	1,890 ~ 1,770	below 50% Ni_2SiO_4	RINGWOOD (1956)
$[MnCo]_2SiO_4$	1,345 ~ 1,415	whole range	BIGGERS (1967)
$[MgMn]SiO_3$	(1,530 ~ 1,325)	between 12 and 85% $MnSiO_3$	GLASSER (1960)
$[FeCo]TiO_3$	1,400 ~ 1,463	whole range	—
$FeTiO_3$–Fe_2O_3	1,400 ~ (1,600)	〃	TAYLOR (1964)

( ) means incongruent melting.

transition metal ions such as Mn, Fe, Co and Ni. In high temperature regions where crystal growth is performed, these ions change valence states depending on partial pressure of oxygen in the surrounding atmosphere. For example, the manganese ion has three stable valence states, of cation, II, III and IV, and the iron ion has two states, II and III, as a function of partial pressure of oxygen. Thus, an atmosphere control during crystal synthesis is essential for transition metal compounds. Atmosphere control can be conveniently performed by introducing a mixed gas flow of CO_2 and H_2, or CO_2 and CO in a growth furnace (DARKEN and GURRY, 1945, 1946). The partial pressure of oxygen ranges from 10^{-20} to 1 atm between 1,600 and 1,000°C (Fig. 1). The chemical reaction of the CO_2–CO system is

$$CO_2 \rightleftarrows CO + \tfrac{1}{2}O_2, \tag{1}$$

and that of the CO_2–H_2 system is

$$\left.\begin{aligned} CO_2 + H_2 &\rightleftarrows CO + H_2O \\ CO_2 &\rightleftarrows CO + \tfrac{1}{2}O_2 \end{aligned}\right\} \tag{2}$$

A phase diagram of Fe–O was first determined by DARKEN and GURRY (1945, 1946). Phase relations of the system MgO–Fe–SiO_2–O have been

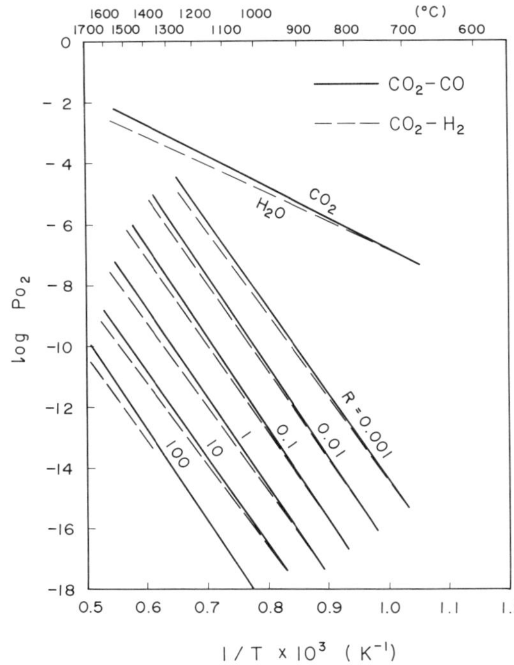

FIG. 1. Temperature dependence on partial pressures of oxygen P_{O_2} in a mixed gas of CO_2–CO or CO_2–H_2. R means the ratio P_{CO}/P_{CO_2} or P_{H_2}/P_{CO_2}.

studied by many investigators. SPEIDEL and NATZIGER (1968) and NITSAN (1974) have determined the stability field of $[Mg_{1-x}Fe_x]_2SiO_4$ where $0 \leq x \leq 1$ represents the functions of temperature and oxygen partial pressure.

Systems of Fe–Ti–O were also investigated by WEBSTER and BRIGHT (1961) under a well controlled atmosphere of oxygen fugacity.

The stability field of growing crystals is generally much different from that in a phase diagram because crystals grow under a non-equilibrium condition. Consequently, such a field must be determined empirically by using the analytical data of crystal composition, X-ray and other information about crystal imperfections.

3.2 The Czochralski-pulling method

Growth by use of the Czochralski-pulling (CZ) method is usually carried out in a furnace arrangement as, illustrated schematically in Fig. 2. It mainly consists of three parts: a melt in a crucible which is thermally and electrically insulated by a ceramic powder, a seed crystal attached to a rotating, pulling shaft, and an induction heating system. In cases of Mg_2SiO_4 and Mn_2SiO_4, the best material for the crucible is Ir metal, which reacts only slightly with the melts under a neutral or reducing atmospheric condition. The ceramic powder is granular zirconia. When the temperature gradient is settled, as indicated by the curve as shown on the right-hand side of Fig. 2, the melt can be solidified by touching and then pulling the seed.

Figures 3a and b are typical as-grown crystals of Mg_2SiO_4 and Mn_2SiO_4. The growth conditions are listed in Table 3 (FINCH and CLARK, 1971; TAKEI and KOBAYASHI, 1974; FINCH *et al.* 1975; TAKEI, 1976). Each crystal has two elongated facets of (010) on the side. The size of the crystals are limited by

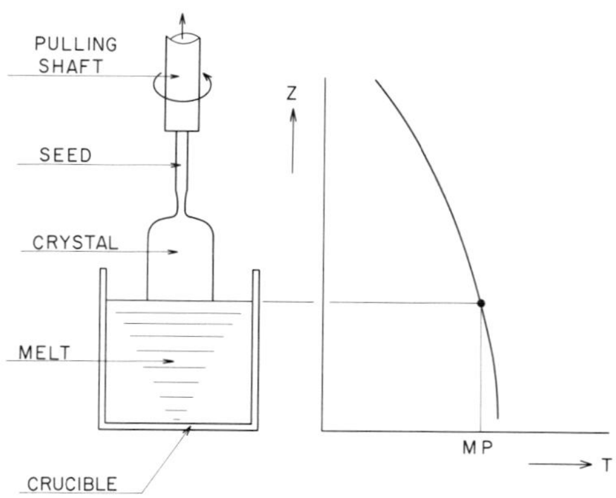

FIG. 2. A schematic illustration of the Czochralski-pulling system and thermal gradient produced.

FIG. 3. Single crystals of silicates grown by the Czochralski-pulling method; (a) Mg_2SiO_4, (b) Mn_2SiO_4.

TABLE 3. Czochralski-growth condition of Mg_2SiO_4 and Mn_2SiO_4 crystals.

	Mg_2SiO_4		Mn_2SiO_4	
	FINCH (1971)	TAKEI (1974)	FINCH (1975)	TAKEI (1976)
Melting pont (°C)	1890		1345	
Pulling rate (mm hr^{-1})	12	3–10	6	3–8
Rotation (rpm)	10–25	20–30	—	10–15
Growth axis	[100] [010], [001]	[100], [001]	[100] [010], [001]	[100], [001]
Atmosphere	—	N_2	Ar + H_2	$N_2 + H_2$
Cooling rate	below 20 min.	>300°C hr^{-1}	—	>200°C hr^{-1}

diameter of the crucible and the volume of the melt. The main defects sometimes observed are light scattering inclusions which are largely gas bubbles and precipitates. These inclusions are observed when the crystal diameter becomes large relative to the crucible size and/or the growth rate is too large.

A flow of inert gas, nitrogen or argon, is usually used as an atmospheric protection from oxidation of an Ir crucible. It was observed in the case of Mn_2SiO_4 growth that a trace amount of oxygen in an inert gas tended to oxidize Mn^{2+} into Mn^{3+} in crystal and melt. So, an addition of a small amount of hydrogen gas into the inert gas was found to be very effective for protection of the oxidation. This effect can be confirmed by measuring the optical absorption of grown crystals in the visible region. Figure 4 apparently shows that the transmittance of the crystal grown in a reducing atmosphere (T–6) is much higher than that grown in a slightly oxidizing condition (T–2) (TAKEI, 1976). These phenomena are attributable to a contamination by Mn^{3+} in the crystal. Table 4 summarizes some physical properties of single crystalline Mg_2SiO_4 and Mn_2SiO_4, grown by the CZ method. The compositions are very close to stoichiometry, and the Archimedian and X-ray densities are nearly equal. The major impurities in the crystals usually come from starting reagents. So, the desired purity of the starting materials should be higher than 99.9%. The crystalline defects in Mg_2SiO_4 will be described by KOMATSU and INOUE in a later section. Single crystals of $CaMgSi_2O_6$ (diopside) have been obtained by the CZ method. As shown in Fig. 5, a cut and polished piece of the crystal is transparent and water-clear. The main defect observed often is cracks which appear during cooling after growth. Details of

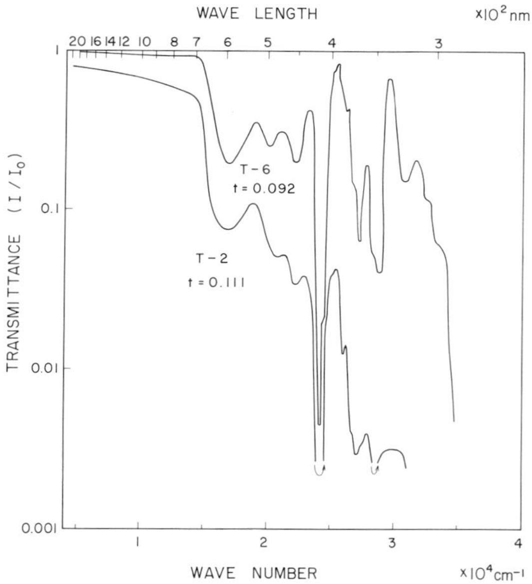

FIG. 4. Absorption spectra of Mn_2SiO_4 crystals. T-2 and T-6 specimens were synthesized in an oxidizing and a reducing conditions, respectively. *t* means specimen's thickness in mm.

TABLE 4. Some physical properties of Mg_2SiO_4 and Mn_2SiO_4 crystals grown by the CZ method.

		Mg_2SiO_4*	Mn_2SiO_4**
Color (1 mm thick)		Colorless	Blueish gray
Lattice constant	a_0	0.47515 [6] nm	0.4904_2 nm
	b_0	1.1914 [7] 〃	1.0567_0 〃
	c_0	0.59781 [6] 〃	0.6254_5 〃
Density $D_{calc.}$		3.228 g cm^{-3}	4.127_7 g cm^{-3}
$D_{obs.}$		3.225 〃	4.12_9 〃
Cleavage		[010]	Not distinct
Hardness (Mohs)		7–6.5	6.5–6

*SYONO *et al.* (1981).
**TAKEI (1976).

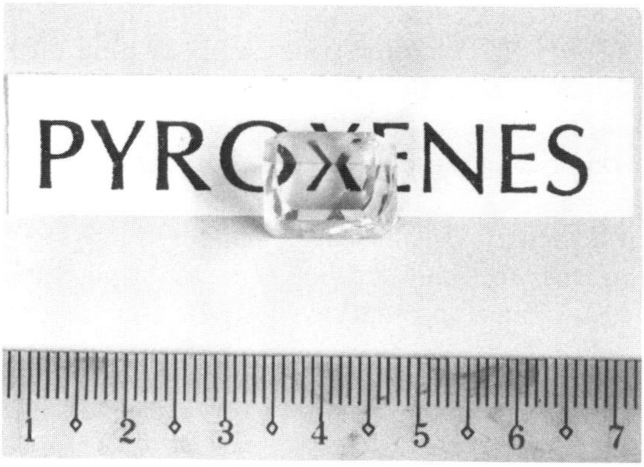

FIG. 5. Cut and polished crystal of $CaMgSi_2O_6$ (diopside). The thickness of the crystal is about 10 mm.

the growth and characteristics of this crystal will be presented in a future article.

3.3 *The floating-zone method*

The principle of the floating zone (FZ) method is illustrated in Fig. 6. At the upper part of the figure there is a nutrient feed which is usually made of a polycrystalline sintered rod. At the center there is a molten-zone, and the lower part is a seed crystal which is cut with a desired orientation. This situation comes about when a sharp temperature gradient is realized, as shown in the left-hand side of the figure. Crystal growth is carried out by moving the assembly of the feed-rod, molten-zone, and seed crystal to the lower side without any change of the zone thickness.

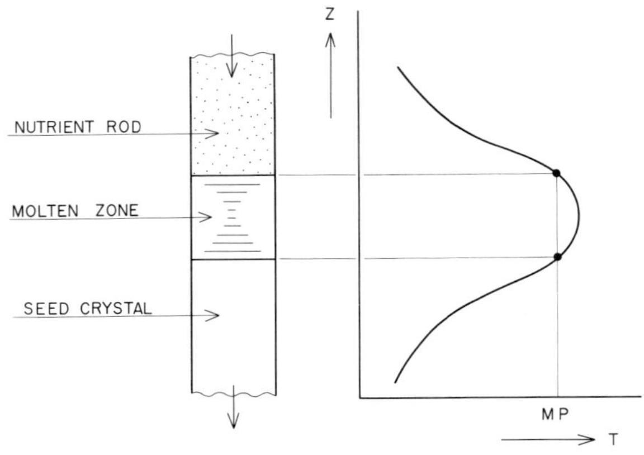

FIG. 6. A schematic illustration of the floating-zone system and thermal gradient produced.

In order to obtain the molten-zone, many heating methods have been developed: resistance heating, r.f. heating, electron beam heating, infrared radiation heating and laser heating. In case of melting silicates or titanates, infrared radiation heating is the best choice because (1) they are insulative or highly resistive and (2) it is easy to construct the growth assembly so as to control oxygen fugacity.

Figure 7 is a cross-sectional drawing of the FZ furnace used by the present authors. It consists of an ellipsoidal cavity, a W-I lamp as an infrared heating

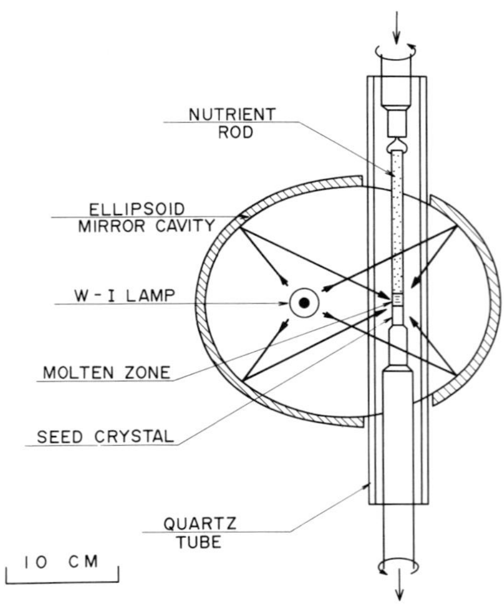

FIG. 7. A cross-section of the infrared-heating floating-zone furnace.

source, and the growth assembly of the feed-rod, molten-zone and seed, which is mounted inside a fused quartz tube. The lamp is located at one focus of the ellipsoid and the molten-zone at the other. The atmosphere control can be achieved by adjusting the flow rates of two kinds of gas, CO_2 and H_2 or CO_2 and CO, in the quartz tube. Figure 8 is a photograph of the infrared-heating FZ furnace. More detailed explanations have already been given in separate papers (TAKEI, 1978a, b). The nutrient rod is made by the usual powder metallurgical technique: starting reagents are weighed in a desired composition and mixed in a blender. Then, the mixture is preheated in a resistant-heating muffle furnace in order to complete a solid-state reaction which forms the desired compound. The powder thus prepared is pressed hydrostatically into the shape of a rod at a pressure of more than 1 ton/cm^2. After that, the rod is sintered in the furnace. The atmosphere should be controlled, if necessary, during pre-heating and sintering.

End member crystals of silicates and titanates synthesized by the FZ furnace are listed in Table 5. It should be noted that the incongruently melting compounds, $MnSiO_3$ and $MnTiO_3$, can be crystallized. This means that the growth condition of TSM (Travelling-Solvent Method) is satisfied between the crystal and the melt. That is, the melt composition changes from the

FIG. 8. Photograph of the infrared-heating floating-zone apparatus. The ellipsoidal cavity shown at the center of the photograph is opened.

TABLE 5. FZ-growth condition of the end member crystals of some silicates and titanates.

	Mn_2SiO_4	Fe_2SiO_4	Co_2SiO_4	$MnSiO_3$	$MnTiO_3$	$FeTiO_3$
Melting point (°C)	1,345	1,205	1,415	(1,450)	(1,400)	1,400
Travelling rate ($mm\ hr^{-1}$)	1	1.2–2	1.5~5	>1	1–3	1.5–5
Totation (rpm)	30	30	30	30	30	30
Travel axis	[100]	[100] [010] [001]	[100] [010]	Spontaneous	$[10\bar{1}0]$ $[11\bar{2}0]$	$[10\bar{1}0]$ $[11\bar{2}0]$
Atmosphere	N_2 or Ar	CO_2–H_2	CO_2	CO_2–H_2	CO_2–H_2	CO_2–H_2
Typical size of boules (mm)	5~8ϕ × 20~50l	8~10ϕ × 50~60l	5~10ϕ × 20~70l	7ϕ × 50l*	6~8ϕ × 80l	7~8ϕ × 50l

*It consists of several grains of crystals.

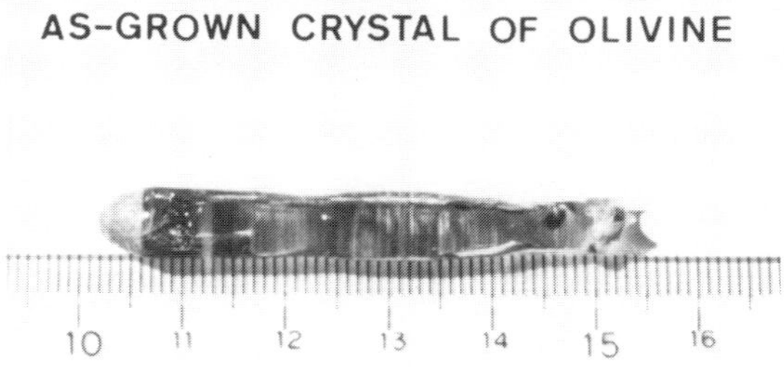

FIG. 9. Single crystal of $[Mg_{0.8}Fe_{0.2}]_2SiO_4$(olivine). The growth direction is [100] (to the left-hand side).

stoichiometry to a non-stoichiometry in which the melt and the crystal can co-exist. Table 5 also lists the growth conditions as well as the maximum size of crystals thus prepared. The physical and chemical properties of these crystals have already been reported in separate papers (TAKEI, 1978a, b, 1981; HOSOYA and TAKEI, 1978). Solid-solution crystals were prepared by use of the FZ method (see Table 12 and Fig. 9). It was found that careful control of oxygen fugacity was essential for obtaining high-quality crystals. In Mg_2SiO_4–Fe_2SiO_4 system, for example, fine particles of metallic iron was found in the crystal when the atmosphere was more reductive than suitable. On the other hand, Fe^{3+} ions appeared when the atmosphere became oxidative. The appearance of Fe^{3+} was confirmed by observing the optical absorption spectra in the visible or near infrared region (HOSOYA and TAKEI, 1982).

TABLE 6. Partition of solute ions in solid-solution crystals of $[MgFe]_2SiO_4$, $[MnMg]_2SiO_4$ and $[MnCo]_2SiO_4$.

System	x	Temperature (°C)	k_0^{obs}*	k_0^{calc}**	$(z/l)_{K_0=1}$
$[Mg_{1-x}Fe_x]_2SiO_4$	0.70	1,470	0.6	0.53	4
	0.80	1,430	0.7	0.60	3
$[Mn_{1-x}Mg_x]_2SiO_4$	0.70	1,570	0.9	0.71	1.5
$[Mn_{1-x}Co_x]_2SiO_4$	0.20	~1,350	1.2	—	2

*Observed distribution coefficient k_0^{obs} is obtained using the Pfann's formula:

$$C_s/C_o = \left[1 - (1 - k_0^{obs}) \cdot \exp\left(-k_0^{obs}\frac{z}{l}\right)\right],$$

where C_s is a solute concentration at travelled length l, C_o is an initial concentration of solute and z is a molten zone length.

**Calculated coefficient k_0^{calc} is taken from the phase diagram of the solid-solution system.

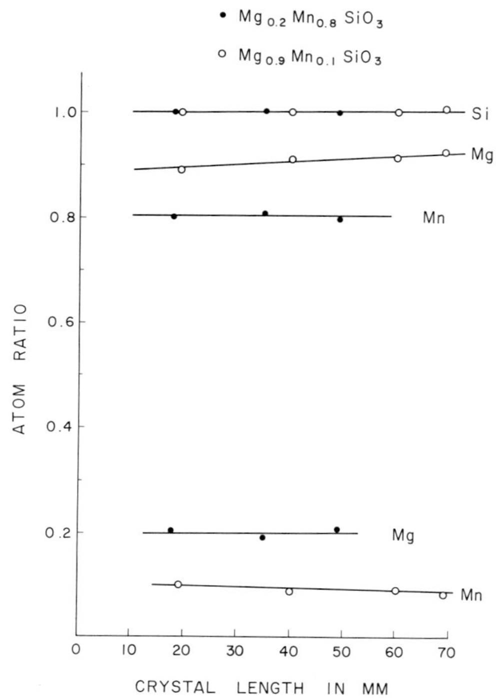

FIG. 10. Mg and Mn distributions along the growth axis [100] in $[MgMn]SiO_3$ crystals.

Cation distributions in solid-solution crystals are very important and numerous studies have been reported. However, very few studies have been done on silicates or titanates. Table 6 lists typical results of the cation distribution coefficient k_0^{obs} in olivine solid-solution crystals obtained by use of the FZ method, where Pfann's theory of single-pass zone refining was used for calculation (PFANN, 1952). It is clear that k_0^{obs} is slightly different from k_0^{calc} estimated from the pseudo-binary phase diagrams. This means that a kinetic effect exists in these FZ conditions. Nevertheless, the homogeneous distributions of cations are realized, as shown in Fig. 10, where the Mg and Mn distributions in the $(MgMn)SiO_3$ crystals are plotted against the crystal length. A typical example of the composition dependence of the unit cell volume, which is a good criteria of the formation of solid-solution crystals, is shown in Fig. 11. The volumes of $[Mg_{1-x}Fe_x]_2SiO_4$, $[Fe_{1-x}Co_x]TiO_3$ and $[Fe^{II}_{1-x}Fe^{III}_x]$ $(Ti^{IV}_{1-x}Fe^{III}_x)O_3$ are appoximately proportional to composition x. This means that the crystals form a complete solid-solution over the whole range of x (Vegard's law).

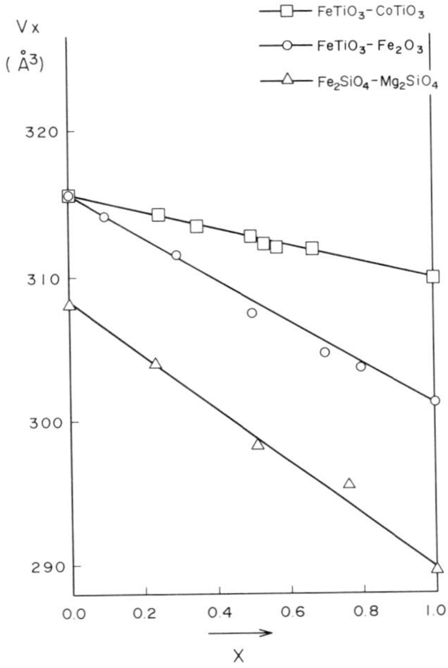

FIG. 11. Unit cell volume Vx of FZ-crystals as a function of x in the systems $[Mg_{1-x}Fe_x]_2SiO_4$, $[Fe_{1-x}Co_x]$ TiO_3, and $[Fe^{II}_{1-x}Fe^{III}_x]$ $[Ti^{IV}_{1-x}Fe^{III}_x]O_3$.

3.4 *Growth of orthoenstatite crystals by the flux method*

As described in the previous sections, CZ and FZ methods are useful for the growth of some silicates or titanates. Nevertheless, the flux method is also important, because there are a large number of minerals which can be grown only by this method. In the 1960s and 1970s, rapid progress was made in the use of the flux method for orthoferrites, rare earth garnets, quartz, and others, and it can be said that the flux method has contributed much to material science in the last two decades. However, few instances of growth of silicate minerals by use of the flux method have been reported so far. For example, only a very small number of papers dealing with the use of the flux method for the growth of silicate minerals is included in a table of the book, *Crystal Growth from High-Temperature Solutions* by ELWELL and SCHEEL (1975), where more than 1,000 papers of flux growth are listed.

The flux method has the following advantages compared with melt-growth technique:

(a) Materials which decompose or vaporize before melting can be crystallized.

(b) Crystals in a low-temperature phase which undergoes a transition before melting can be obtained.

(c) Crystals with well-developed facets can be grown.

(d) In some cases, crystals of high perfection can be synthesized.

On the other hand, this method has the following disadvantages:

(a) Flux materials tend to contaminate the growing crystals.

(b) The growth rate is usually very low, and it commonly takes one week or more.

(c) The size of flux-grown crystals is usually small (mm size) and much technical advancement is needed to obtain larger crystals (cm size).

Also in the case of silicate growth, these points restrict the applicability of the flux method and, consequently, there have been few reports on the use of the flux method for growth of silicate minerals.

3.4.1 Principle of flux growth

Flux growth is simply based on the temperature dependence of the solubility of the solute (nutrient) to the solvent (flux). Usually the solubility increases with increasing temperature. Hence, on cooling the solution, crystal precipitation of the nutrient occurs when the solution reaches a state of supersaturation. Instead of a temperature decrease, evaporation of the solvent leads to the same results. In order to obtain larger crystals, it is essential to control the state of nucleation. For example, a very slow cooling rate at the earliest stage is preferred rather than a constant cooling rate (SCHEEL and ELWELL, 1972), and techniques such as the top-seeded method, stirring the solution (SCHEEL and ELWELL, 1973), accelerated rotation of the crucible (SCHEEL, 1972), and the use of a temperature gradient in the crucible (SHINDO, 1980) are useful for controlling nucleation.

There are many kinds of flux materials. For purposes of silicate growth, the MoO_3-Li_2O system is known to be an excellent flux. It may be partly because SiO_2 can dissolve in this flux very well. In this case, V_2O_5 is necessary as an additive which is effective in suppressing nucleation and reducing the number of the growing crystals. This results in an increase in the size of each crystal.

3.4.2 Growth of orthoenstatite

Enstatite ($MgSiO_3$) which belongs to pyroxene family, is known to be one of the important constituents of the earth's upper-mantle. It melts incongruently above 1,500°C, and has three polymorphs, that is, the high-temperature phase: protoenstatite (PE), the middle phase: orthoenstatite (OE), and the low-temperature phase: clinoenstatite (CE). PE is stable above 950°C at atmospheric pressure. It is unquenchable even by a rapid cooling to room temperature, and turns into CE below 1,000°C by skipping over the OE phase. OE is stable between 950°C and about 600°C, and quenchable under 1 atm. The OE-CE phase transition was only observed when an appreciable amount of shearing stress was applied to the OE specimen (MUNOZ, 1968). Thus, the stability field of OE at 1 atm without any shearing stress is between 950°C and room temperature. This means OE crystals can be obtained using a flux which acts below 950°C.

In order to grow single crystals of OE, ITO (1975) employed a flux "lithiumvanadomolybdate" of which composition in mol % is; MoO_3 : 42.8, Li_2O : 51.2, V_2O_5: 6.0. Figure 12 is a tentative partial diagram for the $MgSiO_3$-lithiumvanadomolybdate join in air (ITO, 1975). In OZIMA's experiment (1980) the flux of MoO_3–Li_2O–V_2O_5 with various compositions was used. Crystal growth was carried out by the following procedure: mixing MoO_3, Li_2CO_3, V_2O_5, MgO, and SiO_2, of guaranteed quality, heating them in a platinum crucible using a muffle furnace, soaking, slow cooling, quenching and finally rinsing with water. The remaining materials were examined by means of X-ray diffraction. The typical results are summarized in Table 7, where it can be seen that the compositional range of enstatite ($MgSiO_3$) in terms of the ratio MoO_3/Li_2O is very narrow. The range is a function of the nutrient concentration as well as the ratio, MgO/SiO_2 (in this experiment, MgO/SiO_2 was fixed at unity).

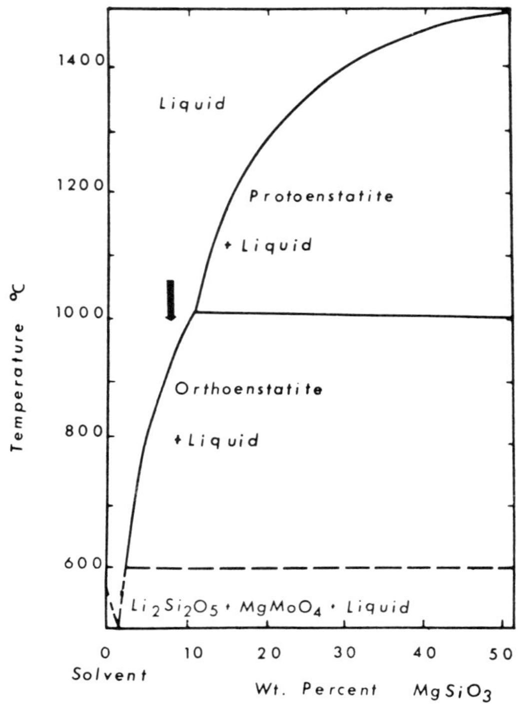

FIG. 12. Partial phase diagram for $MgSiO_3$-lithiumvanadomolybdate (a mixture of Li_2O, V_2O_5, and MoO_3) flux written by ITO (1975).

TABLE 7. Materials obtained by "lithiumvanadomolybdate" fluxes.

Flux	Li_2O	mol %	52.3	52.0	51.7	51.5	51.2	50.0	47.0
	MoO_3		41.9	42.0	42.3	42.5	42.8	44.0	47.0
	V_2O_5		5.8	6.0	6.0	6.0	6.0	6.0	6.0
Nutrient*	MgO	g	1.48	1.45	1.82	1.45	1.45	1.45	1.45
	SiO_2		2.42	2.17	2.72	2.17	2.17	2.17	2.17
Product			Li_2SiO_3	Li_2SiO_3	Li_2SiO_3	Li_2SiO_3	—	—	—
			—	—	—	$Li_2Si_2O_5$	$Li_2Si_2O_5$	—	—
			—	—	$MgSiO_3$	$MgSiO_3$	$MgSiO_3$	$MgSiO_3$	—
			—	—	—	—	—	SiO_2	—
			Mg_2SiO_4	Mg_2SiO_4	—	—	—	—	—
			—	—	—	—	—	—	$Li_2Mg_2Mo_3O_{12}$

*Weight for 70.2 g of MoO_3.

The present experiment shows that the growth condition reported by Ito seems to be incomplete because:

(1) When we put about 9 wt% of nutrient ($MgO + SiO_2$), only a number of small euhedral crystals were grown in the form of aggregate around the upper rim of the crucible together with Li_2SiO_3. On the other hand, when we halved the amount of the nutrient, a small number of big elongated crystals grew separately at the bottom of the crucible.

(2) Single crystals of $Li_2Si_2O_5$ interlocked with OE crystals were often found. This means some part of $Li_2Si_2O_5$ and OE grew in the same temperature range.

Transparent crystals of $Li_2Si_2O_5$ were obtained in the shape of an elongated plate, of which size sometimes became a few cm long (Fig. 13i). Irregularly shaped and opaquely white Li_2SiO_3 crystals were also obtained by a quick cooling of the residual melt after OE growth. A typical growth condition is given in Table 8.

TABLE 8. Crystal growth data for orthoenstatite, $MgSiO_3$.

Charge composition	Li_2CO_3	86.20 g
	MoO_3	140.40
	V_2O_5	24.86
	MgO	2.91
	SiO_2	4.33
Crucible size		100 ml
Mg/Si		1.0
Nutrient/flux		3.6 wt%
Soaking temperature		935°C
Soaking time		14 days
Cooling rate	935°–910°C	0.14°/h
	910°–855°C	0.3°/h
	855°–650°C	1.3°/h
Quenching temperature		650°C
Crystal size		about 20 mm in max length
Product		orthoenstatite crystal and $Li_2Si_2O_5$ crystal

Tables 9 and 10 are lists of the chemical compositions determined by an X-ray microprobe analyser and the lattice parameters of OE crystals, respectively. The crystal compositions are very close to stoichiometry. According to Ito (1975), the crystals contain Li_2O:0.17, V_2O_5:0.27 wt% as major impurities.

The orthoenstatite crystals thus grown are colorless and transparent. The size is typically 1 × 2 × 5 mm. In some cases crystals 20 mm in length were grown. As shown in Figs. 13 a–g, a variety of crystal shapes, that is, euhedral or platy rods (*c*-axis elongation), was observed. The well-developed faces are

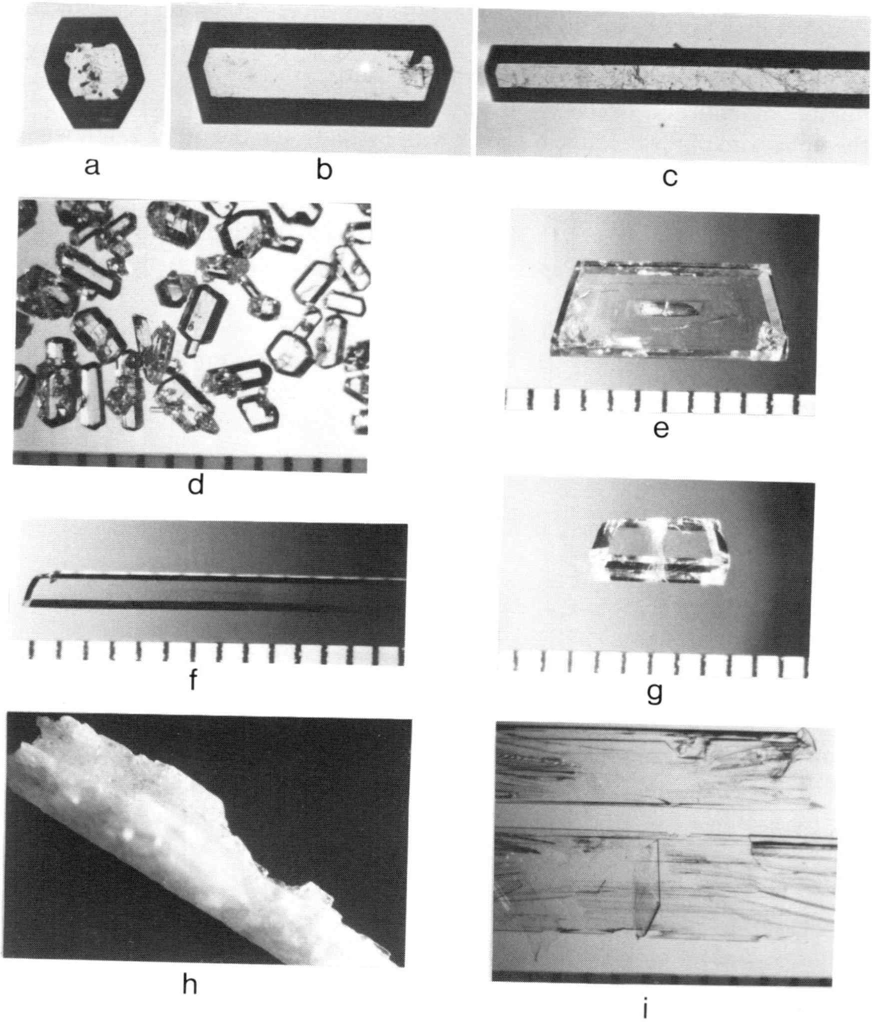

FIG. 13. Crystals grown from the lithiumvanadomolybdate flux. a, b, d, and g: euhedral orthoenstatite (OE) crystals, c, and f: platy rod crystals of OE, e: platy OE crystal, h: proto-inverted-clinoenstatite, and i: $Li_2Si_2O_5$ crystal. (1 division: 1 mm).

(210), (010) and (100). In one particular case, a plate with the (010) face was crystallized. Figure 14 is a top-view of a typical euhedral crystal. The {111} and {211} faces make edges which are perpendicular to [010] or [100].

On cooling from about 1,000°C, a number of white octagonal column crystals were obtained (Fig. 13h). They were confirmed to be a clinoenstatite (CE) aggregate which may have been produced by the transition from

TABLE 9. Chemical composition of orthoenstatite.

	This study	Ito (1975)	Calculated
MgO	40.4 wt%	40.3 wt%	40.16 wt%
SiO_2	59.7	60.0	59.84
total	100.1	100.3	100.00

TABLE 10. Lattice parameter of orthoenstatite.

	This study	JCPDS card Stephenson *et al.* (1966)	Ito (1975)
a	18.2372(20) A	18.2249 A	18.214(4) A
b	8.8176(9)	8.8149	8.818(2)
c	5.1786(8)	5.1746	5.177(2)

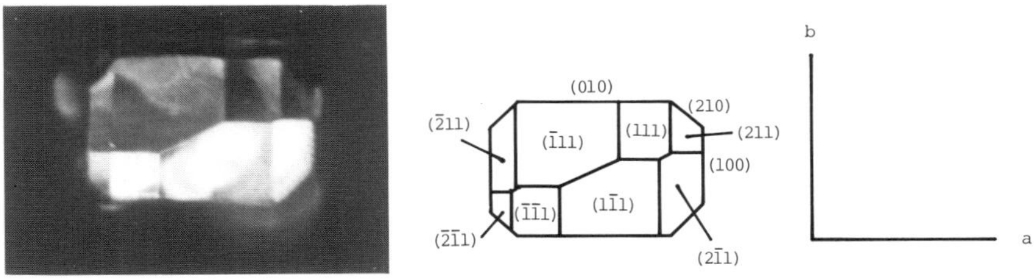

FIG. 14. Top-view of a single crystal of orthoenstatite (*c*-axis is perpendicular to the figure).

protoenstatite (PE). It is very interesting that CE crystals of the aggregate are oriented roughly in a dispersed angle of about 15 degree. This orientation can be explained by a (100) twinning which arises at the phase transition of PE to CE. Space group of this PE-inverted-CE is $P2_1/c$.

4. *Concluding Remarks*

Various methods used to grow single crystals of olivines, pyroxenes, pyroxenoids and ilmenites have been described in this article. The Czochralski-pulling and floating-zone techniques are demonstrated to be the most suitable methods for growing large and high quality crystals of silicates and titanates with congruent melting behaviors. It has also become clear that the floating-zone method is very convenient because solid-solution crystals with a homogeneous composition of the solute can be produced, and because incongruently melting compounds can be crystallized when the traveling-solvent method is applied. Flux growth is obviously another choice for obtaining crystals of silicates and titanates, which decompose or transform into another phase at high temperatures. The growth of orthoenstatite crystals has been described as a successful example of using the flux technique. The methods explained here may be applicable to the growth of a number of other crystals important in the field of geosciences.

The authors wish to acknowledge Professor S. Akimoto of Tokyo University for his encouragement, and Professors H. Komatsu and Y. Syono of Tohoku University for their comments and discussions. Mr. T. Miura is also acknowledged for his assistance of growing diopside crystals.

REFERENCES

Akimoto, S. and Syono, Y. (1972): High pressure transformations in $MnSiO_3$, *Am. Mineral.*, **57**, 76–84.

Baruskava, M. L., Kuznetsov, V. A. and Malinovskaya, E. K. (1972): Crystallization of divalent metal titanates in high temperature solutions, *Kristallografiya*, **17**, 1268–1270.

Darken, L. S. and Gurry, R. W. (1945): The System Iron-Oxygen, I, *J. Am. Chem. Soc.*, **67**, 1398–1412.

Darken, L. S. and Gurry, R. W. (1946): The System Iron-Oxygen, II, *J. Am. Chem. Soc.*, **68**, 798–816.

Elwell, D. and Scheel, H. J. (1975): *Crystal Growth from High-Temperature Solutions*, Academic Press.

Finch, C. B. and Clark, G. W. (1971): Czochralski growth of single crystal Mg_2SiO_4 (forsterite), *J. Crystal Growth*, **8**, 307–308.

Finch, C. B., Clark, G. W. and Kopp, O. C. (1975): Growth of single-crystal Mn_2SiO_4 (tephroite) by Czochralski and edge-defined film-fed (EFG) techniques, *J. Crystal Growth*, **29**, 269–272.

Finch, C. B., Clark, G. W. and Kopp, O. C. (1980): Czochralski growth of single-crystal fayalite under controlled oxygen fugacity conditions, *Am. Mineral.*, **65**, 381–389.

Garton, G., Smith, S. H. and Wanklyn, B. M. (1972): Crystal growth from the flux systems $PbO-V_2O_5$ and $Bi_2O_3-V_2O_5$, *J. Crystal Growth*, **13/14**, 588–592.

Ginley, D. S. and Baughman, R. J. (1976): Preparation and Czochralski crystal growth of the iron titanates, $FeTiO_3$, Fe_2TiO_4, and Fe_2TiO_5, *Mat. Res. Bull.*, **11**, 1539–1544.

Grodkiewcz, W. H. and Van vitert, L. G. (1963): Synthesis of Forsterite, diopside, akermanite, and wollastonite from molten, PbO, *J. Am. Ceram. Soc.*, **46**, 356.

Hofmann, J. and Gruehn, R. (1977): Darrstellung von Ni_2SiO_4-Einkristallen durch chemischen Transport, *J. Crystal Growth*, **37**, 155–158.

Hosoya, S. and Takei, H. (1978): Growth of Co_2SiO_4 crystals by the floating-zone method, *Japan, J. Crystal Growth*, **5**, 101 (in Japanese).

Hosoya, S. and Takei, H. (1979a): Crystal growth of $[Mn,Co]_2SiO_4$ solid-solution by the floating-zone method, *Japan, J. Crystal Growth*, **6**, 73 (in Japanese).

Hosoya, S. and Takei, H. (1979b): Growth of $[Mn,Mg]_2SiO_4$ solid-solution crystals by the floating-zone technique, *Proc. 24th Discussion Meeting of synthetic Minerals*, 107–108 (in Japanese).

Hosoya, S. and Takei, H. (1980): Growth of $[Mg, Fe]_2SiO_4$ solid-solution single-crystals by the floating-zone method, *Collect. Abst. 1980 Meeting, Mineral. Soc. Japan*, 91 (in Japanese).

HOSOYA, S. and TAKEI, H. (1982): Growth of olivine, $[Mg_{1-x}Fe_x]_2SiO_4$, single crystal by the floating-zone method, *J. Crystal Growth*, **57**, 341–346.

ITO, J. (1975): High temperature solvent growth of orthoenstatite, $MgSiO_3$, in air, *Geophys. Res. Lett.*, **2**, 533–536.

JORDAN, W. and NAUGHTON, J. J. (1964): Growth of forsterite crystals in a reactive crucible, *Am. Mineral.*, **49**, 806–808.

KOZIMA, H. and OSADA, S. (1978): Growth of forsterite solid-solution crystals by the floating-zone method, *Proc. 23rd Discussion Meeting of Synthetic Minerals*, 51–52 (in Japanese).

MUNOZ, J. L. (1968): Effect of shearing on enstatite polymorphism, *Carnegie Inst. Wash. Year Book*, **66**, 369–370.

NAKA, S., FURUHASHI, H. and NODA, T. (1968): Crystal growth of cobalt olivine (Co_2SiO_4), *J. Ceram. Soc. Japan*, **76**, 179–183 (in Japanese).

NITSAN, U. (1974): Stability field of olivine with respect to oxidation and reduction, *J. Geophys. Res.*, **79**, 706–711.

OZIMA, M. (1976): Growth of nickel olivine single crystals by the flux method, *J. Crystal Growth*, **33**, 193–195.

OZIMA, M. (1980): Growth of $MgSiO_3$ single-crystals by flux technique, *Proc. 25th Discussion Meeting of Synthetic Minerals*, 13–14 (in Japanese).

PFANN, W. G. (1952): Principles of zone-melting, *Trans AIME*, **194**, 747–753.

SCHEEL, H. J. (1972): Accelerated crucible rotation: a novel stirring technique in high-temperature solution growth, *J. Crystal Growth*, **13/14**, 560–565.

SCHEEL, H. J. and ELWELL, D. (1972): Stable growth rates and temperature programming in flux growth, *J. Crystal Growth*, **12**, 153–161.

SCHEEL, H. J. and ELWELL D. (1973): Stability and stirring in crystal growth from high-temperature solutions, *J. Electrochem. Soc.*, **120**, 818–824.

SCHANKLAND, T. J. (1966): Synthesis and optical properties of forsterite, *Tech. Rept. HP–16*, Gordon McKay Lab., Harvard Univ.

SCHANKLAND, T. J. (1967): Synthetic forsterite, *Ceram. Bull.*, **46**, 1160–1162.

SCHANKLAND, T. J. and HEMMENWAY, K. (1963): Synthesis of forsterite crystals, *Am. Mineral.*, **48**, 200.

SHINDO, I. (1980): On the effect of gravitation in solution growth, *Proc. 25th Discussion Meeting of Synthetic Minerals*, 63–64 (in Japanese).

SOMIYA, S., HIRANO, S. and IWAI, Y. (1978): Hydrothermal growth of fayalite crystals, *Res. Rept. Nagaoka Tech. Coll.*, **14**, 89–96.

SPEIDEL, D. H. and NAFZIGER, R. H. (1968): P–T–fo_2 relations in the system Fe–O–MgO–SiO_2 *Am. J. Sci.*, **266**, 361–379.

STEPHENSON, D. A., SCLAR, C. B. and SMITH, J. V. (1966): Unitcell volumes of synthetic orthoenstatite and low clinoenstatite, *Mineral. Mag.*, **35**, 838–846.

STROBEL, P. and PAGE, Y. L. (1981): Side-reaction in chemical vapor transport: growth of Pt and Co_2SiO_4 crystals, *Mat. Res. Bull.*, **16**, 223–228.

TAKEI, H. (1974): Synthesis of large olivine crystals, *J. Gemol. Soc. Japan*, **1**, 59–64 (in Japanese).

TAKEI, H. (1976): Czochralski growth of Mn_2SiO_4 (tephroite) single crystal and its properties, *J. Crystal Growth*, **34**, 125–131.

TAKEI, H. (1978): Growth of fayalite (Fe_2SiO_4) single crystals by the floating-zone method, *J. Crystal Growth*, **43**, 463–468.

TAKEI, H. (1981): Preferentially orientated precipitation in $MnTiO_3$ single crystals, *J. Materials Sci.*, **16**, 1310–1316.

TAKEI, H. and KOBAYASHI, T. (1974): Growth and properties of Mg_2SiO_4 single crystals, *J. Crystal Growth*, **23**, 121–124.

TAKEI, H. and KITAMURA, K. (1978): Growth of $FeTiO_3$ (ilmenite) crystals by the floating-zone method, *J. Crystal Growth*, **44**, 629–631.

TAKEI, H. and KITAMURA, K. (1978): Synthesis of ilmenite group crystals and their properties, *Proc. 23rd Discussion Meeting of Synthetic Minerals*, 57–58 (in Japanese).

TAKEI, H. and HOSOYA, S. (1982): Growth and properties of olivine single crystals by the floating-zone method, in *High-Pressure Research in Geophysics*, (Proc. 2nd U.S.-Japan Seminar on High-Pressure Research Applications, Hakone 1981), Eds. S. Akimoto and M. H. Manghani, Center for Academic Publications Japan/D. Reidel, Tokyo/Dordrecht, 537–547.

VU TIEN, L., GRANDIN DE L'PREVIER, GABIS, V. and ANTHONY, A. M. (1972): Amèlioration de la méthode des flux, synthèse et purification d'um minéral naturel: la forstérite, *J. Crystal Growth*, **13**/**14**, 601–603.

WANKLYN, B. M. (1970): Two new flux systems, $PbO-V_2O_5$ and $Bi_2O_3-V_2O_5$, *J. Crystal Growth*, **7**, 368–370.

WANKLYN, B. M. (1977): Growth of silicate and germanate crystals from $PbO-SiO_2$ (GeO_2) fluxes, *J. Crystal Growth*, **37**, 51–56.

WEBSTER, A. H. and BRIGHT, N. F. H. (1961): The system, iron-titanium-oxygen at 1,200°C, and oxygen partial pressures between 1 and 2×10^{-14} atm, *J. Am. Ceram. Soc.*, **44**, 110–116.

WILSON, J. (1965): Optical and electrical properties of synthetic olivine crystals, Thesis, Univ. Newcastle-Upon-Tyne, Great Britain, 1965.

WOLFF, G. A. and A. I. MLAVSKY (1974): *Travelling Solvent Technique, Crystal Growth*, Vol. 1, Plenum Press.

Materials Science of the Earth's Interior, edited by I. Sunagawa, pp. 131–148.

SYNTHESIS OF SINGLE CRYSTALS UNDER HIGH PRESSURE—CRYSTAL GROWTH OF Ni_2SiO_4 SPINEL AND BLACK PHOSPHORUS

Syun-iti AKIMOTO, Nozomu HAMAYA* and Ichimin SHIROTANI

The Institute for Solid State Physics, The University of Tokyo, Roppongi, Minato-ku, Tokyo, Japan

1. *Introduction*

High-pressure and high-temperature technique has long been expected as a promising tool for the synthesis of new materials. No doubt a large part of this motivation is due to the successful synthesis of diamonds by the General Electric Company in the 1950s. Actually, pressure-induced phase transformations have produced a great number of denser phases, and many of these new dense phases have been quenched under pressure and preserved as metastable phases at ambient pressure and temperature (see, for example a review by JOUBERT and CHENAVAS, 1979). High-pressure synthesis of dense polymorphs of silicates has first enabled the constructive discussion concerning the mineralogical constitution of the earth's mantle. The synthesis of silicate spinel, stishovite, silicate ilmenite, and silicate perovskite has had a direct influence on our understanding of the mantle transition zone.

In the high-pressure synthesis experiments hitherto attempted for silicates, high-pressure phases have generally been obtained as a polycrystalline aggregate consisting of very small crystals of 1–100 μm in size. Although such small crystals are available for the purpose of single-crystal X-ray diffraction study, they are too small to be used for the detailed characterization study. Recently, necessities of obtaining large single crystals of mantle minerals are increasingly accelerated in various field of mineral physics: well-characterized olivine single crystals are required for the high temperature-pressure creep experiments which are directly related to the large-scale flow in the mantle; ultrasonic measurements on cm-size single crystals of olivine, pyroxene, and garnet have greatly contributed to a better understanding of the

*Present address: Department of Material Physics, Faculty of Engineering Science, Osaka University, Toyonaka, Japan.

elasticity in the upper mantle; Brillouin scattering measurements of mm-size single crystals of silicate spinel, silicate ilmenite, and silicate perovskite are urgent in expectation of refining our knowledge on the mineralogical constitution of the mantle transition zone and the lower mantle.

The possibility of obtaining large single crystals of dense phases has been clearly demonstrated by the successful growth of gem-quality diamonds (STRONG and WENTORF, 1972). However, no positive attempt has been made for growing single crystals of the candidate deep-mantle minerals. Recently, a new cubic-anvil type of high-pressure apparatus with large sample volume was constructed in our laboratory for the purpose of investigating the crystal growth under pressure. In the present paper, typical examples of the crystal growth are demonstrated for Ni_2SiO_4 spinel and black phosphorus. The experimental procedures using a solid-solid phase transformation and a liquid-solid phase transformation are applied for growing single crystals of Ni_2SiO_4 spinel and black phosphorus, respectively.

2. *High-Pressure Apparatus Suitable for the Crystal Growth*

2.1 *A brief review of static high-pressure apparatus*

In the past thirty years, a great variety of static high-pressure apparatuses have been developed. Mechanisms for generating high pressure in these apparatuses can be classified on the basis of the geometrical considerations into five types schematically shown in Fig. 1: (1) piston-cylinder type, (2) anvil-cylinder type, (3) opposed anvil type, (4) multiple-anvil type, and (5) sliding anvil type (see, for example, a review by KUMAZAWA, 1971). It is

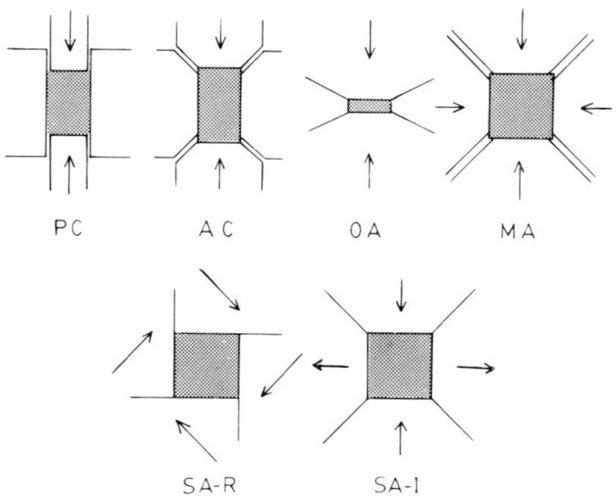

FIG. 1. A classification of high pressure apparatus based on the mechanism of pressure generation. PC: piston-cylinder type, AC: anvil-cylinder type, OA: opposed anvil type, MA: multiple-anvil type, SA-R: sliding anvil type (rotational), SA-I: sliding anvil type (irrotational).

obvious that opposed anvil apparatuses such as Bridgman anvil and diamond anvil are unsuitable for the synthesis of large crystals, because of their extremely limited sample-space. As for the sliding anvil type apparatus, there still remain some problems in the ease of operation and in the reproducibility of pressure-temperature conditions. In the present state of the art, the routine high-pressure and high-temperature runs are performed with the aid of the remaining three types of apparatuses among the above five types.

Large working volume in the pressure cell of piston-cylinder apparatus enables an accurate temperature control under pressure and is convenient for crystal growth. The lower strength of the piston and cylinder materials, however, restrict its applicability within a narrow pressure limit generally below 40 kbar. Anvil-cylinder apparatus and multiple-anvil apparatus are of the most versatile utility in the pressure range 10–300 kbar depending on the sample volume. Typical examples of the anvil-cylinder apparatus is seen in the belt apparatus and the girdle apparatus. Their capability in the crystal growth investigation has already been proved in the synthesis of cm-size diamonds and mm-size cubic BN in a flux. Original type of multiple-anvil apparatus is presented in a tetrahedral-anvil apparatus developed by HALL (1958). Since then, cubic-anvil and octahedral-anvil apparatuses have come into practical use in several high-pressure laboratories, and a design of the multi-stage multi-anvil apparatus has also been presented. Pressures generated in a modern two-stage cubic-octahedral-anvil apparatus reach to nearly 300 kbar at about 1,000°C.

In our laboratory, since 1963, we have been using a tetrahedral-anvil apparatus very similar to the original design by HALL (1958) for phase equilibrium study. Many high-pressure minerals have been recovered with that apparatus as metastable phases and they have been successfully used for the single-crystal X-ray diffraction study. A remarkable success is demonstrated by the discovery of the modified spinel structure through the synthesis of single crystals of βCo_2SiO_4 and βMn_2GeO_4 (AKIMOTO and SATO, 1968; MORIMOTO *et al.*, 1969). Very recently, new facility of a cubic-anvil apparatus with much larger pressure-cell volume compared with that in the tetrahedral-anvil apparatus has been made available for material synthesis research and the crystal growth under pressure. The detailed description of the new cubic-anvil apparatus is given in the next section.

2.2 *A wedge-type cubic-anvil apparatus*

There are several factors to be considered in the effective performance of high-pressure research in areas of phase equilibrium and crystal growth. Those are: (1) ease of operation of high-pressure apparatus, (2) reproducibility of pressure calibration, (3) ease of temperature measurements and control, (4) large volume for the internal furnace assembly, (5) capability of generating ultra-high pressures, and so on. The cubic-anvil system seems to be an excellent apparatus in view of its fulfilment of these requests.

The new cubic-anvil apparatus in our laboratory is the same as a wedge-type apparatus developed in Toshiba Corporation (ICHINOSE *et al.*, 1975; WAKATSUKI and ICHINOSE, 1982). An external view of the present apparatus is shown in Fig. 2. The apparatus is composed of the upper and lower cells. As shown in Fig. 2, each cell consists of three anvils that slide on the wedge systems: one of the small guiding wedge is seen in the figure, but the main wedges formed into shallow V-shaped grooves are masked by the anvils. Complete synchronization of the anvil movement is realized by means of these wedge systems. The new wedge-type cubic-anvil apparatus is driven by a 700-ton uniaxial hydraulic press. Compression axis of the hydraulic ram coincides with a [111] direction of the cube-shaped sample container usually made of pyrophyllite.

Routine runs were made in a combination of cemented tungsten carbide anvils of 16×16 mm^2 top-square face and a 21 mm edge cube of pyrophyllite. Pressure calibration for the combination was made based on the conventional pressure fixed points up to low-Ba transition pressure (55 kbar) at room temperature. The reproducibility of the pressure calibration in the present apparatus was found to be quite good as expected. Ram load (oil pressure)

FIG. 2. A wedge-type cubic-anvil apparatus with cemented tungsten carbide anvils of 16×16 mm^2 cube face.

could be controlled within an accuracy of ± 0.5 kg/cm^2 which corresponds to the uncertainty in pressure determination less than ± 0.2 kbar at 38.0 kbar. Although the limit of the attainable pressure in the present apparatus has not yet been fixed, A. Sawaoka (private communication) has generated pressures up to high-Bi transition point (77 kbar) at the Tokyo Institute of Technology using the same type of apparatus and the same size of anvils.

3. *Growth of Single Crystal Spinel from Single Crystal Olivine in Ni_2SiO_4*

3.1 *Olivine-spinel transformation*

The olivine-(modified spinel)-spinel transformation of orthosilicates has been intensively investigated in view of its great importance in interpreting the rapid increase in seismic wave velocities at the depth range 400–800 km in the mantle. A series of recent investigations in our laboratory on high-pressure phase equilibria in the system Mg_2SiO_4–Fe_2SiO_4 (AKIMOTO *et al.*, 1976; KAWADA, 1977) and in a garnet lherzolite (AKAOGI and AKIMOTO, 1979) strongly suggest that the olivine-modified spinel transformation is responsible for the seismic discontinuity at depths of 380–410 km and that the modified spinel-spinel transformation is related to the origin of the minor seismic discontinuity around 500 km.

In addition to the phase equilibrium data, precise knowledge of the elastic properties of olivine, modified spinel and spinel is indispensable for interpreting the seismic data correctly. Various methods have been developed to determine the elastic properties of crystals. The most accurate method is the measurement of ultrasonic wave velocities using a cm-size single-crystal. This method is applicable for olivine, but not for modified spinel and spinel because of the difficulty in obtaining large single crystals for these high-pressure polymorphs. The only method hitherto applied for silicate spinels is the X-ray diffraction measurements under high pressure using a very small amount of polycrystalline powder sample. The compressibility obtained in this method is still less precise than that obtained by the ultrasonic method, even if the X-ray diffraction measurements were made under hydrostatic pressure condition. It is further difficult to obtain a reliable value for the pressure derivative of compressibility from the X-ray measurements. Recently, successful application of Brillouin scattering technique to the elasticity measurements of sub-mm-size crystals of silicate minerals was reported by Weidner's laboratory in State University of New York at Stony Brook (for example, WEIDNER and CARLETON, 1977; WEIDNER *et al.*, 1978; WEIDNER *et al.*, 1982). This technique is particularly useful for the high-pressure phases such as modified spinel and spinel.

In the present paper, we describe an innovative technique of growing flawless spinel single crystals of 100–500 μm size from olivine single crystals of mm-size. The present experiments also provide an important information on the kinetic features of the olivine → spinel transformation and on the crystallographic orientation relationships between olivine and spinel phases in the transformation. The microscopic mechanism of the olivine → spinel transformation derived from the present data is discussed in detail elsewhere (HAMAYA, 1980; HAMAYA and AKIMOTO, 1982).

3.2 Starting materials and sample assembly

Flux-grown single crystals of Ni_2SiO_4 olivine were kindly provided by Ozima and used as starting materials in the present high-pressure and high-temperature experiments. Details of the flux method developed for the synthesis of mm-size single crystals of Ni_2SiO_4 olivine were reported by OZIMA (1976).

Olivine crystal without inclusions, a few mm in size, were subjected to concentrated nitric acid and cleaned in acetone prior to the high-pressure and high-temperature experiments. A wedge-type cubic-anvil apparatus described in a previous section was used in the present experiments. A 21-mm edge pyrophyllite cube was used as a pressure-transmitting medium and a sample container. A typical example of the sample assembly used for the cubic-anvil apparatus is illustrated in Fig. 3. A tubular graphite heater was used for heating the samples. NaCl was adopted as the pressure-transmitting medium inside the graphite heater. A few olivine crystals were placed at the central part of the NaCl pressure-transmitting medium. Since the temperature-pressure condition in the present experiments, 800° to 1,100°C and 31 kbar to 41 kbar, is close to the melting curve of NaCl, the NaCl insert is softened remarkably although not melted. This produced a uniform stress condition for the samples. Actually olivine crystals have never been broken by the stress even if the phase transformation proceeded considerably.

Temperature was measured with a Pt/Pt · 13%Rh thermocouple, 0.2 mm

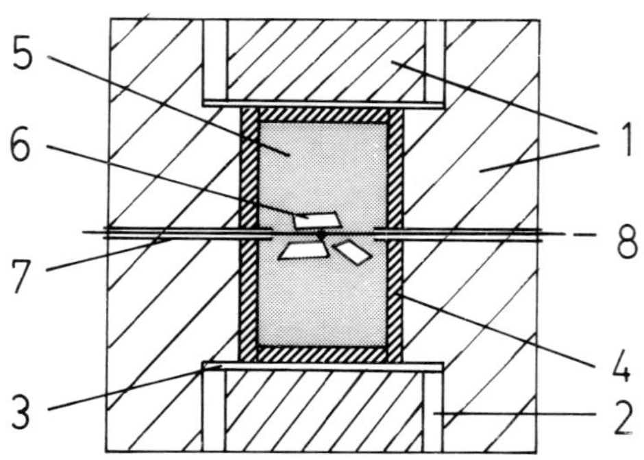

FIG. 3. Sample assembly for the crystal growth of Ni_2SiO_4 spinel. 1: pyrophyllite, 2: steel current ring, 3: Ni foil, 4: graphite heater, 5: NaCl, 6: olivine single crystal, 7: ceramic tube, 8: thermocouple.

in diameter. The hot junction was placed in the central part of the sample space so as to contact with olivine crystals. Thermocouple leads from the hot junction were brought out to the faces of a pair of opposite anvils.

3.3 Pressure and temperature control

In phase equilibrium experiments using a quenching technique the usual procedure is as follows: (1) First, the sample is compressed to a final desired value of pressure at room temperature. (2) Temperature is then increased to a final desired value. (3) The sample is held at the desired pressure-temperature condition for a time sufficient for equilibrium and is quenched isobarically to room temperature. (4) The sample is finally taken out to the ambient condition after releasing the pressure. If the solid-solid phase boundary has a positive slope of dP/dT such as the present olivine-spinel transformation, however, a successful application of this procedure would not be expected for growing large single crystals of high-pressure phase. Since the final synthesis condition is located in the spinel field, the olivine sample has to pass over the stability field of the spinel phase throughout temperature elevation. Considering the kinetics of the phase transformation, it is very probable that the transformation from olivine to spinel inevitably begins to take place at temperatures considerably lower than the finally desired temperature. The simultaneous formation of numerous nuclei of spinel phase in an olivine crystal at a large excess pressure beyond the equilibrium pressure would be inappropriate for growing a limited number of large crystals of spinel phase.

In the present work, a different path of pressure-temperature increase was taken for the purpose of growing large crystals of spinel phase and simultaneously investigating the kinetics of the olivine-spinel transformation. The procedure is as follows: (1) Pressure was elevated to 29 kbar at room temperature, and (2) temperature was then elevated to a desired value above 800°C at a heating rate of about 100°C/min. In advance, the olivine phase had been confirmed to be stable at temperature above 800°C at 29 kbar. (3) Then, pressure was increased rapidly to a final desired value within about 30 sec at the desired temperature. (4) After being maintained at a desired pressure-temperature condition for a desired interval of time the sample was thermally quenched under load. To maintain temperature at a constant value, a power control system was used. Both the graphite furnace and the Pt/Pt · 13%Rh thermocouple were stable enough to keep the accuracy of ±5°C under the present experimental conditions. In quenching the sample, temperature was dropped at a high cooling rate of 60°C/sec after the cutoff of the heating power. (5) Finally pressure was released and the sample was recovered to the ambient condition. In Fig. 4, the present path for the pressure-temperature increase was schematically compared with the usual procedure for the phase equilibrium study.

To estimate the quantity of the transformed spinel phase and to evaluate

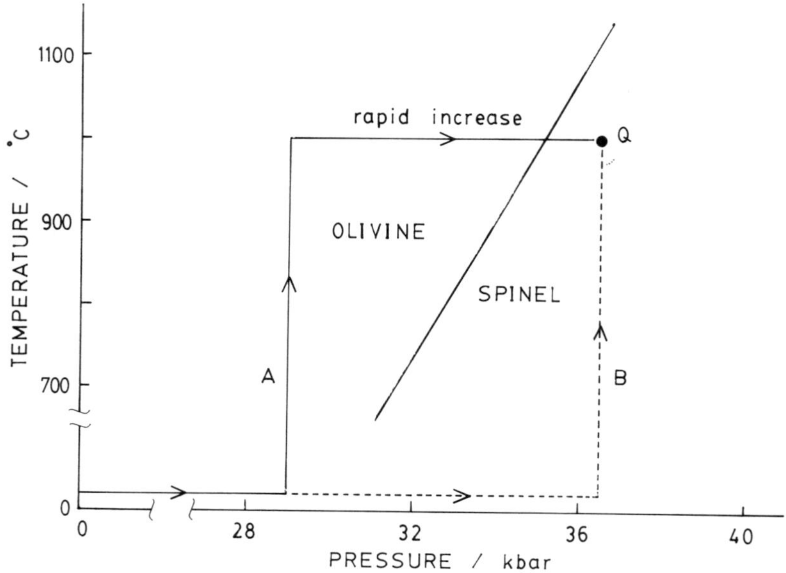

FIG. 4. The path of pressure-temperature treatment for the crystal growth of Ni_2SiO_4 spinel is compared with the usual procedure for the phase equilibrium study. A: present experiment, B: conventional run, Q: final pressure-temperature run-condition.

its quality, all the recovered samples were examined under the ordinary binocular and petrographic microscope. Selected crystals were further analyzed by means of scanning electron microscope and precession X-ray diffraction photography.

3.4 *Synthesis of high-quality Ni_2SiO_4 spinel crystals*

Runs were made in the pressure range from 31 to 41 kbar and over the temperature range from 800° to 1,100°C with the varying time interval of 5 to 180 min. Phases present in the quench product of thirty-two 20-min runs are mapped on the pressure-temperature diagram in Fig. 5. In Fig. 5, the results of three additional runs with different time intervals are also shown for clarifying the phase boundary between olivine field and spinel field. The phase boundary inferred from the present results is in good agreement with the boundary curve given by MA (1974), $P(\text{kbar}) = 23.3 + 0.0118T(°\text{C})$, but that is located in a little higher pressure side than the curve given by AKIMOTO *et al.* (1965), $P(\text{kbar}) = 19 + 0.012T(°\text{C})$.

Microscopic observations of the recovered samples indicate that there is a systematic difference in the growth feature of the spinel crystals in a parent single crystal of olivine depending upon the pressure-temperature condition of the synthesis. At lower temperatures below the chain line in Fig. 5, the olivine → spinel transformation was found to take place only at the surface of the

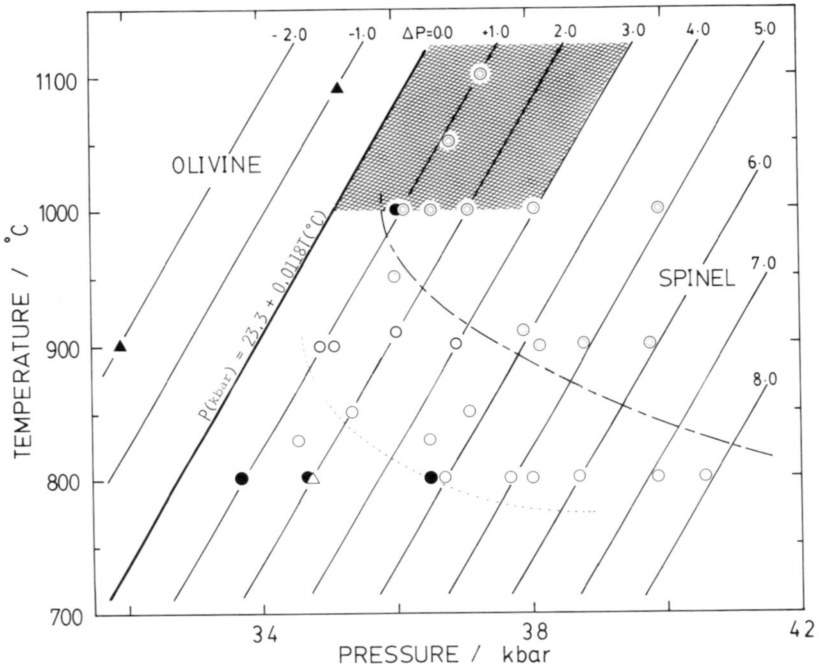

FIG. 5. Synthesis diagram of Ni_2SiO_4 spinel crystals based on the results of 20-min runs (circle) and additional three runs (triangle). Solid symbol: untransformed olivine, open circle: spinel, open triangle: spinel in 180 min-run, double circle: interior nucleation of spinel. The olivine-spinel phase boundary determined by MA (1974) is shown by thick line, and excess pressure ΔP beyond the boundary by thin lines. Dotted line and chain line denote the lowest P–T conditions of the surface and interior nucleations of the spinel phase in 20-min runs, respectively.

olivine single crystals and then the spinel crystals grew into the interior. Figure 6 shows such a hexagonal spinel crystal growing from the surface into the interior of a host olivine crystal.

At higher temperatures above the chain line in Fig. 5, spinel crystals was found to grow not only from the surface but also from the interior of the olivine crystal. Typical example of spinel crystals nucleated and grown in the interior of the olivine crystal is shown in Figs. 7(a) and 7(b). The size of the largest spinel crystal attains to about 500 μm. The spinel crystal usually forms an octahedron with eight {111} faces, and frequently a thin octahedron having a pair of well-developed {111} faces on which a few low steps lie along ⟨110⟩ directions. The faces {100} and {110} are less common and less developed. A number of spinel twins were also observed in these crystals.

High-quality spinel crystals were successfully synthesized under pressure-temperature conditions close to the phase boundary of the olivine → spinel

FIG. 6. Spinel crystal of Ni_2SiO_4 growing from the surface into the interior of a host olivine single crystal. The maximum diagonal dimension of the spinel crystal is approximately 90 μm.

FIG. 7(a). Spinel crystals of Ni_2SiO_4 growing in the interior of a host olivine crystal (scanning electron microphotograph). Spinel twin and some pyramids, probably single crystals of spinel, are seen.

FIG. 7(b). As-grown Ni_2SiO_4 spinel single crystals in the interior of olivine crystals.

transformation. Particularly, the hatched region corresponding to the conditions specified by $T \geq 1{,}000°C$, excess pressure beyond the olivine-spinel phase boundary, $\Delta P \leq 3$ kbar, and reaction time, $t \leq 20$ min, yields excellent quality crystals of up 300 μm in size (Fig. 7b). Here, the boundary curve given by Ma is adopted as a base line for estimating the excess pressure.

To demonstrate the availability of the spinel crystal synthesized for optical measurement, absorption spectrum was measured at one bar in the range 400–750 nm under a microscope combined with a spectroscopy system. An as-grown spinel single crystal, 80 μm in width and 30 μm in thickness, was used for the present measurement. The spectrum obtained are shown in Fig. 8, where three absorption bands are identified. Positions of these bands are in good agreement with a previous report by YAGI and MAO (1976) on the thin sections of Ni_2SiO_4 spinel crystal. Recent measurements by one of the authors (N.H.) at Dr. Weidner's laboratory in State University of New York at Stony Brook have also proved that the quality of the present spinel crystals is high enough for obtaining excellent Brillouin scattering spectrum. A detailed report on the Brillouin scattering measurements will be given in a separate paper (HAMAYA *et al.*, in preparation).

3.5 Crystallographic orientation relationships in the olivine → spinel transformation

Successful synthesis of the spinel crystals from the single crystal olivine made it possible to examine the crystallographic orientation relationships between the olivine and spinel phases in the transformation. A sample

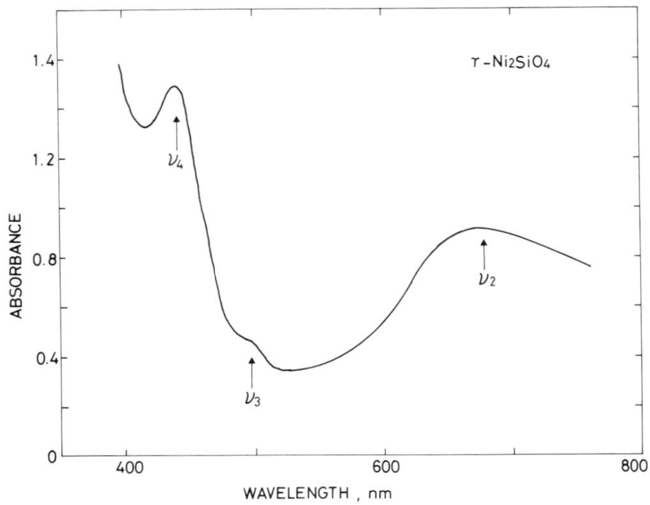

FIG. 8. Absorption spectra of a single crystal of Ni_2SiO_4 spinel.

composed of a 100 μm-size single crystal of spinel enclosed by a fragment of single crystal of olivine was analyzed by means of a precession X-ray diffraction photography.

A typical example of the precession photograph indicating the coincidence of crystallographic directions between the olivine and spinel phases is shown in Fig. 9. The orientational relation, $[100]_{oliv}//[111]_{spinel}$ is demonstrated within an accuracy of a few degrees. The relationships, $[001]_{oliv}//[\bar{1}10]_{spinel}$ and $[010]_{oliv}//[11\bar{2}]_{spinel}$, were also observed on the other precession photographs.

We observed these interphase structural orientation relationships for six samples among eleven samples examined. It is concluded from these X-ray diffraction measurements that the relation represented by $(100)_{oliv}//(111)_{spinel}$ and $[001]_{oliv}//[\bar{1}10]_{spinel}$ is fundamental in the olivine → spinel transformation in Ni_2SiO_4. Recently, the same orientation relationship was found independently by LACAM *et al.* (1980) for a sample prepared by heating Fe_2SiO_4 olivine in a diamond anvil cell in a transition zone between the olivine and spinel structure, using transmission electron microscopy. This leads to the conclusion that the olivine → spinel transformation generally takes place so as to hold the relation $(100)_{oliv}//(111)_{spinel}$ and $[001]_{oliv}//[\bar{1}10]_{spinel}$. This orientation relationship is compatible with a model of a martensitic olivine → spinel transformation, first proposed by HORNSTRA (1960, 1961) and recently developed by POIRIER (1981). In this model, the movement of the partial

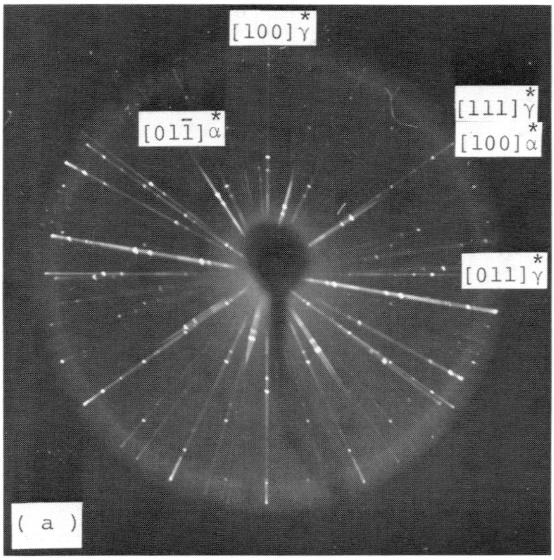

FIG. 9. Precession photograph of a run product subjected to 1,000°C and 38 kbar. The coincidence of crystallographic directions between the olivine and spinel structure, $[100]_{olivine}//[111]_{spinel}$, is observed.

dislocations 1/12[013] and 1/12[0$\bar{1}$3], which are dissociated from the [001] dislocation, on the (100) plane of the olivine structure induces a change in stacking of the oxygen layer of the crystal from olivine-type (hexagonal closest packing) to spinel-type (face-centered cubic), accompanying the rearrangement of cations in synchroshear motion.

Although the presence of the interphase relationships seems to favor the martensitic transformation mechanism, some evidences suggesting a nucleation-and-growth mechanism for the olivine-spinel transformation are also found from the detailed observation of the as-transformed surfaces of Ni_2SiO_4 spinel crystal prepared in this study (H. Komatsu, private communication).

4. *Growth of Single Crystals of Black Phosphorus from Melt*

4.1 *Previous work on the synthesis of black phosphorus*

Bridgman reported in the 1910's that ordinary yellow or white phosphorus are permanently changed into a black-colored solid much like graphite in appearance when they are exposed to pressure above 12 kbar at temperatures above 200°C (BRIDGMAN, 1914, 1916). Since then, a number of investigations have been done on the synthesis of black phosphorus and on its physical properties. Polycrystalline black phosphorus is reported to be a P-type semiconductor with room temperature resistivity of about 1 Ωcm and with an energy gap of about 0.35 eV.

The crystal structure of black phosphorus is determined to be orthorhombic. The atoms are arranged in puckered layers; in a layer each phosphorus atom is joined by covalent bonds to three neighbors, while the layers are held together by Van der Waals forces. The layered structure results in the anisotropic linear compressibilities of polycrystalline black phosphorus.

Recently, some attempts of growing single crystals of black phosphorus were made from a general interest in the physical properties of the layered-structure materials. MARUYAMA *et al.* (1981) have reported the synthesis of black phosphorus single crystals from a solution of white phosphorus in liquid bismuth. They obtained needle-like crystals or thin rod crystal. These crystals are still too small to make detailed measurements of the anisotropy in the electrical resistivity. In this paper, we describe a new technique of growing mm-size single crystals of black phosphorus directly from its melt at high pressures and temperatures. Electrical properties of the black phosphorus crystals prepared by the present method has been reported elsewhere (SHIROTANI, 1981).

4.2 *Preparation of polycrystalline black phosphorus*

A two-step process was adopted to grow large single crystals of black phosphorus. In the first step, polycrystalline black phosphorus was synthesized from white phosphorus or red phosphorus. The polycrystalline black phosphorus thus prepared was used as a starting material for growing the large single crystal in the second-step experiment. The wedge-type cubic-anvil apparatus was used throughout the first-step and the second-step high pressure-temperature experiments. The sizes of tungsten-carbide anvils and pyrophyllite cube are the same as those used for the synthesis of Ni_2SiO_4 spinel crystals.

Figure 10 shows the sample assembly used for the preparation of polycrystalline black phosphorus. Two kinds of sample containers were used inside the graphite furnace depending on the starting materials. As for the starting material of white phosphorus, teflon capsule was adopted. On the other hand, fired-pyrophyllite tube was used as sample container for red phosphorus.

Successful runs for preparing polycrystalline black phosphorus were carried out at 38 kbar and 240–500°C and at 45 kbar and 600°C, using a combination of white phosphorus and teflon capsule. In these runs samples were maintained at these pressure-temperatue conditions for ten minutes and then quenched under working pressure. Black phosphorus was also synthesized from red phosphorus in the 20-min runs at the pressure-temperature conditions of 38 kbar, 600–700°C and 45 kbar, 450–800°C.

4.3 Growth of large single crystals of black phosphorus

We have attempted to grow large single crystals of black phosphorus from its melt by finding a suitable heat treatment under pressure. To determine the melting curve of black phosphorus, change in electrical resistivity of polycrystalline black phosphorus with increasing temperature was measured at three different pressure, 28, 32, and 38 kbar. Sample assembly used for the resistivity measurement has been given in a previous paper (SHIROTANI, 1981).

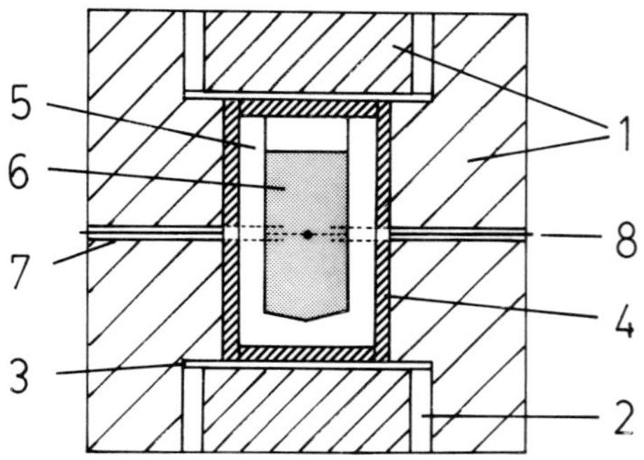

FIG. 10. Sample assembly for the preparation and crystal growth of black phosphorus. 1: pyrophyllite, 2: steel current ring, 3: Ni foil, 4: graphite heater, 5: capsule, 6: sample, 7: ceramic tube, 8: thermocouple.

Resistivity *versus* reciprocal temperature curve at 32 kbar is illustrated in Fig. 11. The resistivity decreased with increasing temperature up to about 900°C. Above this temperature the resistivity slightly increased with increasing temperature. It reached a maximum at about 995°C and then dropped abruptly. This change is reasonably considered to be due to a phase transition of black phosphorus from solid to liquid. In the present study the temperature of the resistivity maximum is assumed to be the melting point. The melting temperature thus determined is plotted as a function of pressure in Fig. 12. Figure 12 suggests that the melting temperature increases almost linearly with pressure at a rate of 12°/kbar in the pressure range 20–40 kbar.

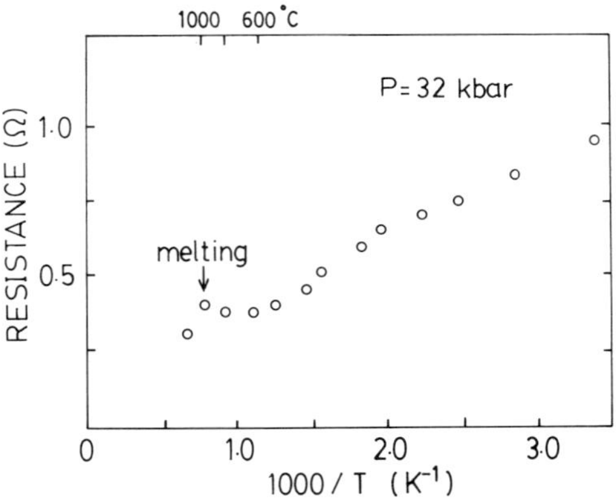

FIG. 11. Resistivity *versus* reciprocal temperature curve of black phosphorus at 32 kbar.

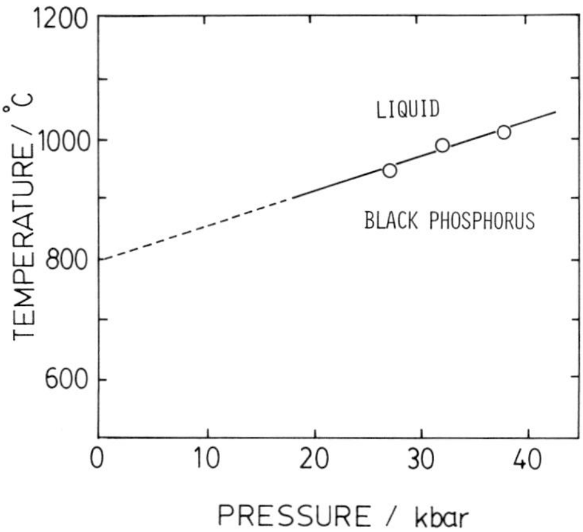

FIG. 12. The melting curve of black phosphorus.

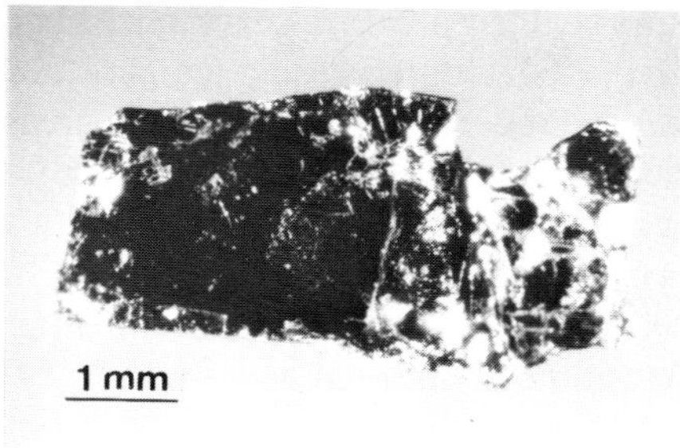

FIG. 13. Cleavage plane of black phosphorus single crystal.

For the purpose of growing single crystals, a similar furnace assembly as shown in Fig. 10 was used. The sample container was made of the pyrophyllite tube baked at 1,100°C for 2–3 hours in air. When polycrystalline black phosphorus is melted in the graphite furnace and slowly cooled at about 23 kbar, large single crystals were successfully grown. The size of the largest one was 4 mm in diameter and 5 mm in length. In this particular run, the sample was cooled from 1,200°C to 850°C at a very slow cooling rate of about 80°C/hour and quenched from 850°C.

The black phosphorus crystal thus synthesized is easily cleaved and exihibits flakiness similar to mica. Figure 13 indicates a cleavage plane of a single crystal of black phosphorus. The flake of black phosphorus crystal is flexible and shows metallic luster like silver. In Fig. 13, a long axis within the plane is parallel to the crystallographic *a*-axis and a short axis is parallel to the *c*-axis. The *b*-axis is perpendicular to the cleavage plane.

As was expected from the crystallographic character, remarkable anisotropy was found in the resistivity measurement along the three axis of the present single crystal. The resistivity along the *a*-, *c*-, and *b*-axis was determined as about 0.1 Ωcm, 1 Ωcm, and 10^2–10^3 Ωcm, respectively at room temperature. An intralayer anisotropy (within *ac* plane) in the resistivity was first observed. More detailed discussion on the resistivity measurement has been given in a previous paper (SHIROTANI, 1981).

This work has been supported by the Ministry of Education, Science and Culture, Japan, under grant 321507, 420906, and 520104.

REFERENCES

AKAOGI, M. and AKIMOTO, S. (1979): High-pressure phase equilibria in a garnet lherzolite, with special reference to Mg^{2+}–Fe^{2+} partitioning among constituent minerals, *Phys. Earth Planet. Inter.*, **19**, 31–51.

AKIMOTO, S. and SATO, Y. (1968): High-pressure transformation in Co_2SiO_4 olivine and some geophysical implications, *Phys. Earth Planet. Inter.* **1**, 498–504.

AKIMOTO, S., FUJISAWA, H. and KATSURA, T. (1965): The olivine-spinel transition in Fe_2SiO_4 and Ni_2SiO_4, *J. Geophys. Res.*, **70**, 1969–1977.

AKIMOTO, S., MATSUI, Y. and SYONO, Y. (1976): High pressure crystal chemistry of orthosilicates and formation of the mantle transition zone, in *The Physics and Chemistry of Minerals and Rocks*, Ed. R. G. J. Strens, Wiley, London. pp. 327–363.

BRIDGMAN, P. W. (1914): Two new modifications of phosphorus, *J. Am. Chem. Soc.* **36**, 1344–1363.

BRIDGMAN, P. W. (1916): Further note on black phosphorus, *J. Am. Chem. Soc.* **38**, 609–612.

HALL, H. T. (1958): Some high-pressure, high-temperature apparatus design considerations: equipment for use at 100,000 atmosphere and 3,000°C, *Rev. Sci. Instrum.*, **29**, 267–275.

HAMAYA, N. (1980): Mechanism of olivine-spinel transformation, Ph. D. Thesis, Univ. of Tokyo, 113 pp.

HAMAYA, N. and AKIMOTO, S. (1982): Experimental investigation on the mechanism of olivine→spinel transformation: Growth of single crystal spinel from single crystal olivine in Ni_2SiO_4, in *High-Pressure Research in Geophysics*, Eds. S. Akimoto and M. H. Manghnani, Center Acad. Publ. Japan/D. Reidel, Tokyo/Dordrecht, pp. 373–389.

HORNSTRA, J. (1960): Dislocations, stacking faults and twins in the spinel structure, *J. Phys. Chem. Solids*, **15**, 311–322.

HORNSTRA, J. (1961): Dislocations, stacking faults and twins in the spinel structure, in *Proc. 4th Int. Symp. Reactivity of Solids*, Ed. J. H. De Boer, Amsterdam, Elsevier, pp. 563–570.

ICHINOSE, K., WAKATSUKI, M. and AOKI, T. (1975): A new sliding type cubic anvil high pressure apparatus, *Atsuryoku Gijutsu*, **13**, 244–253 (in Japanese).

JOUBERT, J. C. and CHENAVAS, J. (1979): New compounds from high pressure, *J. Solid State Chem.*, **27**, 29–39.

KAWADA, K. (1977): The system Mg_2SiO_4–Fe_2SiO_4 at high pressures and temperatures and the earth's interior, Ph. D. Thesis, Univ. of Tokyo, 187 pp.

KUMAZAWA, M. (1971): Multiple-anvil sliding system—a new mechanism of producing very high pressure in a large volume, *High Temp.-High Pres.*, **3**, 243–260.

MORIMOTO, N., AKIMOTO, S., KOTO, K. and TOKONAMI, M. (1969): Modified spinel, beta-manganous orthogermanate: stability and crystal structure, *Science*, **165**, 586–588.

MARUYAMA, Y., SUZUKI, S., KOBAYASHI, K. and TANUMA, S. (1981): Synthesis and some properties of black phosphorus single crystals, *Physica*, **105B**, 99–102.

MA, C. B. (1974): Reinvestigation of the olivine-spinel transformation in Ni_2SiO_4 and the incongruent melting of Ni_2SiO_4 olivine, *J. Geophys. Res.*, **79**, 3321–3324.

LACAM, A., MADON, M. and POIRIER J. P. (1980): Olivine glass and spinel formed in a laser heated, diamond-anvil high pressure cell, *Nature*, **288**, 155–157.

OZIMA, M. (1976): Growth of nickel olivine single crystals by the flux method, *J. Crystal Growth*, **33**, 193–195.

POIRIER, J. (1981): Martensitic olivine-spinel transformation and plasticity of the mantle transition zone, in *Anelastic Properties and Related Processes in the Earth's Mantle*, Am. Geophys. Union Monograph.

STRONG, H. M. and WENTORF, R. H. Jr. (1972): The growth of large diamond crystals, *Naturwissenschaften*, **59**, 1–7.

SHIROTANI, I. (1982): Growth of large single crystals of black phosphorus at high pressures and temperature, and its electrical properties, in *Proc. Inter. Conf. Low-Dimensional Conductors*, Boulder, Colorado; *Molecular Cryst. Liquid Cryst.*, **86** 1943–1951.

WAKATSUKI, M. and ICHINOSE, K. (1982): A wedge-type cubic anvil high-pressure apparatus and its application to material synthesis research, in *High-Pressure Research in Geophysics*, Eds. S. Akimoto, and M. H. Manghnani, Center Acad. Publ. Japan/D. Reidel, Tokyo/Dordrecht, pp. 13–26.

WEIDNER, D. J. and CARLETON, H. R. (1977): Elasticity of coesite, *J. Geophys. Res.*, **82**, 1334–1346.

WEIDNER, D. J., WANG, H. and ITO, J. (1978): Elasticity of orthoenstatite, *Phys. Earth Planet. Inter.*, **17**, 7–13.

WEIDNER, D. J., BASS, J. D. and VAUGHAN, M. T. (1982): The effect of crystal structure and composition on elastic properties of silicates, in *High-Pressure Research in Geophysics*, Eds. S. Akimoto, and M. H. Manghnani, Center Acad. Publ. Japan/D. Reidel, Tokyo/Dordrecht, pp. 125–133.

YAGI, T. and MAO, H. K. (1976): Crystal-field spectra of the spinel polymorph of Ni_2SiO_4 at high pressure, *Carnegie Inst. Washington, Year Book*, **76**, 505–508.

Materials Science of the Earth's Interior, edited by I. Sunagawa, pp. 149–169.

CHARACTERIZATION OF OLIVINE CRYSTALS

Hiroshi KOMATSU and Tetsuo INOUE

The Research Institute for Iron, Steel and Other Metals, Tohoku University, Katahira, Sendai, Japan

1. Introduction

The purpose of the present study is twofold, the one is to characterize the defects in large synthetic crystals of centimeter size and the other is to compare characteristic defect features in crystals of different growth conditions.

The crystals investigated here are as follows:

1. Forsterite (Fo_{100}) grown by the Czochralski methods (CZ-Fo_{100}).
2. Olivine ($Fo_{92}Fa_8$) grown by the floating zone method (FZ-Fo_{92}).
3. Olivine ($Fo_{20}Fa_{80}$) grown by the floating zone method (FZ-Fo_{20}).
4. Natural olivine from San Carlos ($Fo_{91}Fa_9$).
5. Forsterite (Fo_{100}) grown by flux method (FX-Fo_{100}).

The first three crystals were grown from the melt (TAKEI and KOBAYASHI, 1974; HOSOYA and TAKEI, 1982), and the last crystal was grown by Ozima in a high temperature solution using MoO_3 and V_2O_5 flux.

The concrete objects of this study are (1) finding out reliable dislocation etchants, (2) the quantitative measurement of dislocations and sub-boundaries, (3) detecting submicroscopic defects and voids, and (4) the comparison of defect characteristics among the above mentioned crystals.

2. Historical Review

2.1 Methods for detecting dislocations

The modern history of dislocation etching started only recently when SCHOCKLEY and READ (1949) predicted that dislocation etch pits might be seen in sub-boundaries. This was experimentally proved later by VOGEL *et al.* in 1953.

Mineralogists, however, had observed etch figures which were artificially produced on crystal faces as long ago as 1817. Books by BAUMHAUER (1894) and HONES (1927) dealt with these, but their main interests were not in the line of defect characterization. Mineralogists were interested in using the etch figures (note; etch "figures" not etch "pits") for symmetry study.

HORN (1952) intentionally etched artificial silicon carbidge crystals grown by spiral growth mechanism, and proved that (1) screw dislocations could be etched and (2) a one-to-one correspondence existed between etch pits and screw dislocations. VOGEL *et al.* (1953) established a one-to-one correspondence between edge dislocations and etch pits, and this stimulated the application of the etch pit technique to other fields, such as crystal growth, plastic deformations and radiation damages, etc.

HEDGES and MITCHELL (1953) were the first to intentionally decorate dislocations in order to observe them. HIRSCH *et al.* (1956) observed dislocations with an electron microscope for the first time and LANG (1958) observed dislocations within the crystal by X-ray; now known as Lang topography.

Now we have at hand versatile methods of detecting and observing crystal imperfections and numerous reports have been published (NEWKIRK and WERNICK, 1962; JOHNSTON, 1962; AMELINCKX, 1964; ROBINSON, 1968, etc.). A comparison of the methods was described by JOHNSTON (1962) and his summary is shown in Table 1, in which one can choose the best method to detect dislocations. It is clear that etching and decoration methods are suitable for natural crystals which have a dislocation density of $10^6 - 10^8/cm^2$.

The guidelines for finding suitable etchants has not yet been established. However, one may set up the criteria for a good etchant as follows: (1) reproducible, (2) simple in all respects (chemical composition, preparation and not time-consuming), (3) must produce sharp, straight ledges of a low index (symmetry is revealed at once), and (4) must be orientation insensitive (although the reverse is sometimes required).

TABLE 1. Maximum dislocation density that can be resolved by various methods (after JOHNSTON, 1962).

Method	Specimen thickness t	Width of dislocation image, W	Maximum dislocation density
Electron microscopy	500 Å	100 Å	$3 \times 10^{11}/cm^2$
X-ray, transmission	1 mm	5 μ	2×10^4
	0.1 mm	5 μ	2×10^5
X-ray, reflection	50 μ (max.)	2 μ	10^6
	<2 μ (min.)	2 μ	2×10^7
Decoration	~10 μ (depth of focus)	0.5 μ	2×10^7
Etching	Optical microscope	$S = 0.5\ \mu$*	4×10^8
	Electron microscope	$S = 500$ Å	4×10^{10}

*S is the limit of resolution of etch pits.

2.2 Etching of olivine crystals

For characterizing large crystals, such as synthetic boules of several cm in length and 1.5 cm in diameter, etching is the most rapid and suitable technique with reasonable resolution through which one can feed back useful information to crystal growth. Etching is also useful in the study of deformations (Kirby and Wegner, 1978) and fission tracks (Fleischer *et al.*, 1965).

There has been much research into etchants on olivine. Maurrette (1966), Horn and Maurrette (1967) etched olivine with dil. HF or conc. KOH solution and produced etch pits on (010), (100), but not on (001) faces. Young (1969) used conc. HCl or 5HF + $2HNO_3$ (in volume ratio hereafter) for (100) faces and NaOH + KOH + H_2O for (010) faces, but he also failed to etch a (001) face. Grossman *et al.* (1970) reported two etchants: HF + HCl + citric acid and HBF_4 + HCl + citric acid, which were particularly useful to cosmic ray tracks. They could not distinguish these tracks from the dislocations. Wegner and Christie (1974) have made a comprehensive study on etching, confirming a one-to-one correspondence. They found an etchant for a (001) face for the first time and designated this as #6 etchant (conc. HF + conc. HCl + citric acid). Kirby and Wegner (1978) used transmission electron microscopy (TEM) to investigate the properties of dislocations revealed by #6 etchant.

The etchants so far described were for natural olivine, which contained a small amount of iron. Desai and John (1977) etched (010) faces of synthetic forsterite (Fo_{100}) by conc. HCl or KOH melt. Desai and John (1979) discussed the changes in morphology on a (010) face by different etchants.

2.3 Decoration and X-ray topography of olivine

Decoration technique was applied to natural forsterite (Fo_{90-95}) diffusing manganese after Dash's method (Dash, 1956) by Young (1969). Kohlstedt and Vander Sande (1975) studied naturally decorated iron-bearing olivine and identified several minerals nucleated around the dislocations. Artificial decoration of iron-bearing olivine by oxidation was carried out by Kohlstedt *et al.* (1976). Jaoul *et al.* (1979) decorated Fo_{100} in the powder of natural iron-bearing olivine, following oxidation.

Slind and Sørum (1976) made an X-ray topographic study of the defects in natural olivine (Fo_{93}), they found the majority of the dislocations were lying in the (100) or (001) plane with the Burgers vector [001] or [100], being of the edge, screw or mixed type.

In our present investigation, known etchants were used first, and if they failed to reveal dislocation etch pits we modified the etchants using the above criteria. It is time saving to observe surfaces with a differential interference microscope or phase-contrast microscope since they can detect very shallow pits that appear in the early stages of etching (Komatsu and Sunagawa, 1975).

3. *Experimentals*

The composition of etchants used in the present experiments are listed in Table 2, referring to the kind of crystals and crystallographic planes. All synthetic olivine (CZ-Fo_{100}, FZ-Fo_{92}, FZ-Fo_{20}) were grown along a [100] axis, and detailed growth conditions are reported by TAKEI and KOBAYASHI (1974) and HOSOYA and TAKEI (1981). Crystallographic orientation of the surfaces were determined by the Laue method; they were prepared either by cleaving or mirror polishing with diamond paste (<1 μm in grain size).

In addition to etching, X-ray topography (fine focus MoK_α 40KV, 2mA) was used for CZ-Fo_{100} and FX-Fo_{100}. Oxidation decoration (KOHLSTEDT *et al.*, 1976) was also carried out for iron-bearing olivine.

The dark-field light scattering technique is most useful in examining submicroscopic precipitates in the crystal interior and three kinds of instruments which are complementary to each other were used. These were an Olympus gem microscope, a cardioid condenser with biological microscope and laser scanning tomography (MORIYA and OGAWA, 1978).

Local chemical analysis of small precipitates and banded structures were performed with a scanning electron microscope (SEM) equipped with energy dispersive X-ray analyser (EDX).

4. *Results and Discussion*

4.1 *Light scattering particles and voids*

All crystals examined here contained more or less light scattering particles or voids (Figs. 1–3, 15, 19), which are useful for analysing growth conditions. Features of the distribution of scattering centers differ depending on the circumstances of growth. Typical examples are shown in Fig. 1, where contrasting differences are seen between melt-grown crystals and natural

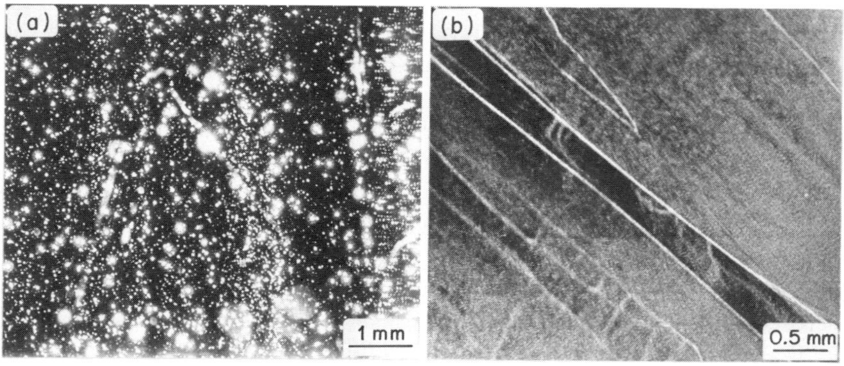

FIG. 1. Light scattering tomograph by an Ar laser; (a) CZ-Fo_{100}, (b) natural olivine.

TABLE 2. Etch Pits and dislocation density.

	010		100		001	
	Etchant	$N, L^{(1)}$	Etchant	N, L	Etchant	N, L
CZ 01. (Fo_{100})	conc. HCl[4] (3.5 hrs) 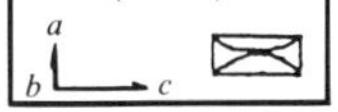	$N = 2 - 8 \times 10^4$ $L = 0$	1HF + 2HCl + 4H_2O (30 min) 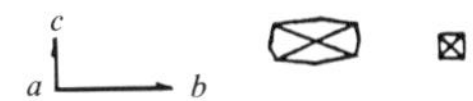	$N = 1 - 2 \times 10^5$ $L = 0$	1HF + 2HCl + 4H_2O (30 min) 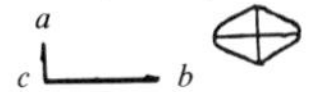	$N = 5 - 6 \times 10^4$ $L = 0$
FZ 01. (Fo_{92})	conc. HCl (3.5 hrs) 	$N_1 = 2 \times 10^5$ $N_2 = 2 \times 10^4$ $L = *(2)$	(a) #6 (1 hr)[5] (b) 20%HF (1 – 2 hrs) (c) 1HF + 2HCl + 4H_2O (30 min) (3) 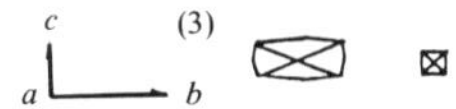	$N = 1 - 2 \times 10^5$ $L = 10$	1HF + 2HCl + 4H_2O (30 min) 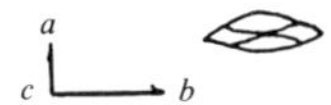	$N_1 = 2 \times 10^5$ $N_2 = 3 \times 10^4$ $L = *$
FZ 01. (Fo_{20})	conc. HCl (7 min) 	$N = 0.7 - 1.2 \times 10^5$ $L = *$	(a) #6 (1 hr) (b) 20%HF (1 hr)  (c) 1HF × 2HCl × 4H_2O (30 min) 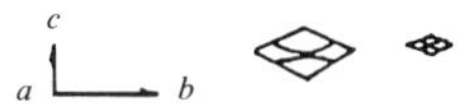	$N = 1 - 3 \times 10^5$ $L = 20$	20%HF (30 min) 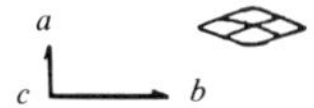	$N = 1.5 - 4.5 \times 10^5$ $L = *$
Natural 01.	conc. HCl (3.5 hrs) #6 (30 min) 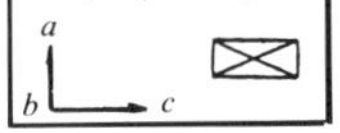	$N = 4 - 8 \times 10^5$ $L = 150$	#6 (30 min) 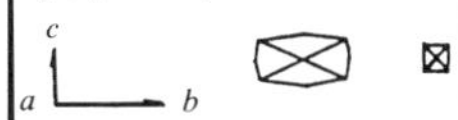	$N = 1 \times 10^6$ $L = 110$	#6 (30 min) 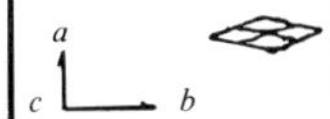	$N = 0.5 - 1 \times 10^6$ $L = *$
Flux 01. (Fo_{100})	$N = 0$ (from X-ray topography); Non-dislocation etch pits were produced with the #6 etchant on the as-grown surfaces, m{110} and k{021}. m{110};  k{021}; 					

Remarks; (1)N(1/cm^2): dislocation density. L(cm/cm^2): subboundary length per unit area. N_1 and N_2 show the dislocation densities in the high and low density region, respectively (see the Fig. 13). (2) L = *: The L was not determined since there were no subboundaries in the area examined. (3) The large rectangular pit and small square one represent screw and edge dislocations, respectively (KIRBY and WEGNER, 1978). (4) The conc. HCl etchant was described by DESAI and JOHN (1977). (5) The #6 etchant was described by WEGNER and CHRISTIE (1974).

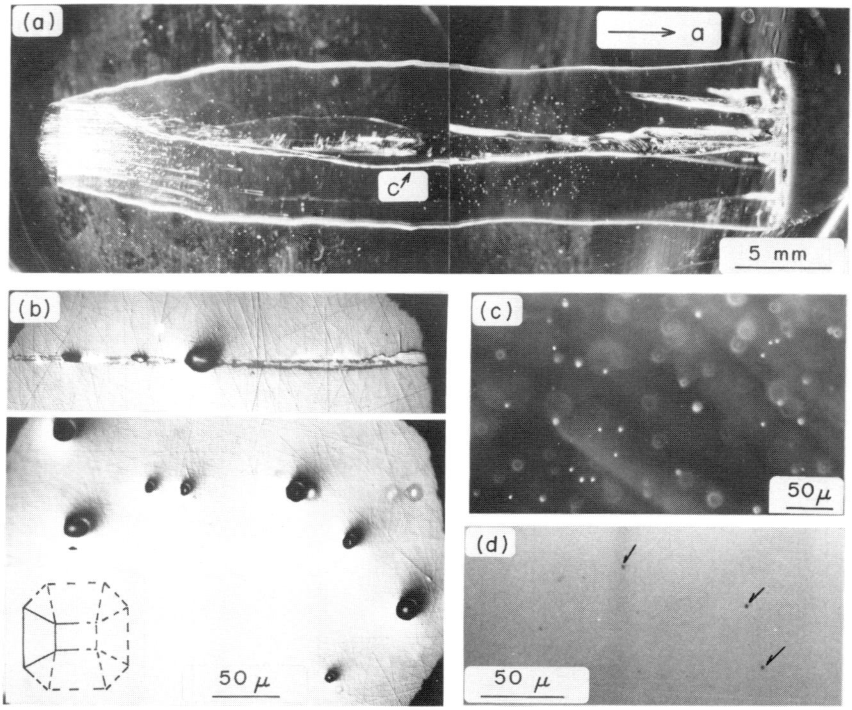

FIG. 2. Scattering centers in FZ-Fo_{92}; (a) Photograph taken by the stereomicroscope under dark-field illumination. The scattering centers are gas bubbles and cracks indicated by arrow "c." Arrow "a" indicates the growth direction [100], (b) Photograph taken by metallurgical microscope showing the scattering centers in (a) at a higher magnification. Some bubbles take the faceted form (negative crystal) as shown in the insert. (c) Microphotograph showing fine scattering centers taken by biological microscope provided with a cardioid condenser, (d) without a cardioid condenser (bright-field illumination). Arrow indicates scattering centers.

olivine. Scattering centers in natural crystals are generally very fine and submicroscopic, whereas melt-grown crystals contain spherical particles of several μm in diameter (Figs. 1a, and 2c). In FZ-Fo_{92} many voids (probably gas bubbles) were presented and some of them showed faceted inner walls (i.e., negative crystals) as shown in Fig. 2b. This suggests that the negative crystals were formed at or near the liquid-solid interface.

In FZ-Fo_{20} many opaque scattering particles were seen and they were disposed in fine bands (Figs. 3a, b). The opaque particles showed metallic luster and EDX analysis showed that they were very rich in iron, but deficient in oxygen and silicon. It is most probably iron particles which precipitated in a reduction atmosphere, since this crystal was grown in an oxygen deficient

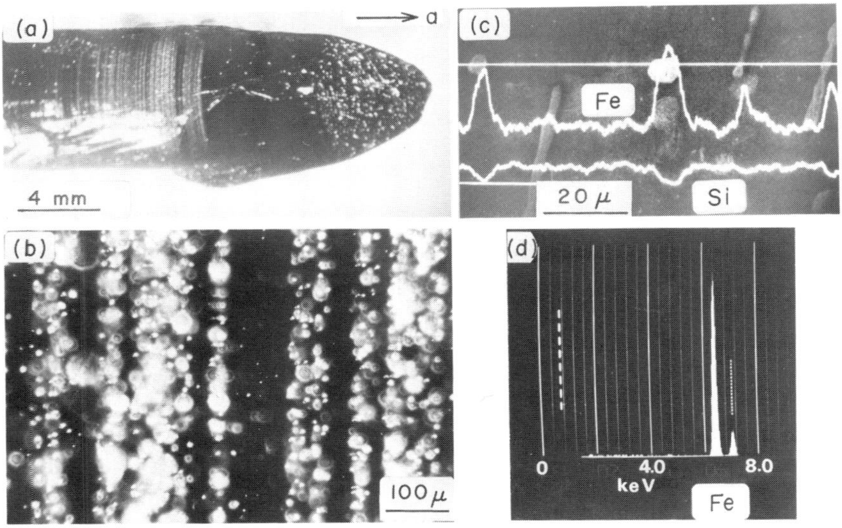

FIG. 3. Scattering centers in FZ-Fo_{20}: (a) Photograph by stereomicroscope under oblique illumination, (b) Microphotograph showing the scattering centers in (a) which was taken by biological microscope provided with a cardioid condenser at higher magnification, (c) The secondary electron image of the scattering centers by SEM and the line profiles of Fe-K_α and Si-K_α along the straight line in the image, (d) X-ray dispersion spectroscopy of the scattering center.

atmosphere in order to control the Fe^{+2}/Fe^{+3} ratio. The fine precipitates in FZ-Fo_{20} crystals generated many dislocation as will be mentioned later.

4.2 Etching

The first necessary step in establishing reliable etchants is to confirm a one-to-one correspondence of etch pits with dislocations. We examined this by matched cleavage and successive etching techniques and the decoration method.

Secondly, orientation effects of the surface and the crystallographic plane in an etch pit formation was examined since the etching behavior critically depends on the inclination of crystallographic planes with respect to the surface. The final results obtained in the present experiments are summarized in Table 2. Those etchants framed by bold lines have been established earlier by other researchers and reconfirmed here.

4.2.1 (010) face

WEGNER and CHRISTIE (1974) have established a reliable etchant called #6 etchant for the (010) plane of natural olivine. This etchant was applied to (010) planes of CZ-Fo_{100} and FZ-Fo_{92} synthetic olivine. Pits of various sizes were produced but no correlation was found between the matched surfaces

(Fig. 4). Obviously this etchant failed to reveal dislocations and the pits produced here may be originated from small inclusions or clustered point defects.

For the (010) plane of synthetic olivine, a single component etchant conc. HCl reported by DESAI and JOHN (1977) was useful to any crystal irrespective of the Fa mole ratio (Fig. 5). In the case of FZ-Fo_{92}, exact correspondence between etch pits and decorated dislocations was also confirmed by using a specimen of 0.2 mm in thickness. After etching, this was heat treated in air at 800°C for 15 hours to decorate the dislocations. A one-to-one correspondence between etch pits and dislocations is clearly shown in Fig. 6. Another important result is that the dislocations were not mobile in these experimental conditions and heat treatment of the olivine did not change the configuration of dislocations.

With respect to the failure of etching by #6 etchant in synthetic olivine crystals, we postulated that the local chemical conditions around the dislocation must be different from those of natural crystals. In order to verify this postulation, we heat-treated the FZ-Fo_{92} crystal in air and then etched by #6 etchant (Figs. 7a, b). It is evident from the results that #6 etchant

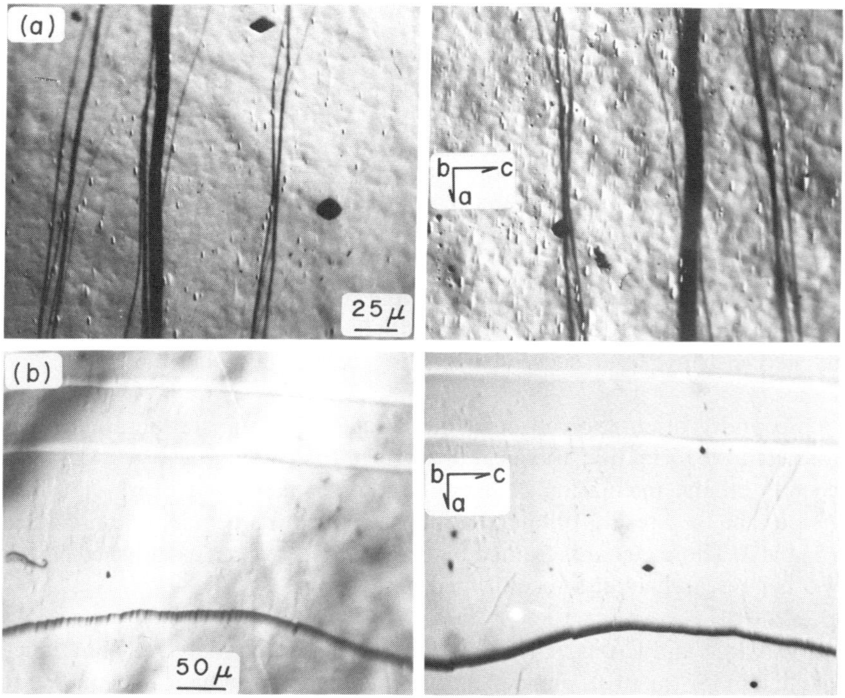

FIG. 4. Etch patterns on the matched {010} cleavage planes etched simultaneously by #6 etchant: (a) CZ-Fo_{100}, (b) FZ-Fo_{92}.

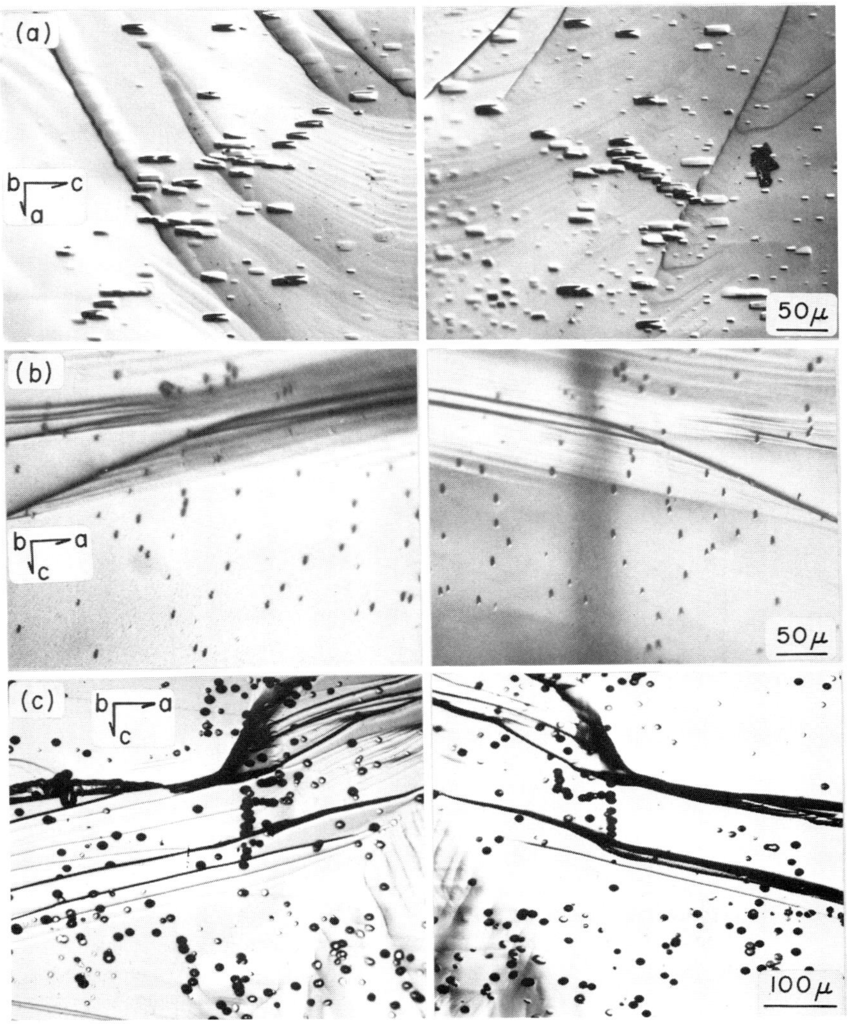

FIG. 5. Etch patterns on the matched {010} cleavage planes etched simultaneously by conc. HCl: (a) CZ-Fo_{100}, (b) FZ-Fo_{92}, (c) FZ-Fo_{20}.

preferentially attacked the dislocations which were entirely inactive before the heat treatment. Many shallow grooves are also revealed by etching. They may be attributed to the dislocations lying nearly parallel to and just below the (010) surface. The first dislocation etch grooves were reported in Si by DASH (1958) and the same phenomenon was already reported in natural olivine by KIRBY and WEGNER (1978).

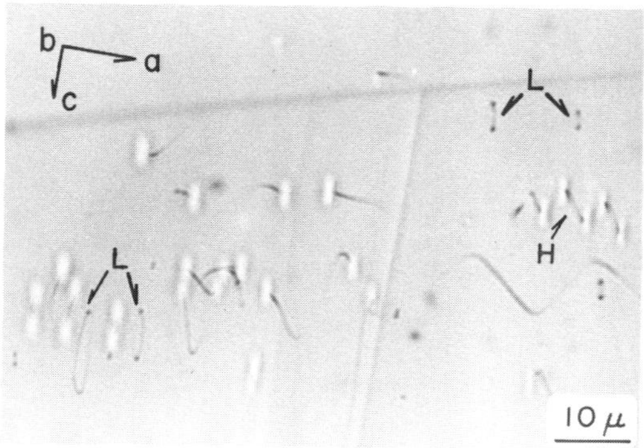

FIG. 6. FZ-Fo_{92}. Microphotograph showing a one-to-one correspondence of the etch pits on (010) plane (Fig. 5b) to dislocations which were decorated. L: dislocation loop, H: helical dislocation.

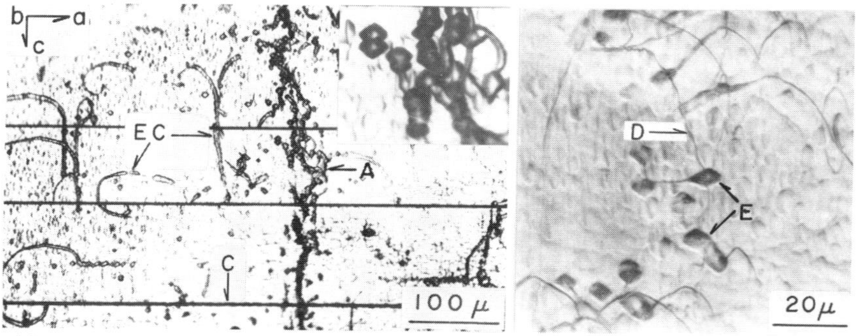

(a) (b)

FIG. 7. FZ-Fo_{92} after annealing for decoration: (a) Etch pattern produced by #6 etchant. A: high density region of etch pits. This corresponds to the high density region of the banded structure (see Fig. 12) (reflection microscope). EC: etch channel, C: crack. Insert: higher magnification, (b) microphotograph showing a one-to-one correspondence of the etch pits of (a) to dislocations (bright-field illumination). E: etch pit, D: decorated dislocation line.

4.2.2 (100) face

The specimen was prepared by mirror polishing the surface with diamond paste of 0.25–1 μm in grain size. Since both #6 etchant and conc. HCl could not produce etch pits on (100) faces of CZ-Fo_{100}, we developed a new etchant (1HF + 2HCl + $4H_2O$) for the (100) surface (KOMATSU and INOUE 1981). Fig. 8 shows a one-to-one correspondence of etch pits between matched (100) faces.

Three dislocation etchants were confirmed for (100) faces using matched faces in the case of iron-bearing olivine, FZ-Fo_{92} and FZ-Fo_{20}, i.e., #6

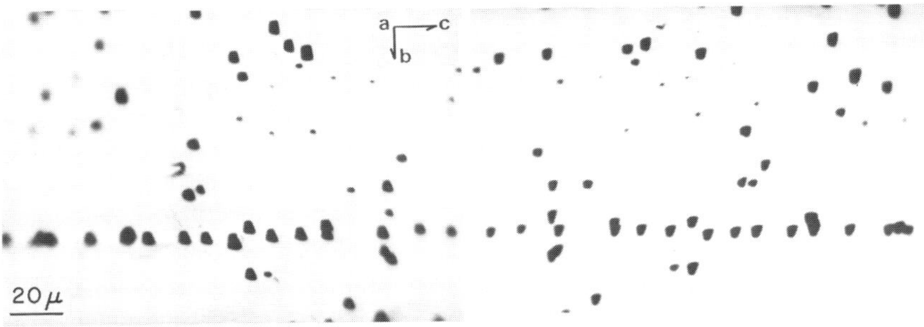

FIG. 8. CZ-Fo_{100}. Etch pattern on matched {100} parted (imperfectly cleaved) planes etched simultaneously by the new etchant, 1HF + 2HCl + $4H_2O$.

etchant, 1HF + $4H_2O$, and 1HF + 2HCl + $4H_2O$. Particularly in FZ-Fo_{92}, a one-to-one correspondence was confirmed by combining the decoration technique with etching by either one of the above three etchants. For an example, a result of #6 etchant was shown in Fig. 9. As to FZ-Fo_{20}, the decoration technique was not successful, because the surface was so heavily oxidized and turned opaque as to be unable to identify dislocations. So we took that procedure of alternate etching in 20% HF and mirror polishing. Good correspondence was found in revealing sub-boundaries in the successive etching. The pits thus produced were point bottomed, which strongly suggests that they were all dislocation etch pits. In addition to this, after removing a surface layer in several tens μm by mirror polishing, the same surface was etched by #6 etchant. Here again we obtained the same pattern as the previous one and a good correspondence in the distribution of sub-boundaries and pits was seen (Fig. 10). The same etching patterns were also observed with a 1HF + 2HCl + $4H_2O$ etchant.

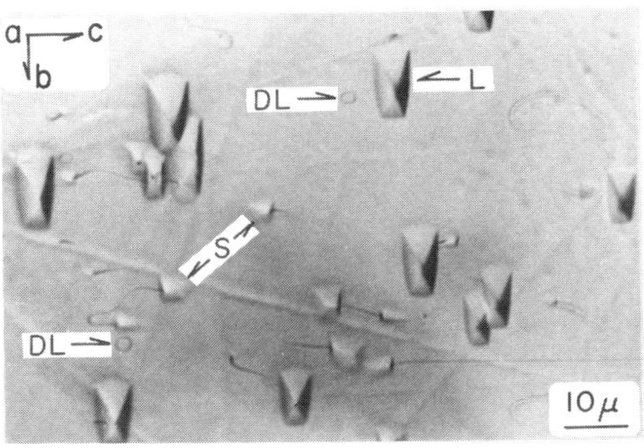

FIG. 9. FZ-Fo_{92}. Microphotograph showing a one-to-one correspondence between dislocations and etch pits on (100) plane produced by #6 etchant. L: large pits, S: small pits, DL: dislocation loops, S: small pit.

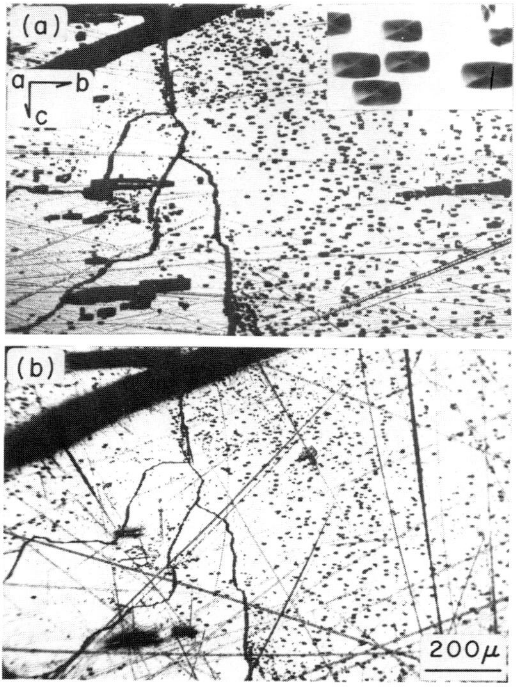

FIG. 10. FZ-Fo_{20}. Etch pattern on polished (100) planes; (a) etched by 20% HF, (b) re-etched by #6 etchant after polishing out the etch pits in (a). A: sub-boundary. Insert: higher magnification.

It should be added here that, as summarized in Table 2, two types of pits (large and small) are seen on the (100) surfaces (Figs. 8, 9, and 10). According to KIRBY and WEGNER (1978), large pits correspond to screw dislocations and small pits to edge dislocations. Further study is needed here to confirm the character of dislocations.

4.2.3 (001) face

In natural olivine, dislocation etch pits were produced on (001) plane (WEGNER and CHRISTIE, 1974), but no dislocation pits were formed on synthetic olivine (CZ-Fo_{100} and FZ-Fo_{92}), again our new etchant was successful for revealing dislocation. Figure 11a shows the matched (001) surfaces, and Figure 11b shows a one-to-one correspondence between dislocations and etch pits. In FZ-Fo_{20} crystal our 1HF + 2HCl + $4H_2O$ etchant was effective in producing dislocation pits, but they were too shallow to be observed in good contrast under microscope, while HF (20%) produced better pits than our new etchant (Fig. 11c).

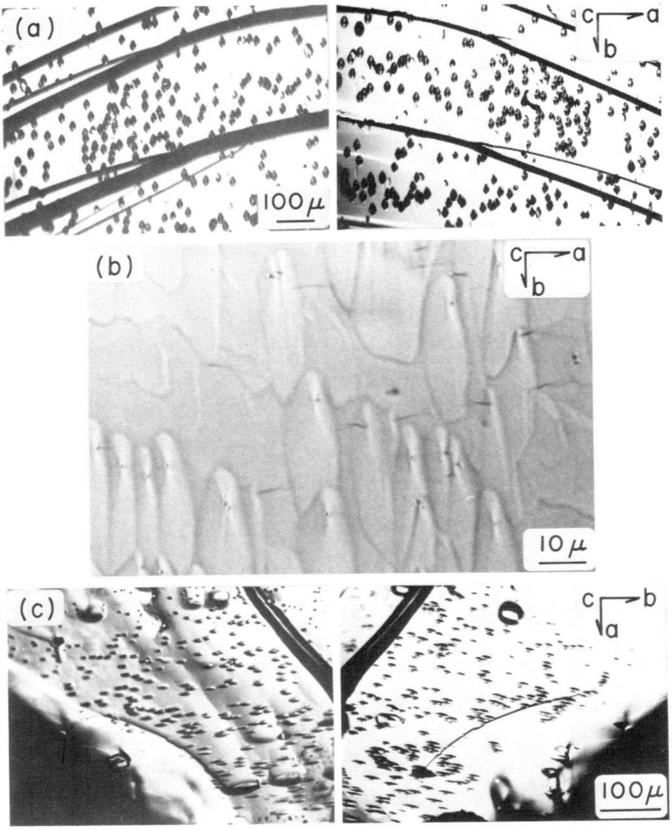

FIG. 11. Microphotographs showing a one-to-one correspondence of the etch pits on (001) planes to the dislocations. (a) CZ-Fo_{100}; Matched {001} parted planes etched by $1HF + 2HCl + 4H_2O$, (b) FZ-Fo_{92}; Etch pits by $1HF + 2HCl + 4H_2O$ and decorated dislocation lines, (c) FZ-Fo_{20}; Matched {001} parted planes etched by 20% HF.

4.3 *Distribution, density, and configuration of dislocations*

Dislocation density N (1/cm^2) and sub-boundary density L (cm/cm^2) of each crystal measured by the above-mentioned etchants are summarized in Table 2. Natural olivine contains dislocations of one order of magnitude higher than synthetic crystals. CZ-olivine showed the lowest dislocation density of the order of 10^4/cm^2 and almost free from sub-boundaries.

An interesting distribution of dislocations was found in FZ-Fo_{92} in which the dislocation density periodically fluctuate at a distance of 0.3–1 mm along the growth axis (Fig. 12). Dark bands correspond to the higher dislocation density area. This banding appears to represent the solid-liquid interface shape during growth.

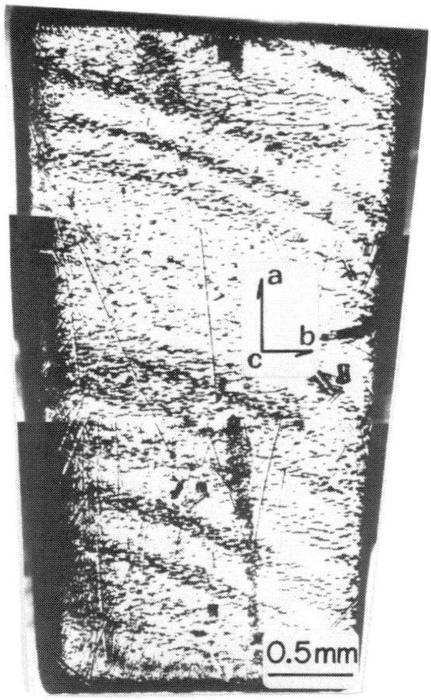

FIG. 12. FZ-Fo_{92}. Microphotograph showing a periodic change of dislocation densities along the growth axis [100] (banded structure). The etched surface is (001).

It can be conjectured that the banded structure was formed due to the periodic variation of the growth rate and, in general, the higher the growth rate the higher the content of impurities and vacancies, which tend to generate and multiply the dislocations. Indeed, many helical dislocations and rows of prismatic loops were observed in the dark band area (Fig. 13). On the other hand, in lower dislocation density area most of the dislocation were straight and elongated parallel to the growth axis (Fig. 14). It is conceivable that helical dislocations are formed when the supersaturation of vacancies is high (HIRTH and LOTHE 1968).

In FZ-Fo_{20} crystals, fine iron rich precipitates were found in thin (001) specimens. Many helical dislocations and loops were seen to originate from the precipitates and arranged in rows along the [100] direction (Figs. 15a, b, c). The Burgers vector of the most prismatic dislocation loops are a_0 [100] and only a small number of dislocation loops were punched out in the [010] direction with a Burgers vector of b_0 [010] (Fig. 15d). Thin slabs of (100) plate which was cut vertical to the growth axis, were also observed after heat treatment in air (Fig. 16). Here again, rows of dislocation loops and helical dislocations were rarely extended to the [010] direction, but were arranged predominantly in the [001] direction. This preferred orientation of the dislocation loops can be explained referring to the unit cell dimension of olivine (approximately $a_0 = 5Å$, $b_0 = 10Å$, $c_0 = 6Å$), and b_0 is about two times larger than a_0 or c_0.

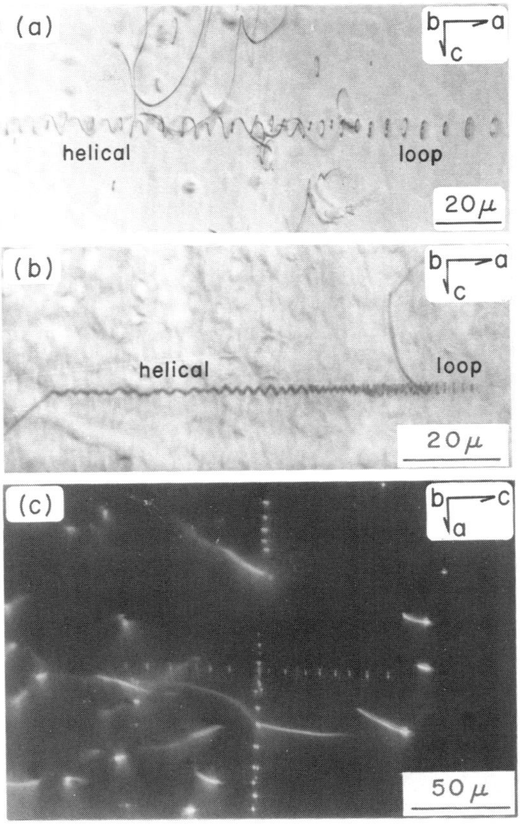

FIG. 13. (010) plate of FZ-Fo_{92}. Microphotographs showing a configuration of the dislocations in a region of high dislocation density in the banded structure (Fig. 12): (a) Helical dislocation followed by a row of closed loops whose axis is parallel to the growth direction [100] (bright-field illumination), (b) the same as the above, (c) rows of dislocation loops punched out from an inclusion along [100] or [001] direction (dark-field illumination).

As the activation energy for nucleating a dislocation loop in a perfect crystal is proportional to the square of the Burgers vector (COTTRELL, 1953), energy needed to nucleate a prismatic dislocation loop in the [010] direction is about four times larger than in the [100] or [001] direction.

Dislocation configuration of CZ-Fo_{100} olivine was observed by Lang topography. Most of the grown-in dislocations were straight and extended to the low index direction (INOUE *et al.*, 1980).

Flux-grown Fo_{100} showed an euhedral shape (Fig. 17). The m (110) face was etched in an as-grown form by #6 etchant with the results shown in Figs. 18a, b. A pair of m (110) faces revealed a great number of pits, but they were

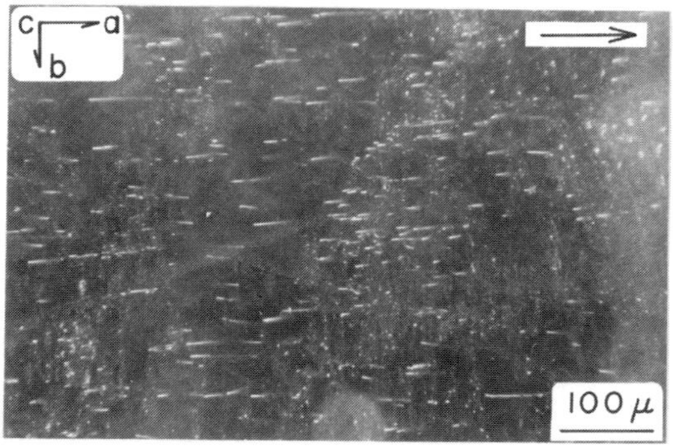

FIG. 14. FZ-Fo_{92}. Microphotograph showing the dislocation structure of the (001) plate in the middle portion of the boule (dark-field illumination). Arrow indicates the growth direction [100].

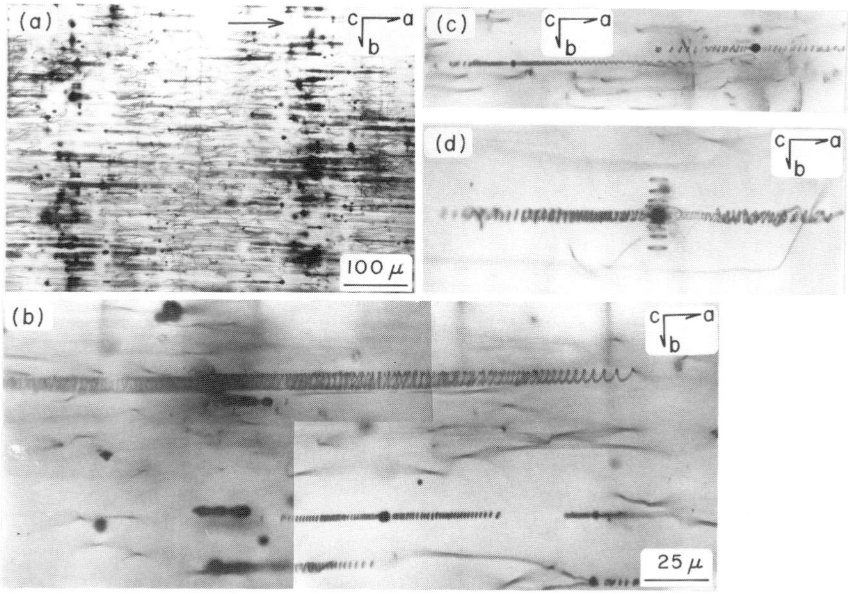

FIG. 15. FZ-Fo_{20}. Microphotograph showing the dislocation structure of (001) plate. Arrow indicates growth direction [100]. (a) dislocation generation from the inclusions. (b), (c) and (d) higher magnification of (a): (b) and (c); row of dislocation loops and helical dislocations, (d); microphotograph showing row of dislocation loops along [010] direction as well as [100] direction.

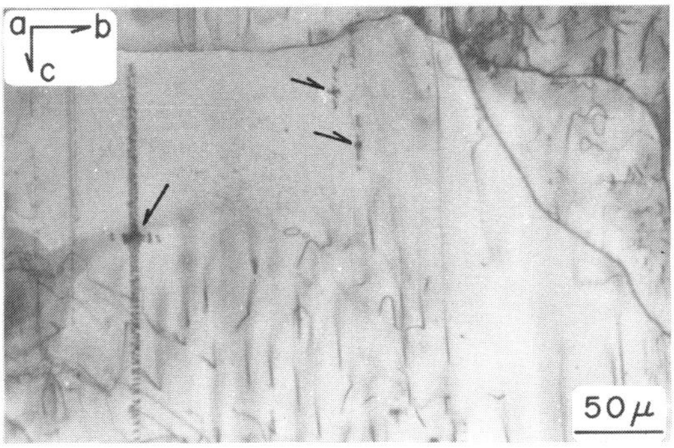

FIG. 16. FZ-Fo_{20}. Microphotograph showing a dislocation structure of the (100) plate (perpendicular to the growth direction). Arrows indicate rows of dislocation loops or helical dislocations.

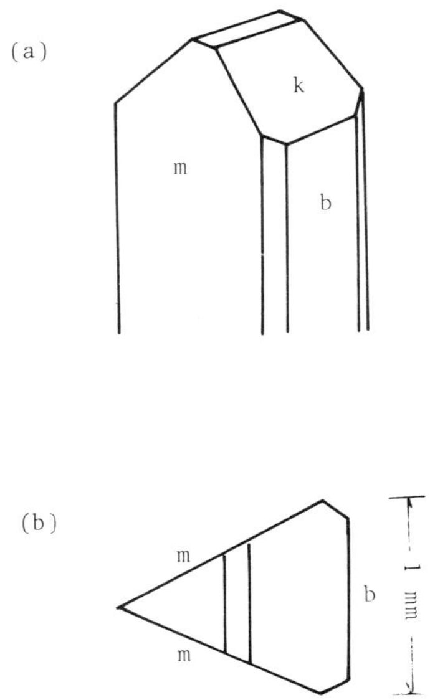

FIG. 17. Typical habit of flux grown forsterite, m {110}, k {021} and b {010}. The orientation of these planes were determined by a two-circle reflection goniometer. (a) typical form, (b) top view.

not dislocation etch pits because (1) most of the pits were flat bottomed, and (2) no dislocation contrast connecting to pits was observed by X-ray topography (Fig. 18c), and the dark contrast here may be the result of an elastic strain due to clustering precipitates. Analysis by SEM with EDX showed that a considerable amount of vanadium derived from flux had been

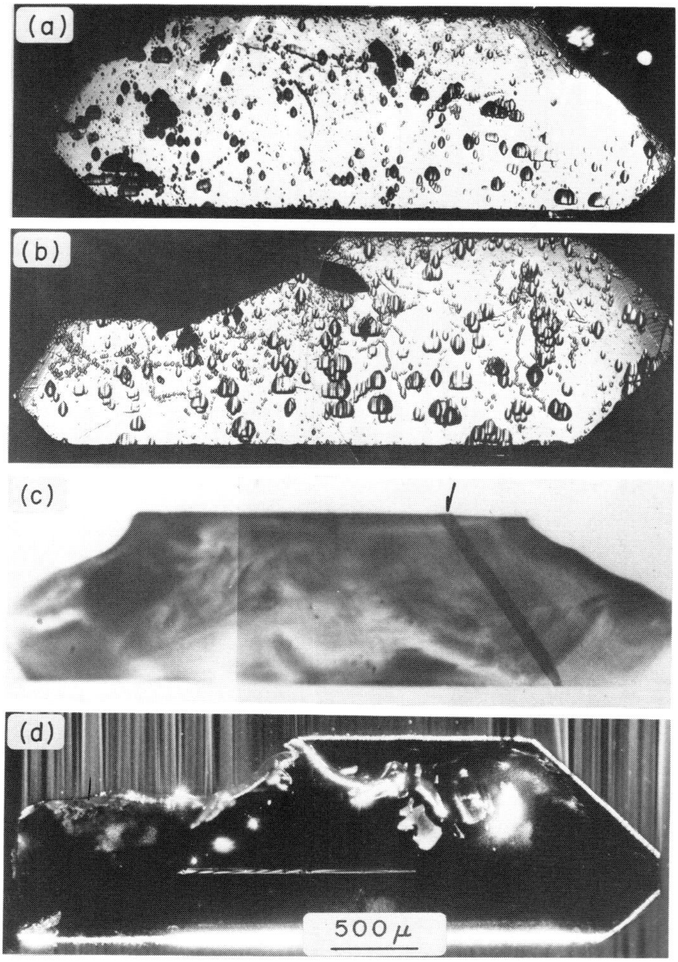

FIG. 18. Flux Fo. All photographs were taken in the same crystal. (a) and (b): etch patterns on a pair of m {110} planes which were produced by #6 etchant. (c) Projection topograph of the 112 reflection. The arrow indicates an artifact which is an over-exposure due to the failure of traverse, but it shows the relative position of the second slit to the crystal. (d) Light scattering tomography by an Ar laser.

captured in the crystal, it may be that many etch pits revealed by #6 etchant might be due to these impurities.

Finally, the dislocation configuration of natural olivine will be mentioned here. Typical features are seen in Fig. 19, which was observed after oxidizing (001) thin slabs in air. Many subboundaries and bowed-out dislocations were developed, as reported earlier by YOUNG (1969) and KOHLSTEDT *et al.* (1976).

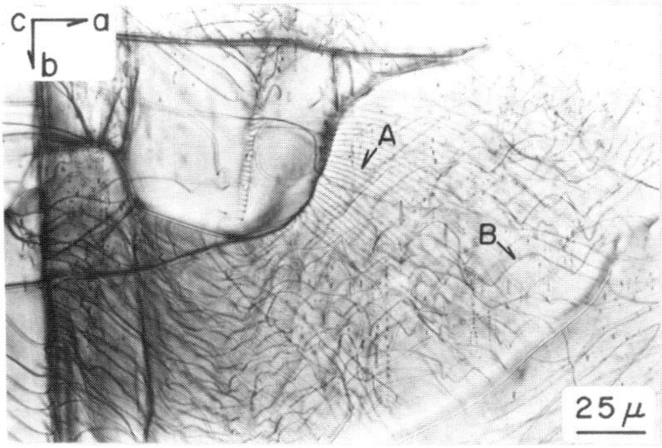

FIG. 19. Dislocation structure (decorated) of (001) plate of a natural olivine (bright-field illumination). "A" indicates the dislocations constituting the sub-boundary and "B" indicates bowed-out dislocations.

5. *Summary*

Defect structures of synthetic and natural olivine were investigated using such techniques as chemical etching, decoration, optical dark-field, X-ray topography, and SEM with EDX. The results are as follows,

(1) A well established etchant #6 for natural olivine (Fo_{71-92}) failed to reveal dislocations on pure forsterite which contains no iron. It was curious enough that the #6 etchant could not etch even FZ-Fo_{92} crystals which are equivalent to natural San Carlos olivine in chemical composition. However, FZ-Fo_{92} become etchable after oxidizing in air. This means that #6 etchant preferentially attacks the oxidized part (presumably Fe^{+3} rich part) in olivine. It can be said that the atmosphere around dislocations may differ between natural and synthetic olivine, and the natural olivine examined here has already sustained oxidation. There have been no etchants for (100), (001) planes of synthetic olivine, and we developed new etchants.

(2) The dislocation density (N) and sub-boundary density (L) were quite different between natural and synthetic olivine. At best, synthetic olivine N was nearly 10^4, while in natural olivine N was about 10^6. The density of subboundary (L) in natural olivine was at least one order of magnitude higher than synthetic olivine, which suggests that natural olivine had experienced plastic deformation.

(3) A distinguished difference in dislocation distribution was found between synthetic (FZ-Fo_{92} and FZ-Fo_{20}) and natural olivine. Many helical dislocations and prismatic loops were punched out from the inclusions in synthetic olivine and most dislocations were straight. On the other hand, natural olivine contained many bowed-out dislocations. These characteristic features of dislocation structures reflect the growth conditions and the deformation history after growth has ceased.

(4) Optical scattering centers were detected in all crystals examined here. The distribution and the individual shape of scattering bodies show remarkable differences among the crystals of different origin and can be used to characterize olivine crystals.

REFERENCES

AMELINCKX, S. (1964): *The Direct Observation of Dislocations, Supplement 6, Solid State Physics*, Academic Press, New York.

BAUMHAUER, H. (1894): *Die Resultate der Äetzmethode*, Wilhelm Engelmann, Leipzig.

COTTRELL, A. H. (1953): *Dislocations and Plastic Flow in Crystals*, Clarendon Press, Oxford.

DASH, W. C. (1958): Evidence of dislocation jogs in deformed silicon, *J. Appl. Phys.*, **29**, 705–709.

DESAI, C. C. and JOHN, V. (1977): Studies on (010) cleavages, etch patterns and dislocations in magnesium orthosilicate single crystals, *Indian J. Pure Appl. Phys*, **15**, 481–484.

DESAI, C. C. and JOHN, V. (1979): Change in morphology and dislocation etch pits on the (010) surface of magnesium orthosilicate crystals, *Krist. Tech*, **14**, 775–782.

FLEISHER, R. L., PRICE, B. P. and WALKER, R. M. (1965): Solid-state track detectors: Applications to nuclear science and geophysics, *Ann. Rev. Nucl. Sci.*, **15**, 1–28.

GROSSMAN, J. J., RYAN, L. A., MUKHERJEE, N. R. and WEGNER, M. W. (1970): Microchemical, microphysical and adhesive properties of lunar material, *Proc. Apollo 11 Lunar Sci. Conf.*, **3**, 2171–2181.

HEDGES, J. M. and MITCHELL, J. W. (1953): The observation of polyhedral sub-structures in silver bromide, *Phil, Mag.*, **44**, 223–224.

HIRSCH, P. B., HORN, R. W. and WHELAN, M. J. (1956): Direct observations of the arrangement and motion of dislocations in aluminium, *Phil. Mag.*, **1**, 677–684.

HIRTH, J. P. and LOTHE, .(1968): *Theory of dislocations*, McGraw-Hill, New York.

HONES, A. P. (1927): *The Nature, Origin and Interpretation of Etch Figures on Crystals*, John Wiley, New York.

HORN, F. H. (1952): Screw dislocations, etch figures, and holes. *Phil. Mag.*, **43**, 1210–1213.

HORN, P. and MAURRETTE, M. (1967): Ein Beitrag zum Ätzverhalten des Olivins, *Neues Jahrb. Mineral Montash*, **1**, 170–178.

HOSOYA, S. and TAKEI, H. (1982): *J. Cryst. Growth*, **57**, 341–346.

INOUE, T., KOMATSU, H., TAKEI, H., OZIMA, M. and SHIMIZU, M. (1980): Internal defect structures of olivine crystals, *Krist. Tech.*, **15**, 1295–1302.

INOUE, T., KOMATSU, H., HOSOYA, S. and TAKEI, H. (1981): Defect structures of synthetic olivine, *J. Crystal Growth*, **55**, 307–316.

JAOUL, O., MICHAUT, M., GUEGUEN, Y. and RICOULT, D. (1979): Decroated dislocations in forsterite, *Phys. Chem. Minerals*, **5**, 15–19.

JOHNSTON, W. G. (1962): Dislocation etch pits in non-metallic crystals, in *Progress in Ceramic Science*, Ed. J. Burke, Pergamon, Vol. 2, pp. 1–75.

KIRBY, S. H. and WEGNER, M. W. (1978): Dislocation substructure of mantle-derived olivine as revealed by slective chemical etching and transmission electron microscopy, *Phys. Chem. Minerals*, **3**, 309–330.

KOHLSTEDT, D. L. and VANDER SANDE, J. B. (1975): An electron microscopy study of naturally occurring oxidation produced precipitates in iron-bearing olivines, *Contrib. Mineral. Petro*, **53**, 13–24.

KOHLSTEDT, D. L., GOETZE, C., DURHAM, W. B. and VANDER SANDE, J. (1976): New technique for decorating dislocations in olivine, *Science*, **191**, 1045–1046.

KOMATSU, H. and INOUE, T. (1981): A New dislocation etchant for the Czochralski grown forsterite, *Cryst. Res. Tech.*, **16**, K50–52.

KOMATSU, H. and SUNAGAWA, I. (1975): Characterization of crystal surfaces by optical microscopy i, ii. in *Crystal Growth and Characterization*, Eds. R. Ueda and J. B. Mullin, North-Holland, pp. 333–359.

LANG, A. R. (1958): Direct observation of individual dislocations by X-ray difflraction, *J. Appl. Phys.*, **29**, 597–598.MAURRETTE, M. (1966): Étude des traces d'ion lourds dans les minéraux naturels d'origine terrestre et extra-terrestre, *Bull. Soc. Franc. Mineral. Crist.*, **89**, 41–79.

MORIYA, K. and OGAWA, T. (1978): Observation of Growth defects in synthetic quartz crystals by light-scattering tomography, *J. Cryst. Growth*, **44**, 53–60.

NEWKIRK, J. B. and WERNICK, J. H. (Ed.) (1962): *Direct Observation of Imperfections in crystals*, Interscience, New York.

ROBINSON, W. H. (1968): Dislocation etch pit techniques, in *Techniques of Metals Research*, Ed. R. F. Bunshah, Interscience Vol. 2, pp. 291–340.

SLIND, T. J. and SØRUM, H. (1976): X-ray topographic study of defects in olivine crystals, *J. Appl. Cryst.*, **9**, 39–41.

SHOCKLEY, W. and READ, W. T. (1949): Quantitative predictions from dislocation models of crystal grain boundaries, *Phys. Rev.*, **75**, 692–693.

TAKEI, H. and KOBAYASHI, T. (1974): Growth and properties of Mg_2SiO_4 single crystals, *J. Cryst. Growth*, **23**, 121–124.

VOGEL, F. L., PFANN, W. G., COREY, H. E. and THOMAS, E. E. (1953): Observations dislocations in lineage boundaries in germanium, *Phys. Rev.*, **90**, 489–490.

WEGNER, M. W. and CHRISTIE, J. M. (1974): Preferential chemical etching, of terrestrial and lunar olivines, *Contrib. Mineral. Petrol.*, **43**, 195–212.

YOUNG III, C. (1969): Dislocations in the deformation of olivine, *Am. J. Sci.*, **267**, 841–852.

Chapter 3: Electron and Crystal Structures

Materials Science of the Earth's Interior, edited by I. Sunagawa, pp. 173–189.

ELECTRON THEORY OF TRANSITION-METAL COMPOUNDS UNDER HIGH PRESSURE

Satoru SUGANO and Shuhei OHNISHI*

Institute for Solid State Physics, The University of Tokyo, Roppongi, Minato-ku, Tokyo, Japan

1. *Introduction*

The electron theory of transition-metal compounds now enjoys a long history starting with the crystalline field model of BETHE (1929). In the early stage, the crystalline field was considered to be an electrostatic field coming from the surroundings of a transition metal ion, which were assumed to be point charges or point dipoles (VAN VLECK, 1939; POLDER, 1942). It was early in 1935 that Bethe's crystalline field theory was shown by VAN VLECK (1935) to be consistent with the more inclusive molecular orbital (MO) picture, but quantitative developments with the use of this picture were started around 1960 when the theory was neatly reformulated as a ligand field theory and successfully applied to the analyses of such experimental observations as nuclear magnetic resonance (NMR), electron spin resonance (ESR), optical spectra, etc. (SUGANO *et al.*, 1970).

The most important physical parameters appearing in the ligand field theory for cubic systems are cubic field splitting parameter Δ (= 10Dq) and Racah parameter B measuring the strength of the Coulomb and the exchange interaction between d-electrons. It is well known that the point-charge model with the simultaneous use of Slater $3d$ wavefunction gives a reasonable value of Δ for the first transition metal ions in cubic crystals. It has, however, been pointed out by KLEINER (1952) that the classical electrostatic model gives a negative value of Δ when the finite spread of the electron clouds at the ligands is taken into account: experimentally Δ should be positive. This is due to imperfect screening of the ligand nuclear charge by ligand electrons. A quantum mechanical treatment of Δ has been given for the first time by TANABE and SUGANO (1956) with an ionic model. In their treatment, the

*Present address: Department of Physics and Astronomy, College of Arts and Sciences, Northwestern University, Evanston, Illinois, U.S.A.

nonorthogonality between the *d* orbital and ligand orbitals is taken into account, and the electrostatic interactions involve exchange interactions. An important result of this study is to show that the contribution of the Heisenberg exchange integral between nonorthogonal orbitals can make Δ positive and of the same order of magnitude as the observed one. It should be noticed again that this study uses an ionic model which is restricted to the Heitler-London framework. In the ionic model, the value of Racah parameter B may be deviated from that of a free ion because of the nonorthogonality between the *d* and ligand orbitals.

The necessity for using the MO picture of VAN VLECK (1935) in the ligand field theory has been clarified by the observation of the hyperfine structure of the ligand nuclei, called "transferred hyperfine structure," associated with the ESR lines of a metal ion (GRIFFITHS *et al.*, 1953; GRIFFITHS and OWEN, 1954) and by the NMR studies of transferred hyperfine interaction (SHULMAN and KNOX, 1960). A quantitative MO treatment of the ligand field theory has been performed by SUGANO and SHULMAN (1963). They have shown that "electron delocalization" or "covalency" plays an important role in giving Δ, as well as in giving the transferred hyperfine interaction. Let us decompose Δ into several contributions hitherto mentioned as follows;

$$\Delta = \Delta_{\text{ion}} + \Delta_{\text{cov}} \tag{1.1}$$

$$\Delta_{\text{ion}} = \Delta_{\text{point}} + \Delta_{\text{K}} + \Delta_{\text{exch}} + \Delta_{\text{nonorth}}, \tag{1.2}$$

where Δ_{cov} is the contribution from the covalency, Δ_{point} from the point charges or dipoles, Δ_{K} Kleiner's correction, Δ_{exch} from the exchange interaction between the *d*-electron and the ligand electron, and Δ_{nonorth} from the nonorthogonality. Each contribution in a $KNiF_3$ crystal has been estimated as follows (SUGANO *et al.*, 1970);

$$\begin{array}{llll} \Delta_{\text{point}} & \sim \ \ 1{,}390 & \Delta_{\text{ion}} & \sim 2{,}000 \\ \Delta_{\text{K}} & \sim -2{,}080 & \Delta_{\text{cov}} & \sim 5{,}000 \\ \Delta_{\text{exch}} & \sim -2{,}880 & & \\ \Delta_{\text{nonorth}} & \sim \ \ 5{,}500 & & (\text{in cm}^{-1}) \end{array}$$

We are now at the stage of discussing pressure effects on Δ and B. These parameters can be determined from the optical spectra of transition-metal compounds in the visible region and Drickamer's school has studied the pressure effects by observing the optical spectra intensively and extensively (DRICKAMER, 1965). It is embarrassing that the pressure experiments show the dependence of Δ upon the distance R between the metal ion and the ligand to be fairly close to that given by the point charge model (DRICKAMER, 1965, 1967), i.e., $\delta(\ln\Delta)/\delta(\ln R) \sim -5$. However, it is now believed that this coincidence is rather accidental and more elaborate quantum mechanical treatments, including the covalency, are necessary to explain the observed pressure dependence. The pressure experiments also show a decrease of the parameter B with a pressure increase, which may be interpreted as being due to an increase of the covelency with a decrease of R. A simple MO theory neglecting all the Coulomb and exchange integrals involving any ligand orbitals has been proposed (KOIDE and PRYCE, 1958) to explain the reduction of B due to the covalency increase, but an elaborate theory on this problem has been reported by ADACHI *et al.* (1979) quite recently. This theory, which also calculate the R dependence of Δ, will be reviewed in Subsection 2.1 and our theory providing a theoretical basis for the high-pressure calibration by the use of ruby will be presented in Subsection 2.2.

Recently, studies of the electronic structure by measuring photoelectron spectra (PES) are becoming popular, and these studies have been applied to transition metal compounds (ASADA *et al.*, 1975). In particular, shake-up satellites in PES observed when a $2p$ electron of a metal ion is photoemitted seem to be quite sensitive to properties of the chemical bond between the metal ion and the ligand (ROBERT, 1975). Our interest concerns an intensity change of the shake-up satellites when the covalency is changed. Since the application of high pressure may change the covalency, we are studying the dependence of the satellite intensities upon R by the use of an elaborate quantum theory described in Subsection 2.1. A simple MO theory for the shake-up satellites (SUGANO, 1978) will be reviewed in Subsection 3.1, and an application of the elaborate theory to a NiO_6 complex will be presented in Subsection 3.2.

Finally, future problems in the electron theory of transition-metal compounds under high pressure will be discussed in Section 4.

2. *Pressure Dependence of the Optical Spectra*

2.1 *Discrete variational (DV)-Xα calculation of Δ and B*

The DV-Xα method is one of the approximations recently developed by ELLIS *et al.* (1977) for the quantum mechanical treatments of a many-electron system with many nuclei. This method has two characteristic features. The first one is to use the exchange potential V_{xc} given by

$$V_{xc}(\boldsymbol{r}) = -(3\alpha/2)[(3/\pi)\rho(\boldsymbol{r})]^{1/3} \qquad (2.1)$$

to take into account the exchange interaction between electrons which takes into account a main part of the electron correlation in a many-electron system. In Eq. (2.1) $\rho(\boldsymbol{r})$ is the density of electrons at $\boldsymbol{r}$, and α is 2/3 in the density functional method of GASPER (1954) and KOHN and SHAM (1965) but sometimes may be left as an adjustable parameter. The use of the exchange potential eliminates difficulties associated with many-body problems, although it is an approximation. In the case of the spin polarization, the different potential given by (2.1) with the spin charge density $\rho_{\uparrow}(\rho_{\downarrow})$ should be applied to electrons of different spins. The second characteristic feature of the

DV-Xα method is to perform numerical integration of many-center integrals in an appropriate way. This allows one to use a non-spherical self-consistent potential which is more accurate than the muffin-tin potential often used.

The cluster approach to the localized electronic system, such as the transition metal ions in a crystal, is known to be very efficient: we can take out a small molecule from the crystal and substitute the Madelung potential for the effect of the ions outside the cluster. The $[NiO_6]^{-10}$ in NiO is a typical example. The calculated result of $\Delta = 8460\ cm^{-1}$ by us with the spin polarization scheme agrees well with the experimental value of 8,500 cm^{-1}. Also, a 5% contraction of the metal-ligand distance causes 1.24 times increase in Δ, which also agrees well with the experimental 1.29 (DRICKAMER, 1967).

ADACHI *et al.* (1979) have applied the DV-Xα method to the studies of the electronic structure of transition-metal compounds. In particular, they have performed the calculation of the electronic structure of MnF_6^{4-} clusters with different Mn-F distances: $R_{Mn-F} = 4.1683$, 4.0643, and 3.9603 a.u. From the results of this calculation, one may derive the pressure-dependence or the R-dependence of Δ and B. The R-dependence of Δ has been derived to be $\delta(\ln\Delta)/\delta(\ln R) = -3.4$ which should be compared with the experimentally determined value of ~ -5 (DRICKAMER, 1967). The R-dependence of B can be derived from the calculated splitting ΔE_2 of the d levels caused by the spin polarization. It has been pointed out (ADACHI *et al.*, 1979) that this splitting can be expressed in terms of Racah parameters B and C as

$$\Delta E_2 = \mu_d(7B/2 + 7C/5), \tag{2.2}$$

where μ_d is the total number of spin-up electrons minus that of spin-down. The calculated values of ΔE_2 for different R_{Mn-F} are as follows;

R_{Mn-F} (a.u.)	ΔE_2 (eV)	
4.1683	5.029	
4.0643	5.016	(2.3)
3.9603	4.997	

Since the ratio C/B is considered to be almost independent of R_{Mn-F}, Eq. (2.2) and Eq. (2.3) shows a decrease of B with pressure increase.

2.2 *Calculation of the R dependence of the U band and the R line spectral shifts in $[CrO_6]^{-9}$—a theoretical contribution to high-pressure calibration*

It is well known that the absorption spectrum of ruby, $Al_2O_3{:}Cr^{3+}$, in the visible region consists of two broad bands, U at 18,000 cm^{-1} and Y at 25,000 cm^{-1}, and three groups of sharp lines, R at 14,000 cm^{-1}, R' at 15,000 cm^{-1}, and B at 21,000 cm^{-1}. The U and Y bands arise from transitions from the $t_{2g}^3\ {}^4A_{2g}$ ground state to the $t_{2g}^2 e_g\ {}^4T_{2g}$ and $t_{2g}^2 e_g\ {}^4T_{1g}$ excited states, respectively, in a cubic field approximation. The peak energy of the U band directly gives the value of cubic field splitting parameter Δ. The R, R', and B lines arise from transitions from the $^4A_{2g}$ ground state to the $t_{2g}^3\ ^2E_g$, $t_{2g}^3\ ^2T_{1g}$, and $t_{2g}^3\ ^2T_{2g}$ excited states, respectively. The inverse transition of the R line gives the laser line of ruby and the spontaneous emission due to this transition is used for high-pressure calibration. The transition energies of these lines are given in terms of Racah parameters in the strong field limit, being independent of Δ: for example, the transition energy of the R line is $9B + 3C$ (SUGANO *et al.*, 1970).

By using the DV-Xα method with a simple approximate self-consistent-field (SCF) potential called the self-consistent-charge (SCC) method (ADACHI *et al.*, 1978) and with spin polarization, spectral positions of the U band peak, ΔE_U, and of the R line, ΔE_R, have been calculated for transition states $(2t_{2g\uparrow})^{2.5}\ (4e_{g\uparrow})^{0.5}$ and $(2t_{2g\uparrow})^{2.5}\ (2t_{2g\downarrow})^{0.5}$, respectively: it has been pointed out (SLATER and JOHNSON, 1972) that the use of the transition state, defined as the state of pseudo configuration $\alpha^{n-\frac{1}{2}}\ \beta^{m+\frac{1}{2}}$ for the transition $\alpha^n\ \beta^m \rightarrow \alpha^{n-1}\ \beta^{m+1}$, for calculating the transition energy takes into account the effects of electron redistribution (or orbital relaxation) in the final state. Orbitals $2t_{2g\uparrow}$, $4e_{g\uparrow}$, and $2t_{2g\downarrow}$ are the spin-polarized Xα eigenstates as shown in Fig. 1. The degree of convergence is less than 0.1%. The peak energy of the U band, ΔE_U, is given by

$$\Delta E_U = \varepsilon(4e_{g\uparrow}) - \varepsilon(2t_{2g\uparrow}), \tag{2.4}$$

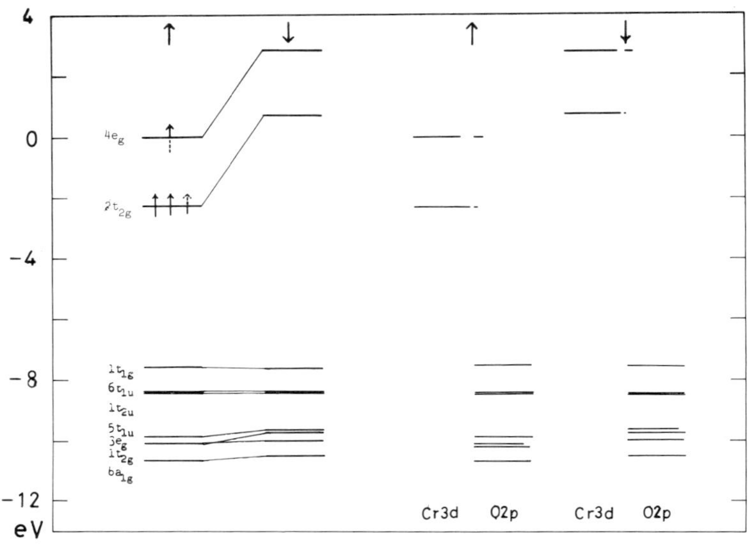

FIG. 1a. The spin-polarized Xα eigenstates and their energies of the $(t_{2g\uparrow})^{2.5}\ (e_{g\uparrow})^{0.5}$ transition state of $(CrO_6)^{-9}$, together with the atomic orbital components.

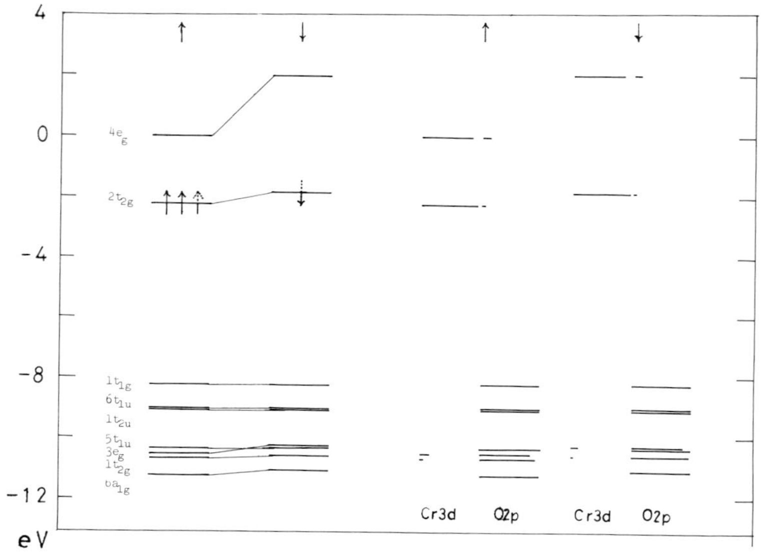

FIG. 1b. A figure for the $(t_{2g\uparrow})^{2.5}\,(t_{2g\downarrow})^{0.5}$ transition state similar to Fig. 1a.

where $\varepsilon(4e_{g\uparrow})$ and $\varepsilon(2t_{2g\uparrow})$ are the Xα eigenvalues for transition state $(2t_{2g\uparrow})^{2.5}$ $(4e_{g\uparrow})^{0.5}$. The peak energy of the R line, ΔE_R, is given by

$$\Delta E_R = \frac{4}{5}\{\varepsilon(2t_{2g\downarrow}) - \varepsilon(2t_{2g\uparrow})\}, \tag{2.5}$$

where $\varepsilon(2t_{2g\downarrow})$ and $\varepsilon(2t_{2g\uparrow})$ are the spin-polarized Xα eigenvalues for transition state $(2t_{2g\uparrow})^{2.5}\,(2t_{2g\downarrow})^{0.5}$. In Eq. (2.5) a numerical factor, 4/5, comes from the fact that the $\{\varepsilon(t_{2g\downarrow}) - \varepsilon(t_{2g\uparrow})\}$ is supposed to give the excitation energy, $15(3B + C)/4$, of the gravity center of the multiplets, $t_{2g}^3\ {}^2E_g$, $t_{2g}^3\ {}^2T_{1g}$, and $t_{2g}^3\ {}^2T_{2g}$, while the excitation energy of the R line in the strong field limit is $3(3B + C)$ (SUGANO *et al.*, 1970). It should be remarked here that the method of ADACHI *et al.* (1979) to determine B by the use of Eq. (2.2) and then $\Delta E_R = 3(3B + C)$ with $C = 4.5B$ is inapplicable to our calculation of ΔE_2 for the transition state $(t_{2g\uparrow})^{2.5}\,(t_{2g\downarrow})^{0.5}$.

The spin-polarized Xα eigenstates and their energies calculated for transition states $(2t_{2g\uparrow})^{2.5}\,(4e_{g\uparrow})^{0.5}$ and $(2t_{2g\uparrow})^{2.5}\,(2t_{2g\downarrow})^{0.5}$ with the Cr-O distance $R = 1.9$ Å are shown in Figs. 1(a) and 1(b), respectively, together with orbital components. It is clear that the $2t_{2g}$ and $4e_g$ states are the antibonding ones. Energies ΔE_U and ΔE_R calculated by using Eqs. (2.4) and (2.5), respectively, for nine points of R, from 2.000 Å through 1.745 Å, are plotted in logarithmic scales in Fig. 2. These points of R are enough to cover the pressure range more than 1 Mbar in Al_2O_3. The results show that

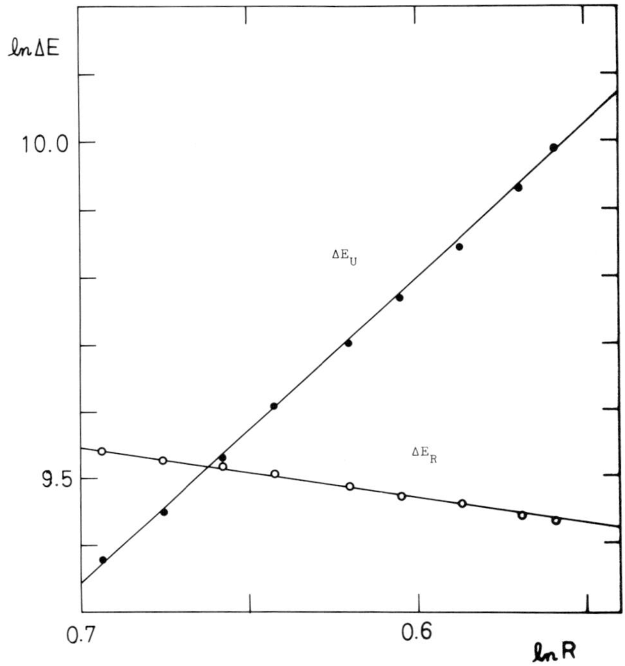

FIG. 2. Log-log plot of the energies of the U band peak (ΔE_U) and the R line (ΔE_R) versus the interatomic distance R.

$$\delta(\ln \Delta E_U)/\delta(\ln R) = -4.5, \tag{2.6}$$

$$\delta(\ln \Delta E_R)/\delta(\ln R) = 0.375. \tag{2.7}$$

By using ruby $Al_2O_3:Cr^{3+}$ the pressure calibration is done by assuming a linear relation (PIERMARINI *et al.*, 1975),

$$p = 2.74\ \Delta\lambda\ \text{(kbar)}, \tag{2.8}$$

where p is the pressure and $\Delta\lambda$ the wavelength shift in Å from $\lambda_0 = 6942$ Å. Assuming the imcompressibility of the $[CrO_6]^{-9}$ cluster to be equal to that of Cr_2O_3, 2.3 Mbar (SATO and AKIMOTO, 1979), one obtains from our result in Eq. (2.7) the relation at $\lambda_0 = 6942$ Å,

$$dp/d\lambda|_{p=0} = 2.65, \tag{2.9}$$

which agrees well with the empirical relation in Eq. (2.8).

The calculated values of ΔE_U and ΔE_R for the nine points of R are listed in Table 1. These absolute values are also in fair agreement with the experimental ones.

TABLE 1. Energies of the U-band peak and the R line.

R(Å)	$\Delta E_U(\times 10^4\ cm^{-1})$	$\Delta E_R(\times 10^4\ cm^{-1})$
2.000	1.443	1.362
1.965	1.552	1.342
1.930	1.680	1.330
1.900	1.820	1.313
1.860	1.996	1.289
1.830	2.136	1.271
1.800	2.301	1.256
1.768	2.511	1.231
1.745	2.670	1.222

3. *Covalency Dependence of the Photoelectron Spectra*

3.1 *Shake-up satellite intensities vs. covalency*

Shake-up satellites in PES represent an energy spectrum of the final state Ψ_E of the photoemission having spectral intensities $|\langle\Psi_E|\Psi_0\rangle|^2$. Here, Ψ_0 is the wavefunction of the initial state where a core electron is omitted. A simple theoretical treatment of the shake-up satellites in transition-metal compounds assumes the two-configuration model for the final state, in which the effect of creating a core hole is represented by the mixing of two states, Φ_1 and Φ_2 (SUGANO, 1978). In state Φ_1 no change of the configuration of valence electrons is induced by the creation of a core hole, and in state Φ_2 an electron is transferred from the bonding to the antibonding orbital of valence electrons. For example, for a simple system of four electrons, Φ_1 and Φ_2 are given by Slater determinants.

$$\Phi_1 = |a\ b\ \bar{b}\ c|,\quad \Phi_2 = |a\, b\, \bar{a}\, c|, \tag{3.1}$$

where a, b, and c are the antibonding, bonding and core orbitals, respectively. The electron in the orbital with the upper bar has a down-spin and that without the bar an up-spin. If one assumes orbitals a and b to be the Hartree-Fock eigenstates in the absence of the core hole, they are expressed in terms of an atomic d-orbital ϕ of the transition-metal ion in the absence of the hole and an appropriate molecular orbital χ of the ligand ions as follows;

$$a = (\cos\sigma)^{-1}\{\phi\cos\theta - \chi\sin(\theta+\sigma)\}, \tag{3.2}$$

$$b = (\cos\sigma)^{-1}\{\chi\cos(\theta+\sigma) + \phi\sin\theta\}, \tag{3.3}$$

where $\sin\sigma = \langle\phi|\chi\rangle \equiv S$. The conventional covalency parameter γ is related to θ as

$$\gamma = \sin\theta/\cos(\theta+\sigma). \tag{3.4}$$

In the presence of a core hole, states Φ_1 and Φ_2 are no longer the eignstates. The eigenstates in the two-configuration model are given by

$$\Phi_m = \Phi_1 \cos\theta' + \Phi_2 \sin\theta', \tag{3.5}$$

$$\Phi_s = -\Phi_1 \sin\theta' + \Phi_2 \cos\theta', \tag{3.6}$$
$$(-\pi/4 \leq \theta' \leq \pi/4)$$

where

$$\tan 2\theta' = -V/\Delta. \tag{3.7}$$

In Eq. (3.7) matrix elements V and Δ of the total Hamiltonian are defined as

$$2\Delta \equiv \langle\Phi_2|\mathscr{H}|\Phi_2\rangle - \langle\Phi_1|\mathscr{H}|\Phi_1\rangle, \tag{3.8}$$
$$V \equiv \langle\Phi_1|\mathscr{H}|\Phi_2\rangle.$$

From Eqs. (3.5) and (3.6), it is evident that the satellite intensity relative to that of the main line is given by

$$I_s = (\tan\theta')^2 \equiv \gamma'^2. \tag{3.9}$$

Since Φ_m and Φ_s in Eqs. (3.5) and (3.6) can be expressed by Slater determinants, $|a'\,b'\,\bar{b}'\,c|$ and $|a'\,b'\,\bar{a}'\,c|$, respectively, in which a' and b' are a and b in Eqs. (3.2) and (3.3) with θ replaced by $\theta + \theta'$, parameter γ' in Eq. (3.9) may be regarded as the increment of the covalency parameter induced by the creation of a core hole. In the limit of small initial covalency, $\gamma \ll 1$, the relative satellite intensity has been pointed out to be proportional to $(\gamma + S)^2$. It is highly probable that the observed increase of I_s in the series of MnF_2 ($I_s = 0.05$), $MnCl_2$ (0.30), and $MnBr_2$ (0.55), and also in that of MnF_2 (0.05), FeF_2 (0.15), CoF_2 (0.35) and NiF_2 (0.45) is due to the increase of the initial covalency, γ (ROBERT, 1975; ASADA and SUGANO, 1978).

On the other hand, ROBERT (1975) has pointed out that the shake-up satellites are absent in $K_2Ni(CN)_4$ and $Fe(C_5H_5)_2$ and the absence of the satellites is due to screening of the core-hole potential by electrons in the strong covalent bond between the metal ion and the ligand. It, however, is unclear what the screening really means. At this stage, it seems worthwhile theoretically examining the R-dependence of the satellite intensities, as the initial covalency could be increased with the decrease of R. Our expectation is that the intensity would increase with the decrease of R at first, then it would steeply decrease at a certain stage of the R decrease.

3.2 *Calculation of the R-dependence of the electronic structures of* $[NiO_6]^{-10}$ *in both the presence and the absence of a core hole*

In order to examine the covalency dependence of the shake-up satellite intensity of an octahedral system of $[NiO_6]^{-10}$, we have applied the spin-restricted DV-Xα cluster method to calculate the electronic structures in the presence and the absence of a core hole. We have dealt with four electronic configurations as shown in Fig. 3, where system-0 corresponds to the ground state in the absence of a core hole, system 1 to Φ_1, system 2 to Φ_2 in the presence of a Ni $2p$ hole, and system T to the transition state for systems 1 and 2: this transition state is used for evaluating the energy difference between systems 1 and 2. For each system, the self-consistent calculations in the SCC scheme have been performed by assuming five cases of the Ni-O distance: $R = 1.2\,R_0$, $1.1\,R_0$, R_0, $0.9\,R_0$, and $0.8\,R_0$ ($R_0 = 2.1$ Å) where R_0 is the distance in the unpressed and unexpanded normal state. This means that we have performed twenty series of the self-consistent calculations. The $X\alpha$ eigenstates and eigenvalues thus calculated for systems 0, 1, and 2 are shown in Figs. 4, 5, and 6, respectively. In each figure, energy shifts of the eigenstates for five values of R are indicated. Further, atomic orbital components of Mulliken charges in the $X\alpha$ eigenstates for systems with different R are shown in Figs. 7–11. Total atomic orbital components of Mulliken charges are also listed in Table 2. The energies of the transition from system 1 to system 2 as a function of R/R_0 calculated by using system T are shown in Fig. 12.

As was done in Subsection 3.1, let us denote the wavefunction of system 1 as Φ_m and that of system 2 as Φ_s. Denoting the wave-function of the initial state, in which a core electron is emitted, as Ψ_0, the relative intensity of the shake-up satellite, I_s, is given as

$$I_s = |\langle\Phi_s|\Psi_0\rangle|^2/|\langle\Phi_m|\Psi_0\rangle|^2. \tag{3.10}$$

Figures 10 and 11, in particular Fig. 11, show that the atomic orbital components in the antibonding $4e_g$ and bonding $3e_g$ orbitals of system 2, respectively, are quite close to those of system 0. This result, when inserted in Eq. (3.10), shows that the relative intensity of the shake-up satellite is almost vanishing in the compressed cases of D and E. On the other hand, Fig. 9 for the

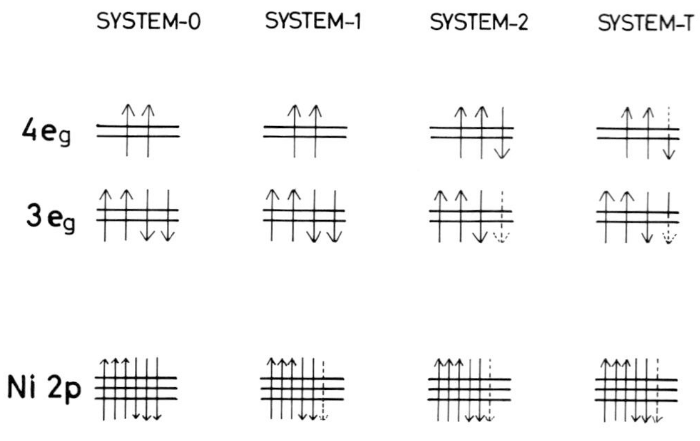

FIG. 3. Electron configurations of a $(NiO_6)^{-10}$ cluster. Orbitals $4e_g$ and $3e_g$ are the antibonding and bonding orbitals, respectively.

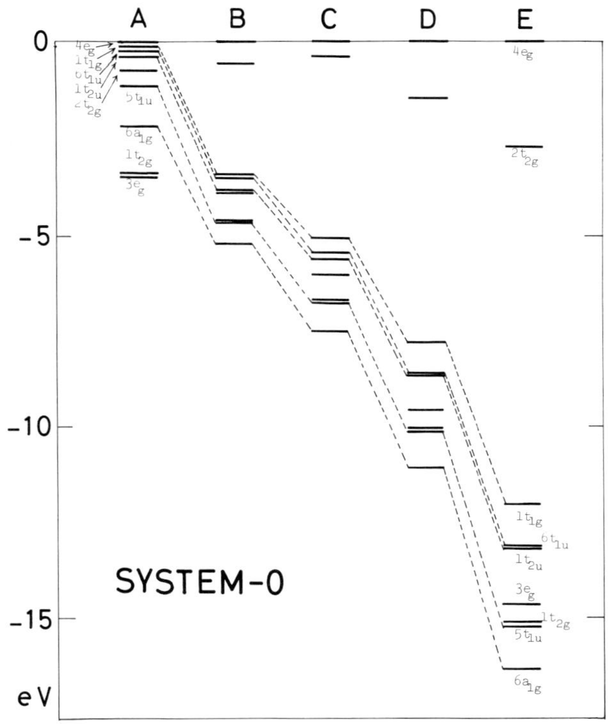

FIG. 4. The Xα eigenstates and their energies of system 0: *A*, *B*, *C*, *D* and *E* represent the interatomic distances between Ni and 0 to be $1.2R_0$, $1.1R_0$, R_0, $0.9R_0$ and $0.8R_0$ respectively. R_0 is the normal distance of 2.1 Å.

TABLE 2. Atomic orbital components calculated for $(NiO_6)^{-10}$ clusters.

	Atomic orbital	*A**	*B*	*C*	*D*	*E*
SYSTEM-0	Ni (3*d*)	8.10	8.26	8.26	8.33	8.31
	O (2*p*)	6.02	5.98	6.00	6.04	6.14
SYSTEM-1	Ni (3*d*)	9.37	9.18	8.97	8.81	8.69
	O (2*p*)	5.72	5.78	5.86	5.94	6.02
SYSTEM-2	Ni (3*d*)	9.00	9.00	9.01	9.01	9.02
	O (2*p*)	5.76	5.80	5.85	5.91	5.97

**A*, *B*, *C*, *D* and *E* are clusters of the interatomic distances $1.2R_0$, $1.1R_0$, R_0, $0.9R_0$, and $0.8R_0$, respectively.

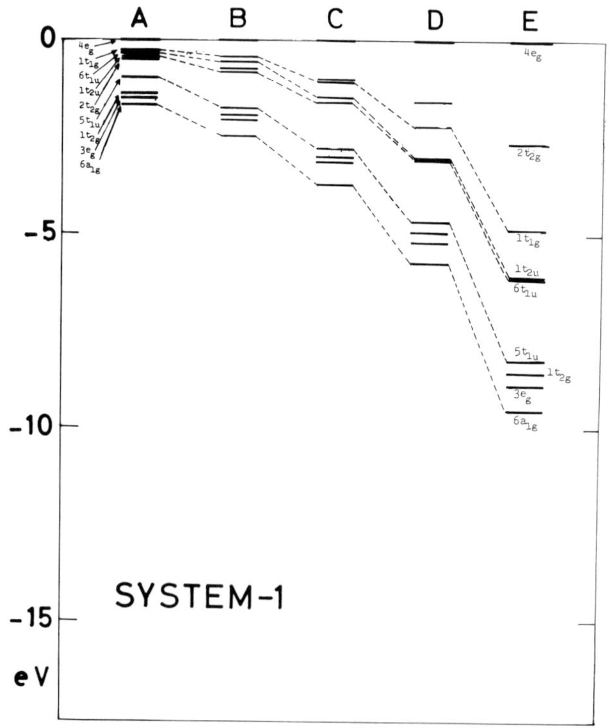

FIG. 5. A figure for system 1 similar to Fig. 4.

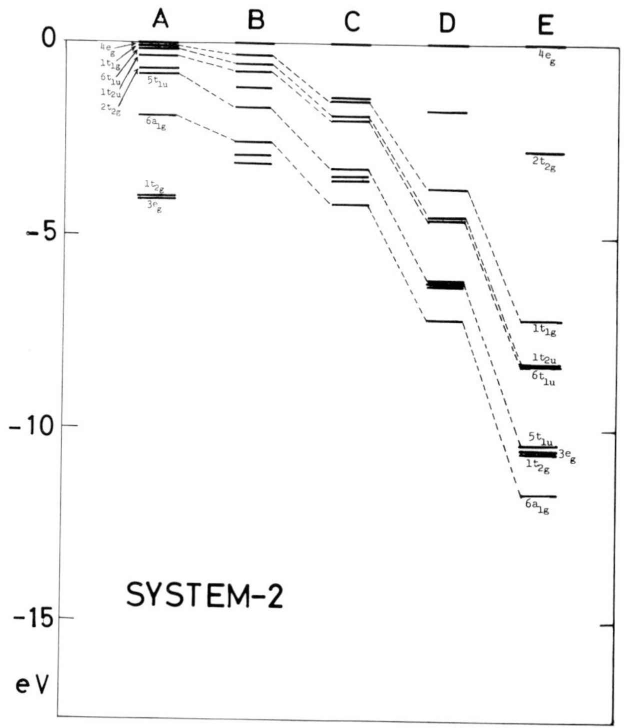

FIG. 6. A figure for system 2 similar to Fig. 4.

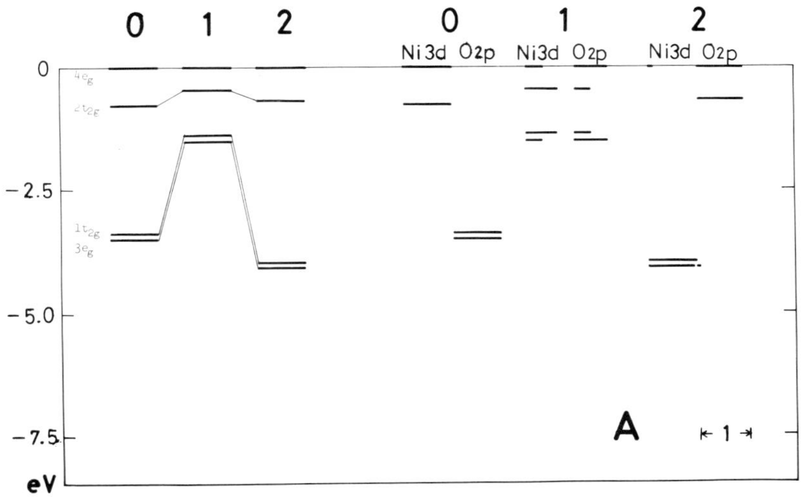

FIG. 7. Comparison of the orbital energies and orbital components between different systems with $R = 1.2R_0$.

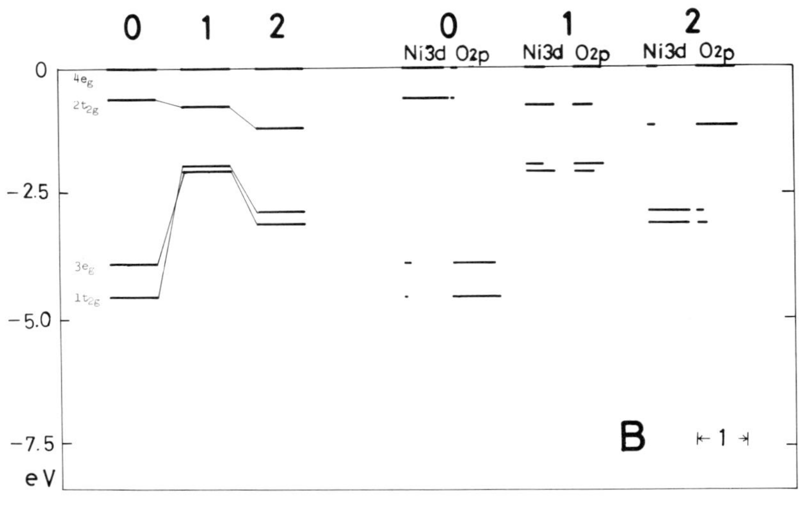

FIG. 8. A figure for $R = 1.1R_0$ similar to Fig. 7.

the normal case of C shows that the atomic orbital components in $4e_g$ and $3e_g$ of system 2 are somewhat different from the corresponding ones of system 0, that indicates the presence of a nonvanishing shake-up satellite in the normal state. The vanishing of the satellite in the strongly compressed cases agrees with our expectation just mentioned at the end of Subsection 3.1.

It has been pointed out (ASADA *et al.*, 1975) that the multiplet structure in

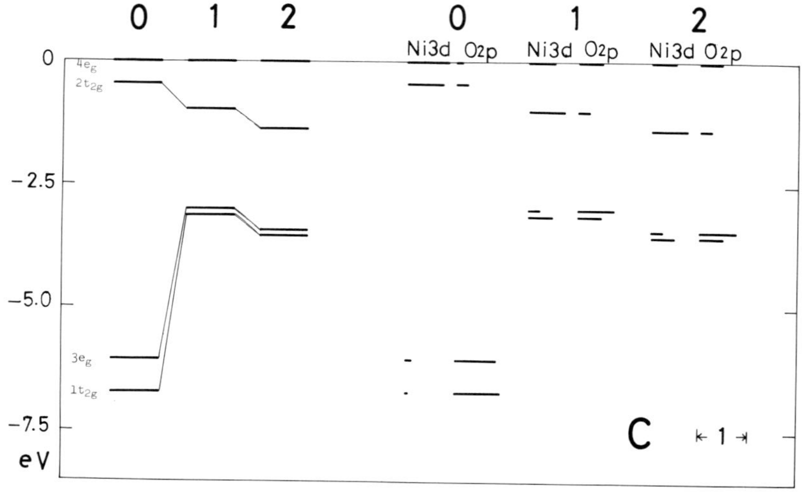

FIG. 9. A figure for $R = R_0$ similar to Fig. 7.

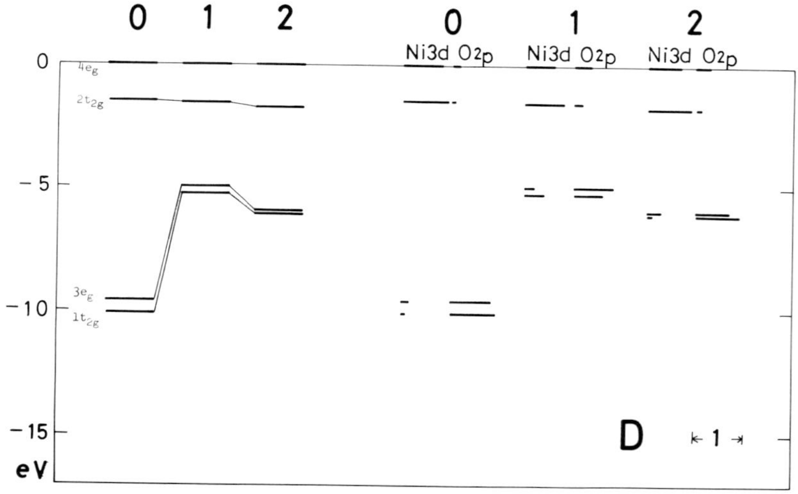

FIG. 10. A figure for $R = 0.9 R_0$ similar to Fig. 7.

PES of transition-metal compounds may be analyzed by assuming cubic field splitting parameter Δ to be about 1 eV larger than that of the initial state and the p-hole-d-electron interaction parameter reduced almost down to 50–60% of that of the free ion. The results shown in Fig. 9 for the normal case seem to explain the increase of Δ, the energy difference between $4e_g$ and $2t_{2g}$, of system 1 as compared with that of system 0 and the decrease of the p-d interaction

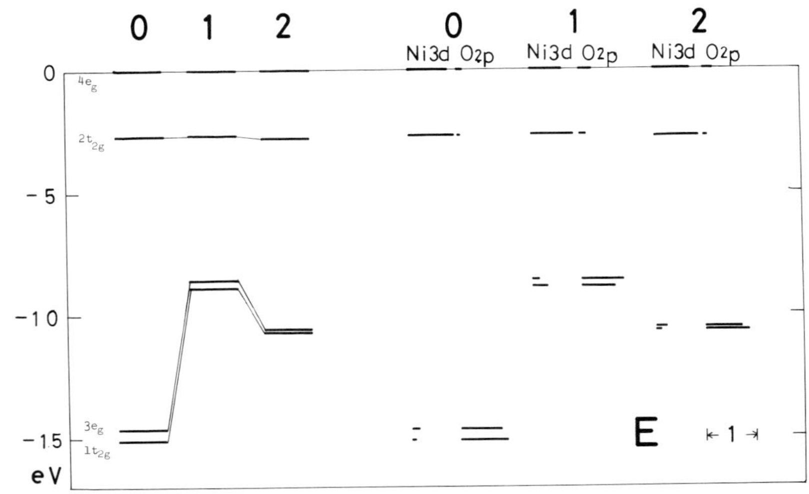

FIG. 11. A figure for $R = 0.8\,R_0$ similar to Fig. 7.

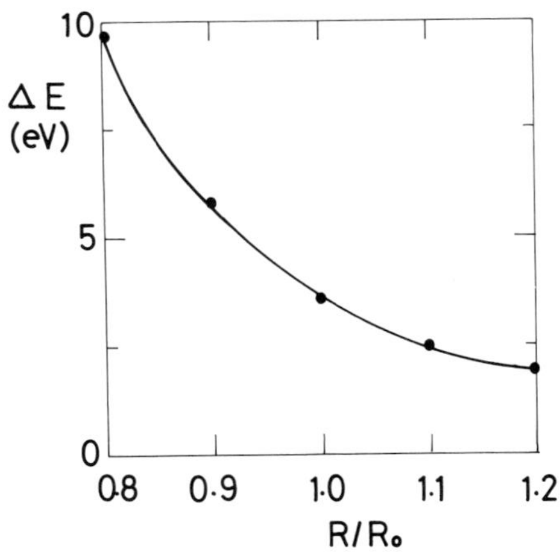

FIG. 12. Transition energies from system 1 to system 2 calculated for various R by the use of system T in Fig. 3.

parameter: the $3d$ orbital components in $4e_g$ and $2t_{2g}$, in particular $4e_g$, of system 1 seem to be smaller than those of system 0. In the compressed cases, however, both the cubic field parameter and the p-d interaction parameter do not seem to change in systems 0, 1, and 2 as shown in Fig. 11. This fact could be naturally explained by assuming the screening of the core-hole potential by electrons in the strong covalent bond formed by compression, and supports ROBERT's speculation (1975) on the absence of shake-up satellites in strongly covalent transition-metal compounds.

4. *Some Future Problems*

In concluding the present article we should like to point out some future problems concerning the electron theory of transition-metal compounds under high pressure besides future developments of the topics already mentioned.

One of the important problems to be developed further in the present field is the temperature- and pressure-induced high-spin–low-spin transition of the $3d^4$–$3d^7$ transition-metal ions in crystals. The transition has been investigated by taking into account the elastic interaction between the high-spin (or low-spin) clusters produced in a low-spin (or high-spin) crystal (OHNISHI and SUGANO, 1981). The critical temperature at which the discontinuous phase transition takes place has been evaluated as a function of cubic field splitting parameter Δ in units of spin-paring energy π which is given in terms of Racah parameter. It has been shown that the "crossover range" where this critical temperature exists is restricted to a finite range of Δ/π around the crossover point. Since Δ may be varied by pressure, a system outside the crossover range may be brought into it by applying high pressure. Future experimental studies on such a problem of the high-spin-low-spin transition are strongly desired.

Another interesting problem is the insulator-metal transition of transition-metal compounds. A good review on this problem has been given by ADLER (1968), so that we are not discussing any details. It might be interesting to point out in connection with this topic that the effect of electron transfer or itinerancy of d-electrons on the shake-up satellites in PES of transition-metal compounds has been studied by using a simple model (ASADA and SUGANO, 1978).

REFERENCES

ADACHI, H., TSUKADA, M. and SATOKO, C. (1978): Discrete variational Xα cluster calculations I, Application to metal clusters, *J. Phys. Soc. Japan*, **45**, 875–883.

ADACHI, H., SHIOKAWA, S., TSUKADA, M., SATOKO, C. and SUGANO, S. (1979): Discrete variational Xα cluster calculations III, Application to transition metal complexes, *J. Phys. Soc. Japan*, **47**, 1528–1537.

ADLER, D. (1968): Insulating and metallic states in transition metal oxides, in *Solid State Physics*, Vol. 21, Eds. F. Seitz, D. Turnbull, and H. Ehrenreich, Academic Press, New York, pp. 1–113.

ASADA, S., SATOKO, C. and SUGANO, S. (1975): Multiplet structure in X-ray P-shell photoelectron and K-emission spectra of nickel compounds, *J. Phys. Soc. Japan*, **38**, 855–865.

ASADA, S. and S. SUGANO, (1976): Satellite in X-ray photoelectron spectra of transition-metal compounds, *J. Phys. Soc. Japan*, **41**, 1291–1299.

ASADA, S. and SUGANO, S. (1978): A model calculation for studying the effects of electron transfor on shake-up satellites in transition-metal compounds, *J. Phys. C*, **11**, 3911–3918.

BETHE, H. (1929): Termaufspaltung in Kristallen, *Ann. Physik.*, **3**, 133–208.

DRICKAMER, H. G. (1965): The effect of high pressure on the electron structure of solids, in *Solid*

State Physics, Vol. 17, Eds. F. Seitz, and D. Turnbull, Academic Press, New York, pp. 1–133.

DRICKAMER, H. G. (1967): Effect of interionic distance on the ligand field in NiO, *J. Chem. Phys.*, **47**, 1880–1880.

ELLIS, D. E., BENESH, G. A. and BYRON, E. (1977): Molecular cluster studies of binary alloys: LiAl, *Phys. Rev. B*, **16**, 3308–3313.

GÁSPÁR, R. (1953): Über eine Approximation des Hartree-Fockschen Potentials durch eine Universelle Potentialifunction, *Acta Phys. Hung.*, **3**, 263–268.

GRIFFITHS, J. H. E., OWEN, J. and WARD, I. M. (1953): Paramagnetic resonance in palladium and platinum group compounds, *Proc. R. Soc. Lond., A*, **219**, 526–542.

GRIFFITHS, J. H. E. and OWEN, J. (1954): Complex hyperfine structures in microwave spectra of covalent iridium compounds, *Proc. R. Soc. Lond., A*, **226**, 96–111.

KLEINER, W. H. (1952): Crystalline field in chrome alum, *J. Chem. Phys.*, **20**, 1784–1791.

KOHN, W. and SHAM, L. J. (1965): Self consistent equations including exchange and correlation effects, *Phys. Rev. A*, **140**, 1133–1138.

KOIDE, S. and PRYCE, M. H. L. (1958): Intensity calculation of some optical absorption lines in hydrated manganous salts, *Phil. Mag.*, **3**, 607–624.

OHNISHI, S. and SUGANO, S. (1981): Strain interaction effects on the high-spin-low-spin transition of transition-metal compounds, *J. Phys. C*, **14**, 39–55.

PIERMARINI, G. J., BLOCK, S., BARNETT, J. D. and FORMAN, R. A. (1975): Calibration of the pressure dependence of the R_1 ruby fluorescence line to 195 kbar, *J. Appl. phys.*, **46**, 2774–2780.

POLDER, D. (1942): On the theory of the paramagnetic anisotropy of some hydrated cupric salts, *Physica*, **9**, 709–718.

ROBERT, T. (1975): Correlation between chemical bonding and satellite lines in X-ray photoelectron spectra of transition metal compounds, *Chem. Phys.*, **8**, 123–135.

SATO, Y. and AKIMOTO, S. (1979): Hydrostatic compression of four corundum-type compounds: α–Al_2O_3, V_2O_3, Cr_2O_3, and α–Fe_2O_3, *J. Appl. Phys.*, **50**, 5285–5291.

SHULMAN, R. G. and KNOX, K. (1960): Nuclear magnetic resonance in $KMnF_3$, *Phys. Rev.*, **119**, 94–101.

SLATER, J. C. and JOHNSON, K. H. (1972): Self-consistent-field Xα cluster Method for polyatomic molecules and solids, *Phys. Rev. B*, **5**, 844–853.

SUGANO, S. and SHULMAN, R. G. (1963): Covalency effects in $KNiF_3$. III. Theoretical studies, *Phys. Rev.*, **130**, 517–530.

SUGANO, S., TANABE, Y. and KAMIMURA, H. (1970): *Multiplets of Transition-Metal Ions in Crystals*, Academic Press, New York.

SUGANO, S. (1978):Theoretical studies of shake-up satellites in transition-metal compounds, *Japan J. Appl. Phys.*, **17**, Suppl. 17-2, 38–43.

TANABE, Y. and SUGANO, S. (1956): On the absorption spectra of complex ions, III. The calculation of the crystalline field strength, *J. Phys. Soc. Japan*, **11**, 864–877.

VAN VLECK, J. H. (1935): Valence strength and the magnetism of complex salts, *J. Chem. Phys.*, **3**, 807–813.

VAN VLECK, J. H. (1939): The Jahn-Teller effect and crystalline stark splitng for cluster of the form XY_6, *J. Chem. Phys.*, **7**, 72–84.

Materials Science of the Earth's Interior, edited by I. Sunagawa, pp. 191–231.

HIGH-TEMPERATURE CRYSTALLOGRAPHY OF OLIVINES AND SPINELS

Yoshio TAKÉUCHI, Takamitsu YAMANAKA, Nobuhiko HAGA, and Masahiro HIRANO

Mineralogical Institute, Faculty of Science, The University of Tokyo, Hongo, Tokyo, Japan.

1. *Introduction*

With the use of a gas flame as a heat source for temperatures above 1,000°C and an electric resistance furnace for those below 1,000°C, the diffraction intensity data of forsterite and spinel were collected at various temperatures up to around 1,700°C. Using the data thus cellected, we have refined the crystal structures of these minerals at high temperatures. In order to study the effect of temperature on the olivine-type structures in general, the structures of Mg_2GeO_4, Ca_2GeO_4, Al_2BeO_4, $AlMgBO_4$, $LiFePO_4$ (NAKANISHI, 1981), and $LiScSiO_4$ (SATO, 1980) have also been studied at temperatures up to around 800°C ~ 1,200°C. The present paper presents the results of our studies mainly on forsterite and spinel and discusses the thermal response of these crystal structures. Details of these olivine-type structures will be reported elsewhere at a later date. To begin with, reported high-temperature studies of olivine and structural studies of spinel are briefly reviewed in subsequent paragraphs.

1.1 *Olivine*

The thermal expansion of olivine has been studied by KOZU *et al.* (1934), RIGBY *et al.* (1946), SKINNER (1962), and SOGA and ANDERSON (1967). In the latest paper, to our knowledge, SUZUKI (1975) measured thermal expansion of natural forsterite (92.3% Fo) at temperatures from room temperature to 1,000°C and, based on Grüneisen's theory (GRÜNEISEN, 1926), provided the calculated thermal expansion coefficients in a temperature range from −250°C to 1,500°C.

Studies of crystal structures at high temperature have been reported for pure forsterite and $Fo_{38}Fa_{62}$ (SMYTH and HAZEN, 1973), $Fo_{69}Fa_{31}$ (BROWN and PREWITT, 1973), pure fayalite (SMYTH, 1975), synthetic forsterite (HAZEN,

1976) and ferromagnesian olivine (HAZEN, 1977). More recently, LAGER and MEAGHER (1978) reported a high temperature structural study of Ni_2SiO_4, monticellite, and glaucochroite. Through these works, a number of basic and important features of the thermal response of the olivine structure have been brought out. HAZEN and PREWITT (1977) compiled the variations of metal-oxygen bond distances with temperature and pressure, providing two empirical relations between bond expansion, compression, and other bonding variables. An equation of state for ferromagnesian olivine has also been given by HAZEN (1977). An extensive review of olivines and silicate spinels has been provided by BROWN (1980). The following olivine-type structures have been refined at room temperature: chrysoberyl (FARRELL *et al.*, 1963), sinhalite (FANG and NEWNHAM, 1965), triphylite (FINGER and RAPP, 1970), $LiScSiO_4$ (Steele *et al.*, 1978), and Na_2BeF_4 (HANKE, 1965).

Temperatures covered by the reported studies are, however, limited up to around 1,000°C, which seems to be too low, particularly for forsterite, to extrapolate observed structural variations to the melting point at 1,890°C. Although mean Si–O bond lengths in olivine remain nearly the same up to around 1,000°C, SMYTH (1975) observed a slight negative expansion of Si–O bonds for fayalite (melting point, 1,205°C), and HAZEN (1976) reported an anomalous thermal response of a specific Si–O bond in forsterite. In order to gain knowledge of the structure of forsterite at around its melting point, it is of substantial importance to clarify the nature of thermal response of Si–O bonds in wider temperature range covering those higher than 1,000°C.

1.2 Spinel

The geometrical aspect of the spinel structure, which was first studied by BRAGG (1915) and NISHIKAWA (1915), is primarily characterized by the coordinates, u, u, u, of an oxygen atom on the cubic body diagonal. NISHIKAWA (1915) estimated that the oxygen u parameter took a value of between 0.389 ~ 0.396. It is interesting to observe that the parameter has since been improved upon with intervals of about 15 ~ 20 years. By X-ray diffraction, BARTH and POSNJAK (1932) obtained $u = 0.390(6)$. Then, by neutron powder diffraction of a synthetic spinel, BACON (1952) determined $u = 0.387 \pm 0.001$. Through this work, Bacon established that the cation arrangement was of the normal type with Mg at the tetrahedral positions (A sites) and Al at the octahedral positions (B sites). Based on a comparison between observed and calculated intensities of reflections, Bacon further commented that the idealized normal structure was not completely achieved and effectively there was a small movement toward the inverse form with Al at A, and 1/2 Mg, 1/2 Al at B. This Bacon's observation was later confirmed by FISCHER (1967). Using synthetic crystal, annealed below 800°C for 14 days, he showed by neutron powder diffraction, that the oxygen u parameter had a value of 0.3873 $\pm$ 0.0001 and that the cation arrangement was of the 8 $\pm$ 4% inverse type. In his paper, Fischer mentioned that a natural spinel $MgAl_{1.9}Fe_{0.1}O_4$ was a normal type free from the inverse type of cation arrangement. As had been inferred by BRAGG (1937), the manner of distribution of the atoms amongst the positions in the spinel structure depends upon the thermal treatment of a spinel when synthetically prepared. In the above crystallographic studies, the space group was referred to *Fd*3*m*. However, JAGODZINSKI and SAALFELD (1958) pointed out the possibility of a lower symmetry for spinel by means of an elaborated analysis of X-ray Fourier peaks of oxygen atoms; spinel might be a twin consisting of individuals having a possible space group *Fddd*.

There are indeed quite a few arguments about the space group of the cubic spinel-type structures; the space group of this structure type would be lower than *Fd*3*m*. Various arguments dealing with the space group of the spinel group of materials may be divided into two categories. Those which fall in one category, lay stress upon crystal field distortions of the coordination polyhedra. For example, GRIMES (1971, 1972) suggested that the conventional site within the octahedral hole is not a potential minimum, producing off-center displacement of the cation. Then, the acentric space group $F\bar{4}3m$ is more probable than *Fd*3*m*. He showed that this idea, originally applied to $MgCr_2O_4$, may be more or less applicable to other spinel-type structures. Those which fall in the other category are based on the fact that there are certain types of cation ordering other than the normal or inverse type. In fact, JOUBERT and DURIF (1964) showed that, depending upon the annealing temperature, $Mn_3Li_3ZnO_8$ takes a symmetry of either $P4_332$ or $P2_13$ in addition to *Fd*3*m* at high temperature. In the structures of these low symmetries, different cations are distributed in an ordered fashion over the octahedral positions as well as tetrahedral.

So far as the routine diffraction studies of spinel $MgAl_2O_4$ are concerned, however, there is no evidence that rules out *Fd*3*m* and no definite evidence that supports $F\bar{4}3m$. What is clear is that the distribution of diffraction intensities is consistent with *Fd*3*m* (SMITH, 1978; TOKONAMI and HORIUCHI, 1980). Using the ψ scan technique and MoKα radiation, we have, in fact, confirmed that the forbidden reflections, such as 200,600 and 10, 0, 0, for *Fd*3*m* are due to double reflections. In a very recent work, GRIMES *et al.* (1981) measured about 4,000 X-ray diffraction intensities of a spinel. Based on these, they reported that the space group $F\bar{4}3m$, compared to *Fd*3*m*, was found to give a significantly superior fit to the experimental data. Even if the space group $F\bar{4}3m$ (or even lower symmetry) is really the case, the possible atomic shifts from the positions in the *Fd*3*m* structure are expected to be exceedingly small. It is therefore anticipated that the basic features of the thermal response of the spinel structure can be studied in terms of the *Fd*3*m* structure.

Spinel undergoes an order-disorder transition at around 800°C (HAFNER, 1960; HAFNER and LAVES, 1961; BRUN and HAFNER, 1962). They showed that the inversion parameter i_A (defined as the fraction of A sites occupied by Al^{3+}

ions (Schmocker and Waldner, 1976)) of spinel varies depending upon the heat treatment of the crystal. According to Schmocker and Waldner (1976), the disorder in natural spinels starts at around 750°C ~ 850°C and terminates at around 900°C with $i_A \approx 30\%$. They showed that the value of i_A of some of the synthetic spinels attained was as large as around 60%. Order-disorder in spinel was also discussed by Datta and Roy (1967, 1968).

In recent papers, Suzuki and Kumazawa (1980) reported an anomalous thermal expansion which suggests the existence of a second-order phase transition in spinel at 660 ± 10°C, Weeks and Sonder (1980) discovered that the electrical conductivity was temperature-dependent and Hill *et al.* (1979) presented systematic trends of the group of spinel-type structures based on structural data of 149 oxide and 80 sulfide spinels. Baumgartner *et al.* (1981) carried out neutron diffraction studies of natural spinel at temperatures from 690 to 880°C and reported that the space group of spinel is *Fd*3*m* above 750°C and $F\bar{4}3m$ below that temperature; there are differences in the atomic occupancy between the structures above and below that temperature. Reported structural works on silicate spinels include those by Yagi *et al.* (1974), Marumo *et al.* (1974), Marumo *et al.* (1977), and Finger *et al.* (1979).

2. *Olivine*

2.1 *Characteristics of the tetrahedral bonds in the olivine-type structures*

The oxygen atoms in the olivine structure type are arranged in hexagonal closest packing. One-half of the octahedral voids found in the closest packing are occupied by cation M, and one-eighth of the tetrahedral voids by cations T, thus giving a general formula M_2TO_4. The orthorhombic unit cell, having space group *Pbnm* (D_{2h}^{16}), contains four formula units. The olivine-type structures other than silicate olivines are often described with the crystallographic axes whose orientations are different from those for silicate olivines. Throughout the present paper we adopt the orientation which was originally used by Bragg and Brown (1926) to describe the forsterite structure and is consistent with *Pbnm*.

The oxygen layers found in the hexagonal-closest packing of this structure type are then parallel to (100). The location of each tetrahedral cation is such that one T–O bond (according to conventional notation, T–O(1)) is perpendicular to the layer, while the remaining three, T–O(2), T–O(3) and T–O(3)′, are oblique to the layer; O(1) may be called an apical oxygen and others by basal oxygen atoms. This situation inherently gives rise to the following characteristic difference in the distance between the tetrahedral cation and neighboring cations.

Among nine octahedral cations surrounding the tetrahedron (Fig. 1), the set of three cations, M(1)′ × 2 and M(2)′, in the top layer in Fig. 1 are bound to O(1), while the set of three cations, M(1) × 2 and M(2), in the bottom are

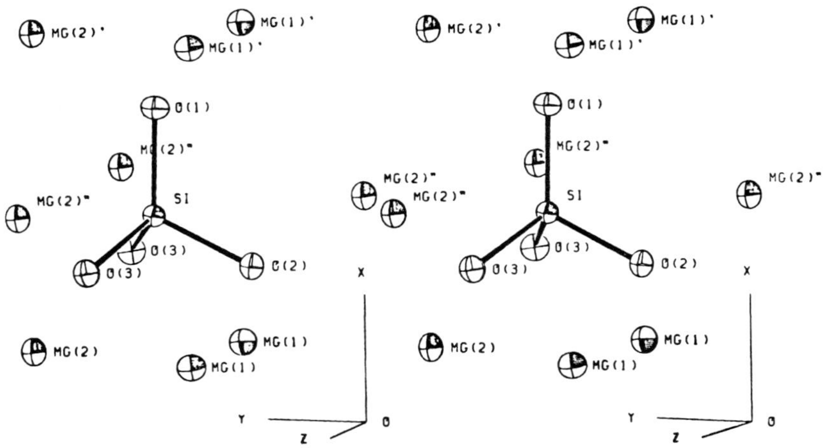

FIG. 1. A stereographic drawing, showing a silicate tetrahedron and surrounding Mg atoms in the structure of forsterite.

bound to the basal oxygen atoms (the edges of the basal triangle of the tetrahedron are shared with the octahedra formed about M's). Let $\{M'\}$ denote the first set of three cations and $\{M\}$ the second. Then we find that the $\{M\}-T$ distances are 0.797 times as short as the $\{M'\}-T$ distances in the ideal hexagonal-closest packing. For some olivine-type structures, the mean value of $\{M\}-T$ ($= \overline{\{M\}-T}$) and those of $\{M'\}-T$ ($= \overline{\{M'\}-T}$) are evaluated and listed in Table 1, in which the set of three cations in the middle layer of Fig. 1 are denoted by M″. As will be observed in this table, the $\overline{\{M\}-T}$ distances in actual cases are shorter than $\overline{\{M'\}-T}$ by around an average of 16%, 19% for sinhalite, $AlMgBO_4$, being the maximum.

Consequently, if T were located at the center of the tetrahedron, the repulsive forces between $\{M\}$ and T would be stronger than those between $\{M'\}$ and T. The tetrahedral cation T would then be shifted toward O(1). Thus, the fact that the T–O(1) bond in the olivine type is unexceptionally the shortest, would be the result of a delicate balance between the effect of the local potential gradient and the repulsive interactions between T and the octahedral cations. In Table 2, we give T–O bond lengths for some olivine-type structures. In the difference Fourier map of forsterite (Fig. 2) (FUJINO *et al.*, 1980), the residual peak on the Si–O(1) bond, among those in the tetrahedron, has a highest peak height in accordance with the fact that Si–O(1) is the shortest silicate bond in the structure.

As observed in Fig. 1, each oxygen atom is tetrahedrally coordinated by four cations, including T. In the event that the formal charge of M(1) is larger than that of M(2) in a given olivine-type structure, the O(1) atom is apparently excessive in negative charge because O(1) is bound to $2 \times M(1)$ and $1 \times M(2)$.

TABLE 1. Distances between tetrahedral cation, T, and surrounding octahedral cations.

Material	T	Surrounding cation		Distance	Material	T	Surrounding cation		Distance
Mg_2SiO_4	Si	Mg(1)′	× 2	3.2513 Å	$MgCaSiO_4$[a]	Si	Mg′	× 2	3.3827 Å
		Mg(2)′		3.2698			Ca′		3.4871
		Mg(1)	× 2	2.6943			Mg	× 2	2.7017
		Mg(2)		2.7856			Ca		3.0137
		Mg(2)″		3.2487			Ca″		3.4305
		Mg(2)″	× 2	3.2767			Ca″	× 2	3.5740
Fe_2SiO_4	Si	Fe(1)′	× 2	3.2981	Al_2BeO_4	Be	Al(1)′	× 2	2.9876
		Fe(2)′		3.2879			Al(2)′		3.0093
		Fe(1)	× 2	2.7717			Al(1)	× 2	2.5144
		Fe(2)		2.8751			Al(2)		2.5806
		Fe(2)″		3.3497			Al(2)″		3.0249
		Fe(2)″	× 2	3.3121			Al(2)″	× 2	3.0297
Ca_2GeO_4	Ge	Ca(1)′	× 2	3.6021	$AlMgBO_4$	B	Al′	× 2	3.0478
		Ca(2)′		3.6157			Mg′		3.1121
		Ca(1)	× 2	3.0267			Al	× 2	2.4236
		Ca(2)		3.1118			Mg		2.6133
		Ca(2)″		3.6302			Mg″		3.1042
		Ca(2)″	× 2	3.6744			Mg″	× 2	3.157

[a]Calculated from atomic coordinates given by LAGER and MEAGHER (1978).

TABLE 2. Tetrahedral bond lengths in some olivine-type structures. In the pair of columns for each bond, the first gives the bond length and the second deviation from the average bond length.

	Mg_2SiO_4		Fe_2SiO_4[a]		Mn_2SiO_4[a]		Ni_2SiO_4[b]	
T–O(1)	1.614 Å	−1.22%	1.625 Å	−0.67%	1.619 Å	−1.22%	1.620 Å	−1.16%
T–O(2)	1.654	+1.22	1.653	+1.04	1.658	+1.16	1.660	+1.28
T–O(3)	1.635	+0.06	1.633	−0.18	1.640	+0.06	1.637	+0.12
Average	1.634		1.636		1.639		1.639	

	$MgCaSiO_4$[b]		Mg_2GeO_4		Ca_2GeO_4		Al_2BeO_4	
T–O(1)	1.617	−1.22	1.726	−1.70	1.741	−1.14	1.5680	−4.20
T–O(2)	1.656	+1.16	1.783	+1.54	1.783	+1.04	1.6907	+3.30
T–O(3)	1.637	0	1.757	+0.06	1.760	−0.06	1.6442	+0.46
Average	1.637		1.756		1.761		1.6367	

	$AlMgBO_4$		$LiScSiO_4$[c]		$LiFePO_4$	
T–O(1)	1.4387	−3.85	1.601	−1.96	1.519	−1.48
T–O(2)	1.5824	+5.75	1.616	−1.04	1.538	−0.26
T–O(3)	1.4821	−0.94	1.658	+1.53	1.555	+0.84
Average	1.4963		1.633		1.542	

[a]FUJINO *et al.* (1981), [b]LAGER and MEAGHER (1978), [c]STEELE *et al.* (1978).

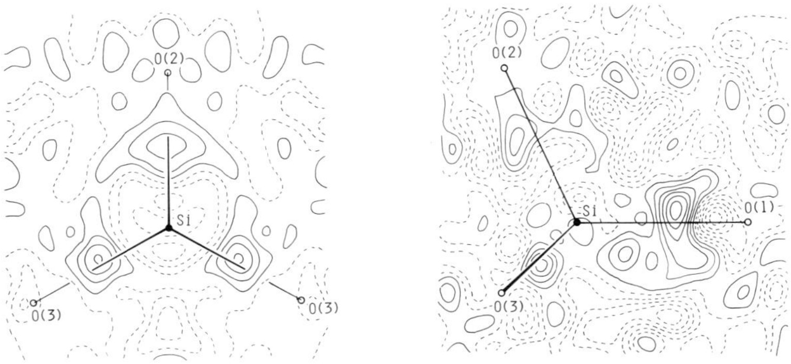

FIG. 2. Difference Fourier synthesis of forsterite, showing the residual peaks around Si (after FUJINO *et al.* (1981) but with some modifications). Left: the $x = 0.305$ section, showing peaks on the Si–O(2) and Si–O(3) bonds. Right: the $z = 1/4$ section, showing peaks on the Si–O(1) and SiO–(2) bonds. Contours are at intervals of 0.1 eA^{-3}; zero and negative contours being broken.

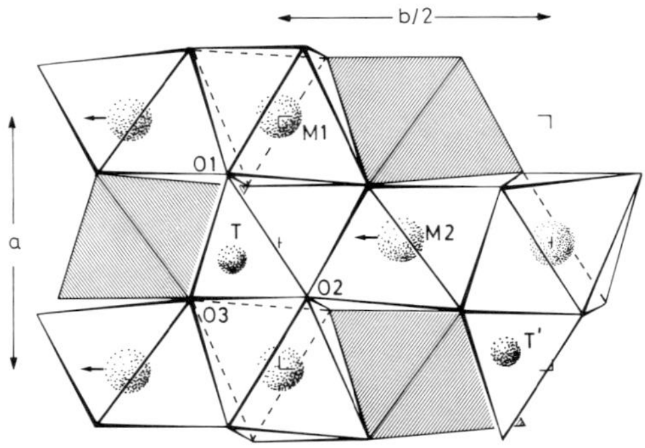

FIG. 3. The links of polyhedra in the forsterite structure viewed down the *c* axis. Cations T (and T′) and M2 are at the $z = 3/4$ level and M1 at $z = 1/2$. The octahedra about M2 at $z = 1/4$ are ruled. Arrows indicate the approximate direction of shifts of the M2 cations with raising temperature.

Even in such a case, the T–O(1) bond length is shorter than others, as in the case of sinhalite (Table 2).

Another notable feature in the T–O bonds is that the T–O(2) is always the longest bond (Table 2). This is probably due to the situation that the repulsive force between T′ and M(2) across the shared edge, as shown in Fig. 3, tends to shorten the bond between M(2) and O(2) associated with the tetrahedron T. The M(2)–O(2) bond is, in general, the shortest octahedral bond in the olivine type.

2.2 *Structure of forsterite at temperatures up to around* 1,600°*C*

2.2.1 *Experimental procedure*

Our study was carried out using a large transparent, colorless single crystal synthesized by TAKEI (1982). Several spherically ground specimens (r = 0.15 ~ 0.2 mm) were prepared out of the crystal and used for X-ray study. Unit cell dimensions and diffraction intensities were measured with a single crystal diffractometer, which is combined with a rotating-anode X-ray generator (200 mA/60 kv), using graphite-monochromatized Mo $K\alpha_1$ radiation, $\lambda = 0.70926$ Å. For temperatures below 1,000°C, we used, as a heating device, a miniature electric-resistance furnace similar to the one reported by BROWN *et al.*, (1973). For those above 1,000°C, a gas-flame (YAMANAKA *et al.*, 1981) was used as the heat source. In both cases, samples were heated in the open air. For details of the high-temperature single-crystal technique, refer to YAMANAKA (1982). The cell dimensions were obtained by the least-squares procedure applied to sin 2θ values of 15 ~ 25 reflections. The $\omega - 2\theta$ scan technique was used to collect diffraction intensities of reflections. Intensities were corrected for Lorentz and polarization effects, transmission and isotropic extinction effects (COPPENS and HAMILTON, 1970) and reduced to structure factors.

One of the experimental difficulties we encountered was that the samples underwent severe sublimation at temperatures above 1,500°C. For this case, we corrected diffraction intensities for volume loss in terms of decreasing in intensity of a set of three reference reflections which were measured every 30 reflections. The data for the intensity study and final values of R are listed in Table 3.

2.2.2 *Unit-cell expansion*

The unit cell dimensions at temperatures under which diffractions intensities were collected are listed in Table 4. We also obtained, from single-crystal diffractometry, the cell dimensions at temperatures other than these; they are plotted all together in a diagram versus temperature, as shown in Fig. 4. Data for thermal expansion, particularly of the *b* edge, is closely similar to that reported by SUZUKI (1975) and HAZEN (1976). The expansion rates for *a* and *c* edges are, however, slightly smaller than those given by SUZUKI (1975)

TABLE 3. Intensity study data of forsterite.

Temperature (°C)	25	310	530	700	950	1,050	1,210	1,400	1,600
Scan speed (ω°/sec)	0.17	0.10	0.13	0.17	0.17	0.17	0.20	0.20	0.20
Maximum 2θ (°)	90	80	70	70	80	80	80	70	60
No. of reflexions measured	1,277	974	702	707	997	998	1,005	727	482
No. of reflexions used	1,020	733	627	619	721	681	591	478	286
R(%)	3.89	4.34	2.03	2.27	4.56	5.12	4.28	3.89	5.76
wt. R(%)	4.12	5.41	2.51	2.87	4.61	4.97	4.06	3.76	5.76

TABLE 4. Cell dimensions of forsterite.

T(°C)	Number of reflexions used	Cell dimension a	b	c	Volume
25	25	4.7503(4) Å	10.187(1) Å	5.9771(7) Å	289.22(5) Å^3
310	24	4.758(1)	10.219(3)	5.9899(9)	291.2(1)
530	23	4.767(1)	10.248(3)	6.0059(8)	293.4(1)
700	25	4.777(1)	10.269(3)	6.0184(9)	295.2(1)
950	22	4.789(1)	10.321(3)	6.045(1)	298.8(1)
1,050	22	4.7951(5)	10.343(2)	6.0571(5)	300.40(7)
1,210	17	4.808(1)	10.378(4)	6.077(1)	303.2(1)
1,400	17	4.819(3)	10.424(5)	6.099(3)	306.4(3)
1,600	15	4.834(1)	10.455(2)	6.1147(9)	309.1(1)

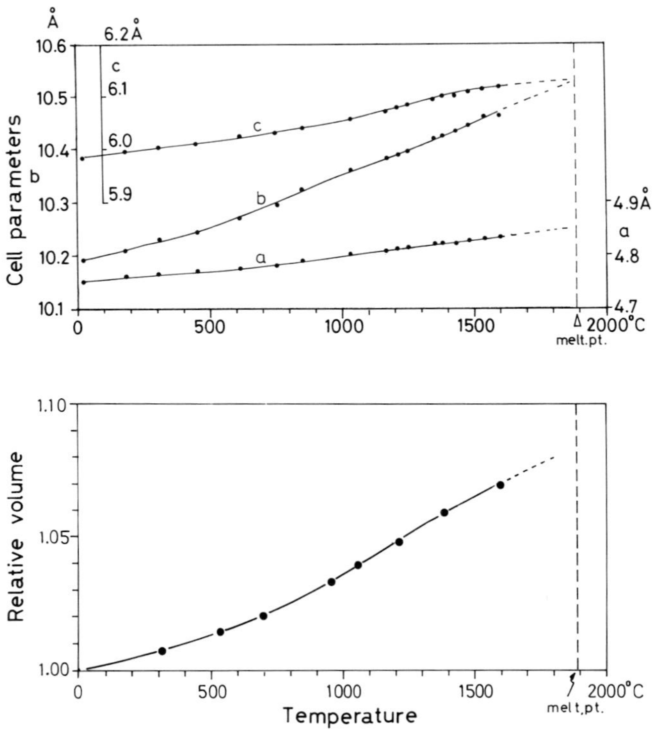

FIG. 4. Top: The unit-cell dimensions of forsterite versus temperature. Bottom: Relative unit-cell volume of forsterite versus temperature.

and slightly larger than those reported by HAZEN (1976). Excepting b, we find that the thermal expansions of a and c exhibit a near linear relationship with the temperature in the temperature range between 25° and 1,000°C. At temperatures above 1,000°C, the rates of expansion, particularly of b and c, increase compared to those at temperatures below 1,000°C. Above around 1,400°C, however, the rate of expansion in the specific case of c is decreased. The mode of expansion of these cell edges appears to be correlated with the mode of M–O expansion, as will be shown later.

The observed variation mode in the cell parameter may now be extrapolated to the melting point to estimate the parameters at the extreme temperature. The obtained values are: $a = 4.86$ Å, $b = 10.53$ Å, $c = 6.13$ Å and $V = 314$ Å^3, giving a calculated density of ~2.98 g/cm^3. The unit cell volume is about 8.4% larger (Fig. 4) than that of the room temperature volume, 289.58(7) Å^3 (FUJINO *et al.*, 1981). The above cell volume at the melting point is about 1.6% smaller than that obtained by HAZEN (1976) by extrapolating his expansion data in the temperature range from 23°C to around 1,400°C.

2.2.3 Variation of Si–O bond lengths

As will be observed in Table 5, which shows the interatomic distances and polyhedral volumes, our study of forsterite shows that the mean Si–O length appears to remain nearly the same up to a temperature of around 1,250°C. At around this temperature, the mean length appears to start decreasing, provided that corrections of bond lengths for thermal vibrations are not considered for the moment. In Fig. 5, which shows the variation of the individual Si–O bond lengths with temperature, we observe the following characteristics: (1) In the temperature range between ~300 and ~1,200°C, the Si–O(2) bond, the longest one, is slightly longer in relation to the room

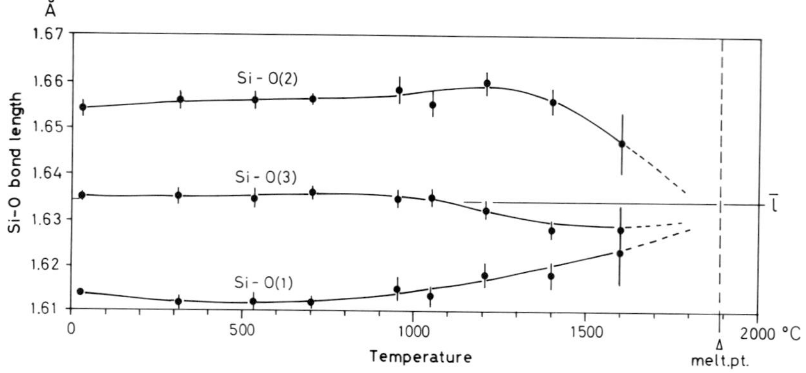

FIG. 5. A plot of individual Si–O bond lengths of forsterite versus temperature, $\bar{l}$ indicating the mean value at room temperature (for thermal correction, see Fig. 10(bottom)).

temperature length. The anomalous negative expansion of this bond, as reported by HAZEN (1976), between 300 ~ 1,000°C was not observed. At around 1,200°C, the bond starts decreasing. (2) The shortest Si–O(1) bond exhibits a trend of increasing with a rising temperature at around 1,000°C. (3) The Si–O(3) bond, whose length is closely similar to the mean tetrahedral bond length, remains constant up to around 1,000°C and is then, slightly contracted at temperatures above 1,000°C. Thus at around 1,000°C, the modes of thermal expansion of the three crystallographically independent Si–O bonds start to vary in such a way that they would give the same length at around the melting temperature of 1,890°C. This situation may be regarded as a realistic consequence because it is conceivable that increased cation-oxygen distances at the extreme temperature would reduce the effect of repulsive interactions between the tetrahedral cation and the surrounding octahedral cations, as mentioned before. As the result, the rigid nature of the tetrahedra would be enhanced. The key experimental data which yield the result, as shown in Fig. 5 are those obtained at 1600°C. We carefully repeated the collection of diffraction intensities at around the temperature several times; they gave nearly the same result.

The above-mentioned thermal response of Si–O(2) between ~ 300 and ~ 1,200°C appears to be insignificant by itself. As will be shown later, however, we can point out that the T–O(2) bonds in the olivine-type structures in general are liable to exhibit an anomalous thermal response.

2.2.4 Expansion of octahedral bonds

Since each oxygen atom in forsterite is coordinated by three Mg atoms and one Si atom, any change in the Mg–O bond lengths would, in principle, be reflected in the Si–O bond lengths or *vice versa*. Therefore, in order to study the thermal response of Mg–O bonds in connection with those of individual Si–O bonds, we are here interested in an average, like $\overline{\text{Mg–O1}}$, which denotes an average of the bond lengths between O(1) and the nearest three Mg atoms, Mg(1) × 2 and Mg(2). For O(2) and O(3), the quantities $\overline{\text{Mg–O2}}$ and $\overline{\text{Mg–O3}}$ are likewise defined, respectively. For forsterite at room temperature, their values are respectively 2.113(1), 2.060(1), and 2.123(1) Å. They vary with a raising of temperature, as shown in Fig. 6, in which we observe a marked difference in the rate of expansion between these mean values, particularly between 25 and 500°C. In that temperature range, the expansion rate of $\overline{\text{Mg–O2}}$ is about $1.4 \times 10^{-5}\text{C}^{-1}$, while that of $\overline{\text{Mg–O3}}$ is as large as about $5.5 \times 10^{-5}\text{C}^{-1}$; that of $\overline{\text{Mg–O1}}$ is $3.8 \times 10^{-5}\text{C}^{-1}$ and between the two but closer to the latter. Since Mg–O(2) bonds include the bonds which have large components to the *b* axis, the small expansion rate of $\overline{\text{Mg–O2}}$ below 500°C would be related to the relatively small expansion rate of *b* in that temperature range.

Between 500°C ~ 1,250°C, however, the expansion rates are approximately $4.9 \times 10^{-5}\text{C}^{-1}$ for $\overline{\text{Mg–O1}}$, $4.5 \times 10^{-5}\text{C}^{-1}$ for $\overline{\text{Mg–O3}}$, and

TABLE 5. Interatomic distances (Å) in forsterite at room tmeperature up to around 1,600°C.

Interatomic distances	25°C	310°C	530°C	700°C	950°C	1,050°C	1,210°C	1,400°C	1,600°C
Tetrahedron									
Si–O(1)	1.614(1)	1.612(2)	1.611(1)	1.612(2)	1.615(3)	1.613(3)	1.618(3)	1.618(3)	1.623(7)
Si–O(2)^p	1.654(2)	1.656(2)	1.656(1)	1.656(2)	1.658(3)	1.655(3)	1.660(3)	1.656(3)	1.647(7)
Si–O(3)	1.635(1)	1.635(2)	1.6340(9)	1.636(1)	1.634(2)	1.635(2)	1.632(2)	1.628(2)	1.628(5)
Average	1.634(1)	1.634(2)	1.634(1)	1.635(1)	1.636(2)	1.635(3)	1.635(2)	1.633(3)	1.632(6)
O(1)–O(2)^p	2.744(2)	2.747(3)	2.746(2)	2.747(2)	2.751(4)	2.744(4)	2.748(4)	2.744(5)	2.74(1)
O(1)–O(3)	2.757(2)	2.752(3)	2.750(2)	2.752(2)	2.752(3)	2.751(3)	2.751(4)	2.747(4)	2.752(8)
O(2)^p–O(3)	2.553(2)	2.555(3)	2.555(2)	2.557(2)	2.561(3)	2.560(3)	2.563(3)	2.561(4)	2.554(8)
O(3)–O(3)^r	2.592(2)	2.593(2)	2.593(1)	2.593(2)	2.593(3)	2.597(3)	2.595(3)	2.587(4)	2.578(7)
Average	2.659(2)	2.659(3)	2.658(2)	2.660(2)	2.661(3)	2.660(3)	2.662(3)	2.658(4)	2.656(8)
Volume (Å^3)	2.204(4)	2.205(7)	2.203(4)	2.206(5)	2.211(7)	2.208(9)	2.212(8)	2.202(9)	2.20(2)
Mg(1) Octahedron									
Mg(1)–O(1)	2.081(1)	2.089(2)	2.096(1)	2.101(1)	2.110(2)	2.114(2)	2.121(2)	2.130(2)	2.134(5)
Mg(1)–O(2)^p	2.067(1)	2.068(2)	2.0724(9)	2.076(1)	2.083(2)	2.087(2)	2.094(2)	2.101(2)	2.106(5)
Mg(1)–O(3)	2.131(1)	2.139(2)	2.1470(9)	2.154(1)	2.166(2)	2.172(2)	2.181(2)	2.196(2)	2.205(5)
Average	2.093(1)	2.099(2)	2.105(1)	2.110(1)	2.119(2)	2.125(2)	2.132(2)	2.142(2)	2.148(5)
O(1)–O(2)^t	3.0218(5)	3.0283(7)	3.0373(5)	3.0439(5)	3.0577(8)	3.0648(7)	3.0751(8)	3.088(2)	3.098(2)
O(1)–O(3)^s	3.103(2)	3.114(3)	3.125(2)	3.133(2)	3.149(3)	3.155(4)	3.167(3)	3.186(4)	3.200(8)

$O(2)^t$–O(3)	3.333(2)	3.343(3)	3.358(2)	3.370(2)	3.391(3)	3.405(3)	3.423(3)	3.451(4)	3.474(8)
O(1)–$O(2)^p$	2.842(2)	2.847(3)	2.855(2)	2.861(2)	2.868(4)	2.874(4)	2.884(4)	2.892(5)	2.89(1)
O(1)–O(3)	2.848(2)	2.861(3)	2.871(2)	2.880(2)	2.892(3)	2.902(3)	2.913(3)	2.926(4)	2.931(8)
$O(2)^p$–O(3)	2.553(1)	2.555(3)	2.555(2)	2.557(2)	2.561(3)	2.560(3)	2.563(3)	2.561(4)	2.554(8)
Average	2.950(2)	2.958(2)	2.967(2)	2.974(2)	2.987(3)	2.994(3)	3.004(3)	3.017(4)	3.025(7)
Volume (Å^3)	11.75(2)	11.83(3)	11.93(2)	12.01(2)	12.15(3)	12.22(4)	12.34(4)	12.48(5)	12.55(10)
Mg(2) Octahedron									
Mg(2)–O(1)	2.177(1)	2.190(3)	2.200(2)	2.210(2)	2.226(3)	2.237(3)	2.249(3)	2.265(3)	2.283(7)
Mg(2)–O(2)	2.045(1)	2.050(2)	2.055(1)	2.060(2)	2.068(3)	2.073(3)	2.075(3)	2.086(3)	2.090(7)
Mg(2)–$O(3)^q$	2.065(1)	2.068(2)	2.074(1)	2.079(1)	2.088(2)	2.091(2)	2.099(2)	2.110(2)	2.120(5)
Mg(2)–O(3)	2.210(1)	2.219(2)	2.2259(9)	2.230(1)	2.242(2)	2.248(2)	2.258(2)	2.265(2)	2.271(5)
Average	2.129(1)	2.136(2)	2.143(1)	2.148(1)	2.159(2)	2.165(2)	2.173(2)	2.183(3)	2.193(6)
O(1)–$O(3)^q$	3.021(2)	3.030(3)	3.040(2)	3.048(2)	3.068(3)	3.076(3)	3.091(3)	3.109(4)	3.122(8)
O(2)–$O(3)^q$	2.927(2)	2.935(3)	2.942(1)	2.950(2)	2.960(3)	2.964(3)	2.971(3)	2.982(4)	2.997(8)
O(2)–O(3)	3.182(2)	3.195(3)	3.208(2)	3.217(2)	3.237(3)	3.250(3)	3.263(3)	3.283(4)	3.302(8)
O(3)–$O(3)^q$	2.989(1)	2.997(2)	3.005(1)	3.010(2)	3.024(2)	3.028(3)	3.041(3)	3.051(3)	3.062(7)
$O(3)^q$–$O(3)^t$	3.385(1)	3.397(2)	3.413(2)	3.425(2)	3.452(3)	3.461(3)	3.482(2)	3.513(4)	3.536(7)
O(1)–O(3)	2.848(2)	2.861(3)	2.871(1)	2.880(2)	2.892(3)	2.902(3)	2.913(3)	2.926(4)	2.931(8)
O(3)–$O(3)^r$	2.592(2)	2.593(2)	2.593(1)	2.593(2)	2.593(3)	2.597(3)	2.595(3)	2.587(4)	2.578(7)
Average	2.993(2)	3.002(3)	3.012(1)	3.019(2)	3.034(3)	3.042(3)	3.053(3)	3.067(4)	3.079(8)
Volume (Å^3)	12.39(2)	12.50(3)	12.61(2)	12.70(2)	12.87(4)	12.96(4)	13.09(4)	13.25(5)	13.38(10)

For symmetry code, refer to APPENDIX II.

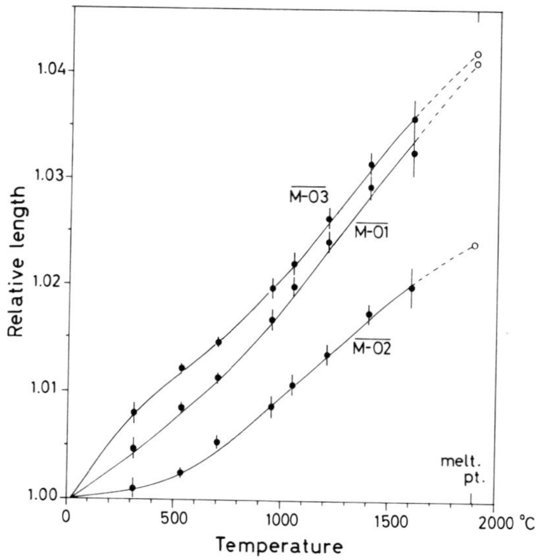

FIG. 6. A plot of $\overline{M-O1}$, $\overline{M-O2}$, and $\overline{M-O3}$ lengths (relative to room-temperature values) of forsterite versus temperature. Open circles for the melting point denote the values which were calculated from the atomic coordinates obtained by extrapolation.

$3.4 \times 10^{-5}C^{-1}$ for $\overline{Mg-O2}$. This decreasing order of three values are parallel to the increasing order of the Si–O bond lengths, namely Si–O(1) = 1.614 Å, Si–O(3) = 1.635 Å, and Si–O(2) = 1.654 Å.

A salient feature is found when we evaluate a ratio such as $\overline{Mg-O1}/\overline{Mg-O}$for each temperature and plot it in a diagram, as shown in Fig. 7, where $\overline{Mg-O}$ is an over all average of octahedral bond lengths at the same temperature for which $\overline{Mg-O1}$ is calculated. As observed in this diagram, $\overline{Mg-O2}/\overline{Mg-O}$ markedly decreases with an increasing temperature in contrast to $\overline{Mg-O1}/\overline{Mg-O}$ or $\overline{Mg-O3}/\overline{Mg-O}$. This implies that $\overline{Mg-O2}$ relatively decreases with the raising of temperature, in spite of the fact that $\overline{Mg-O2}$ itself increases with temperature. It is then conceivable that this situation would impose an certain effect upon the Si–O(2) bond; this specific tetrahedral bond would possibly show a trend of lengthening to balance the effectively shortened distances between Mg atoms and O(2). The slight positive expansion of Si–O(2) observed at the temperatures between ~300 and ~1,300°C, would thus be significant and it occurs as the consequence of such a decreasing effect of $\overline{Mg-O2}$. The shifting of Mg(2) almost along *b* with an increasing temperature (Fig. 3) (LAGER and MEAGHER, 1978) will largely be responsible for the effective decreasing in $\overline{Mg-O2}$.

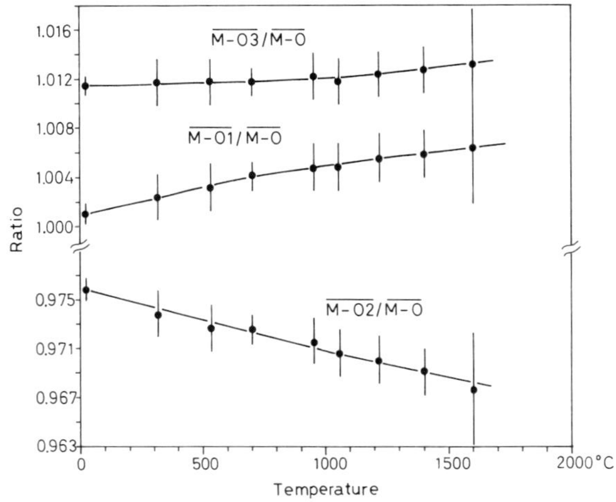

FIG. 7. A plot of $\overline{M-Oi}/\overline{M-O}$ ratios of forsterite versus temperature ($i = 1, 2, 3$).

2.3 *Anomalous expansion of T–O(2) bonds in non-silicate olivines*

The above trend of thermal response of M–O and T–O bonds seems to be substantiated by the study of non-silicate olivines. As an example, the case of sinhalite $AlMgBO_4$ is given here. In this material, $\overline{M-O1} = 2.0489(6)$ Å, $\overline{M-O2} - 1.9107(5)$ Å, and $\overline{M-O3} = 2.0078(7)$ Å; for tetrahedral B–O lengths, see Table 2. The variations of the B–O bond lengths with a raising of temperature are plotted in Fig. 8, which shows a marked expansion of the B–O(2) bond; at 880°C the bond length is 4.7% larger than the room temperature length. In conformity with such a conspicuous rate of thermal expansion of B–O(2), the value of $\overline{M-O2}/\overline{M-O}$ shows a striking rate of decreasing with a raising of temperature (Fig. 9). Although the sum of Pauling bond strengths about O(2) (also that about O(1)) in this structure exceed two, the anomalous expansion of B–O(2) can be primarily accounted for in terms of the decreasing effect of the $\overline{M-O2}/\overline{M-O}$ ratio with an increasing temperature. An anomalous expansion of T–O(2) bond, which is similar to the above, is also observed for chrysoberyl Al_2BeO_4. We may thus decide that the T–O(2) bonds in the olivine-type structures in general potentially have a trend of expansion with a raising of temperature.

2.4 *Structure of forsterite at around the melting point*

Extrapolation, by means of polynomial approximation, of the variation in atomic coordinates, with a raising of temperature, to the melting point yields the atomic coordinates for that extreme temperature as given in Table 6. In order to depict a rough view of the forsterite structure at around the melting temperature, the atomic coordinates thus derived were used to calculate

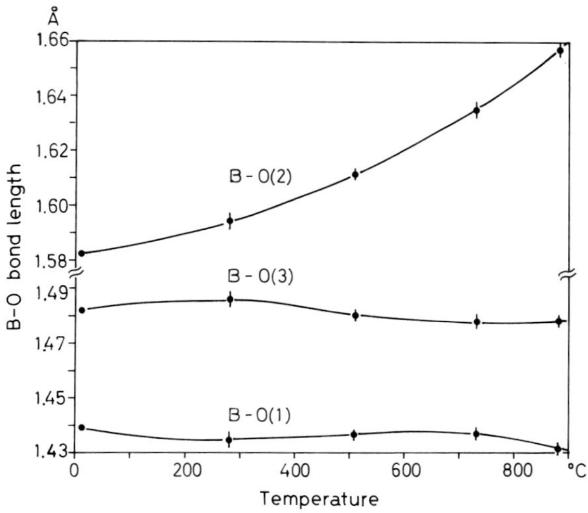

FIG. 8. A plot of tetrahedral B–O bond lengths of sinhalite versus temperature.

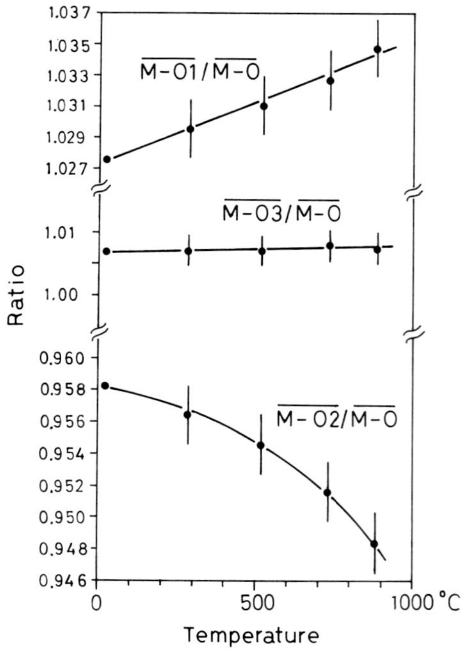

FIG. 9. A plot of $\overline{M-Oi}/\overline{M-O}$ ratios of sinhalite versus temperature ($i = 1, 2, 3$).

interatomic distances of the extreme structure (Table 6). The Si–O bond lengths, which were considerably different from one another at room temperature, are now almost the same within the limited accuracy of the procedure used to derive the values. In spite of such an 'equalization' of bond lengths which takes place in the silicate tetrahedra, the elongation of tetrahedra along *a* remains nearly the same as those at lower temperatures. The variance of tetrahedral angles, however, decreases with a raising of temperature and takes an estimated value of 24($^{\circ 2}$) at around the melting point, which should be compared to 49.7(3) at room temperature.

In contrast to such a relatively steady configuration of the silicate tetrahedra, thermal expansion of Mg octahedra are conspicuous (Fig. 10); the volume of Mg(2) octahedron, in particular, is about 10% larger than the room temperature volume. The variance of octahedral bond angles in fact increases with a raising of temperature and attains ~136 and ~129 (respectively for Mg(1) and Mg(2) octahedra) at around the melting point, which are to be compared with 96.3(3) and 90.7(3) at room temperature, respectively.

The above-mentioned shape of the silicate tetrahedra, which is expected at the melting point, seems to suggest that the interactions between Si and the

TABLE 6. Structural data of forsterite at around the melting temperature.

Extrapolated atomic coordinates

	Mg(1)	Mg(2)	Si	O(1)	O(2)	O(3)
x	0	0.9969	0.4275	0.7626	0.2175	0.2870
y	0	0.2830	0.0943	0.0908	0.4545	0.1631
z	0	1/4	1/4	1/4	1/4	0.0410

Interatomic distances

	Si tetrahedron		Mg(1) octahedron		Mg(2) octahedron	
Bond lengths (Å)						
	Si–O(1)	1.63	Mg(1)–O(1)	2.14	Mg(2)–O(1)	2.32
	–O(2)	1.63	–O(2)	2.11	–O(2)	2.10
	–O(3)	1.62	–O(3)	2.23	–O(3)	2.13
					–O(3)	2.29
	Average	1.63 [−0.2%]	Average	2.16 [3.2%]	Average	2.21 [3.8%]
Average edge distances (Å)	2.65 [−0.3%]		3.04 [3.1%]		3.10 [3.6%]	
Polyhedral Volume (Å^3)	2.18 [−1.2%]		12.70 [8.1%]		13.64 [10.1%]	

Numbers in brackets indicate expansion rate relative to room-temperature values.

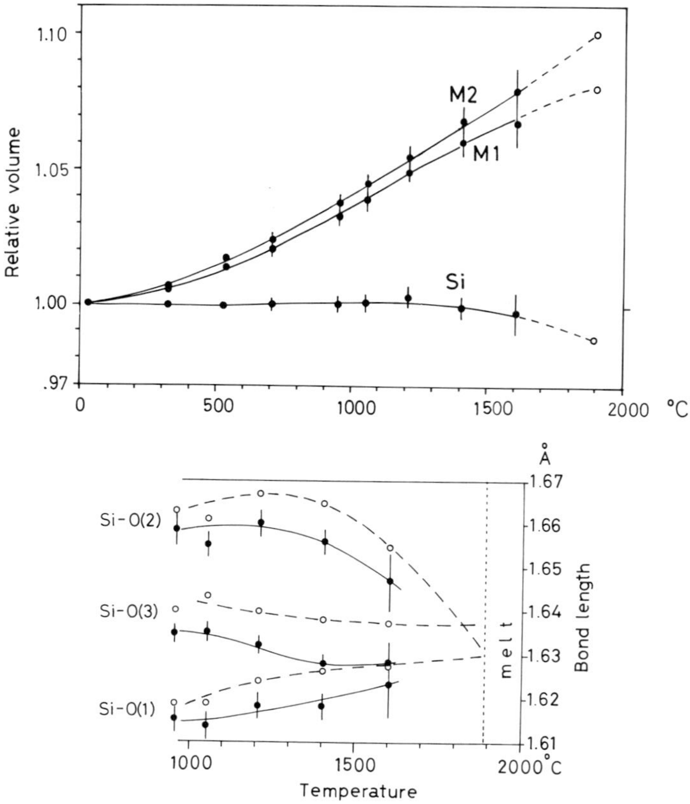

FIG. 10. Top: Relative polyhedral volumes of forsterite versus temperature. Open circles indicate extrapolated values. Bottom: A plot of Si–O bond lengths corrected for rigid-body thermal motion (open circles) versus temperature. Uncorrected lengths are indicated by solid circles.

surrounding Mg atoms are still significant. Since, however, the Mg–O bondings are evidently weakened, particularly at temperatures higher than around 1,500°C, the silicate tetrahedra may be treated approximately as rigid-bodies, as mentioned before. Then a trial was made of correcting the Si–O bond lengths for rigid-body thermal motions (CRUICKSHANK, 1956; BUSING and LEVY, 1964) by using a computer program provided by SAKURAI *et al.* (1980). As an example, the bond lengths at 1,600°C are compared below with corrected bond lengths

	Si–O(1)	Si–O(2)	Si–O(3)
Uncorrected	1.623(7) Å	1.647(7) Å	1.628(5) Å
Corrected	1.628	1.655	1.637.

The corrected bond lengths at temperatures higher than around 1,000°C are plotted in Fig. 10 (bottom) versus temperature. If these data are extrapolated to the melting point, they nearly give the same length as shown in this figure. Although the accuracy is quite unknown in the Si–O bond lengths thus obtained for the melting temperature, we may at least conclude that the Si–O bond lengths at around the melting point would be nearly the same and have values of around 1.63 ~ 1.64 Å.

In general, melting is classified as a nucleation and growth phenomenon (CRISTIAN, 1965), resulting from the formation and movement of the solid-liquid interface. The movement is interpreted in terms of the spontaneous generation of dislocations (COUCHMAN and JESSER, 1977; COTTERILL, 1975). In the case of the breakdown of forsterite to a melt, the formation (nucleation) of such a solid-liquid interface will immediately follow the instance when the 'equalization' of the Si–O bond lengths takes place.

2.5 *Anomalous expansion of tetrahedra in* Ca_2GeO_4 *(γ form)*

As a final remark on the thermal expansion of olivine-type structures, we present here the case of Ca_2GeO_4 which undergoes the phase transition to a hexagonal form (α form) at 1,450°C (BOIKOVA and DOMANSKY, 1974). Unlike the case of forsterite (and also Mg_2GeO_4), the thermal expansion of the unit cell of this material is characterized by a great expansion rate of the *c* edge. The thermal expansion coefficient α_c in the temperature range from 25°C to 500°C has a value of about $1.9 \times 10^{-5}C^{-1}$ which is about two times as large as that of forsterite in the same temperature range. The value α_a is roughly $1.0 \times 10^{-5}C^{-1}$, and that of α_b $0.3 \times 10^{-5}C^{-1}$.

One of the striking features of this material is that the germanate tetrahedra start expanding from around 700°C, although up to that temperature the volume retains nearly the same as the room temperature volume (Fig. 11). The expansion from around 700°C is such that the volume at the transition temperature would become about 9% larger than the room temperature volume if the expansion data are extrapolated. Such an expansion of tetrahedra appears to be explainable in terms of the expansion of Ca octahedra in the following fashion. As observed in Table 7, in the temperature range 25 ~ 500°C, the unshared edge, O(1)–O(2), which is nearly parallel to *c*, expands with an approximate rate of $1.9 \times 10^{-5}C^{-1}$ which is about two times as large as that of forsterite. Such a conspicuous lengthening of the unshared edge would primarily be due to the relatively large thermal vibration of Ca(1) atoms ($B_{eq} = 2.1$ Å^2 at 300°C, and 2.6 Å^2 at 600°C). Note that the edge, O(1)–O(2), shared by two Ca(1) octahedra even significantly shows a negative expansion.

To match with the expansion of the Ca(1) octahedra in the direction of the *c* axis, the unshared edge, $O(3)^t$–$O(3)^q$ of the Ca(2) octahedra (Fig. 12), which is parallel to *c*, exhibits a marked expansion with an approximate rate of 2.8

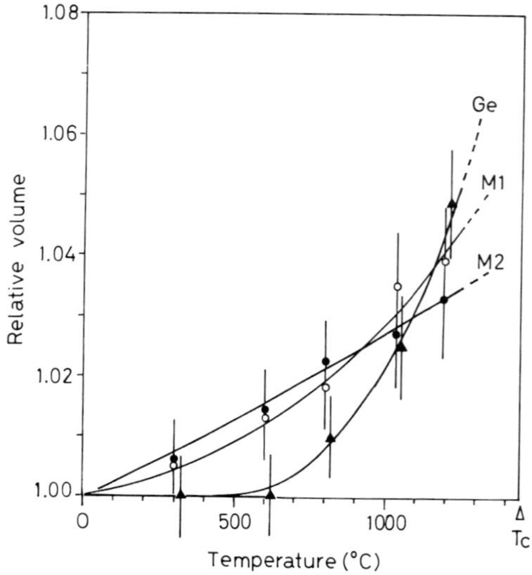

FIG. 11. Relative polyhedral volume of Ca_2GeO_4 versus temperature. Triangle: germanate tetrahedra, open circle: M1 octahedra and solid circle: M2 octahedra. The entries of the germanate tetrahedra are slightly shifted along the temperature axis.

$\times 10^{-5}C^{-1}$ (Fig. 13). Because of such a conspicuous exapnsion which has selectively taken place in the specific edges, the Ca(2) octahedra, with a raising of temperature, become highly distoreted and would reach an extreme limit at around 700°C, where the germanate tetrahedra start expanding. At that temperature, the mean Ca(1)–O bond length has a value of 2.376(6) Å, and mean Ca(2)–O 2.418 Å, suggesting that these are still strong bondings between Ca and O. Then, with raising the temperature further, the expansion of the octahedral framework (consisting of Ca(1) and Ca(2) octahedra) would force the tetrahedra to expand because Ca(2) octahedra, as mentioned above, would no longer be distorted by lengthening their O(3)–O(3) edges. When the volume expansion of the tetrahedra, in turn, reach a critical limit at the transition temperature, they would abruptly shrink to take a smaller volume proper to the tetrahedra, accompanying a rearrangement of Ca atoms. The mechanism of phase transition in Ca_2GeO_4 thus speculated would probably be the case. Mean lengths of the tetrahedral edges shared with octahedra are plotted in Fig. 13 versus temperature.

3. *Spinel*

3.1 *Geometrical characteristics of the spinel structure type*

The spinel structure type consists of an anion X which forms a cubic closest-packed array. One-eighth of the tetrahedral voids are basically

TABLE 7. Approximate thermal expansion coefficients of edges and volume of octahedra in Ca_2GeO_4 in the temperature range between 25°C and 500°C as compared with those of Mg_2SiO_4.

Edges and volume			Expansion coefficient	
			Ca_2GeO_4	Mg_2SiO_4
M(1) octahedron				
$O(1)–O(2)^t$		×2	$1.9 \times 10^{-5}C^{-1}$	$1.0 \times 10^{-5}C^{-1}$
$O(1)–O(3)^s$		×2	1.4	1.4
$O(2)^t–O(3)$		×2	1.0	1.5
$O(1)–O(2)^p$	*	×2	−1.0	0.9
O(1)–O(3)	*	×2	1.6	1.6
O(2)–O(3)	**	×2	0.2	0.2
Volume			2.5	3.1
M(2) octahedron				
$O(1)–O(3)^q$		×2	0.8	1.3
$O(2)–O(3)^q$		×2	1.6	1.0
O(2)–O(3)		×2	0.2	1.6
$O(3)–O(3)^q$		×2	−0.1	1.1
$O(3)^q–O(3)^t$			2.8	1.6
O(1)–O(3)	*	×2	1.6	1.6
$O(3)–O(3)^r$	**		0.7	0
Volume			2.9	3.4

Each edge shared with octahedron is indicated by a single star, while that shared with tetrahedron by a double stars. For symmetry code, refer to Appendix II.

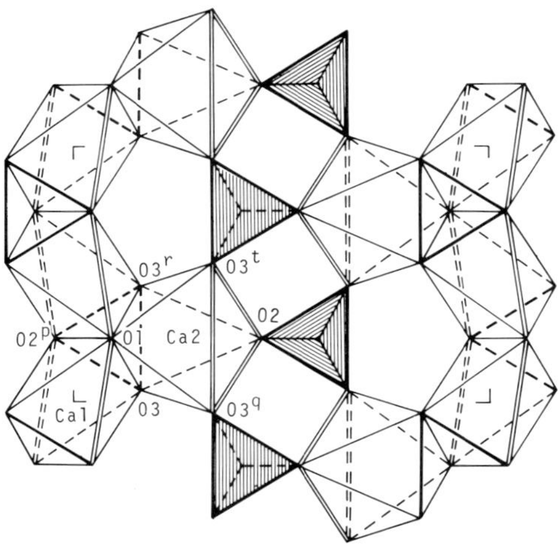

FIG. 12. The crystal structure of Ca_2GeO_4 bounded approximately by $x = -1/4$ and $x = 1/4$. The cell corners are indicated; the origin is taken at the lower left corner, the c axis vertical and b horizontal. Double lines (and double broken lines) indicate unshared octahedral edges which show great expansion rates. Heavy lines indicate the octahedral edges shared with tetrahedra.

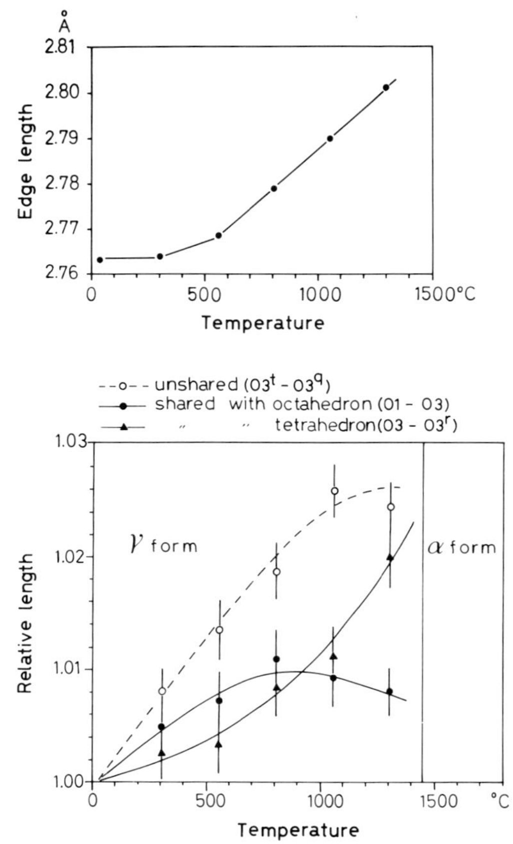

FIG. 13. Top: Mean length of the tetrahedral edges shared with octahedra of Ca_2GeO_4 versus temperature. Bottom: A plot of three selected edge lengths (relative to room temperature value) of the Ca(2) octahedra versus temperature.

occupied by a cation A, while one-half of the octahedral voids by a cation B, thus giving a basic chemical formula AB_2X_4. The equivalent point positions and point symmetries of these atoms in the space group *Fd*3*m* are: the A atom: 8(a), $\bar{4}3m$; B atom: 16(d), $\bar{3}m$; X atom: 32(e), 3*m*.

The crystal structure can be looked upon as consisting of two kinds of layers perpendicular to [111], one being composed of edge sharing octahedra, and the other composed of both octahedra and tetrahedra, by sharing corners. There are three pairs of the two kinds of layers in the cubic unit cell. The A cation at the conventional origin of *Fd*3*m* in the first layer is at the level of 1/8 $(=3/(4 \times 6))$ from the bottom (Fig. 14). Accordingly, the oxygen atom at the top level of the third layer (Fig. 14) is at the height of $3/6 - 1/8 = 0.375$ from the origin level. Among anions in this level, one located on the cubic body

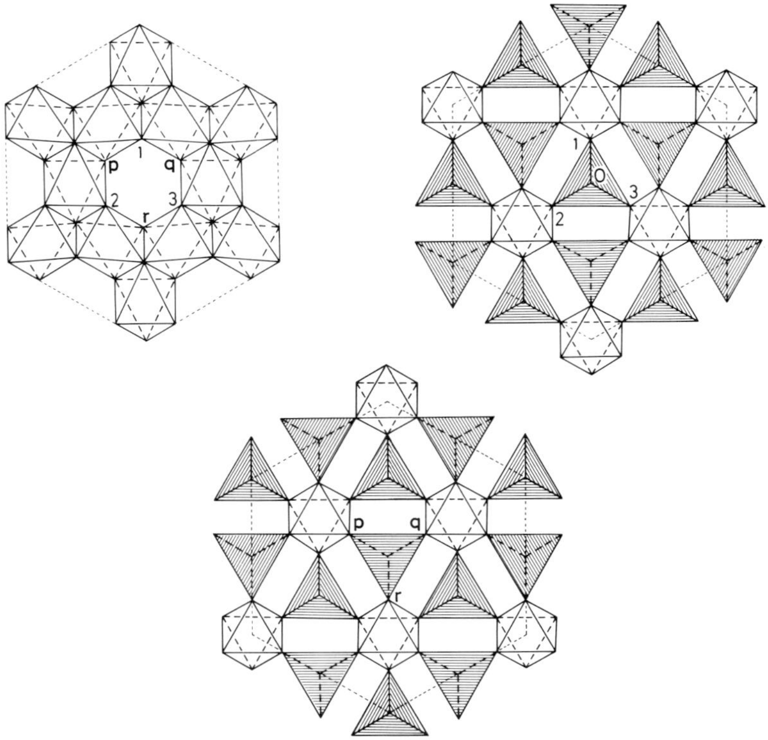

FIG. 14. The crystal structure of spinel viewed down the [111] direction, showing three polyhedral layers (among six layers in the cubic unit cell): the bottom layer with the origin at the center of the central tetrahedron (*bottom*), the second layer (*left in the top*), and the third layer (*right in the top*). They are combined with each other in such a way that the corners of the polyhedra indicated by letters or numbers are joined together with those indicated by the same letters or numbers. Dotted lines trace the cubic cell edges (the cell edges which fall into the figures are not shown).

diagonal is at an apex of a tetrahedron in the same layer. The detailed value of the coordinates, u, u, u, of the anion at this specific position is commonly used to characterize the structure. When the origin of the structure is taken not at the standard position of $Fd3m$ but at the center of symmetry at the level 0.125, the oxygen parameter u for an undistorted spinel structure takes a value of 0.250 (=0.375 − 0.125). In the present paper we adopt u referred to as the standard space-group origin.

In that case, the relations between interatomic distances and u for a spinel AB_2X_4, as given by HILL *et al.* (1979) are rearranged as follows:

$$A - X: a\sqrt{3}(u - 1/4)$$

$$B - X: a(3u^2 - 11u/4 + 43/64)^{1/2}$$

Tetrahedral edge: $a\sqrt{2}(2u - 1/2)$

Shared octahedral edge: $a\sqrt{2}(1 - 2u)$

Unshared octahedral edge: $a(4u^2 - 3u + 11/16)^{1/2}$

Here, a represents the cubic cell edge. Noting that the tetrahedron at the center of the third layer (Fig. 14) has a point symmetry $\bar{4}3m$ with the point at the center of the tetrahedral cation at the level of 0.25, we readily observe that with increasing u, the volume of the tetrahedron increases, accompanying a contraction of the octahedral framework if a is kept constant. Or, if the octahedral framework expands, the tetrahedron must necessarily expand likewise. Such a geometrical relation between octahedra and tetrahedra in the spinel-structure type is of substantial importance in studying the expansion or compression of spinel structures. As pointed out by HILL *et al.* (1979), the values of the oxygen parameter u are in the range $0.375 < u < 0.399$.

3.2 The crystal structure of spinel at temperatures up to around 1,700°C

3.2.1 Thermal expansion of cell parameter for spinel

In subsequent paragraphs, we present and discuss the effect of temperature on the spinel structure based on the work carried out by YAMANAKA *et al.* (1979). For their work, single crystals were used which were provided by Dr. S. Takasu and had a maximum dimension of about 3 mm. They were grown by a flux method with the cooling rate of 3.6°C per hour to 900°C. A spherically-ground crystal with a radius of 0.3 mm was prepared and used for X-ray single crystal work. The cell parameter and diffraction intensities were measured with a Syntex $P2_1$ automated four-circle diffractometer, using graphite-monochromatized MoKα radiation ($\lambda = 0.71069$ Å). The length of the cubic cell edge obtained by the least-squares procedure applied to the sin 2θ values of 15 reflections, is compared in Table 8 with some reported lengths. The value of 8.0806 Å obtained by YAMANAKA *et al.* (1979) agrees well with those reported by HENDERSON and TAYLOR (1975), GRIMES and HILLEARD (1970), and by SUZUKI and KUMAZAWA (1980). The considerably larger value for natural spinels are probably due to the fact that they commonly contain cations like Fe and Zn. DATTA and ROY (1968) obtained a small value of 8.074 ± 0.002 Å on hydrothermal samples.

From single-crystal diffractometry, cell parameters at various temperatures were determined with the results as shown in Fig. 15; for temperatures below 1,000°C, the crystal was heated with a miniature electric furnace in the open air, while for those above that temperature, by gas flame (see p. 621). As observed in this diagram, the variation of cell parameter exhibits a discontinuity at around a temperature between 600°C and 700°C in conformity with the anomalous thermal expansion reported by SUZUKI and KUMAZAWA (1980). Among cell parameters, those for the temperatures at which col-

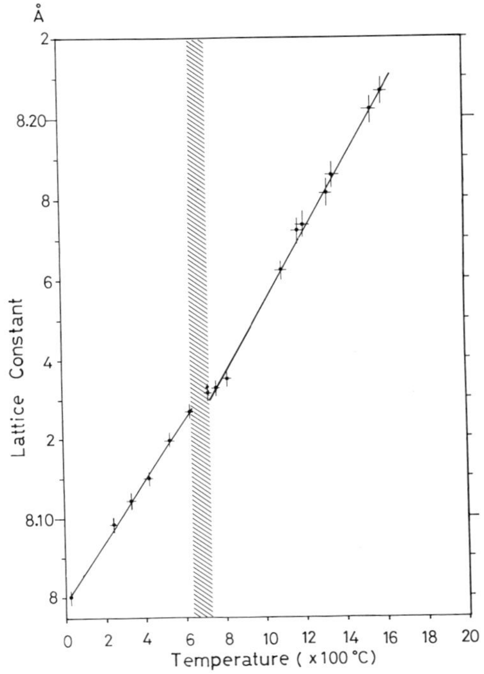

FIG. 15. A plot of cell parameter of spinel versus temperature.

lections of diffraction intensities were made are listed in Table 9.

3.2.2 The oxygen u parameter of spinel at room temperature

With the $\omega - 2\theta$ scan mode, the graphite-monochromatized MoKα diffraction intensities were measured on the four-circle diffractometer. The intensities were corrected for Lorentz and polarization effects, transmission factors, and isotropic-extinction effects (COPPENS and HAMILTON, 1970). The atomic parameters were refined with the least-squares program LINUS (COPPENS and HAMILTON, 1970). The final values of R (and weighted R) and u parameter obtained are listed in Table 9, together with the cell parameter, maximum 2θ values of reflections covered by intensity measurements, and the number of reflections used for refinement.

The oxygen u parameter at room temperature is identical with the one obtained by FISCHER (1967) although our cell parameter is slightly smaller than his value (Table 8).

3.2.3 Thermal response of u parameter

From each of the sets of diffraction intensities collected at high temperatures, the oxygen u parameter (Table 10) was obtained in a way which is similar to the above procedure for the room-temperature case. In Fig. 16, which shows the plots of u parameters versus temperature, we observe a

TABLE 8. Comparison of reported cell parameters for spinel

Preparation or source of crystal	Cell parameter	T(°C)	Reference
Flux growth Very slow cooling	8.0806(5) Å	20	YAMANAKA *et al.* (1979)
Flux growth	8.0804(12)	20	HENDERSON and TAYLOR (1975)
Solution growth Sintered at 1,550°C and slow cooling	8.080		GRIMES and HILLEARD (1970)
Synthetic Cryst. Tech. Inc U.S.A.	8.081(1)		SUZUKI and KUMAZAWA (1980)
Synthetic	8.0832(5)		GUBSER and SCHEEL in FISCHER (1967)
Synthetic	8.0832	26	SINGH *et al.* (1975)
Melt Quenched from 1,300°C	8.084		DATTA and ROY (1968)
Synthetic	8.086		ROMEIJIN in HAFNER (1960)
Natural Antsiriki, Madagascar	8.0877	22	SINGH *et al.* (1975)
Natural	8.089(1)		HAFNER and LAVES (1961)

striking feature that the value of u abruptly drops at a temperature between 600 and 700°C in accordance with the discontinuity observed in thermal expansion of the cell parameter (Fig. 15). This abrupt change in u appears to support the existence of the second-order transition in spinel with a lambda temperature located at 660 ± 10°C, as evidenced by SUZUKI and KUMAZAWA (1980) in their study of the thermal expansion of spinel by means of dilatometry. Both the octahedral bond length, B–O, and the tetrahedral bond length, A–O, then exhibit anomalous expansions as shown in Fig. 16 (also see Table 10). The rates of expansion, α(A–O) and α(B–O), of these two kinds of bonds after the transition are markedly different from those before the transition. The approximate expansion rates are

	20°C – 420°C	720°C – 1,660°C
α(A–O)	$1.4 \times 10^{-5}C^{-1}$	$0.2 \times 10^{-5}C^{-1}$
α(B–O)	0.8	1.4

Before the transition, the value of α(A–O) is similar to an average of the expansion rate of Mg–O for forsterite, and that of α(B–O) is close to the expansion rates of the Al–O bonds in chrysoberyl. The situation that the

TABLE 9. Unit cell dimension and atomic parameters for spinel.

T(°C)	20	190	420	600	720	1,230	1,390	1,660
a(Å)	8.0806(5)	8.0948(7)	8.1136(5)	8.1251(11)	8.1311(9)	8.1740(13)	8.1860(10)	8.2065(12)
u	0.3873(1)	0.3872(2)	0.3876(1)	0.3875(1)	0.3867(2)	0.3862(2)	0.3860(1)	0.3857(2)
B_{eq}(Å^2)								
A	0.563	1.159	0.829	1.139	1.717	1.625	1.822	2.115
B	0.614	1.179	0.854	1.112	1.755	1.678	1.856	2.181
O	0.848	1.357	1.035	1.310	1.987	2.083	2.312	2.621
Anisotropic temp. factor coef.								
A β_{11}	230(4)	422(12)	315(8)	431(8)	649(14)	618(14)	680(8)	786(10)
B β_{11}	248(3)	450(7)	324(8)	421(5)	664(7)	628(10)	693(5)	810(6)
β_{12}	−7(7)	−17(24)	−15(14)	−29(22)	−29(22)	−52(18)	−61(14)	−72(18)
O β_{11}	333(4)	518(12)	393(7)	496(8)	751(13)	779(14)	863(8)	974(12)
β_{12}	−5(4)	−11(13)	−6(8)	−20(9)	−17(13)	−32(12)	−25(9)	−27(12)
R (%)	1.97	2.48	2.77	2.96	3.65	4.13	4.32	4.31
wt. R (%)	1.91	3.23	3.11	3.02	3.68	4.67	4.11	4.14
Number of reflexions used	145	65	88	87	82	91	130	89
$2\theta_{max}$ (°)	100	80	80	80	80	80	100	80

Anisotropic temperature factor coefficients are × 10^5.

TABLE 10. Interatomic distances of spinel at high temperatures up to 1,660°C.

T (°C)	20	190	420	600	720	1,230	1,390	1,660
Tetrahedron								
A–O (Å)	1.9221(8)	1.9267(9)	1.9327(9)	1.935(1)	1.926(2)	1.928(2)	1.928(1)	1.929(1)
O–O	3.1388(8)	3.1463(13)	3.1566(13)	3.160(2)	3.144(2)	3.154(2)	3.149(2)	3.150(2)
Octahedron								
B–O	1.9257(8)	1.9284(9)	1.9319(9)	1.935(1)	1.942(2)	1.956(2)	1.961(1)	1.968(1)
O–O [*s*]	2.5751(11)	2.5776(13)	2.5806(13)	2.585(2)	2.605(2)	2.626(2)	2.640(2)	2.653(2)
O–O [*u*]	2.8639(11)	2.8690(13)	2.8758(13)	2.880(2)	2.881(2)	2.896(2)	2.900(2)	2.907(2)
O–B–O	83.92°	83.88°	83.83°	83.83°	84.24°	84.52°	84.62°	84.78°
B–O to A–O ratio	1.0019	1.0009	0.9996	1.0000	1.0087	1.0120	1.0168	1.0202

The octahedral angle listed is one subtended to the shared edge. Octahedral shared and unshared edges are indicated by *s* and *u* in parentheses, respectively.

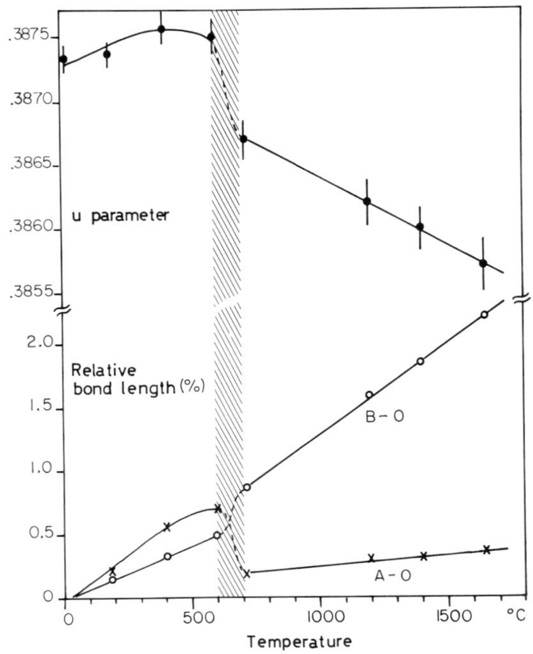

FIG. 16. The oxygen *u* parameter and per cent expansion of A–O and B–O bonds of spinel versus temperature.

expansion rate of the tetrahedral bond is significantly larger than that of the octahedral bond is, however, reversed after the transition.

Thus the anomalous change in bond length that takes place at the transition temperature, is characterized by an abrupt shrinkage of the tetrahedral bonds and a lengthening of octahedral bonds. Such a mode of change in bond length itself well assumes a feature that the interchange of tetrahedral and octahedral cations is taken place at the temperature (TAKÉUCHI *et al.*, 1981). If this is the case, the transition temperature of 660°C is significantly lower than the reported temperature of 750° ~ 850°C at which cation disorder in natural spinel starts (SCHMOCKER and WALDNER, 1976). This difference, however, would be ascribed to the fact that natural spinel is well ordered while synthetic spinel tends to have invert type of cation arrangement to some extent. In any case, the mechanism of transition at 660°C for synthetic spinel should be reserved for future study.

3.2.4 Structure of spinel as a function of temperature

Instead of speculating the physical significance of the transition at 660°C, further details in the mode of structural change involved in the transition are discussed in subsequent paragraphs. Since, as mentioned previously, there are simple geometrical relations between the oxygen u parameter, polyhedral edge lengths and cell parameter of spinel, we can predict the modes of variation in u for various assumed modes of polyhedral expansions. Some simple modes of polyhedral exapnsion are: (1) Expansion with the length of shared edge being kept constant, (2) both unshared and shared edges expand with an equal rate, (3) expansion with the length of the unshared edge being kept constant, and (4) expansion with the length of tetrahedral edge being kept constant. For each of these cases, the mode of variation in the oxygen u parameter is illustrated in a u' versus a' diagram (Fig. 17), here u' denotes u parameter relative to a room temperature value of 0.3873, and a' the cubic cell edge relative to a room temperature value of 8.0806 Å. It is notable that u' (or u) sensibly decreases if shared edges of the octahedra expand with the length of unshared edges being kept unchanged.

Now, observed variation in u parameter of spinel, which is given in the same diagram (Fig. 17), shows an abrupt drop at around $a' = 1.006$ (corresponding to the value at the transition temperature), suggesting that a

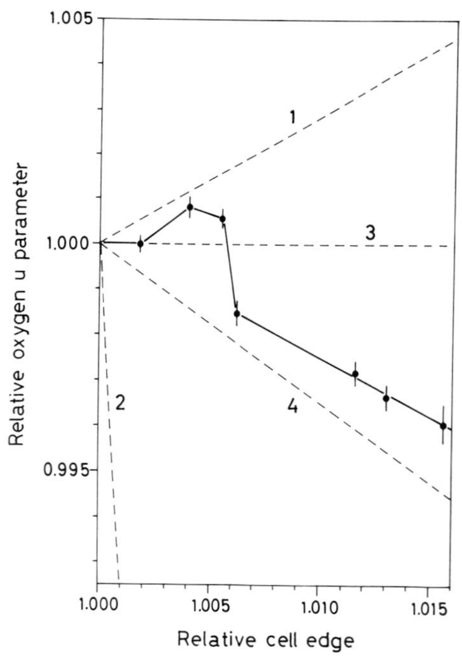

FIG. 17. Relative u parameter versus relative cell edge of spinel, showing the modes of change in u for several assumed modes of polyhedral expansions: 1, the length of shared edges are kept constant; 2, the length of unshared edge are kept constant; 3, both unshared and shared edges expand with the same rate, and 4, tetrahedral edge lengths are kept constant.
The heavy lines indicate the observed variation of u.

sudden expansion of unshared edges takes place at the transition. We find that this is indeed the case when the variations, with raising temperature, of shared- and unshared-edge lengths are plotted versus temperature (Fig. 18). Thus, the lambda transition observed by SUZUKI and KUMAZAWA (1980) may structurally be accounted for in terms of abrupt displacements of oxygen atoms which have taken place in such a way that the shared edges are suddenly lengthened. Since, in general, shared edges tend to resist expansion, their expansion, if it occurs, with a raising of temperature, often shows peculiar features, as demonstrated previously in the case of Ca_2GeO_4. The sudden expansion of the shared edges in spinel may be due to a delicate balance of the bondings between cations and oxygen atoms involved in the formation of the shared edges. Or, such a change might perhaps be indicative of the existence of a structure in spinel having a space group lower than *Fd*3*m* and related to the transition between the low-symmetry-structure and the *Fd*3*m* structure. In either case, however, it is difficult to explain the shrinkage of tetrahedra that takes place at the transition.

The diagram given in Fig. 17 shows that before transition, the expansion of the unshared edges is dominant, while after the transition, the polyhedral expansion is such that it keeps the length of the tetrahedral edges nearly constant. The latter evidence may suggest a gradual increase of Al in the tetrahedral site (or a gradual increase of the inversion parameter).

Difference Fourier syntheses based on sets of structure factors for 1,230°C, 1,390°C, and 1,660°C exhibit salient features. As observed in Fig. 19 (ii–iv), residual peaks like the one observed at room temperature, are no

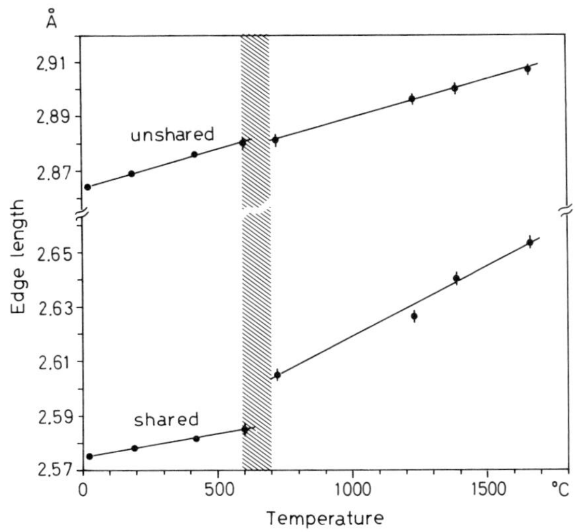

FIG. 18. A plot of shared and unshared edge lengths of spinel versus temperature.

longer existent but well-defined four residual peaks occur around the tetrahedral site and in the opposite direction of each tetrahedral bond. The distances between the peaks and the center of the tetrahedron are 0.45 Å (at 1,230°C), 0.47 Å (at 1,390°C), and 0.51Å (at 1,660°C); the peak height significantly increases with a raising of temperature. The occurrence of these residual peaks may be interpreted in two alternative ways: (1) each tetrahedral cation vibrates across the A site in an asymmetric manner which arises from

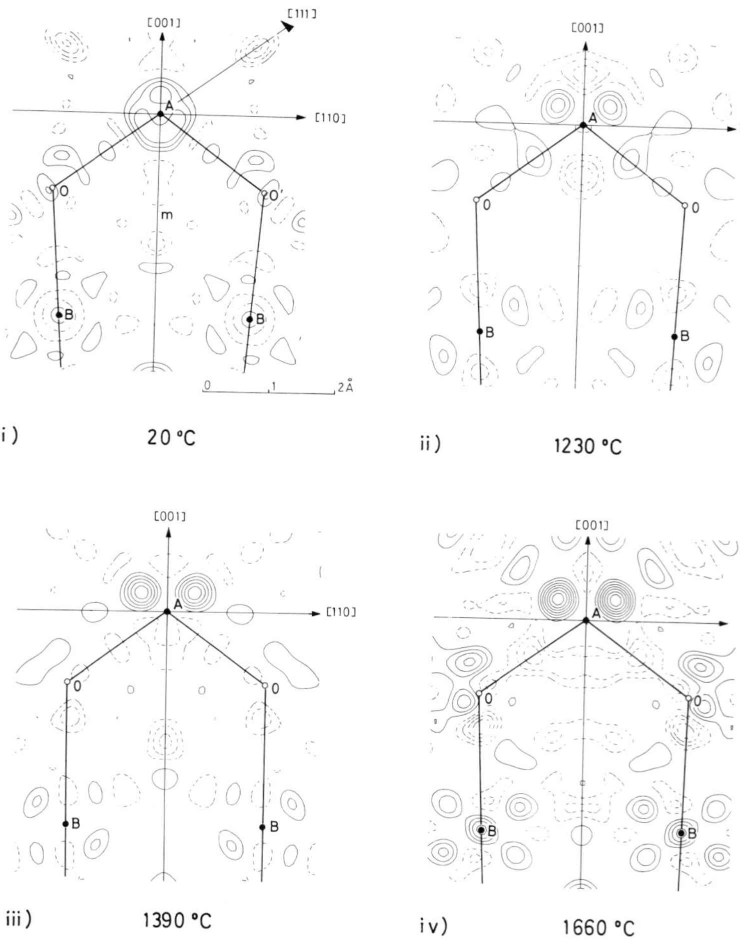

FIG. 19. Difference Fourier synthesis of spinel based on structure factors at room temperature (i), 1,230°C (ii), 1,390°C (iii) and 1,660°C (iv), each showing the $(1\bar{1}0)$ section. Contours are at the intervals of 0.2 $e\text{Å}^{-3}$; zero contours being omitted and negative contours broken. A threefold axis and mirror plane, m, are indicated in (i).

anharmonic contributions to the Debye-Waller factor, or (2) the displacement of the cation from the geometrical center of the tetrahedron. Since the radius of Al, in particular, is small for the size of the tetrahedron in spinel, they may be displaced from its center when they partially substitute for Mg at high temperatures. Although conclusive interpretation for the residual peaks should be reserved for further study, the temperature dependence of the peaks appears to favor the anharmonic interpretation. Such a displacement effect of atoms has been reported for tetrahedrally coordinated fluorine atoms in fluorite (WILLIS, 1965).

3.3 *Effect of temperature on the spinel form of* Fe_2SiO_4

We present here a typical case in which thermal expansion of a spinel-type structure is dominantly controlled by that of the octahedral framework. The spinel-type structure, now consider, is the spinel form of Fe_2SiO_4. The discussions given below are based on the structural data provided by YAMANAKA *et al.* (1981).

The crystals used were those synthesized by Prof. Akimoto in a tetrahedral anvil high pressure apparatus; the stable form of this material at ordinary pressure is of the olivine form (i.e., fayalite). The collection of diffraction intensities and structure refinements were made with the procedures which are similar to those in the case of spinel, except the crystal sample in this case was sealed with a silica tube (it took about three hours to collect a set of diffraction intensity data at each temperature). The crystallographic data obtained at several temperatures up to 692(5)°C are listed in Table 11. Those at room condition are also reported by YAGI *et al.* (1974), MARUMO *et al.* (1977), and FINGER *et al.* (1979).

The octahedral bond length Fe–O of 2.138 Å at room temperature is smaller than the average Fe–O of 2.168 Å for fayalite (SMYTH, 1975), While, the Si–O bond length of 1.653 Å for the spinel form is significantly longer than the average Si–O of 1.628 Å for fayalite (SMYTH, 1975). With a raising of temperature, the octahedra about Fe expands in such a mode that the expansion rate of the shared edges is nearly the same as that of the unshared edges (the expansion rate of Fe–O is $9.0 \times 10^{-6}C^{-1}$ for 20°C ~ 383°C). Consequently, as mentioned previously, the oxygen u parameter remains almost the same in the temperature range extended over 692°C (Fig. 20). Because of such an expansion mode of octahedra, the silicate tetrahedra are necessarily expanded (Fig. 20) with the expansion rate of Si–O of 8.3 $\times 10^{-6}C^{-1}$ for 23°C ~ 383°C. Such a thermal response of the spinel form of Fe_2SiO_4 then suggests that the transformation (by heating) from the spinel form to fayalite would be executed by an abrupt shrinkage of the silicate tetrahedra (after expansion to a certain limit) accompanying a rearrangement of Fe atoms as in the case of the transition in Ca_2GeO_4, which we discussed previously.

TABLE 11. Crystallographic data for silicate spinel, Fe_2SiO_4.

T(°C)		20	191(5)	385(5)	585(5)	692(5)
Number of reflexions used		100	87	83	86	76
R%		2.06	1.60	2.92	2.24	4.22
a(Å)		8.237(1)	8.251(1)	8.264(1)	8.311(1)	8.340(1)
u		0.3658(2)	0.3658(2)	0.3658(2)	0.3657(3)	0.3654(6)
Anisotropic temp. factor coef. $\times 10^5$						
Si	β_{11}	118(8)	354(7)	476(11)	654(16)	760(36)
Fe	β_{11}	121(3)	402(3)	529(8)	687(6)	712(24)
	β_{12}	−5(3)	−13(4)	−6(5)	−19(9)	−68(22)
O	β_{11}	137(9)	472(7)	558(13)	763(18)	854(59)
	β_{12}	−5(12)	−10(10)	−11(17)	−87(30)	−102(55)
Interatomic distances						
Tetrahedron						
Si–O		1.653(1)	1.655(1)	1.658(2)	1.665(2)	1.668(5)
O–O		2.699(2)	2.702(2)	2.707(2)	2.719(3)	2.723(7)
Volume (Å^3)		2.32	2.32	2.34	2.37	2.38
Octahedron						
Fe–O		2.138(1)	2.142(1)	2.145(2)	2.158(2)	2.168(5)
O–O [s]		3.126(2)	3.133(2)	3.137(2)	3.158(3)	3.174(7)
O–O [u]		2.916(2)	2.921(2)	2.926 (3)	2.943(3)	2.953(7)
Volume (Å^3)		12.92	12.99	13.05	13.30	13.47

Up to 585°C, the same crystal, having the shape of a 0.2 mm cube, was used; total duration of heating was about 9 hrs. A new crystal was used for the collection of diffraction intensities at 692°C; heating duration was about 3 hrs. [s] and [u] denote shared and unshared edges, respectively.

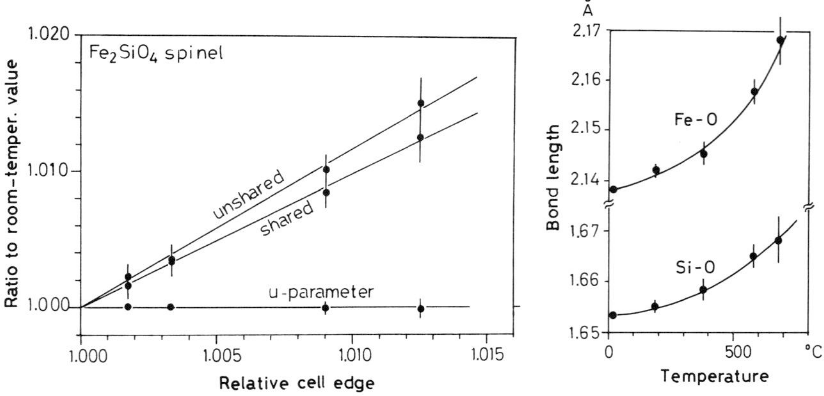

FIG. 20. *Left*: A plot of relative lengths of octahedral edges and relative u parameter for Fe_2SiO_4 spinel versus the relative cell edge. *Right*: A plot of the bond lengths of Fe_2SiO_4 spinel versus temperature.

4. *Concluding Remarks*

The crystals we have discussed in this paper fall in the category of *heterodesmic crystals* (MEGAW, 1939), in which some or all of the atoms are bound to certain of their neighbors by forces of a stronger kind than those which link them to the others. In order to fully interpret thermal expansion of a bond, say a strong bond, in these crystals, the present study suggests that both the bond strength of the strong bond and the variations (according to temperature) of those of weak bonds should be taken into account on a quantitative basis.

We thank Dr. S. Takasu, Prof. F. Takei and Prof. S. Akimoto for specimens, and Dr. T. Sakurai for his kind help in computation of the correction of bond lengths for thermal motion. We are indebted to Prof. R. Sadanaga for his continued encouragement throughout this work. The first two authors have benefitted from discussions on the spinel structure with Professors H. Jagodzinski, N. W. Grimes and A. Preisinger on the occasion of the XIIth Congress of the International Union of Crystallography held at Ottawa.

APPENDIX I. Atomic coordinates and thermal anisotropic parameters ($\times 10^3$) of forsterite

Atom	x	y	z	$B_{eq.}$	β_{11}	β_{22}	β_{33}	β_{12}	β_{13}	β_{23}
[25°C]										
Mg(1)	0	0	0	0.53	6.3(2)	1.64(5)	2.5(1)	−0.14(9)	−0.4(1)	−0.33(7)
Mg(2)	−0.0084(2)	0.27750(8)	0.25	0.53	7.0(2)	1.23(5)	3.2(1)	0.06(9)	0	0
Si	0.4266(1)	0.09413(6)	0.25	0.37	4.2(2)	1.06(4)	2.1(1)	0.04(6)	0	0
O(1)	0.7664(3)	0.0914(2)	0.25	0.52	4.7(4)	1.69(9)	3.0(2)	0.1(2)	0	0
O(2)	0.2215(3)	0.4472(1)	0.25	0.49	6.4(4)	1.13(8)	3.0(2)	0.1(2)	0	0
O(3)	0.2778(2)	0.1631(1)	0.0332(2)	0.54	6.6(3)	1.54(6)	2.6(2)	0.0(1)	−0.2(2)	0.41(8)
[310°C]										
Mg(1)	0	0	0	0.93	9.8(4)	2.83(9)	5.0(2)	−0.1(1)	−0.9(2)	−0.7(1)
Mg(2)	−0.0081(3)	0.2780(1)	0.25	0.95	12.3(4)	2.05(8)	6.1(2)	0.1(2)	0	0
Si	0.4266(2)	0.09406(8)	0.25	0.67	7.4(3)	1.77(6)	4.2(2)	0.0(1)	0	0
O(1)	0.7653(5)	0.0915(2)	0.25	0.86	7.2(6)	2.7(2)	5.7(4)	0.4(3)	0	0
O(2)	0.2223(5)	0.4476(2)	0.25	0.88	10.9(8)	1.9(1)	6.1(4)	0.5(3)	0	0
O(3)	0.2788(3)	0.1630(2)	0.0336(3)	0.94	10.6(5)	2.6(1)	5.3(3)	0.3(2)	−0.5(3)	0.7(1)
[530°C]										
Mg(1)	0	0	0	1.13	10.6(2)	3.52(6)	6.5(2)	−0.23(8)	−1.2(1)	−0.96(7)
Mg(2)	−0.0077(1)	0.27838(7)	0.25	1.11	13.1(3)	2.36(5)	8.0(2)	0.19(9)	0	0
Si	0.4267(1)	0.09419(5)	0.25	0.71	6.2(2)	1.91(4)	5.4(1)	0.04(6)	0	0
O(1)	0.7647(3)	0.0916(1)	0.25	1.01	6.7(4)	3.3(1)	7.2(3)	0.2(2)	0	0
O(2)	0.2219(3)	0.4481(1)	0.25	0.98	10.9(5)	1.96(9)	7.7(3)	0.0(2)	0	0
O(3)	0.2796(2)	0.16302(9)	0.0341(2)	1.10	11.3(3)	3.02(7)	7.0(2)	0.1(1)	−0.4(2)	1.1(1)
[700°C]										
Mg(1)	0	0	0	1.37	12.8(3)	4.25(7)	7.9(2)	−0.2(1)	−1.5(2)	−1.13(9)
Mg(2)	−0.0075(2)	0.27867(8)	0.25	1.34	15.7(3)	2.80(7)	9.4(2)	0.3(1)	0	0
Si	0.4268(1)	0.09419(6)	0.25	0.86	7.5(2)	2.31(5)	6.4(1)	0.04(7)	0	0
O(1)	0.7642(3)	0.0915(2)	0.25	1.20	8.1(5)	3.8(1)	8.7(3)	0.2(2)	0	0

O(2)	0.2218(3)	0.4485(1)	0.25	1.16	12.7(5)	2.2(1)	9.4(3)	0.0(2)	0	0
O(3)	0.2799(2)	0.1631(1)	0.0346(2)	1.32	13.2(4)	3.56(9)	8.6(2)	0.2(1)	−0.4(2)	1.4(1)
[950°C]										
Mg(1)	0	0	0	2.00	19.5(5)	5.5(1)	12.9(3)	−0.3(2)	−1.9(3)	−1.5(2)
Mg(2)	−0.0074(3)	0.2791(1)	0.25	1.97	22.0(5)	3.8(1)	15.6(3)	0.5(2)	0	0
Si	0.4268(2)	0.09419(9)	0.25	1.33	11.8(3)	3.17(7)	10.6(2)	0.2(1)	0	0
O(1)	0.7640(5)	0.0914(3)	0.25	1.84	14.1(8)	4.7(2)	15.1(6)	0.5(4)	0	0
O(2)	0.2216(6)	0.4490(2)	0.25	1.73	19.5(9)	2.9(2)	14.7(6)	−0.1(3)	0	0
O(3)	0.2813(4)	0.1630(2)	0.0355(3)	1.89	18.6(6)	4.7(1)	13.3(4)	0.2(3)	−0.4(5)	1.7(2)
[1050°C]										
Mg(1)	0	0	0	2.29	22.2(6)	6.2(1)	14.9(4)	−0.7(2)	−2.3(4)	−1.9(2)
Mg(2)	−0.0070(3)	0.2795(2)	0.25	2.25	26.0(6)	4.2(1)	17.6(4)	0.5(2)	0	0
Si	0.4270(2)	0.0944(1)	0.25	1.47	13.1(3)	3.42(8)	11.8(2)	0.2(1)	0	0
O(1)	0.7633(6)	0.0912(3)	0.25	2.07	14.6(9)	5.3(3)	17.7(8)	0.8(4)	0	0
O(2)	0.2208(6)	0.4497(3)	0.25	2.00	22.(1)	3.2(2)	17.5(7)	0.1(4)	0	0
O(3)	0.2820(4)	0.1631(2)	0.0357(4)	2.19	21.7(8)	5.3(2)	15.9(5)	0.2(3)	0.0(6)	2.2(3)
[1210°C]										
Mg(1)	0	0	0	2.64	25.0(5)	7.3(1)	16.7(4)	−0.6(2)	−2.9(4)	−2.3(2)
Mg(2)	−0.0068(3)	0.2799(1)	0.25	2.61	29.7(6)	4.6(1)	21.0(4)	0.6(2)	0	0
Si	0.4268(2)	0.0946(1)	0.25	1.68	15.0(3)	3.82(7)	13.5(2)	0.1(1)	0	0
O(1)	0.7631(6)	0.0911(3)	0.25	2.40	17.3(9)	6.1(3)	20.0(8)	0.3(4)	0	0
O(2)	0.2204(6)	0.4499(3)	0.25	2.24	24(1)	3.6(2)	20.0(8)	0.0(4)	0	0
O(3)	0.2829(4)	0.1629(2)	0.0365(4)	2.52	24.8(8)	6.0(2)	18.1(5)	0.7(3)	0.0(6)	2.6(3)
[1400°C]										
Mg(1)	0	0	0	3.21	30.1(6)	8.9(2)	19.9(4)	−0.9(3)	−3.3(5)	−2.4(2)
Mg(2)	−0.0063(4)	0.2803(2)	0.25	3.15	36.1(7)	5.5(1)	24.7(5)	0.4(3)	0	0
Si	0.4269(2)	0.0947(1)	0.25	2.03	17.5(4)	4.76(9)	16.0(3)	0.1(2)	0	0

(continued)

APPENDIX I. (continued)

Atom	x	y	z	$B_{eq.}$	β_{11}	β_{22}	β_{33}	β_{12}	β_{13}	β_{23}
O(1)	0.7625(6)	0.0911(3)	0.25	2.89	22(1)	7.3(3)	23.2(9)	0.3(5)	0	0
O(2)	0.2195(7)	0.4510(3)	0.25	2.72	32(1)	4.2(2)	22.9(9)	0.0(5)	0	0
O(3)	0.2845(5)	0.1630(2)	0.0380(4)	3.02	30(1)	7.1(2)	21.1(6)	0.8(4)	−0.6(7)	3.1(3)
[1600°C]										
Mg(1)	0	0	0	3.53	36(2)	11.6(4)	14.5(9)	−0.8(7)	−2(1)	−2.9(5)
Mg(2)	−0.0054(8)	0.2814(4)	0.25	3.39	44(2)	7.6(4)	18(1)	1.1(7)	0	0
Si	0.4274(5)	0.0945(3)	0.25	2.16	18(1)	5.8(2)	15.3(7)	−0.1(4)	0	0
O(1)	0.763(1)	0.0910(7)	0.25	3.00	29(3)	9.7(9)	13(2)	0(1)	0	0
O(2)	0.219(1)	0.4523(7)	0.25	2.78	29(3)	6.3(7)	19(2)	0(1)	0	0
O(3)	0.285(1)	0.1631(5)	0.0392(8)	3.17	34(2)	9.3(6)	15(1)	0.1(9)	0(2)	3.8(7)

APPENDIX II. Symmetry code for the olivine type.

p: $(1/2) - x, (1/2) + y, z$
q: $(1/2) + x, (1/2) - y, (1/2) + z$
r: $x, y, (1/2) - z$;
s: $-x, -y, -z$
t: $(1/2) + x, (1/2) - y, -z$

REFERENCES

BACON. G. E. (1952): A neutron-diffraction study of magnesium aluminium oxide, *Acta Crystallogr.*, **5**, 684–686.

BARTH, T. F. W. and POSNJAK, E. (1932): Spinel structures with and without variate atom equipoints, *Z. Kristallgr*, **82**, 325–341.

BAUMGARTNER, O., PREISINGER, A., HEGER. G. and GUTH, H. (1981): Low and high temperature spinel ($MgAl_2O_4$): studies on natural single crystals by neutron diffraction, *Collected Abst: C–187 Twelfth IUCr Congress*, Ottawa.

BRAGG, W. H. (1915): The structure of spinel group of crystals, *Philos, Mag.*, **30**, 305–315.

BRAGG. W. H. and BROWN, G. B. (1926): Die Structure des Olivines, *Z. Kristallogr.*, **63**, 538–556.

BRAGG, W. L. (1937): *Atomic Structure of Minerals*, Cornell University Press, 101.

BOIKOVA, A. I. and COMANSKY, A. I. (1974): High temperature transformation of pure calcium germanate, *Cement Concrete Res.*, **4**, 773–784.

BROWN, G. E. and PREWITT, C. T. (1973): High-temperature crystal chemistry of hortonolite, *Am. Mineral.*, **58**, 577–587.

BROWN, G. E. SUENO, S. and PREWITT, C. T. (1973): A new single-crystal heater for the precession camera and four-circle diffractometer, *Am. Mineral.*, **58**, 698–704.

BROWN, G. E. (1980): Olivines and silicate spinels. in *Orthosilicates, Reviews in Mineralogy*, Vol. 5, Ed. P. H. Ribbe, Mineral Soc. Am., pp. 275–381.

BRUN, E. and HAFNER, S. (1962): Die electrische Quadrupolaufspaltung von Al^{27} in Spinel $MgAl_2O_4$ und Korund Al_2O_3, *Z. Kristallogr.*, **117**, 37–78.

BUSING, W. R. and LEVY, H. A. (1964): The effect of thermal motion on the estimation of bond lengths from diffraction measurements, *Acta Crystallogr.*, **17**, 142–146.

CHRISTIAN, J. W. (1965): *The Theory of Transformations in Metals and Alloys*, Pergamon Press, London.

COPPENS, P. and HAMILTON, W. C. (1970): Anisotropic extinction corrections in the Zachariasen approximation, *Acta Crystallogr. Sect. A*, **26**, 71–83.

COTTERILL, R. M. J. (1975): Melting-point depression in very thin Lannard-Jones crystals: A molecular dynamics study, *Philos. Mag.*, **32**, 1283–1288.

COUCHMAN, P. R. and JESSER, W. A. (1977): Comments on the melting mechanism for crystalline species, *Philos. Mag.*, **35**, 787–790.

CRUICKSHANK, D. W. J. (1956): The analysis of the anisotropic thermal motion of molecules in crystals, *Acta Cryst.*, **9**, 754–756.

DATTA, R. K. and ROY, R. (1967): Equilibrium order-disorder in spinel, *J. Am. Ceram. Soc.*, **50**, 578–583.

DATTA, R. K. and ROY, R. (1968): Order-disorder in $MgAl_2O_4$. The system: $MgAl_2O_4$–$LiAl_5O_8$, $MgAl_2O_4$–$NiCr_2O_4$, $MgAl_2O_4$–$NiAl_2O_4$, and $NiAl_2O_4$–$ZnAl_2O_4$, *Am. Mineral.*, **53**, 1456–1475.

FARRELL, E. F., FANG, J. H. and NEWNHAM, R. E. (1963): Refinement of the chrysoberyl structure, *Am. Mineral.*, **48**, 804–810.

FANG, J. H. and NEWNHAM, R. E. (1965): The crystal structure of sinhalite, *Minral. Mag.*, **35**, 196–199.

FINGER, L. W. and RAPP, G. R. (1970): Refinement of the crystal structure of triphylite, *Carnegie Inst. Year Book*, **68**, 230–232.

FINGER, L. W., HAZEN, R. M. and YAGI, T. (1979): Crystal structures and electron densities of nickel and iron silicate spinels at elevated temperature or pressure, *Am. Mineral.*, **64**, 1002–1009.

FISCHER, P. (1967): Neutronenbeugungsuntersuchung der Strukturen von $MgAl_2O_4$–und $ZnAl_2O_4$–Spinellen, in Abhangigkeit von der Vorgeschichite, *Z. Kristallogr.*, **124**, 275–302.

FUJINO, K., SASAKI, S., TAKÉUCHI, Y. and SADANAGA, R. (1981): X-ray determination of electron distributions in forsterite, fayalite and tephroite, *Acta Crystallogr. Sect. B*, **37**, 513–518.

GRIMES, N. W. and HILLEARD, R. J. (1970): X-ray diffraction studies of the spinel series $Mg(Cr_xAl_{2-x})O_4$ I: Lattice parameters and structures, *J. Phys. C. (Ser 2)*, **3**, 866–871.

GRIMES, N. W. (1971): Structural distortions in $MgCr_2O_4$, *J. Phys. C.* **4**, L342–L344.

GRIMES, N. W. (1972): 'Off-centre' ions in compounds with spinel structure, *Philos. Mag.*, **26**, 1217–1226.

GRIMES, N. W., THOMPSON, P. and KAY, H. F. (1981): Space-group symmetry and structure-parameter refinement for magnesium aluminium spinel, *Collected Abst: C–187 Twelfth IUCr Congress*, Ottawa.

GRÜNEISEN, E. (1926): Zustand der festen Körper, in *Handbuch der Phisik Bd 10 Berlin*, Springer, Heidelberg, New York, pp. 1–52.

HAFNER, S. and LAVES, F. (1961): Ordung/Unordung und Ultrarotabsorption. III. Die Systeme $MgAl_2O_4$–Al_2O_3 und $MgAl_2O_4$–$LiAl_5O_8$, *Z. Kristallogr.*, **115**, 321–330.

HAFNER, S. (1960): Metalloxyde mit Spinellstructur, *Schwiz Miner. Petrogr. Mitt.*, **40**, 207–242.

HANKE, K. (1965): Beiträge zu Kistallstrukturen von Olivin-type, *Beitr. Mineral Petrogr.*, **11**, 535–558.

HAZEN, R. M. (1976): Effects of temperature and pressure on the crystal structure of forsterite, *Am. Mineral.*, **61**, 1280–1293.

HAZEN, R. M. (1977): Effects of temperature and pressure on the crystal structure of ferromagnesian olivine, *Am. Mineral.*, **62**, 286–295.

HAZEN, R. M. and PREWITT, C. T. (1977): Effects of temperature and pressure on interatomic distances in oxygen-based minerals, *Am. Mineral.*, **62**, 309–315.

HENDERSON, C. M. B. and TAYLOR, D. (1975): Thermal expansion by X-ray diffraction: Use of specimen holder as internal standard and expansion of MgO (periclase) and $MgAl_2O_4$ (spinel), *Trans. Brit. Ceram. Soc.*, **74**, 55–57.

HILL, R. J., CRAIG, J. R. and GIBBS, G. V. (1979): Systematics of the spinel structure type, *Phys. Chem. Minerals*, **4**, 317–339.

JAGODZINSKI, H. and SAALFELD, H. (1958): Kationenverteilung and Strukturbeziehungen in Mg-Al Spinellen, *Z. Kristallogr.*, **110**, 197–218.

JOUBERT, JEAN-CLAUDE and DURIF, A. (1964) Étude de deus types d'ordre dans le spinelle $Mn_3Li_2ZnO_8$, *C. R. Acad. Sc. Paris.*, **t258**, 4482–4485.

KOZU, S., UEDA, J. and TSURUMI, S. (1934): Thermal expansion of olivine, *Imp. Acad. Japan. Proc.*, **10**, 83–86.

LAGER, G. and MEAGHER, E. P. (1978): High-temperature structural study of six olivines, *Am. Mineral.*, **63**, 365–377.

MARUMO, F., ISOBE, M. and AKIMOTO, S. (1977): Electron-density distributions in crystals of γ–Fe_2SiO_4 and γ–Co_2SiO_4, *Acta Crystallogr. Sect. B*, **33**, 713–716.

MARUMO, F., ISOBE, M., SAITO, Y. and AKIMOTO, S. (1974): Electron-density distribution in crystals of γ–Ni_2SiO_4, *Acta Crystallogr. Sect. B*, **30**, 1904–1906.

MEGAW, H. D. (1939): The thermal expansion of crystals in relation to their structures, *Z. Kristallogr.*, **100**, 58–76.

NAKANISHI, K. (1981): The crystal structure of triphylite at elevated temperatures up to around 700°C, Graduate work, Mineralogical Institute, Univ. of Tokyo.

NISHIKAWA, S. (1915): Structure of some crystals of the spinel group, *Proc. Math. Phys. Soc. Tokyo.*, **8**, 199–209.

RIGBY, G. R., LOVELL, G. H. B. and GREEN, A. T. (1946): The reversible thermal expansion and other properties of some magnesian ferrous silicates, *Trans. Brit. Ceram. Soc.*, **45**, 237–250.

SAKURAI, T., ITO, T. and KOBAYASHI, K. (1980): *UNICS III (3) system*, Inst. Phys. Chem. Res. (Japan).

SATO, G. (18): Effects of temperature or pressure on the crystal structure of $LiScSiO_4$, D. Sc. Thesis, Faculty of Science, Univ. of Tokyo.

SCHMOCKER, U. and WALDNER, F. (1976): The inversion parameter with respect to the space group of $MgAl_2O_4$ spinels, *J. Phys. C., Solid. State. Phys.*, **9**, L235–L237.

SINGH, H. P., SIMMONS, G. and MCFARLIN, P. F. (1975): Thermal expansion of natural spinel, ferroan gahnite, magnesiochromite and synthetic spinel, *Acta crystallogr. Sect. A*, **31**, 820–822.

SKINNER, B. J. (1962): Thermal expansion of ten minerals, *U. S. Geol. Surv. Prof. Pap.*, **450D**, 109–112.

SMITH, P. R. J. (1978): Note on the space group of spinel minerals, *Philos. Mag. B*, **38**, 99–102.

SMYTH, J. R. and HAZEN, R. M. (1973): The crystal structure of forsterite and hortonolite at several temperatures up to 900°C, *Am. Mineral.*, **58**, 588–593.

SMYTH, J. R. (1975): High temperature crystal chemistry of fayalite, *Am. Mineral.*, **60**, 1092–1097.

SOGA, N. and ANDERSON, O. L. (197: High-temperature elasticity and expansivity of forsterite and steatite, *J. Am. Ceram. Soc.*, **50**, 239–242.

STEELE, I. M., PLUTH, J. J. and ITO, J. (1978): Crystal structure of synthetic $LiScSiO_4$ olivine and comparison with isotypic Mg_2SiO_4, *Z. Kristallogr.*, **147**, 119–127.

SUZUKI, I. (1975): Thermal expansion of periclase and olivine, and their anharmonic properties, *J. Phys, Earth.*, **23**, 145–159.

SUZUKI, I. and KUMAZAWA, M. (1980): Anomalous thermal expansion in spinel $MgAl_2O_4$, *Phys. Chem. Minerals*, **5**, 279–284.

TAKEI, H., HOSOYA, S., and KOJIMA, H. (1982): Synthesis of large and high-quality single crystals of olivines, pyroxenes, pyroxenoids and ilmenites, this volume, pp. 107–130.

TAKÉUCHI, Y., YAMANAKA, T., HAGA, N. and HIRANO, M. (1981): High temperature structural study of the olivine type structures and spinels, *Collected Abst: C–156 Twelfth IUCr Congress*, Ottawa.

TOKONAMI, M. and HORIUCHI, H. (1980): On the space group of spinel, $MgAl_2O_4$, *Acta crystallogr. Sect. A*, **36**, 122–126.

WEEKS, R. A. and SONDER, E. (1980): Electrical conductivity of pure and Fe-doped magnesium-aluminium spinel, *J. Am. Ceram. Soc.*, **63**, 92–95.

WILLIS, B. T. M. (1965): The anomalous behaviour of the neutron reflexions of fluorite, *Acta Crystallogr.*, **18**, 75–76.

YAGI, T., MARUMO, F. and AKIMOTO, S. (1974): Crystal structures of spinel polymorphs of Fe_2SiO_4 and Ni_2SiO_4, *Am. Mineral.*, **59**, 486–490.

YAMANAKA, T. (1982): Technique and gas-flame heater for single-crystal X-ray diffraction studies at high temperatures up to around 2,500°C, this volume, pp. 621–631.

YAMANAKA, T., TAKÉUCHI, Y. and SADANAGA, R. (1979): The crystal structure of spinel at elevated temperature up to 1,800°C, in *Crystalloger. Soc. Japan, 1979 Annual Meeting Abst: 3B–7.*

YAMANAKA, T., TAKÉUCHI, Y. and SADANAGA, R. (1981): Gas-flame high-temperature apparatus for single-crystal X-ray diffraction studies, *Z. Kristallogr.*, **154**, 147–153.

YAMANAKA, T., TAKÉUCHI, Y., SADANAGA, R. and ISHII, N. (1981): Dynamics of silicate spinels in their unstable region, in *Mineral Soc. Japan, Annual Meeting Abst: A–10.*

Materials Science of the Earth's Interior, edited by I. Sunagawa, pp. 233–247.

IN-SITU OBSERVATIONS OF CRYSTAL STRUCTURE AND PHASE TRANSFORMATION UNDER PRESSURE

Takehiko YAGI and Syun-iti AKIMOTO

The Institute for Solid State Physics, The University of Tokyo, Roppongi, Minato-ku, Tokyo, Japan

1. *Introduction*

Pressure is one of the important parameters in determining a crystal's structure. When pressure is applied to a crystal, interatomic distances and, hence, the volume of the crystal, decrease first. Pressure-dependence of the interatomic distance depends on many factors, such as the nature of the bond, the sort of ion, and the structure type of the crystal. Even in a single crystal, pressure response of a different bond has a wide variation. Complicated responses of many bonds in the crystal can be simplified by using bulk moduli of constituent coordination polyhedra (HAZEN and PREWITT, 1977). When the pressure is further increased, it becomes difficult to reduce the interatomic distances because of the repulsive forces between the nearest atoms. In order to reduce the volume of the crystal under these circumstances, rearrangement of the atoms occurs. This is called a phase transformation and is accompanied with a discontinuous volume decrease. In many cases, ions in a high-pressure phase have a higher coordination number and longer interatomic distances than in a low-pressure phase. Many solids have more than one high pressure polymorphs and packing of the structure becomes more denser with pressure. In extremely high pressure, electronic orbitals of the component ions are compressed and electronic transitions such as high spin-low spin transition, and metallic transition occur.

In order to study the behaviors of crystal under pressure, X-ray diffraction study gives us the most fundamental and important information. It gives us knowledge of the equations of state (P-V-T relations) of each phase, the locations of the phase boundary between two polymorphs in P-T space, and the crystal structures of high-pressure polymorphs. Unfortunately, however, the experimental technique for high pressure-temperature *in situ* X-ray observation is much more difficult compared to that of conventional X-ray work at one bar.

In order to generate high pressure, specimens must be sealed in a high-pressure vessel that is made of mechanically-strong materials such as tungsten carbide. On the other hand, X-ray diffraction study requires an X-ray path around the specimen. However, most of the materials with low X-ray absorption coefficients are mechanically very weak. In 1938, JACOBS first succeeded in obtaining a diffraction pattern of AgI under high pressure using He as a pressure transmitting medium.

Since then new apparatuses have been developed to solve these contradictory problems. Among them are Brigdman anvil-type (JAMIESON and LAWSON, 1962) and Drickamer anvil-type (PEREZ-ALBUERNE *et al.*, 1964) apparatuses which were most extensively used to study the behavior of solids up to 200 kbar at room temperature. Diamond anvil cell (PIERMARINI and WEIR, 1962) has also been used to study the compression curves of many oxides and silicates up to 300 kbar (BASSETT and TAKAHASHI, 1974). An attempt to perform simultaneous high-pressure and high-temperature *in situ* observation was made using a tetrahedral press (BARNETT and HALL, 1964) and split-belt type apparatus (FREUD and SCLAR, 1969). Only very simple materials, such as metal, alkali halide, and monoxide, with high symmetry, could be studied by these apparatuses because the high pressure X-ray study had many problems, including the following: (1) Resolution was very low because nonhydrostatic stress and pressure inhomogeneity in the specimen broadened the diffraction lines. (2) The signal-to-noise ratio was very low because the X-ray beam was scattered by a pressure transmitting medium and increased the background considerably. (3) Exposure time was very long. The amount of the specimen in the high-pressure vessel is very small and the diffracted beam is weak. The intensity is further reduced by the absorption of the pressure-transmitting medium. This very long exposure time (100 to 300 hours or more) was unfavorable for a detailed X-ray diffraction study and made it difficult to perform an experiment at an elevated temperature. (4) The accuracy of the measurement was low, mainly because it was difficult to determine the distance from sample to film accurately. (5) The high temperature limit was too low to study the transition in silicate because most of the pressure-transmitting media for X-ray work could not sustain high temperatures.

Since the 1970's, various new developments were made, not only in high pressure apparatus but also in X-ray diffraction systems, and many of the difficulties described above have been overcome. Developments of the cubic anvil-type apparatus and the intense X-ray source extended the high-pressure-temperature limit and made it possible to perform *in situ* observations of the phase transformations in silicates and oxides (INOUE, 1975; YAGI and AKIMOTO, 1976). These authors also contributed much to increase the accuracy of the measurement and precise *P-V-T* relations were obtained (YAGI, 1978). Development of the diamond anvil cell made it possible to extend the pressure to the megabar range (MAO *et al.*, 1978) and clarified the existence of spin transition in iron oxide in this extremely high-pressure region (YAGI and AKIMOTO, 1982). A single crystal analysis technique under pressure was developed (MERRILL and BASSETT, 1974) and much data on the pressure responses of polyhedra has been accumulated (HAZEN and FINGER, 1979). Efforts were also made to establish an experimental technique to perform single crystal analysis under combined high pressure and high temperature (HAZEN and FINGER, 1981). An intense X-ray source combined with new detectors, such as an energy dispersive-type solid state detector and a position sensitive detector (FUJII *et al.*, 1980), contributed much to shorten the exposure time. Improvement of the slit system also considerably reduced the exposure time in the diamond anvil cell (YAGI and AKIMOTO, 1982). This short exposure time made it possible to study the kinetics of the transition under pressure (HAMAYA and AKIMOTO, 1981).

In this section the recent advance of high pressure and high temperature X-ray diffraction study will be described. Direct determination of coesite-stishovite transformation in SiO_2, kinetics study and the equilibrium phase boundary in B1-B2 transition in KCl, and *in situ* observation of the orthorhombic-cubic transition in PbO_2 are the main topics. Experimental techniques of the high pressure and temperature X-ray diffraction study using cubic anvil-type apparatus will also be described briefly.

The diamond anvil cell played a very important role in a recent high-pressure X-ray diffraction study. The latest research using this apparatus can be found in the work edited by AKIMOTO and MANGHNANI (1982) and will not therefore be described here.

2. *High Pressure X-Ray Diffraction Technique*

2.1 *Cubic anvil-type high-pressure apparatus*

Cubic anvil-type high-pressure apparatus was successfully combined with an X-ray diffraction system (INOUE and ASADA, 1973). This apparatus consists of six tungsten carbide anvils and is driven by a uniaxial 250 ton press (Fig. 1). By applying load, the six anvils are advanced simultaneously and compress a cube-shaped pressure-transmitting medium. The position of the specimen is well controlled during the compression because the strokes of the six anvils are precisely synchronized. This fact contributed much to increase the accuracy of the X-ray measurement. The pressure-transmitting medium is made of a mixture of amorphous boron and epoxy resin which is a good heat-resistant material. Specimens are imbedded in the center of the pressure-transmitting medium and, for high temperature experiments, a small graphite furnace was installed together with a specimen (Fig. 2). Various improvements were made in the sample assembly to increase the hydrostaticity of the pressure (YAGI, 1978). The X-ray beam goes into the sample chamber through

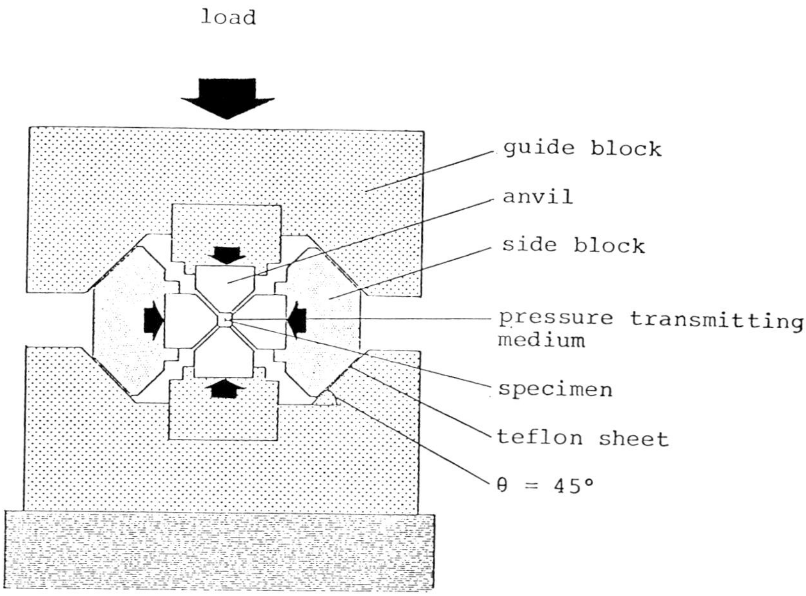

FIG. 1. Schematic diagram of cubic anvil-type high pressure apparatus.

a narrow gap between two anvils, which is filled with a self-formed gasket. A diffracted X-ray passed through an opposite gasket is measured by a detector. A proper design of X-ray slits improved the S-N ratio of the diffraction pattern. The diffracted X-ray in 2θ angles up to $\pm 35^\circ$ can be measured by this system. Development of this cubic anvil-type high-pressure X-ray diffraction system, combined with a high-power X-ray source, considerably improved the quality of the diffraction data and made it possible to perform precise X-ray diffraction studies up to 100 kbar and 1,100°C (YAGI and AKIMOTO, 1976; YAGI, 1978).

2.2 X-ray diffraction system

Both energy dispersive-type and angle dispersive-type X-ray diffraction systems were used in the present studies. An intense X-ray was generated by a rotating-anode type Mo target, which is operated at 55 kV and 160 mA. The use of this high-power X-ray source contributed much to shorten the exposure time and hence to increase the accuracy of the measurement. In the angle dispersive technique, a conventional scintilation counter was used to detect the diffracted X-ray. It was moved along a vertical plane and a step-scanning method was adopted for measurement. In the energy dispersive technique, a pure germanium detector was fixed to a desired 2θ angle. Detector signals were collected with a 1,024 channel pulse-height analyzer. In this system, a 300

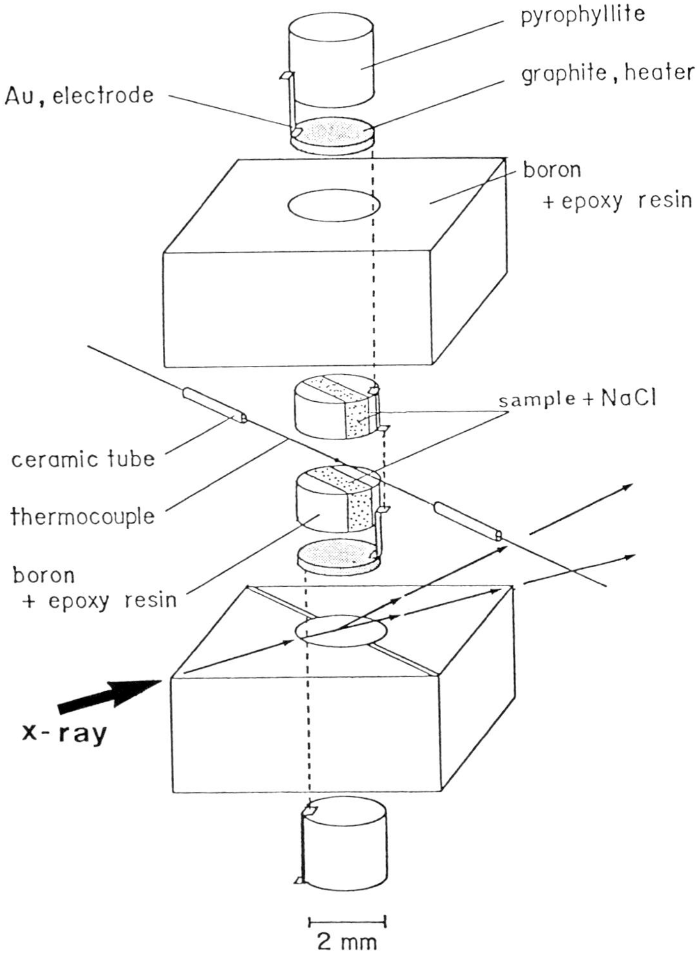

FIG. 2. Sample assembly for high pressure and high temperature *in situ* X-ray diffraction study. The powdered specimen is mixed with NaCl which works as an internal pressure standard.

sec exposure is enough to get a good diffraction pattern at high pressure (HAMAYA and AKIMOTO, 1981).

3. *Phase Transformations and Transformation Rate in SiO_2, KCl and PbO_2*

Phase transformations in silicates and oxides have long been studied by a quenching method. Phase diagrams obtained by this method have the following problems: (1) In some materials, retrogressive transition occurs during a quench process and, in that case, an analysis based on the quench product leads to serious error. (2) Pressures in a high-pressure vessel change with temperature, but the pressure values in quench experiments have generally been determined at room temperature on the basis of conventional-pressure fixed points. (3) Experimental time is not so long and there is a possibility that the specimen is not in equilibrium with given *P-T* conditions.

Most of these problems can be solved by adopting a high pressure-temperature *in situ* X-ray diffraction method to study the phase diagrams.

3.1 SiO_2

Silicon dioxide has many polymorphs. Among them stishovite (rutile-type polymorph) is known to be stable in the most high-pressure region. Silicon in stishovite has six coordination numbers and, consequently, the density of stishovite is more than 1.6 times larger than that of α-quartz. Its stability and high density suggest that this can be a candidate constituent of the lower mantle.

Since the first synthesis of stichovite (STISHOV and POPOVA, 1961), various studies have been made to determine its pressure-temperature stability field. All of these studies, however, were based on quench experiments. The development of the high pressure-temperature X-ray diffraction technique made it possible to perform *in situ* observations of this phase transformation. An attempt was made to determine coesite-stishovite phase boundary using this technique (YAGI and AKIMOTO, 1976).

In this experiment NaCl was mixed with a starting material (amorphous silica) and a simultaneous observation of the X-ray diffraction from both SiO_2 and NaCl were made (Fig. 3). From the diffraction pattern of SiO_2, the stable phase can be determined in a given *P-T* condition. Temperature was measured by a thermocouple and pressure was calculated from the lattice parameter of NaCl through equations of state of NaCl (DECKER, 1971). Results are summarized in Fig. 4 together with the boundary determined by the quenching method (AKIMOTO and SYONO, 1969). A comparison of this result with previous works indicates that the position of the boundary does not differ much but the slope of the boundary is different. This result can be understood in the following way: In SiO_2 there was no discrepancy between *in situ* observation and the quench product. Pressure value, however, was different between room temperature and elevated temperature even when the press load was kept constant. Up to approximately 800°C, pressure in the sample chamber increased with temperature, probably because of the thermal expansion of the pressure transmitting medium. Above 800°C pressure began to decrease, probably because pressure transmitting medium becomes soft and the gasket flows out slightly. These pressure variations with temperature can well explain the discrepancy between quench experiment and *in situ* observation (Fig. 4).

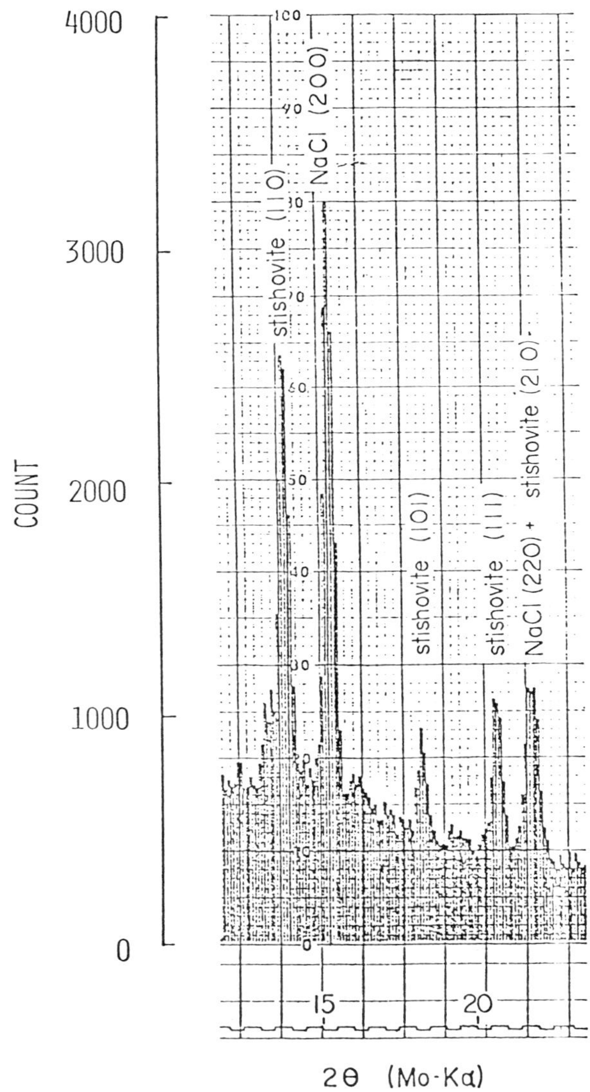

FIG. 3. X-ray diffraction pattern of SiO_2 + NaCl obtained at 89.5 kbar and 700°C by the angle dispersive method. Stishovite is a stable phase of silica in this *P-T* condition. Data were collected by the step-scanning method and exposure time was 40 sec for each step.

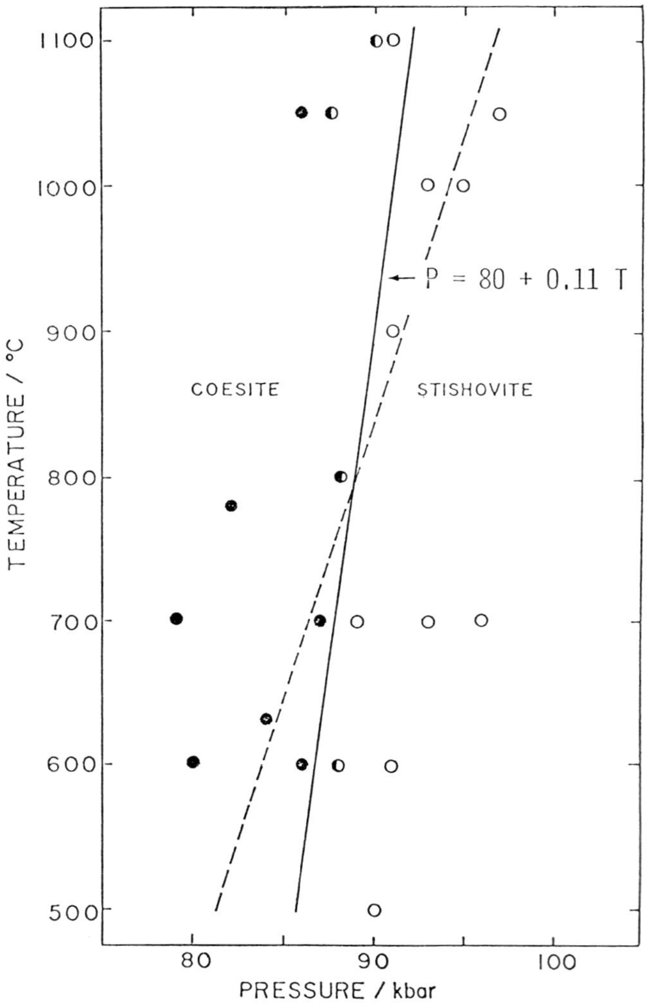

FIG. 4. Pressure-temperature stability diagram for the coesite-stishovite transformation obtained by *in situ* X-ray measurement. Open and solid circle represent stishovite and coesite, respectively, and the solid line is the boundary determined by this experiment. The broken line is the phase boundary determined by the quenching method (AKIMOTO and SYONO, 1969). Discrepancy between these two lines can be explained by the change of pressure in the high-pressure vessel at an elevated temperature.

The first two problems in quench experiments, uncertainty of the quenching and the pressure calibration at elevated temperature, were solved by this *in situ* X-ray observations. The third equilibrium problem, however, remained unsolved. In order to solve this, consideration is required which is based on the kinetics of the transformation. High pressure *in situ* X-ray diffraction, however, require a long exposure time for each measurement and it was very difficult to follow the time dependence of the transformation. Development of the energy dispersive-type X-ray diffraction technique shortened the exposure time considerably and made it possible to study the kinetics under pressure. Phase transformations of KCl and PbO_2 were studied by this method.

3.2 *Equilibrium phase boundary in KCl*

Potassium chloride transforms from the NaCl structure to the CsCl structure at about 19 kbar at room temperature. Many works have been done to determine this transition pressure precisely. HAMAYA and AKIMOTO (1981) determined the equilibrium transition pressure based on the kinetics of this transition.

A series of X-ray diffraction patterns obtained by energy-dispersive technique are shown in Fig. 5. In this experiment, pressure was kept constant by making a fine adjustment to the external load to compensate for the pressure variation caused by the volume change of the specimen. Exposure time for each measurement is only 300 sec. From these data at several different pressures, the transformation rate in a both increasing and decreasing pressure cycle was determined as a function of pressure (Fig. 6). The equilibrium phase boundary is defined as a pressure where the transformation rate becomes zero for both directions. From Fig. 6, the pressure at which the growth speed of the B1 and B2 phase becomes zero is estimated to be 18.9 kbar and 19.3 kbar, respectively. By taking the arithmetic mean value of these two, the equilibrium pressure for the B1 $\rightleftarrows$ B2 transformation in KCl was determined to be 19.1 kbar at room temperature.

3.3 *Orthorhombic-cubic transformation in PbO_2*

Phase transformations in AX_2-type compounds are predominantly controlled by the ionic radius ratio of cation to anion (r_C/r_A). Cations are less compressible than anions and this ratio is expected to increase with pressure. Among many oxides with the rutile structure, lead dioxide has the highest value of this ratio. Therefore, lead dioxide can be regarded as a high-pressure model material of stishovite. The phase transformations of lead dioxide has been studied in conjunction with the possible phase transformations of stishovite under a lower mantle condition.

WHITE *et al.* (1961) found that the rutile-type phase (PbO_2 I) transforms into an orthorhombic phase (PbO_2 II) with the columbite-type structure at

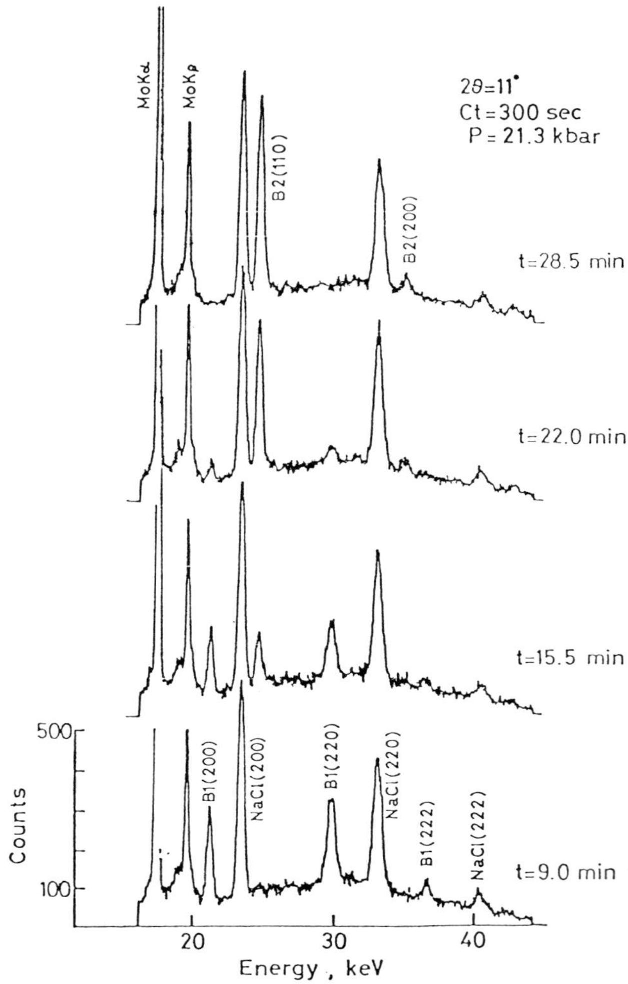

FIG. 5. Change in the energy dispersive X-ray diffraction pattern of KCl with time. At first, a pressure of 21.3 kbar was applied to KCl in the low-pressure phase (B1) at room temperature. Since this *P-T* condition is a stable field of high-pressure phase (B2), it gradually transformed into B2 phase. Times in the right-hand column are measured from when the pressure was elevated to 21.3 kbar. Exposure time for each measurement was 5 min.

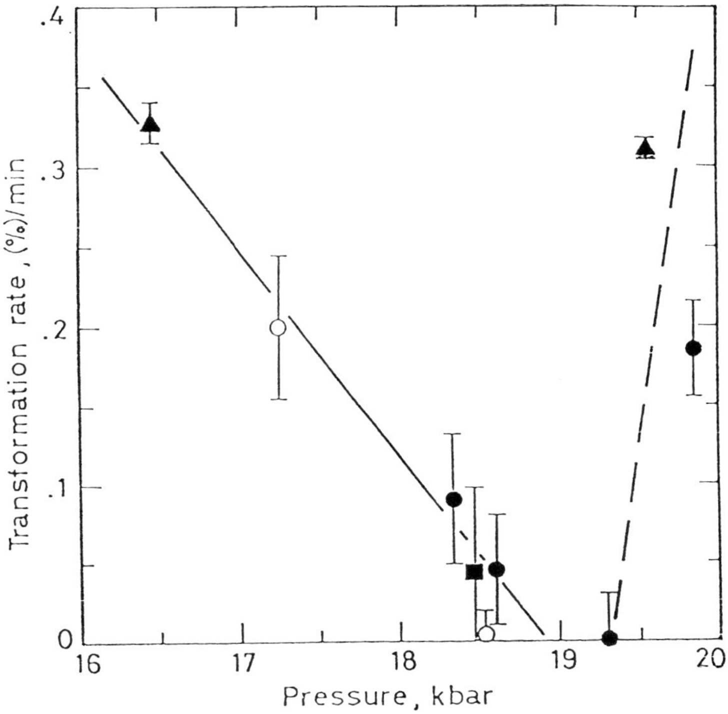

FIG. 6. Pressure-dependence of the transformation rate of the B1-B2 transformation of KCl. Below 18.9 kbar, B2 → B1 transformation takes place, while above 19.3 kbar, transformation from B1 to B2 occurs. By taking the arithmetic mean value of these two, the equilibrium phase boundary was determined to be 19.1 kbar at room temperature.

pressures higher than 10 kbar. SYONO and AKIMOTO (1968) reported the synthesis of a cubic fluorite-type polymorph (PbO_2 III) above 60 kbar and determined the phase boundary between PbO_2 II and III by a conventional quenching technique. The quenched samples of PbO_2 III, however, were quite unstable and undergo a retrogressive transformation to PbO_2 II, even at room temperature and they could never get a single phase of PbO_2 III. The phase boundary determined in their experiment is uncertain. The phase diagram of PbO_2 was reinvestigated using the *in situ* X-ray diffraction technique (YAGI and AKIMOTO, 1980).

Figure 7 is an energy dispersive X-ray diffraction pattern of PbO_2 III observed at 49 kbar and 25°C. This sample was first kept at 63 kbar and 300°C and then pressure was lowered at room temperature. LiF was mixed with PbO_2 as an internal pressure standard. Twelve diffraction peaks of the cubic phase are clearly observed together with diffraction lines of LiF. It is obvious from this figure that cubic fluorite-type PbO_2 exists as a stable single phase at

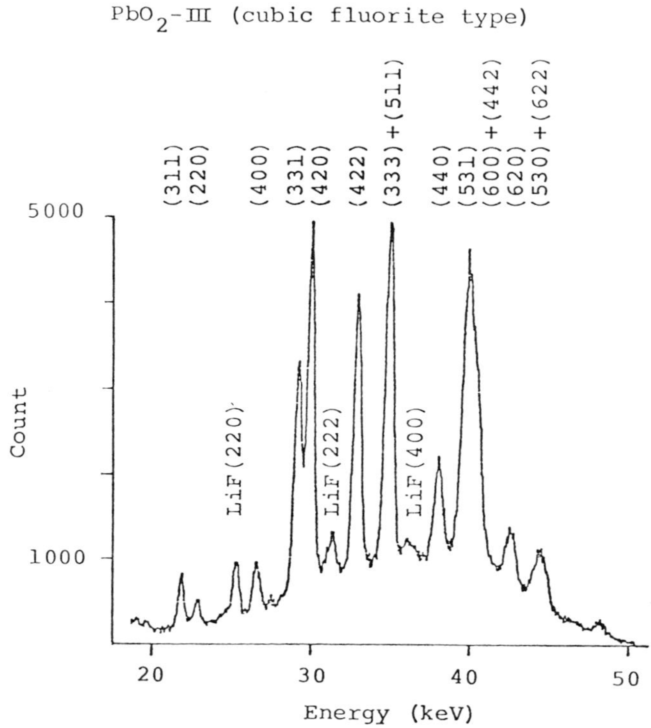

FIG. 7. Energy dispersive X-ray diffraction pattern of PbO_2 III (cubic fluorite type) observed at 49 kbar and 25°C. LiF was mixed with PbO_2 as an internal pressure standard.

high pressure. The transformation rate between the orthorhombic phase and the cubic phase was studied in a similar manner as that in KCl at 300, 400, and 500°C. These experimental results are summarized in Fig. 8. Open circles and solid circles in Fig. 8 represent the place where PbO_2 II has formed from PbO_2 III and *vice versa*, respectively. Dots represent the points where no phase transformation was observed during the experiment. Dashed-dotted lines are the contour lines where 30 min was required to accomplish 50% conversion. In the region between these two lines, it took much longer to undergo phase transformation. The region of no change increased rapidly at low temperatures, making it difficult to determine the phase boundary below 400°C. In the present study, however, transformation from PbO_2 I to II or I to III occurred very rapidly, even at 300°C, and these data were also used as complementary information to determine the phase boundary between PbO_2 II and III. The boundary curve thus determined was expressed as follows: P (kbar) = 69 − 0.019 T(°C). SYONO and AKIMOTO (1968) reported a boundary curve of P(kbar) = 52.5 + 0.030 T(°C) based on their quench experiment. The present

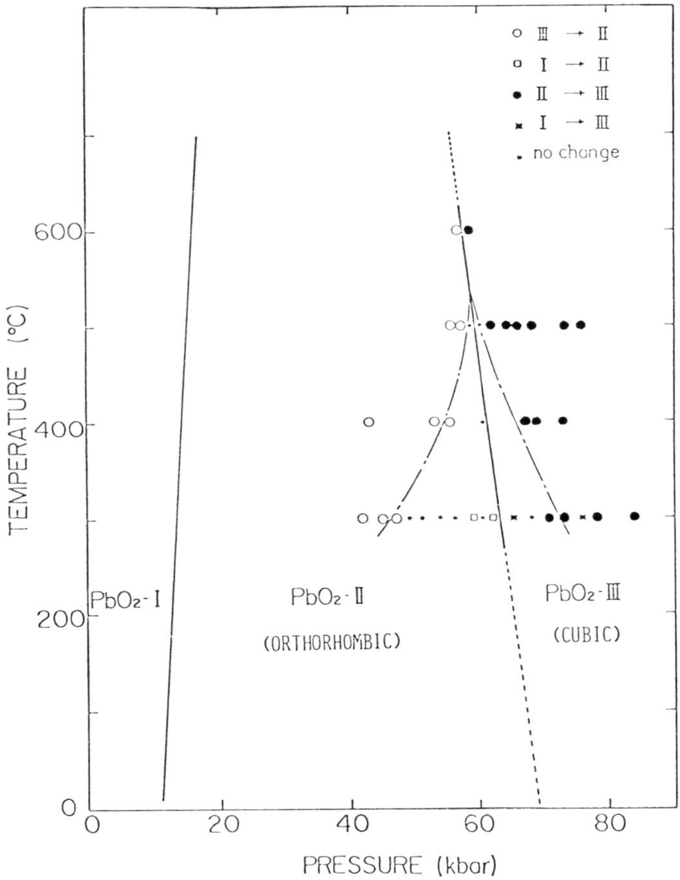

FIG. 8. Pressure-temperature phase diagram of PbO_2. The phase boundary between I and II are from WHITE *et al.* (1961) and other results are obtained by *in situ* X-ray measurement. Thick solid line represents the boundary between PbO_2 II and PbO_2 III. The dashed-dotted lines are the contour lines where 30 min is required to accomplish 50% conversion. In the region between these two lines, it takes a much longer time to undergo phase transformation.

data clearly indicate a negative slope, while they reported a positive slope. This discrepancy may be caused by both an instability of the quenched PbO_2 III phase and the lack of full information on the transformation rate in a previous quench experiment.

4. *Conclusion*

Most of the phase diagrams of geophysically important materials such as olivine and pyroxene, have been so far obtained by the quenching method. The quenching method, however, has some problems in identifying stable phase and pressure measurement at an elevated temperature. *In situ* X-ray observations under high pressure and temperature can eliminate these problems and reliable phase boundaries were determined by this method on coesite-stishovite transformation in SiO_2, and on orthorhombic-cubic transformation in PbO_2.

The knowledge of the kinetics of the transformation is very important in many respects. A time scale of geophysical phenomena is very long and it is believed that in many cases a thermodynamical equilibrium state is achieved in the earth. In order to obtain an equilibrium phase boundary in a finite experimental time, a consideration of the kinetics of the transformation is indispensable. The knowledge of kinetics is also very important for clarifying the microscopic process of the phase transformation and for discussing the dynamic phenomena in the earth, like a phase transition in the subduction zone and the mechanism of a deep earthquake (HAMAYA, 1980). Development of the rapid X-ray diffraction system made it possible to perform a kinetics study under high pressure, and the transitions in KCl and PbO_2 were studied using this system.

Our knowledge of the kinetics of the transformation of silicates and oxides are still very limited and further studies are required to construct a reliable dynamic model of the earth's interior. Construction of a new high pressure X-ray laboratory is now in progress which combines cubic anvil-type high-pressure apparatus with an intense coherent X-ray beam from synchrotron radiation. In this new system, an exposure time required for each measurement is estimated to be less than one minute. Therefore, it is expected to contribute much in advancing kinetics study under high pressure and temperature.

REFERENCES

AKIMOTO, S. and MANGHNANI, M. H. (Eds.) (1982): *High-Pressure Research in Geophysics*, The Center for Academic Publications Japan/D. Reidel, Tokyo/Dordrecht, 632 pp.

AKIMOTO, S. and SYONO, Y. (1969): The coesite-stishovite transition, *J. Geophys. Res.*, **74**, 1653–1659.

BARNETT, J. D. and HALL, H. T. (1964): High pressure-high temperature, X-ray diffraction apparatus, *Rev. Sci. Instrum.*, **35**, 175–182.

BASSETT, W. A. and TAKAHASHI, T. (1974): X-ray diffraction studies up to 300 kbar, in *Advances in High Pressure Research* 4, Ed. R. H. Wentorf, London, pp. 165–247.

DECKER, D. L. (1971): High-pressure equation of state for NaCl, KCl, and CsCl, *J. Appl. Phys.*, **42**, 3239–3244.

FREUD, P. J. and SCLAR, C. B. (1969): Apparatus for high-pressure high-temperature X-ray powder diffraction studies, *Rev. Sci. Instrum.*, **40**, 434–437.

FUJII, Y., SHIMOMURA, O., TAKEMURA, K. and HOSHINO, S. (1980): The application of a position-sensitive detector to high-pressure X-ray diffraction using a diamond-anvil cell, *J. Appl. Cryst.*, **13**, 284–289.

HAMAYA, N. (1980): Mechanism of olivine-spinel transformation, Ph. D. Thesis, University of Tokyo, 113 pp.

HAMAYA, N. and AKIMOTO, S. (1981): Kinetics of pressure-induced phase transformation in KCl at room temperature, *High Temp.-High Press.* **13**, 347–358.

HAZEN, R. M. and PREWITT, C. T. (1977): Effects of temperature and pressure on interatomic distances in oxygen-based minerals, *Am. Mineral.*, **62**, 309–315.

HAZEN, R. M. and FINGER, L. W. (1979): Bulk modulus-volume relationship for cation-anion polyhedra, *J. Geophys. Res.*, **84**, 6723–6728.

HAZEN, R. M. and FINGER, L. W. (1981): High-temperature diamond-anvil pressure cell for single-crystal studies, *Rev. Sci. Instrum.*, **52**, 75–79.

INOUE, K. (1975): Development of high temperature and high pressure X-ray diffraction apparatus with energy dispersive technique and its geophysical applications, Ph. D. Thesis, University of Tokyo, 139 pp.

INOUE, K. and ASADA, T. (1973): Cubic anvil X-ray diffraction press up to 100 kbar and 1,000°C, *Jpn. J. Appl. Phys.*, **12**, 1786–1793.

JACOBS, R. B. (1938): X-ray diffraction of substances under high pressures, *Phys. Rev.*, **54**, 325–331.

JAMIESON, J. C. and LAWSON, A. W. (1962): X-ray diffraction studies in the 100 kilobar pressure range, *J. Appl. Phys.*, **33**, 776–780.

MAO, H. K., BELL, P. M., SHANER, J. W. and STEINBERG, D. J. (1978): Specific volume measurements of Cu, Mo, Pd, and Ag and calibration of the ruby R_1 fluorescence pressure gauge from 0.06 to 1 Mbar, *J. Appl. Phys.*, **49**, 3276–3283.

MERRILL, L. and BASSETT, W. A. (1974): Miniature diamond anvil pressure cell for single crystal X-ray diffraction studies, *Rev. Sci. Instrum.*, **45**, 290–294.

PEREZ-ALBUERNE, E. A., FORSGREN, K. F. and DRICKAMER, H. K. (1964): Apparatus for X-ray measurements at very high pressure, *Rev. Sci. Instrum.*, **35**, 29–33.

PIERMARINI, G. J. and WEIR, C. E. (1962): A diamond cell for X-ray diffraction studies at high pressures, *J. Res. Natl. Bur. Std.*, **A66**, 325–331.

STISHOV, S. M. and POPOVA, S. V. (1961): A new dense modification of silica, *Geochimiya*, **10**, 837–839.

SYONO, Y. and AKIMOTO, S. (1968): High pressure synthesis of fluorite type PbO_2, *Mater. Res. Bull.*, **3**, 153–158.

WHITE, W. B., DACHILLE, F. and ROY, R. (1961): High-pressure high-temperature polymorphism of the oxides of lead, *J. Am. Ceram. Soc.*, **44**, 170–174.

YAGI, T. (1978): Experimental determination of thermal expansivity of several alkali halides at high pressures, *J. Phys. Chem. Solids.*, **39**, 563–571.

YAGI, T. and AKIMOTO, S. (1976): Direct determination of coesite-stishovite transition by in-situ X-ray measurements, *Tectonophysics*, **35**, 259–270.

YAGI, T. and AKIMOTO, S. (1980): Phase boundary and transition rate of orthorhombic-cubic transformation in PbO_2, *J. Geophys. Res.*, **85**, 6991–6995.

YAGI, T., and AKIMOTO, S. (1982): X-ray measurements to 100 GPa range and static compression of α-Fe_2O_3, in *High-Pressure Research in Geophysics*, Eds. S. Akimoto and M. H. Manghnani, The Center for Academic Publications Japan/D. Reidel, Tokyo/Dordrecht, pp. 81–90.

Chapter 4: Analysis of Thermal and Stress Histories

Materials Science of the Earth's Interior, edited by I. Sunagawa, pp. 251–269.

APPLICATION OF TRANSMISSION ELECTRON MICROSCOPY TO THE STUDIES OF DECOMPOSITION AND EXSOLUTION OF MINERALS

Nobuo MORIMOTO and Masao KITAMURA

Department of Geology and Mineralogy, Faculty of Science, Kyoto University, Sakyo, Kyoto, Japan

1. *Introduction*

One of the main purposes of mineralogy is to elucidate the geological history which the minerals experienced. To attain this purpose, four important sources of information must be utilized. They are the chemical compositions, crystal structures, microtextures and occurrences of minerals.

The minerals crystallized in the earth interior have experienced transformations, in general, in wide ranges of temperature and pressure conditions. These minerals usually keep the traces of decomposition and exsolution as microtextures. Therefore, the study of the microtextures and their chemical inhomogeneity in the minerals are indispensable for deriving the information on the parameters describing the geological history of the minerals.

The textures due to decomposition and exsolution change in size from a macroscopic order to a unit cell scale order. They have been studied by optical (or petrographic) microscopy (OM), scanning electron microscopy (SEM), and transmission electron microscopy (TEM).

In the course of the study of the microtextures of minerals, it has become increasingly evident that the textures of the scale of 10 ~ 100 nm are especially important for elucidation of the cooling history of minerals and transformation kinetics which the minerals experienced. During the last ten years, transmission electron microscopes have shown that they are indispensable for investigating this range of the microstructures, and have added numerous important new data to mineralogy and considerably changed its outlook.

An electronprobe microanalyser (EPMA) has been used as a useful method for the determination of chemical composition. The thickness of the specimens, considered to be infinite for EPMA, results in the spatial resolution of about

2 μm. However, specimens thin enough to be transparent for an electron beam allow us to achieve the spatial resolution of less than 100 nm for the chemical analysis. Using these characteristics, an analytical transmission electron microscope (ATEM) has been constructed by combining the transmission electron microscope and X-ray detectors. Thus, the high resolution transmission electron microscope with the analytical mode, or analytical transmission electron microscope, has been proved to be the ideal instrument to investigate the microtextures, and chemical inhomogeneity of the microtextures in minerals.

In this review, the investigation of the microtextures of minerals by analytical transmission electron microscopy, most of which have been carried out in our institute in the last three years, are described. In the first section, the development and technical aspect of analytical transmission electron microscopy is explained. In the last section, the actual examples of the microtextures of minerals are described.

2. *Application of Electron Microscopy in Mineralogy*

2.1 *Historical*

Since the transmission electron microscopic research of minerals, such as mica (AMELINCKX, 1952) and clay minerals (HONJO and MIHAMA, 1954), extensive literature on clay minerals has been published. RIBBE (1962) and McCONNELL and FLEET (1963) were among the first mineralogists to use transmission electron microscopy. YADA (1967) studied the microstructures of chrysotile and confirmed its concentric and spiral tube structures proposed from the previous X-ray diffraction studies.

The study of lunar materials by transmission electron microscopy and the advent of reliable ion-thinning devices stimulated a tremendous boom of using transmission electron microscopes for the study of minerals in the beginning of 1970's. Their results quickly demonstrated electron microscopy as a powerful tool and changed the aspect of modern mineralogy.

The many-beam lattice imaging technique has been applied to the structure of tourmaline by IIJIMA *et al.* (1973), whose result indicates a close correspondence between the lattice image and the crystal structure. BUSECK and IIJIMA (1974) published a series of beautiful lattice images of various types of silicate minerals. They called their images structure images rather than lattice images, because the contrasts of the images represent the structural features. The success accomplished by BUSECK and IIJIMA (1974) accelerated the application of transmission electron microscopy, especially lattice imaging, to the study of minerals. Until now, new data obtained by transmission electron microscopy have had a most significant impact on the understanding of the important rock-forming minerals and revealed many unexpected features.

Since the use of EMMA-4 by CLIFF and LORIMER (1972), the validity of analytical transmission electron microscopy has been confirmed in quantitative analysis of small areas of minerals. NOBUGAI *et al.* (1978, 1979) studied the microtextures of Skaergaard pyroxenes by 100 kV analytical transmission electron microscopy. BUCHANAN (1978), MORIMOTO and KITAMURA (1981) and KITAMURA *et al.* (1981a) described the use of 100 kV and 200 kV analytical transmission electron microscopes to Bushveld pyroxenes, respectively. Many other recent studies by analytical transmission electron microscopy have indicated that it is indispensable for the study of microtextures of minerals.

2.2 *Technical aspect of high-resolution transmission electron microscopy*

By 1970, new advancements had been accomplished in high-resolution electron microscopy (HREM) as follows:

a) *Instrumental development of electron microscope*: The interpretable resolution has become 3.0 to 3.5 Å for 100 kV transmission electron microscope. With high accelerating voltages in the electron microscope, this limit of interpretable resolution has been reduced to 1.5 to 2.0 Å. Because of the increase of resolution capacity, the many-beam lattice imaging has become common procedure in high-resolution electron microscopy.

For lattice imaging, the crystal specimen must be oriented with respect to the incident beam with an accuracy of 1 milliradian (1×10^{-3} radians, or 0.06°). This has become possible by use of the modern goniometer stage.

b) *Sample preparation*: To obtain the lattice images representing the crystal structures, the thickness of the crystals should be less than 150 Å. The ion-thinning devices of the early 1970's solved the problem of preparing thin foils of definite orientations for minerals. However, for the lattice images representing the lattice relations or superlattice structures, specimens of several hundred Å can be used in high-resolution electron microscopy.

c) *Confirmation of the lattice images by computation*: Because the image of a structure is very sensitive to many factors reflecting both specimen and instrument (e.g., thickness, orientation, defocussing and lens aberration), it is dangerous to assume that the lattice image has any direct relationship to the crystal structure. It is, therefore, necessary to confirm theoretically the intensity distribution in lattice images and diffraction patterns for postulated structures. This computation is now possible by using the many-beam dynamical diffraction theory of electrons based on the formulation derived originally by COWLEY and MOODIE (1957). However, as mentioned above, to investigate only the change of periodicity of the unit translation, such as the superlattice relation or domain relation, thick specimens are often used, for which the computation is not generally needed.

d) *Advancement of analytical transmission electron microscopy*: The transmission electron microscope with energy dispersive microanalysis has made semi-quantitative chemical analysis possible for the range of a few hundred Å in diameter. Based on this new means, it has become possible to observe directly the relationship between the microtextures and chemical inhomogeneity.

3. *Analytical Transmission Electron Microscopy*

The energy dispersive-microanalysis has been attached to high-resolution electron microscopes of 100 kV and 200 kV, to perform quantitative chemical analysis of a few hundred Å in diameter. The explanation of the analytical transmission electron microscope has been given by Nobugai *et al.* (1978) and Morimoto and Kitamura (1981). The main features of the 200 kV analytical transmission electron microscope will be given briefly below. A large take-off angle (68°) for emerging X-rays has been assured by using the side-entry sample holder. To keep a high signal/noise ratio of the X-ray peaks, special shielding has been designed for the transmission electron microscope.

If the specimen is thin enough to transmit a few hundred kV electrons it is also transparent to most of the primary X-rays produced by the electrons, so that the X-ray absorption and fluorescence effects can be neglected. Although the absolute intensity of the characteristic X-rays for elements will depend on the specimen thickness, the ratio of the X-ray intensity for any two elements will be independent of thickness (Cliff and Lorimer, 1972), resulting in the relation of $I_1/I_2 = k_{12} \cdot (C_1/C_2)$, where I_1 and I_2 are the observed characteristic X-ray intensities, C_1 and C_2 are concentrations of the elements, and k_{12} is a factor that can be calculated or determined experimentally from a thin-film standard.

The linear relationship between the counts of the characteristic X-rays from each element and the total counts from all elements has been established for fragments thin enough for the conventional 100 kV analytical transmission electron microscope (Cliff and Lorimer, 1972; Nobugai *et al.*, 1978) and for the 200 kV analytical transmission electron microscope (Morimoto and Kitamura, 1981). One of the typical examples showing the linear relationship is given in Fig. 1, where the intensities of characteristic X-rays have been plotted for various thicknesses of specimens or total counts under 200 kV. With an increase of thickness, the intensities deviate downwards for Si and Mg and upwards for Fe and Ca from the linear relation. This is considered to be due to the effect of absorption and fluorescence. The corrections for these effects have been proposed in various ways.

Two disadvantages are clear for the 200 kV analytical transmission electron microscope compared with the 100 kV. They are the decrease of signal/noise ratio and the relative decrease of X-ray intensities for light elements from those for heavy elements. However, the 200 kV analytical transmission electron microscope results in larger areas possible for trans-

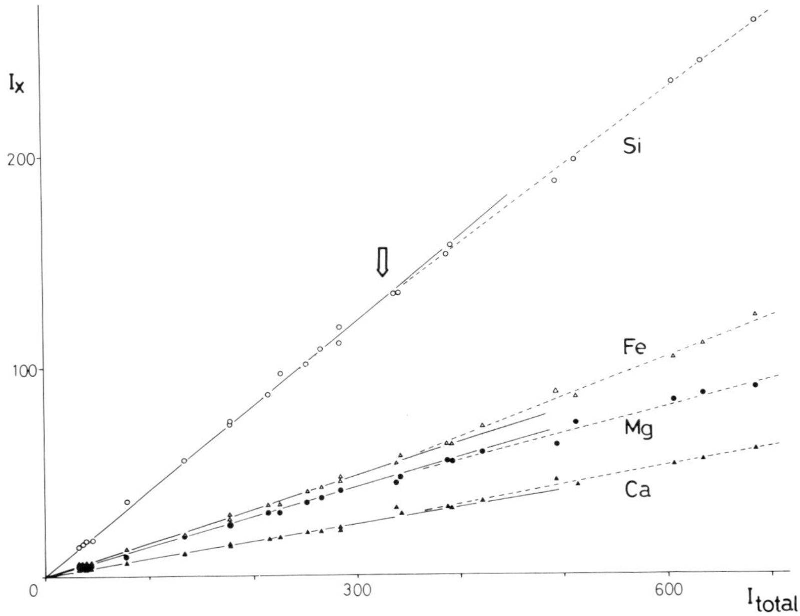

FIG. 1. The intensities of the characteristic X-rays for Si, Fe, Mg and Ca are compared with the total count. The mineral is a garnet from kimberlite. A linear relationship between the X-ray intensity and the total count is followed in the thinner regions. The arrow indicates the thickness at which the simple linear relation is no longer valid (MORIMOTO and KITAMURA, 1981).

mission than the 100 kV analytical transmission electron microscope. This effect is very important for the study of the mineral specimens.

4. *Microtextures in Pyroxenes*

The microtextures due to decomposition and exsolution in the silicate minerals are full of variety. The important factors controlling them are the degree of deviation from the equilibrium condition in the reaction, diffusion of elements, heat transfer, interface kinetics, and strain energy.

Decomposition by the reactions between the grains occurs mainly around the grain boundaries or crystal surfaces. However, the exsolution is a special case where decomposition takes place within a crystal. One of the most important differences in microtextures between these two processes is the effect of the strain energy. The exsolution textures are seriously affected by the strain fields where exsolution takes place, but the textures, due to the reactions among the crystals, are by the reaction rate.

The exsolution and decomposition of pyroxenes in the pyroxene quadrilateral have been one of the most important problems in mineralogy and have been extensively studied, because these were thought to be able to provide the information on the phase relations of the pyroxenes in a subsolidus condition and on the cooling history of the rocks which contain the pyroxenes (BUSECK *et al.*, 1980).

4.1 Exsolution textures of augite and pigeonite

The crystal structures of augite and pigeonite belong to the same space group, $C2/c$, at high temperatures, but the space group of pigeonite changes from $C2/c$ to $P2_1/c$ when temperature decreases.

The nucleation growth process must take place in the exsolution which takes place near an equilibrium state. Crystal defects of the host must be the prefered sites for the nucleation of the exsolved phase. In a further growth process, the elastic strain energy will play an important role and affect the morphology of the exsolved phase.

In the system of augite and pigeonite, the change of the unit cell dimensions of both host and exsolved phases will affect the morphology or the orientation of the interfaces. Based on the precise work by the optical microscope and the X-ray method, ROBINSON *et al.* (1977) have found that the exsolution lamellae in augite from Bushveld gabbro consist of '100' hypersthene, '100' pigeonite and three kinds of '001' pigeonite. The '001' pigeonite lamellae were divided into three types, type I, II and III, based on their thickness (Fig. 2). Because the orientations of the phase boundaries between the '001' pigeonite lamellae and the host augite are different, depending on the thickness of lamellae, ROBINSON *et al.* (1977) have concluded that these different orientations were caused by the difference of the unit cell dimensions of both phases at different temperatures, and that thicker lamellae were precipitated at higher temperatures than thinner lamellae.

Recent analytical transmission electron microscopic study on the Bushveld specimen (KITAMURA *et al.*, 1981a) shows that type III lamellae are richer in iron than the other two types, though type I and II lamellae have insignificant difference in composition. These results are consistent with those of ROBINSON *et al.* (1977) and indicate the re-equilibration of type I lamellae.

By strain energy produced by changes of the unit cell dimensions of both phases during the cooling, pigeonite of type I and II lamellae suffers the lattice distortion which causes the stacking faults in (100), or the unit cell scale twinning representing a unit cell scale hypersthene structure (Fig. 3) (CHAMPNESS and COPLY, 1976; KITAMURA *et al.*, 1981a). This fault acts as a nucleation site for the hypersthene lamellae in a process from augite + pigeonite to augite + hypersthene.

A clear example of the effect of the strain energy produced at the interfaces is the transformation of hypersthene to a pigeonite-type structure, or clino-hypersthene, in the interfaces between augite and hypersthene (Fig.

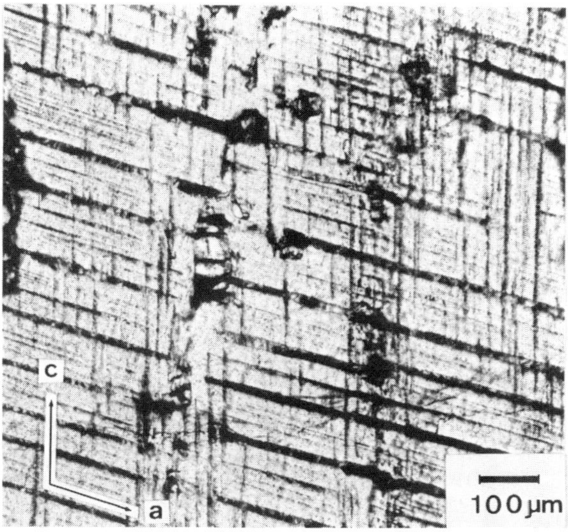

FIG. 2. Optical micrograph of augite from Bushveld gabbro (SA-1019). Note large coarsened '001' lamellae, type I, and narrower '001' type II and III between them. '100' lamellae can also be observed (KITAMURA *et al.*, 1981a).

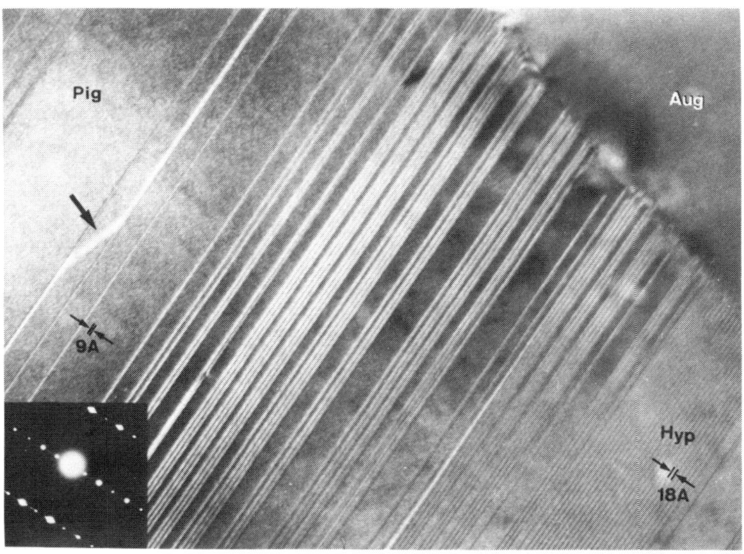

FIG. 3. Direct lattice image of the '001' pigeonite lamella in the augite host. The stacking fault occurs separately in the upper-left, but some clusters in the central part. The continuous repetition of the stacking faults in the lower-right apparently results in hypersthene with the spacing of 18 Å (KITAMURA *et al.*, 1981a).

4). The spacing among these plane defects with the structure of clino-hypersthene becomes greater by going further from the interfaces, as shown by the intervals between the small arrows. This feature also indicates the decrease of the strain energy by the increase in distance from the interface.

An example showing a drastic change of the morphology of the exsolved phases due to the strain energy, will be described from the Bushveld augite. Coexistence of two types augite lamellae, '001' and '100', in one pigeonite crystal was first described by MORIMOTO and TOKONAMI (1969). Based on the calculation of the strain energies of the coherent interfaces between pigeonite and augite for the two types, they concluded that the '001' lamellae must be common. According to SMITH's suggestion (1969), NAKAZAWA and HAFNER (1977) proposed the existence of a certain temperature where the best-fitting plane between pigeonite and augite changes from '001' to '100.' This temperature has been called MCT (morphology change temperature) (KITAMURA *et al.*, 1981b). The texture given in Fig. 5 (KITAMURA *et al.*, 1981b) indicates that the '001' lamellae which had appeared at a higher temperature, changed their morphology to the assemblage of '100' lamellae to reduce the strain energy. This relation can be expressed in a simple way using the diagram of coherent solvi (Fig. 6). The shape of the coherent solvi for '100' and '001' is different. At higher temperatures than MCT, '001' interfaces become more

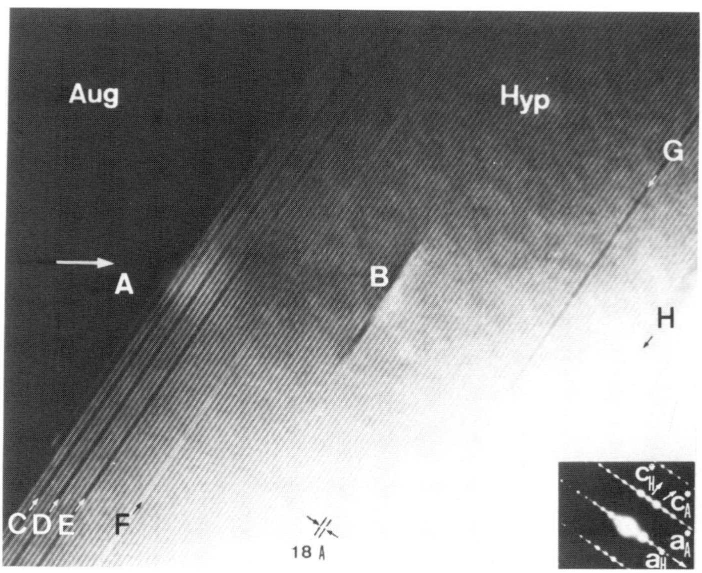

FIG. 4. Direct lattice-fringe image of the interface between augite and hypersthene. The (100) 18 Å hypersthene planes are imaged. The plane defects are also observed at C, D, E, . . . , which are related to the clino-hypsersthene structure. A dislocation at B and the special contrast at A are observed (KITAMURA *et al.*, 1981a).

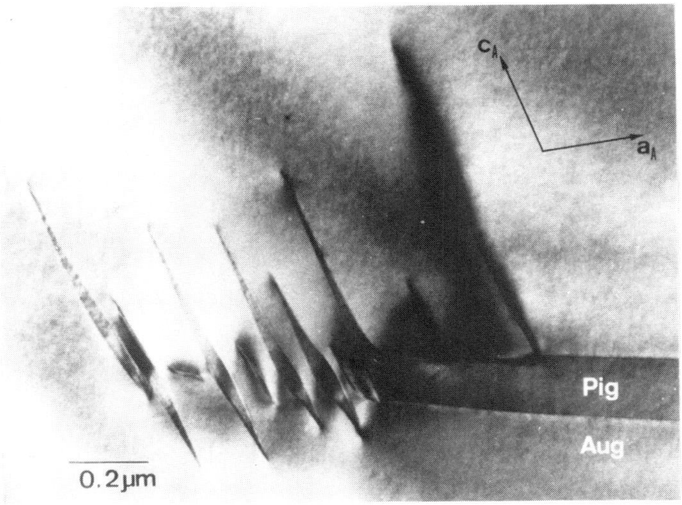

FIG. 5. Electron micrograph of fragmental '100' pigeonite lamellae located in the area extrapolated from the stem portion of the '001' lamellae in Bushveld augite (KITAMURA *et al.*, 1981b).

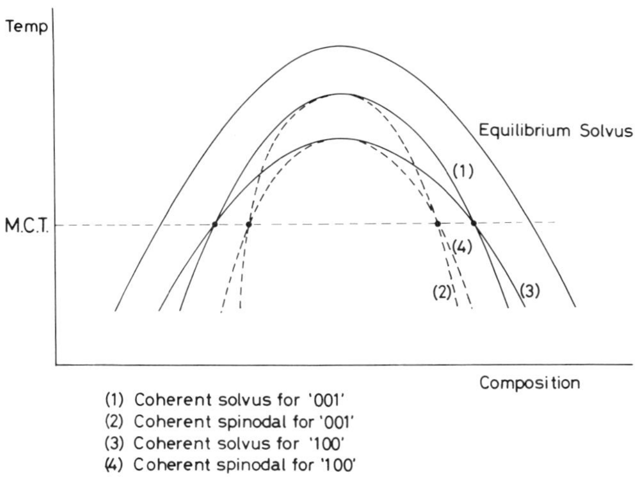

FIG. 6. Schematic diagram of equilibrium solvus, coherent solvi, and spinodal's for '001' and '100' in the augite-pigeonite system. At higher temperatures than morphology change temperature (MCT), '001' interfaces become more stable than '100,' while this situation is reversed below MCT.

stable than '100,' while this situation is reversed below MCT. This relationship explains the coexistence of the '001' and '100' lamellae in the same specimen.

4.2 Spinodal decomposition in pyroxene

Spinodal decomposition occurs by rapid cooling of clinopyroxene, because both pigeonite and augite (or diopside) have the same space group of *C*2/*c* at high temperatures. Because the wavelength of the compositional change is too small to be observed under the optical microscope in the spinodal decomposition, a transmission electron microscope has been employed to study the textures by the spinodal decomposition in the specimens from lunar basalts and kimberlite. The textures due to the spinodal decomposition have also been experimentally produced (McCallister and Yund, 1977). The most important characteristics of the textures are the periodic arrangement of the lamellae and the modulation of the interfaces, where the host and exsolved phases are completely coherent. The high resolution electron micrograph (Fig. 7, Kitamura and Aoki, 1980) shows the curved fringes indicating that the interfaces are coherent and a gradual change of the structural state or composition wave had existed. This texture is considered to be formed by an eruption of kimberlite magma (Kitamura and Aoki, 1980).

In other specimens from lunar basalts (cf. Nord *et al.*, 1976), the tweed textures consisting of two waves of (001) and (100), have been observed. These textures are considered to be formed at around MCT because the wavelengths of the two orientations are almost same reflecting the similar strain energy along the two orientations.

5. *Microtextures of the Plagioclase Solid Solution*

Most of the rock-forming minerals form wide ranges of the solid solutions in nature. Plagioclase forms a complete solid solution between albite, $NaAlSi_3O_8$, and anorthite, $CaAl_2Si_2O_8$, at high temperatures. At very low temperatures, however, only two end-members, albite and anorthite are considered to be stable. In between, complex immiscibility relations take place. All forms of plagioclase have the three-dimensional framework of SiO_4 and AlO_4 tetrahedra. Alkali or Ca atoms are accommodated in the cavities of the framework. The crystal structures of the two end-members, albite and anorthite, are compared schematically with the high-temperature plagioclase structure in Fig. 8. The study of plagioclase by the X-ray diffraction study by Chao and Taylor (1940) revealed the incommensurate-type *e*-satellites in the composition range of An_{25}–An_{75}. Since then, the superstructure of the *e*-plagioclase has been intensively studied (Smith, 1974).

A complete structure model of the *e*-plagioclase was presented based on the X-ray structure analysis of bytownite (An_{73}) by Kitamura and Morimoto

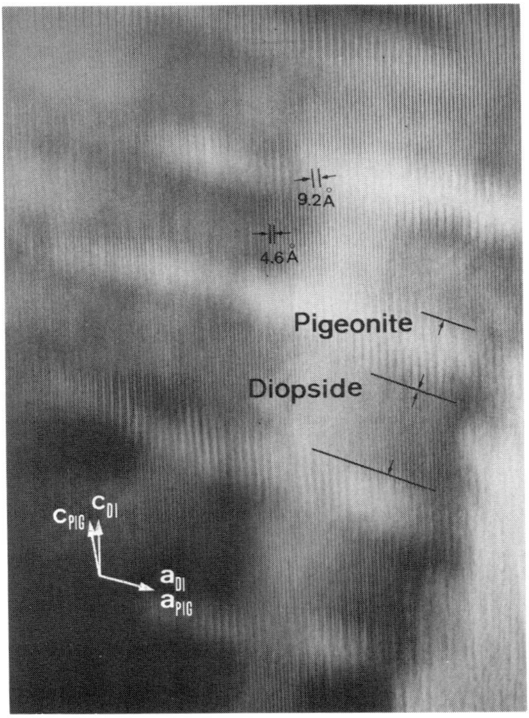

FIG. 7. Lattice image of the coherent intergrowth in diopside megacryst. The lattice fringes of 4.6 and 9.2 Å are the spacings of (200) for diopside and (100) for pigeonite, respectively (KITAMURA and AOKI, 1980).

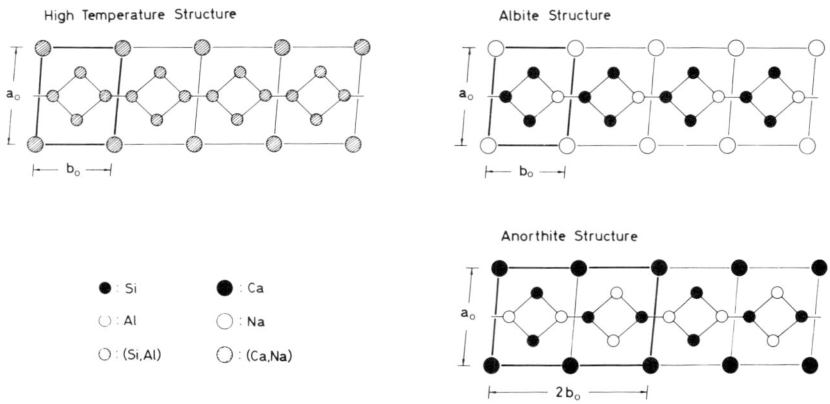

FIG. 8. Schematic representation of the structures of plagioclase feldspars. Unit cells are followed by the definition by KITAMURA and MORIMOTO (1977).

(1975, 1977). This model has been developed by the electron microscopic study of labradorite (An_{52}) from Labrador (MORIMOTO *et al.*, 1975a, b; NAKAJIMA *et al.*, 1977).

Figure 9 is the electron diffraction and the lattice image of the labradorite. A micro-lamella structure with an interval of about 30 Å is clearly observed running approximately in a N-S direction, perpendicular to the satellite pairs of the diffraction pattern. A part of Fig. 9 is enlarged and shown schematically in Fig. 10. The supercell given by KITAMURA and MORIMOTO is compared with the anorthite and albite cells.

The broader set of the lamellae is characteristic, where the white columns alternate with black ones. This alternation represents the reversed Si–Al distribution in the anorthite structure. The broader Ca-rich anorthite bands alternate with the narrower Na-rich albite bands. The contrast of the black and white subcells in each anorthite band is reversed across the albite band, resulting in an antiphase relation with the displacement vector of b_0.

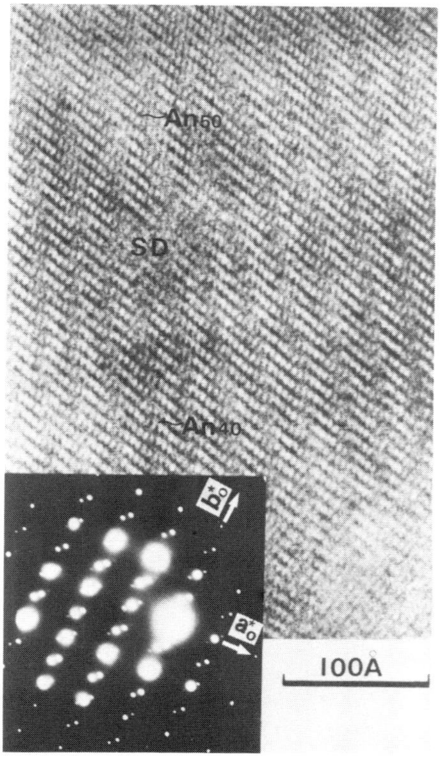

FIG. 9. High magnification lattice image from labradorite. The electron diffraction pattern used for image formation is also presented (MORIMOTO *et al.*, 1975b).

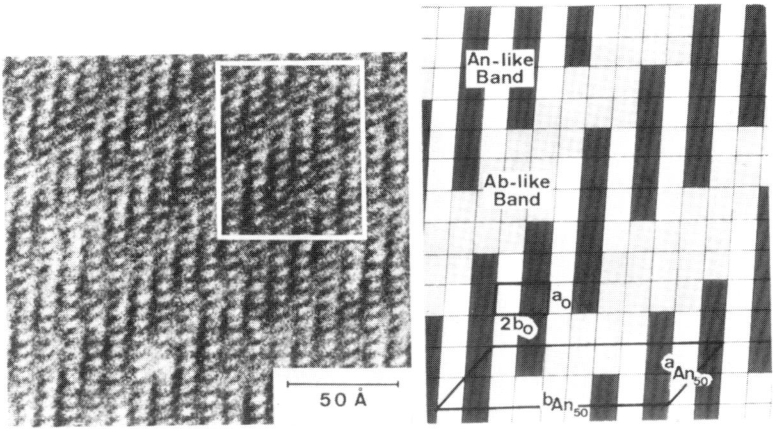

FIG. 10. Magnified portion of Figure 9, and its schematic representation. Only the part enclosed by white lines is shown. The supercell by KITAMURA and MORIMOTO (1975) is compared with that of anorthite.

The change in orientation and interval of the *e*-satellite pairs in plagioclase indicates that the superstructure of the *e*-plagioclase changes as shown schematically in Fig. 11. The orientation and the width of the two lamellae sets change with composition.

The modulated structures of plagioclase in nature are intermediate products in the process of the transformation to the two end-members from the high-temperature phase, and tell the geological history which the plagioclase experienced.

6. *Microtextures of Other Reaction Products*

6.1 *Reaction rims of garnet in kimberlite*

After the formation of rocks, further reactions proceeded among the constituent minerals in most rocks. These reactions sometimes left the reaction rims around crystals, when the reaction did not proceed throughout the crystals. Thus, the reaction rims give information on the subsolidus relations among the constituent minerals.

One of the examples of these reaction rims is the kelyphitic rims or rinds which appear around garnet crystals in kimberlite, eclogite, and other rocks.

Kelyphitic rims around garnet in kimberlite have been studied by the optical microscope and electron probe microanalysis (see NIXON *et al.*, 1963) and 200 kV analytical transmission electron microscopy (MORIMOTO and KITAMURA, 1981). The main constituent minerals are orthopyroxene, clinopyroxene, and spinel less than 2 μm in size. Fibrous crystallites of these minerals are observed adjacent to the garnet, and rather larger euhedral

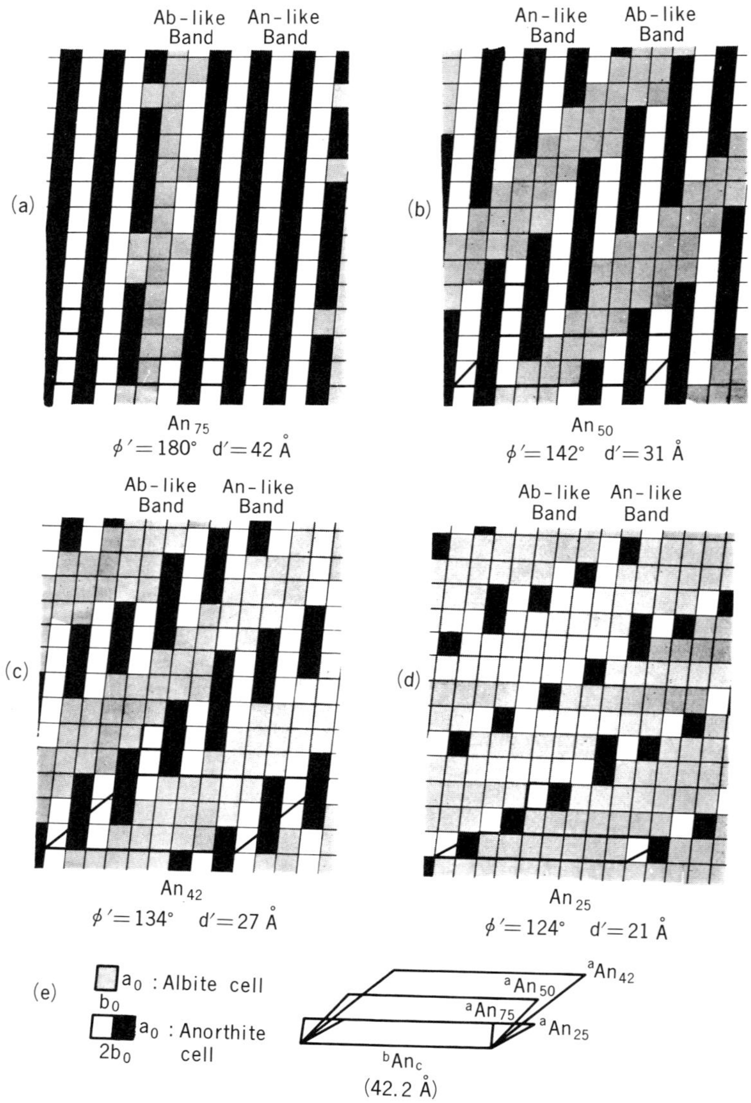

FIG. 11. The superstructures of the *e*-plagioclase of An_{75} (a), An_{50} (b), An_{42} (c) and An_{25} (d) projected along the c_0 axis (MORIMOTO *et al.*, 1975a). The anorthite-like band shown by white and black subcells, and the albite-like band by the faint subcells.

crystals adjacent to the olivine (Fig. 12a, b). These textures indicate a rapid reaction followed by an annealing process to form these minerals.

The mineralogy of the rims indicates the existence of the following reaction between garnet and olivine:

Garnet + Olivine → Orthopyroxene + Clinopyroxene + Spinel.

The chemical composition of these pyroxenes indicates that this reaction occurred at around 1000°C from the WOOD and BANNO's (1973) method. The textures and estimated temperature indicate that the reaction is considered to have occurred rapidly during the ascent of the kimberlite.

6.2 *Morphology and chemical heterogeneity in the minerals synthesized in subsolidus conditions*

To elucidate the reaction kinetics of the formation of minerals under subsolidus conditions, synthetic methods must play an important role. High pressure and high temperature synthesis has been intensively carried out by many workers to investigate the equilibrium mineral assemblages in the earth interior. However, only very few studies have been conducted with the intention of understanding the reaction kinetics. A study for this purpose has been carried out with the 200 kV analytical transmission electron microscope (KITAMURA *et al.*, 1982).

The solubility of Al_2O_3 in orthopyroxene in the system of orthopyroxene-pyrope was studied to understand the pressure and temperature conditions where natural orthopyroxene was formed (BOYD and ENGLAND, 1964; MACGREGOR, 1974; ARIMA and ONUMA, 1977). The solubility change of Al_2O_3 in orthopyroxene in this system has been clearly demonstrated by the study of synthetic crystals from the gel by means of the 200 kV analytical transmission electron microscope (KITAMURA *et al.*, 1982). The fine textures of the synthetic crystals grown from the gel mixture at 2.86 GPa and 1,500°C have shown that the content of Al_2O_3 in orthopyroxene increases from the core to rims (Fig. 13a, b).

Although this experiment was carried out with the gel mixture as a starting material, the reaction kinetics act as one of the important factors governing the solubility of elements in natural crystal formation under subsolidus conditions.

7. *Conclusion*

To elucidate the history of the minerals and rocks, the study of the fine textures of minerals has become very important. To this end, analytical transmission electron microscopy is an indispensable tool, because it can give the information on the textural characteristics and chemical heterogeneity.

The textures of minerals formed under the subsolidus conditions are full of variety. In case of exsolution, which occurs within a crystal, the most

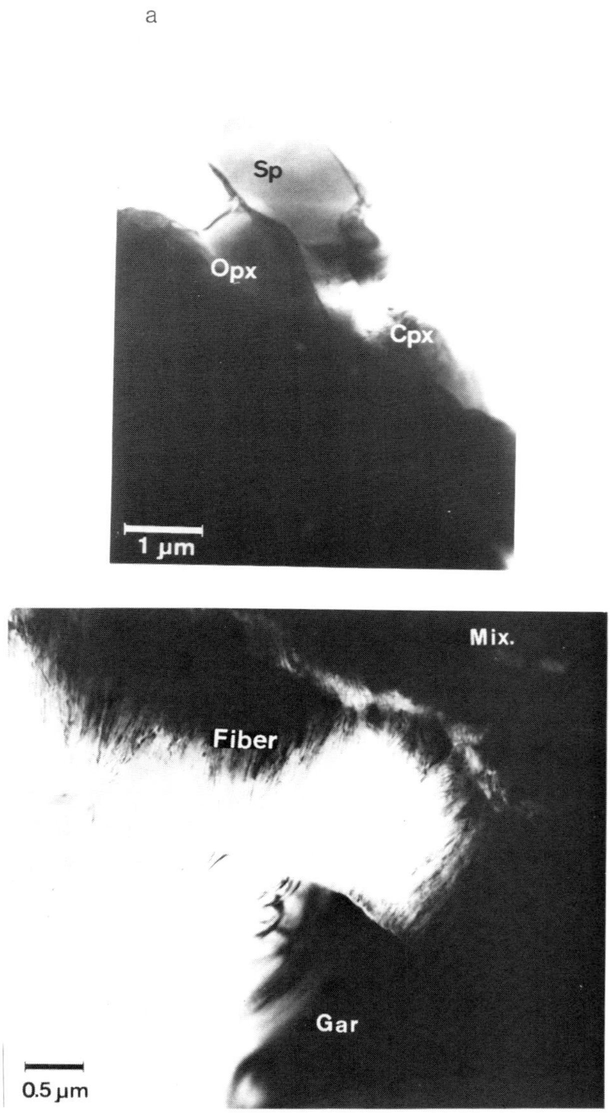

FIG. 12. Electron micrographs of the constituent minerals of the kelyphitic rims from kimberlite. (a) the assemblage of spinel, orthopyroxene and clinopyroxene observed in the region near the olivine. (b) fibrous crystallites adjacent to the garnet (MORIMOTO and KITAMURA, 1981).

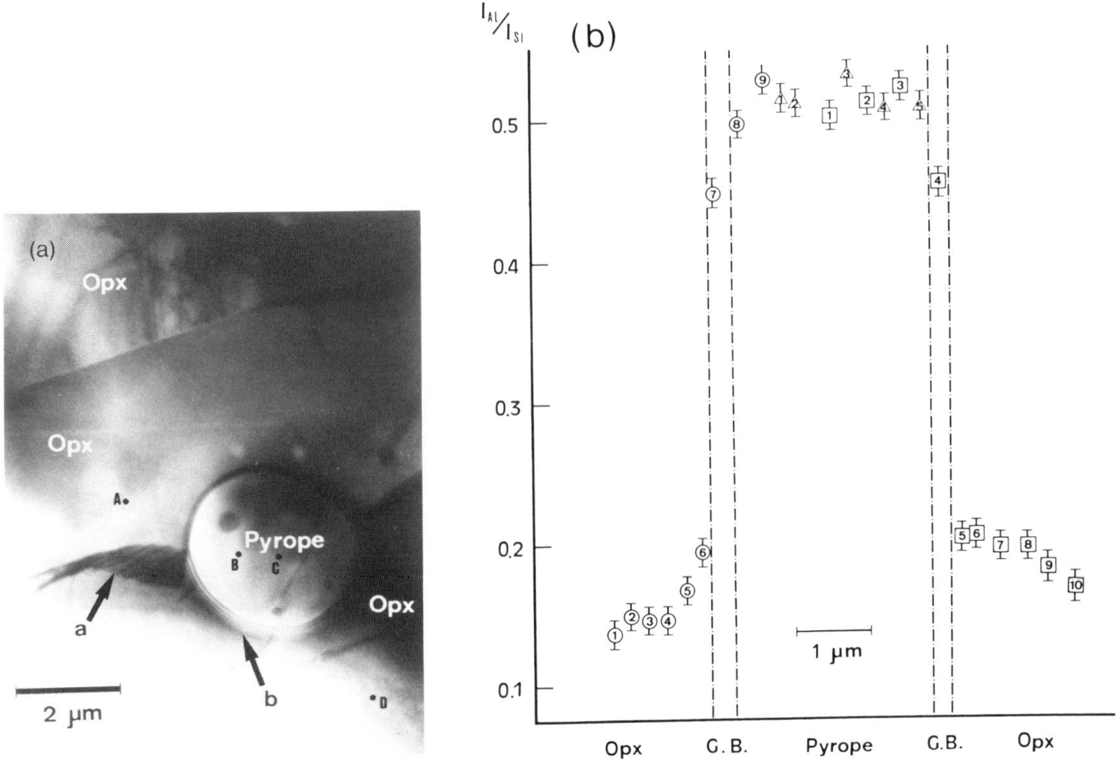

FIG. 13. (a) Electron micrograph of orthopyroxene and pyrope synthesized from the gel mixture at 2.86 GPa and 1,500°C. (b) Variation of the ratios of the characteristic X-ray intensities of Al and Si, I_{Al}/I_{Si} in orthopyroxene. The analytical points are in the lines of A-B, B-C, and C-D in (a).

important factor governing the textures is the strain. The reaction rates also play an important role to form the complex textures and chemical heterogeneity within the crystal. In case of the spinodal decomposition, characteristic textures usually appear and can be recognized by transmission electron microscopy.

We thank Dr. T. Kawasaki of Kochi University, Miss M. Yasuda in our institute and Dr. Y. Nakajima of the State University of Stony Brook, New York for their collaboration in our studies, and Prof. K. Aoki of Tohoku University for providing us with several specimens used in our studies. This work was supported by Grant in Aid for Scientific Research from the Ministry of Education of Japan, No. 420903. We also thank to Prof. I. Sunagawa of Tohoku University and Prof. M. Tokonami of Tokyo University for many discussions and encouragement.

REFERENCES

AMELINCKX, S. (1952): Screw dislocations in mica, *Nature*, **169**, 580.

ARIMA, M. and ONUMA, K. (1977): The solubility of alumina in enstatite and the phase equilibria in the join $MgSiO_3$–$MgAl_2SiO_6$ at 10–25 kbar, *Contrib. Mineral. Petrol.*, **61**, 251–265.

BOYD, F. R. and ENGLAND, J. L. (1964): The system enstatite-pyrope, *Carnegie Inst. Washington Yearbook*, **63**, 157–161.

BUCHANAN, D. L. (1979): A combined transmission electron microscope and electron microprobe study of Bushveld pyroxenes from the Bethal area, *J. Petrol.*, **20**, 327–354.

BUSECK, P. R. and IIJIMA, S. (1974): High resolution electron microscopy of silicates, *Am. Mineral.*, **59**, 1–21.

BUSECK, P. R., NORD, G. L. and VEBLEN, D. R. (1980): Subsolidus phenomena in pyroxenes, in *Reviews in Mineralogy, Vol. 7, Pyroxenes*, Ed. C. H. Prewitt, Mineral. Soc. Am., pp. 117–211.

CHAMPNESS P. E. and COPLEY, P. A. (1976): The transformation of pigeonite to orthopyroxene, in *Electron Microscopy in Mineralogy*, Ed. H. -R. Wenk, Springer, Berlin-Heidelberg-New York, pp. 228–233.

CHAO, S. H. and TAYLOR, W. H. (1940): Isomorphous replacement and superlattice structures in the plagioclase feldspares, *Proc. R. Soc. Lond. A*, **176**, 76–87.

CLIFF, G. and LORIMER, G. W. (1972): The quantitative analysis of thin metal foils using EMMA-4, the ratio technique, in *Proc. Fifth European Congr. Electron Microscopy*. Institute of Physics, London, pp. 140–141.

COWLEY, J. M. and MOODIE, A. F. (1957): The scattering of electrons by atoms and crystals, I. A new theoretical approach, *Acta Cryst.*, **10**, 609–619.

HONJO, G. and MIHAMA, K. (1954): A study of clay minerals by electron diffraction diagrams due to individual crystallites, *Acta Cryst.* **7**, 511–513.

IIJIMA, S., COWLEY, J. M. and DONNAY, G. (1973) High resolution electron microscopy of tourmaline crystals, *Tschermaks Mineral. Petrog. Mitt.*, **20**, 216–224.

KITAMURA, M. and MORIMOTO, N. (1975): The superstructure of intermediate plagioclase, *Proc. Japan Acad.*, **51**, 419–424.

KITAMURA, M. and MORIMOTO, N. (1977): The superstructure of plagioclase feldspars (A modulated coherent structure of the *e*-plagioclase), *Phys. Chem. Minerals*, **1**, 199–212.

KITAMURA, M. and AOKI, K. (1980): Coherent intergrowth in subcalcic diopside megacryst, *Mineral. J.*, **10**, 35–42.

KITAMURA, M., YASUDA, M. and MORIMOTO, N. (1981a): A study of fine textures of bushveld augite by 200 kV analytical electron microscopy, *Bull. Minéral.*, **104**, 278–284.

KITAMURA, M., YASUDA, M. and MORIMOTO, N. (1981b): Morphology change of exsolution lamellae of pigeonite in Bushveld augite—an electron microscopic observation—, *Proc. Japan Acad.*, **57**, 183–187.

KITAMURA, M., KAWASAKI, T. and MORIMOTO, N. (1982): Study of synthetic materials of the orthopyroxene-pyrope system by analytical transmission electron microscopy, in *High-Pressure Research in Geophysics*, Eds. S. Akimoto and M. Manghnani, Center for Academic Publications Japan (CAPJ)/D. Reidel, Tokyo/Dordrecht, pp. 499–509.

MACGREGOR, I. D. (1974): The system MgO–Al_2O_3–SiO_2: solubility of Al_2O_3 in enstatite for spinel and garnet peridotite compositions., *Am. Mineral.*, **59**, 110–119.

MCCALLISTER, R. H. and YUND, R. A. (1977): Coherent exsolution in Fe-free pyroxenes, *Am. Mineral.*, **62**, 721–726.

MCCONNELL, J. D. C. and FLEET, S. G. (1963): Direct electron-optical resolution of antiphase domains in a silicate, *Nature*, **199**, 586.

MORIMOTO, N. and TOKONAMI, M. (1969): Oriented exsolution of augite in pigeonite, *Am. Mineral.*, **54**, 1101–1117.

MORIMOTO, N., KITAMURA, M. and NAKAJIMA, Y. (1975a): Antiphase relation in superstructure of the *e*-plagioclase, *Proc. Japan Acad.*, **51**, 729–732.

MORIMOTO, N., NAKAJIMA, Y. and KITAMURA, M. (1975b): Direct observation of the superstructure of labradorite by electron microscopy, *Proc. Japan Acad.*, **51**, 725–728.

MORIMOTO, N. and KITAMURA, M. (1981): Applications of 200 kV analytical electron microscopy to the study of fine textures of minerals, *Bull. Minéral.*, **104**, 241–245.

NAKAJIMA, Y., MORIMOTO, N. and KITAMURA, M. (1977): The superstructure of plagioclase feldspar. Electron microscopic study of anorthite and labradorite, *Phys. Chem. Minerals*, **1**, 213–225.

NAKAZAWA, H. and HAFNER, S. S. (1977): Orientation relationships of augite exsolution lamellae in pigeonite hosts, *Am. Mineral.*, **62**, 79–88.

NIXON, P. H., VON KNORRING, O. and ROOKE, J. M. (1963): Kimberlites and associated inclusions of Basutoland: A mineralogical and geochemical study, *Am. Mineral.*, **48**, 1090–1132.

NOBUGAI, K., TOKONAMI, M. and MORIMOTO, N. (1978): A study of subsolidus relations of the Skaergaard pyroxenes by analytical electron microscopy. *Contrib. Mineral. Petrol.*, **67**, 111–117.

NOBUGAI, K. and MORIMOTO, N. (1979): Formation mechanism of pigeonite lamellae in Skaergaard augite, *Phys. Chem. Minerals.*, **4**, 361–371.

NORD, G. L. Jr., HEUER, A. H. and LALLY, J. S. (1976): Pigeonite exsolution from augite, in *Electron Microscopy in Mineralogy*, Ed. H. -R. Wenk, Springer, Berlin, Heidelberg, New York, pp. 228–233.

RIBBE, P. H. (1962): Observations on the nature of unmixing in peristerite plagioclases, *Norsk. Geol. Tidsskr.*, **42**, 138–151.

ROBINSON, P., ROSS, M., NORD, G. L. JR., SMITH, J. R. and JAFFE, H. W. (1977): Exsolution lamellae in augite and pigeonite: fossil indicators of lattice parameters at high temperature and pressure, *Am. Mineral.*, **62**, 857–873.

SMITH, J. V. (1969): Crystal structure and stability of the $MgSiO_3$ polymorphs; physical properties and phase relations of Mg, Fe pyroxenes, *Mineral. Soc. Am. Spec. Pap.*, **2**, 3–29.

SMITH, J. V. (1974): *Feldspar Minerals, 1, Crystal Structure and Physical Properties*, Springer, Verlag, Heidelberg.

WOOD, B. J. and BANNO, S. (1973): Garnet-orthopyroxene and orthopyroxene-clinopyroxene relationships in simple and complex systems, *Contrib. Mineral. Petrol.*, **42**, 109–124.

YADA, K. (1967): Study of chrysotile asbestos by a high resolution electron microscope, *Acta Cryst.*, **23**, 704–707.

Materials Science of the Earth's Interior, edited by I. Sunagawa, pp. 271–280.

OXYGEN SELF-DIFFUSION COEFFICIENT IN SINGLE-CRYSTAL FORSTERITE

Yasumichi OISHI and Ken ANDO

Department of Nuclear Engineering, Kyushu University, Fukuoka, Japan

1. *Introduction*

In clarifying diffusion-controlled phenomena of a mineral such as creep, sintering, crystal growth, phase transition, and other solid-state reactions, it is essential to previously know the details of the diffusion mechanisms of constituent ions. If the activation energies of diffusion for the respective ions are known with precise differences, they will provide a criterion for determination of the controlling species in the phenomenon of interest.

Forsterite has an orthorhombic structure the same as that of olivine but involved lattice defects should be simpler than in olivine which contains Fe ions as an additional constituent. Information of the diffusion mechanisms in forsterite will be useful for clarification of diffusion characteristics common to olivine.

Stable isotope, ^{18}O (or ^{17}O), is utilized as the tracer for determination of the oxygen self-diffusion coefficient, since oxygen has no convenient radioactive isotope, as well as is the case for silicon. Several techniques are used for determination of the diffusion amount of the tracer. The first reported oxygen self-diffusion coefficient in single crystal forsterite was determined by SOCKEL and HALLWIG (1977), using secondary ion mass spectroscopy. JAOUL *et al.* (1980) employed ion probe microanalysis and REDDY *et al.* (1980) proton activation analysis. The present work describes the results obtained by a mass spectrometer technique (ANDO *et al.*, 1981).

2. *Determination of Oxygen Self-Diffusion Coefficient*

The oxygen self-diffusion coefficient is determined by an isotope exchange technique; a solid sample (forsterite) is diffusion-annealed at a selected temperature for a scheduled time in an O_2 atmosphere enriched with ^{18}O (50% ^{18}O enrichement and 2×10^4 Pa). The diffusion coefficient to be determined is not high enough to change noticeably the oxygen isotope ratio of the atmosphere, as in the present case, so that the isotope ratio change is determined for the solid sample (OISHI and KINGERY, 1960). The forsterite sample enriched with ^{18}O by diffusion annealing is crushed and reacted with purified carbon powder at approximately 1,800°C in a closed furnace to gasify the oxygen in the solid sample as CO. The collected CO is then converted to CO_2, in accordance with Eq. (1), with the aid of a pure iron catalyser at approximately 700°C; a liquid-nitrogen bath is employed so as to facilitate the conversion by cold-trapping the formed CO_2.

$$2CO = CO_2 + C \tag{1}$$

The obtained CO_2 is submitted to determination of the oxygen isotope ratio by a mass spectrometer.

From the ^{18}O uptake into the solid sample, the oxygen diffusion coefficient is calculated for the boundary condition of the diffusion from a limited-volume, well-stirred solution to a plane sheet as described by Eq. (2) (CRANK, 1975).

$$\frac{M_t}{M_\infty} = (1 + \alpha)\left[1 - e\,\mathrm{erfc}\left(\frac{S}{\alpha}\sqrt{Dt}\right)\right] \tag{2}$$

where M_t and M_∞ are the total amounts of ^{18}O diffused into the solid in times t and infinity, respectively, t is the diffusion time, D is the diffusion coefficient, α is the oxygen mole ratio in the gas and solid phases, S is the specific surface area for the isotope exchange, and e erfc (Z) represents $\exp(Z^2)\,\mathrm{erfc}(Z)$.

The present determination is conducted with an assumption that the isotope exchange at the gas-solid interface is fast enough relative to the diffusion in the solid phase, since the oxygen diffusion coefficient in forsterite is relatively low.

3. *Effect of Surface Condition on the Diffusion Coefficient Determination*

The fractional diffusion amount, M_t/M_∞ is a function of S, for a sheet of a given D, as described by Eq. (2). Consequently, if the specific surface area, S, is simply estimated from the macroscopic dimensions of the sample, the diffusion coefficient calculated by Eq. (2) from an experimentally obtained M_t/M_∞ is not always the true diffusion coefficient. The S must be estimated for the microscopic surface area in the case where the penetration depth of the tracer diffusion cannot be sufficiently large, owing to the diffusion coefficient being small, as is the case for the oxygen ion in forsterite. The S value for an actual sample can be increased, from that for the ideal geometry, by surface roughness, microcracks, and surface-emerging dislocations, which vary with sample preparation. This effect of the surface condition was examined for a forsterite single crystal by preparing the samples with three different surface finishes from a single boule.

The sample employed was a single crystal forsterite synthesized by the Czochralski technique by TAKEI and KOBAYASHI (1974). A similar single crystal synthesized by them contained a dislocation density of 2 to 8 $\times 10^4/cm^2$ and no subgrain boundaries.

(1) A plate of 3 by 3 by 8 mm was carefully sawed from a single crystal boule to expose (100) planes by a diamond-impregnated wire saw, using abrasive slurry.

(2) The wire sawed sample was chemically polished, using a $2HF-HNO_3-2HCl$ solution to remove the 30 μm-thick surface layer. The polished surface was not flat, under an optical microscope, but jagged with crystallographically-oriented pyramids $\approx$ 50 μm which apparently increased its surface area.

(3) The wire-sawed sample was mechanically polished to 570 μm thick, using successively finer diamond pastes of 15 to 1 μm. Subsequently, the polished sample was ion-milled to remove a 10 to 15 μm thickness, by employing a 6.0 kV argon-ion beam at grazing angle of 30°. The ion-milled surface appeared flawless under an optical microscope (280 ×).

The oxygen diffusion coefficients determined at 1,650°C for the samples with different surface finishes are compared in Fig. 1. The chemically polished

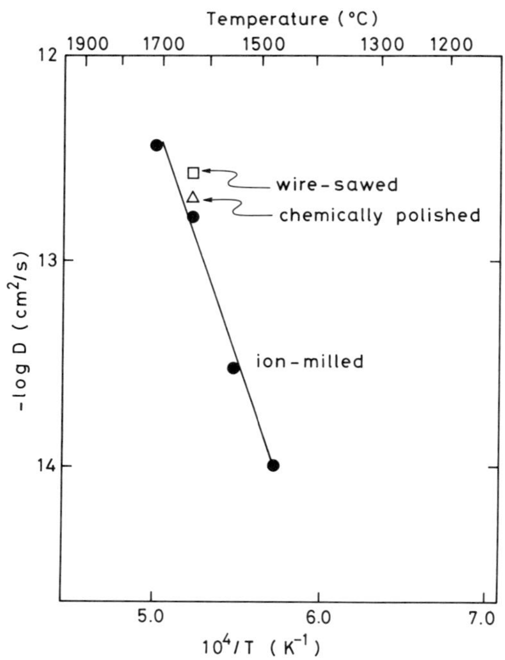

FIG. 1. Oxygen self-diffusion coefficient in single-crystal forsterite as a function of temperature.

sample showed a diffusion coefficient lower than that of the wire-sawed sample, and the ion-milled sample showed the lowest diffusion coefficient. A sample separately prepared by chemical polishing followed by ion-milling exhibited a diffusion coefficient similar to that of the above mechanically polished and ion-milled sample. These results indicated that ion-milling gave the least defective surface to the diffusion sample. The high diffusion coefficient exhibited by the wire-sawed sample is interpreted as due to surface roughness, microcracks, and/or surface-emerging dislocations introduced in the sample by wire sawing. Chemical polishing should be a useful technique to prepare the sample free from introduced microcracks and dislocations, if proper condition is provided so as not to form the above-mentioned pyramids. Ion-milling is a preferable polishing technique for forsterite to prepare the diffusion sample with flat surface without introducing defects.

4. *Oxygen Self-Diffusion Coefficient in Single Crystal Forsterite*

Although it has not been experimentally confirmed as yet, the presence of grain boundaries in forsterite is assumed to enhance diffusion of the oxygen ion, which is slower than the Mg ion (Sockel and Hallwig, 1977), by analogy with the oxygen ions in corundum (Al_2O_3) (Oishi and Kingery, 1960), periclase (MgO) (Shirasaki and Oishi, 1971), and spinel ($MgO \cdot Al_2O_3$) (Oishi and Ando, 1975). This grain boundary effect can be avoided by employment of single-crystal forsterite in the present determination of oxygen self-diffusion coefficients. The diffusion samples were prepared by mechanical polishing followed by ion-milling as mentioned above.

The oxygen self-diffusion coefficients determined in the *a*-direction, shown in Fig. 1, were represented by

$$D = 2.85 \times 10^{-2} \exp\left(-\frac{99.4\ \text{kcal/mol}}{RT}\right) \text{cm}^2/\text{s} \qquad (3)$$

for the temperature range of 1,472° and 1,734°C.

From the magnitude of the determined activation energy, the results are interpreted that the activation energy is not only the migration energy, but also involves the formation energy for oxygen defects.

The results are compared in Fig. 2 with previously reported self-diffusion coefficients of oxygen (Sockel and Hallwig, 1977; Jaoul *et al.*, 1980; Reddy *et al.*, 1980), magnesium (Sockel and Hallwig, 1977; Morioka, 1981), and silicon (Jaoul *et al.*, 1981). With respect to the oxygen self-diffusion coefficients, all these results should be understood to be in good agreement, taking into consideration the different samples and techniques employed for the determination, though the activation energy of the present work is somewhat higher than the others. Carefully polished samples are necessary to

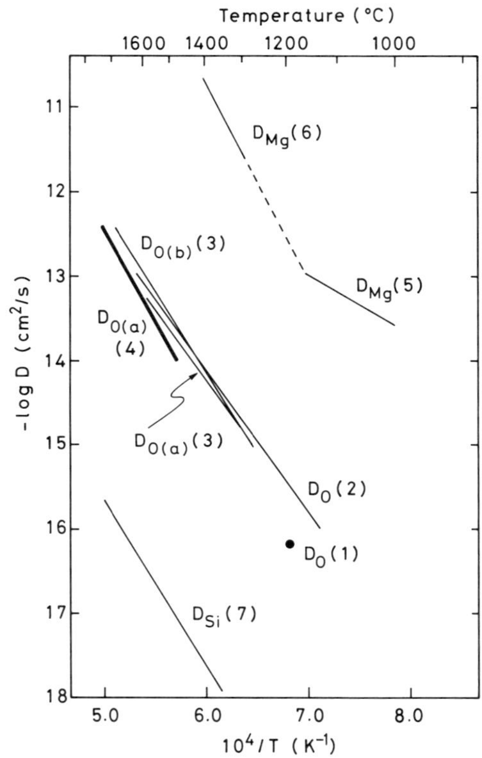

FIG. 2. Comparison of self-diffusion coefficients of oxygen, silicon and magnesium in single-crystal forsterite: (1) D_O, SOCKEL and HALLWIG, 1977; (2) D_O, JAOUL *et al.*, 1980; (3) $D_{O(a)}$ and $D_{O(b)}$, REDDY *et al.*, 1980; (4) $D_{O(a)}$, ANDO *et al.*, 1981; (5) D_{MK}, SOCKEL and HALLWIG, 1977; (6) D_{MK}, MORIOKA, 1981; (8) D_{Si}, JAOUL *et al.*, 1981.

determine whether the small differences between these in magnitude and activation energy are due to impurity effect or anisotropy of the diffusivity. Effect of crystallographic orientation on introduction of the defects by polishing must be distinguished from the anisotropy of the intrinsic diffusivity. If the contribution of pipe diffusion through surface-emerging dislocations becomes appreciable, this will result in a slightly decreased activation energy with an increased magnitude from the true diffusion coefficient.

As seen in Fig. 2, the oxygen diffusivity is several orders of magnitude lower than the Mg diffusivity. The activation energy for the Mg diffusion (106 kcal/mol) recently determined by MORIOKA (1981) is close to that for the oxygen diffusion in the present work (99.4 kcal/mol). The oxygen diffusivity is three orders of magnitude higher than the silicon diffusivity determined by JAOUL *et al.* (1981), but the activation energies are similar (99.4 kcal/mol for oxygen and 90 ± 10 kcal/mol for silicon). If this relativity between the oxygen

TABLE 1. Activation energies of diffusion and creep for oxides.

Oxide	Sample	Activation energy (kcal/mol)			Reference
		Diffusion		Creep	
Mg_2SiO_4	s.c.	Mg	(approx. 35)		Sockel and Hallwig, 1977
	s.c.	Mg	106		Morioka, 1981
	s.c.	Si	90		Jaoul *et al.*, 1981
	s.c.	O	77		Jaoul *et al.*, 1980
	s.c.	O	89 (*b*-direction)		Reddy *et al.*, 1980
	s.c.	O	77 (*a*-direction)		Reddy *et al.*, 1980
	s.c.	O	99.4 (*a*-direction)		Ando *et al.*, 1981
	s.c.			120–160	Jaoul *et al.*, 1981
Olivine	p.c.			100	Raleigh and Kirby, 1970
	p.c.			120	Carter and Ave'Lallemant, 1970
	p.c.			126	Post, 1977
	s.c.			125	Kohlstedt and Goetze, 1974
$MgO-Al_2O_3$ (spinel)	s.c.	Mg	86		Lindner and Åkerström, 1958
	s.c.	O	105		Ando and Oishi, 1974
	s.c.			127	Doukhan *et al.*, 1973

α-Al_2O_3	p.c.	Al	114		PALADINO and KINGERY, 1962
	s.c.	O	152		OISHI and KINGERY, 1960
	s.c.	O	188		REED and WUENSCH, 1980
	s.c.	O	155		OISHI *et al.*, 1981
	p.c.			130	FOLWEILER, 1961
	p.c.			130	WARSHOW and NORTON, 1962
	s.c.			180	CHANG, 1960
ThO_2	s.c.	Th	149.5		KING, 1971
	p.c.	O	65.8		EDWARDS *et al.*, 1963
	s.c.	O	49.9		ANDO *et al.*, 1976
	p.c.			112	POTEAT and YUST, 1966
UO_2	s.c.	U	98.3		REIMANN and LUNDY, 1969
	s.c. p.c.	O	59.3		MARIN and CONTAMIN, 1969
	p.c.			95	SCOTT *et al.*, 1959
	p.c.			91	ARMSTRONG *et al.*, 1962

and silicon diffusivities is true, the results may indicate that diffusion of silicon requires motion of oxygen ions, while diffusion of oxygen does not necessarily require motion of silicon ions. The oxygen diffusivity higher than that of silicon with similar activation energies are known for a liquid silicate (OISHI *et al.*, 1975), in which all the oxygen ions are not necessarily in the same state, but can be existent in different states with different diffusivities (WAKABAYASHI and OISHI, 1978; OISHI and NANBA, 1982).

5. *Comparison of Diffusion and Creep Data*

Listed in Table 1 are activation energies for self-diffusion of constituent ions and creep for forsterite and several other oxides. The creep mechanism for single crystal forsterite has been discussed in connection with the diffusion data by JAOUL *et al.* (1981). In their discussion, since the diffusion activation energy for Mg was far smaller than the creep activation energy, the Mg ion was ruled out of the possibility to be the rate-controlling species in creep of forsterite. However, since Morioka's activation energy for the Mg diffusion is now close to those for the oxygen and silicon diffusion, the Mg ion cannot be ruled out from the possibility, only on the basis of the activation energies.

While the creep activation energy (160 ~ 120 kcal/mol) is higher than the diffusion activation energies, the creep activation energy is temperature-dependent in Jaoul *et al.*'s. determination. Extrapolation of their data toward the low temperature side suggests the activation energy lower than 120 kcal/mol and closer to the diffusion activation energies. This supports that the low-temperature creep of forsterite is diffusion-controlled. Creep activation energies determined for olivines are also in the range of 100 ~ 126 kcal/mol.

The diffusion activation energy is not a clear-cut criterion for forsterite in determining which ion is the rate-controlling species for creep at the present stage, since diffusion activation energies determined for magnesium, silicon, and oxygen appear to be similar. One may think silicon is rate-controlling because of the magnitude of D_{Si} which is smallest of the three constituent ions. However, the creep rate should be controlled by the slowest diffusing species along its fastest path. The comparison of relative diffusivities may have to be made for short-circuit diffusions of the respective species, such as pipe diffusion through dislocations, in the case where the short-circuit diffusion takes part in the rate-controlling step of the deformation mechanism. In the case of α-Al_2O_3, the lattice self-diffusion coefficient of the oxygen ion is smaller than that of the Al ion and the O ion should be the rate-controlling species for the single-crystal creep (CANNON and COBLE, 1975). However, in the case of the polycrystal creep, grain-boundary enhancement of the oxygen diffusion reverses the effective diffusivity of the oxygen ion relative to that of the Al ion; consequently, the Al ion becomes rate-controlling in creep and sintering for small-grain-sized polycrystalline alumina (COBLE, 1963).

In $MgO-Al_2O_3$ spinel, in contrast to α-Al_2O_3, the lattice self-diffusion coefficient of the oxygen ion is far smaller than that of the cation, so that the grain-boundary enhancement of the oxygen diffusion does not reverse the relative diffusivities of the oxygen ion and the cation. Consequently, the oxygen ion remains as the rate-controlling species in sintering (ANDO and OISHI, 1974) as in the single-crystal creep (DOUKHAN *et al.*, 1973).

The difference between the magnitudes of D_O and D_{Si} is relatively small in forsterite (by 10^3); so that information on grain-boundary enhanced diffusion of the oxygen ion relative to that of the silicon ion will be useful in clarification of the creep mechanism. Comparison of P_{O_2}-dependences of creep, D_{Si}, and D_O in olivine will also give a clue to solve this problem (JAOUL *et al.*, 1981).

REFERENCES

ANDO, K. and OISHI, Y. (1974): Self-diffusion coefficients of oxygen ion in single crystals of $MgO \cdot nAl_2O_3$ spinels, *J. Chem. Phys.*, **61**, 625–629.

ANDO, K., OISHI, Y. and HIDAKA, Y. (1976): Self-diffusion of oxygen in single crystal thorium oxide, *J. Chem. Phys.*, **65**, 2751–2755.

ANDO, K., KUROKAWA, H., OISHI, Y. and TAKEI, H. (1981): Self-diffusion of oxygen ion in single crystal forsterite, *J. Am. Ceram. Soc.*, **64**, C–30.

ARMSTRONG, W. M., IRVINE, W. R. and MARTINSON, R. H. (1962): Creep deformation of stoichiometric uranium dioxide, *J. Nucl. Mater.*, **2**, 133–141.

CANNON, R. M. and COBLE, R. L. (1974): Review of diffusional creep of Al_2O_3, in *Deformation of Ceramic Materials*, Eds. R. C. Bradt, and R. E. Tressler, Plenum Press, New York, pp. 61–100.

CARTER, N. L. and AVE'LALLMENT, H. G. (1970): High temperature flow of dunite and peridotite, *Geol. Soc. Am. Bull.*, **81**, 2181–2202.

CHANG, R. (1960): Creep of Al_2O_3 single crystals, *J. Appl. Phys.*, **31**, 484–487.

COBLE, R. L. (1963): A model for boundary diffusion controlled creep in polycrystalline materials, *J. Appl. Phys.*, **34**, 1679–1682.

CRANK, J. (1975): *The Mathematics of Diffusion*, 2nd ed., Oxford University Press, New York: p. 58

DOUKHAN, N., DUCLOS, R. and ESCAIG, B. (1973): Structural and mechanical study of creep in Al_2O_4Mg single crystals, *J. Physique*, **34**, C9 379–387.

EDWARDS, H. S., ROSENBERG, A. F. and BITTEL, J. T. (1963): Thorium oxide-diffusion of oxygen, compatibility with borides, and feasibility of coating borides by pyrohydrolysis of metal halides, Aeronautical Systems Division, Wright-Patterson Air Force Base, OH, Report No. ASD–TDR–63–635, July 1963.

FOLWEILER, R. C. (1961): Creep behavior of pore-free polycrystal-line aluminum oxide, *J. Appl. Phys.*, **32**, 773–778.

JAOUL, O., FROIDEVAUX, C., DURHAM, and MICHAUT, M. (1980): Oxygen self-diffusion in forsterite: implications for the high temperature creep mechanism, *Earth Planet. Sci. Lett.*, **47**, 391–397.

JAOUL, 0. (1981): private communication.

KING, A. D. (1971): Thorium diffusion in single crystal ThO_2, *J. Nucl. Mater.*, **38**, 347–349.

KOHLSTEDT, D. L., GOETZE, C. (1974): Low-stress high-temperature creep in olivine single crystal, *J. Geophys. Res.*, **79**, 2045–2051.

Lindner, R. and Åkerström, Å. (1958): Diffusion von Ni–63 in nickel-spinellen, *Zeit. Phys. Chem.*, **18**, 303–307.

Marin, J. F. and Contamin, P. (1969): Uranium and oxygen self-diffusion in UO_2, *J. Nucl. Mater.*, **30**, 16–25.

Morioka, M. (1981): Cation diffusion in olivine II. Ni–Mg, Mn–Mg, Mg and Ca, *Geochim. Chsmochim. Acta*, in printing.

Oishi, Y. and Kingery. W. D. (1960): Self-diffusion of oxygen in single crystal and polycrystalline aluminium oxide, *J. Chem. Phys.*, **33**, 480–486.

Oishi, Y. and Ando, K. (1975): Self-diffusion of oxygen in polycrystalline $MgAl_2O_4$, *J. Chem. Phys.*, **63**, 376–378.

Oishi, Y., Terai, R. and Ueda, H. (1975): Oxygen diffusion in liquid silicates and relation to their viscosity. in *Mass Transport Phenomena in ceramics.*, Eds. A. R. Cooper, and A. H. Heuer, Plenum Press, New York, pp. 297–310.

Oishi, Y., Ando, K. and Kubota, Y. (1980): Self-diffusion of oxygen in single crystal alumina, *J. Chem. Phys.*, **73**, 1410–1412.

Oishi, Y. and Nanba, M. (1982): Analysis of interdiffusion in liquid silicates by multiatomic ion model, *Sem-Ammual Report, The Asahi-Glass Foundation for the Contribution to Industrial Technology* (in printing).

Paladino, A. E. and Kingery, W. D. (1962): Aluminum-ion diffusion in aluminum oxide, *J. Chem. Phys.*, **37**, 957–962.

Post, R. L., Jr. (1977): High-temperature creep of Mt. Burnet Dunite, *Tectonophysics*, **42**, 75–110.

Raleigh, C. B. and Kirby, S. H. (1970): Creep in upper mantle, *Mineral Soc. Am. Spec, Pap.*, **3**, 113–121.

Reddy, K. P. R., Oh, S. M., Major, L. D., Jr. and Cooper, A. R. (1980): Oxygen diffusion in forsterite, *J. Geophys. Res.*, **85**, [B1] 32–326.

Reed, D. J. and Wuensch. B. J. (1980): Ion-probe measurement of oxygen self-diffusion in single-crystal Al_2O_3, *J. Am. Ceram. Soc.*, **63**, 88–92.

Reimann, D. K. and Lundy, T. S. (1969): Diffusion of ^{233}U in UO_2, *J. Am. Ceram. Soc.*, **52**, 511–512.

Scott, R., Hall, A. R. and Williams, J. (1959): Plastic deformation of uranium oxides above 800°C, *J. Nucl. Mater.*, **1**, 39–48.

Shirasaki, S. and Oishi, Y. (1971): Role of grain boundaries in oxygen self-diffusion in polycrystalline MgO, *Jpn. J. Appl. Phys.*, **10**, 1109.

Sockel, H. G. and Hallwig, D. (1977): Ermittelung kleiner Diffusionskoeffizienten mittels SIMS in oxydischen Verbindungen (SIMS determination of small diffusion coefficients in oxidic compounds), *Mikrochim. Acta*, Suppl. **7**, 95–107.

Takei, H. and Kobayashi, T. (1974): Growth and properties of Mg_2SiO_4 single crystals, *J. Cryst. Growth*, **23**, 121–124.

Wakabayashi, H. and Oishi, Y. (1978): Liquid-state diffusion of Na_2O–CaO–SiO_2 system, *J. Chem. Phys.*, **68**, 2046–2052.

Warshaw, S. I. and Norton, F. H. (1962): Deformation behavior of polycrystalline aluminum oxide, *J. Am. Ceram. Soc.*, **45**, 479–486.

Materials Science of the Earth's Interior, edited by I. Sunagawa, pp. 281–300.

TRANSIENT AND STEADY STATE CREEP OF OLIVINE

Mitsuhiro TORIUMI[1], Shun-ichiro KARATO[2], and Toshitsugu FUJII[3]

[1]*Department of Earth Sciences, Faculty of Science, Ehime University Matsuyama, Japan*
[2]*Ocean Research Institute, University of Tokyo, Minamidai, Nakano, Tokyo, Japan*
[3]*Department of Geological Sciences, University of Alberta, Alberta, Canada*

1. *Introduction*

It is believed that dislocation climb is a rate-controlling process of stationary deformation of the mantle (WEERTMAN, 1970; GOETZE and BRACE, 1972; GREEN and RADCLIFFE, 1972; GOETZE and KOHLSTEDT, 1973, KOHLSTEDT and GOETZE, 1974). The recent accumulation of data on creep of single and polycrystalline olivine raises important questions about the simple dislocation climb model for steady-state creep. POIRIER and VERGOBBI (1978) have pointed out a dissociation of [100] screw dislocation, which had been found by VANDER SANDE and KOHLSTEDT (1976), and have proposed a cross-slip controlled model. On the other hand, KARATO (1977), MINSTER and ANDERSON (1981), and KARATO *et al.* (1982) have suggested a cell-wall climb model which has been theoretically proposed by GITTUS (1976) and experimentally on MgO by HÜTHER and REPPICH (1973).

Previous experimental works on single and polycrystalline olivine have been focused on the mechanical properties concerning steady-state flow and the microstructure of dislocations, subgrain and grain under several ten-percent strains. In these studies, it is clarified that the stationary dislocation density is achieved only at a few percent strains in an olivine single crystal (DURHAM and GOETZE, 1977). The microstructure of dislocations, which form a network of cell walls in the steady-state creep, has been proposed by TORIUMI and KARATO (1978) for olivine and HÜTHER and REPPICH (1973) and GITTUS (1976) have stressed the formation and role of cell walls during steady-state creep.

Theoretical work interpreting cell formation has been carried out by HOLT (1970) in which he derived the relationship between average dislocation spacing and cell size. The [100] organization of dislocation, which shows an abundant tilt wall of (100), was reproduced experimentally by DURHAM and GOETZE (1977). They considered that the [100] organization is a stationary dislocation structure during creep, for it appears at a few percent strains with nearly constant strain rates. In naturally deformed mantle olivines, [100] organization of dislocations is common (GUEGUEN, 1977; GREEN, 1976; BUISKOOL TOXOPEUS and BOLAND, 1976), but slightly different from the cellular network dislocation structure of TORIUMI and KARATO (1978). The cellular structure of MgO named by HÜTHER and REPPICH (1973) and of olivine, named by TORIUMI and KARATO (1978) is outlined as an equant subgrain by low-angle tilt and twist boundaries.

Experimental reproduction of cellular networks of dislocation walls has been made by HÜTHER and REPPICH (1973) on MgO, by POIRIER (1972) on NaCl, and by ZEUCH and GREEN (1979), and KARATO *et al.* (1980) on olivine. The dislocation structures obtained by KARATO *et al.* (1980) are quite similar to those in mantle-deformed olivine. Such dislocation structures are revealed after several ten-percent strains , such as in MgO and NaCl, and they are preserved for larger strains during creep.

It has been noticed that recrystallization occurs during the creep process. Two recrystallization processes are often recognized in metals: one is grain-boundary migration (GBM) and the other is subgrain rotation (SGR). In naturally-deformed peridotites, POIRIER and NICOLAS (1975) have stressed SGR recrystallization of the texture during the deformation process of rocks. Recent works on NaCl by GUILLOPÉ and POIRIER (1979) and on olivine by KARATO *et al.* (1980) have revealed a dynamic recrystallization in high-temperature creep at large strains. Dynamic recrystallization of single-crystalline minerals is important for giving a possible way of stabilizing grain geometry during creep at applied stress. MERCIER (1980), on the other hand, noticed a discontinuous grain-size distribution in the upper mantle, and suggested that the SGR-grain size is larger than the GBM-grain size in natural deformation conditions. However, GUILLOPÉ and POIRIER (1979) have proposed a contrary relationship between these two grain sizes for NaCl, as against Mercier's conclusion for olivine. GUILLOPÉ and POIRIER (1979) also reported a successive recrystallization from SGR to GBM, and the grain which had been formed by SGR grew during the GBM process, due to a reduction of total interfacial energy. They, however, did not propose a mechanism by which grain size is determined as a function of the applied stress in GBM. Relations between grain size and applied stress for largely strained olivine polycrystals, suggest the stability of grain geometry during the creep of polycrystals. What is the process counteracting GBM that makes a grain fine? The SGR process most likely counteracts GBM, if the GBM velocity is very high during creep. It seems that grain stability is derived from a balancing of SGR and GBM.

The stability of grain geometry during creep is also satisfied by a grain-switching model for superplastic flow (ASHBY and VERALL, 1973; GITTUS, 1978). Dislocation-controlled creep and diffusional creep models, on the other hand, may not involve the conservation of grain geometry during creep. These mechanisms initially make the equant grain very flat for very large strains. Inasmuch as long-term stress is considered, the actual process of such a mantle flow seems to be dislocation and diffusional creep involving the dynamic recrystallization mentioned above and/or superplastic flow, as suggested by TWISS (1976).

The possibility of superplastic flow in the mantle was discussed by GUEGUEN and BOULLIER (1976). They noticed that sheared garnet lherzolite in Lethoto kimberlites went through a superplastic deformation in the upper mantle diapir, for sheared xenoliths are composed of fine-grained equigranular olivine, which displays a cubo-octahedral outline and a very fine-grained orthopyroxene around porphyroclastic orthopyroxene. The outline of olivine and fine-grained orthopyroxene is approximately the equilibrium in shape, that is, the system of these shaped-grains have no curved interfaces which have been readjusted by GBM at a rate which attaims textural equilibrium. GBM must occur in many interfaces if the deformation of polycrystals is performed by intragrain dislocation-glide and/or intragrain diffusion of vacancy.

As seen previously, creep experiments and discussions have been concentrated on steady-state deformation and microstructure, but transient creep has not been investigated. According to KARATO *et al.* (1980, 1982), creep behaviour of olivine below several ten percent strains corresponds to transient creep. Therefore, the time-dependent creep law at small strains will be discussed in this paper. There are two types of creep equations in the transient state (except for the incubation stage): one is logarithmic creep and the other is parabolic creep (including Andrade and exponential creep) (GAROFALO, 1965). In this paper, we discuss these two types of creep in olivine single crystals as well as dynamic recrystallization, secondary creep (steady-state creep), and the change of dislocation structure during high-temperature creep, as discussed in the papers of KARATO *et al.*, (1980, 1982).

2. *Experimental Procedures*

A detailed description of experimental procedures and data has been written in previous papers (KARATO *et al.*, 1980, 1982). In this paper, brief comments on high-temperature creep experiments are made.

Olivine single crystals of San Carlos, Arizona (Fo_{91}) were used for starting materials, for their dislocation density is very low (10^3–10^4 mm^{-2})

and their dislocation structure is of a cellular type (TORIUMI and KARATO, 1978). Oriented samples ($[110]_c$ and $[101]_c$) cut from a single olivine grain, were deformed in different creep conditions. Creep experiments were carried out using a W-Mo wound furnace at 0.1 MPa. A H_2/CO_2 gas mixture (ratio is 2.5) was flowed into the furnace for performing the oxygen fugacity range of olivine stability.

Compressional stresses were operated using an iron block weight with a DWL (dead weight lever) system. In order to keep the constant stress for large strains, the load was increased in 0.5 or 1.0 kg steps for a small strain estimated to assume a constant sample volume and a uniform change of sample-geometry.

The experimental techniques are as follows: the specimen was heated under no-differential stress. After relaxation of thermal expansion of the system, the load was applied. After the desired strain of the sample was achieved, the load was set off and then the temperature was lowered. The cooling rate in the high-temperature range is large enough to maintain the dislocation structure and the texture of the sample.

To obtain time-dependent equations of the transient creep, small strain creep experiments were carried out for a single olivine before large strain experiments at low-stress conditions. In order to avoid displacement by a collapse of asperities between the sample interface and the W-plate, a 3 MPa prestress was operated for 1 hour at the same temperatures as the creep experiments.

3. *Results and Discussions*

3.1 *Transient creep (microcreep)*

During the early stage of creep, transient creep appeared in the experiments. The transient creep in low temperatures is commonly approximated by the logarithmic creep (GAROFALO, 1965)

$$\varepsilon = \alpha \log(1 + vt) \tag{1}$$

in which ε is the creep strain, t is the time, and α and v are temperature and stress-dependent terms. Further, in high temperature conditions, the common creep equation in a transient state becomes a parabolic creep (which involves the Andrade creep) as follows;

$$\varepsilon = (k \cdot t)^m, \tag{2}$$

in which k is dependent on temperature and applied stress. The experimental creep curves, therefore, were plotted in a log t versus strain which is defined by ln (l/l_0) (l = sample length) as shown in Fig. 1, and log t versus log (strain) (Fig. 2) diagrams for stressing evidence of the transient creep.

It is clearly seen that two different segments compose the experimental

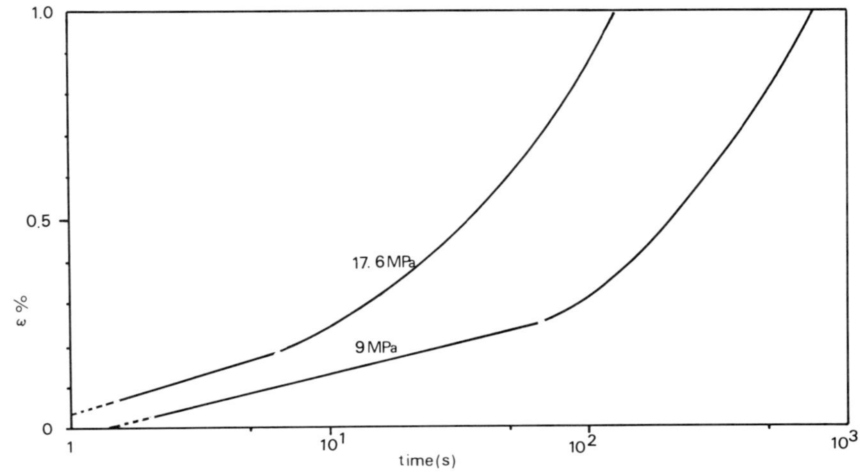

FIG. 1. Time-dependent transient creep curves of $[101]_c$ olivine single crystals for 17.6 and 9 MPa at 1,923 K. The linear segments in the early stage are the logarithmic creep and the curved ones are the parabolic creep.

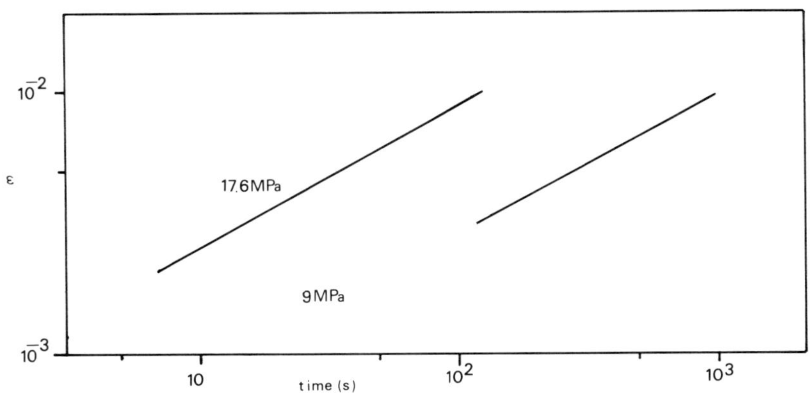

FIG. 2. Log strain versus log t plots of parabolic creep of $[101]_c$ single crystals for 17.6 and 9 MPa at 1,923 K.

creep curves: in the early stage the logarithmic creep appears and later the creep curve seems to be fitted by a parabolic creep. Transition from the logarithmic to parabolic creep was also reported on single crystals of Cu by WYATT (1960) (ref., GAROFALO, 1965). As compiled by GAROFALO (1965), the logarithmic creep of metal is common below $0.3\,T_m$ (T_m = the melting temperature). In the range from 0.3 to $0.7\,T_m$, the transient creep is usually parabolic. Combining these two facts, it is possible that the strain rate of the parabolic creep depends on temperature more than on logarithmic law. In this model, the schematic changes of creep curves with increasing temperature in the log t versus strain diagram are illustrated in Fig. 3. As pointed out by GAROFALO (1965), low stress conditions favour parabolic creep as well as the low temperature conditions in Fig. 3.

The difference between logarithmic and parabolic creep may be due to a change in dislocation density and structure. In logarithmic creep, dislocation motion due to kink-movement, even under relative constant dislocation density, controls the net strain of creep (the exhaustion theory of logarithmic creep by Mott (ref. COTTRELL, 1956)). On the other hand, parabolic creep is possibly due to a creep resistance from a dislocation multiplication and tangling (so-called work-hardening) (GAROFALO, 1965).

In Eq. (1), v is dependent up on temperature, as reviewed by GAROFALO (1965) for metals;

$$v = v_0 \cdot \exp(-U_1(\sigma)/RT) \tag{3}$$

and α is nearly constant. In Eq. (3), $U_1(\sigma)$ is the activation energy for dislocation glide. Therefore, we obtain

$$\varepsilon = \alpha \cdot \log t + C(T, \sigma) \tag{4}$$

with the conditions of $1 \ll vt$, in which $C(T, \sigma)$ is the stress and temperature-dependent term. As shown in Fig. 1, parabolic creep is predominant in transient creep at high temperatures. According to GAROFALO (1965), the k in Eq. (2) of parabolic creep holds,

$$k = k_0 \exp(-U_2(\sigma)/RT), \tag{5}$$

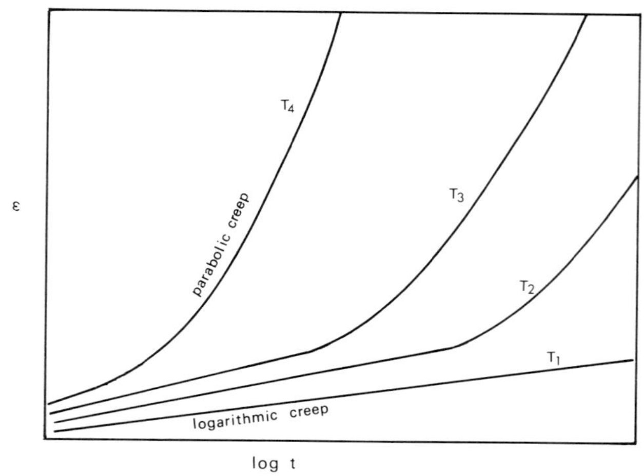

FIG. 3. Schematic change of transient creep with increasing creep temperature from T_1 to T_4. At low temperatures, the logarithmic creep appears, but it disappears and the parabolic one is dominant at high temperatures (T_4).

in which U_2 is the activation energy for parabolic creep. Applying Eqs. (1) and (2) to obtain the creep curves for 9 and 17.6 MPa experiments, time-dependent equations for transient creep at 1,923 K of olivine single crystal hold as follows;

$$\varepsilon = 0.18 \log (1.46 \cdot t) \qquad \text{at 17.6 MPa,} \qquad (6)$$

and

$$\varepsilon = 0.14 \log (0.72 \cdot t) \qquad \text{at 9 MPa} \qquad (7)$$

for logarithmic creep, but for parabolic creep they are

$$\varepsilon = (10^{-5.6} \cdot t)^{0.56} \qquad \text{at 17.6 MPa,} \qquad (8)$$

and

$$\varepsilon = (10^{-6.4} \cdot t)^{0.56}. \qquad \text{at 9 MPa.} \qquad (9)$$

To estimate the temperature and stress-dependence on creep equations, relations between the strain rate at a constant strain and stress and/or temperature were investigated. In Fig. 4, the relations between the stress and strain rate at 1% strain are shown in which data were obtained in such ways that the applied load is added step by step at the desired stresses and after set-off, the load is again operated in the same manner as a primary load operation. This figure clearly displays the strain-rate hysteresis at the cyclic operation of loading, thereby showing that dislocations multiplied at high stress were locked (work-hardening). The relationship between the stress and strain rate at 1% strain are shown in Fig. 5.

On the other hand, the activation energy of the creep at a 1% strain is estimated to be 668 kJ/mol in the stress range of 40–100 MPa (KARATO *et al.*, 1982). This value for transient creep is surprisingly larger than the activation energies for the steady-state creep of polycrystalline olivine and oxygen self-diffusion, as discussed later.

The relationship of the strain rate at a 1% strain and the stress and/or temperature obtained here can be introduced into the time-dependent creep equations previously mentioned.

Assuming the parabolic creep for a 1% strain, we obtain a strain rate from differentiation of Eq. (2) by time as follows;

$$\dot{\varepsilon} = m \cdot k^m \cdot t^{m-1} \qquad (10)$$

with $k = k_0 \exp(-U_2/RT)$.

Using $\varepsilon = (k \cdot t)^m$, and eliminating the time, we have

$$\dot{\varepsilon} = m \cdot k \cdot \varepsilon^{(m-1)/m}. \qquad (11)$$

Assuming $m = 0.56$ in Eq. (8) for experimentally-obtained parabolic creep of olivine single crystals, we take

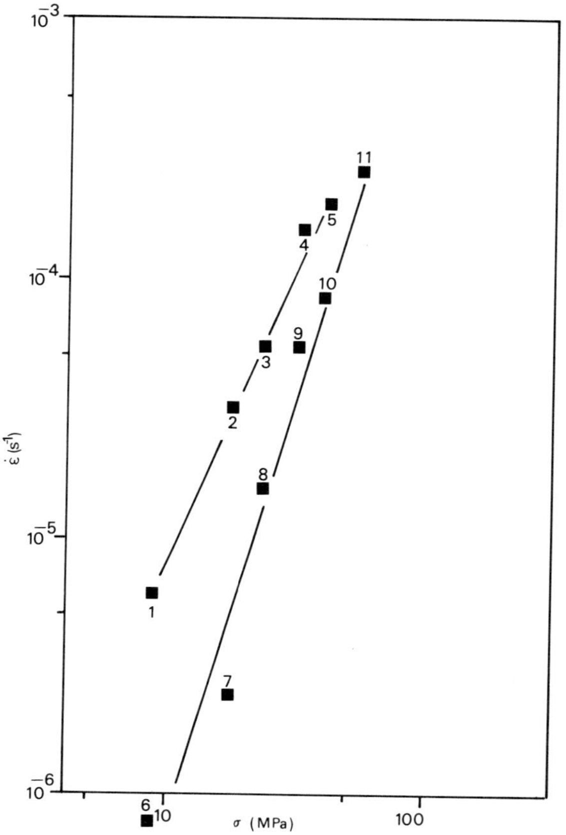

FIG. 4. Stress-strain rate relations at 1% strain showing the strain rate-hysteresis due to work-hardening. Numbers indicate the order of operation of loading from step 1 to step 11. A load was added from step 1 to step 5, and after set-off of the load, stress was applied again, the increasing from step 6 to step 11.

$$\dot{\varepsilon} = 0.56 \cdot k \cdot \varepsilon^{-0.79} \tag{12}$$

and at a 1% strain, we have

$$\dot{\varepsilon} = 20.4 \cdot k. \tag{13}$$

Therefore, the activation energy for microcreep can be obtained from the Arrhenius plotting of the strain rate at a constant applied stress, as investigated by KARATO *et al.* (1982). The relationship of the strain rate and stress are also indicated in Fig. 5 at a 1% strain.

Therefore, using Eq. (13) we obtain

$$k(s^{-1}) = 1 \times 10^9 \cdot \sigma^{2.5} \cdot \exp(-668{,}000/RT) \tag{14}$$

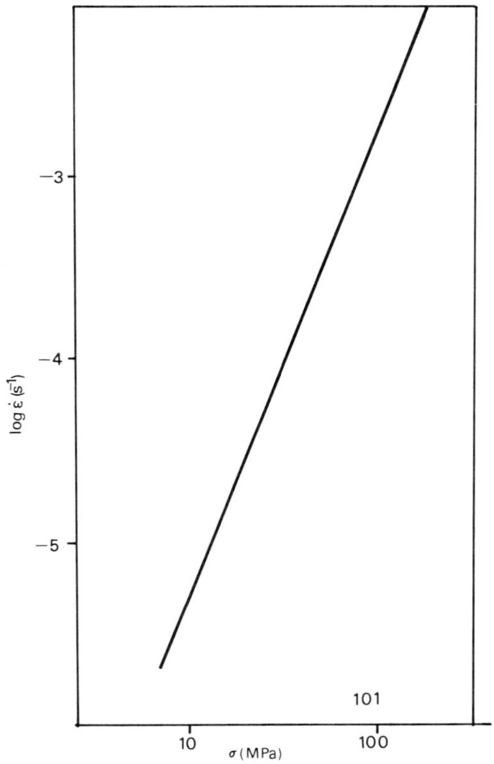

FIG. 5. Experimental stress-strain rate relations of $[101]_c$ olivine single crystals at 1% strain and 1,923 K.

in which σ is stress (MPa) and R is the gas constant (8.31 $JK^{-1}mol^{-1}$).

3.2 Dislocation structure

Change in dislocation density, as shown in Fig. 6, also supports the transient creep model under several ten-percent strains. In the range from 0.1 to 5% strains, dislocation density increases with creep strain, but at 5–10% strains, it reaches to saturation density, which depends on the applied stress. Therefore, it is concluded that recovery is not effective below 5–10% strains, but beyond these, recovery contributes towards a reduction of dislocation density.

Dislocation density is a function of applied stress above a few percent strains as compiled in Fig. 7. In this figure, results of previous works by many authors are shown together with the data obtained by this study.

Our data are fitted into the following equation;

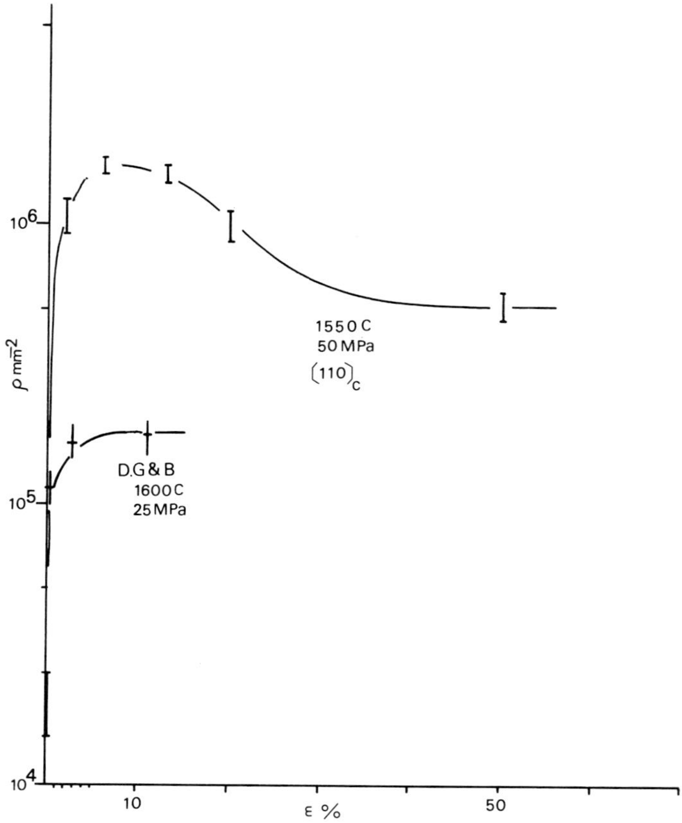

FIG. 6. Change of dislocation density with a increasing strain of $[110]_c$ single crystals at 50 MPa and 1,823 K. A change of dislocation density at 25 MPa and 1,873 K was obtained by DURHAM *et al.* (1977) (D G & B).

$$\rho(\mathrm{mm}^{-2}) = 7.0 \times 10^2 \sigma^{1.67}(\mathrm{MPa}). \qquad (15)$$

Dislocation structures are shown in Fig. 8. Although DURHAM *et al.* (1977) suggested that [100] organization of dislocations is a typical stationary dislocation structure, our data indicate the change from the [100] organization to the cellular-network dislocation structure at large strains (KARATO *et al.*, 1980) as shown in Fig. 8d. Therefore, it is suggested that [100] organization of dislocations is also in a transieent state. The transition of the texture of [100] organization to an equant cellular structure of the dislocation wall is illustrated in Figs. 8b and 8c. Incipient (001) dislocation arrays appear in the bands sandwitched by the (100) dislocation walls (Fig. 8b), suggesting the initiation of an equant cell formation during creep.

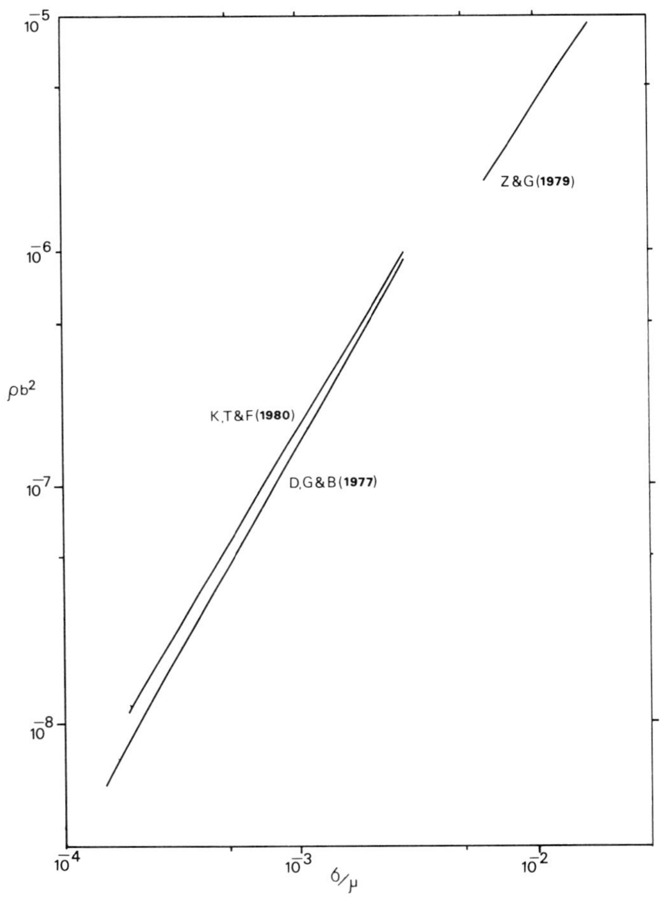

FIG. 7. Relations of dislocation density and stress. Abbreviations are as follows: D G & B = DURHAM *et al.* (1977), Z & G = ZEUCH and GREEN (1979), K T & F = this study.

3.3 *Steady-state creep*

During experiments of large strains, KARATO *et al.* (1980, 1982) indicated that olivine single crystals change to polycrystals, due to a dynamic recrystallization during creep. Further, at strains above 60–80%, creep curves display nearly constant slopes against time and creep is considered to be in a steady state.

As discussed in a previous paper, although strain rates are slightly fluctuated during creep, average strain rates have physical meanings within the accuracy of an order. As compiled in Fig. 9, the empirical creep law obtained by KARATO *et al.* (1982) holds,

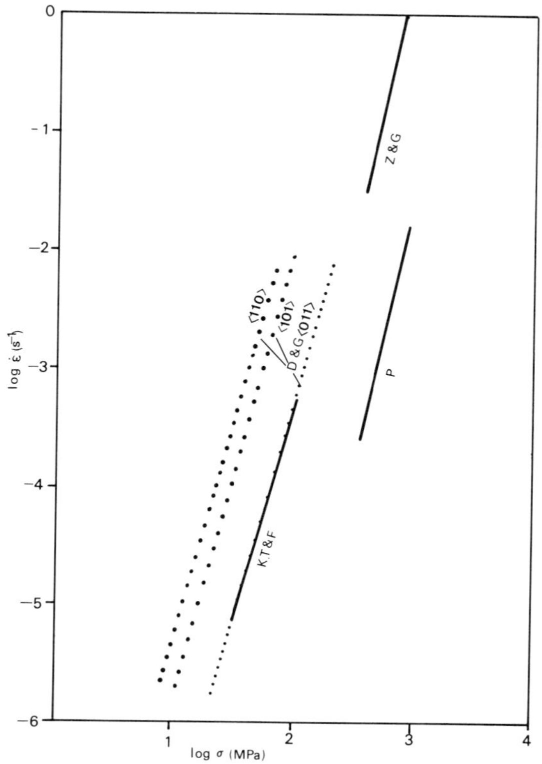

FIG. 8. Relations of stress and strain rate of polycrystalline olivine (full lines) and single-crystalline olivine (dotted lines). Abbreviations are as follows: P = POST (1977), D & G = DURHAM and GOETZE (1977), Z & G = ZEUCH and GREEN (1979), K T & F = KARATO *et al.* (1982).

$$\dot{\varepsilon} = 3.4 \times 10^4 \cdot \sigma^{3.5} \cdot \exp(-543{,}000/RT). \quad (16)$$

It is greatly stressed that the activation energy for the creep process obtained here has good accuracy, for it is estimated by means of a temperature-cyclic determination method described in a previous paper. The creep law in Eq. (16) is consistent with those of ZEUCH and GREEN (1979) at high pressure conditions (Fig. 9). The strain rates of polycrystals obtained here are smaller than those of single crystals of olivine obtained by DURHAM and GOETZE (1977) and KARATO *et al.* (1982).

It is noticed that the activation energy for steady-state creep of polycrystalline olivine is larger than that for oxygen and silicon self-diffusion in olivine, as pointed out by JAOUL *et al.* (1980), MINSTER and ANDERSON (1981), and KARATO *et al.* (1982). This is very important in determining the creep mechanism of silicates and will be discussed in later sections.

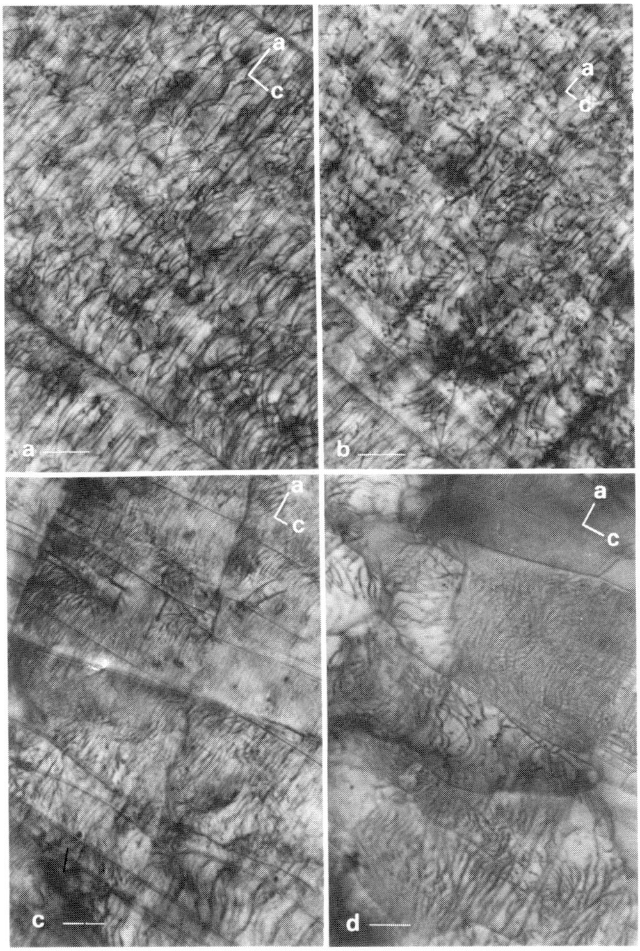

FIG. 9. Dislocation structures of parabolic creep stage (a and b) (transient creep) and steady-state creep stage (c, d) a and b : [100] organization in olivine deformed at 1,823 K, 50 MPa and 9% strains ; c : cellular network of subgrain boundaries in olivine deformed at 1,923 K, 60 MPa and 20% strains ; d : cellular network of subgrain boundaries in olivine deformed at 1,923 K, 40 MPa and 50% strains.

3.4 Dynamic recrystallization

In previous papers, KARATO *et al.* (1980, 1982), gave evidence of a dynamic recrystallization at large strains. The high-angle subgrains (HA-subgrains) denoted by KARATO *et al.* (1982) are common in deformed olivine over 40% strains. As seen in Fig. 10, HA-subgrains are outlined by sharp-planer interfaces which had a nearly low-angle tilt and twist boundary in the earlier stages of creep. Mutual angles of HA-subgrain boundaries meeting at a

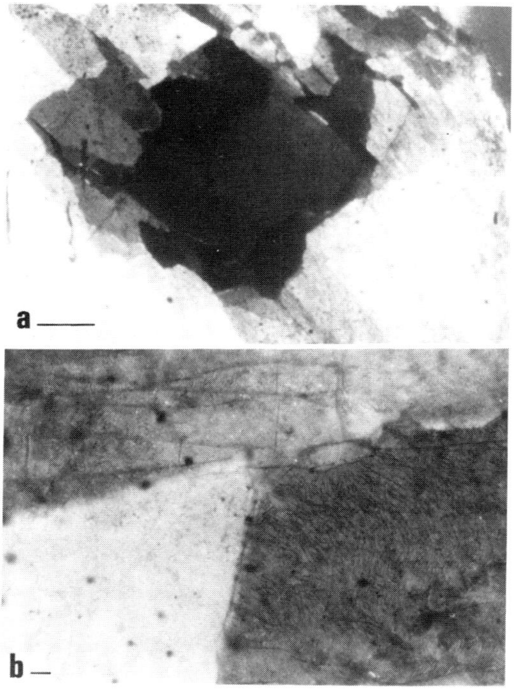

FIG. 10. HA-subgrains in olivine deformed at 1,923 K, 40 MPa and 60% strains, showing planer HA-subgrain boundaries. Bars represent 200 μm in a and 20 μm in b.

triple points are different from those determined from the balance of interfacial tension (that is, the textural dis-equilibrium), as shown in Fig. 10b.

It is stressed that the HA-subgrain size is about 3–5 times larger than the subgrain size and it suggests that some of the subgrain boundaries develop into a high-angle subgrain boundary (under 5°) and eventually into a grain boundary (over 20°). Therefore, it is concluded that high-angle subgrains have been formed due to subgrain rotation which was theoretically proposed by LI (1962) and described in natural mantle-lherzolite by POIRIER and NICOLAS (1975).

Subgrain rotation (hereafter called SGR), first proposed by LI (1962), is part of a recrystallization process of annealing; it occurs in such a way as to reduce the total subgrain-boundary energy in the sample. Further, SGR is controlled by the cooperatative climb of cell-wall dislocations.

Following results of LI (1962), the time required for the formation of a HA-subgrain becomes

$$t_{sg} = L^2 kT/3DE_0 b^3 j, \tag{18}$$

with

$$j = j_0 \exp(-Q_j/RT) \tag{19}$$

and

$$D = D_0 \exp(-Q_s/RT), \tag{20}$$

in which L is the size of the subgrain, b is the magnitude of Burger's vector, E_0 is the strain energy around the dislocation which is equal to $Gb/4\pi(1 - v)$, j is the jog density of the dislocation, D is the diffusion constant of oxygen, and k is Boltzmann's constant.

Thus, SGR requires the activation energy of the sum of activation energy for oxygen diffusion (Q_s) and jog-formation energy (Q_j).

Grain boundary migration (hereafter called GBM) seems to take place during the process of forming grains from HA-subgrains, thereby the triple points meeting three grain boundaries attain an approximate textural equilibrium. In this case, GBM occurs to reduce the interfacial excess energy resulting from the disequilibrium interfacial angle.

A bulging of the grain boundary in largely strained polycrystals is common. A typical feature of bulge GBM is shown in Fig. 11b. As noticed in

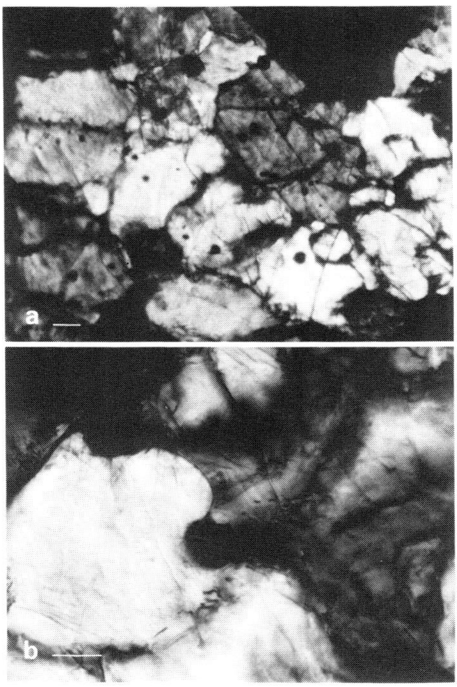

FIG. 11. Recrystallized grains outlined by wavy-undulated grain boundaries in a sample deformed at 1,923 K, 65 MPa and 60% strains.

our previous papers, grains at very large strains are often outlined by wavy interfaces, thereby showing that most grain boundaries at large strains are believed to have experienced GBM.

This type of GBM may differ from, GBM recrystallization of NaCl reported by GUILLOPÉ and POIRIER (1979). They reported that very large grains grow by GBM consuming the fine grains derived from SGR.

Grain size: GUILLOPÉ and POIRIER (1979) pointed out that the grain size of NaCl induced by GBM is much larger than that determined by SGR. On the other hand, MERCIER (1980) proposed that the SGR-grain size is larger than the GBM-grain size in olivine.

There is little difference in grain size between the polyhedral-outlined grain induced by SGR and the wavy-shaped grain induced by GBM, suggesting that the SGR-grain size is not very different from the GBM one in our experimental conditions. Relations between stress and grain size, as

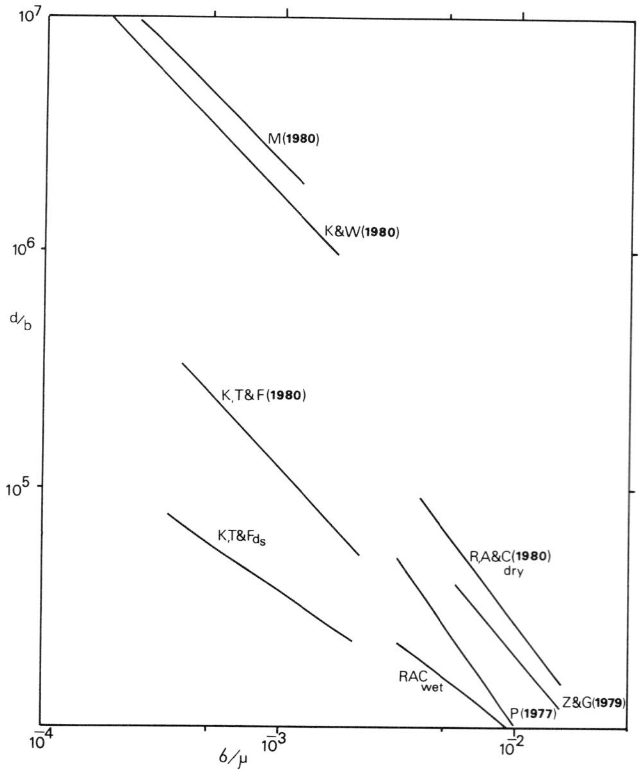

FIG. 12. Relations of grain-size (d), subgrain size (d_s), and stress (σ) of olivine. Abbreviations are as follows: P = POST (1977), Z & G = ZEUCH and GREEN (1979), M = MERCIER (1980), K & W = KOHLSTEDT and WEATHERS (1980), R A & C = ROSS *et al.* (1980), K T & F = KARATO *et al.* (1980).

illustrated in Fig. 12, indicate that many data obtained from creep experiments on polycrystalline olivine in various conditions are in good agreement with each other.

3.5 Mechanism of steady-state creep

It is worth noting that activation energy for steady-state creep in olivine polycrystals is larger than that for oxygen self-diffusion (REDDY *et al.*, 1980; JAOUL *et al.*, 1980). MINSTER and ANDERSON (1981) and KARATO *et al.* (1982) considered that the steady-state creep in olivine is rate-controlled by a dislocation climb of the intracell wall where dislocations are straight and without athermal jogs. This type of creep model has been examined by GITTUS (1976) and KARATO (1977). Following them, activation energy for creep is equivalent to the sum of that of the diffusion of the slowest atom and jog-formation energy, for a non-conservative climb is performed by jog-migration. MINSTER and ANDERSON (1981) estimated jog-formation energy in olivine as within range the 146 to 242 kJ/mol, using data of activation energy for creep as 523 kJ/mol, and for oxygen diffusion as 376 kJ/mol. The difference between the core and bulk diffusion of oxygen was zero. More reliable data of activation energy for steady-state creep of olivine polycrystals obtained by KARATO *et al.* (1982) is 540 $\pm$ 40 kJ/mol, and then we calculate

$$Q_s + Q_j = 540 \text{ kJ/mol}. \tag{22}$$

According to REDDY *et al.* (1980), activation energy for self-diffusion of oxygen, Q_s, is 376 kJ/mol. Therefore, we obtain the jog-formation energy, Q_j, as

$$Q_j = 167 \text{ kJ/mol}. \tag{23}$$

The cell wall climb model for steady-state creep was examined by KARATO and OGAWA (1982) and the difference of activation energy between steady-state creep and dislocation recovery was noted by TORIUMI and KARATO (1978). Their data (388 kJ/mol), which is similar to activation energy of diffusion of oxygen, is explained in the following way: Dislocations in a cell have many athermal jogs which are produced by an intersection of dislocations, and thus they can climb through a pre-existing jog-movement which emits vacancies. Dislocations in cell walls are straight and without athermal jogs, and therefore the climb motion of dislocations requires a formation of thermal jogs. Therefore, it is suggested that the difference of activation energy between these two processes, that is, the cell wall climb (creep) and intracell climb (recovery), is just that of the formation energy of a thermal jog.

Further, there is a clear difference in strain rates between polycrystalline and single crystalline olivine, as shown in Fig. 9. This can be attributed to a lack of Von Mise's condition for homogeneous deformation in polycrystals (PATERSON, 1969; KARATO, 1977). Deformation of polycrystalline olivine requires strains due to dislocation climb and/or diffusion. Discrepancy of activation energy between diffusion and creep indicates that diffusion and intercell wall climb are not rate-controlling processes for creep. Intracell wall climb is, therefore, a possible rate-controlling process of creep which can account for both the discrepancy of activation energy and for the strain rate mentioned above. We think that SGR due to the intracell wall climb, as discussed in the previous sections, is responsible for the deformation of polycrystalline olivines. In addition, this model can account for the conservation of grain geometry during deformation as like, which is similar to superplastic flow (ASHBY and VERALL, 1973).

4. Conclusion

We obtained following conclusions from high-temperature creep experiments of olivine at 0.1 MPa and 1,773–1,923 K. Transient creep continues at about 40% compressive strains and its time-dependent equation is composed of two segments: one is the logarithmic creep which is dominant in the early stage, and the other the parabolic one which is dominant in the later stage. The logarithmic creep is interpreted by the so-called exhaustion theory.

The activation energy for parabolic creep is 670 ± 20 kJ/mol, and stress-dependence is expressed as a power law of $\sigma^{2.5}$. After transient creep, olivine single crystals change to polycrystal due to a dynamic recrystallization at about 50% strains, and creep is approximately in a steady state. Activation energy for steady-state creep of polycrystalline olivine is 540 ± 40 kJ/mol, suggesting a good agreement with previous creep experiments of polycrystals (500–590 kJ/mol).

The discrepancy in activation energy of steady-state creep and oxygen self-diffusion can be well-interpreted by a cell-wall climb model of creep. In this model, activation energy is approximated to be the sum of oxygen diffusion and jog-formation energy. Jog-formation energy, therefore, is calculated to be about 167 kJ/mol for olivine.

A dynamic recrystallization involves subgrain rotation (SGR) and grain boundary migration (GBM) processes. HA-subgrains (high-angle subgrain) are produced and are changed to grains by SGR. A sharp and planer grain boundary is converted by GBM to wavy-undulated interfaces, and equilibrium geometry of the sample is introduced.

Relations appear between grain size and applied stress in dynamically recrystallized samples. Grain sizes induced by SGR do not largely differ from those by GBM.

We thank Prof. I. Kushiro of University of Tokyo for continuous encouragement and Dr. H. Fukuyama of University of Tokyo for many discussions. This work was supported by Grant in Aid for Scientific Research from the Ministry of Education of Japan No. 420903, and in part by funds from Cooperative Program (No. 81105) provided by the Ocean Research Institute, University of Tokyo.

REFERENCES

ASHBY, M. F. and VERALL, R. A. (1973): Diffusion accommodated flow and superplasticity, *Acta Metal.*, **21**, 149–163.

BUISKOOL TOXOPEUS, J. M. S. and BOLAND, J. N. (1976): Several types of natural deformation in olivines, an electron microscope study, *Tectonophysics*, **32**, 207–233.

COTTRELL, A. H. (1953): *Dislocations and Plastic Flow in crystals*, Clarendon Press, Oxford.

DURHAM, W. B., and GOETZE, C. (1977): Plastic flow of oriented single crystals of olivine 1. Mechanical data, *J. Geophys. Res.*, **82**, 5737–5753.

DURHAM, W. B., GOETZE, C. and BLAKE, B. (1977): Plastic flow of oriented single crystals of olivine 2. Observations and interpretations of the dislocation structures, *J. Geophys. Res.*, **82**, 5755–5770.

GAROFALO, F., *Fundamentals of creep and Creep-Rupture in Metals*, Macmillan, 1965.

GITTUS, J. H. (1976): Theoretical equation for steady-state dislocation creep; Effects of jog-drag and cell-formation, *Phil. Mag.*, **34**, 401–411.

GITTUS, J. H. (1978): High temperature deformation of two phase structures, *Phil. Trans. R. Soc. Lond., A,* **288**, 121–146.

GOETZE, C. and BRACE, W. F. (1972): Laboratory observations of high temperature rhelogy of rocks, *Tectonophysics*, **12**, 583–600.

GOETZE, C. and KOHLSTEDT, D. L. (1973): Laboratory study of dislocation climb and diffusion in olivine, *J. Geophys. Res.*, **78**, 5961–5971.

GREEN, H. W. II. (1972): Plasticity of olivine in periodotites, in *Electron Microscopy in Mineralogy*, Ed. H. R. Wenk, pp. 443–464.

GREEN, H. W. II and RADCLIÉE, S. V. (1972): Dislocation mechamisms ni olivine and flow in the upper mantle, *Earth Planet. Sci. Lett.*, **15**, 239–247.

GUEGUEN, Y. (1977): Dislocations in mantle peridotite nodules, *Tectonophysics*, 39, 231–251.

GUEGUEN, Y. and BOULLIER, A. M. (1976): Evidence of superplasticity in mantle peridotites, in *The Physics and Chemistry of Minerals and Rocks*, Ed. R. J. G. Strens, Wiley-Interscience, New York, pp. 19–33.

GUILLOPE, M. and POIRIER, J. P. (1979): Dynamic recrystallization during creep of single-crystalline halite: An experimental study, *J. Geophys. Res.*, **84**, 5557–5567.

HOLT, D. L. (1970): Dislocation cell formation in metals, *J. appl. Phys.*, **41**,3197–3201.

HÜTHER, W. and REPPICH, B. (1973): Dislocation structures during creep of MgO single crystals, *Phil. Mag.*, **28**, 363–371.

JAOUL, O., FROIDEVEAUX, C., DURHAM, W. B. and MICHANT, M. (1980): Oxygen self-diffusion in forsterite: implications for the high-temperature creep mechanism, *Earth Planet. Sci. Lett.*, **47**, 391–397.

KARATO, S. (1977): Rheological properties of materials composing the Earth's mantle, Ph.D Thesis, Univ. Tokyo.

KARATO, S. and OGAWA, T. (1982): High pressure recovery of olivine: implications for creep mechanisms and creep activation volume, *Phys. Earth. Planet. Inter*, in press.

KARATO, S., TORIUMI, M. and FUJII, T. (1980): Dynamic recrystallization of olivine single crystals during high temperature creep, *Geophys. Res. Lett.*, **7**, 649–652.

KARATO, S., TORIUMI, M. and FUJII, T. (1982): Dynamic recrystallization and high temperature rheology of olivine, in *High Pressure Researches in Geophysics*, Eds. S. Akimoto and M. Manghnani, Center for Acadamic Pulications Japan/D. Reidel, Tokyo/Dordrecht, pp. 171–189.

KOHLSTEDT, D. L. and GOETZE, C. (1974): Low-stress and high temperature creep in olivine single crystals, *J. Geophys. Res.*, **79**, 2045–2051.

KOHLSTEDT, D. L. and WEATHERS, M. S. (1980): Deformation-induced microstructures paleopiezometers and differential stresses in deeply eroded fault zones, *J. Geophys. Res.*, **85**, 6269–6285.

LI, J. C. M. (1962): Possibility of subgrain rotation during recrystallization, *J. Appl. Phys.*, **3**, 2958–2965.

MERCIER, J-C. C. (1980): Magnitude of the continental lithospheric stresses inferred from rheomorphic petrology, *J. Geophys. Res.*, **85**, 6293–6303.

MERCIER, J-C. C., ANDERSON. D. A. and CARTER, N. L. (1977): Recrystallization and stress in the lithosphere, *Pure Appl. Geophys.*, **115**, 199–226.

MINSTER, J. B. and ANDERSON, D. L. (1980): Dislocations and nonelastic processes in the mantle, *J. Geophys. Res.*, **85**, 6347–6352.

MINSTER, J. B. and ANDERSON, D. L. (1981): A model of dislocation-controlled rheology for the mantle, *Phil. Trans. R. Soc. Lond. A*, **299**, 319–356.

PATERSON, M. S. (1969): The ductility of rocks, in *Physics of Strength and Plasticity*, Ed., A. S. Argon, MIT Press, pp. 377–392.

POIRIER, J. P. (1972): High temperature creep of single crystalline sodium chloride, *Phil. Mag.*, **26**, 701–725.

POIRIER, J. P. and NICOLAS, A. (1975): Deformation induced recrystallization by progressive misorientation of subgrain boundaries with special reference to mantle periodotites, *J. Geol.*, **83**, 707–720.

POIRIER, J. P. and VERGOBBI, B. (1978): Splitting of dislocations in olivine: cross-slip-controlled creep and mantle rheology, *Phys. Earth Planet. Inter.*, **10**, 370–378.

POST, R. L. (1977): High Temperature creep of Mt. Burnet dunite, *Tectonophysics*, **42**, 75–110.

REDDY, K. P. R., OH, S. M. MAJOR, L. D.and COOPER, A. R. (1980): Oxygen diffusion in forsterite, *J. Geophys. Res.*, **85**, 322–326.

ROSS, J. V., AVELALLEMENT, H. G. and CARTER, N. L. (1980): Stress dependence of recrystallized-grain and subgrain size in olivine, *Tectonophysics*, **70**, 39–61.

SELLARS, C. M. (1978): Recrystallization of metals during hot deformation, *Phil. Trans. R. Soc. Lond., A*, **288**, 147–158.

TORIUMI, M., (1982) Grain boundary migration in olivine, *Phys, Earth Planet. Inter.*, in press.

TORIUMI, M. and KARATO, S. (1978): Experimental studies on the recovery processes of deformed olivine and the mechanical state of the upper mantle, *Tectonophysics*, **49**, 79–95.

TWISS, R. J., Structural superplastic creep and linerar viscosity in the Earth's mantle, *Earth Planet. Sci. Lett.*, **33**, 86–100.

VANDER SANDE, J. B. and KOHLSTEDT, D. L. (1976): Observation of dissociated dislocations in deformed olivine, *Phil. Mag.*, **34**, 653–658.

WEERTMAN, J. (1970): The creep strength of the Earth's mantle, *Rev. Geophys. Space Phys.*, **8**, 145– 168.

Chapter 5: Solid Materials in the Earth's Interior

Materials Science of the Earth's Interior, edited by I. Sunagawa, pp. 303–330.

MORPHOLOGY OF NATURAL AND SYNTHETIC DIAMOND CRYSTALS

Ichiro SUNAGAWA

Institute of Mineralogy, Petrology and Economic Geology, Tohoku University, Aoba, Sendai, Japan

1. *Introduction*

The morphological aspects of natural diamond crystals have been a subject of great interests and dispute for many years. Extensive works have been carried out since as early as the 18th Century and various controversial theories have been put forward. Most natural diamond crystals show rounded corners, edges, and curved faces, and even foot ball-like forms, in contrast to ordinary crystals whose corners are pointed and whose edges are straight. The origin of rounded crystals has long been the subject of controversy. One argument stands on growth theory and another on dissolution theory. There is also an argument for a combined theory.

Octahedral faces of natural diamonds almost invariably exhibit triangular depressions with negative orientation in respect to the outline of a {111} face. These are called trigons. The origin of trigons has been another controversial subject. One theory considers that they are formed by a failure in layer growth on octahedral faces. Another theory considers their origin as being due to etching at the outcrops of dislocations or to the defects of other types.

When internal morphologies (growth bandings, sector structures, etc.) of natural diamonds are revealed by etching, X-ray topography or other physical techniques, many crystals exhibit crystallographic growth bandings parallel to octahedral faces. Occasionally, however, there are crystals which show irregular agate-like bandings (SEAL, 1965), or centre cross-patterns consisting of crystallographic {111} and curved {100} bandings. The origin of this centre cross pattern has also been the subject of strong arguments. One argument explains it in terms of plastic deformation, an other mentions the combined growth of rough {100} and smooth {111} interfaces.

The morphology of crystals is a direct result of growth and/or dissolution and thus reflects the mechanism of growth and dissolution, conditions of growth and dissolution, and the history that the crystal has experienced. This is the reason why the morphological aspect of natural diamond crystals has been a subject of such great interest and severe controversy. Since we cannot see actual processes of growth and dissolution of natural diamond crystals, people have been very anxious to solve the problem through morphology. The crystal's external forms (habits) will tell us under what conditions they grew and what sort of history have they experienced after the growth was completed. The surface microtopography of crystal faces will tell in molecular order how growth or dissolution proceeded. Internal morphology (growth bandings or sector structures) will offer information relating to the whole growth and dissolution histories of the crystal.

Therefore, it is worthwhile studying morphological aspects of natural diamond crystals, as well as synthetic crystals. It is the purpose of this chapter to discuss the present understandings of these problems, and to compare natural and synthetic diamonds.

2. *Classification of Natural Diamonds*

Natural diamonds occur not only as single crystalline forms, but also in polycrystalline aggregates. Various variety names, often not well defined and sometimes given only as trade names, have been given to polycrystalline aggregates; in most text books of mineralogy, variety names of bort, ballas, and carbonado are mentioned, and in Dana's System of Mineralogy, varieties are classified into bort, short bort, hailstone bort, framesite, stewartite, and carbonado (Palache *et al.*, 1944). Orlov (1977) classified natural diamonds into ten different varieties; varieties I to V are single crystalline types, whereas varieties VI to X are polycrystalline types, and variety VI corresponds to ballas, varieties VII to IX to bort, and variety X to carbonado. Jeynes (1978) critically reviewed the works and terminology of polycrystalline diamonds.

The present author proposes the following classification.

(1) Single crystalline diamonds. Octahedral and dodecahedral crystals, twinned crystals, as well as aggregates of a small number of single crystals which are coalesced.

(2) Coated diamonds, consisting of a single crystalline core and thin rims with fibrous textures.

(3) Cuboid, having a radiated structure with or without a single crystalline core.

(4) Polycrystalline aggregates, consisting of cryptocrystalline diamonds. These may be further divided into the following varieties depending on the mode of aggregations, i.e., internal structures.

a.	Framesite	bort-containing minute crystals of single crystalline type
b.	Bort	granular to cryptocrystalline aggregates

c.	Stewartite	magnetic bort
d.	Carbonado	massive, granular and at random aggregates of cryptocrystalline diamonds, individual sizes being in the order of microns
e.	Short bort	bort with a radiated structure
f.	Ballas	radiated structure, the individual fiber lengths being 30 to 40 μ, width 1 μ
g.	Hailstone bort	with concentric structure

The above classification is based on the considerations of the morphological changes of crystals, depending on supersaturation σ, or the chemical potential difference between solid and liquid phases $\Delta\mu/kT$ (refer to SUNAGAWA, 1977, 1981, 1982). As discussed in these references, there are two transitional supersaturations σ^* and σ^{**}. Above σ^{**}, the solid-liquid interface is rough and continuous or an adhesive-type growth occurs. Thus, the morphology of crystals will be dendritic to spherulitic. When going into a higher supersaturation region, spherulites will appear. Below σ^*, polyhedral crystals, bounded by flat surfaces, are expected, since, growth of crystals under this condition is principally controlled by interface kinetics and dislocations. Between σ^* and σ^{**}, hopper crystals are expected. Thus, single crystalline-type diamonds represent the crystals grown under the condition below σ^*, where a spiral growth mechanism principally operates. This was witnessed by a combined study of surface microtopographs and X-ray topographs of octahedral crystals from Siberia (SUNAGAWA *et al.*, 1982). In contrast, polycrystalline aggregates of cryptocrystalline diamonds represent the crystals formed under the conditions above σ^{**}. These may be regarded as having been formed by abnormal growth, in contrast to normal growth for single crystalline diamonds. Depending on growth conditions and histories, the micro or cryptocrystals may form granular, radiated or concentric structures, resulting in different varieties, but all these polycrystalline aggregates are considered to have been formed under the conditions above σ^{**}. The cuboid also belongs to this category, but has a slightly lower supersaturation condition. According to LANG and MOORE (1972), LANG (1974a) and KOLOMYTSEV and YAKUBOVA (1976), most cuboids have radiated structures with or without a single crystalline core. Coated stones represent those which have experienced changes in environmental conditions during their growth histories. At first they grew as single crystalline diamonds, but due to a change of environmental condition (probably due to uprift of the magma, and, thus, an increase in effective supersaturation), dendritic overgrowth occurred which formed rim portions with a fibrous structure (LANG, 1974b). Some cuboid diamonds which contain a single crystalline core also belong to this category. Therefore, natural diamonds can be classified into three different categories: a single crystalline-type grown by layer by layer growth mechanism under stable conditions, polycrystalline aggregates and most cuboid which are formed under unstable conditions and by abnormal growth (above σ^{**}), and those experiencing two-growth conditions, like coated stones.

Synthetic diamonds take dendritic, hopper, and polyhedral morphologies depending on the growth conditions. When they are synthesized under not well-controlled conditions, dendritic crystals are formed. Among earlier synthetic crystals, dendritic and hopper crystals are commonly found and are used for resin bond purposes (TOLANSKY and SUNAGAWA, 1959, 1960). When they are synthesized under well-controlled conditions, crystals become polyhedral bounded by flat faces, and these are good for metal bond purposes. Synthetic carbonados are produced by sintering diamond powders under high temperature and high pressure conditions, in the presence of such binders as Co.

3. *Origin of Rounded Crystals*

The origin of rounded crystals of natural diamonds of a single crystalline-type has been the subject of controversy between growth and dissolution theories.

The growth theories consider that the rounded morphology is derived by a concretion-like collection of ultrafine individuals (ROME DE L'ISLE, 1783), stepped superposition of growth layers on {111} (HAÜY, 1817; VAN DER VEEN, 1913; WILLIAMS, 1932; ANSHELES, 1954; TOLANSKY, 1955; VARMA, 1967a–d), by combination of a large number of vicinal faces (SCACCHI, 1862; KOKSHAROV, 1869; FEDROV, 1901), and also due to the crystallization of droplet of carbon melt (RAMAN and RAMASESHAN, 1946). ORLOV (1973, 1977) has made a lengthy critical review on this subject. Russian references cited here are due to his book (1977).

The dissolution theories were initially put forward based on goniometric investigations of many diamond crystals by FERSMAN and GOLDSCHMIDT (1911). These were then supported by KUKHARENKO (1954, 1955), KUKHARENKO and TITOVA (1957), and ORLOV (1963). ORLOV (1977) gave ten evidences to support dissolution theory based on the observations of both external morphology and surface microtopography.

SEAL (1965) observed that crystallographic growth bandings parallel to {111} revealed by etching are often truncated by the rounded outline of the dodecahedroidal crystal, and suggested that the crystal was originally much bigger than the present size, had octahedral morphology, and was dissolved to the present size, thus the rounded forms of dodecahedroidal crystals are due to dissolution. This was further supported by MOORE and LANG (1974) through optical and X-ray topographic investigations of rounded dodecahedroidal crystals. A similar relation as that observed by Seal was found in their study as well. MENDELSSOHN and MILLEDGE (1974) made computer simulations of how dissolution transformed the morphology of a diamond.

A combined theory was put forward by SCHAFRANOVSKI (1948), who considered that the rounded forms were principally due to dissolution, but followed by a brief regrowth.

The origin of trigons has also been a subject of controversy. This problem has a close connection with the arguments about the origin of rounded forms. Thus, those standing on the growth theory for rounded crystals believe growth mechanism as the origin of trigons, whereas those standing on dissolution theory believe dissolution mechanism. There are many scientists who supported growth theory in which trigons are formed by a failure in layer growth on octahedral faces. VAN DER VEEN (1913) was the first to put forward the growth theory. This was followed by WILLIAMS (1932), HALPERIN (1954), TOLANSKY (1955, 1963, 1965), ANSHELES (1954, 1957), VARMA (1967 a–d). The growth theory is based on the analysis of morphological characteristics of trigons and on the observations that artificial etching can produce only triangular etch pits of positive orientation, i.e., parallel to the triangle of the octahedral face.

The dissolution theory was initially put forward by FERSMAN and GOLDSCHMIDT (1911). This was further supported by KUKHARENKO (1954, 1955), FRANK (1958, 1966, 1969), FRANK *et al.* (1958), FRANK and LANG (1959), ORLOV (1963), LANG (1964, 1965), PATEL and AGARWAL (1968), MOORE and LANG (1975), KANDA *et al.* (1977), and YAMAOKA *et al.* (1980). The theory is based on the morphological analysis of trigons (FRANK, 1958; FRANK *et al.*, 1958; Computer simulation by PONTON *et al.*, 1974), experiments to artificially reproduce trigons in kimberlite melt (FRANK *et al.*, 1958), in the oxidizing agent (PATEL and AGARWAL, 1968; PATEL and PATEL, 1968a; PATEL and RAMANATHAN, 1962, 1964), and more recently in high pressure water (KANDA *et al.*, 1977; YAMANAKA *et al.*, 1980), and on the correlation between trigons and the outcrops of dislocations (FRANK and LANG, 1959; LANG 1964, 1965; SUNAGAWA *et al.*, 1982).

Although there has been a long-holding controversy as to the origin of rounded forms and of trigons, it is now reasonable to assume that both morphologies are due to the dissolution processes which natural diamond crystals have sufferend after the completion of their growth. The following observations provide a firm basis to support the dissolution theory:

1. When a crystal with pointed corners, straight edges and flat faces, and containing no lattice defects, is dissolved, the corners are the easiest sites for dissolution, followed by the edges and faces. Thus, rounded morphology is a natural result of dissolution. All stages of this process have been observed in natural diamond crystals.

2. Crystallographic growth bandings parallel to octahedral faces are often truncated by the outline of rounded dodecahedroidal crystals (SEAL, 1965; MOORE and LANG, 1974). This is a definite proof that the original crystals were much bigger and they were dissolved to the present size.

3. Smaller diamond crystals included in a larger diamond exclusively exhibit morphology with pointed corners, straight edges, and flat faces, while the host crystal is rounded.

4. When a crystal containing lattice defect is subjected to dissolution, etch pits are formed. At the outcrop of line defects, point-bottomed etch pits are formed, whereas at point defects or line defects issuing very close to the surface, flat-bottomed etch pits will appear. Morphology and orientations of etch pits can vary depending on the etching condition. Trigons of both point-bottomed and flat-bottomed types are observed on natural diamond octahedral faces. Triangular etch pits of both positive and negative orientations have been artificially produced. Trigons can be produced easily in the laboratory, particularly under the presence of water.

5. Morphological analysis of detailed features of trigons indicated that they are etch pits and not due to growth failure (FRANK, 1958; MENDELSSOHN and MILLEDGE, 1974; PONTON *et al.*, 1974).

6. Some of the trigons are related to the outcrops of dislocations (LANG, 1964, 1965; SUNAGAWA *et al.*, 1982).

These observations support dissolution theory for the origin of rounded crystals and trigons. Growth theory has lost its basis. Thus, we may proceed our discussion on this basis. Namely, natural diamonds have experienced dissolution after the completion of growth.

Starting from simple octahedral crystals with pointed corners, straight edges and flat faces, crystals will become more rounded and transform to an hexoctahedral form with rounded edges and faces as dissolution proceeds, and eventually become dodecahedroidal forms. This series of morphological changes due to dissolution is schematically shown in Fig. 1. Octahedral

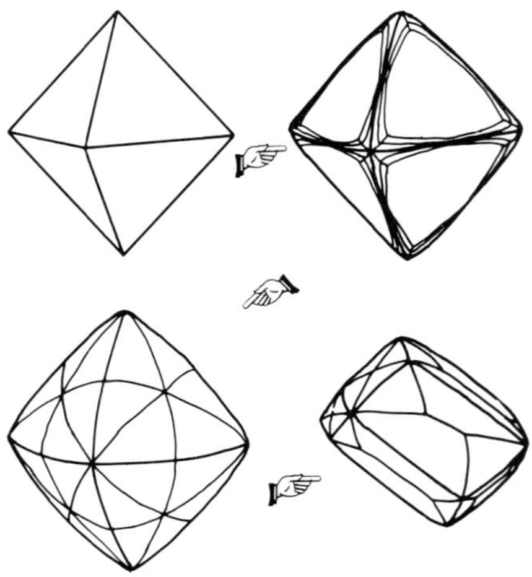

FIG. 1. Schematical drawing to show changes of morphologies as dissolution proceeds, starting from a simple octahedral crystal and ending at a dodecahedroidal crystal.

morphology with slightly rounded corners and edges represents the morphology of crystals which received only a slight dissolution, whereas dodecahedroidal or foot ball-like morphology represents the morphology that is expected for the crystals which experienced heaviest dissolution.

The degree of dissolution may be different depending on the conditions and the perfections of the crystals, and thus vary from locality to locality, as well as from crystal to crystal (BOBRIEVICH, 1959; ORLOV, 1977; KOMATSU, 1978; and many other observations). In Fig. 2, the frequencies of appearance of octahedral, intermediate and dodecahedral crystals at different kimberlite pipes are shown. These are cited from the data of BOBRIEVICH *et al.* (1959) and those by TOLANSKY (1973), both of which were obtained from diamond crystals of known localities. There is a marked difference in the frequencies of octahedral and dodecahedroidal crystals in different localities. Such differences may be related to or may represent the differences in the degree of dissolution in different kimberlite magmas. Judging from the recent experiments on the formation of trigons with the presence of H_2O (KANDA *et al.*, 1977; YAMANAKA *et al.*, 1980), the contents of H_2O or probably CO_2 as well, in kimberlite magmas, will have an important implication in the dissolution of

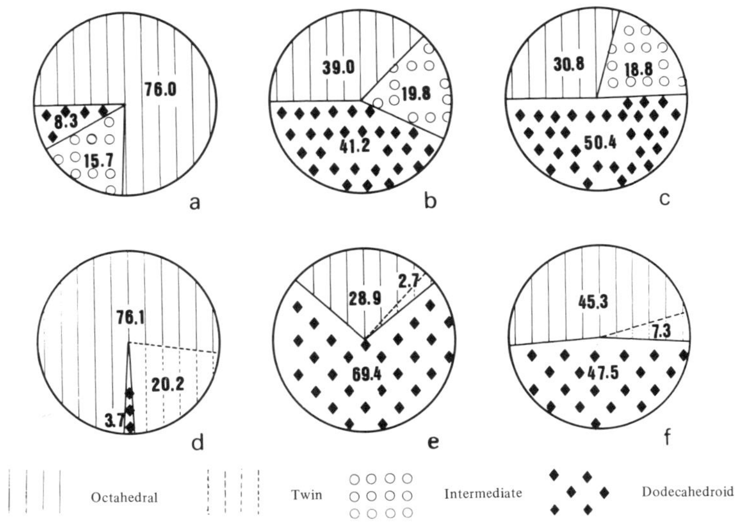

FIG. 2. Graphs showing frequencies of appearance of different habits of diamond crystals of a single crystalline type in different kimberlite pipes. (a) Mir, (b) Udachnaya, (c) Zarnitsa, in Yakutiya, USSR, and (d) Premier, (e) Finch, (f) De Beers in South Africa. (a)–(c) from the data in BOBRIEVICH *et al.* (1959), and (d)–(f) from the data of TOLANSKY (1973). In (a)–(c) habits are classified into octahedral, intermediate and dodecahedroidal, and in (d)–(f) into octahedral, twin, and dodecahedroidal. The statistics were made in the former three for naked-eye size crystals, whereas in the latter three, for micro-crystals only.

diamonds. Thus, the above differences may be taken as representing the differences in H_2O and CO_2 contents in different kimberlite magmas. The kimberlite pipes showing a higher percentage of octahedral crystals may represent the magma with lower H_2O and CO_2 contents.

It is worthwhile noting that diamond crystals occurring in eclogite nodules (and also peridotite nodules) in kimberlite show, in general, less dissolved morphological characteristics than those occurring in kimberlite. It has been well established through the studies of inclusion mineralogy of diamond that the chemical environments are different in diamond growth between ultra-mafic and eclogite suites (HARRIS, 1979; HARRIS and GURNEY, 1979). There are also other differences between diamonds of ultra-mafic and eclogite suites. As compared to diamonds in kimberlite (ultra-mafic suites), those occurring in eclogite nodules in kimberlite are higher in concentration, smaller in size, and show less dissolution, more marked stepped octahedral faces, larger number of growth centres on {111} faces, and steeper growth pyramids, and exhibit striated {110} faces, which are rarely seen on diamond crystals in kimberlite.

In addition to the dissolution process within kimberlite magmas, corrosion during the processes of weathering, transportation and sedimentation will have additional influences upon the appearance of rounded crystals. It is reported in BOBRIEVICH *et al.* (1959) that the percentage of rounded crystals increases apart from the source kimberlite pipes. It is also well known that diamonds occurring in placer deposits, particularly in Pre-Cambrian placers, exhibit very rounded morphology, as compared to diamond crystals in original rocks.

4. *Crystal Habits of Single Crystalline Type*

Natural diamond crystals of a single crystalline type exhibit a wide range of forms and a different degree of roundness. According to PALACHE *et al.* (1944), diamond crystals are predominantly octahedral, less commonly dodecahedral, rarely cubic, and occasionally tetrahedral. Often flattened {111}, they are built of successive flattened octahedral plates and are variously distorted. Their faces are commonly much curved and are often striated. FERSMAN and GOLDSCHMIDT (1911) reported forms of {001}, {011}, {111}, {112}, {223}, {122}, {133} and {233}, the latter five forms being very tiny and rare.

Among crystals of octahedral habit, there is a wide range of forms depending on the degree of roundness. There are crystals with almost flat octahedral faces and only slightly rounded corners and edges. Such crystals are regarded as representing crystals which have received only a very slight dissolution and showing their almost original as-grown forms (refer to SUNAGAWA *et al.*, 1982). Octahedral faces, corners and edges can be more rounded, eventually resulting in a hexoctahedral habit. On such crystals, hexoctahedral faces are not crystallographically-flat faces, and cannot be given proper indices. These represent crystals which received heavy dissolution. The degree of roundness vary from locality to locality. Siberian and Chinese diamonds of kimberlite origins show a higher percentage of less-rounded octahedral habit than in other localities. This is well demonstrated by the huge pyramids constructed by numerous Siberian diamond crystals, which are exhibited in the Diamond Museum in the Kremlin, Moscow.

Crystals of dodecahedral habit are, in a majority of cases a very much rounded, with curved faces, edges, and corners. They are far more rounded than crystals of octahedral habit, and are more deformed or elongated than the latter. In extreme cases, dodecahedral crystals having foot ball-like forms are seen. The dodecahedral faces are, in most cases, not crystallographically flat faces and are divided into two rounded $\{hk0\}$ and $\{kh0\}$ faces or into four $\{hk1\}$ faces. Thus they are not dodecahedron but dodecahedroid. Dodecahedroidal faces usually exhibit elongated ridges, and not straight striations. It should be noted, however, that almost flat dodecahedral faces are occasionally found, which exhibit straight striations among very small crystals (microdiamonds) and also among diamond crystals occurring in eclogite and peridotite nodules in kimberlites, which show evidence of a very slight dissolution. Growth bandings revealed by etching (SEAL, 1965) sometimes exhibit the presence of stepped dodecahedral faces consisting of repeated small steps of $\{111\}$ during the course of growth. This is an indication that dodecahedral faces appear through growth process, and not only through dissolution.

Diamond crystals of cubic habit occur rarely and are not generally bounded by flat crystallographic $\{100\}$ faces. Cubic faces are always rugged and curved and often depressed at the centre of a face. Thus, these crystals should be called cuboid, rather than cube. According to optical and X-ray topographic studies by Lang and MOORE (1972), LANG (1974a, 1979), natural cuboid diamonds exhibit internal structures consisting of radiating columns, suggesting space-filling by branching columnar crystal growth. Thus, natural cuboid crystals are not formed by a layer-by-layer growth mechanism, but under the conditions above σ^{**}. It is also observed that growth bandings seen inside a crystal with a centre-cross pattern (FRANK, 1958; LANG, 1974, 1979; SUZUKI and LANG, 1976) are irregularly curved on $\{100\}$ growth sectors in contrast to crystallographic straight bandings on $\{111\}$ growth sectors. This indicates that under the same growth conditions, $\{100\}$ takes a rough interface, whereas $\{111\}$ takes a smooth interfaces (refer to SUNAGAWA, 1982). This will be described in more detail in Section 6.

Crystals of tetrahedral habit are rare and, in general, exhibit curved surfaces. This morphology may be formed by an anisotropic development of octahedral faces, followed by dissolution. Most flattened crystals with triangular forms are spinel twinned crystals. The malformation of morphology from octahedral to triangular platy forms is due to a preferential growth along three re-entrant corners which appeared by spinel twinning, as schematically illustrated in Fig. 3. Dislocations are concentrated in the composition plane of such twins, and act as preferential growth sites (KITAMURA *et al.*, 1979).

From the above observations, we may assume that {111} is the only flat face on which free development of growth layers is seen, and is thus, morphologically, most important face of natural diamond crystals. The second important face is {110}, which, in general, appears as a dissolution form (dodecahedroid), but in certain cases can appear as a growth form with a stepped nature consisting of alternating small {111} steps. {100} face does not appear as a crystallographic flat face. Therefore, in the case of natural diamond, layer-by-layer growth (including spiral growth) occurs only on {111} faces, and never on {110} or {100} faces.

The morphology of synthetic diamond crystals has also been a subject of interests, and it has been observed that the morphology varies as a function of thermodynamic conditions (BOVENKIRK, 1961; GIARDINI and TYDINGS, 1962; LITVIN and BUTUZOV, 1968; YAMAOKA *et al.*, 1977; and other many works). The morphodrom of synthetic diamond crystals in relation to temperature and pressure may be schematically illustrated as in Fig. 4, based on available data. A recent work by Chinese scientists provides a slightly different morphodrom, but we shall use Fig. 4 for our present discussion. Octahedral habit appears in the region closer to the equilibrium line and, as apart from the line cubo-octahedral followed, and then simple cubic habits appear. YAMAOKA *et al.* (1977) reported that {110} appears only in a very narrow region close to the equilibrium line, and shows a flat surface.

It should be noted here that {111} and {100} faces of synthetic diamonds are macroscopically flat faces on which growth spirals are observed, about which a detailed description will be given in the following section. This makes a strong contrast with the observations on the morphology of natural

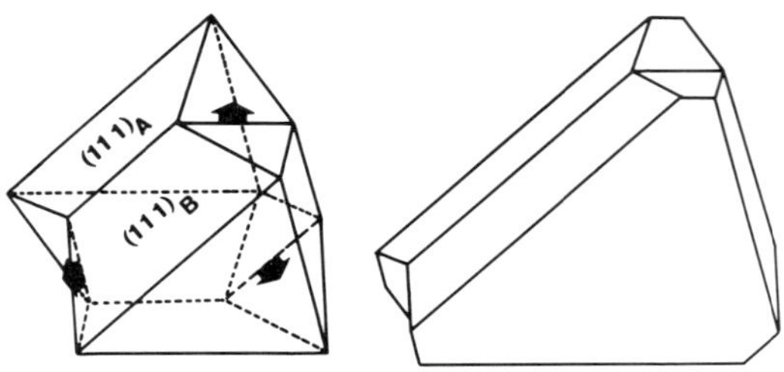

FIG. 3. Schematically showing how octahedral crystals twinned after spinel twin law change their morphology to triangular platy habit due to the preferential growth at the re-entrant twin junctions indicated by arrows.

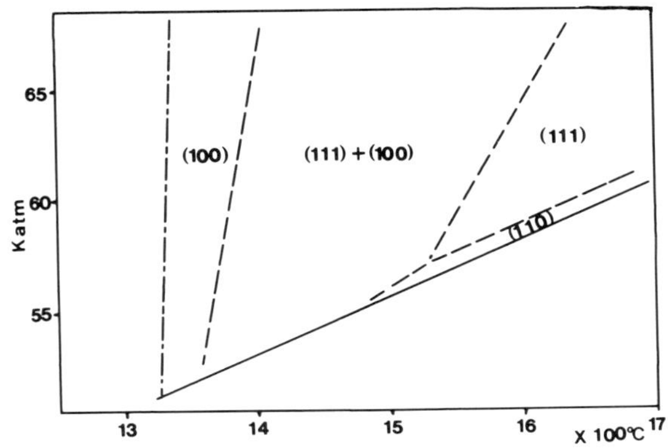

FIF. 4. Morphodrom of synthetic diamond.

diamond crystals, and indicates a big difference in growth mechanisms between natural and synthetic diamond crystals.

5. *Surface Microtopography*

The surface microtopographs of natural diamond crystals, particularly of octahedral faces, have been a subject of extensive studies for many years. A vast amount of detailed observations, using very sensitive methods, such as phase contrast microscopy and multiple-beam interferometry, which can reveal and measure the steps of molecular heights, has been accumulated. As representative works of the earlier days, where only ordinary microscopy was used, we may mention the observations by FERSMAN and GOLDSCHMIDT (1911), SUTTON (1928), and WILLIAMS (1932), and of more recent days, where sophisticated methods have been used, the works of TOLANSKY (1955, 1960, 1961, 1963, 1965).

The observations that have been reported on the surface microtopographs of natural diamond crystals are;

(1) growth layers and pyramids with triangular or modified forms, which have the same orientation as the triangle of the octahedral face on which they are seen,

(2) trigons of various forms, which have the opposite orientation to the triangle of the octahedral face,

(3) slip lines parallel to $\langle 110 \rangle$,

(4) ring and net work patterns on dodecahedroidal faces,

(5) ridges on dodecahedroidal faces,

(6) mottled or blocky $\{100\}$ surfaces,

(7) percussion marks, etc.

Since natural diamond crystals have received dissolution, it is not easy to find original as-grown surfaces. As a result, earlier works have been mainly concerned with the surface microtopographs which appeared due to dissolution processes, such as trigons, although these features were often explained in terms of growth failure. SUNAGAWA *et al.* (1982) have recently made a detailed observation of growth layers and pyramids on some octahedral diamond crystals which have received a very slight dissolution. The growth layers and pyramids are believed to show almost original as-grown features. An example of these growth pyramids is shown in Fig. 5. They have triangular forms consisting of triangular layers with step heights of less than 10Å, and the triangle have the same orientation as that of the triangle of the octahedral face, i.e., opposite orientation to the trigons. The step separation is variable, ranging from less than 10 μ to more than 50 μ, i.e., the ratio of step height and step separation ranges 10^3 to 10^4. At the summits of these growth pyramids are small P-type trigons develop. SUNAGAWA *et al.* (1982) have demonstrated that there are one-to-one correlations between outcrops of dislocations of the screw type revealed by X-ray topography and the summits of growth pyramids, as well as small P-type trigons appearing at the summits. This definitely proves that the growth pyramids are formed by the spiral growth mechanism, and the growth layers having step heights less than 10 Å originate

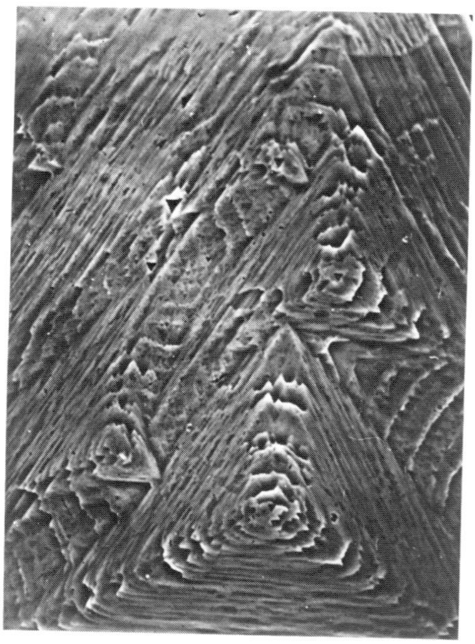

FIG. 5. Phase contrast photomicrograph of growth pyramids observed on a (111) face of an octahedral diamond crystal from Siberia, which has received only a very slight dissolution. Note P-type trigons at the summits of the growth pyramids. Photographed by K. Tsukamoto.

from such dislocations, and that some of the P-type trigons appear at the outcrops of the dislocations. It is almost exceptional that such original as-grown features are well preserved on the octahedral faces, though among some photomicrographs published earlier, one sometimes find similar growth pyramids. Most octahedral faces of natural diamond crystals have received much heavier dissolution, so that the surfaces exhibit either very much deformed growth pyramids or only a large number of trigons, preserving almost none of as-grown features.

Trigons are observed on almost all octahedral faces of natural diamond crystals. It is not an exaggeration to say that all octahedral faces of natural diamond crystals exhibit trigons, either only a few on one face or almost

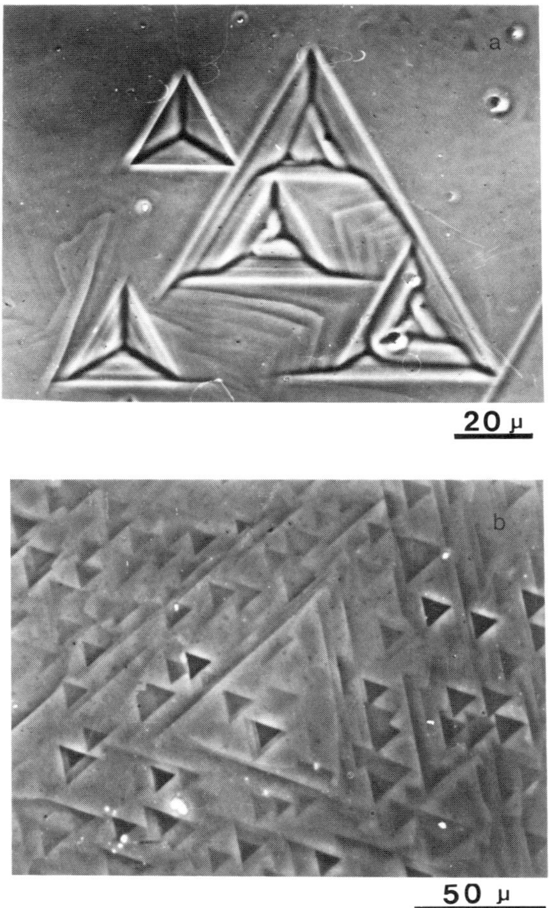

FIG. 6. Phase contrast photomicrographs of representative trigons. (a) P-type trigons, (b) F-type trigons.

uncountable numbers on one face. The number of trigons on one face varies from locality to locality and from crystal to crystal, or from face to face, showing the difference in the degree of dissolution and the perfection of crystals among different crystals and localities. In Fig. 6, representative examples of trigons are shown.

There are two major types of trigons; point-bottomed (or pyramidal) trigons are called P-type, flat-bottomed trigons F-type. These are further divided by LANG (1965) according to the slope of the side faces. F-type trigons have very steep side faces, probably octahedral faces. F-type trigons may be divided into FD (deep F-type) and FS (shallow F-type) types. The former is easily visible under ordinary reflected light, whereas the latter needs phase contrast for recognition. P-type trigons have two different slopes; one group exceeds the slope of the other by a factor of 5 to 10. An average value of the gentler slope is about 15′ and of the steeper slope about 100′ (FRANK *et al.*, 1958; WILKS, 1961). The former is called PS-type (shallow P-type) and the latter PD-type (deep P-type).

As to the size of trigons, LANG (1965) described P-type trigons as showing a rather narrow size distribution, particularly on the more perfect stones (a typical value of the edge length of an isolated P-type trigons being about 100μ), though when they occur in clusters their sizes are variable and can fall below 10μ. On crystals which received a very slight dissolution, the sizes of P-type trigons can be much smaller, even when they appear isolated, around 10μ in edge length (SUNAGAWA *et al.*, 1982). As compared to P-type trigons, the sizes and the depths of F-type trigons are variable. In addition to the above four types, Lang mentioned a sail-boat pattern, a larger number of very small trigons, edge length being down to 3μ, and depth down to 10Å.

Morphologies of trigons are also variable. There are perfect trigons in a sense that three edges close by themselves, thus forming regular triangular morphology. Trigons whose one, two or three corners are truncated, and in an extreme case resulting in a hexagonal depression, are also observed. There are trigons whose one base is extended and sometimes aligned on one slip line. F-type trigons often show many much smaller trigons on the base surface.

Combined studies of X-ray topography and surface microtopography (LANG, 1964, 1965, and other papers; SUNAGAWA *et al.*, 1982) have demonstrated the existence of one-to-one correlation between the outcrops of dislocations and P-type trigons. Laboratorical etching under high pressure water (KANDA *et al.*, 1977; YAMANAKA *et al.*, 1980) has also been successful in easily reproducing trigons in laboratories. These indicate that at least P-type trigons are formed at dislocation outcrops by the dissolution process. For F-type trigons, surface point defects like impurities and dislocations issuing not very far from the surface are suggested as possible sites for the formation through dissolution processes. It is interesting to note that the densities of trigons vary from locality to locality, suggesting the difference of the degree of dissolution. Although growth theory has been put forward for the origins of trigons, it is safe to assume that trigons are the result of dissolution, and not due to growth failure.

Ring and network patterns observed on dodecahedroidal faces are also considered to have been formed by dissolution processes, although some authors believe their origins as due to growth (PANDEYA and TOLANSKY, 1961). Ridges observed on the dodecahedroidal faces are also due to dissolution. In case that dodecahedral faces appear by the pile up of steps of growth layers developing on the neighbouring octahedral faces, the dodecahedral faces exhibit straight striations. Such dodecahedral faces with well developed striations are observed on natural diamond crystals which have received only a slight dissolution, like those occurring in eclogite or peridotite nodules in kimberlite.

Mottled or blocky {100} surfaces are due to the radiated columnar structure of cuboid crystals, which received dissolution. It is to be noted that no growth layers have been observed on cuboid faces of natural diamond crystals.

The most characteristic and commonly observed surface microtopographs of synthetic diamond crystals are dendritic patterns on both octahedral and cubic faces. These are shown in Fig. 7. Although their morphologies vary considerably from a rounded and more or less irregular form to a polygonal and more crystallographic form, their symmetry follows the symmetry of the face on which they appear. On octahedral faces, branches of the dendritic pattern follow the trigonal symmetry of the face and, on a cubic face, they have crossed branches at 90° following the symmetry of the cubic face. On crystals grown under a well controlled condition, like the crystals of gem quality, dendritic patterns exhibit more crystallographic morphology as compared to the crystals synthesized for abrasive purposes under not well controlled conditions. The dendritic patterns are considered as due to the imprint of solution film from which crystallization occurs. They may be diamond itself or metal carbide.

Underneath these dendritic patterns, growth features of diamond itself are observed, among which most striking feature is growth spirals. These were observed for the first time by TOLANSKY and SUNAGAWA (1959, 1960) on {100} faces, and then by PATEL and RAMACHANDRAN (1968) on {111} faces, followed by numerous observations on {100} faces (TOLANSKY, 1963, 1965).

The spirals observed on {100} faces of synthetic diamond are rectangular bounded by ⟨011⟩. These spirals have characteristics which are different from ordinary growth spirals commonly observed on crystal faces, which have mono-molecular step heights, smooth steps and surfaces. The step heights of growth spirals observed on synthetic diamonds range from 200Å to 1,500Å, and molecularly thin (less than 10Å) growth spirals have never been observed on synthetic diamonds. In the case of growth spirals commonly observed on

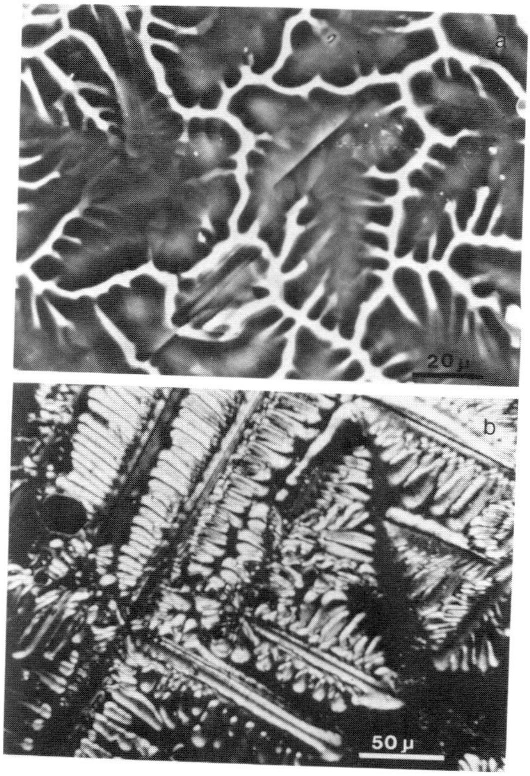

FIG. 7. Dendritic patterns observed on (a) (111) and (b) (100) faces of a synthetic diamond.

the crystals grown from vapour phase, the step heights are mono-molecular and steps are not very much rugged, the surface of spiral layers are molecularly smooth. In such cases, bunched thick layers may occur due to bunching of thinner layers, but they always occur in association with mono-molecular layers. This is not so in the case of growth spirals observed on synthetic diamond crystals, which always have big step heights, without association of mono-molecular growth spirals. Therefore, they are not formed by bunching of mono-molecular growth spirals, but have a high-step height from their origin. They are always denticulated, and the surface of the spiral layers are wavy, not molecularly smooth, as in the case of ideal growth spirals. Figure 8b is a representative example of growth spirals observed on cubic faces.

Growth spirals on {111} faces of synthetic diamonds reported by PATEL and RAMACHANDRAN (1968) have similar morphological characteristics to those observed on {100} faces. They have triangular forms with the same orientation as the triangle of the face, have denticulated steps and wavy surface,

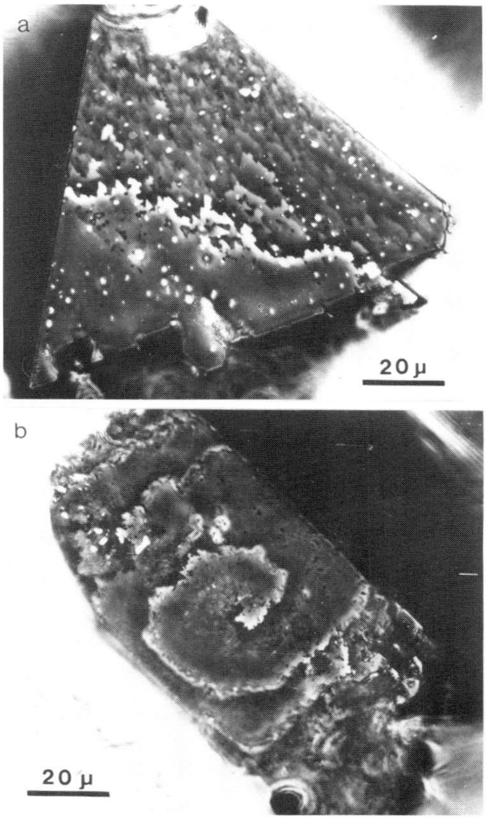

FIG. 8. Hackly growth layers on (111) (a) and growth spiral on (100) faces of a synthetic diamond.

and have high step heights. Growth spirals are less commonly observed on {111} faces than on {100} faces. Commonly observed surface microtopographs of {111} faces of synthetic diamonds are of dendritic pattern, or of a triangularly denticulated pattern consisting of a large number of triangular layers. The triangular layers composing the denticulated hackly pattern have high step heights, being more than a few hundred Angstrom (Fig. 8). Growth layers having mono-molecular step heights have not been observed on both {100} and {111} faces of synthetic diamonds, which is a marked difference from natural diamonds. If Fig. 5 and 8 are compared, the big difference of surface microtopographs between natural and synthetic diamonds is evident.

Trigons are observed only on synthetic crystals whose experimental conditions indicate dissolution. Thus, the formation of trigons is consistent with the dissolution theory.

6. *Internal Morphology*

Internal microstructures of polycrystalline-type diamonds have been studied by several workers, by means of X-ray topography or electron microscopy. Trueb and Butterman (1969) investigated carbonado by SEM, TEM, EPMA analysis and auto-radiographic technique and showed that carbonado consists of anhedral diamond grains with diameters ranging from 1 to 20 μ. Trueb and Barrett (1972) investigated ballas by SEM and X-ray topography and observed that it has radiated structure consisting of fibrous diamonds elongated parallel to [110], with lengths from 30 to 40 μ. Lang and Moore (1972), Lang (1974a), and Kolomyitsev and Yakubova (1976) studied cuboid crystals by X-ray topography and found that they have a radiated structure formed by space-filling columnar growth. Lang (1974b) also applied X-ray topography to coated stone and observed that the rim portion consists of fibrous structure, whereas the core portion has the same internal structure as a single crystalline type. Very recently, Moriyoshi *et al.* (1980) have applied high voltage, high resolution transmission electron microscopy to polycrystalline diamonds and observed dislocation structures in individual grains and grain boundaries. They found that atomic bondings exist between the neighbouring grains in these polycrystalline-type diamonds.

Internal structures of single crystalline type diamonds were at first revealed by Seal (1963, 1965) and Harrison and Tolansky (1964) who etched in potassium nitrate the sawn and polished (100) surface of loupe-clean gem-quality diamond crystals. Various internal microstructures were revealed by this technique. Later, X-ray topography (Lang, 1964, 1965, 1974c, 1979), ultra-violet trnsmission topography, birefringence topography (Takagi and Lang, 1964), cathodoluminescence topography (Moore and Lang, 1974, 1975), and other techniques have been applied to study internal microstructures of single crystalline type diamonds.

The most commonly encountered internal microstructures of natural diamond of a single crystalline type are stratigraphic bandings parallel to {111}. These bandings become visible in loup clean crystals, because etch pit densities vary alternatively and stratigraphically. Alternative stratigraphic bands appear since there are defect-free zones (low etch pit densities and roughly correspond to Type II) and the zones with many defects (high etch pit densities, and roughly correspond to Type I, which contains Nitrogen platelets). The bandings are parallel to the outline of the crystal in case of octahedral crystals, but are often truncated by the outline of rounded or dodecahedroidal crystals (Seal, 1965; Moore and Lang, 1974), definitely suggesting that the rounded and particularly dodecahedroidal morphology are due to dissolution of larger crystals. The stratigraphic bandings sometimes show step-wise bandings consisting of small steps of {111}, and the general orientation is parallel to {110}, indicating the appearance of striated dodecahedral faces during the course of growth, due to a pile up of small {111} steps. The common occurrence of {111} stratigraphic bandings indicate that the growth of natural diamond of a single crystalline type mainly takes place by layer-by-layer growth mechanism (including spiral mechanism) on the octahedral faces and that the growth rates as well as the rate of impurity incorporation have changed alternatively during the process of growth, and that {111} interface was smooth under the conditions of natural diamond growth of a single crystalline type.

Another commonly encountered microstructures is dislocation bundles, which are revealed by X-ray topography. Dislocations usually originate at a central point in a crystal and radiate in bundles nearly perpendicular to {111} surfaces. One-to-one correlations are seen between the outcrops of dislocations and the summits of growth pyramids on the {111} faces (SUNAGAWA *et al.*, 1982), or the area where trigons concentrate on the {111} surfaces (LANG, 1964, 1965). Such dislocation configurations are expected only when crystals grow from vapour or solution phases under unconstrained environments, and are not expected for the crystals formed in metamorphic environments.

A most striking, though not very often seen, internal structure observed in single crystalline diamonds is the so-called centre-cross pattern. This pattern appears only at the central portion of a crystal and was observed for the first time by SEAL (1963, 1965) and HARRISON and TOLANSKY (1964), and aroused arguments as to the origin of such peculiar morphology. The centre-cross pattern consists of sectors of faceted straight bandings parallel to {111} and curved bandings parallel to a non-faceted mammilary cuboid. It was later observed that impurity nitrogen platelets are more concentrated in octahedral growth sectors, whereas in cuboidal growth sectors, solid inclusions appear more frequently (LANG, 1974c). HARRISON and TOLANSKY gave an explanation for the origin of the centre-cross pattern as due to the process of growth followed by dissolution and further regrowth. There was also a theory to explain its origin as due to plastic deformation that diamond crystals experienced during growth. FRANK (1966) explained the origin in terms of combined habit, i.e., faceted octahedral and non-faceted cuboidal growths. SUZUKI and LANG (1976) have made further detailed studies of the centre-cross pattern by X-ray topography and supported Frank's explanation. Non-faceted growth corresponds to the growth on a rough interface, i.e., abnormal growth, where inclusions will be trapped more frequently than in the facet-growth sectors. Facet growth corresponds to growth on a smooth interface, i.e., normal growth, and impurities with $K_{eff} < 1$ will concentrate more in such growth sectors than in non-faceted growth sectors (KITAMURA and SUNAGAWA, 1977).

The observations of the centre-cross pattern in single crystalline type diamonds indicate that at the earliest stage of a growth of single crystalline type diamond, there was a stage of combined faceted and non-faceted growths, and that $\{100\}$ always takes a non-faceted rough interface, whereas $\{111\}$ takes a faceted and smooth interface under the condition of growth of natural diamonds of single crystalline type. This makes a strong contrast to the case of synthetic diamonds, where $\{100\}$ can appear as faceted face.

X-ray topographic (LANG, 1965, 1979) and etching studies (SEAL, 1965) of synthetic diamonds have been also carried out. Synthetic diamond crystals do not usually show stratigraphic bandings and other internal microstructures, except solidified metal solution inclusions. The difference between natural and synthetic crystals in internal microstructures is probably due to the fact that synthetic crystals are much smaller than natural crystals and are formed in much shorter duration, during which small variations in environmental conditions are assumed.

7. *Discussions*

As described in the preceding sections, there are marked differences in crystal morphologies and surface microtopographies between natural and synthetic diamond crystals. The important differences are as follows:

(1) On natural diamonds, $\{111\}$ is the only face on which layer-by-layer (including spiral) growth can take place. This face maintains a smooth interface throughout the whole history of growth of single crystalline diamonds. Such growth processes are proved to have not taken place on $\{110\}$, $\{100\}$, nor on other faces. $\{110\}$ can appear sometimes through growth process as nearly flat faces, but this is due to the pile up of steps of growth layers on the neighbouring $\{111\}$ faces, and not due to free development of growth layers on $\{110\}$. In majority of cases, dodecahedroidal faces are formed due to dissolution. Cuboid faces are in most cases due to abnormal growth, but sometimes can appear as non-faceted growth interfaces in association with faceted octahedral interfaces in the earlier stage of growth of single crystalline diamonds. In any case, $\{100\}$ does not appear as a face on which free development of growth layers can occur. Growth layers on $\{111\}$ are thinner than 10Å and have wide step separations, suggesting a slow growth rate and also small, probably atomic entities as growth units.

(2) On synthetic crystals, both $\{111\}$ and $\{100\}$, and also probably $\{110\}$ faces show macroscopically flat surfaces on which growth layers develop. The growth layers on $\{111\}$ and $\{100\}$, however, always have macro step heights usually in the order of a few hundreds Ångstroms and rugged steps, and no mono-molecular thin growth layers have been observed. These suggest higher growth rates and larger growth units than in the case of natural diamonds.

In the following, we shall analyse why these morphological characteristics and the differences between natural and synthetic diamonds appear.

The morphology (crystal habit) of a crystal is determined by the relative growth rates of faces concerned. The growth rates of crystal faces are determined by the roughness of the interfaces and the driving force for crystal growth. Under the condition of the same driving force, the growth rate of the rougher interface is faster than that of the smoother interface, and thus the former face will disappear faster from the crystal. The roughness of the interfaces is determined by α-factor, which is related to bonding energy ϕ_{ss}, interaction energy between solid and fluid ϕ_{sf}, between fluid and fluid ϕ_{ff}, and growth temperature T, and also by chemical potential differences between solid and liquid phases $\Delta\mu/kT$ (refer to SUNAGAWA, 1982). If these values can be evaluated for actual cases, they will provide a firm basis for the analysis of morphological characteristics of both natural and synthetic diamond crystals. However, such evaluation is not easy. Thus, we will use an indirect approach.

In evaluating the morphological importance of crystal faces, HARTMAN (1953) and HARTMAN and PERDOK (1955) proposed the PBC method. This method predicts the relative morphological importance of crystal faces for a particular crystal structure, although the morphology derived from this analysis is a kind of abstracted morphology, not related to thermodynamic functions. However, this method is very useful in that it can predict the relative importance of crystal faces, as well as their nature. On an F-face (flat face, which contains more than two PBCs, and is, morphologically, the most important face), one can expect growth layers to develop freely. On an S-face (stepped face containing only one PBC, and morphologically less important than an F-face), we expect striations, and not growth layers. On a K-face (kinked face containing no PBC and a morphologically least-important face), neither growth layers nor striations are expected. A K-face corresponds to a rough interface, whereas an F-face corresponds to a smooth interface, and an S-face is an interface of an intermediate type. Thus, F-faces always appear and develop on a crystal, but a K-face will not usually appear. S-faces will develop much smaller than F-faces.

According to the PBC analysis of diamond by HARTMAN (1953), diamond structure contains [1/2, 1/2, 0] PBC vectors, which determines the morphological importance of faces. He concluded that {111} is the only F-face, whereas {110} and {100} are S-faces. Between {110} and {100}, the former is more important because the face contains one PBC, whereas the latter is an S-face on two identical zones [110] and [1$\bar{1}$0]. Thus {100} has closer characteristic to a K-face than {110}.

It follows from this analysis that:

(1) When diamond crystals grow near the equilibrium condition, they will take on the simple octahedral habit bounded by flat {111} faces only, or with the association of smaller {110} faces. Cubic habit bounded by {100} faces only will not be expected.

(2) Free development of growth layers can be expected only on {111} faces. No growth layers are expected to develop on both {110} and {100} faces.

(3) Only when the factors which are not involved in the PBC analysis, such as impurity or solvent effect, become effective, deviation from the above morphology will occur.

(4) Since morphology of growth layers follows the symmetry of the face, when they are polygonized, and this is controlled by strong PBCs, we can expect that {111} faces will show triangular growth layers parallel to the edges of the triangle of the face, {110} will exhibit striations parallel to [110] direction, and {100}, rough surface.

We have seen that both crystal morphology and surface microtopography of natural diamond crystals of a single-crystalline type are very close to the expected morphology summarized above. In contrast, morphological characteristics of synthetic diamond crystals deviate greatly from the expected morphology, in that {111}, {100}, and probably {110} faces appear due to layer-by-layer growth mechanism, and that layer growth through the incorporation of larger growth units is assumed from their surface microtopographic characteristics.

Morphological differences between natural and synthetic diamond crystals may be summarized as follows:

	{111}	{110}	{100}	growth unit
PBC analysis	smooth	rough	rough	atomic
natural	smooth	rough	rough	atomic
synthetic	smooth	smooth	smooth	larger

The essential differences in growth conditions between natural and synthetic diamonds are as follows;

(1) Both grow from a solution phase, but the solvents are different. Natural crystals grow from a silicate solution in which solubility of carbon is very low. Synthetic crystals grow from a metal solution, which has much higher solubility for carbon than silicate solution. The higher the solubility, the smaller is the α value (refer to SUNAGAWA, 1982; also SUNAGAWA and BENNEMA, 1982). Thus the solid-liquid interface should be rougher and so facet growth is less expected for synthetic crystals.

(2) Growth temperature is lower in natural diamond growth (estimated to be around 1,000 ~ 1,300°C, HARRIS, 1979) than synthetic diamond growth (1,350 ~ 1,700°C). The higher the growth temperature, the smaller is the α value and, thus, a rougher interface is expected for synthetic diamonds.

(3) Higher supersaturation is assumed for synthetic crystal growth than for natural crystal growth. The higher the supersaturation, the rougher is the interface through the kinetic process. Thus the rougher interface is expected for synthetic crystals than for natural crystals.

In any cases, a rougher interface is expected for synthetic crystals. Yet, the observed morphology indicates smoother interfaces for {111}, {110}, and {100} faces of synthetic crystals, and only for {111} faces of natural crystals.

Any existing theories on crystal growth mechanism and morphology of crystals assume atomic entities as growth units. Indeed, such growth units can be assumed for natural diamond crystals, as judged from the surface microtopographs. On synthetic crystals, the observed growth layers on {111} and {100} faces have always high-step heights. No mono-molecular growth layers have been observed on these crystals. Probably, the difference in morphological characteristics between natural and synthetic crystals and the discrepancy between theoretically-expected and experimentally-observed morphologies are due to such differences in the sizes of growth units.

The observed dependency of crystal habits of synthetic diamond upon thermodynamic functions indicate the temperature dependence of growth rates of {111} and {100} should be as schematically shown in Fig. 9. The growth rate of {111} face has less dependence upon temperature than that of {100}. This indicates that the α value of {100} decreases more sharply than that of {111} with the increase of growth temperature. In other words, roughening transition of {100} occurs at a much lower temperature than that of {111}. In the case of natural diamond of single crystalline type, {100} interface is always rough even when it appears together with {111}, which always takes smooth interface. The roughening transition temperature of {100} should be much lower than 1,000°C and that of {111} much higher than 1,500°C in a silicate solution system, where the growth unit is assumed to be atomic. The roughening transition temperature of {100} is around 1,600°C, and of {111} higher than 1,700°C in metal solution system, where larger growth unit is assumed.

8. *Conclusions*

Based on the available data of external and internal morphologies and surface microtopography of natural and synthetic diamonds, and also on the

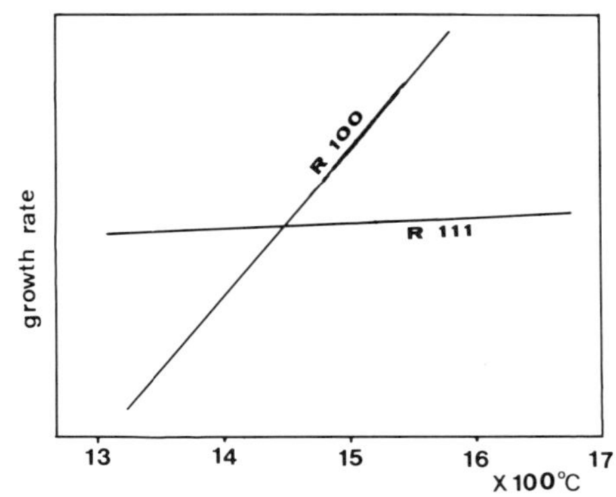

FIG. 9. A schematic drawing to show temperature dependency of growth rates of (111) and (100) faces.

present understandings of the theories and experiments on crystal growth mechanisms and the morphology of crystals, we may assume the following:

(1) Natural diamonds are classified into three categories; (a) single crystalline diamonds which are formed by normal growth under the supersaturation below σ^*, (b) polycrystalline diamonds which are formed by abnormal growth under the supersaturation conditions above σ^{**}, and (c) those which have experienced two-growth conditions.

(2) Natural diamonds have invariably experienced dissolution while they passed through the region unstable for diamond. The degree of dissolution varies from crystal to crystal due to the difference in crystal perfections, and from locality to locality, probably due to the differences in H_2O and CO_2 contents in different kimberlite magmas.

(3) Rounded morphology of single crystalline-type diamonds, particularly of dodecahedroidal morphology, and trigons on octahedral faces, are due to the dissolution process.

(4) Both natural and synthetic diamonds are formed in solution phases, although the solvents and the supersaturations are different between the two. Natural diamonds are not formed in metamorphic environments.

(5) Growth units in the formation of natural single crystalline diamonds are assumed to be of atomic sizes, whereas larger growth units are assumed for the formation of synthetic diamonds.

(6) The growth conditions, particularly the concentrations of carbon in the solutions, are assumed to be different between the diamonds occurring in ultra-mafic suites and eclogite suites. A higher carbon concentration is assumed for the latter.

(7) In the growth of natural diamond of single crystalline type, {111} is the only face which maintains smooth interface and on which layer-by-layer growth (including spiral growth) can occur. {110} may sometimes appear during growth by pile up of small steps of {111}, and {100} appears only as non-faceted mammilarly cuboid interface, i.e., rough interface. Thus, in a silicate solution system, the roughening transition temperature of {111} is higher than the highest estimated temperature for natural diamond growth (ca. 1,500°C), whereas that of {100} is lower than the lowest estimated temperature (ca. 1,000°C).

(8) In the growth of synthetic diamonds, {111}, {100}, and possibly {110} exhibit growth layers and spirals, indicating smooth interfaces under the conditions of synthesis. The roughening transition temperatures for these interfaces are higher than those for natural diamonds.

(9) The differences in the morphologies between natural and synthetic diamonds are considered to have been derived by the differences in solvents and in the sizes of growth units.

REFERENCES

ANSHELES, O. M. (1954): On the nature of rounded forms of diamond, *Uchen. Zap. Leningrad, Gos. Univ.*, No. 178, Issue 4 (in Russian).

ANSHELES, O. M. (1956): On the nature of rounded forms of diamond, *Zap. Vses Mineralog. Obshch*, Part 85, No. 2 (in Russian).

BOBRIEVICH, A., BONDARENKO, M. N., GUEVNSHEV, M. A., KRASOV, L. M., SMIRNOV, G. I. and YURKEVICH, R. K. (1959): *The Diamond Deposits of Yakutiya*, Ed. V. S. Sobolev, Dept. Geol. Conserv. Mineral Resour., Moscow (in Russian).

BOVENKIRK, H. P. (1961): Some observations on the morphology and physical chracteristics of synthetic diamond, *Am. Mineral.*, **46**, 952–963.

EMARA, S. H. and TOLANSKY, S. (1957): The microstructure of dodecahedral faces of diamond, *Proc. R. Soc. Lond., A*, **239**, 289–295.

FEDROV, E. S. (1901): *Course of Crystallography*, St Petersburg (in Russian).

FERSMAN, A. E. (1955): The *Crystallography of Diamond*, Acad. Sci. USSR, Moscow (in Russian).

FERSMAN, A. and GOLDSCHMIDT, V. (1911): *Der Diamant*, Winter, Heiderberg.

FRANK, F. C. (1958): On the kinetic theory of crystal growth and dissolution processes, in *Growth and Perfection of Crystals*, Eds. R. H. Doremus, B. W. Roberts, and D. Turnbull, John Wiley, New York, pp. 411–419.

FRANK, F. C. (1966): Defects in diamond, in *Proc. Intern. Indust. Diamond Conf., Oxford*, Ed. R. Berman, Industrial Diamond Information Bureau, London, pp. 119–135.

FRANK, F. C. (1969): Diamond and deep fluids in the mantle, in *The Application of Modern Physics to Earth and Planetary Interiors*, Ed. S. K. Runcorn, John Wiley, New York, pp. 247–250.

FRANK, F. C. and LANG, A. R. (1959): Observation by X-ray diffraction on dislocations in a diamond, *Phil. Mag.*, **4**, 383.

FRANK, F. C., PUTTICK, K. E. and WILKS, E. M. (1958): Etch pits and trigons on diamond, I, *Phil. Mag.*, **3**, 1262–1272.

GIARDINI, A. A. and TYDINGS, J. E. (1962): Diamond synthesis; Observations on the mechanism of formation, *Am. Mineral.*, **47**, 1393–1421.

HALPERIN, A. (1954): The formation of trigons on diamond, *Proc. Phys. Soc.*, **B67**, 538–545.

HARRIS, J. W. (1979): Physical and chemical constraints on the formation of natural diamond in the upper mantle, in *Diamond Research, 1979.*

HARRIS, J. W. and GURNEY, J. J. (1979): Inclusions in diamond, Chapter 18 in *The Properties of Diamond*, Ed. J. E. FIELD, Academic Press, London, pp. 555–591.

HARRISON, E. R. and TOLANSKY, S. (1964): Growth history of a natural octahedral diamond, *Proc. R. Soc. Lond., A*, **279**, 490–496.

HARTMAN, P. (1953): Relation between structure and morphology of crystals, Ph. D. Thesis, Groningen Univ., The Netherlands, 158 pp.

HARTMAN, P. and PERDOK, W. G. (1955): On the relation between structure and morphology of crystals, *Acta Cryst.*, **8**, 49–52, 521–524, 525–529.

HAÜY, R. J. (1817): *Traité des caractéres physiques des pierres précieuses pour servir à leur détermination lorsquelles ont été tailleés*, Paris.

JEYNES, C. (1978): Natural polycrystalline diamond, *Indust. Diamond Rev.*, Jan., 14–23.

KANDA, H., YAMAOKA, S., SETAKA, N. and KOMATSU, H. (1977): Etching of diamond octahedron by high pressure water, *J. Cryst. Growth*, **38**, 1–7.

KITAMURA, M. and SUNAGAWA, I. (1977): The effective distribution of solute and the anisotropy of a crystal, *Collected Abstract*, *ICCG–5*, Boston, p. 91.

KITAMURA, M., HOSOYA, S. and SUNAGAWA, I. (1979): Re-investigation of the re-entrant corner effect in twinned crystals, *J. Cryst. Growth*, **47**, 93–99.

KOKSHAROV, N. I. (1869): On the origin of rounded crystal faces, *Zap. Vses. Mineralog. Obshch.*, Part 4 (in Russian).

KOLOMYITSEV, A. I. and YAKUBOVA, S. A. (1976): On the columnal mechanism of growth of natural diamond crystals of cubic habit, *Zap. Vses. Mineralog. Obshch.*, **105**, 469–472 (in Russian).

KOMATSU, H. (1978): Growth, dissolution and destruction of diamond, in *Collected Abstract, 1st, Meeting of Res. Proj. "Materials Sci. Earth's Interior"*, pp. 9–12.

KUKHARENKO, A. A. (1954): On rounded diamond crystals, *Uchen. Zap. Leningrad, Gos. Univ.*, No. **178**, Issue 4 (in Russian).

KUKHARENKO, A. A. (1955): *Diamonds of the Urals*, Gosgeoltekhizdat, Moscow (in Russian).

KUKHARENKO, A. A. and TITOVA, V. M. (1957): New data on dissolution of diamond crystals, *Uchen. Zap. Dalnevostoch Gos. Univ. Ser. Geol*, No. **215**, Issue 8 (in Russian).

LANG, A. R. (1964): Dislocations in diamond and the origin of trigons, *Proc. Roy. Soc., A*, **278**, 235–242.

LANG, A. R. (1965): X-ray topography of diamond, in *Physical Properties of Diamond*, Ed. R. BERMAN, Clarendon Press, Oxford, pp. 69–115.

LANG, A. R. (1974a): Space-filling by branching columnar single-crystal growth; an example from crystallization of diamond, *J. Cryst. Growth*, **23**, 151–153.

LANG, A. R. (1974b): Glimps into the growth history of natural diamonds, *J. Cryst.* Growth, **24/25**, 108–115.

LANG, A. R. (1974c): On the growth sectorial dependence of defects in natural diamonds, *Proc. R. Soc. Lond., A*, **340**, 233–248.

LANG, A. R. (1979): Internal structure, in *The Properties of Diamond*, Chapter 14, Ed. J. E. Field, Academic Press, London, pp. 425–469.

LANG, A. R. and MOORE, M. (1972): On the internal structure of natural diamonds of cubic habit, *Phil. Mag.*, **26**, 1313–1325.

LITVIN, YU. A. (1969): On the problem of the origin of diamond, *Zap. Veses. Mineralog. Obshch.*, Part 98, No. 1 (in Russian).

LITVIN, YU. A. and BUTUZOV, V. P. (1968): On the growth of synthetic diamond crystals, *Dokl. AN SSSR, ser. mat fiz.*, **181**, No. 5 (in Russian).

MENDELSSOHN, M. J. and MILLEDGE, H. J. (1974): Computing the consequences of different growth, etch and abrasion rates in diamond, in *Diamond Research, 1974*, pp. 26–32.

MOORE, M. (1979): Optical studies of diamonds and their surfaces: A review of the late Professor Tolansky's work, in *The Properties of Diamond*, Ed. J. E. Field, Academic Press, London, pp. 245–277.

MOORE, M. and LANG, A. R. (1974): On the origin of the rounded dodecahedral habit of natural diamond, *J. Cryst. Growth*, **26**, 133–139.

MOORE, M. and LANG, A. R. (1975): Growth and dissolution features on diamond octahedron faces; Studies by X-ray topography, cathodoluminescence and optical micrography, *Acta Cryst.* **A31**, S 213 only.

MORIYOSHI, Y., SETAKA, N. and KAMO, M. (1980): Microstructures of natural single and polycrystal diamonds, *J. Gemmol. Soc. Japan*, **7**, 45–54 (in Japanese).

MUNCKE, G. (1979): Physics of diamond growth, in *The Properties of Diamond*, Ed. J. E. Field, Academic Press, London, pp. 473–499.

ORLOV, YU. L. (1963): *Morphology of Diamond*, Akad. Nauk SSSR, Moscow (in Russian).

ORLOV, YU. L. (1977): *The Mineralogy of the Diamond*, John Wiley, New York, 235 pp.

PALACHE, C., BERMAN, H. and FRONDEL, C. (1944): *The System of Mineralogy*, Vol. 1 7th Ed., John Wiley, New York, pp. 146–151.

PANDYA, D. C. and TOLANSKY, S. (1961): Micro-disk patterns on diamond dodecahedra, *Proc. Phys. Soc.*, **78**, 12.

PATEL, A. R. and AGARWAL, M. K. (1968): Studies of dodecahedral form of diamonds, *J. Cryst. Growth*, **2**, 377–379.

PATEL, A. R. and PATEL, M. M. (1969): Studies on the dodecahedral faces of diamonds, *Am. Mineral.*, **54**, 1324–1329.

PATEL, A. R. and PATEL, S. M. (1968a): Observations on trigons on natural octahedral faces of diamond, *Indust. Diamond Rev.*, June, 272–275.

PATEL, A. R. and PATEL, S. M. (1968b): Etching of (100) faces of natural diamond, *Indust. Diamond Rev.*, Nov., 498–499.

PATEL, A. R. and RAMACHANDRAN, N. (1968): Growth spirals on octahedral faces of synthetic diamond, *Indust. Diamond Rev.*, March, 127–129.

PATEL, A. R. and RAMANATHAN, S. (1962): Etchpits on diamond surfaces, *Phil. Mag.* **7**, 1305–1314.

PATEL, A. R. and RAMANATHAN, S. (1964): Dislocation energy and etch pit orientation on diamond octahedral faces, *Physica*, **30**, 2003–2004.

POTON, J. W., REEKIE, T. D. R. and ANGUS, J. C. (1974): Computer simulation of etching of diamond surfaces, in *Diamond Research, 1974*, pp. 33–38.

RAMAN, C. V. and RAMANASESHAN (1946): *Proc. Indian Acad. Sci.* **A24**, 1.

ROME DE L'ISLE (1783): *Cristallographie ou Description des Formes Propres à Touts les Corps du Régne Minéral*, Paris.

SCACCHI, A. (1862): Sulla poliedria della facce dei cristali, *Mem. Acad. Sci. Torino*, **21**.

SCHAFRANOVSKI, I. I. (1948): Crystallography of rounded diamonds, *Leningrad Gos. Univ.* (in Russian).

SEAL, M. (1963): The growth of natural diamonds as revealed by etching experiments, in *Proc. 1st Intern. Congr. Diamond in Industry, Paris, 1962*, Ed. P. Greene, Industrial Diamond Information Bureau, London, pp. 361–374.

SEAL, M. (1965): Structure in diamonds as revealed by etching, *Am. Mineral.*, **50**, 105–123.

SPITSYN, B. V., BOUILOV, L. L. and DERJAGUIN, B. V. (1981): Vapour growth of diamond on diamond and other surfaces, *J. Cryst. Growth*, **52**, 219–226.

SUNAGAWA, I. (1977): Natural crystallization, *J. Cryst. Growth*, **42**, 214–223.

SUNAGAWA, I. (1981): Characteristics of crystal growth in nature as seen from the morphology of mineral crystals, *Bull. Minéral. (France)*, **104**, 81–87.

SUNAGAWA, I. (1982a): Growth of crystals in nature, this volume, pp. 61–103.

SUNAGAWA, I. (1982b): Morphology of diamond crystals, *J. Jpn. Assoc. Mineral. Petrol. Econ. Geol.*, Special Issue **3**, 129–142 (in Japanese).

SUNAGAWA, I. and BENNEMA, P. (1982): Morphology of growth spirals, theoretical and experimental, in *Preparation and Properties of Solid State Materials, Vol. 7*, Ed. W. R. Wilcox, Marcel Dekker Inc., New York, pp. 1–129.

SUNAGAWA, I., TSUKAMOTO, K. and YASUDA, T. (1982): Surface microtopographic and X-ray topographic study of octahedral crystals of natural diamonds from Siberia, this volume, pp. 331–349.

SUTTON, J. R. (1928): *Diamond, A Descriptive Treatise*, Murby, London.

SUZUKI, S. and LANG, A. R. (1976): Internal structures of natural diamonds revealing mixed-habit growth, in *Diamond Research, 1976*, pp. 39–47.

TAKAGI, M. and LANG, A. R. (1964): X-ray Bragg reflection, spike reflection and ultra-violet absorption topography of diamonds, *Proc. R. Soc. Lond., A*, **281**, 310–322.

TOLANSKY, S. (1955): *Microstructures of Diamond Surfaces*, NAG Press, London.

TOLANSKY, S. (1960): *Surface Microtopography*, Longmans, London.

TOLANSKY, S. (1961): Comparative optical and interferometric studies on synthetic diamonds, *Proc. R. Soc. Lond., A*, **263**, 31–38.

TOLANSKY, S. (1963): Surface topography of natural and synthetic diamonds, *Proc. 1st. Intern. Congr. Diamond in Industry, Paris,1962*, Ed. P. Greene, Industrial Diamond Information Bureau, London, pp. 343–359.

TOLANSKY, S. (1965): Optical studies on diamond, in *Physical Properties of Diamond*, Ed. R. Berman, Clarendon Press, Oxford, pp. 135–173.

TOLANSKY, S. (1973): Distribution of Type I and Type II in South African diamonds, in *Diamond Research, 1973*, pp. 28–31.

TOLANSKY, S. and SUNAGAWA, I. (1959): Spirals and other growth forms of synthetic diamonds; A distinction between natural and synthetic diamonds, *Nature*, **184**, 1526–1527.

TOLANSKY, S. and SUNAGAWA, I. (1960): Interferometric studies on synthetic diamonds, *Nature*, **185**, 203–204.

TRUEB, L. F. and BUTTERMAN, W. C. (1969): Carbonado; A microstructural study, *Am. Mineral.*, **54**, 412–425.

TRUEB, L. F. and BARRETT, C. S. (1972): Microstructural investigation of ballas diamond, *Am. Mineral.*, **57**, 1664–1680.

VARMA, C. K. R. (1967a): Trigons on diamond, *Phil. Mag.*, **16**, 959–974.

VARMA, C. K. R. (1967b): Studies on natural diamonds of the dodecahedral form, *Phil. Mag.*, **16**, 621–634.

VARMA, C. K. R. (1967c): Studies on the structure of natural diamond crystals of the cube form, *Phil. Mag.*, **16**, 657–667.

VARMA, C. K. R. (1967d): Etch pits and trigons on diamond, *Phil. Mag.*, **16**, 611–620.

VAN DER VEEN A. L. (1913): Die Symmetrie des Diamanten, *Z. Krist. Miner.* **15**, No. 4.

WILKS, E. M. (1961): Steep-sided trigons on diamonds, *Phil. Mag.*, **6**, 701.

WILLIAMS, A. F. (1932): *The Genesis of Diamond*, Benn, London.

YAMANAKA, S., KANDA, H. and SETAKA, N. (1980): Etching of diamond octahedrons at high pressure with controlled oxygen partial pressure, *J. Materials Sci.*, **15**, 332–336.

YAMAOKA, S., KOMATSU, H., KANDA, H. and SETAKA, N. (1977): Growth of diamond with rhombic dodecahedral faces, *J. Cryst. Growth*, **37**, 349–352.

Note added in proof

SPITSYN *et al.* (1981) have recently reported that they could grow diamond crystals with habit faces on various substrates including Cu, by chemical transport technique, under the metastable conditions for diamond, i.e. at atmospheric pressure and temperatures ranging from 800 to 1,000°C. According to their report, the diamond crystals are around 50 μm in diameter and octahedral to cubo-octahedral in habit. On both $\{111\}$ and $\{100\}$ faces, growth layers develop, but judging from the SEM photographs in their paper, their morphology is the reverse that observed on natural and synthetic crystals grown under stable conditions, i.e. triangular growth layers on $\{111\}$ faces have opposite orientations to the triangle of the $\{111\}$ face. It is also striking that habit changes of the morphologies of these crystals depending on thermodynamic parameters are the quite reverse the tendency exhibited by diamonds synthesized under stable conditions, i.e. $\{111\}$ faces develop more as apart from the equilibrium. Thus, the morphologies of diamond crystals synthesized under metastable conditions deviate most greatly from the theoretical morphology of diamond crystals. This problem is discussed in a paper by SUNAGAWA (1982b).

Materials Science of the Earth's Interior, edited by I. Sunagawa, pp. 331–349.

SURFACE MICROTOPOGRAPHIC AND X-RAY TOPOGRAPHIC STUDY OF OCTAHEDRAL CRYSTALS OF NATURAL DIAMOND FROM SIBERIA

Ichiro SUNAGAWA[1], Katsuo TSUKAMOTO[1], and Toshikazu YASUDA[2]

[1]*Institute of Mineralogy, Petrology and Economic Geology, Tohoku University, Aoba, Sendai, Japan*
[2]*Geological Survey of Japan, Yatabe-cho, Tsukuba-gun, Ibaragi Prefecture, Japan*

1. *Introduction*

Since natural diamond crystals were brought up from the depth of the Earth and passed through the region unstable for diamond, which is estimated to be less than 120 to 150 km in depth, they have invariably experienced dissolution (FRANK, 1967). Rounded corners, edges and curved faces, which most natural diamonds exhibit, are part of the evidence of dissolution. Trigons, i.e., triangular depressions with negative orientations in respect to the triangle of an octahedral face, are observed universally on octahedral faces of natural diamond crystals, and are another testimony to support the dissolution theory (LANG, 1964). SEAL (1962, 1965) observed that the outline of natural diamond crystals often truncates crystallographic growth bandings revealed by etching in the crystal, and interpreted the fact in terms of dissolution.

Due to this, the surface microtopographs of diamond crystals have been modified to such an extent that the details of the original growth features become obscure. Therefore, it has not been possible to analyse in detail how diamonds grew in nature, though it was suggested through the studies of growth bandings (SEAL, 1965), and also X-ray topographic studies (e.g., LANG, 1965) that natural diamond crystals grew by the layer-by-layer growth mechanism mainly on octahedral faces and that screw dislocation might not play important roles as the sources of growth layers. The occurrence of diamond crystals well preserving their original as-grown surfaces are really exceptional.

Although most natural diamond crystals have suffered dissolution, there are two possibilities of finding crystals on which original as-grown surfaces may be seen; namely, the diamond crystals included in larger diamond crystals and the crystals which suffered only very slight dissolution through various reasons. Taking out diamond crystals, without destroying their surfaces, from the host crystal is extremely difficult, which one may not dare to do unless he does not care for the value of the stone. The only way to investigate the original growth features is, therefore, to find diamond crystals of the latter type. Such diamond crystals may be found in quenched kimberlite dykes. Fortunately, we could obtain, through De Beers Diamond Research Laboratory in Johannesburg, diamond crystals of this type which were found in a very thin kimberlite dyke in Siberia whose details of geologic occurrence are not known. Altogether 12 octahedral crystals of 1 to 2 mm in diameter were granted; six crystals suffered dissolution, the other six suffered only a very slight dissolution.

We made surface microtopographic studies of seven selected stones using a highly sensitive phase contrast microscope and prepared mosaic photographs of each face, with which crystal models with surface microtopographs were constructed. We then selected three representative crystals which suffered only slight dissolution, and made X-ray topographic studies using the Lang method. Both results are correlated so that the growth mechanism of these crystals can be elucidated. Such a correlation between surface microtopographs and dislocations has been previously made only on one crystal by LANG (1964), who showed that pyramidal trigons are always associated with dislocation outcrops, but no dislocations outcrop at flat-bottomed trigons. His correlations were made mainly between trigons and dislocation outcrops, and not between growth features and dislocation outcrops. We have paid special attention to finding a correlation between the summits of growth pyramids and dislocation outcrops. One-to-one correlation has been established. Based on this, we arrived at a conclusion that natural diamonds of a single crystalline type grew by the spiral growth mechanism from a low supersaturated solution phase.

2. *Experimental*

Crystal surfaces were cleaned with a H_2O_2 solution, detergent solution and distilled water. They were silvered in a vacuum evaporation plant to secure a high reflectivity suitable for observation under reflecting light. Photomicrographs were taken with the Olympus PMF-type and Reichert MeF_2-type microscopes equipped with phase plates of 95% absorption, with which steps of a few Angstrom can be detected (SUNAGAWA, 1961, 1975). High contrast film and high contrast processing were used so that the thinnest steps could be detected. Since nearly ten frames are required to cover the whole surface of an octahedral face, mosaic photographs were prepared for each face, with which an octahedral model was constructed for each crystal. We could therefore investigate each crystal enlarged about 100 times, with detailed surface microtopographs of each face revealing topographs down to a few Ångstroms. Figure 1 is a photograph of such octahedral models. Step heights of growth layers were measured by multiple-beam interferometry (TOLANSKY, 1948).

X-ray topography was applied to three representative crystals, which have experienced only a very slight dissolution. Crystal No. 2 (see Fig. 1) shows triangular growth pyramids on its octahedral faces, and No. 3 and 4 extremely flat octahedral faces on which only very thin growth layers and shallow growth pyramids develop. Since crystals are less than 2 mm in diameter, which is an adequate thickness for X-ray topography, they were investigated without slicing or cutting.

X-ray topographs were taken by transmission, using a Rigaku Denki Lang camera. All traverse topographs were taken with $AgK\alpha_1$ radiation at 50 kV and 1 mA. Images were recorded on Ilford Nuclear Plates emulsion L4, 50 μ thick. Each topograph was subsequently enlarged by a photographic procedure through an intermediate contrast inversion step. Nearly one hundred topographs of reflections {220} and {111} were taken. From these data, models of octahedral crystals showing the dislocations and stacking faults were constructed for crystals Nos. 2 and 3 (Fig. 2). By correlating the

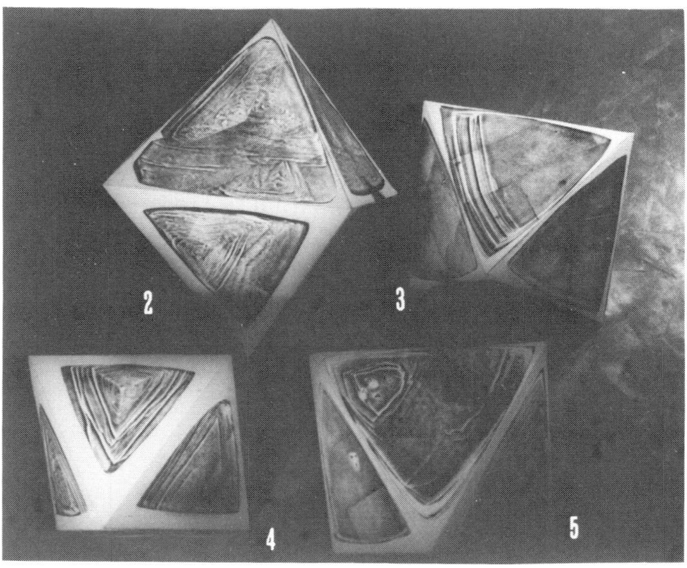

FIG. 1. Models of octahedrons with surface microtopographs revealing steps down to 5Å in height. Numerals indicate the specimen numbers.

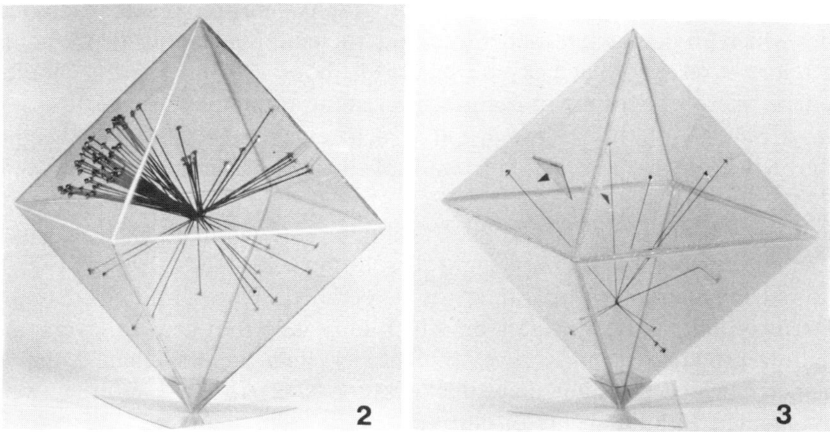

FIG. 2. Models of two octahedrons (left, crystal No. 2; right, No. 3) indicating distribution of dislocations and stacking faults inside the crystals.

two models for surface microtopography and X-ray topography, it was possible to precisely correlate each surface microtopograph with an X-ray topograph in the case of crystal No. 3. With crystal No. 2, however, such a one-to-one correlation was not possible because of the large number of dislocations, but a general correlation was possible with satisfactory results.

3. *Surface Microtopography*

A wide range of surface microtopographs has been observed. Not a single crystal was observed whose history could be interpreted to be entirely free from dissolution. Surface microtopographs of octahedral faces vary considerably from crystal to crystal and even from face to face on one crystal (see Fig. 1). There are crystals whose {111} faces are characterized by the development of triangular growth pyramids, others by extremely flat surfaces, on which only very thin growth layers are seen. Even on the latter type crystals, there are faces which show development of growth pyramids. These differences are due to the difference in growth and dissolution histories, as well as in crystal perfections.

In the following subsections the representative surface microtopographs will be described separately.

3.1 *Outline of the {111} faces*

A regular triangular outline bounded by straight edges of the {111} faces has not been observed. The three corners are rounded off to a varying extent. As schematically shown in Fig. 3, when dissolution proceeds two-

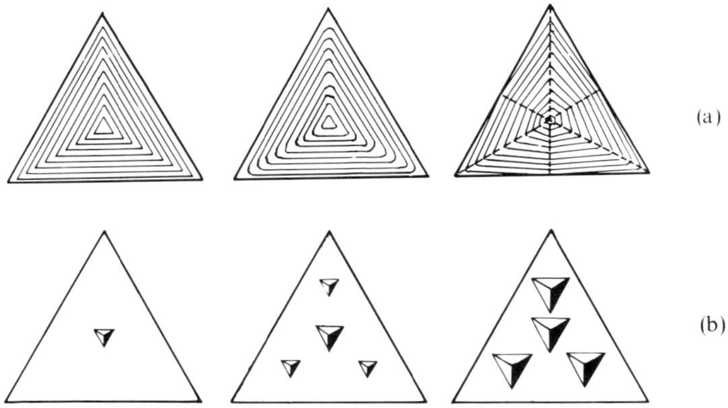

FIG. 3. A schematic drawing to show morphological changes of a (111) face as two-dimensional dissolution proceeds from the corners (a) and dislocation outcrops on the surface (b).

dimensionally starting from a regular triangular octahedral face, the three corners of the face are preferentially dissolved, eventually resulting in a hexa-octahedral form. In addition to this, dissolution also starts from the outcrops of lattice defects on the surface. Combination of the two dissolution processes results in a rather complicated appearance of the {111} faces, such as shown in Fig. 4b. However, it is not impossible to judge the degree of dissolution that individual crystals have experienced, if careful judgement is made based on the outline of the octahedral faces, the morphology of growth pyramids and growth layers, densities of well-defined trigons, etc. The three categories which

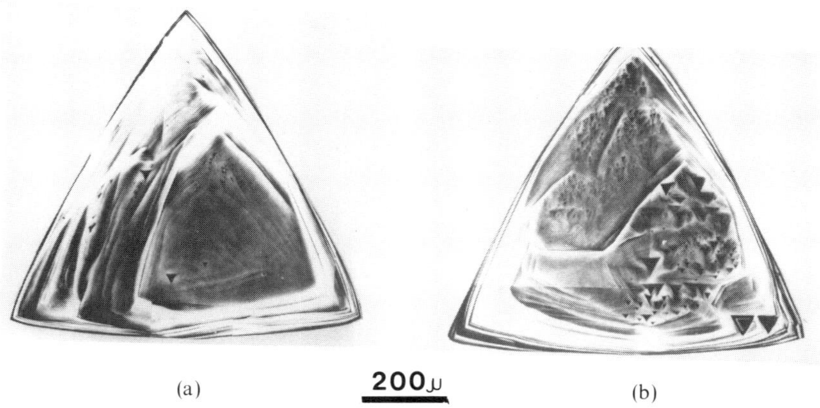

FIG. 4. Positive phase contrast photomicrographs representing the octahedral faces which received a very slight dissolution (a) and heavy dissolution (b). All surface microtopographs in this chapter are taken with positive phase contrast.

will be used in the following description, are judged mainly on this basis. Crystals of category A have suffered only a slight dissolution, thus their original surface microtopographs have not been modified very much. Crystals of category C have suffered heavy dissolution, and their original surface microtopographs have been strongly modified. Crystals of category B are in an intermediate state. In Fig. 4a and b, two extreme examples are shown.

3.2 Growth pyramids and growth layers

A wide range of morphologies of growth pyramids (hillocks) has been observed, starting from almost regular triangular to hexa-octahedral pyramids. Regular triangular pyramids (Fig. 5) are observed only on crystals of category A, whereas hexa-octahedral pyramids on the crystals of category C. Slightly malformed pyramids are observed on the crystals of category B. It is evident that the regular triangular pyramids represent the growth pyramids which suffered a very little dissolution and preserve the morphology of original as-grown growth pyramids, whereas the malformed pyramids, those received much heavier dissolution.

FIG. 5. Well developed steep regular triangular growth pyramids on a (111) face of crystal No. 2. Note trigons at the summits of the pyramids.

There is also a range in the steepness of the pyramids. The pyramid shown in Fig. 5 belongs to a steeper one, yet the height of the pyramid is less than 3,000 Å, and the inclination of the side faces to the octahedral face is far less than 1°. The pyramid consists of a pile up of thin triangular layers, with narrow step separations. The total height of these pyramids were measured by multiple-beam interferometry and the number of layers constituting the pyramids were counted on phase contrast photomicrographs. The average step height of growth layers is measured from these values to be less than 50 Å. The real step height should be smaller than this value, since these layers are often bunched.

Crystals belonging to category A (e.g. No. 3 and 4) exhibit extremely flat octahedral faces to such an extent that only straight parallel interference fringes are produced. On these octahedral faces, one finds widely separated triangular steps (step separation being 10 to 90 μ), with step heights of less than 5Å (Figs. 6 and 11). These steps originate from one or a few centres on the face, and the step systems correspond to very shallow growth pyramids, as

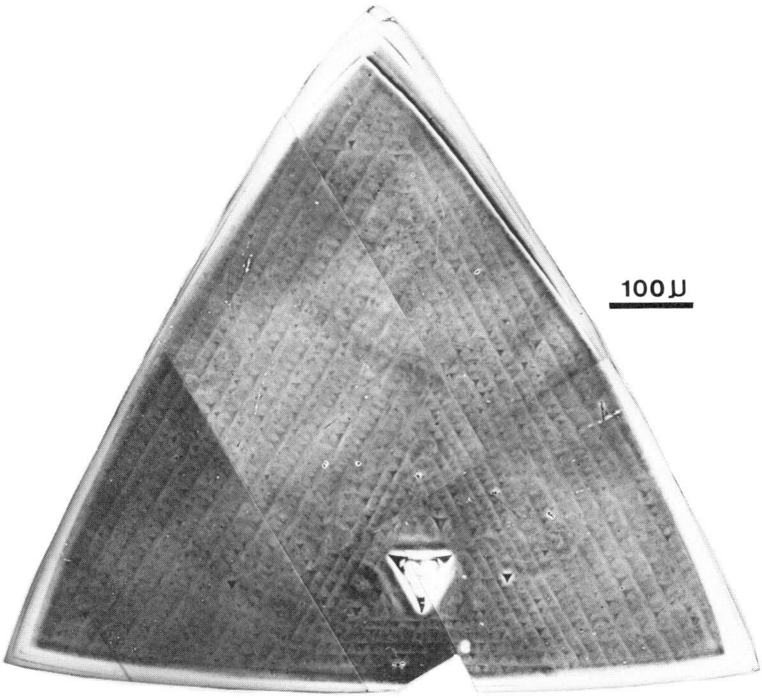

FIG. 6. Mosaic phase contrast photomicrograph of a very flat (111) surface of crystal No. 4. Note triangular growth layers with step heights less than 5Å, shallower growth pyramids, numerous sail-boat type trigons and a large trigon at the centre of the growth layers.

compared to the steeper pyramids shown in Fig. 5. The difference between the two is simply due to the difference in step separations.

On one octahedral face, one to more than ten growth pyramids are observed. Some are dominant and show a regular triangular form, others are dominated and show only a portion of the triangle. The number of pyramids occurring on one octahedral face is smaller for shallower pyramids as compared to the case of steeper pyramids. Shallower and steeper pyramids do not co-exist on one octahedral face.

No step system to be called spirals has been observed on both steeper and shallower pyramids. However, it should be noted that at the summits or centres of these pyramids, one usually, though not always, finds small P-type (pyramidal or point-bottomed type) trigons, as clearly seen in Figs. 5 and 6.

3.3 Trigons

Trigons (TOLANSKY, 1955) have been observed universally on all crystals investigated in this study, though their sizes, types and densities vary significantly from crystal to crystal. From the depth of trigons, three groups have been distinguished. There are also three different categories in the sizes of trigons.

Deep trigons can easily be seen under the ordinary reflection microscope, and they appear as white trigons under a positive phase contrast microscope. There are two categories; one of which has an edge length ranging from 90 to 140 μ and occurs only on crystals of category C. From their sizes, this type corresponds to FD (deep flat bottomed) trigon of LANG's (1964, 1965) classification. Under higher magnification of a phase contrast microscope, it is seen that the trigons of this type have stepped side faces and with a large number of tiny trigons, 1 to 10 μ across, at the bottom (Fig. 7). One or three

FIG. 7. Photomicrograph of a large FD-type trigon. Note microstructure of the trigon.

corners of the trigons are, in some cases, truncated with a gentler slope than those of the three principal edges, resulting in a hexagonal depressions in an extreme case.

Smaller deep trigons whose edge length ranges from 15 to 50 μ are P-type (point-bottomed or pyramidal type) trigons with step-wise side slopes (Fig. 8). Trigons of this type are observed on all three categories of crystals. Although they are smaller than the P-type trigons reported by LANG (1965), both are of the same type.

The second group of trigons are shallow trigons with the edge length ranging from 3 to 15 μ. These cannot be recognized under an ordinary reflection microscope, but can be easily seen under a phase contrast microscope. They are observed on all crystals and appear as dark trigons under the positive phase contrast microscope, suggesting that their depth is much smaller than the previous type. Their densities are much higher on that crystals of category C than those of category A. On crystals of category A, they appear only rarely and isolatedly and sometimes selectively at the summits of growth pyramids or growth centres of shallow growth pyramids (Figs. 5 and 9). They are mostly of the PS-type (shallow point-bottomed type) trigons according to Lang's classification. On the crystals of category C, they often appear in clusters. The cluster may be in a regular triangular form or in a trapezoidal form.

The trigons of the third group are much smaller in size and much shallower in depth. They are hardly seen even under a phase contrast microscope, showing that their depths are less than 10 Å. They are detectable only when special care is taken in phase contrast photographing. Their edge lengths are around 1 μ. Trigons of this type are invariably of the FS-type (shallow flat-bottomed type) in Lang's classification and universally occur on all crystals. They appear densely on crystals of category A, having flat

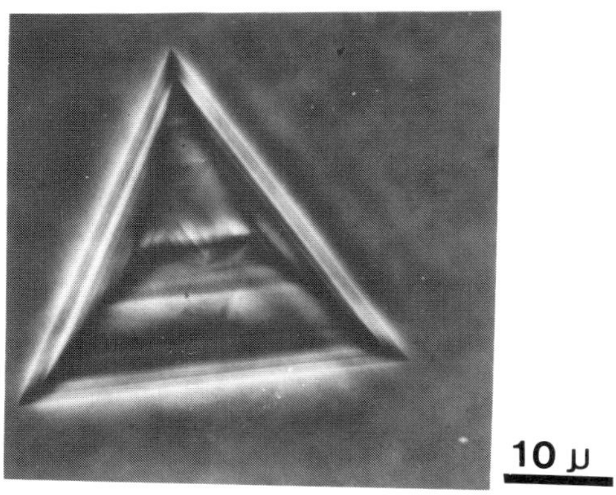

FIG. 8. Photomicrograph of a PD type trigon.

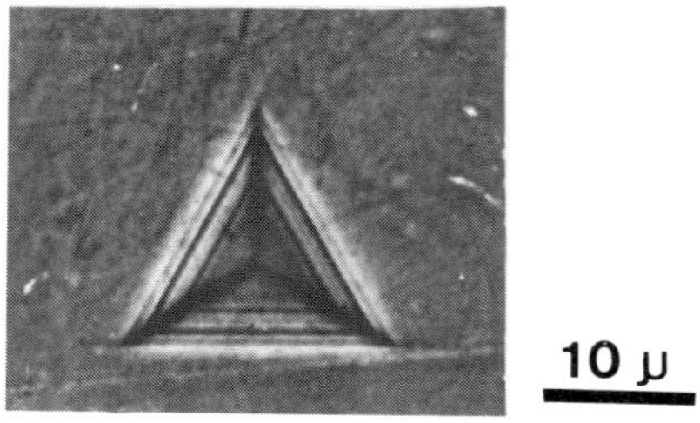

FIG. 9. Photomicrograph of a smaller P-type trigon appearing at the centre of growth layers.

octahedral faces, or on the flat area of the octahedral faces showing growth pyramids. A typical example is seen in Fig. 6. In this photograph, it is clearly seen that there are a wide range of sizes and that smaller ones appear on the surface of, and larger ones preferentially along the steps of thin growth layers. It is seen that the larger trigons aligned along the steps are, in reality, trigons consisting of coalesced smaller trigons and, therefore, edge lengths exceed 1 μ. Many of these trigons are completed in a sense that they show triangles, but there are some which are not completed, i.e., triangles are not closed. This group corresponds to the sail-boat patterns of Lang's classification.

The results of observations are summarized in Table 1, together with the number of dislocations detected by X-ray topographic investigations.

4. *X-Ray Topography*

X-ray topographs of crystals No. 2, 3 and 4 show various types of contrast images; triangular planer type, irregularly curved plane type, fringe patterns, spotty type, and dislocation bundles radiating from a point in the crystal.

Triangular planer contrast images correspond to a stacking fault parallel to $\{111\}$, as discussed by LANG (1974). Irregularly-curved plane-type contrast images may be due to an internal strain field around the solid inclusions, about which further study is required. The fringe pattern is thickness fringes, and spotty type contrast images are due to surface damage. Since these are not directly related to the growth process, we shall mainly focus our attention on dislocation bundles.

Crystal No. 2, whose surface exhibits well-developed triangular growth pyramids, shows bundles consisting of a large number of dislocations which radiate from the centre of the crystal to the surface (Fig. 2a). As a general tendency, it is seen that the bundles outcrop at the portions on the surface where triangular growth pyramids occur and/or where P-type trigons concentrate. However, there are so many dislocations in the bundles that it is

TABLE 1. Summary of observations.

Category			A			B		C	
Crystal No.			3	4	2	7	1	10	9
Corners and edges of (111) face			Triangular	Triangular	Triangular	Triangular	Hexa-octahedral	Hexa-octahedral	Hexa-octahedral
Growth centres									
Morpgology			F	F + P	P	P + F	P	P	P
Number			1–2	1–2	Several	1–2	1–2	Several	1–2
Growth layers									
Step height			$\lesssim$10Å	$\lesssim$5Å	$\simeq$10Å	$\gtrsim$10Å	$\gtrsim$10Å	$\gtrsim$10Å	$\gtrsim$10Å
Step separation			$\simeq$90 μ	$\simeq$20 μ	<10 μ	$\simeq$10 μ	<10 μ	<10 μ	<10 μ
Mode			Single	Single	Bunch	Bunch	Bunch	Bunch	Bunch
Trigons, number/face	Deep	PD	—	1–2* (25 μ)	1–2 (10 μ) On 1 face only	—	—	—	—
		FD	—	1–2 (50 μ)	Few (15 μ) On 1 face only	—	—	Many (40–100 μ)	Few (70–140 μ)
	Shallow	PS	1–2*	1–2 (3–10 μ)	Few* (3 μ)	1–2* (3–10 μ)	Few (3–10 μ)	Many (3–15 μ)	Many (3–20 μ)
		FS	1–2 (1–10 μ)	Many (1–3 μ)	Many ($\approx$3 μ)	1–2 (3–7 μ)	—	Many (3–10 μ)	Many (3–10 μ)
	Sail-boat		Many	Numerous	Numerous on flat area	Many on flat area	Many on flat area	Many	Many
Number of dislocations			11	$\gtrsim$20	$\gtrsim$100	n.d.	n.d.	n.d.	n.d.

Morphology of growth centres: F, shallow growth pyramid, P steep growth pyramid. Trigons: PD, deep pyramidal, FD deep flat-bottomed, PS, shallow pyramidal, FS shallow flat-bottomed, *indicate trigons appearing at the centre of growth pyramid.

difficult to distinguish individual dislocations and to correlate them with individual trigons or summits of growth pyramids.

In contrast, crystal No. 3 (Fig. 2b), whose surface is extremely flat and shows only a small number of trigons (see Fig. 1), exhibits only a small number of dislocations, radiating from a somewhat off-centred point in the crystal. Dislocations are seen individually on X-ray topographs, and it is possible to correlate them with surface microtopographs, and also to decide the nature of individual dislocations. Thus, we selected this crystal for further detailed X-ray topographic study. Crystal No. 4, which also contains only a small number of dislocations, was used as a reference stone, on which one-to-one correlation between surface and X-ray topographies was also possible.

By taking many X-ray topographs of crystal No. 3, we see that altogether this crystal contains eleven dislocations, all directly connected with trigons on the surface. Figure 10 schematically shows dislocations and the trigons, together with a denotation of the two dislocations cropping out on the H face. Since the H face exhibits clearly visible growth layers and two shallow growth hillocks and two trigons only, we selected this face for the identification of dislocation types. From Figs. 11a and b, surface microtopograph and X-ray topograph, respectively, it is safely concluded that the two dislocations H-1 and H-2 outcrop at the two trigons on the H face. It is also safe to conclude that the two trigons on the H face are formed at the two shallow growth centres on the face, one fully developed, the other dominated by the former and only partly developed. We therefore conclude that the growth layers originate from

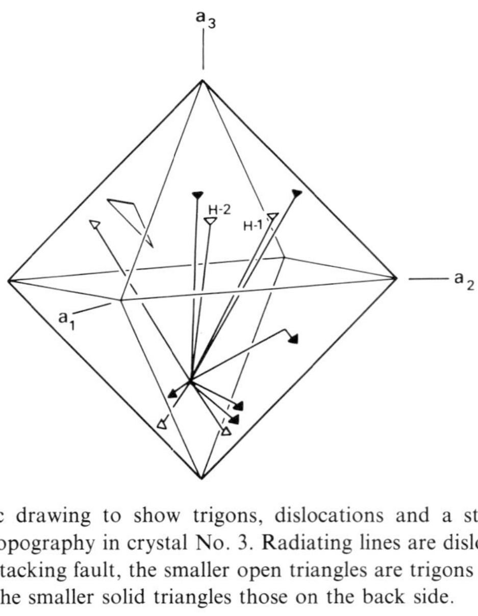

FIG. 10. Schematic drawing to show trigons, dislocations and a stacking fault revealed by X-ray topography in crystal No. 3. Radiating lines are dislocations, the larger triangle is a stacking fault, the smaller open triangles are trigons on the (111) faces of front side, the smaller solid triangles those on the back side.

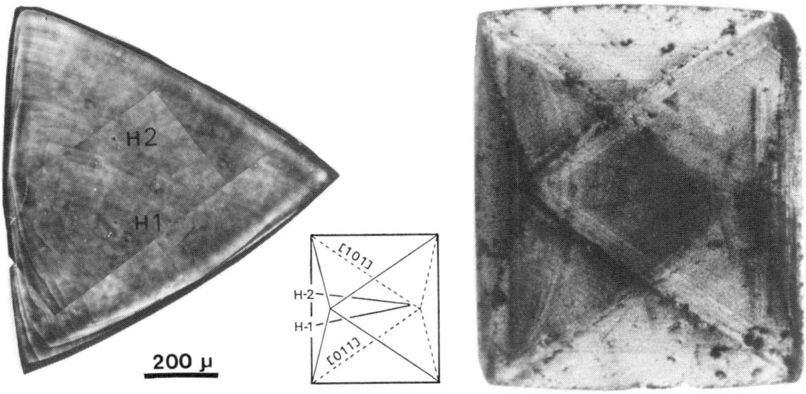

(a) (b)

FIG. 11. Surface microtopograph of H face on crystal No. 3 (a) and the corresponding X-ray topograph (b).

and the trigons are formed at the points where the two dislocations emerged. On the H-face, no other growth centres are seen and all growth layers seen on this face originate from either of the two points.

The nature of the two dislocations H-1 and H-2 are investigated both by LANG's method (1965) and on the basis of general extinction rules. According to Lang, edge and screw dislocations in diamond can be distinguished, provided that the Burgers' vector is 1/2 ⟨110⟩, by taking X-ray topographs of three sets of reflections (220), (202), (022). If a contrast image appears on the three topographs, a dislocation outcropping on a (111) face is a screw type, whereas it is an edge type if contrast images appear only on two topographs. When a dislocation is a screw type, its Burgers' vector is either [110], [101] or [011], whereas if it is an edge type, the Burgers' vector is either $[1\bar{1}0]$, $[0\bar{1}1]$ or $[\bar{1}01]$.

In Figs. 12a, b, c, X-ray topographs, taken by the reflections (022), $(\bar{2}0\bar{2})$, $(\bar{2}\bar{2}0)$, are shown, respectively. The two dislocations H-1 and H-2 are visible in all of the three reflections, showing that they are a screw type. Among the three topographs, the contrast of the two dislocations H-1 and H-2 are rather weak on two topographs and strong on one. Although their contrast becomes weak in two reflections, they do not entirely vanish in any of the three reflections. The dislocation H-1 shows a strong contrast in Fig. 12a, (022) reflection and H-2 in Fig. 12b, $(\bar{2}0\bar{2})$ reflection. This shows that the direction of the Burgers' vector of the two dislocations are perpendicular to the respective reflection planes, i.e., the direction of the Burgers' vector of H-1 is 1/2[011], that of H-2 is 1/2[101]. This is in accordance with the above consideration, based on Lang's method.

According to the general extinction rules of dislocations on X-ray topographs, a screw-type dislocation should lose its image on the topograph

FIG. 12. X-ray topographs of crystal No. 3. (a) (022), (b) ($\bar{2}0\bar{2}$), (c) ($\bar{2}\bar{2}0$).

taken by the reflection which includes the Burgers' vector of the dislocation. Figures 13a, b, c show three X-ray topographs taken by the reflections $(\bar{1}1\bar{1})$, $(1\bar{1}\bar{1})$ and $(11\bar{1})$, respectively. On the topograph of $(\bar{1}1\bar{1})$ reflection (Fig. 13a) which includes [011], the image of the dislocation H-1 is not seen at all, whereas H-2 is visible. On the topograph of $(1\bar{1}\bar{1})$ reflection which includes [101] (Fig. 13b), the image of H-1 is visible, but H-2 disappears. On the topograph of $(11\bar{1})$ reflection which includes both [011] and [101], both images are not seen at all. These observations again show that the direction of the Burgers' vector of the dislocation H-1 is 1/2[011], and that of the dislocation H-2 is 1/2[101].

From both results, we may safely conclude that the two dislocations H-1 and H-2 outcropping at the trigons on the summits of the two growth centres, are screw-type dislocations with the direction of Burgers' vector indicated above.

Although no trial was made to identify the remaining nine dislocations in crystal No. 3, it is safe to extend the above result to these dislocations and to consider that they are also screw-type dislocations. These dislocations outcrop on the remaining seven {111} faces at P-type trigons appearing at the centre of growth layers, and all {111} faces of this crystal have at least one outcrop of dislocations. A similar conclusion was also obtained for crystal No. 4. It was concluded that screw dislocations outcrop at the trigons in Fig. 6, a (111) face of crystal No. 4. The only probable conclusion obtainable from this is that crystals No. 3 and No. 4 grew by the spiral growth mechanism throughout their whole growth histories, although spiral patterns are not recognized on their surfaces. The spiral must have originated from isolated screw dislocations and not from a group of dislocations in these crystals.

Although the types of dislocations in crystal No. 2 were not identified individually, we may infer that at least some are screw-type, since they emerge at the summits of growth pyramids, whereas others may not be screw-type, but are most probably edge-type since they emerge at the areas where P-type trigons concentrate but are not related with any growth centres.

It is interesting to note that diamond crystals containing a smaller number of dislocations exhibit a smaller number of growth centres on one face and shallower growth pyramids than the crystals containing a larger number of dislocations. The latter-type crystals exhibit several steeper growth pyramids on one octahedral face, and several P-type trigons occur at the summit of the one growth pyramid. The steeper growth pyramids seen on the surface of the latter type crystal, are therefore composite spirals originating from a group of screw dislocations.

No connection was observed whatsoever between dislocations and some FD-type and smaller FS-type (sail-boat type) trigons. These trigons must have been formed at defect points other than line defects, most probably at micro defects, such as impurity sites, on the surface (sail-boat type), or at the

FIG. 13. X-ray topographs of crystal No. 3. (a) $(\bar{1}1\bar{1})$, (b) $(1\bar{1}\bar{1})$, (c) $(11\bar{1})$.

outcrops of dislocations issuing from points not far below the surface (FD-type), as suggested by LANG (1964).

Where stacking faults outcrop on the surface, it is observed that growth steps show kinks, suggesting that stacking faults were formed during growth.

5. *Discussion*

5.1 *Growth mechanism*

On the basis of one-to-one correlation between screw-type dislocations and the centres of growth layers, we can safely conclude that natural diamond crystals are formed by the spiral growth mechanism, though actual spiral patterns have not been observed on their surfaces. The number of screw dislocations active for growth vary from face to face and from crystal to crystal. Some crystals exhibit growth layers originating from an isolated screw dislocation, while in other crystals, their layers originate from a group of screw dislocations. When the number of dislocations cropping out on the surface is very small, the {111} face shows an extremely flat surface with widely spaced thin-growth layers. When a large number of dislocations emerge on the surface, the (111) face exhibits well-developed growth pyramids. In any case, it is a characteristic of surface microtopographs of natural diamonds that molecularly-thin growth layers with step heights less than 5Å develop on the octahedral faces. Thicker growth layers are formed by bunching these thinner growth layers originating from screw dislocations. This shows that natural diamond crystals (single crystalline) are formed by the addition of atomic entities and not by the accumulation of larger, macro-molecular entities. This makes a good contrast with the observations on the surface microtopographs of synthetic diamonds, which usually show growth layers thicker than 1,000Å and rugged steps and wavy surfaces (TOLANSKY, 1965; SUNAGAWA, 1982b).

The validity of the above statement that single crystalline natural diamonds are formed by the incorporation of atomic entities, is naturally limited to the diamond crystals grown under the conditions where dislocation-controlled growth predominates (i.e., the supersaturation region below σ^*, refer to SUNAGAWA, 1977, 1981, 1982a), and not to those formed above the critical conditions (i.e., above σ^{**}, such as cuboid, coated diamond, or microcrystalline aggregates of diamond, refer to SUNAGAWA 1982b, and also LANG (1974).

Another point to be mentioned is that these diamond crystals show dislocations radiating from the centre of a crystal, which run nearly perpendicular to the as-grown octahedral faces. Such dislocation configurations are typical for solution-grown crystals (TANNER, 1976) and are not expected for the crystals grown from the melt, neither for the crystals formed in metasomatic or metamorphic environments (SUNAGAWA, 1982a). Therefore, it follows from these observations that natural single crystalline diamonds grew from the low supersaturated solution phase, through the incorporation of atomic entities, by the spiral growth mechanism. It is highly improbable that these crystals grew under metamorphic conditions.

5.2 *Formation of trigons*

It has been established in the present study that P-type trigons at the summits of growth pyramids are directly related to the outcrop of screw dislocations. In crystal No. 2, some dislocations outcrop at the summits of growth pyramids, others at areas where P-type trigons concentrate, but are not directly related with growth centres. The latter dislocations are believed to be edge dislocations. It is therefore safe to assume that P-type trigons are dislocation etch pits, and not formed by a growth failure, as previously suggested by TOLANSKY (1955). In LANG's observation (1964, 1965), it is described that P-type trigons show, narrow distribution of sizes, being about 100 μ when they are isolated, and the sizes can fall below 10 μ when they occur in clusters. In contrast, P-type trigons observed in the present study are invariably much smaller than Lang's P-type trigons, being 3 to 15 μ (shallower one) and 15 to 50 μ (deeper one), in spite of that, they appear isolatedly. This difference is inferred to the difference in the duration of dissolution between the two cases. The present crystals have received dissolution for a much shorter duration or weaker dissolution.

The FD-type trigons on the present specimens whose sizes are much larger compared to the P-type trigons directly related to dislocations, invariably show microstructures and consists of many smaller trigons. No one-to-one correlation has been observed between this type of trigons and dislocations. Lang also observed similar FD-type trigons which are not related with any dislocations, and suggested that they might be formed at the outcrops of dislocations issuing from points not far below the surface.

In both observations by Lang and by the present study, it was found that smaller FS type trigons, which are equivalent to the sail-boat type trigons in Lang's description, are not related with any types of dislocations. The characteristics of these trigons are; (1) their depths are less than 10Å, (2) the sizes of isolated trigons of this type are smaller than 10 μ, and only when they coalesce together can FS trigons larger than this size occur, (3) incomplete forms of trigons of various degrees are seen, (4) along the molecularly-thin growth steps, coalesced trigons larger than the above value preferentially develop, and on the surface of growth layers smaller isolated trigons are distributed randomly. These trigons cannot be interpreted in terms of growth failure. A more plausible explanation is two-dimensional dissolution originating from micro defects on the surface.

Impurities may be another plausible defect. If dissolution starts from such micro defects on the surfaces, the dissolution fronts advance more easily in lateral direction than in a vertical direction, creating flat-bottomed depressions.

6. *Conclusion*

Surface microtopographic and X-ray topographic studies were made of octahedral crystals of diamond from Siberia which suffered only a slight dissolution and whose almost as-grown original surfaces are preserved. One-to-one correlation between the growth centres and the outcrops of screw-type dislocations was established for one crystal, which altogether contains only 11 dislocations. Similar, but less precise, correlations were also found for other crystals containing a larger number of dislocations. Based on these correlations and the observations on the step height and step separations of growth layers, it is concluded that natural diamond crystals of a single crystalline type grew by the spiral growth mechanism from the solution phase of low supersaturation, through the incorporation of atomic growth units. It is highly improbable that they grew in metamorphic environments.

We are grateful to Prof. A. R. Lang, Bristol University, for critically reading the original manuscript and giving many valuable comments. Thanks are also due to Japan Ministry of Education for a Grant-in-Aid for Fundamental Researches.

REFERENCES

FRANK, F. C. (1962): Defects in diamond, in *Science and Technology of Industrial Diamonds*, Vol. 1, Ed. J. Burls, Industrial Diamond Information Bureau, London, pp. 119–135.

LANG, A. R. (1964): Dislocations in diamond and the origin of trigons, *Proc. R. Soc., A*, **278**, 234–242.

LANG, A. R. (1965): X-ray topography of diamond, in *Physical Properties of Diamond*, Ed. R. Berman, Clarendon Press, Oxford, pp. 69–115.

LANG, A. R. (1974): Glimps into the growth history of natural diamonds, *J. Cryst. Growth*, **24/25**, 108–115.

LANG, A. R. (1978): Internal structure, in *The Properties of Diamond*, Chapter 14, Ed. J. E. Field, Academic Press, London, pp. 425–469.

SEAL, M. (1962): The growth history of natural diamonds as revealed by etching experiments, *Proc. 1st Internl. Cong. on Diamonds in Industry, Paris*, pp. 361–375.

SEAL, M. (1965): Structure in diamond as revealed by etching, *Am. Mineral.*, **50**, 105–123.

SUNAGAWA, I. (1961): Step height of spirals on natural hematite crystals, *Am. Mineral.*, **46**, 1216–1226.

SUNAGAWA, I. (1975): Characterization of crystal surfaces by optical microscopy II, in *Crystal Growth and Characterization*, Eds. R. Ueda, and J. B. Mullin, North Holland Pub. Co., Amsterdam. pp. 347–359.

SUNAGAWA, I. (1977): Natural crystallization, *J. Cryst. Growth*, **42**, 214–223.

SUNAGAWA, I. (1981): Characteristics of crystal growth in nature as seen from the morphology of mineral crystals, *Bull. Mineral. (France)*, **104**, 81–87.

SUNAGAWA, I. (1982a): Growth of crystals in nature, this volume, pp. 61–103.

SUNAGAWA, I. (1982b): Morphology of natural and synthetic diamond crystals, this volume, pp. 303–330.

TANNER, B. K. (1976): *X-ray Diffraction Topography*, Pergamon Press, London, 174 pp.

TOLANSKY, S. (1948): *Multiple-beam Interferometry of Surfaces and Films*, Clarendon Press, Oxford.

TOLANSKY, S. (1955): *The Microstructure of Diamond Surfaces*, N. A. G. Press, London.

Materials Science of the Earth's Interior, edited by I. Sunagawa, pp. 351–374.

SYNTHESIS RESEARCHES OF DIAMOND

Masao WAKATSUKI

Institute of Materials Science, The University of Tsukuba, Sakura-mura, Ibaraki-ken, Japan

1. Introduction

The study of diamond may be one of the way of getting information about thermodynamical and geochemical situations in the depth of the earth, along with discovering the history of their changes. Such a study will be effected by a detailed characterization of natural diamonds. The relation between characters of the crystal and the condition of its formation can be made by the use of crystals synthesized at known and carefully controlled conditions. Synthesis research is also required because of the substantial importance of this material to industry. The mechanisms of formation should be understood in detail for controlling the quality of the product, as well as for analysing the situation of the depth of the earth.

In this report a general knowledge of the synthesis and characteristics of synthesized diamonds will be briefly summarized (Section 2). Then new knowledge obtained through the present researches, especially relating to the nucleation stage will be described (Section 3). Section 4 deals with inclusions in synthetic diamond crystals from the viewpoint of the growth stage.

2. Present Status of Knowledge about the Synthesis of Diamond

A very early trial of synthesizing diamond (PARSONS, 1919) showed that this modification of carbon was already believed to be formed at high pressure. Pressure and temperature conditions of the thermodynamical stability of diamond are indeed of primary importance in the problems of synthesis and are important in discussions about the depths of the earth at which natural diamond is formed or occasionally dissolved, as well. Thermodynamical estimations of the equilibrium pressure and temperature were made by several researchers (BRIDGMAN, 1947; LILJEBLAD, 1955; BERMAN and SIMON, 1955).

Berman and Simon carefully examined published data on thermodynamical properties of graphite and diamond and calculated the equilibrium pressures at temperatures from absolute zero to 1,200 K. The linear extrapolation of the calculated curve in a higher temperature range was later shown to be in reasonable agreement with an experimentally-determined equilibrium curve of between 1,500–2,700 K (BUNDY *et al.*, 1961). The calculation was revised by BERMAN (1965a), especially with an estimation of the equilibrium curve in the higher temperature region (1,100–3,000 K). Estimated equilibrium pressures are slightly higher at temperatures above 2,000 K than those given by simple extrapolation. The equilibrium relations revised by Berman are believed to be the most reliable. Figure 1 shows the phase diagram of carbon proposed at present, on the basis of thermodynamical calculation and various experiments (WENTORF, 1974).

A reproducible method of synthesis was first realized by the advent of suitable high pressure-high temperature apparatus and appropriate metallic solvents for carbon, so called catalysts (BOVENKERK *et al.*, 1959; HALL, 1960; BUNDY *et al.*, 1961). The solvents were claimed to be those elements which belong to group 8 in the periodic table, Mn, Cr and Ta, and their alloys. Since then, there have been very many reports on synthesizing diamond in many countries.

In actual synthesis work, graphite and solvents are placed in a reaction cell, as shown in Fig. 2. They are pressurized and then heated in high pressure-high temperature apparatus. The pressure and the temperature should be in

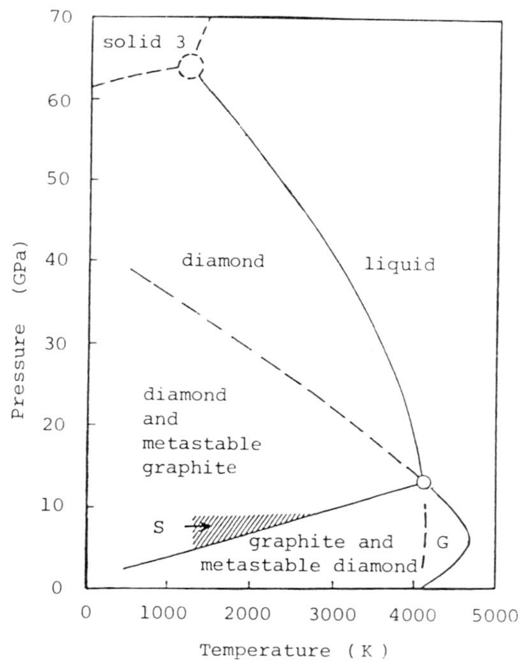

FIG. 1. A phase diagram for carbon (after WENTORF, 1974).

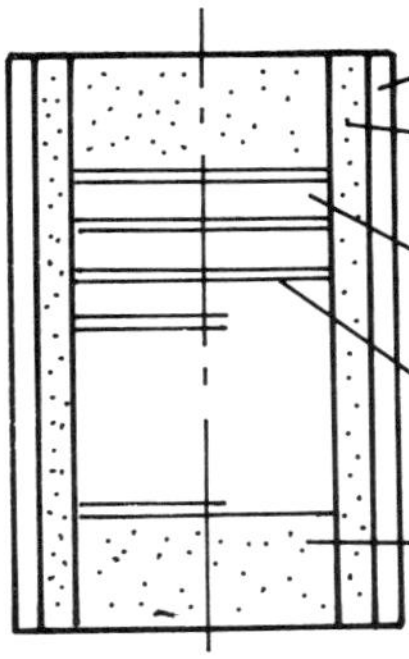

FIG. 2. A typical sample assembly for the synthesis of diamond.

the stable region of diamond in the phase diagram, and in the liquid region of the solvent material-carbon system. The lowest temperature limit for the reaction is different, according to the solvent materials used. The pressure-temperature area explored so far with various solvent materials, is shown as a hatched zone in Fig. 1. General descriptions of methods for synthesis of this type can be found in some text books (WENTORF, 1974; ELWELL and SCHEEL, 1975).

Besides the solvent materials mentioned above, copper-base alloys, including transition elements of groups 4a, 5a and 6a in the periodic table, were also found to effect the formation of diamond (WAKATSUKI, 1966). A difference in the type of solvents seems to produce different characteristics in synthesized diamond (TAKASU *et al.*, 1968; WAKATSUKI *et al.*, 1975).

At first, the solvents were introduced as catalysts. It is, however, known today that the catalyst is melted at the synthesis condition and acts as a solvent for carbon, that is, graphite (carbon supply) is dissolved into the molten metallic medium, then the dissolved carbon is transported through it and finally crystallized as diamond, according to thermodynamical stability. The driving force for these reactions is the supersaturation of carbon in the solvent due to the difference in chemical potentials of graphite and diamond, which, in turn, is determined by the excess pressure-the difference between synthesis pressure and equilibrium pressure at synthesis temperature.

STRONG and HANNEMAN (1967) and STRONG and CHRENKO (1971) reported details of observed phenomena in diamond formation with typical solvents, Ni, Fe and Fe–Ni alloys, on the basis of phase diagrams for these metals and carbon. In the former, are described the growth kinetics, such as transportation of carbon and the nucleation behavior in the molten Ni–C medium, as well as a slight shift of equilibrium pressure ($\simeq 0.03$ GPa) due to the existence of the solvent material. Observed diffusion coefficients of carbon in the molten solvent lie in the range of $2\text{–}4 \times 10^{-5}\ cm^2\ sec^{-1}$, at a pressure of 5.4 GPa and temperatures from 1,660–1,800 K, typical for the synthesis condition. The diffusion is the rate-controlling mechanism in the growth at moderate driving forces. The nucleation density is also described as having a strong dependence on the excess pressure.

Seeded growths of large single crystals up to 5 mm have been realized with very low and carefully controlled driving forces (WENTORF, 1971; STRONG and WENTORF, 1972). In this case, the carbon supply is not graphite but diamond itself, and the driving force for the growth is given by a temperature difference between the supply and the seed. Figure 3 shows a reaction cell for this type of growth. With pressures around 5.5–6 GPa, and temperatures of, say, 1,450 and 1,420°C for the supply and the seed, respectively, sound crystals of 5 mm were grown in one week. The maximum growth rate allowed to get a sound crystal is also discussed. According to the discussion by WENTORF (1972), the rate of growth normal to a growing facet should be inversely proportional to the diameter of the facet, and thus should be lowered as the facet is increased in diameter. Impurity concentrations are reduced or controlled by doping some elements to the solvent, as will be described in the following.

Special dopants added to the solvent can vary the character of a synthesized diamond. Be, B and Al impart the p-type semiconductivity to the synthesized diamond (WENTORF and BOVENKERK, 1962). B and Al seems to prompt the nucleation of diamond from graphite, and Si reduces it but increases appearance of twinned crystals (CANNON and CONLIN, 1964). Nitrogen, the unintended but very common impurity in synthesized diamond, can be excluded or drastically reduced by adding dopants to the solvent, such as Al, Ti and Zr (STRONG and CHRENKO, 1971; WENTORF, 1971).

Diamonds are classified into four types: type Ia, type Ib, type IIa and type IIb, according to the concentrations of nitrogen incorporated. (BERMAN, 1965b; FIELD, 1979). Nature produces crystals of the three types except for

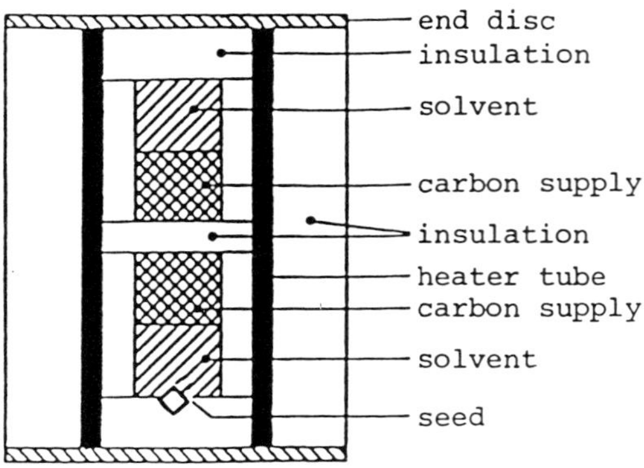

FIG. 3. A high-pressure reaction cell for the growth of diamond crystals in temperature gradients (after WENTORF, 1974).

type Ib. Those of type Ia and type Ib contain large amounts of nitrogen up to several times 10^{20} cm^{-3}. Nitrogen atoms in the two types are further specified as forming aggregates (platelets, 600–1,000 Å in diameter) in type Ia, and as dispersed atomically in type Ib. On the other hand, nitrogen concentrations in type IIa and IIb diamonds are less than 10^{17} cm^{-3}. Type IIb is different from type IIa in that the crystal shows semiconductivity, occasionally accompanied by a blue color.

Synthesized diamonds are never of type Ia, but always of type Ib if they contain nitrogen in sufficient concentrations. According to STRONG (1977), dispersed nitrogen can be, at least partially, corverted to that of the aggregated state by annealing the crystal for 1–50 hours, at 1,500–2,200°C and 4.8–7 GPa. This implies a hypothetical mechanism for the origin of natural type Ia diamonds, in which crystals containing dispersed nitrogen may be annealed in the mantle (MAL'KOV and ASKHABOV, 1979).

Diamonds with least nitrogen concentrations, synthesized using a solvent with nitrogen getted off by the dopants already mentioned, shows optical and thermal properties the same as those of natural type IIa diamonds (STRONG and CHRENKO, 1971). Copper-based alloy solvents, including groups 4a, 5a and 6a transition elements together with Mg, were also reported to produce diamonds showing an optical behavior without strong absorption in the 220–330 nm range of wavelength, which is one of the features of type IIa diamonds (WAKATSUKI *et al.*, 1975). Boron-doped diamonds correspond to those of type IIb.

It may appear that sufficient knowledge has been accumulated about the artificial formation of diamond. Indeed, all types of natural diamonds can now be simulated by synthesis, as described above. But it should be noted that the simulation is limited only to the bulk properties of diamond. We are, really, at the entrance of, but with effective methods for analysing, many other problems of the synthesis and characteristics of diamond, such as problems of impurities or inclusions, surface structures and morphology, especially relating to the mechanisms and histories of the growth and the post-growth stages. On this point of view, a few, but interesting works can be cited. For example, nitrogen in the growing environment seems to have a tendency of favoring the appearance of {111}-growth surfaces (STRONG and CHRENKO, 1971). Such an effect may give a substantial stand point of understanding the growth mechanisms, since this element is usually incorporated into both the environment of the synthesis and the natural formation of diamond. Etching of octahedral faces of natural diamond by water at high pressure and temperature (KANDA *et al.*, 1977) was revealed to create triangular pits almost the same as "trigons." Trigons are small recesses seen on the octahedral face of natural diamond and bounded by epiangular triangles with vertices pointing oppositely to those of the face. The bottom is shaped either as a negative pyramid, or is flat. Experiments by Kanda *et al.* gave evidence of etching for their natural origin (LANG, 1964; FRANK and LANG, 1965), and implied surface dissolution of the natural diamond by water included in silicate magma, during transportation to the earth's surface after its formation.

Diamond particles of sizes of up to several hundred micrometers are produced for industrial use, from graphite using the solvent method. Knowledge about the formation mechanisms are also needed to control their quality. In general, interference among crystals being grown, which results in the forming of irregularly-shaped particles, will be prevented by controlling the nucleation density. It will also relate to the particle size obtained. Incorporation of impurities or inclusions, which may affect their chemical, mechanical and thermal characteristics in the use, should be understood and controlled on the basis of knowledge of the growth stage. And their characterizations will help to understand that stage.

3. *Nucleation Stage and Its Control in The Synthesis of Diamond*

3.1 *Some problems relating to the nucleation stage*

Diamond particles of high quality in industrial use (for metal-bonded tools) are specified as having at least a blocky and euhedral shape without large recesses on the surface, without large inclusions, and with the size controlled within an appropriate range. These features are coincident with the general crystallographic qualities.

But the synthesis of diamond from graphite contains a problem—to grow thousands of crystals at the same time—quite different from that in many other cases of crystal growth, in which a single seed is grown. Therefore, the control of nucleation density, thus understanding the mechanism governing the nucleation stage, is of substantial importance. There have been, however, few reports on analyses or on observations of the nucleation stage, compared to those on the growth stage. STRONG and HANNEMAN (1967) observed the nucleation behavior at a temperature of 1,455°C, and excess pressures of up to 0.3 GPa over the equilibrium pressure (liquid-diamond-graphite eutectic) in the presence of Ni solvent. According to their results, the spontaneous nucleation begins at the equilibrium (l-d-g eutectic) pressure and shows a very remarkable dependence on the excess pressure. The spontaneous nucleation can be suppressed to some extent if a seed crystal coexists at lower excess pressures. But the former greatly attenuates the growth of the latter as the excess pressure is increased, at least at 0.22 GPa. The remarkable dependence of nucleation density on pressure is described by BEZRUKOV *et al.* (1972).

Besides, Bezrukov *et al.* report a fact that the main part of the nucleated crystals are formed in a rather short period (within two minutes) at the beginning of the reaction. On the other hand, HORTON and PETERSON (1978) assume a nucleation rate kept constant throughout the reaction time, to explain their experimental results in terms of yield as a function of crystal size and reaction time. The experiments were made at 5.5 GPa and 1,450 K ($\simeq$1,180°C) with a Ni–Mn alloy solvent. The maximum reaction time, however, is limited to within two minutes.

Further observations of the nucleation behavior, especially the period for its main part in a sufficiently long reaction time, will give a standpoint for understanding the mechanism controlling the nucleation stage.

3.2 *Method of synthesis experiments*

The strong dependence of nucleation density on the excess pressure requires performing experiments at moderate excess pressures which are controlled with sufficient reproducibility. In the present studies, a cubic anvil apparatus (WAKATSUKI and ICHINOSE, 1981) was used. The pyrophyllite cube, 24 mm on an edge, contained a heating element (tube) of graphite, 10 mm in diameter and 13.5 mm in length. Although accurate absolute values for generated pressures at high temperature were not evaluated, pressures were reproduced within a width of 0.04 GPa at around 5 GPa if every part of the sample arrangement was carefully prepared.

The sample arrangement was the same as that shown in Fig. 2, in which the carbon supply was made of a graphite for spectroscopic analysis. Two kinds of solvent materials were used; the one was Co and the other was an alloy of Fe (55 wt%), Ni (29 wt%) and Co (16 wt%), the both being very similar in solvent characteristics. The reaction temperature was always selected so that almost all crystals were grown in a cubo-octahedron consisting of square ({100}) and triangular ({111}) facets. The absolute temperature was evaluated with a Pt-Pt13%Rh thermocouple, but the pressure effect on its emf was not corrected.

The solvent material and/or residual graphite could be selectively removed from the recovered sample by boiling in appropriate acids, according to requirements from observation.

3.3 *Number of nucleated crystals in relation to reaction time*

Cobalt was used as the solvent and the synthesis runs were made for various reaction times from 2 to 10 minutes, at a temperature of 1,350°C and a pressure of 5.3 GPa (about 0.1 GPa higher than the threshold of diamond formation, experimentally determined in this system.

In order to evaluate the number of crystals obtained in each synthesis run, recovered crystals were sieved into three classes according to their sizes: class (1) < 149 μm, class (2) 149–297 μm and class (3) > 297 μm. Then the yield of each class was weighed. It was observed under the microscope that crystals smaller than 100 μm were not included in class (1) and neither were those exceeding 420 μm in class (3). Therefore, representative sizes were assumed as 105 μm (corresponding to 140 US mesh), 210 μm (70 US mesh) and 354 μm (45 US mesh) for the classes (1), (2) and (3), respectively. The number of each class in each sample was then estimated on the basis of the yield and the representative size, and the total number was calculated for each sample.

The results are shown in Fig. 4, in terms of the number of crystals as a function of reaction time. Yields are also shown. The number does not vary essentially, in spite of an increase in the yield with increasing reaction time. This behavior agrees with that reported by BEZRUKOV *et al.* (1972). Further, the period for the increase in the number of crystals (≤ 2 minutes) is also consistent between the two cases. One can, thus, reasonably conclude that the nucleation stage is limited to an early short period in the reaction. The next problem was seeking a mechanism for realizing such a behavior of nucleation.

3.4 *A model for the nucleation stage competing with a recrystallization of graphite*

Diamond always nucleates on the interface of solvent and graphite, or in the solvent, but very closely adjacent to the interface, therefore the surfaces of the graphite discs were examined after the synthesis run. Figure 5 shows a graphite disc with the solvent material removed with aqua regia. Many small (a few tens of μm) recrystallized graphite particles cover the entire area, except for the diamonds. Such a phenomena is always observed unless the whole surface of the graphite is covered with diamond crystals heavily impinging on each other.

This kind of the change on graphite surface is presumably caused by the unstableness of pure graphite in contact with the solvent. According to a simple theory of regular solution, the chemical potential difference, $\Delta\mu(g_0 \rightarrow g)$, between pure graphite (g_0) and that saturated with the solvent material as a solid solution (g) is written as

$$\Delta\mu(g_0 \rightarrow g) = \mu_g - \mu_{g_0} = RT\ln(1 - X_M^g) = -RTX_M^g \qquad (1)$$

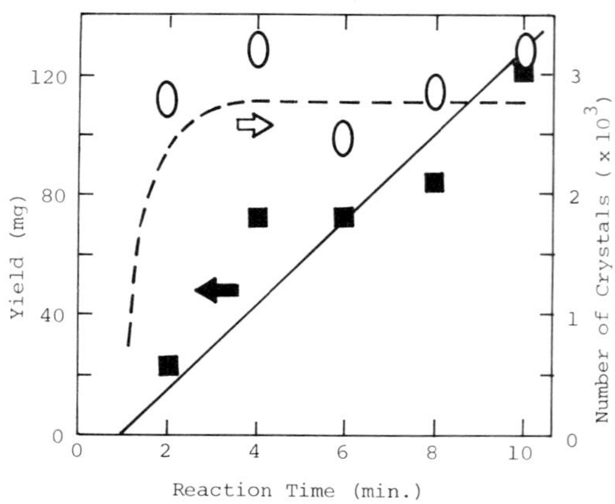

FIG. 4. The number of nucleated crystals and yield versus reaction time.

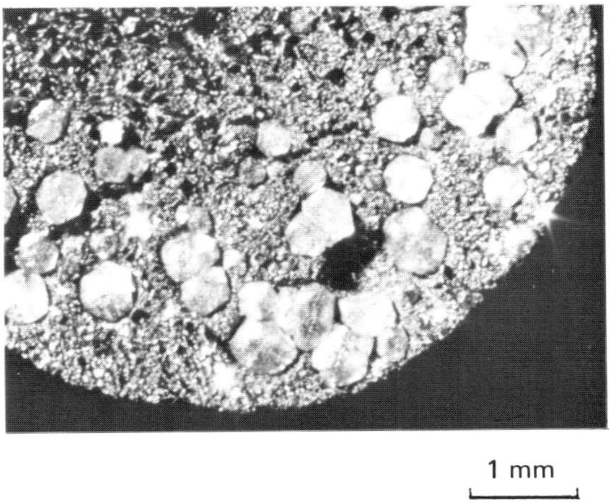

FIG. 5. Recrystallized graphite particles covering the surface of a graphite disc.

where μ means chemical potential and suffixes show the phases. X_M^d is the atom fraction of the solvent material in the graphite as a solid solution and thought to be much smaller than unity. STRONG and HANNEMAN (1967) measured this quantity in the case of Ni solvent, and gave a value as $X_{Ni}^g = 0.003$ (0.3 at%). Similarity of Ni and elements used in the present case (Fe, Co and Ni) allows us to presume a similar value for X_M^g.

The chemical potential difference, $\Delta\mu\,(g_0 \rightarrow g)$, is always negative. So that the recrystallized graphite can appear together with diamond, if the former is not consumed by some kinetics to further nucleate (or grow) the latter. And survival of the former is evident in the present experiment. Thus, it seems that the nucleation of diamond from the recrystallized graphite is much more difficult than from pure (and fresh) graphite.

Such an idea can lead us to a kinetic model of the nucleation stage, in which it competes with the recrystallization of graphite. The diamond nucleation rate, dN/dt, can be expressed as

$$\frac{dN}{dt} = K_d \cdot S \tag{2}$$

where N is the number of diamond nuclei, t expresses time and S is the surface area of the graphite disc that has not yet been covered with either recrystallized graphite or with diamond nuclei. In order to get an expression for K_d, the homogeneous nucleation theory (ELWELL and SCHEEL, 1975) can be modified by expressing ΔG_v, decrease in Gibbs' free energy per unit volume of formed diamond phase, as a function of excess pressure

$$\Delta G_v \cdot v_d = -(\mu_d - \mu_l)$$

where: μ_{d} = chemical potential of carbon in diamond phase,
μ_{l} = chemical potential of carbon in solution phase, and
v_{d} = molar volume of diamond.

In the present case, the solution can be thought to be nearly in equilibrium with graphite. This is a possible situation since the energy barrier for nucleation is generally much larger than that for dissolution. Thus we get

$$\Delta G_{\mathrm{v}} = -\frac{\mu_{\mathrm{d}} - \mu_{\mathrm{g0}}}{v_{\mathrm{d}}} = \left(\frac{-\Delta v}{v_{\mathrm{d}}}\right)\delta P \tag{3}$$

where: μ_{g0} = chemical potential of carbon in pure graphite, $\mathrm{g_0}$,
$\Delta v (= v_{\mathrm{d}} - v_{\mathrm{g0}})$ = difference in molar volume of diamond and graphite,
$\delta P (= P - P_0)$ = excess pressure,
P = synthesis pressure, and
P_0 = pressure of equilibrium between pure graphite and diamond.

Using this relation in the expression of the nucleation rate, K_{d} can be written as

$$K_{\mathrm{d}} = \frac{3kTZ_{\mathrm{d}}^{*}n}{\sigma_{\mathrm{d}}}\left(\frac{\Pi}{\delta P}\right)^2 \exp\left\{-\left(\frac{\Pi}{\delta P}\right)^2\right\}$$

where: $$\Pi^2 = \frac{16\pi\sigma_{\mathrm{d}}^3}{3kT(-\Delta v/v_{\mathrm{d}})^2}$$

k = Boltzmann's constant,
T = temperature,
Z_{d}^{*} = the number of solute atoms striking the surface unit area of embryos in unit time,
n = the number of solute atoms in the unit volume of the solution, and
σ_{d} = the interface energy between diamond and the solution.

The quantity S of Eq. (2) can be written as

$$S = S_0\{1 - a_{\mathrm{g}}(t) - a_{\mathrm{d}}(t)\} \tag{4}$$

where S_0 is the initial surface area of the graphite disc, the quantities $a_{\mathrm{g}}(t)$ and $a_{\mathrm{d}}(t)$ are ratios of the graphite disc surface area already covered with recrystallized graphite and with diamond, respectively, to the initial surface area (S_0). $a_{\mathrm{g}}(t) \gg a_{\mathrm{d}}(t)$, at least within the nucleation stage of diamond, since nuclei should not be crowded in the present case. Then, Eq. (2) is reduced to

$$\frac{\mathrm{d}N}{\mathrm{d}t} = K_{\mathrm{d}} \cdot S_0\{1 - a_{\mathrm{g}}(t)\}. \tag{5}$$

The area factor of this equation should become zero at some instance, $t = t_0$, and determines the period allowed for diamond crystals to nucleate, which is mostly governed by the covering of the graphite disc surface with the recrystallized graphite. This period (t_0) may depend upon temperature but should have little dependence on pressure, as suggested by Eq. (1). On the other hand, K_d does depend strongly on pressure.

According to the above arguments, changes of characteristic quantities, S, K_d, dN/dt and N, in the nucleation stage can be deduced qualitatively; these effects are illustrated schematically in Fig. 6. If one considers the isothermal case, the number of diamond nuclei can be related to pressure as

$$N = N_0\left(\frac{\Pi}{\delta P}\right)^2 \exp\left\{-\left(\frac{\Pi}{\delta P}\right)^2\right\}. \tag{6}$$

3.5 *Experimental results according to the nucleation model*

Figure 7 shows experimental results of the pressure dependence of the density of diamond crystals formed. The synthesis runs were made with a solvent of the Fe–Ni–Co alloy, at a temperature of 1,280°C for a reaction time of 5 minutes. The oil pressure for actuating the apparatus is also indicated for each photograph. The minimum oil pressure (220 kg/cm^2) corresponds to about 5.1 GPa, and the difference between neighboring cases (5 kg/cm^2) corresponds to 0.075 GPa. At an oil pressure of 215 kg/cm^2, no crystal was confirmed to be formed. The number of crystals formed on one graphite disc surface was counted, and the counted numbers were averaged for all discs in one synthesis run. The results are plotted in terms of density versus pressure in Fig. 8. The observed relation is consistent with one that is expected from Eq. (6).

Both nucleation and growth stages are governed by a single excess pressure in the simple synthesis process. But it will be now possible, according to the above model, to vary the balance of the competition between the nucleation and recrystallization of graphite, while the excess pressure for the

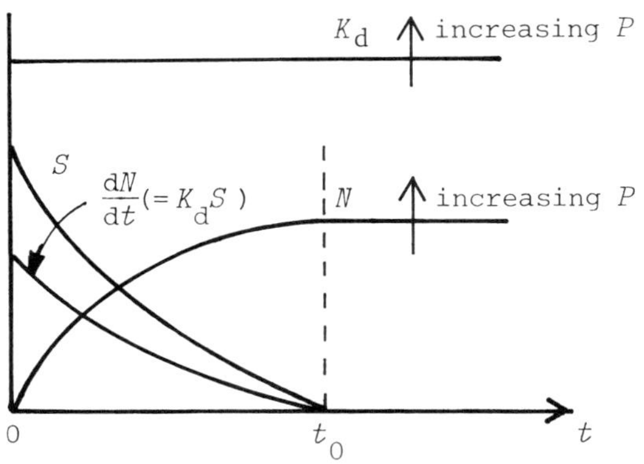

FIG. 6. Schematic changes with time of characteristic quantities in the nucleation stage of diamond.

FIG. 7. The effect of pressure on the nucleation density of diamond.

growth is independently selected. Figure 9 shows a pressure diagram for such a synthesis process. The expected changes of the characteristic quantities for the nucleation stage are also shown. The pressure should be kept at first at an appropriate level, P_1, for some period, t_1, typically 1 or 2 minutes. In this period, the nucleation stage (or the covering of the graphite disc surface with the recrystallized particles) should be almost completed. P_1 is selected so as to give a desired nucleation density. Then the pressure is raised and kept at P_2 for a sufficient period, t_2, to give a sufficient excess pressure to the growth. Temperature may be kept constant for the whole reaction time.

In Fig. 10 are shown changes of the nucleation density with t_1. The oil pressures for the first stage (P_1) were selected as 210 kg/cm^2 (A), 225 kg/cm^2 (B) and 232.5 kg/cm^2 (C). The second-stage pressure (P_2) was 250 kg/cm^2 for all cases. Temperature was the same as that in the case of Fig. 7 and the growth time, t_2, was 10 minutes. These results seems to support the present model of the nucleation stage. Numbers of crystals formed at conditions of $t_1 \geq 3$ minutes are understood as corresponding to those nucleated at $P = P_1$ (first stage), while those for $t_1 \leq 2$ minutes, including those nucleated at $P = P_2$ (second stage). The numbers for $t_1 \geq 3$ minutes are in good agreement with those expected from Fig. 8. The transition value for t_1, 2–3 minutes, is also consistent with that deduced from Fig. 4. Thus, it is concluded that the present model explains very well the experimental behavior of nucleation observed so far.

The results obtained in the present study are applicable to the production technology of superior quality particles for industry, as well. If the pressure had been kept at P_2 for the whole reaction period, as shown with broken lines in Fig. 9, crystals should have been formed in a state of heavy impingement or

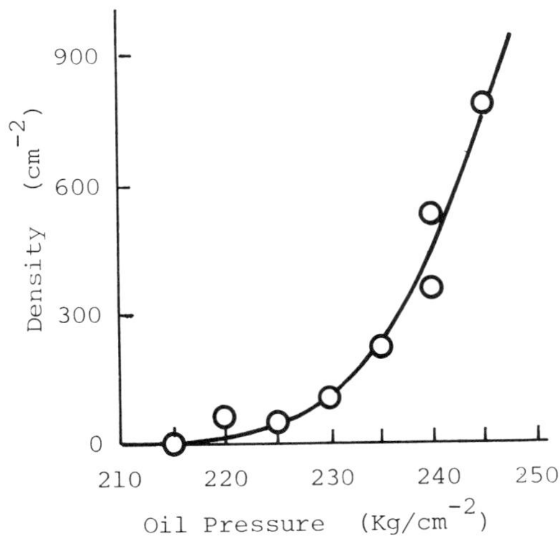

FIG. 8. Nucleation density of diamond as a function of pressure.

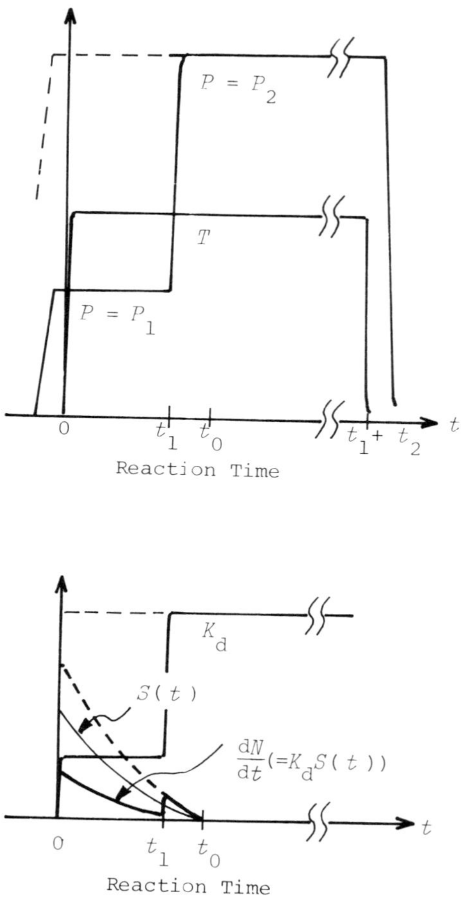

FIG. 9. A schematic illustration for synthesis process with a modified pressure program.

intergrowth, as expected from the results of Fig. 7 and 8. Crystals obtained in the modified process are very well crystallized, as shown in Fig. 11, in spite of a rather high growth pressure.

4. *Inclusions in Synthesized Crystals of Diamond*

4.1 *Inclusions in synthesized diamond*

Well crystallized diamonds, synthesized with the solvent method, usually contain a great number of very small inclusions distributed in a peculiar manner: the inclusions are distributed on narrow lines, or on bands, radially directed from the central part of a crystal to its apices, or edges, respectively.

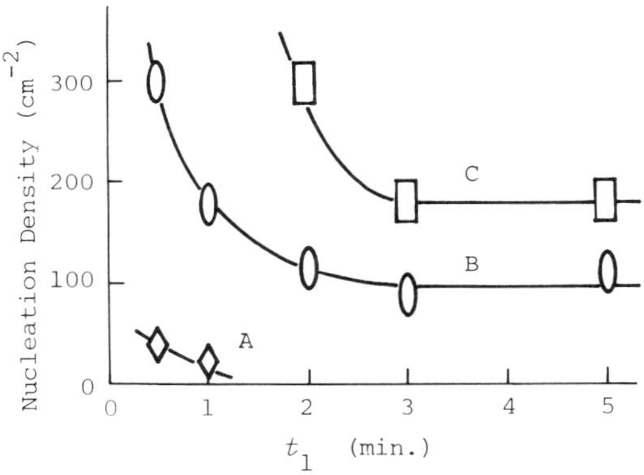

FIG. 10. Changes of nucleation density with an increase in t_1.

FIG. 11. Crystals obtained through a modified pressure program.

The characterization of these inclusions will give a key to understanding the growth stage of diamond. Such knowledge will be useful in the application of synthetic diamond to industry, since those inclusions, as well as atomically-dispersed impurities, may affect chemical, mechanical and thermal properties of the host crystal. Reports of systematic observation of this type of inclusions are very scarce.

In the present study, detailed observations have been made on their distribution, shape and size, chemical compositions and magnetic properties. The results are related to the growth stage of diamond. Specimens (about 400 μm) were synthesized by the method previously described, with the alloy

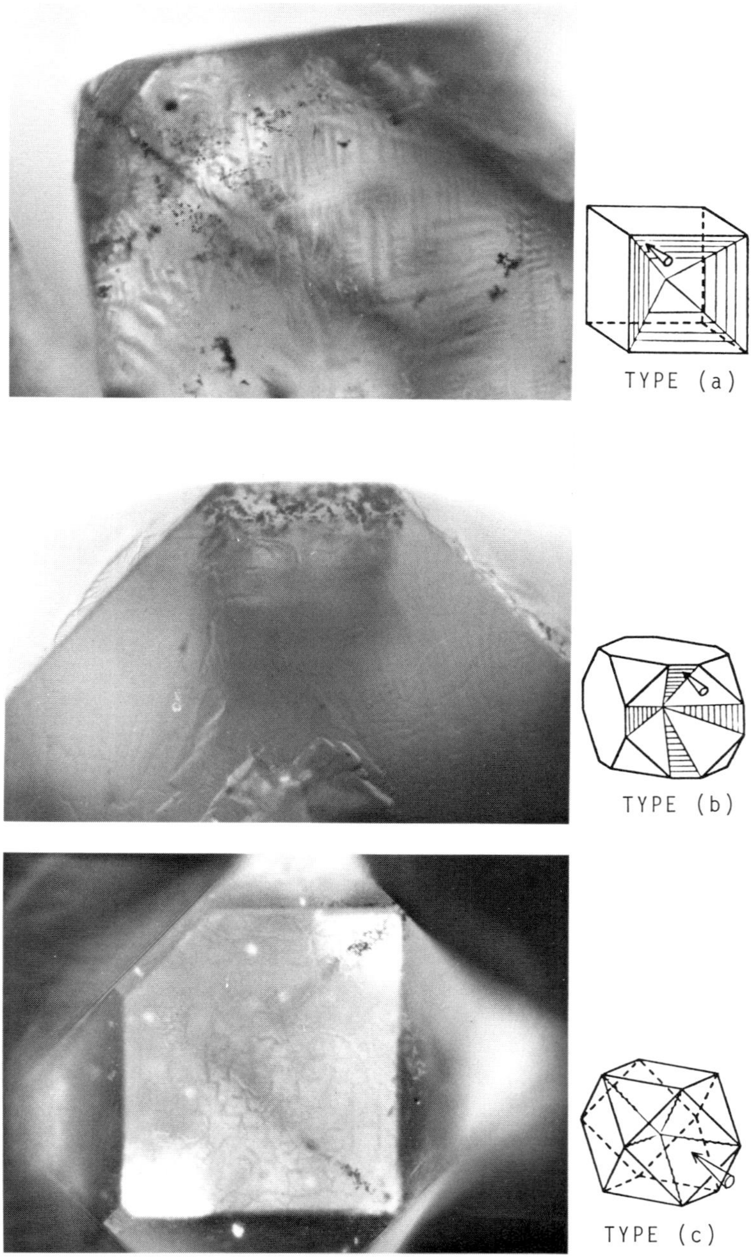

FIG. 12.

FIG. 12. Distribution of arrayed inclusions as related to crystal habit.

of Fe, Ni and Co. Commercially-supplied crystals (G. E., MBS-type, about 600 μm) were also used. Essentially, no difference exists between the two kinds of specimens regarding the inclusions.

4.2 Distribution of the inclusions

Microscopic observations revealed that distribution depends on the crystal's habit, and can be classified into five types. Photographs and schematic illustrations for the five types are shown in Fig. 12. The distribution rule can be expressed as follows: (1) If the crystal has the edge at which two facets of the same kind of the crystallographic index ({100} or {111}) meet together, inclusions are distributed on sectorial planes (not necessarily flat) defined by the nucleus site and the said edges (types (a), (b) and (d)). (2) If the crystal has an apex at which four facets meet together, inclusions are distributed on lines (not necessarily straight) connecting the nucleus site and the said apex (type (c) and (e)). It is an interesting fact that the edge, where a {100} facet and a {111} facet meet, does not concern the existence of inclusions arrayed in the plane.

In order to get more detailed information about the location of included particles in relation to the growth history of the crystal, observations were made on their sectional surfaces. Growth sectors could be clarified by means of cathodoluminescence after WOOD and LANG (1975). In Fig. 13 a luminescence photograph and an illustration of a {111} section of a crystal of the type (c) are shown. Growth regions (each grown by one particular facet) are labelled with g, b and o, according to the color of the luminescence. The labels g and b, and o correspond to {100} and {111} growth regions, respectively. Irregular contours surrounding them are caused by the following reason. The different kinds of growth regions are ideally opposing each other on both sides of this section in this type of habit, but their actual interface is not completely flat-it is curved-and the section intersects this actual interface to result in the irregular contours.

As shown in Fig. 13, included particles are arrayed on boundaries of growth regions. Growth sectors were also detected by relief on polished sections of a few more crystals, and the inclusions were again confirmed to be really arrayed on the interface between neighboring two growth regions, both of which correspond to the same kind of index.

4.3 *Shape and size of the inclusions*

Included particles appearing just on carefully polished sections were selected and observed with a scanning electron microscope. Figure 14 shows a typical micrograph taken on a {111} section. It is revealed that the particles are enclosed in holes having the same morphological symmetry as that of the host crystal (negative crystals of diamond). Their sizes seem to be within 10 μm.

4.4 *Composition and Curie temperature of the included material*

X ray microanalyses were made on the inclusions of the section of crystals which were synthesized in the present study with the Fe (55 wt%)-Ni (29 wt%)-Co (16 wt%) alloy solvent previously described. Quantitative results were deduced by means of ZAF correction. Figure 15 shows relative compositions between the three elements, determined for many inclusions. For sufficiently small inclusions (completely immersed in the electron beam), determined compositions for 25 particles agree well with that of the solvent alloy, with a very small scattering, as shown with an ellipse. Those scattered to a large extent (out of the ellipse)are compositions determined on one part of the large inclusions. The total content of the three elements was estimated as 90–95 wt%, the remaining part, 5–10 wt%, being mainly expected to be carbon.

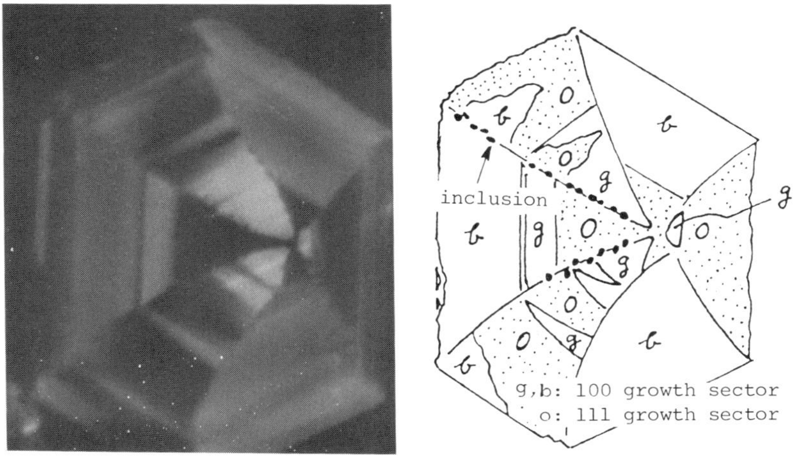

FIG. 13. Location of inclusions: cathodoluminescence (a) and illustration (b) of a {111} section.

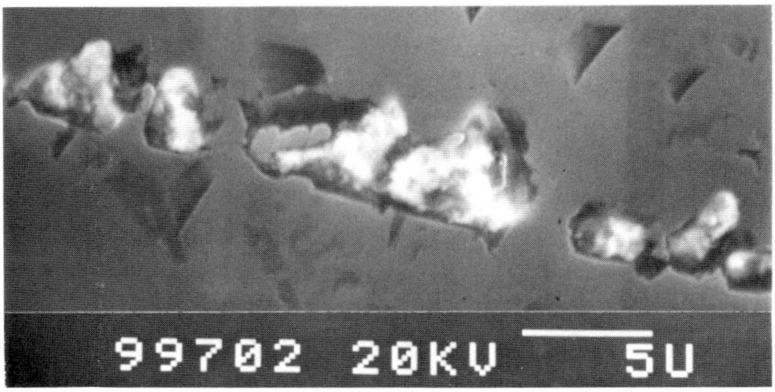

FIG. 14. The shape and size of included particles.

Solubility of carbon in the present alloy at the synthesis condition has not been made clear. It may, however, be estimated as 2–6 wt% on the basis of published data for the three elements (HANSEN and ANDERKO, 1958). A value of 6–7 wt% may also be implied by assuming the formation of carbide of M_3C-type, which cannot be deleted simply by the determined content of carbon.

Fortunately, the synthesized diamond shows an apparent ferromagnetism caused by the included material. Therefore, Curie temperature measurement will be useful in knowing whether the remaining carbon forms carbide. The diamonds were roughly classified into two parts, B and C, according to susceptibility to a magnet, and saturation magnetization was measured as a function of temperature. As shown in Fig. 16, the temperature

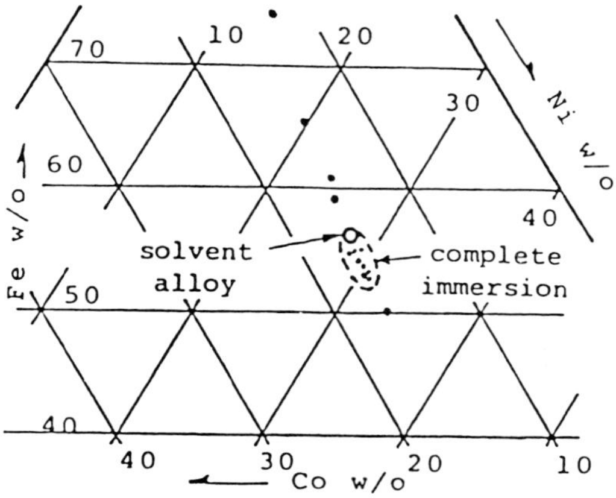

FIG. 15. The composition of inclusions.

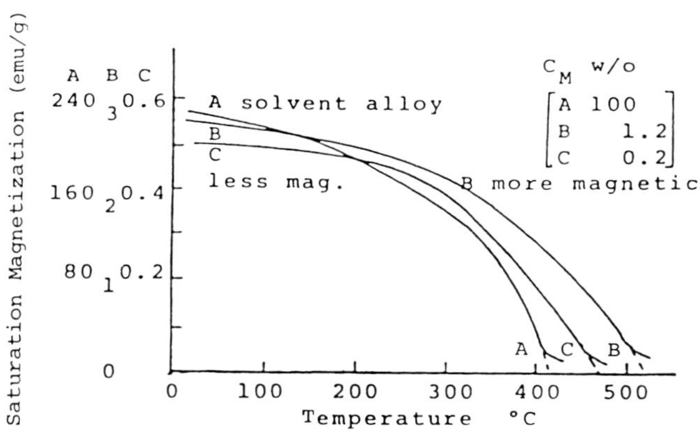

FIG. 16. Saturation magnetization versus temperature relation for solvent alloy and inclusions.

dependences for both specimens B and C are similar to that of the solvent alloy. Thus it is supposed that the included material is a solidified alloy concerning the growth of diamond and contain carbon as a solid solutions and very small diamonds in the state of a eutectic, instead of forming carbide. This supposition is supported by comparing the Curie temperatures of the specimens with those of relating materials (Ni 358°C, Fe 770°C, Co 1,131°C, $FeNi_3$ > 502°C, FeCo 727°C and Fe_3C 212°C).

4.5 *Enclosing process for the inclusions*

The shape, the composition and the location of the included particles revealed so far suggest the following process for their formation. A carbon-containing molten metallic solvent, working as the growing medium for diamond, at first fills a recess which is formed by some growth instability at the particular edge or apex described. Then overhang is made at the periphery of the recess, since the periphery gathers more nutrient than the bottom. And finally, the recess is closed to form a single particle of inclusion. Another recess should again be formed soon after, as long as the situation inducing the instability is not changed. Thus the enclosing process is repeated on the edge or the apex to result in an array of the inclusions along the boundary of neighboring growth regions. The repeated enclosing process, presumed as above, is shown in Fig. 17. The included material solidifies at the cooling stage.

Crystals grown by the method of temperature difference (STRONG and CHRENKO, 1971; KANDA and FUKUNAGA, 1981), in which a thick (more than 1 mm) layer of solvent exists between the supply and the growing crystal, do not seem to contain arrayed inclusions of interest, even when they contain large inclusions enclosed at the central region of the growing facet (KANDA and FUKUNAGA, 1981). According to a recent observation by KANDA *et al.* (1981), the radially-arrayed inclusions are especially developed towards the side of a crystal facing to graphite as shown in Fig. 18. On this side, dissolved carbon is supplied through a thin ($\simeq 0.1$ mm) layer of solvent. In any way, growth instabilities are always caused by a high growth rate. But the growth instability responsible for inclusions of the present interest may be characteristic of the case in which the thin layer of solvent concerns the growth.

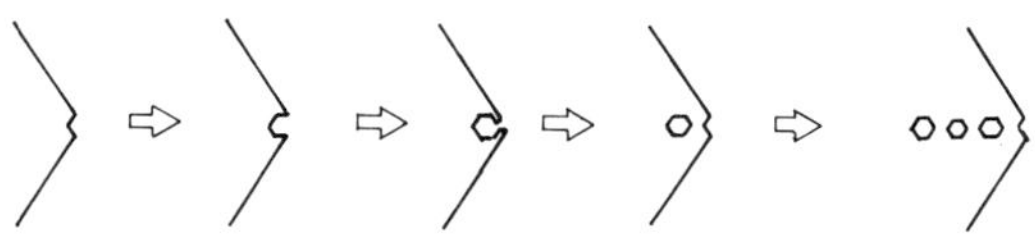

FIG. 17. An enclosing process of arrayed inclusions.

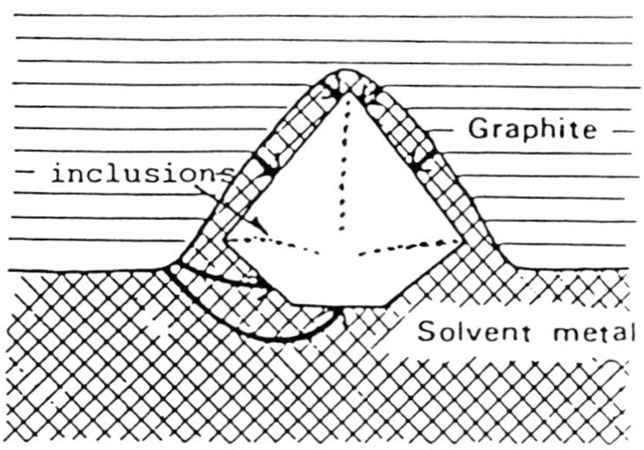

FIG. 18. Schematic illustration for carbon transport to the diamond surface from graphite (after KANDA *et al.*, 1981).

As for the site of the instability occurrence, growing facets of the same kind of index meet together at this site. Thus, growing steps sweeping on the both facets and meeting at this site have the same height and the same sweeping velocity. Such a situation may be susceptible to the growth instability. The mechaism for this type of instability should also explain the different features, caused by the thickness of the working layer of solvent. However, it, remains unclear at present.

5. *Concluding Remarks*

Among the problems of synthetic diamond, a model for the nucleation stage has been developed. The model can explain behaviors of nucleation very well. Further, nucleation density and growth rate have become separately controllable on the basis of the model. But there remains the problem of understanding the mechanism enabling the recrystallized graphite to inhibit the nucleation. The nucleation should occur, in general consideration, even from the recrystallized graphite at a sufficiently large excess pressure. Confirmation, thereof, is left for future research.

In the field of characterizing crystals, small and arrayed inclusions, inherent to the synthetic diamond, were revealed to be the solvent material enclosed during growth. One meaning of such a characterization makes it clear that some growth instability occurs at the special edges or apices of the crystal, though the mechanism is unclear. The other meaning is related to the industrial application. Since metallic inclusions will decrease the mechanical strength, thermal conductivity and chemical stability of this material, the obtained knowledge will be used in evaluating synthetic particles for industry. The present study, however, is incomplete on the other side; a synthesis condition to eliminate enclosing the solvent seems to be required.

The author would like to express his sincere thanks to Mr. Kazuaki Ichinose and Mr. Toshio Aoki of the Toshiba Corporation, with whom the author collaborated for the study of the nucleation of diamond. This portion of the present article is included in a presentation at the U.S.-Japan Seminar on "High pressure Research in Geophysics" (1981, Hakone). The author's sincere thanks are also due to Dr. Kaoru Takano of The University of Tsukuba and Dr. Katsuhiko Inoue of Kobe Steel Co., Ltd., the author's colleagues in the study of the inclusion. The study of the inclusion has been presented at the International Conference on "High Pressures in Research and Industry" (1981, Uppsala).

REFERENCES

BERMAN, R. (1965): Thermal Properties, in *Physical Properties of Diamond*, Ed. R. Berman, Clarendon Press, Oxford, pp. 371–393.

BERMAN, R. (Ed.) (1965): *Physical Properties of Diamond*, Clarendon Press, Oxford.

BERMAN, R. and SIMON, F. (1955): On the graphite-diamond equilibrium, *Z. Electrochem*, **59**, 333–338.

BEZRUKOV, G. N., BUTUZOV, V. P. and LAPTEV, V. A. (1972): Growth kinetics of artificial diamond crystals of various crystal habits, *Soviet Phys. Dokl*, **16**, 800–804.

BOVENKERK, H. P., BUNDY, F. P., HALL, H. T., STRONG, H. M. and WENTORF, R. H., JR. (1959): Preparation of diamond, *Nature*, **184**, 1094–1098.

BRIDGMAN, P. W. (1947): An experimental contribution to the problem of diamond synthesis, *J. Chem. Phys*, **15**, 92–98.

BUNDY, F. P., BOVENKERK, H. P., STRONG, H. M. and WENTORF, R. H., JR. (1961): Diamond-graphite equilibrium line from growth and graphitization of diamond, *J. Chem. Phys.*, **35**, 383–391.

CANNON, P. and CONLIN, E. T., II. (1964): Formation of diamond III, Effects of silicon on the nucleation and growth of diamond. Comparison of effects with other light atom impurities, *J. Chem. Soc.*, **86**, 4540–4544.

ELWELL, D. and SCHEEL, H. J. (1975): *Crystal Growth from High-Temperature Solutions*, Academic Press, London.

FIELD, J. E. (Ed.) (1979): *The Properties of Diamond*, Academic Press, London.

FRANK, F. C. and LANG, A. R. (1965): X-Ray Topography of Diamond, in *Physical Properties of Diamond*, Ed. R. Berman,.Clarendon Press, Oxford, pp. 69–115.

HALL, H. T. (1960): Ultra-high-pressure, high-temperature apparatus: the belt, *Rev. Sci. Instr.*, **31**, 125–131.

HANSEN, M. and ANDERKO, K. (1958): *Constitution of Binary Alloys*, McGraw-Hill, New York.

HORTON, M. D. and PETERSON, G. R. (1978): Nucleation and growth of diamond, *High Temp. High Press.*, **10**, 399–405.

KANDA, H., YAMAOKA, S., SETAKA, N. and KOMATSU, H. (1977): Etching of diamond octahedrons by high pressure water, *J. Cryst. Growth*, **38**, 1–7.

KANDA, H., SETAKA, N., OHSAWA, T. and FUKUNAGA, O. (1981): Asymmetries of the crystal shapes of synthetic diamonds, *J. Cryst. Growth*, **51**, 629–631.

KANDA, H. and FUKUNAGA, O. (1982): Growth of large diamond crystal, in *High-Pressure Research in Geophysics*, (Proc. U.S.-Japan Seminar on High Pressure Research in Geophysics, 1981, Hakone), Eds. S. Akimoto and M. H. Manghnani, Center for Academic Publications Japan/D. Reidel, Tokyo/Dordrecht, pp. 525–536.

LANG, A. R. (1964): Dislocations in diamond and the origin of trigons, *Proc. Roy. Soc. A.*, **278**, 234–242.

LILJEBLAD, R. (1955): Equilibrium pressure for different temperatures between graphite and diamond, *Arkiv för Kemi*, **8**, 423–432.

MAL'KOV, B. A. and ASKHABOV, A. M. (1979): Nitrogen platelets in diamond—a product of mantle annealing, *Dokl.* Acad. Nauk SSSR, **248**, 1420–1423.

PARSONS, C. (1920): Experiments on the artificial production of diamond, *Trans. Roy. Soc. A*, **220**, 67–107.

STRONG, H. M. (1977): Annealing Synthetic Diamond Type IIb, U.S. Patent No. 4,174,380.

STRONG, H. M. and CHRENKO, R. M. (1971): Further studies on diamond growth and physical properties of laboratory-made diamond, *J. Phys. Chem.*, **75**, 1838–1843.

STRONG, H. M. and HANNEMAN, R. E. (1967): Crystallization of diamond and graphite, *J. Chem. Phys.*, **46**, 3668–3676.

STRONG, H. M. and WENTROF, R. H., JR. (1972): The growth of large diamond crystals, *Naturwiss*, **59**, 1–7.

TAKASU, S., WAKATSUKI, M. and WAKAMATSU, N. (1968): Influence of solvent types on the characteristics of synthetic diamonds, in *Science and Technology of Industrial Diamonds*, Vol. 2, Ed. J. Burls, Industrial Diamond Information Bureau London, pp. 369–378.

WAKATSUKI, M. (1966): New catalysts for synthesis of diamond, *Japan. j. Appl. Phys.*, **5**, 337.

WAKATSUKI, M., ICHINOSE, K. and AOKI, T. (1975): Formation of diamond with copper-base alloy catalysts. in *Proc. 4th Intern. Conf. on High Pressure Research, Kyoto*, Physico-Chemical Soc. Japan, 1975, pp. 413–417.

WAKATSUKI, M. and ICHINOSE, K. (1982): A wedge-type cubic anvil apparatus and its application to material synthesis research, in *High-Pressure Research in Geophysics* (Proc. U.S.-Japan Seminar on High Pressure Research in Geophysics, 1981, Hakone), Eds. S. Akimoto and M. H. Manghnani, Center for Academic Publications Japan/D. Reidel, Tokyo/Dordrecht, pp. 13–26.

WENTORF, R. H., JR. (1971): Some studies of diamond growth rates, *J. Phys. Chem.*, **75**, 1833–1837.

WENTROF, R. H., JR. (1965): Diamond Formation at High Pressures, in *Advances in High Pressure Research*, Ed. R. H. Wentorf, Jr. Academic Press, London, pp. 249–281.

WENTORF, R. H. JR. and BOVENKERK, H. P. (1962): Preparation of semiconducting diamonds, *J. Chem. Phys.*, **36**, 1987–1989.

WOOD, G. S. and LANG, A. R. (1975): Cathodoluminescence, optical absorption and X-ray topographic studies of synthetic diamonds, *J. Cryst. Growth*, **28**, 215–226.

Materials Science of the Earth's Interior, edited by I. Sunagawa, pp. 375–386.

AR ISOTOPIC RATIOS AND K, Na AND OTHER TRACE ELEMENT CONTENTS IN PREMIER AND FINSCH MINE DIAMONDS

M. OZIMA[1], N. TAKAOKA[2], O. NITO[3], and S. ZASHU[1]

[1]*Geophysical Institute, University of Tokyo, Tokyo, Japan*
[2]*Department of Earth Sciences, Yamagata University, Yamagata, Japan*
[3]*Faculty of Engineering, Tokyo University of Agriculture and Technology, Tokyo, Japan*

1. *Introduction*

A noble gas state in the mantle, especially the isotopic ratios such as $^{40}Ar/^{36}Ar$ and $^{3}He/^{4}He$, imposes crucial constraints on the degassing history of the mantle and the evolution of the atmosphere (e.g., OZIMA, 1975). The usual approach in obtaining the information on the noble gas state in the mantle is to examine noble gases trapped in some mantle-derived materials such as volcanic rocks and volcanic xenoliths. However, it is generally difficult to ascertain that noble gases trapped in the mantle-derived materials are faithful representatives of a mantle noble gas state, since volcanic rocks and volcanic xenoliths often show an exchange of the noble gases with the atmosphere (BERNATOWICZ, 1980). In this regard, diamond is a unique mantle-derived material; its inertness against any known chemicals and high temperature and pressure stability suggest that diamond would retain original mantle noble gases.

To examine noble gases in diamonds, however, is a formidable task; firstly because of the enormous high temperature stability, diamond must be heated to very high temperature (above 2,000°C), around which graphitization takes place in high vacuum (TAKAOKA and OZIMA, 1978) to release the noble gases. Secondly, a high temperature furnace must be kept under ultra-high vacuum condition, since the noble gas content in diamond is, generally, extremely small and even a slight air contamination would totally mask the original trapped component. Thirdly, in order to obtain geologically useful information on diamonds, the geological settings in which diamonds were sampled must be known. The first two difficulties were solved with the use of a ultra-high vacuum gas extraction system, which was built by one of the authors (TAKAOKA, 1976). In the present study, earlier parts of the noble gas experiments (Premier 1 ~ 4) were carried out at Osaka University with the extraction line built by Takaoka and later experiments were done at University of Tokyo with an extraction line which was built on the basis of Takaoka's design. With regards to the third problem, we are very grateful to De Beers Consolidated Mines Ltd. for the donation of 12 carat diamond stones, of which the geological sources are well traced.

In addition to the noble gas study, we also studied K and Na and other trace elements contents by neutron activation analysis. The major objective of this was to measure the K content to obtain the K–Ar age of the diamond stones with a K–Ar isochron method (e.g., HAYATSU and PALMER, 1975). As will be stated below, because of the extremely small amount of radiogenic ^{40}Ar, this attempt was not very successful.

2. *Samples*

Two stones (one 'with-inclusion' and one 'without-inclusion') from Premier mine (Sample numbers 1 and 2), eight stones (four 'with-inclusion' and four 'without-inclusion') from Finsch mine and six Premier unidentified stones (sample numbers 3–8) were studied (Hawthorne, written communication, 1979). Among them, the Premier mine and Finsch mine stones are identified to come from a respective single pipe. However, the genetic origin of the Premier unspecified stones are not well known. Each stone is about 1 carat (0.2 g) in weight. The samples are specified as either 'with-inclusion' or 'without-inclusion' in Table 1, in that the conventionally accepted criterion for 'with-inclusion' or 'without inclusion' is whether or not to observe inclusions in a diamond stone with a magnifying glass (magnification power × 10). The inclusions were identified as mostly graphite under a binocular (× 20), but olivine and other inclusions may also be present (Komatsu, private communication, 1979). The eruption age of the Premier pipe is estimated to be 1.25 Ga and that of Finsch pipe 94 ~ 104 Ma (KRAMERS, 1979).

3. *Experimental Results*

3.1 *Noble gas analyses*

Noble gas extraction was made at two temperature steps, 800°C and 2,050°C for the Finsch samples and at three steps 800°C, 2,000°C and 2,100°C for the four Premier samples (Premier 1 ~ 4). Before the experiment, the diamonds were pre-heated in vacuum at 150°C for 8 hours. On heating at 2,000°C, the diamond stones were completely graphatized, in which an original single diamond stone (about 1 carat each) was found to be converted to a fine graphite powder. In our previous experiments (TAKAOKA and OZIMA,

TABLE 1. Ar and He in diamonds.

T(°C)	$(^4He)_b$ $\times 10^8$ cm³ STP/g	$(^{36}Ar)_b$ $\times 10^{11}$ cm³ STP/g	$(^{40}Ar)_b$ $\times 10^9$ cm³ STP/g	^4He $\times 10^8$ cm³ STP/g	^{36}Ar $\times 10^{11}$ cm³ STP/g	^{40}Ar $\times 10^9$ cm³ STP/g
	Finsch 1, without inclusion, $m = 0.13719$ g					
800	1.96	2.16	6.0	2.21	3.12	8.62
2,050	3.39, 2.95	6.95, 7.86	19.2 , 21.9	7.73	11.4	32.5
$(2,050)_c$				4.56 ± 0.12	4.50 ± 0.04	12.0 ± 1.4
	Finsch 2, without inclusion, $m = 0.15533$ g					
800	2.21	2.45	6.8	1.76	2.16	6.09
2,050	2.06, 2.82	5.63, 6.15	15.8, 17.0	5.20	7.47	20.5
$(2,050)_c$				2.77 ± 0.38	1.58 ± 0.25	4.1 ± 0.6
	Finsch 6, with inclusion, $m = 0.13895$ g					
800	12.30	2.59	7.00	5.89	2.10	5.68
2,050	2.27, 2.30	5.90, 6.30	16.4, 17.7	21.8	8.42	23.6
$(2,050)_c$				19.6 ± 0.1	2.32 ± 0.2	6.56 ± 0.06
	Finsch 8, with inclusion, $m = 0.17809$ g					
800	1.80	2.02	5.65	1.44	2.02	5.45
2,050	1.77	7.37, 4.55	20.7, 12.6	18.4	9.43	26.3
$(2,050)_c$				16.7	3.5 ± 1.5	9.7 ± 3.3

(continued)

TABLE 1. (continued)

T(°C)	$(^4He)_b$ $\times 10^8$ cm^3 STP/g	$(^{36}Ar)_b$ $\times 10^{11}$ cm^3 STP/g	$(^{40}Ar)_b$ $\times 10^9$ cm^3 STP/g	^4He $\times 10^8$ cm STP/g	^{36}Ar $\times 10^{11}$ cm^3 STP/g	^{40}Ar $\times 10^9$ cm^3 STP/g
	Premier 1, without inclusion, $m = 0.13509$ g					
800	0.44, 0.59	4.4, 2.96	13.3, 6.7	0.59	3.33	9.63
2,000	0.52, 0.67	37.0, 7.4	10.4, 20.7	111	15.6	45.2
2,100	0.52, 0.67	111.0, 20.0	333.0, 57.8	0.67	46.7	133
$(2{,}000)_c$				11.05 ± 0.5	n.m	n.m
	Premier 2, with inclusion, $m = 0.15047$ g					
800	0.4, 0.53	4.0, 2.7	12.0, 6.0	0.67	3.07	8.67
2,000	0.47, 0.60	33.0, 6.7	93.3, 18.7	86.7	8.67	28.0
2,100	0.47, 0.60	100.0, 18.0	300.0, 52.0	0.93	20.0	59.3
$(2{,}000)_c$				86.2 ± 0.4	n.m	n.m
	Premier 3, (unspecified) with inclusion, $m = 0.16067$ g					
800	0.37, 0.50	3.7, 2.5	11.2, 5.6	0.63	0.10	87.5
2,000	0.43, 0.56	31.0, 6.2	87.0, 17.4	3630.0	53.8	700
2,100	0.43, 0.56	93.2, 16.8	280.0, 48.4	0.44	61.9	181
$(2{,}000)_c$				3630.0 ± 0.3	34.5 ± 12.5	640 ± 40
	Premier 4, (unspecified) with inclusion, $m = 0.17215$ g					
800	0.35, 0.47	3.5, 2.3	10.5, 5.2	1.10	3.49	14.5
2,000	0.41, 0.52	29.0, 5.8	81.4, 16.3	128	13.4	308.0
2,100	0.41, 0.52	87.2, 15.7	261.6, 45.3	0.52	26.7	75.6
$(2{,}000)_c$				127.6 ± 0.3	n.m	260 ± 30

	Premier 5, (unspecified) with inclusion, $m = 0.13114$ g					
800	—	2.7	7.0	—	7.33	30.6
2,050	—	25.5, 10.5	63.5, 29.1	—	26.8	85.4
$(2{,}505)_c$				—	10.0 ± 6	39.5 ± 17.5
	Premier 6, (unspecified) with inclusion, $m = 0.15646$ g					
800	2.50	2.3	6.2	1.84	1.96	5.29
2,050	1.78	5.91, 6.89	15.8, 18.7	63.1	8.24	31.5
$(2{,}050)_c$				61.3	1.8 ± 0.6	14.5 ± 1.3
	Premier 7, (unspecified) with inclusion, $m = 0.16678$ g					
800	2.2	2.4	6.6	1.36	1.75	4.78
2,050	16.7	6.17, 15.6	16.7, 42.6	36.9	39.1	153
$(2{,}050)_c$				20.2	27.9 ± 8	123 ± 13
	Premier 8, (unspecified) with inclusion, $m = 0.17086$ g					
800	—	2.5	6.8	1.72	2.22	5.33
2,050	—	6.0, 5.4	16.5, 14.4	153	10.4	130
$(2{,}050)_c$					4.6 ± 0.3	114 ± 1

Suffix b indicates the amount of hot blank.
Two figures given for blanks correspond to the blanks measured before and after noble gas analysis.
Suffix c indicates the values corrected for the hot blanks.
Errors in the noble gas amounts correspond to the corrections for the minimum and maximum amounts of the hot blanks.

1978), we showed that almost all of the noble gases were degassed from the diamond during graphitization. 800°C fractions were generally less than 10% of 2,000° or 2,050°C fractions and are likely to be surface contaminated air. Hence, for the later discussion, we only use the 2,000 and 2,050°C fractions. Hot blank is negligibly small for the 800°C fractions, and is about 10 ~ 50% of the 2,000° or 2,050°C fractions (Table 1).

The extracted noble gases were measured with a 12-inch radius, single focusing, 60 degree mass spectrometer, with a Faraday cap for Ar and with an electron multiplier for the other noble gases. The noble gas amount was determined by means of a peak height method, in which the known amount of noble gases extracted from the air were occasionally measured as calibration. Reproducibilities are generally about 10% for He and less than 3% for Ar.

3.2 *Neutron activation analyses*

Using the technique of instrumental neutron activation analysis, K, Na, and other trace elements in diamonds were determined. Diamond stones were first irradiated for six hours with a total thermal neutron flux of about 3×10^{16} by a TRIGA Mark-II reactor at Rikkyo University. Before γ-ray counting, the individual irradiated samples were completely washed by aqua regia, distilled water and ethyl alcohol in an ultrasonic bath. In the case of K-determination, three different standard samples with varied K-contents were employed, but only one standard sample for Na. For the determination of other trace elements, a standard basalt powder (JB–1; prepared by Geological Survey, Japan) was used as a standard. The results are shown in Table 2.

4. *Results*

4.1 *K–Ar systematics*

Analytical data were plotted in an $^{40}Ar/^{36}Ar$–$K/^{36}Ar$ correlation diagram (Fig. 1). Due to high ^{36}Ar blanks, $^{40}Ar/^{36}Ar$ ratios were not determined for the two Premier mine stones (No. 1 and No. 2) and for one Premier unspecified sample (No. 4). Also, because of the extremely small content of K, no attempt was made to construct the $^{40}Ar/^{36}Ar$–$K/^{36}Ar$ plot for the Finsch mine samples. Hence, in $^{40}Ar/^{36}Ar$–$K/^{36}Ar$ plot only data for the Premier unspecified samples were plotted. In Fig. 1, reference isochrons for 4.55 and 1.25 Ga (the latter being an eruption age of the Premier pipe) are also shown. The data are largely scatterd in Fig. 1, showing no correlation trend. This result indicates either that the Premier unspecified samples are genetically not related to each other, i.e., each having a different age, or that the scatter is simply due to the analytical uncertainty. The error bars shown in Fig. 1 indicate the maximum possible values, and analytical errors are likely to be much smaller. Hence, we feel that the scatter of the data in Fig. 1 is genuine, that is, the Premier unspecified diamond stones are not cogenetic.

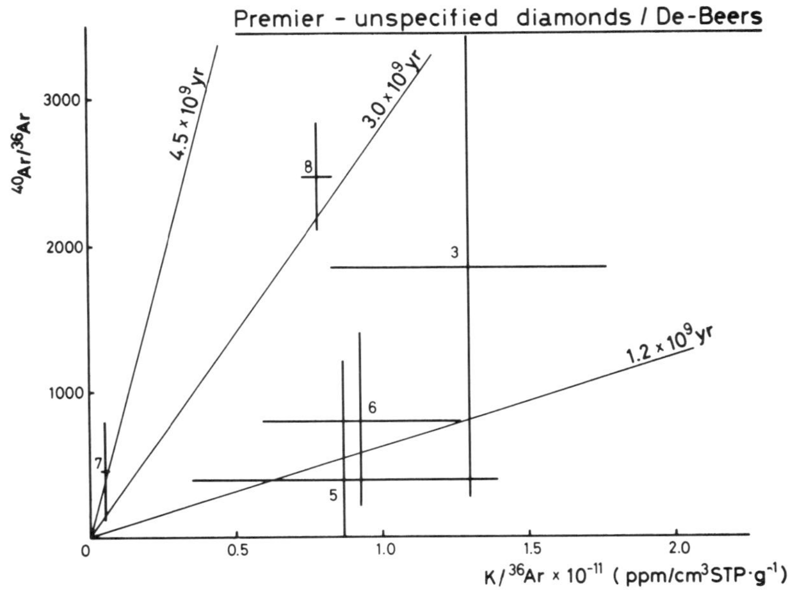

FIG. 1. $K/^{36}Ar$–$^{40}Ar/^{36}Ar$ isochron plot for Premier unspecified diamonds. Reference isochrons of 4.55 G.a. and 1.25 G.a. are also shown, 1.25 G.a. corresponds to the age of kimberlite magma eruption in Premier Mine. Error bars indicate 1σ. Numerals indicate sample number listed in Table 1.

4.2 *Ar isotopic ratios*

Ar in Finsch mine diamonds are characterized by a rahter uniform $^{40}Ar/^{36}Ar$ isotopic ratio of about 290 (Fig. 2). The ratio is very close to the value of the present air Ar (295.5). Such uniformity in the isotopic ratio is consistent with the fact that the Finsch mine diamonds are all from the same pipe, possibly being of cogenetic origin, and Ar is essentially a primary trapped component without the addition of any significant amount of radiogenic ^{40}Ar (see Section 5 for further discussions). As will be discussed below, the origin of the trapped Ar may be attributed to the ancient air Ar introduced by subducted materials in the diamond source region.

In the case of the Premier unspecified stones, the $^{40}Ar/^{36}Ar$ ratio varies considerably (Fig. 2). The scatter in the isotopic ratio must be primarily a reflection of the different origin of each stone, in that the trapped Ar had a different Ar isotopic ratio and there was also the addition of varied amounts of radiogenic ^{40}Ar, due to the different age and K content of each stone. At any rate, it is not possible to make a reasonable estimation of the isotopic ratio of the primary trapped Ar in the Premier unspecified diamond stones.

Here, it is important to note that the amount of ^{36}Ar in the Premier unspecified diamonds is almost ten times as large as in the Finsch mine stones.

TABLE 2.

	Premier stones		Premier unspecified stones					
	No. 1	No. 2	No. 3	No. 4	No. 5	No. 6	No. 7	No. 8
	Without inclusions	With inclusions						
	135.09 mg	150.47 mg	160.67 mg	172.15 mg	131.14 mg	156.46 mg	166.78 mg	170.86 mg
K (ppm)	<0.013	0.43	45.1	4.00	8.72	1.68	1.55	3.62
Na (ppm)	0.072	4.86	55.2	1.25	8.01	1.96	1.03	1.29
Ca (ppm)	<10	<27	522	49	130	44	<8	<11
Cr (ppm)	<0.0057	0.544	1.82	<0.0098	0.718	<0.016	<0.0090	0.020
Fe (ppm)	<1.3	49.6	301	26.1	336	32.3	10.2	22.3
Co (ppm)	<0.0024	0.040	0.386	0.024	0.176	<0.0077	0.028	0.022
Ni (ppm)	<0.044	0.18	2.58	<0.11	10.0	0.75	0.34	0.22
Sc (ppb)	0.14	17.9	23.9	1.2	19.7	3.0	0.25	1.7
La (ppb)	<0.058	0.62	334	7.9	62.3	20.0	5.5	12.3
Sm (ppb)	<0.0045	0.199	73.7	0.993	8.77	4.42	0.46	1.05

Finsch stones	No. 1	No. 2	No. 3	No. 4	No. 5	No. 6	No. 7	No. 8
	Without inclusions				With inclusions			
	137.19 mg	155.33 mg	155.55 mg	173.31 mg	135.24 mg	139.14 mg	145.72 mg	178.09 mg
K (ppm)	0.0212	0.0091	0.0032	0.0035	0.039	0.112	0.034	0.200
Na (ppm)	0.00254	0.00073	0.00019	0.00038	0.206	0.203	0.079	0.258
Ca (ppm)	<6	<6	<3	<3	<10	<9	26	<9
Cr (ppm)	<0.0044	<0.0064	<0.0045	<0.0036	0.378	0.306	0.038	10.6
Fe (ppm)	0.8	<0.6	<0.3	0.3	26.4	27.8	5.1	22.6
Co (ppm)	0.036	0.048	0.0023	0.0013	0.059	0.047	0.012	0.046
Ni (ppm)	1.49	2.94	<0.032	<0.017	1.24	1.10	0.13	0.69
Sc (ppb)	<0.064	<0.071	<0.039	<0.030	0.47	0.72	0.12	39.6
La (ppb)	0.23	<0.026	0.057	<0.058	0.19	1.31	1.52	0.51
Sm (ppb)	<0.0087	<0.0033	<0.0036	<0.0084	<0.0055	0.080	0.056	0.17

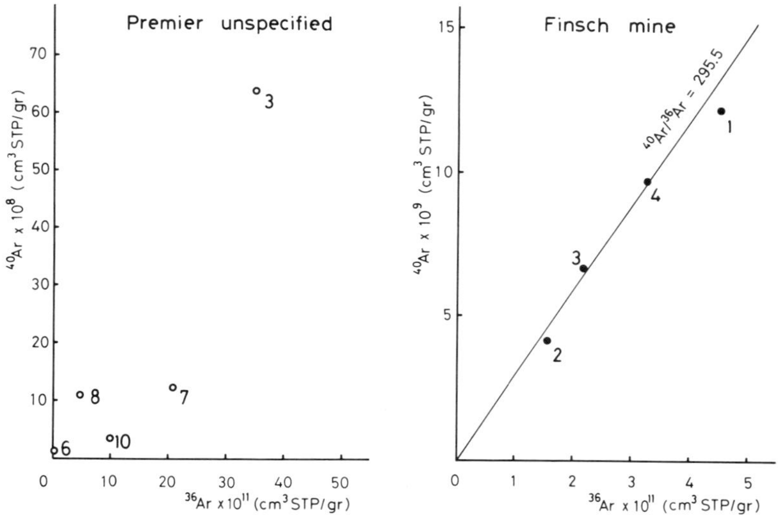

FIG. 2. ^{40}Ar content is plotted against ^{36}Ar content for Premier unspecified diamonds (left) and Finsch mine diamonds. A straight line ($^{40}Ar/^{36}Ar = 295.5$) corresponds to the present atmospheric Ar isotopic ratio.

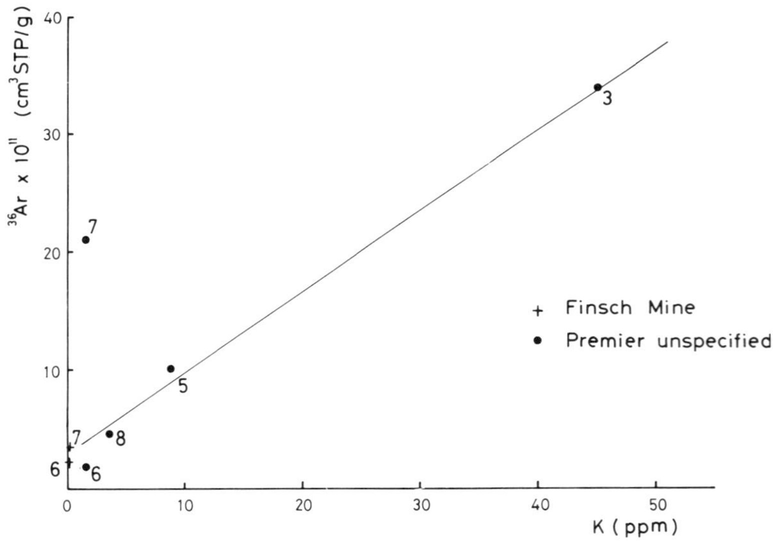

FIG. 3. ^{36}Ar content is plotted against K content for Premier unspecified diamonds and Finsch mine diamonds. A correlation (except for sample No. 7) suggests that both K and ^{36}Ar reside in the same inclusion.

The Premier unspecified samples are also characterized by the large content of trace elements such as K, Na (Table 2). There is a nice positve correlation in the Premier unspecified samples between the K content and the amount of ^{36}Ar (Fig. 3), suggesting that ^{36}Ar is essentially carried by K-bearing inclusions, but not by the diamond matrix. However, because of a too small K content in the Finsch mine diamonds, it is difficult to conclude such a correlation.

5. *Discussion*

FRANK (1966) has proposed that carbon, a source material of diamond, was originally derived from subducted oceanic crust materials, which must have abundant carbonaceous matters. Argon isotopic ratio $^{40}Ar/^{36}Ar$ in Finsch 'without-inclusion' diamonds is not inconsistent with Frank's hypothesis, as discussed below.

$^{40}Ar/^{36}Ar$ isotopic ratio in the Finsch 'without-inclusion' diamonds is very close to the present atmospheric Ar isotopic ratio. Both the theoretical model of Ar isotope evolution in the earth (HAMANO and OZIMA, 1978) and the observed Ar isotopic ratio in various mantle-derived materials (e.g., ALEXANDER and OZIMA, 1978) indicate that the $^{40}Ar/^{36}Ar$ ratio in the present mantle must be larger than a few thousands and the ratio should not have changed significantly for the last several hundred millions years. The observed nearly-atmospheric Ar isotopic ratio strongly suggests that the Ar in the diamonds cannot be of mantle origin, but more likely to be from atmospheric Ar contamination. However, it is important to note that the contaminating Ar is not necessarily the present one, but can be as old as a few hundred millions years, since the terrestrial Ar isotopic ratio evolution model (HAMANO and OZIMA, 1978) shows that the $^{40}Ar/^{36}Ar$ in the atmosphere, as well as in the mantle, has remained almost constant for the last several hundred millions years. Judging from the very tight binding of Ar in the diamond structure, post-eruption surface contamination is very unlikely; Ar was degassed essentially at 2,000°C, but only less than 10% was degassed at 800°C. It is more likely that air Ar contamination had occurred during the process of the formation of the diamonds in such a way that air Ar was trapped intrinsically within the structure of the diamonds. One plausible process to introduce air Ar into the mantle when the diamonds crystallized is subduction; that is, oceanic crust materials subducted shortly before the eruption of kimberlite magma had introduced the air Ar into the mantle, which was then incorporated in the diamonds. This then is consistent with Frank's speculation.

We are very grateful to De Beers Consolidated Mines Ltd., who very generously offered the diamond stones for our study. The arrangement was made possible by Professor H.L. Allsopp and Drs. F.A. Raal and B. Hawthorne, to whom we are much indebted.

REFERENCES

ALEXANDER, E. C. JR. and OZIMA, M. (ed.) (1978): *Terrestrial Rare Gases*, 229 pp., Cent. Acad. Publ. Japan, Tokyo.

BERNATOWICZ, T. J. (1980): Noble gases in ultramafic xenoliths from San Carlos, Arizona, Ph. D. Thesis, Washington University.

FRANK, F. C. (1967): Defects in diamond, in *Proc. Industrial Diamond Conference, Oxford, 1966*, Industrial Diamond Information Bureau, London, pp. 119–135.

HAMANO, Y. and OZIMA, M. (1978): Earth-atmosphere evolution model based on Ar isotopic data, in *Terrestrial Rare Gases*, Eds. E. C. Alexander, JR. and M. Ozima, Cent. Acad. Publ. Japan, Tokyo

HAYATSU, A. and PALMER, H. C. (1975): K–Ar isochron study of the Tudor Gabbro, Grenville Province, Ontario, *Earth Planet. Sci. Lett.*, **25**, 208–215.

KRAMERS, J. D. (1979): Lead, uranium, strontium, potassium, and rubidium in inclusion-bearing diamonds and mantle-derived xenoliths from southern Africa, *Earth Planet. Sci. Lett.*, **42**, 58–70.

OZIMA, M. (1975): Ar isotopes and earth-atmosphere evolution models, *Geochim. Cosmochim. Acta*, **39**, 1127–1134.

TAKAOKA, N. (1976): A low-blank, metal system for rare-gas analysis, *Mass Spectros.*, **24**, 73–86.

TAKAOKA, N. and OZIMA, M. (1978): Rare gas isotopic compositions in diamonds, in *Terrestrial Rare Gases*, Eds. E. C. Alexander, JR. and M. Ozima, Cent. Acad. Publ. Japan, Tokyo.

Materials Science of the Earth's Interior, edited by I. Sunagawa, pp. 387–394.

ULTRA-HIGH PRESSURE PHASE RELATIONS OF THE SYSTEM MgO–FeO–SiO_2 AND THEIR GEOPHYSICAL IMPLICATIONS

Eiji ITO

Institute for Thermal Spring Research, Okayama University, Misasa, Tottori-ken, Japan

1. *Introduction*

The depth region between about 400 and 1,000 km in the earth's mantle, the transition zone, was characterized by the large gradients of seismic velocities, which could be mainly explained by a series of phase transformations of the upper mantle minerals such as olivine, pyroxenes, and garnet, to the denser lower mantle minerals (BIRCH, 1952). Recent models of the velocity-depth distributions of the mantle, however, show that the transition zone is marked by two and/or three abrupt increases. Among them, the discontinuity found around the depth of 670 km has attracted special interest since the discontinuity is very large and sharp (e.g., FUKAO, 1977) and the interpretation of the discontinuity would provide strong constraint for the constitution of the lower mantle.

Recent high pressure experiments, on the other hand, have revealed several phase transformations which are regarded as of basic importance for the consideration of the state of the deep mantle. The most significant findings are the stable existence of the ilmenite and perovskite modifications of $MgSiO_3$ and the dissociation of spinel into perovskite plus periclase in Mg_2SiO_4 (LIU, 1976a, b; ITO, 1977; ITO and MATSUI, 1978). In the cases of iron silicates, however, no high pressure compound was found for the $FeSiO_3$ stoichiometry and Fe_2SiO_4 spinel decomposes into wüstite plus stishovite (e.g., ITO and MATSUI, 1977). Under these circumstances, phase relations of the MgO–FeO–SiO_2 ternary system at pressures higher than 20 GPa are of urgent need for the detailed consideration of the nature of the 670 km discontinuity and the formation of the lower mantle.

We have carried out systematic experiments to clarify the stability relations of the spinel, ilmenite, and perovskite phases in the MgO–FeO–SiO_2 system. It is the purpose of the present article to summarize the results obtained so far and to apply them to the constitution of the middle and lower mantle.

2. *Experimental Procedure*

High pressure and high temperature experiments were performed using a uniaxial split-sphere apparatus (cf. KAWAI *et al.*, 1973) installed in our laboratory in 1979, in which the cubic anvil assembly of tungsten carbide is compressed with the aid of a 5,000-ton press. The cubic anvil assembly and the specimen configuration are schematically illustrated in Fig. 1. The pressure medium was pyrophyllite backed at 700°C for 30 min in air. The powdered sample was put directly into a cylindrical heater of nichrome or tantalum embedded in the center of pyrophyllite octahedron. The sample temperature was monitored by the Pt/Pt13%Rh thermocouple without regard to the pressure effect on emf.

The generated pressure was calibrated against the applied load, adopting the transition pressures of the following phase transformations at 1,000°C as the fixed points:

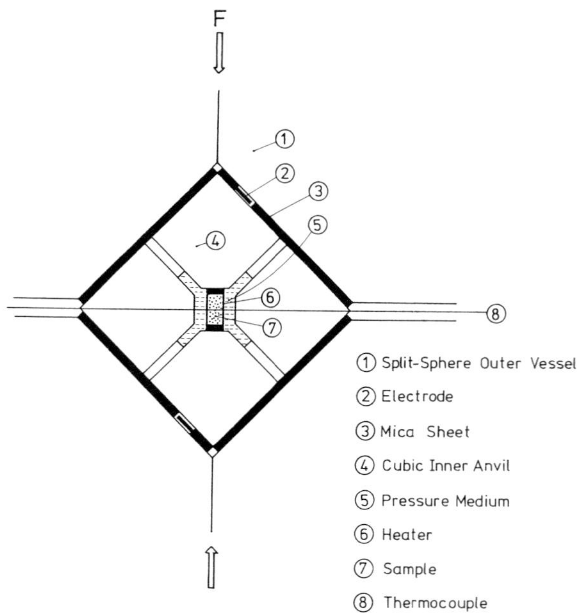

FIG. 1. Schematical drawing of the uniaxial split-sphere high-pressure vessel and the speciman assembly for high temperature experiments.

$ZnSiO_3$(clinopyroxene)
→ $ZnSiO_3$(ilmenite) at 11 GPa
(AKIMOTO *et al.*, 1977)

$MgAl_2O_4$(spinel) → MgO(rocksalt) + Al_2O_3 (corundum) (LIU, 1980) at 15 ± 1 GPa

and

Ni_2SiO_4(spinel) → $2NiO$(rocksalt) + SiO_2(stishovite) at 18 GPa
(NAVROTSKY *et al.*, 1979).

The pressure value for $ZnSiO_3$ was determined according to the NaCl scale and others were calculated on the basis of available thermochemical data for the reactants and products. The calibration curve thus made was carefully expraporated up to 27 GPa.

The starting materials were the synthetic olivine solid solutions of $(Mg_{1-x}Fe_x)_2SiO_4$ ($x = 0$–0.5) and synthetic pyroxene solid solutions of $(Mg_{1-x}Fe_x)SiO_3$ ($x = 0$–0.5). After the sample was held under certain pressure and temperature conditions for a desired time interval, it was quenched by shutting off the electric power supply. Then the perssure was released slowly and the product was recovered in ambient conditions.

Identification of phases were mostly performed by the powder X-ray diffraction method, in which only the central portion of the product near the junction of thermocouple was picked up and examined. Chemical compositions of the phases were determined on the basis of their unit cell parameters. Some products were made into polished sections and examined under the microscope and by the X-ray probe micro-analyzer.

3. *Results*

3.1 *Negative slopes in the $MgSiO_3$ perovskite formation reactions*

Stability relations of $MgSiO_3$ and Mg_2SiO_4 were investigated at pressures from 24 to 27 GPa and temperatures from 900° to 1,400°C. The experimental results indicated that the sign of the slopes are negative for the phase boundary curves for transformations of the ilmenite-perovskite in $MgSiO_3$ and the spinel-perovskite + periclase in Mg_2SiO_4 (ITO and YAMADA, 1982). The overhung of transformation pressure for the latter to that for the former was found to be quite small, probably less than 1 GPa at 1,100°C, and both the phase boundary curves can be tentatively expressed as $P(\mathrm{GPa}) = 27 - 0.002T(^\circ\mathrm{C})$. The feature suggests that the stabilization of perovskite would play some decisive role in realizing the post-spinel transformation since the perovskite phase has small volume and presumably large entropy.

3.2 Phase relations in the MgO–FeO–SiO$_2$ system at very high pressures

The detailed phase relations in the $MgSiO_3$–$FeSiO_3$ system has been described elsewhere (ITO and YAMADA, 1982). Stability relations obtained at 1,100°C are shown in Fig. 2 as the ternary phase diagrams for the MgO–FeO–SiO$_2$ system in the sequence of increasing pressure. Relations indicated by dotted line remain to be of reconnaissance. Special remarks for each diagram are as following:

At 19 GPa (a), spinel phase is destabilized for compositions of iron content higher than $(Mg_{0.35}Fe_{0.65})_2SiO_4$, which decomposes into magnesiowüstite $(Mg_{1-x}Fe_x)O$ plus stishovite and any compound with the $(Mg_{1-x}Fe_x)SiO_3$ stoichoimetry is nonexistent. At 23 GPa (b), the stability region of spinel phase diminishes to $(Mg_{0.6}Fe_{0.4})_2SiO_4$ composition and the

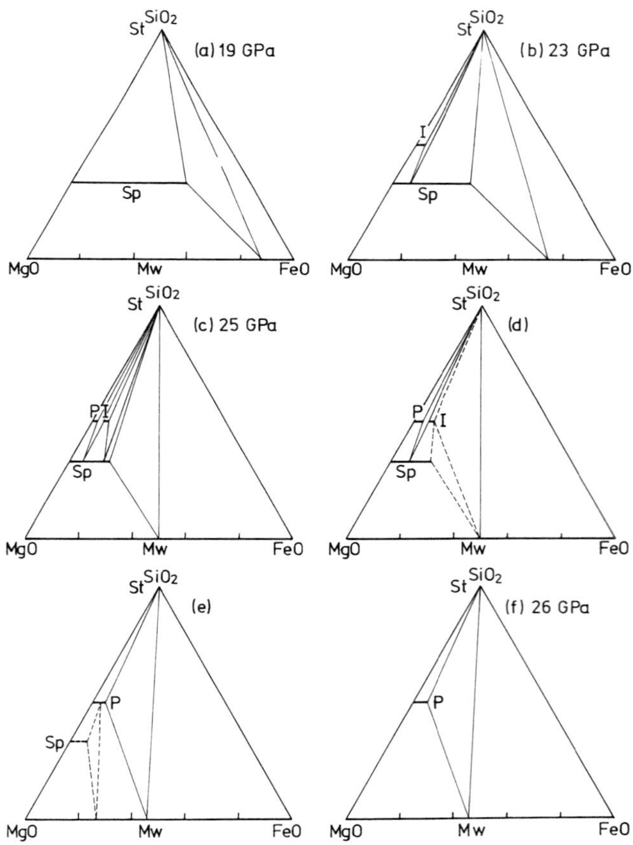

FIG. 2. Phase diagrams for the MgO–FeO–SiO$_2$ system at 1,100°C and various pressures. Sp–$(Mg_{1-x}Fe_x)_2SiO_4$ spinel; Mw–$(Mg_{1-x}Fe_x)O$ magnesiowüstite; I–$(Mg_{1-x}Fe_x)SiO_3$ ilmenite; P–$(Mg_{1-x}Fe_x)SiO_3$ perovskite.

ilmenite phase is stabilized for composition close to $MgSiO_3$. The ilmenite region expands toward a $FeSiO_3$ end member with increasing pressure, which, however, does not largely exceed the $(Mg_{0.9}Fe_{0.1})SiO_3$ composition. At 25 GPa (c), the spinel region becomes narrower to the composition of some $(Mg_{0.8}Fe_{0.2})_2SiO_4$ and the ilmenite, with composition very close to $MgSiO_3$, transforms into perovskite. Under further increasing pressure, the system reach the equilibrium state of (f) in a very small pressure interval, probably less than 1 GPa, passing through the transitional states like (d) and (e). At 26 GPa (f), the ilmenite is completely replaced by the perovskite and the survived spinel is dissociated into the assemblage of perovskite plus magnesiowüstite. It should be noted that the solubility of $FeSiO_3$ component into the perovskite is limited to nearly 10 mol percent. The pressure needed to dissociate spinel is inferred to decrease slightly with an increasing iron content, as suggested in the diagram (e).

The iron contents of the coexisting phases are plotted in Fig. 3 as the Roozeboom diagrams for the coexisting pairs of perovskite-magnesiowüstite and spinel-magnesiowüstite. The apparent Fe–Mg partition coefficient, K', were calculated from the plotted data to be

$$K'_{\text{P-Mw}} = 0.13(2) \text{ for perovskite-magnesiowüstite}$$

and

$$K'_{\text{Sp-Mw}} = 0.27(3) \text{ for spinel-magnesiowüstite.}$$

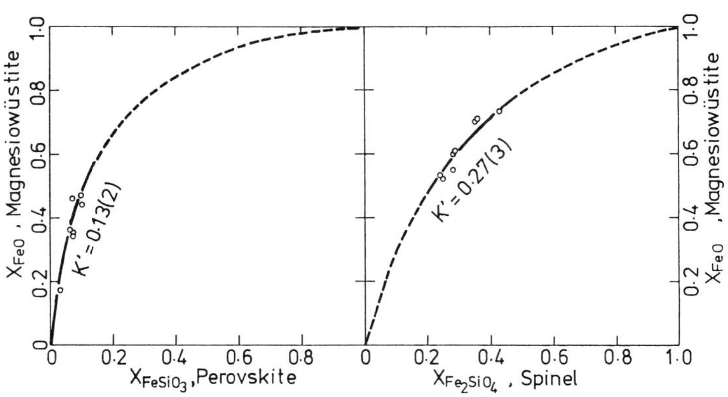

FIG. 3. The Roozeboom diagrams for the coexisting pairs of perovskite-magnesiowüstite (left) and spinel-magnesiowüstite (right).

4. *Geophysical Implications*

4.1 *The 670 km discontinuity*

It has been generally considered that MgO, FeO, and SiO_2 are the most dominant components of the mantle, the sum of these three components being greater than 90 percent of the total mass of the mantle, and that the molar ratios of (MgO + FeO)/SiO_2 and FeO/(MgO + FeO) are in the ranges of 1–2 and 0.1–0.2, respectively (e.g., MASON, 1966; RINGWOOD, 1975). According to the stability relations shown in Fig. 2, any material with such composition exists as the assemblage of spinel + magnesiowüstite + stishovite at 23 GPa, which successively transforms under increasing pressure and is completely replaced by that of perovskite + magnesiowüstite (+ stishovite) at 26 GPa. It is highly likely that the series of transformations would be the dominant cause of the 670 km discontinuity. Considering the situation that the temperature around 670 km should be largely higher than the experimental temperature of 1,100°C (STACEY, 1977) and the negative slopes of the phases boundaries of the perovskite formation reactions (cf. Section 3.1), the transformation pressures could be well reconciled with those near the discontinuity, approximately 24 GPa (DZIEWONSKI and ANDERSON, 1981). The accompaning density increase is estimated to be about 10 percent, which is also in concordance with recent density models of the mantle (DZIEWONSKI and ANDERSON, 1981). It should be noted that the density increase is donimantly due to the stabilization of the perovskite phase, which is completely in a quite small pressure interval, as mentioned in last chapter. The feature is in harmony with the seismological characteristics of the 670 km discontinuity that the velocity increase is sharp, especially near its base (FUKAO, 1977).

As frequently discussed, the 670 km discontinuity of the phase boundary with negative slope has special implication with respect to the mantle dynamics that the discontinuity act as a barrier to the mantle-wide convection and a sinking litospheric plate meets a strong mechanical brake when passing through the depth.

4.2 *Comments on the constitution of the lower mantle*

As decribed in last section, the dominant phases immediately below the 670 km discontinuity are inferred to be perovskite, magnesiowüstite, and possible trace stishovite. The amount of stishovite depends not only on the SiO_2 content but also on the FeO/(MgO + FeO) ratio of the mantle chemistry since the solubility of iron into the perovskite phase is limited to nearly ten percent (cf. Fig. 3 (f)).

DZIEWONSKI *et al.* (1975) recommended 4.15 g/cm^3 as the zero pressure and room temperature density for the lower mantle, indicating that the density distribution in the lower mantle follow the Adams-Williamson equation very closely. The density value, however, is strongly affected by the iron contents of the constituent minerals. It is requisite, therefore, for the lower mantle not only to account for the zero pressure and room temperature density of 4.15 g/cm^3 but also to satisfy the Fe–Mg partitioning between perovskite and magnesiowustite, i. e., $K'_{P\text{-}Mw} = 0.13$. The derived conclusions on the mineralogy and chemistry of the lower mantle (ITO and TAKAHASHI, 1982) are as follows:

The dominant phases are the perovskite and magnesiowüstite, possessing the compositions of some $(Mg_{0.95}Fe_{0.05})SiO_3$ and $(Mg_{0.75}Fe_{0.25})O$, respectively, and stishovite would not be stably existent. The MgO/(MgO + FeO) ratio is less than 0.14, assuming that the $(MgO + FeO)/SiO_2$ ratio is less than 2. The lower mantle dominantly composed of metasilicate (e.g., LIU, 1979) is required to be extremely magnesian $(FeO/(MgO + FeO) \sim 0.05)$. The SiO_2 content should rapidly decrease with an increasing FeO content and, if we assume $FeO/(MgO + FeO) \sim 0.1$, the $(MgO + FeO)/SiO_2$ ratio is close to 1.5.

It should be pointed out that the Fe–Mg partitioning between perovskite and magnesiowüstite largely diminishes the density difference between both phases, and the chemical zoning of the deep mantle due to the gravitational instability of both phases (LIU, 1979) would be unlikely.

The author is grateful to Y. Matsui and E. Takahashi for their stimulating discussion. He thanks H. Yamada for his help. This work was financially supported by the Grant-in-Aid for Scientific Research from the Ministry of Education, Science, and Culture of the Japanese Government (No. 321515 to E. Ito and No. 510109 to M. Kumazawa).

REFERENCES

AKIMOTO, S., YAGI, T. and INOUE, K. (1977): High temperature-pressure phase boundaries in silicate systems using in situ X-ray diffraction, in *High-Pressure Research: Applications in Geophysics*, Eds. M. H. Manghnani and S. Akimoto, Academic Press, New York, pp. 585–562.

BIRCH, F. (1952): Elasticity and constitution of the earth's interior, *J. Gephys. Res.*, **57**, 227–286.

DZIEWONSKI, A. M. and ANDERSON, D. L. (1981): Preliminary reference earth model, *Phys. Earth Planet. Inter.*, **25**, 297–356.

DZIEWONSKI, A. M., HALES, A. L. and LAPWOOD, E. R. (1975): Parametrically simple earth models consistent with geophysical dada, *Phys. Earth Planet. Inter.*, **10**, 12–48.

FUKAO, Y. (1977): Upper mantle *P* structure on the ocean side of the Japan-Kurile Arc., *Geophys. J. R. Astr. Soc.*, **50**, 621–642.

ITO, E. (1977): The absence of oxide mixture in high-pressure phases of Mg-silicates, *Geophys. Res. Lett.*, **4**, 72–74.

ITO, E. and MATSUI, Y. (1977): Silicate ilmenites and the post-spinel transformations, in *High-Pressure Research: Applications in Geophysics*, Eds. M. H. Manghnani and S. Akimoto, Academic Press, New York, pp. 193–208.

ITO, E. and MASTUI, Y. (1978): Synthesis and crystal-chemical characterization of $MgSiO_3$ perovskite, *Earth Planet Sci. Lett.*, **33**, 443–450.

ITO, E. and TAKAHASHI, E. (1982): Mineralogy and chemistry of the lower mantle, in preperation.

ITO, E. and YAMADA, H. (1982): Stability relations of silicate spinels, ilmenites, and perovskite, in *High-Pressure Research in Geophysics*, Eds. S. Akimoto and M. H. Manghnani, Center for Academic Publication Japan/D. Reidel, Tokyo/Dordrecht, pp. 405–419.

KAWAI, N., TOGAYA, M. and ONODERA, A. (1973): A New device for high pressure vessels, *Proc. Japan. Acad.*, **49**, 623–626.

LIU, L. (1976a): The high-pressure phases of $MgSiO_3$, *Earth Plant. Sci. Lett.*, **31**, 200–208.

LIU, L. (1976b): The post-spinel phases of forsterite, *Nature*, **262**, 770–772.

LIU, L. (1979): On the 650–km discontinuity, *Earth Planet. Sci. Lett.*, **42**, 202–208.

LIU, L., (1980): The equilibrium boundary of spinel $\rightleftarrows$ corudum + periclase: a calibration curve for pressure above 100 kbar, *High Temp. High Press.*, **12**, 217–220.

MASON, B., (1966): Composition of the earth, *Nature*, **211**, 616–618.

NAVROTSKY, A., PINCHOVSKI, F. S. and AKIMOTO, S. (1979): Calorimetric study of the stability of high pressure phases in the system CoO–SiO_2 and "FeO"–SiO_2, and calculation of phase diagrams in MO–SiO_2 system, *Phys. Earth Plant. Inter.*, **19**, 275–292.

RINGWOOD, A. E., (1975): *Composition and Petrology of the Earth's Mantle*, McGraw-Hill, Inc., New York.

STACEY, F. D. (1977): A thermal model of the earth, *Phys. Earth Planet. Inter.*, **15**, 341–348.

Materials Science of the Earth's Interior, edited by I. Sunagawa, pp. 395–414.

SHOCK-INDUCED PHASE TRANSITION IN OXIDES AND SILICATES

Yasuhiko SYONO

The Research Institute for Iron, Steel and Other Metals, Tohoku University, Katahira, Sendai, Japan

1. *Introduction*

Generation of ultrahigh pressures by shock wave is characterized by the inertial compression of the material behind a shockfront travelling with a supersonic speed. Let us consider the case where a plane shock wave propagates in solids with a shock velocity, U_1, thereby accelerating the material to a particle velocity, u_1 (Fig. 1). Although strain is of a uniaxial fashion, the material behind the shockfront is compressed hydrostatically, due to the applied stress much greater than the yield strength. Based on the conservation relations of mass, momentum and energy across the shockfront, the compressed state is to be computed by the following equations,

$$\rho_0(U_1 - u_0) = \rho_1(U_1 - u_1) \tag{1}$$

$$P_1 - P_0 = \rho_0(U_1 - u_0)(u_1 - u_0) \tag{2}$$

$$E_1 - E_0 = \frac{1}{2}(V_0 - V_1)(P_0 + P_1) \tag{3}$$

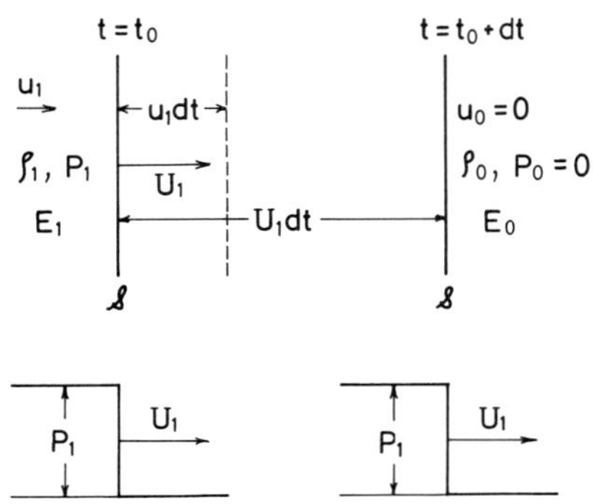

FIG. 1. Propagation of plane shock wave.

where subscripts 0 and 1 indicate the initial and compressed state respectively. The pressure-volume relation can be determined from a combination of Eqs. (1) and (2) using measured shock-velocity, U_1, and particle velocity, u_1, and is called Hugoniot. The temperature of the shocked state will be estimated with the aid of Eq. (3), and shock compression data are to be reduced to the isothermal case (WALSH and CHRISTIAN, 1957; MCQUEEN *et al.*, 1970). When a shock wave induces a phase transition, a double shock wave propagates in the material, as illustrated in Fig. 2. Measurements of shock parameters of each wave lead to the determination of both the phase transition state and the final deformational state with a single experiment. Details of the measurement technique of shock parameters and the reduction method of shock data will be described in (GOTO and SYONO, 1982).

Shock compression experiments have been extensively carried out for various materials (VAN THIEL *et al.*, 1967; MARSH, 1980), and good agreement between shock and static compression data has been established, in particular for various metals and alkali halides, up to a few tens GPa, warranting the validity of shock data for the equation-of-state (EOS) study at very high pressures. As the static high pressure apparata to be used above 10 GPa generally lack a reliable means for pressure calibration, shock data have provided pressure standards for static experiments with the aid of EOS considerations. Furthermore, frequently observed phase transition phenomena in these materials under shock loading, have also been found to compare favorably with the observations at static high pressures, again lending support for the validity of shock data (JONES and GRAHAM, 1971). It is noteworthy that α-ε transition in iron at 13 GPa was first found in shock experiments (BANCROFT *et al.*, 1956), and later confirmed by static experiments (TAKAHASHI and BASSETT, 1964).

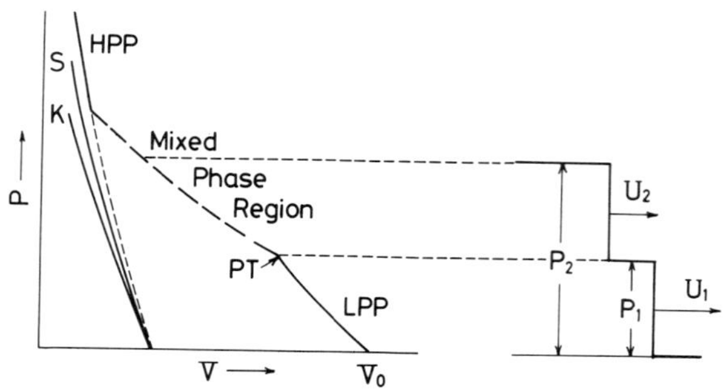

FIG. 2. Generation of double shock wave due to the phase transition. Subscripts 1 and 2 represent the phase transition state and final deformational state respectively. S: Isentrope. K: Isotherm.

As the shock compression technique had been almost a unique means for realizing the pressures corresponding to the lower mantle and the core of the earth, a considerable part of shock investigations was motivated with the elucidation of the internal structure and composition of the earth (AHRENS, 1980). Success in the analysis of shock data of metals and alkali halides encouraged the application of similar EOS study to earth materials such as oxides and silicates. Particularly, shock-induced phase transitions frequently observed in these substances (DUVALL and GRAHAM, 1977) have stimulated the interest of the earth scientists, and pressure-volume relations of the high pressure phase and its possible structure have been extensively argued on the basis of EOS study (AHRENS *et al.*, 1969). However, recent elaborate investigations have revealed that the shock response of these brittle materials differs remarkably from that of more ductile metals, and the validity of simplified EOS discussion based on the assumption of equilibrium thermodynamic state should be reconsidered (GRAHAM, 1974; GRADY, 1977).

Characteristic features of shock compression of these brittle materials are first observed in a very large value of Hugoniot elastic limit (HEL), where dynamic yielding takes place. Above HEL, shock compression curve of these brittle materials, particularly remarkable in the case of single crystals, was found to collapse to hydrostat, indicating a substantial loss of shear strength, in contrast to the case of metals which shows a relatively small value of HEL and an almost complete retaining of shear strength (FOWLES, 1961). Disappearance of shear components above HEL was directly confirmed by the sound velocity measurements in the shock compressed state (GRADY *et al.*, 1975). Measured sound velocity was substantially lower than the longitudinal sound velocity, and rather close to the bulk sound velocity. This strongly suggested that a partially-melted zone is formed in the shock compressed state above HEL, although the average temperature estimated from the conservation relations is far below the melting temperature. Evidence for such partially-melted textures was first inferred in shocked quartz monocrystals (ANAN'IN *et al.*, 1974). Recovered materials shocked above HEL were found to consist of finely divided crystal grains of 10–100 μm, which were surrounded by glassy layers. These textures were considered to be evidence for a remarkable temperature distribution in the shocked materials.

These observations are most comprehensively explained by a heterogeneous yielding concept, in which stress concentrates along the shear zone, which is often close to the direction of maximum resolved stress, as schematically shown in Fig. 3. These sheared zones are considered to be brought to high temperatures, often exceeding the melting temperature, while the region between the sheared zone should be only isentropically compressed and its temperature remains low (Fig. 3). Such temperature distribution will not be equilibrated within a time scale of microsecond duration of shock pulse because of the low thermal conductivity of these materials.

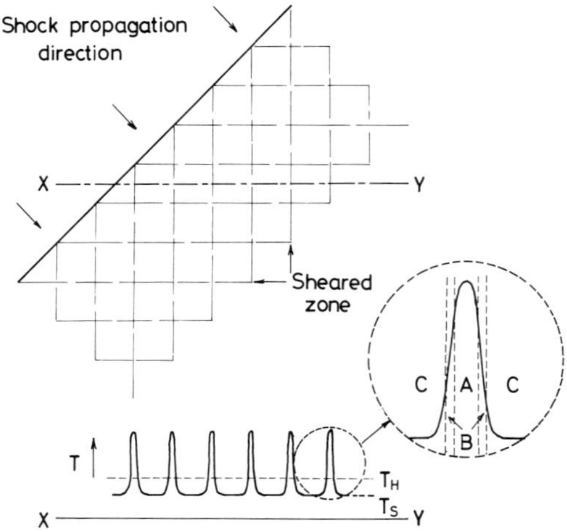

FIG. 3. Schematic illustration of heterogeneous yielding in the shock-loaded brittle substances with low thermal conductivity. Temperature profile along the line X-Y is schematically shown in the lower part of the figure, where T_H is average Hugoniot temperature calculated from conservation relations and T_S is the temperature expected for isentropic compression. The enlarged section in the circle demonstrates how the phase transition proceeds under shock compression in accordance with temperature increase. A: High temperature shear zone where melting may occur. B: Transition zone where solid-state reconstructive phase transition is possible. C: Low temperature region in which materials are compressed only isentropically and remain relatively undamaged.

When the high pressure phase transition of the reconstructive type is involved, reactions will preferentially occur along the sheared zone where temperature is high enough to promote the transition, resulting in extraordinarily heterogeneous textures consisting of high pressure reaction products and a relatively undamaged portion in the shock recovered materials. Dissociation of forsterite recently found in shock recovery experiments (SYONO *et al.*, 1981a) is presumably a typical example, as will be discussed later. Unusually wide mixed phase regions often encountered in the shock compression curve of various silicates are also reasonably understood to be due to the high-pressure disequilibrium condition realized with the shock-induced phase transitions in brittle substances with low thermal conductivity.

This review is intended to provide a clearer understanding of shock-induced phase transitions in oxides and silicates, which have a significant meaning for the application of shock data to the problems of the internal structure of the earth. Results of measurements of the shock compression curve and the observation of shock residual effects, which are complementary approaches with each other, are compared for several oxides and silicates, mostly based on new results obtained in the author's laboratory.

2. *Measurements of Shock-Induced Phase Transition*

Shock compression experiments so far have revealed that most oxides and silicates, with a notable exception for MgO and Al_2O_3, undergo shock-induced phase transitions in the pressure range up to about 200 GPa (AHRENS *et al.*, 1969; AHRENS, 1980). The phase transition induced by a shock wave can be classified into two major categories, reconstructive type and non-reconstructive type, depending upon whether it requires atomic diffusion or not. The phase transition of the non-reconstructive type may be completed within a microsecond time scale of shock compression experiments. Martensitic or electronic transition belongs to this category. The martensitic case is apparently demonstrated in the B1 (NaCl)-B2 (CsCl) transition recently observed in CaO (JEANLOZ *et al.*, 1979). As shown in Fig. 4, shock

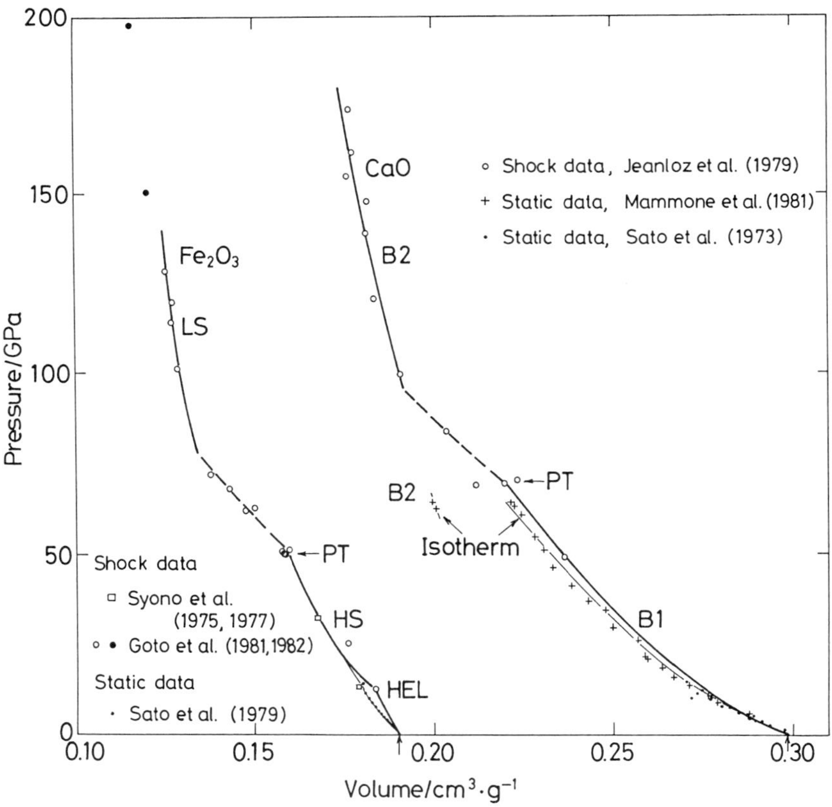

FIG. 4. Shock compression curve of CaO and α-Fe_2O_3, which show non-reconstructive type phase transitions. Shock data shown by full circles were obtained by the shock reflection technique. Arrows indicate zero-pressure volume. B1: NaCl structure. B2: CsCl structure. LS: Low-spin state. HS: High spin state. PT: Onset of phase transition.

compression data clearly exhibited a discontinuous volume decrease at about 70 GPa, indicating a phase transition in which the low pressure phase with the B1 structure is converted to the high pressure phase. The structure of the high pressure phase was confirmed by the high-pressure X-ray experiments using a diamond anvil cell to be the B2 structure. Both the dynamic and static compression data agree fairly well both in the B1 and B2 phases, if the thermal pressure due to temperature increase in the case of dynamic compression is taken into considerations. The martensitic nature of the B1-B2 transition has been discussed by many investigators (FRAZER and KENNEDY, 1974; WATANABE *et al.*, 1977). The fact that the B1-B2 transition in NaCl shocked parallel to [111] occurs at a lower pressure by 4 GPa than that shocked parallel to [100] (FRITZ *et al.*, 1971) further lends support to the above interpretation.

The other example of instantaneously-driven phase transition under shock loading is an electronic one. Phase transitions observed in shock compression of Fe_3O_4 and α-Fe_2O_3 (MCQUEEN and MARSH, 1966; SYONO *et al.*, 1975, 1977a) have been suggested to be due to the change in the electronic configuration in Fe^{3+} ions from a high-spin state to a low-spin state (SYONO *et al.*, 1971; CHENAVAS *et al.*, 1971; DAVIES and GAFFNEY, 1973). Mössbauer spectroscopy in diamond cell experiments revealed that these substances were transformed into a non-magnetic state at about 50 GPa, where shock-induced phase transitions were observed (MAO *et al.*, 1977). A recent high-pressure X-ray study by YAGI and AKIMOTO (1982) also supported the interpretation based on the high-spin low-spin transition.

Recently, OHNISHI (1978) developed an elaborate theory of pressure-induced high-spin low-spin transition in transition metal oxides, in which formulation to determine the transition point in terms of crystal field splitting, Δ, and spin-pairing energy, Π, is given. A precise redetermination of a shock compression curve of hematite, using high purity natural single crystals, was carried out (GOTO *et al.*, 1981; 1982), as shown in Fig. 4. The phase transition pressure was determined to be 50.5 GPa, which is consistent with the static high pressure experiments. The zero-pressure volume of the high pressure phase was estimated to be 0.161 $cm^3 \cdot g^{-1}$ from the extrapolation of the shock compression curve above 100 GPa. The volume change accompanied by the phase transition was estimated to be 15.2%, which is in good agreement with empirically predicted value of 13%, based on the density-cation radius systematics in the corundum structure. Ohnishi's theory was extended to the general case where both the crystal-field splitting and the spin-pairing energy depend on the volume of the crystal (GOTO *et al.*, 1982). The phase transition pressure and the pressure-volume relation of the low-spin state calculated from the theory can be well fitted with the observed transition pressure and compression curve of the high pressure phase, if $\Delta \propto r^{-3.0}$ and $\Pi \propto r^{1.3}$ are assumed, where r is the interatomic distance.

As is shown in the above examples, the agreement between the dynamic and static compression experiments has been satisfactory in the case of non-reconstructive type phase transition. However, if the phase transition is of a reconstructive nature in which time sufficient for atomic diffusion is required to complete the transition, the situation is no longer simple. Special features for the reconstructive type phase transition observed under shock loading can be summarized as follows: a substantial amount of pressure overdriving is necessary for the onset of the phase transition and the wide mixed phase region is observed before entering the high-pressure phase region.

Most comprehensive investigations of shock-induced phase transition of reconstructive type were made for quartz (WACKERLE, 1962; FOWLES, 1967). Phase transition in quartz started at about 13 GPa, slightly higher than the equilibrium transition pressure to the stishovite phase at about 9 GPa, although with no indication of the coesite phase. A wide mixed-phase region persisted up to around 50 GPa. The observed Hugoniot of the high-pressure phase was successfully explained to be due to stishovite. Extrapolated zero-pressure density coincides very well with the measured density of stishovite. Detailed EOS discussion of the high pressure phase based on the elastic parameters of stishovite, has been proved very successful (MCQUEEN *et al.*, 1963; GRAHAM, 1973).

The phase transition observed in complex silicates such as olivine and pyroxene is essentially of the reconstructive type. As olivine is the most important constituent in the earth's mantle, shock compression studies were extensively carried out on polycrystalline olivine and olivine rocks (TRUNIN *et al.*, 1965; MCQUEEN *et al.*, 1967; MCQUEEN, 1968; AHRENS *et al.*, 1971; SIMAKOV *et al.*, 1974; JEANLOZ and AHRENS, 1977). Indication of the phase transition was noted above about 40 GPa, which is substantially higher than the equilibrium phase transition pressure observed in static high pressure experiments, and the nature of the high pressure phase has been disputed (AHRENS *et al.*, 1969; KALININ *et al.*, 1972; DAVIES and GAFFNEY, 1973).

Recently, a more comprehensive study of shock compression measurements of single-crystal forsterite (Mg_2SiO_4) was performed up to 170 GPa (JACKSON and AHRENS, 1979; SYONO *et al.*, 1981b). Results are summarized in Fig. 5. Extraordinary high values of HEL up to 12 GPa were observed, and they were found to depend on the shock direction with respect to crystallographic orientations, reflecting large elastic anisotropies in the olivine structure. However, no anisotropy was observed in the hydrostatic regime of the low pressure olivine phase. The shock compression curve was found to be consistent with the high pressure X-ray study by OLINGER (1977) and ultrasonic measurements (GRAHAM and BARSCH, 1969; KUMAZAWA and ANDERSON, 1969). The onset of the phase transition was clearly demonstrated at about 50 GPa, which is remarkably higher than those of known high pressure transformations in static experiments, e. g., below 20 GPa for olivine-modified spinel or -spinel transition (AKIMOTO *et al.*, 1976) and below 30 GPa

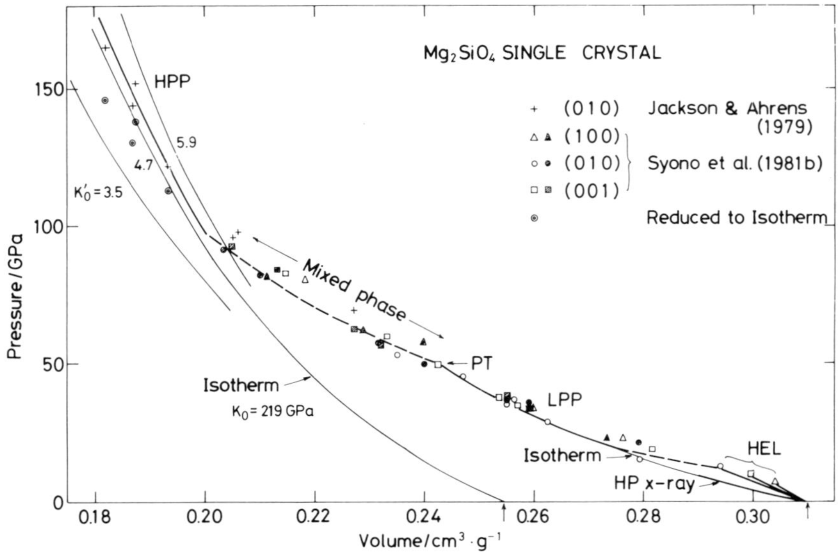

FIG. 5. Shock compression curve of single-crystal forsterite (Mg_2SiO_4). HEL: Hugoniot elastic limit. LPP: Low pressure phase with the olivine structure. HPP: High pressure phase presumably consisting of $MgSiO_3$ perovskite plus MgO. PT: Onset of phase transition. Arrows indicate zero-pressure volume. High pressure X-ray data are after OLINGER (1977).

for post-spinel dissociation to ilmenite or perovskite type $MgSiO_3$ plus MgO (ITO *et al.*, 1982), and also slightly higher than that observed in shock compression of polycrystalline olivines. The mixed phase region persisted up to about 100 GPa, where the shocked state merged into the high pressure regime. JACKSON and AHRENS (1979) showed that Hugoniot of the high pressure phase was possibly due to the dissociation product of $MgSiO_3$ perovskite plus MgO, using the measured zero pressure density and the estimated bulk modulus of $MgSiO_3$ perovskite (ITO and MATSUI, 1978; YAGI *et al.*, 1978; LIEBERMANN *et al.*, 1977).

3. *Observation of Shock Residual Effects*

Residual effects in minerals caused by the hypervelocity impact of a body such as meteorite has drawn much attention of earth scientists, and are termed as shock metamorphism. Both naturally and experimentally shocked minerals have been extensively investigated. Various fractures, deformation lamellae, planar features, mosaicism, formation of glass or kink bands have been identified to be specific features of the shock deformation process (CHAO, 1967; STÖFFLER, 1972, 1974).

Observation of shock residual effects should provide direct informations on the phase transition, if any. Very few examples of the formation of high-pressure phases, however, have been reported in shock experiments, in contrast to many successful results in new material synthesis at static high pressures. This can be understood, at least partially, as that shock duration of the pressure pulse is not sufficiently long for the reconstructive reactions. On the other hand, the high pressure form of the non-reconstructive type phase transition may not be retained after the passage of shock waves because of its rapid retrogressive process. Scarce number of experimental confirmations of shock-induced phase transformations is also partly due to insufficient efforts, because the yield of high pressure phases under shock compression is generally very low. Modern diagnostic means such as observation with a transmission electron microscope (TEM) or various spectroscopic analyses will be useful for identification of a small amount of the high pressure phases in the shock recovered products.

In this section, several successful cases in obtaining evidence for phase transformations by the observation of shock residual effects will be selectively demonstrated for quartz, feldspar, olivine and ilmenite.

3.1 Quartz

Shock residual effects in quartz have been most extensively investigated for both naturally and experimentally shocked specimens. Stishovite, a high pressure form of SiO_2 with the rutile structure, which was first synthesized at static high-pressure experiments (STISHOV and POPOVA, 1961), was later found in the shocked quartz rocks in meteorite crater in Arizona (CHAO *et al.*, 1962). Subsequently, both stishovite and coesite were observed in many crater sites, and taken for indicators of strong shock events in nature. Experimental confirmation of stishovite in shocked SiO_2 was first made by DE CARLI and MILTON (1965), and the synthesis field was found to coincide with the phase equilibrium study in static high-pressure experiments (KLEEMAN and AHRENS, 1973). Formation of coesite under dynamic experiments was rather rare (DERIBAS *et al.*, 1966), although it shows much higher yield in shocked rocks of natural impact craters such as Ries (STÖFFLER, 1971). The yield of these high-pressure forms in shocked SiO_2 was generally as low as 0.1 percent.

Recently, GERMAN *et al.* (1974) reported shock synthesis of a new high-pressure phase of SiO_2 above 70 GPa, which was interpreted to be of the α-PbO_2 type structure, denser than stishovite by about 4 percent. This would be likely because an α-PbO_2 type TiO_2 was formed under shock compression (LINDE and DE CARLI, 1969), although reconfirmation of German's experiments is needed. Indication of post-stishovite transformations in SiO_2 were obtained from flash temperature measurements in shock compressed SiO_2 in a similar pressure region (MCQUEEN *et al.*, 1979; LYZENGA and AHRENS, 1980).

Apart from these crystalline high-pressure phases, more general features in shocked quartz, as well as other framework silicates, are the formation of diaplectic glass (DE CARLI and JAMIESON, 1959; ENGERHARDT *et al.*, 1967; CHAO, 1967) which has been generally accepted to be formed by a solid-state disordering process, because it shows no signs of melting such as flow structures and retains morphology of original crystals. The diaplectic SiO_2 glass often shows higher density, or higher refractive index, than that of fused silica (ENGERHARDT *et al.*, 1967; STÖFFLER and HORNEMANN, 1972; MASHIMO *et al.*, 1980). Similar densification of SiO_2 glass under shock loading was also reported (ARNDT *et al.*, 1971). The densification in the diaplectic glass was explained to be due to local formation of six-coordinated silicon and called the short-range ordered phase. The dense glass was observed only for the pressure range of 10–30 GPa (GIBBONS and AHRENS, 1973). Probably high residual temperatures accompanied by more intense shock loading may easily erase the high density glass structure in the shock released state.

As the glass phase was detected in the shocked products recovered from above the phase transition pressure, it has been generally thought to be a reversion product of unquenchable high pressure phases. However, alternatively it could be of the melted high density phase where freezing took place prior to pressure release. A short duration of shock loading would not be sufficient for smearing out the morphology of the original crystals, even by the liquid diffusion process. Experimental evidence such as glassy layers around fragmented grains (ANAN'IN *et al.*, 1974) and lamellae textures of glass and crystal (MÜLLER, 1969) suggests that the melting due to heterogeneous yielding could be a more direct explanation for the formation mechanism of diaplectic glass in shocked quartz.

3.2 Feldspar

Shock residual effects in feldspar group minerals have also been investigated in detail and found to be very remarkable, as in quartz. Shocked glass having plagioclase composition was first found in meteorite and supposed to be caused by impact and named maskelynite (BUNCH *et al.*, 1967). Artificial verification of formation of feldspar glass under shock loading was first reported by MILTON and DE CARLI (1963). Remarkable densification of shocked feldspar glass, compared with a fused one, was also found by means of refractive index measurements. Contrary to the case of shocked quartz, no indication of high-pressure phases was observed in shocked feldspar. Experimental results so far were critically reviewed by STÖFFLER (1972, 1974).

Controlled shock loading experiments on single crystal anorthites, a calcium end member of plagioclase, revealed interesting intergrowth textures of glass and crystals with TEM observation (KITAMURA *et al.*, 1977). Intergrowth textures of shocked anorthite observed with TEM is reproduced in Fig. 6. It is noteworthy to mention that the principal direction of lamellae textures was $(\bar{1}\bar{2}1)$, which makes approximately an angle of 45° with the shock propagation direction [010]. The lamellae texture was considered to be caused by activation of particular slip system close to the direction of maximum resolved stress. The subsidiary lamellae were observed as parallel to (010), which is a remarkable cleavage plane of plagioclase.

The whole particle was proved to hold an identical crystallographic orientation, as shown in the diffraction pattern, in spite of the formation of submicron-size lamellae. Furthermore, the diffuseness of the *b*-reflection originated from the ordering of aluminum and silicon ions in the *p*-anorthite (NAKAJIMA *et al.*, 1976), showed no substantial change upon shock loading, indicating that the atomic arrangements in the crystalline part of the lamellae were little affected by shock loading. It is surprising that such an undamaged crystalline part coexists with a completely disordered glass part in the TEM scale. These observations support the heterogeneous yielding of the shock deformation process in brittle substances.

Close correspondence was noticed between these TEM observations of shock residual effects (SYONO *et al.*, 1977b) and the results of Hugoniot measurements (AHRENS *et al.*, 1969b). The intergrowth textures were only found in the mixed phase region ranging from 15 to 40 GPa. A complete glass formation was observed above about 40 GPa, corresponding to the high pressure regime, while no indication of glass was detected below about 15 GPa

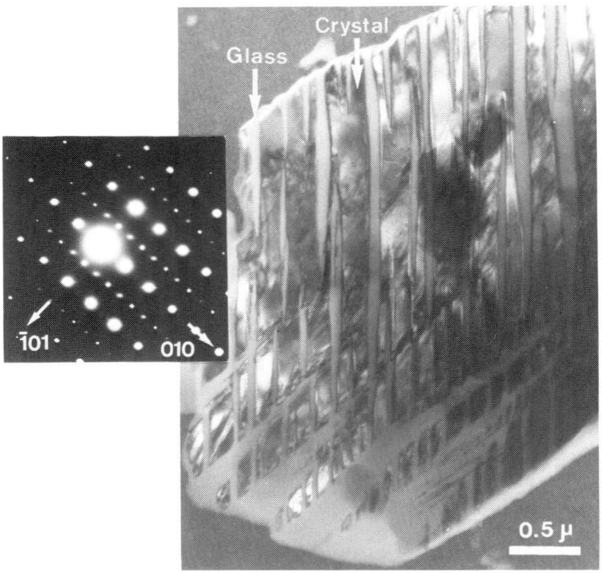

FIG. 6. TEM photograph of natural p-anorthite shocked parallel to [010] to 30 GPa, indicating the intergrowth textures of glass and crystals, and its electron diffraction pattern consisting of halo and diffraction spots due to glass and crystals, respectively. Weaker spots are *b*-, *c*- and *d*-reflections due to superstructure.

of the low-pressure phase.

As in the case of shocked quartz, for the formation mechanism of feldspar glass under shock loading, the conventional interpretation that the shocked glass is the reversion product of unquenchable high pressure phases (AHRENS *et al.*, 1969b) may be applicable. However, an alternative explanation that the glass is directly formed by melting along the sheared zone introduced by shock loading may be equally possible. As the structure of melt at high pressures would be similar to that of the virtual high-pressure phases, the discrepancy between these two explanations should be very subtle.

3.3 Olivine

Olivine crystals have been known to be rather little affected by shock effects. Only planar features and mosaicism were notable for the shock loading below about 50 GPa (CARTER *et al.*, 1968; MÜLLER and HORNEMANN. 1969; SNEE and AHRENS, 1975; REIMOLD and STÖFFLER, 1978; JEANLOZ, 1980). The anisotropic behavior of olivine with respect to the HEL was reflected in the orientation and frequency of planar deformation structures in shocked olivine. Some minor indications of shock recrystallization and formation of diaplectic glass became notable above about 50 GPa (JEANLOZ *et al.*, 1977), but was not sufficient to explain the shock-induced phase transition in olivine, as was demonstrated in Hugoniot measurements. This inconsistency had been a serious difficulty in the application of shock data to geophysical problems.

An evidence for dissociation reaction in single crystal forsterite was first observed in the shock recovered products for the pressure range of 78–92 GPa (SYONO *et al.*, 1981a). TEM observation of shocked crystals revealed very heterogeneous textures, i.e., coexistence of relatively undamaged crystals, completely recrystallized polycrystalline forsterite, dissociation products of microcrystalline MgO plus glass, and further differentiated glass, as demonstrated in Fig. 7. The composition of the glass was determined to be close to $MgSiO_3$, indicating that Mg_2SiO_4 dissociates into $MgO + MgSiO_3$ under shock loading. Most probable explanation for this observation is that forsterite incongruently melts into MgO plus liquid with $MgSiO_3$ composition during shock compression. The possibility of incongruent melting of forsterite was suggested by molecular dynamics calculation of forsterite and enstatite melts under high pressures (MATSUI *et al.*, 1982). An alternative interpretation is due to a reconstructive type solid-state process of the breaking down of forsterite to MgO plus $MgSiO_3$ perovskite (JACKSON and AHRENS, 1979) which may revert to a glassy state upon shock unloading.

These observations are completely compatible with the current heterogeneous yielding model for brittle substances with low thermal conductivity (GRADY, 1977). Evidence that dissociation reaction only takes place along the sheared zones where temperature is expected to be high, has been obtained for a TEM observation of the shocked specimens prepared by the ion thinning

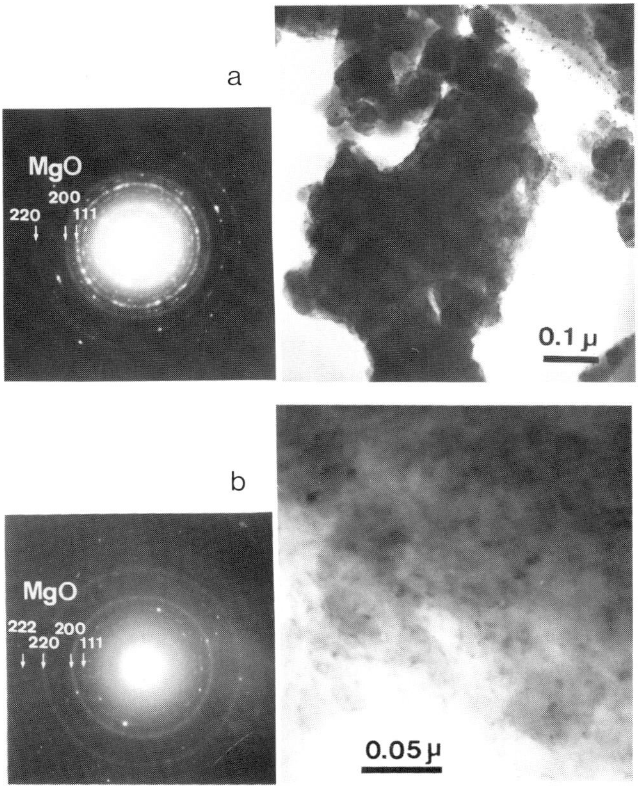

FIG. 7. TEM photographs of single-crystal forsterite shocked to 92 GPa. (a) Observation with 200 kV TEM shows very heterogeneous textures consisting of crystal particles, recrystallized polycrystalline forsterite and dissociation products. (b) Enlarged portion of (a) observed with 1,000 kV TEM. The electron diffraction pattern consists of broad-ring patterns due to microcrystalline MgO and the halo due to glass, indicating the dissociation of Mg_2SiO_4 under shock compression. The size of MgO precipitates is estimated to be about several nanometers.

technique (Nobugai *et al.*, unpublished data). Considering the relative amount of highly disordered parts in the shocked crystal in these pressure range, the extraordinarily wide mixed phase region observed in the shock compression curve of forsterite (SYONO *et al.*, 1981b) is reasonably understood.

The inconsistency that no evidence for the phase transition was found in the shock recovered products up to 75 GPa (JEANLOZ, 1980), well above the phase transition pressure inferred from shock compression measurements, can be removed, if one considers that the temperature achieved by multiple shock reflection in the shock recovery experiments should be significantly lower than that attained by a single-shock process in the shock compression measurements, even if the final pressure is apparently the same. The kinetics of the phase transition under shock loading is principally governed by temperature rather than by pressure itself (SYONO and GOTO, 1982).

It is noteworthy to mention that a similar dissociation reaction under shock loading was also observed in andalusite and sillimanite (SCHNEIDER and HORNEMANN, 1977, 1981). Since the fraction of the sheared region will be small around the phase transition pressure, only a careful examination of shock recovered products will lead to confirmation of the reaction due to the phase transition.

3.4 Ilmenite

Ilmenite has been known to be a sensitive indicator of shock events in nature, as was revealed in detailed studies of lunar ilmenites (MINKIN and CHAO, 1971; SCLAR *et al.*, 1973). Meanwhile, shock compression measurements of ilmenite revealed that anomalous compression suggesting a phase transition occurred above about 30 GPa together with considerable compression anisotropy (KING and AHRENS, 1976).

Controlled shock-loading experiments have been carried out for single crystal ilmenite up to peak pressure of 80 GPa, and examined by single-crystal X-ray study as well as Mössbauer spectroscopy (SYONO *et al.*, 1981c). Shock residual effects were found to be controlled by the anisotropic structure of ilmenite. Remarkable deformations were observed at as low as 30 GPa in the crystal shocked parallel to the *c*-axis, whereas little effects were seen to 55 GPa when the crystal was shocked parallel to the *c*-plane. A typical example of X-ray precession photograph of ilmenite shocked to 33 GPa along the *c*-axis is shown in Fig. 8. The greatest deformation was introduced in the planes containing the *c*-axis, as represented by pronounced concentric spreadings and anomalous polysynthetic twinning (Fig. 8a), while less-remarkable effects were observed in the plane perpendicular to the *c*-axis (Fig. 8b). These results can be interpreted as that the shock-loading process in ilmenite occur in an anisotropic fashion. The introduction of a sheared zone in ilmenite by shock loading may lead to the formation fragmented domains consisting of *c*-platelets, which are susceptible to a relatively free rotational motion, so as to give misorientation of each domain as well as specific twinning (SYONO *et al.*, 1981c).

Mössbauer measurements of shock loaded ilmenite showed an increase in average quadrupole splitting as well as line profile broadening in the paramagnetic spectrum. The results can be interpreted to be due to the increase in the fraction of the sheared region, which may have a short-range configuration similar to the disordered ilmenite (SYONO *et al.*, 1981d) under shock loading. All these observations are again consistent with the heterogeneous yielding picture.

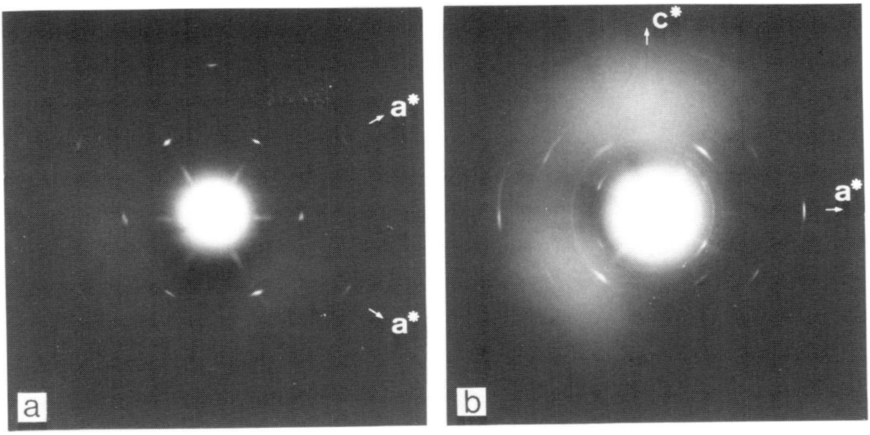

FIG. 8. X-ray precession photographs of ilmenite shocked along the *c*-axis to 33 GPa. X-rays are parallel to the *c*-axis for (a) and perpendicular to the *c*-axis for (b).

4. *Concluding Remarks*

Although shock wave experiments have demonstrated that phase transformations are possible on a sub-micosecond time scale, shocked states realized in oxides and silicates should be considered to be highly disequilibrium. Reconstructive-type phase transformations under shock compression preferentially occur along the shear zone where temperature is high enough to promote the transition, resulting in very heterogeneous textures in the shock recovered materials. These specific features can best be explained on the basis of a heterogeneous yielding model developed for brittle, low thermal conductivity materials, in which dynamic yielding occurs along the stress-concentrated sheared zone. High onset pressure of the phase transition and wide mixed phase region can also be reasonably understood with this model. Under these circumstances, simplified EOS discussion of the shock-induced high pressure phases, based on the assumption of the equilibrium thermodynamic state, should be carefully examined. Elaborate investigations of shock-induced phase transitions, from both precise *in-situ* measurements of shock compressed state and detailed observation of shock residual effects in recovered products, are required so that shock data could provide useful information for the discussion of the earth's internal structure.

REFERENCES

AHRENS, T. J. (1980): Dynamic compression of earth materials, *Science*, **207**, 1035–1041.
AHRENS, T. J., ANDERSON. D. L. and RINGWOOD, A. E. (1969a): Equations of state and crystal

structures of high-pressure phases of shocked silicates and oxides, *Rev. Geophys.*, **7**, 667–707.

AHRENS, T. J., PETERSEN, C. F. and ROSENBERG, J. T. (1969b): Shock compression of feldspar, *J. Geophys. Res.*, **74**, 2727–2746.

AHRENS, T. J., LOWER, J. H. and LAGUS, P. L. (1971): Equation of state of forsterite, *J. Geophys. Res.*, **76**, 518–528.

AKIMOTO, S., MATSUI, Y. and SYONO, Y. (1976): High-pressure crystal chemistry of orthosilicates and formation of the mantle transition zone, in *The Physics and Chemistry of Minerals and Rocks*, Ed. R. G. J. Strens, Wiley Intersciene London, pp. 327–363.

ANAN'IN, A. V., BREUSOV, O. N., DREMIN, A. N., PERSHIN, S. V. and TATSII, V. F. (1974): The effect of shock waves on silicon dioxide. I. Quartz, *Combust. Expl. and Shock Waves*, **10**, 372–379.

ARNDT, J., HORNEMANN, U. and MÜLLER, W. F. (1971): Shock-wave-densification of silica glass, *Phys. Chem. Glasses*, **12**, 1–7.

BANCROFT, D., PETERSON, E. L. and MINSHALL, S. (1956): Polymorphism of iron at high pressure, *J. Appl. Phys.*, **27**, 291–298.

BUNCH, T. E., DENCE, M. R. and COHEN, A. J. (1967): Natural terrestrial maskelynite, *Am. Mineral.*, **52**, 244–253.

CARTER, N. L., RALEIGH, C. B. and DE CARLI, P. S. (1968): Deformation of olivine in stony meteorites, *J. Geophys. Res.*, **73**, 5439–5461.

CHAO, E. C. T. (1967): Shock effects in certain rock-forming minerals, *Science*, **156**, 192–202.

CHAO, E. C. T., JAMES, O. B., MINKIN, J. A., BOREMAN, J. A., JACKSON, E. D. and RALEIGH, C. B. (1970): Petrology of unshocked crystalline rocks and evidence of impact metamorphism in Apollo 11 returned lunar sample, *Proc. Apollo 11 Lunar Sci. Conf., Geochim. Cosmochim. Acta*, Suppl 1, Vol 1, 287–314.

CHENAVAS, J., JOUBERT, J. C. and MAREZIO, M. (1971): Low-spin→high-spin state transition in high pressure cobalt sesquioxide, *Solid State Commun.*, **9**, 1057–1060.

DAVIES, G. F. and GAFFNEY, E. S. (1973): Identification of high-pressure phase of rocks and minerals from Hugoniot data, *Geophys. J. R. Astr. Soc.*, **33**, 165–183.

DE CARLI, P. S. and JAMIESON, J. C. (1959): Formation of an amorphous form of quartz under shock conditions, *J. Chem. Phys.*, **31**, 1675–1676.

DE CARLI, P. S. and MILTON, D. J. (1965): Stishovite: Synthesis by shock wave, *Science*, **147**, 144–145.

DERIBAS, A. A., DOBRETSOV, N. L., KUDINOV, V. M. and ZYNZIN, N. I. (1966): Shock compression of SiO_2 powders, *Dokl. Akad. Nauk SSSR*, **168**, 127–130.

DUVALL, G. E. and GRAHAM, R. A. (1977): Phase transitions under shock wave loading, *Rev. Mod. Phys.*, **49**, 523–579.

ENGERHARDT, W. V., ARNDT, J., MÜLLER, W. F., STÖFFLER, D., JEZIORKOWSKI, H. and GUBSER, R. A. (1967): Diaplektische Gläser in der Breccien des Ries von Nördlingen als Anzeichen für Sto*β*wellenmetamorphose, *Contrib. Mineral. Petrol.*, **15**, 91–100.

FOWLES, G. R. (1961): Shock wave compression of hardened and annealed 2024 aluminum, *J. Appl. Phys.*, **32**, 1475–1487.

FOWLES, G. R. (1967): Dynamic compression of quartz, *J. Geophys. Res.*, **72**, 5729–5742.

FRÄZER, W. L. and KENNEDY, S. W. (1974): The crystal-structure transformation NaCl-type→CsCl-type; Analysis by martensitic theory, *Acta Cryst. A.*, **30**, 13–22.

FRITZ, J. N., MARSH, S. P., CARTER, W. J. and MCQUEEN, R. G. (1971): The Hugoniot equation of state of sodium chloride in the sodium chloride structure, in *Accurate Characterization of the High Pressure Environment*, Ed. E. C. Lloyd, U. S. National Bureau of Standards, Special Publ. 326, pp. 201–208.

GERMAN, V. N., ORLOVA, N. N., TARASOVA, L. A. and TRUNIN, R. F. (1974): Synthesis of the orthorhombic phase of silica in dynamic compression, *Izv. Acad. Sci. USSR, Earth Phys.*, No. 7, 431–434.

GIBBONS, R. V. and AHRENS, T. J. (1971): Shock metamorphism of silicate glasses, *J. Geophys. Res.*, **76**, 5489–5498.

GOTO, T., SYONO, Y., SATO, J. and NAKAGAWA, Y. (1981): High pressure phase transition observed in Fe_2O_3 under shock compression, in *Ferrites* (Proc. ICF 3), Eds. H. Watanabe, S. Iida, and M. Suginoto, Center for Academic Publication Japan, Tokyo, pp. 250–253.

GOTO, T. and SYONO, Y. (1982): Technical aspects of shock compression experiments using the gun method, this volume, pp. 605–619.

GOTO, T., SATO, J. and SYONO, Y. (1982): Shock-induced spin-pairing transition in Fe_2O_3 due to the pressure effect on the crystal-field, in *High-Pressure Research in Geophysics*, Eds. S, Akimoto. and M. H. Manghnani, Center for Academic Publication Japan/D. Reidel, Tokyo/Dordrecht, pp. 595–610.

GRADY, D. E. (1977): Processes occurring in shock wave compression of rocks and minerals, in *High-Pressure Research: Applications in Geophysics*, Eds. M. H. Manghnani and S. Akimoto, Academic Press, New York, pp. 389–438.

GRADY, D. E., MURRI, W. J. and DE CARLI, P. S. (1975): Hugoniot sound velocities and phase transformations in two silicates, *J. Geophys. Res.*, **80**, 4857–4861.

GRAHAM, E. K. (1973): On the compression of stishovite, *Geophys. J. R. Astr. Soc.*, **32**, 15–34.

GRAHAM, E. K. and BARSCH, G. R. (1969): Elastic constants of single-crystal forsterite as a function of temperature and pressure, *J. Geophys. Res.*, **74**, 5949–5960.

GRAHAM, R. A. (1974): Shock-wave compression of *x*-cut quartz as determined by electrical response measurements, *J. Phys. Chem. Solids*, **35**, 355–372.

ITO, E. and MATSUI, Y. (1978): Synthesis and crystal-chemical characterization of $MgSiO_3$ perovskite, *Earth Planet. Sci. Lett.*, **38**, 443–450.

ITO, E., MATSUI, Y. and YAMADA, H. (1982): Stability relations of silicate ilmenites and perovskites, in *High Pressure Research in Geophysics*, Eds. S. Akimoto and M. H. Manghnani, Center for Academic Publication Japan D. Reidel, Tokyo/Dordrecht, pp. 405–420.

JACKSON, I. and AHRENS, T. J. (1979): Shock wave compression of single-crystal forsterite, *J. Geophys. Res.*, **84**, 3039–3048.

JEANLOZ, R. (1980): Shock effects in olivine and implications for Hugoniot data, *J. Geophys. Res.*, **85**, 3163–3176.

JEANLOZ, R. and AHRENS, T. J. (1977): Pyroxenes and olivines: Structural implications of shock wave data for high pressure phases, in *High-Pressure Research: Application in Geophysics*, Eds. M. H. Nanghnani and S. Akimoto, Academic Press, New York, pp. 439–461.

JEANLOZ, R., AHRENS, T. J., LALLY, J. S., NORD, G. L., JR. CHRISTIE, J. M. and HAUER, A. H. (1977): Shock-produced olivine glass: First observation, *Science*, **197**, 457–459.

JEANLOZ, R., AHRENS, T. J., MAO, H. K. and BELL, P. M. (1979): B1–B2 transition in calcium oxide from shock-wave and diamond-cell experiments, *Science*, **206**, 829–830.

JONES, O. E. and GRAHAM, R. A. (1971): Shear strength effects on phase transition "pressure" determined from shock compression experiments, in *Accurate Characterization of the High Pressure Environment*, U. S. National Bureau of Standards, Special Publ. 326, pp. 229–242.

KALININ, V. A., PAN'KOV, V. L. and ZHARKOV, V. N. (1972): Equations of state of dunites and bronzitites undergoing polymorphic transformations under pressure, *Izv. Acad, Sci. USSR, Phys. Solid Earth*, No. 7, 472–482.

KING, D. A. and AHRENS, T. J. (1976): Shock compression of ilmenite, *J. Geophys. Res.*, **81**, 931–935.

KITAMURA, M., GOTO, T. and SYONO, Y. (1977): Intergrowth textures of diaplectic glass and crystal in shock-loaded p-anorthite, *Contrib. Mineral. Petrol.*, **61**, 299–304.

KLEEMAN, J. D. and AHRENS, T. J. (1973): Shock-induced transition of quartz to stishovite, *J. Geophys. Res.*, **78**, 5954–5960.

KUMAZAWA, M. and ANDERSON, O. L. (1969): Elastic moduli, pressure derivatives and temperature derivatives of single crystal olivine and single crystal forsterite, *J. Geophys. Res.*, **74**, 5961–5972.

LIEBERMANN, R. C., JONES, L. E. and RINGWOOD, A. E. (1977): Elasticity of aluminate, titanate, stannate and germanate compounds with perovskite structure, *Phys. Earth Planet. Inter.*, **14**, 165–178.

LINDE, R. K. and DECARLI, P. S. (1969): Polymorphic behavior of titania under dynamic loading, *J. Chem. Phys.*, **50**, 319–325.

LYZENGA, G. A. and AHRENS, T. J. (1980): Shock temperature measurements in Mg_2SiO_4 and SiO_2 at high pressures, *Geophys. Res. Lett.*, **7**, 141–144.

MAMMONE, J. F., MAO, H. K. and BELL, P. M. (1981): Equations of state of CaO under static pressure conditions, *Geophys. Res. Lett.*, **8**, 140–142.

MAO, H. K., VIRGO, D. and BELL, P. M. (1977): High-pressure ^{57}Fe Mössbauer data on the phase and magnetic transitions of magnesioferrite ($MgFe_2O_4$), magnetite (Fe_3O_4) and hematite (Fe_2O_3), *Ann. Rept. Geophys. Lab. Yearbook*, **76**, 522–525.

MARSH, S. P. (Ed) (1980): *LASL Shock Hugoniot Data*, Univ. of Calif. Press.

MASHIMO, T., NISHII, K., SŌMA, T. and SAWAOKA, A. (1980): Some physical properties of amorphous SiO_2 synthesized by shock compression of α-quartz, *Phys. Chem. Miner.*, **5**, 367–377.

MATSUI, Y., KAWAMURA, K. and SYONO, Y. (1982): Molecular dynamics calculation applied to silicate system: molten and vitreous $MgSiO_3$ and Mg_2SiO_4 under low and high pressures, in *High-Pressure Research in Geophysics*, Eds. S. Akimoto and M. H. Manghnani, Center for Academic Publication Japan/D. Reidel, Tokyo/Dordrecht, pp. 511–524.

MCQUEEN, R. G. (1968): The equation of state of mixtures, alloys and compounds, in *Seismic Coupling, Advanced Research Project Agency Meeting*, Ed. G. Simmons, Nat. Tech. Inform. Serv., Springfield, pp. 53–106.

MCQUEEN, R. G., FRITZ, J. N. and MARSH, S. P. (1963): On the equation of state of stishovite, *J. Geophys. Res.*, **68**, 2319–2322.

MCQUEEN, R. G. and MARSH, S. P. (1966): Unpublished data, *Handbook of Physical Constants*, Ed. Clark, S. P. Jr., Geological Society of America, New York, pp. 153–159.

MCQUEEN, R. G., MARSH, S. P. and FRITZ, J. N. (1967): Hugoniot equation of state of twelve rocks, *J. Geophys. Res.*, **72**, 4999–5036.

MCQUEEN, R. G., MARSH, S. P., TAYLOR, J. W., FRITZ, J. N. and CARTER, W. J. (1970): The equation of state of soilds from shock wave studies, in *High-Velocity Impact Phenomena*, Ed. R. Kinslow, Academic Press, New York, pp. 293–417.

MCQUEEN, R. G., FRITZ, J. N. and HOPSON, J. W. (1979): The temperature of shocked SiO_2, *EOS*, **60**, 952 (Abstract).

MINKIN, J. A. and CHAO, E. C. T. (1971): Single crystal X-ray investigation of deformation in terrestrial and lunar ilmenite, *Proc. 2nd Lunar Sci. Conf., Geochim. Cosmochim. Acta*, Suppl. 2, Vol. 1, 237–246.

MILTON, D. J. and DE CARLI, P. S. (1963): Maskelynite: Formation by explosive shock, *Science*, **140**, 670–671.

MÜLLER, W. F. (1969): Elektronenmikroskopischer Nachweis amorpher Bereiche in stoβwellen-beanspruchtem Quarz, *Naturwiss*, **56**, 279.

NAKAJIMA, Y., MORIMOTO, N. and KITAMURA, M. (1977): The superstructure of plagioclase feldspars. Electron microscopic study of anorthite and labradorite, *Phys. Chem. Miner.*, **1**, 213–226.

OHNISHI, S. (1978): A theory of the pressure-induced high-spin low-spin transition of transition metal oxides, *Phys. Earth Planet. Inter.*, **17**, 130–139.

OLINGER, B. (1977): Compression studies of forsterite (Mg_2SiO_4) and enstatite ($MgSiO_3$), in *High-Pressure Research: Application in Geophysics*, Eds. M. H. Manghnani and S. Akimoto, Academic Press, New York, pp. 325–334.

REIMOLD, W. U. and STÖFFLER, D. (1978): Experimental shock metamorphism of dunite, in

Proc. Lunar Planet Sci. Conf. 9th, pp. 2805–2824.

SATO, Y., IDA, Y. and AKIMOTO, S. (1973): Equation of state determined with the Bridgman-anvil high-pressure apparatus, *High Temp.-High Press.* **5**, 679–688.

SATO, Y. and AKIMOTO, S. (1979): Hydrostatic compression of four corundum-type compounds: α–Al_2O_3, V_2O_3, Cr_2O_3 and α–Fe_2O_3, *J. Appl. Phys.*, **50**, 5285–5291.

SCHNEIDER, H. and HORNEMANN, U. (1977): The disproportionation of andalusite (Al_2SiO_5) to Al_2O_3 and SiO_2 under shock compression, *Phys. Chem. Miner.*, **1**, 257–264.

SCHNEIDER, H. and HORNEMANN, U. (1981): Shock-induced transformation of silimanite powders, *J. Mater. Sci.*, **16**, 45–49.

SCLAR, C. B., BAUER, J. F., PICKART, S. J. and ALPERIN, H. A. (1973): Shock effects in experimentally shocked terrestrial ilmenite, lunar ilmenite of rock fragments in 1–10 mm fines (10085, 19), and lunar rock 60015, 127, *Proc. 4th Lunar Sci. Conf. Geochim. Cosmochim. Acta*, Suppl. 4, Vol. 1, 841–859.

SIMAKOV, G. V., PAVLOVSKII, M. N., KALASHNIKOV, N. G. and TRUNIN, R. F. (1975): Shock compressibility of twelve minerals, *Izv. Acad. Sci. USSR, Phys. Solid Earth*, No. 8, 488–492.

STISHOV, S. M. and POPOVA, S. V. (1961): A new dense modification of silica, *Geochemistry*, No. 10, 923–926.

STÖFFLER, D. (1971): Coesite and stishovite in shocked crystalline rocks, *J. Geophys, Res.*, **76**, 5474–5488.

STÖFFLER, D. (1972): Deformation and transformation of rock-forming minerals by natural and experimental shock process. I. Behavior of minerals under shock compression, *Fortschr. Miner.*, **49**, 50–113.

STÖFFLER, D. (1974): Deformation and transformation of rock-forming minerals by natural and experimental shock process. II. Physical properties of shocked minerals, *Fortschr. Miner.*, **51**, 256–289.

STÖFFLER, D. and HORNEMANN, U. (1972): Quartz and feldspar glasses produced by natural and experimental shock, *Meteoritics*, **7**, 371–394.

SYONO, Y., AKIMOTO, S. and ENDOH, Y. (1971): High pressure synthesis of ilmenite and perovskite type $MnVO_3$ and their magnetic properties, *J. Phys. Chem. Solids*, **32**, 243–249.

SYONO, Y., GOTO, T., NAKAI, J. and NAKAGAWA, Y. (1975): Shock compression study of transition metal oxides, in *Proc. 4th Intern'l Conf. on High Pressure*, Ed. J. Osugi, The Physico-Chemical Society of Japan, Kyoto, pp. 466–472.

SYONO, Y., GOTO, T. and NAKAGAWA, Y. (1977a): Phase-transition pressures of Fe_3O_4 and GaAs determined from shock-compression experiments, in *High Pressure Research: Application in Geophysics*, Eds. M. H. Manghnani and S. Akimoto, Academic Press, New York, pp. 463–476.

SYONO, Y., GOTO, T., NAKAGAWA, Y. and KITAMURA, M. (1977b): Formation of diaplectic glass in anorthite by shock-loading experiments, in *High Pressure Research: Application in Geophysics*, Eds. M. H. Manghnani and S. Akimoto, Academic Press, New York, pp. 477–489.

SYONO. Y., GOTO, T., TAKEI, H., TOKONAMI, M. and NOBUGAI, K. (1981a): Dissociation reaction in forsterite under shock compression, *Science*, **214**, 177–179.

SYONO, Y., GOTO, T., SATO, J. and TAKEI, H. (1981b): Shock compression measurements of single-crystal forsterite in the pressure range 15–93 GPa, *J. Geophys. Res.*, **86**, 6181–6186.

SYONO, Y., TAKEI, H., GOTO, T. and ITO, A. (1981c): Single crystal X-ray and Mössbauer study of shocked ilmenite to 80 GPa, *Phys. Chem. Miner.*, **7**, 82–87.

SYONO, Y., YAMAUCHI, H., ITO, A., SOMEYA, Y., ITO, E., MATSUI, Y., AKAOGI, M. and AKIMOTO, S. (1981d): Magnetic properties of the disordered ilmenite $FeTiO_3$ II synthesized at very high pressure, in *Ferrites* (Proc ICF3), Eds. H. Watanabe, S. Iida, and M. Sugimoto, Center for Academic Publication Japan, Tokyo, pp. 192–195.

SYONO, Y. and GOTO, T. (1982): Behavior of single-crystal forsterite under dynamic compression, in *High-Pressure Research in Geophysics*, Eds. S. Akimoto and M. H. Manghnani, Center for Academic Publication Japan/D. Reidel, Tokyo/Dordrecht, pp. 563–578.

TAKAHASHI, T. and BASSETT, W. A. (1964): High pressure polymorph of iron, *Science*, **145**, 483–486.

TRUNIN, R. F., GON'SHAKOVA, V. I., SIMAKOV, G. V. and GALDIN, N. E. (1965): A study of rocks under the high pressures and temperatures created by shock compression, *Izv. Acad. Sci. USSR, Phys. Solid. Earth*, No. 9, 1–12.

VAN THIEL, M., KUSUBOV, A. S. and MITCHELL, A. C. (1967): Compendium of shock wave data, UCRL–50108, Vol. 1–3, National Technical Information Service, Spring field, Revised version (1978).

WACKERLE, J. (1962): Shock-wave compression of quartz, *J. Appl. Phys.*, **33**, 922–937.

WALSH, J. M. and CHRISTIAN, R. H. (1955): Equation of state of metals from shock wave measurements, *Phys. Rev.*, **97**, 1544–1556.

WATANABE, M., TOKONAMI, M. and MORIMOTO, N. (1977): The transition mechanism between the CsCl-type and NaCl-type structures in CsCl, *Acta Cryst. A.*, **33**, 294–298.

YAGI, T., MAO, H. K. and BELL, P. M. (1978): Structure and crystal chemistry of perovskite-type $MgSiO_3$, *Phys. Chem. Miner.*, **3**, 97–110.

YAGI, T. and AKIMOTO, S. (1982): Rapid X-ray measurements to 100 GPa range and static compression of α–Fe_2O_3, in *High-Pressure Research in Geophysics*, Eds. S. Akimoto and M. H. Manghnani, Center for Academic Publication Japan/D. Reidel, Tokyo/Dordrecht, pp. 81–90.

Materials Science of the Earth's Interior, edited by I. Sunagawa, pp. 415–444.

PETROLOGY OF MATERIALS DERIVED FROM THE UPPER MANTLE

Ken-ichiro AOKI

Institute of Mineralogy, Petrology and Economic Geology, Tohoku University, Sendai, Japan

1. Introduction

The part of the mantle 400 km deep from the Mohorovicic discontinuity is called the upper mantle. It is one of the most important subjects in the field of solid-state earth sciences in elucidating the physico-chemical nature of this mantle. Especially, to make clear the evolution of the crust and the origin of basaltic magmas, it is also indispensably important in investigating and understanding the upper mantle materials.

The upper mantle has higher pressures and temperatures than the crust, and the pressures and temperatures directly under the M-discontinuity are at 10 kbar and 450°C respectively, and those of 400 km deep from it are estimated at about 130 kbar and at 1,500°–1,800°C respectively. The pressure-temperature condition is so varied in these ways that the mineral assemblage is assumed to vary according to depth, although the chemical composition of the upper mantle material is rather homogeneous, and the composition is also assumed to vary though the minerals are of the same kind.

The major mineral assemblages estimated for the petrological, geophysical, and geochemical data are the following;

(a) olivine + pyroxene + spinel,
(b) olivine + pyroxene + garnet,
(c) pyroxene + garnet.

Naturally, such materials are found as xenoliths in alkali basalts and kimberlites.

2. Xenolith and Megacryst

2.1 Distribution and occurrence

The lower crustal and upper mantle materials and megacrysts crystallized from magmas under a high-pressure condition are usually found in the volcanic rocks produced from magmas which arrived at the surface from a deeper part in a short period. In most cases, it is contained in kimberlite, assumed to have come into being at the deepest part, and in alkali basalt. It is not found in other rock types. When xenoliths and megacrysts in kimberlites and alkali basalts are investigated, the former contain rock fragments and minerals derived from a deeper part and have larger varieties than the latter.

Kimberlite, a kind of ultramafic volcanic rock which occurs very rarely or in almost negligible amounts, is found only in stable continents. It can be found in central and southern Africa, Siberia, India, the U.S.A., and Canada, and more recently, it has been found in Greenland and China. Although kimberlite is the host rock, it does not necessarily contain diamond. For example, 15 localities of kimberlite are known in the U.S.A., but only in two localities has diamond been found.

Kimberlite shows an appearance of tuff breccia or tuff in most cases, and it outcraps as a dyke of less than 1 m in width and less than 50 m in length, but sometimes it occurs as a pipe or a diatreme. Even the largest pipe is of about 500 m in diameter and 20 ha in square measure. Possibly, it may have given off ashes, lapilli and blocks at the time of eruption, but they seem to have been removed by erosion afterwards. Kimberlite is rarely intruded as a liquid containing abundant crystals and volatiles, and in the Igwisi Hills in Tanzania, a small quantity of lava flow can be recognized in it.

Kimberlite generally contains phenocrystic minerals with large amounts of olivine and small amounts of enstatite, chromian diopside, pyrope, ilmenite, and phlogopite, and its groundmass is composed of olivine, pyroxene, spinel, apatite, perovskite, magnetite, and carbonate. It has generally undergone intense alteration, producing large amounts of serpentine and secondary carbonate, and it is not easy to determine the original composition of its minerals.

Kimberlite is assumed to have arrived at the surface as crystal and volatile-rich magma at one stretch through the upper mantle and the crust from an underground depth of 100–200 km. The speed of eruption at the surface is calculated to be 400 m per second and, at the upper mantle, to be 10–20 m per second and 30–60 km per hour.

From stratigraphic and radiometric evidence, it is apparent that kimberlite activity has occurred several times during a long geologic period. The oldest kimberlite found in the world is in West Africa, and it occurred 2.1–2.3 billion years ago, while that in South Africa (at Premier) occurred about 1.5 billion years ago, that at Colorado Plateau in the U.S.A., 30 million years ago, and that at Igwisi in Tanzania occurred recently. However, kimberlite activity on a largest scale occurred about 100 million years ago. The most typical and representative examples are found in Lesotho, South Africa and Siberia.

The geographic distribution of ultramafic xenolith-bearing basaltic rocks in the world was first investigated by FORBES and KUNO (1965). According to them, xenolith-bearing basaltic rocks were recognized to be distributed in most continents, including Antarctica, volcanic islands in the Pacific, Atlantic, and Indian oceans and island arcs. More than 200 localities were recognized at the time of Forbes and Kuno's investigation. Later, they have been found in various other places in the world and the number of locations seems now to be more than 1,000. In Japan, the number is more than 30, and in the Hawaiian Islands, about 50, and in the West of U.S.A., more than 50.

Xenolith-bearing basaltic rocks are alkali basalts in a broad sense and, on close examination, they are alkali basalt, basanite, nephelinite, melilitite and leucitite, and their hypabyssal equivalents. They are rarely contained in differentiated mugearite, benmoreite and mafic phonolite belonging to the alkali series, but are not found in trachyte or alkali rhyolite. They are very rarely contained in olivine tholeiite (and high-alumina basalt) and calc-alkali andesite. The location of the former is in Tasmania, the Hawaiian Islands, and Sannome-gata, Japan, and the latter is in Itinome-gata, Japan.

Xenolith-bearing basaltic rocks occur as lava flow, dyke, small intrusion, pyroclastic flow, vent breccia, and air fall scoria. Their occurrences are more varied than those of kimberlite.

The geologic time of activities of xenolith-bearing basaltic rocks are all in the Cenozoic era, with few exceptions, contrasting with that of kimberlite. ENGEL *et al.* (1965) demonstrated that in the first three quarters of geologic time, tholeiites were active as basalts and alkali basalts were rarely seen. Alkali basalt did not erupt so much in the Mesozoic and Paleozoic eras. It may be very natural that ultramafic xenolith should be found in the volcanic rock of the Cenozoic era in astonishingly great amount, for it is contained only in alkali basalt. Volcanic rocks in the olden times were often subjected to erosion, alteration, deformation, and metamorphism, and this fact may be one of the reasons why ultramafic xenolith was not found. Ultramafic xenolith is exceptionally contained in the complex of carbonatite and ultramafic alkalic rocks, and in carbonaceous volcanic rocks, as well as in kimberlites and alkali basalts. The locality of the former case is the Fen alkali complex (Paleozoic), Norway, and that of the latter is Lashaine volcano (Recent), Tanzania.

The volume of ultramafic xenolith is generally very small, both in kimberlite and alkali basalt, so that it is very hard to gather. Its size is usually less than 5 cm, and that of 30 cm is very rare. Ultramafic xenolith found in the Matsoku kimberlite pipe, Lesotho, is exceptionally large and it occupies 20 to 30% of the pipe and its weight is rarely more than 400 kg.

Megacryst is sometimes found in kimberlite and alkali basalt and the frequency of its occurrence is higher than that of ultramafic xenolith. Its size is from several millimeters to 20 cm and it is usually less than 2 cm. It sometimes shows as euhedral in form but, in most cases, it shows as anhedral after its crystal margin has undergone magmatic corrosion or has been crushed down.

Table 1 shows the chemical composition of the representatives of xenolith-bearing kimberlite, basalt, and andesite. Almost all kimberlites have undergone a remarkable hydration after emplacement, and most of primary minerals have been altered to hydrous minerals so that, in most cases, it is very difficult to estimate precisely the composition of the original rocks.

2.2 Type and classification of upper mantle rocks

There are more than 100 rock types of the upper crust, and the range of their chemical composition is very wide. It is said that more than 2,000 mineral species can be found there. On the contrary, the chemical range of the upper mantle is remarkably narrow, its average composition is considered to be ultramafic and of a more reduced condition. Here, the rock types and minerals are very limited and it is very ambiguous to distinguish igneous rocks from metamorphic ones, owing to the high temperature. Most of the rocks derived from the upper mantle show a metamorphic texture and structure.

The classification shown in Fig. 1 has been recently applied to ultramafic rocks. These are mainly divided into peridotites, of which the main is olivine, and pyroxenite, of which the main is pyroxene. They are further subdivided by the volume of olivine, orthopyroxene, and clinopyroxene. And they contain garnet, chromian spinel or plagioclase as their accessories to the major components (0–20%). For example, when harzburgites contain garnet they are called garnet harzburgite, and, when websterites contain chromian spinel they are called chromian spinel websterite. The stability fields of these three accessory minerals at high pressure are different respectively, and so the depth of their formation can be estimated roughly from the rock names.

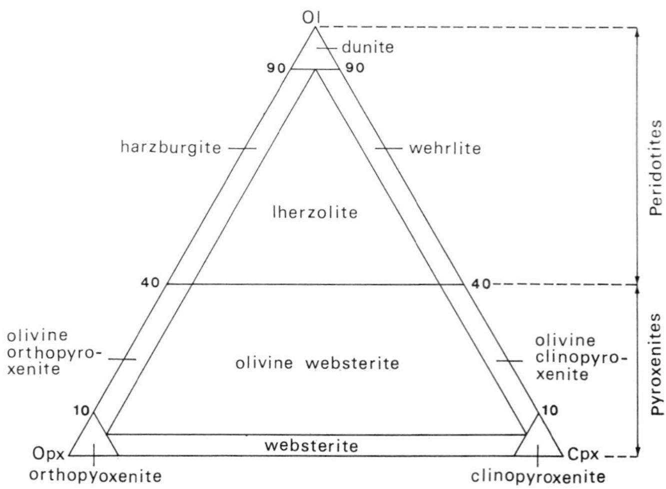

FIG. 1. Classification of ultramafic rocks. Ol = olivine; Opx = orthopyroxene; Cpx = clinopyroxene (IUGS SUBCOMMISION, 1973).

Eclogite is defined as a high-pressure phase of gabbro or basalt, though it belongs to mafic rocks. It is chiefly composed of garnet and clinopyroxene. The stability fields of rocks with this mineral assemblage against temperature and pressure are widely varied, and such an assemblage of minerals is recognized in xenoliths in kimberlites and eclogite facies and a part of granulite facies of metamorphic rocks. In order to emphasize the formation under high-pressure and high-temperature conditions, garnet websterite and garnet clinopyroxenite from Hawaii were once named eclogite. But in this paper, eclogite refers only to that contained in kimberlites.

The main component of eclogite is garnet and clinopyroxene, and garnet is usually 28% or more in volume. Sometimes corundum or kyanite is contained, and eclogite containing corundum as an accessory mineral is called corundum eclogite, eclogite containing kyanite eclogite. Grospydite is a kind of kyanite eclogite, and garnet contained in it is characterized to have 50–80% of grossular components (garnet in eclogite belongs to the pyrope-almandine series, its grossular component being usually less than 30%).

Genetically, deep-seated xenoliths are divided into the following two groups: (1) Cognate, and (2) Accidental. Cognate xenoliths are fragments of cumulative rocks and rocks gradually cooled in the depth which were formed and generated during a course of igneous activities of magmas. These are characterized by the presence of igneous textures and structures. Minerals formerly crystallized from magmas are euhedral and those crystallized later are subhedral or anhedral in form. Especially in rocks showing poikilitic and ophitic textures, the order of crystallization of minerals from magmas can sometimes be determined. Accidental xenoliths are further subdivided into two groups. The first is rocks composing the upper mantle itself, and the second is cumulative rocks formed and generated during the former igneous activity, which are independent of the activity of the magmas containing them. They sometimes show igneous textures, but in most cases they have undergone deformation and metamorphism afterwards. Table 2 shows the summary of characteristics of ultramafic xenoliths.

Megacryst refers to minerals crystallized from magmas under such high-pressure conditions as in the lower crust or in the upper mantle, and it is distinguished from porphyritic minerals crystallized from magmas under lower pressure condition (such as in the upper crust) (Binns, 1969). Though it has the name of "mega," megacryst is not necessarily a large crystal, and in most cases it is isolated. It is not rare that it occurs as a glomeroporphyritic aggregate of more than one kind of mineral. In the latter case, it is sometimes treated as cognate xenolith. Among kimberlites the intergrowth of clinopyroxene-ilmenite, orthopyroxene-ilmenite or clinopyroxene-garnet is recognized. This is composed of two kinds of minerals, but it is treated as a kind of megacryst. Fragments which magmas took from the upper mantle and from the lower crust when rising, are sometimes crushed and become

TABLE 1. Chemical analyses of representative xenolith-bearing kimberlite, basalt, and andesite.

	1	2	3	4	5	6	7	8	9	10	11	12
SiO_2	17.54	36.36	33.21	27.64	24.15	37.37	42.78	44.74	50.25	52.76	50.55	60.65
TiO_2	4.34	0.98	1.97	1.65	6.46	3.12	2.78	2.67	1.93	1.90	1.06	1.12
Al_2O_3	5.01	5.13	4.45	3.17	2.58	11.67	13.64	15.67	13.90	11.19	17.20	16.41
Cr_2O_3		0.22	0.17	0.14		0.01						
Fe_2O_3	12.56	7.71	6.78	5.40	7.67	6.19	5.29	3.76	2.21	3.91	2.58	1.73
FeO	3.60		3.43	2.75	8.36	9.50	4.43	7.21	7.80	3.61	5.85	2.81
MnO	0.32	0.16	0.17	0.13	0.16	0.23	0.21	0.19	0.15	0.14	0.19	0.15
MgO	19.80	17.43	22.78	24.31	24.03	11.30	8.29	9.83	10.76	7.25	6.82	3.36
NiO					0.02	0.03						
CaO	17.05	11.16	9.36	14.13	10.27	11.20	14.44	8.80	6.87	7.18	9.59	5.57
Na_2O	0.32	0.42	0.19	0.23	0.25	5.18	4.07	3.31	3.43	2.50	2.46	3.38
K_2O	0.07	1.52	0.79	0.79	1.02	1.47	2.31	1.96	1.54	6.83	1.43	2.29
H_2O^+	7.17		8.04	7.89	4.98	2.02	0.92	1.56	0.40	1.35	1.80	2.05
H_2O^-	0.87		2.66		0.90	0.19	0.16	0.06	0.20	0.49		
P_2O_5	5.11	0.55*	0.65	0.55	0.23	1.30	0.89		0.45	0.87	0.21	0.29

	1	2	3	4	5	6	7	8	9	10	11	12
CO_2	6.33		4.58	10.84	9.02							
S			0.28	0.24								
$O \equiv S$			0.14	0.12								
Total	100.09		99.37	99.74	100.10	100.78	100.21	99.76	99.89	99.98	99.74	99.81

1. Kimberlite, Benfontein sill, South Africa (DAWSON and HAWTHORNE, 1973).
2. Average of 80 South African kimberlite (GURNEY and EBRAHAIM, 1973).
3. Average of 25 Lesotho kimberlite (GURNEY and EBRAHAIM, 1973).
4. Average of 623 Siberian kimberlite (ILUPIN and LUTZ, 1971).
5. Kimberlite, Bachelor Lake, Quebec (WATSON, 1955).
6. Olivine nephelinite, Salt Lake Crater, Hawaii (FORBES and KUNO, 1965).
7. Olivine titanaugite basanite, Dreiser Weiher, Germany (AOKI and KUSHIRO, 1968).
8. Olivine alkali basalt, Ludlow, California (FORBES and KUNO, 1965).
9. Titanaugite olivine alkali basalt, Dogo, Japan (AOKI, 1977).
10. Minette, Shiprock, New Mexico (AOKI, unpubl.).
11. Groundmass of olivine tholeiite, Sannome-gata, Japan (AOKI and FUJIMAKI, 1982).
12. Hornblende biotite andesite, Itinome-gata, Japan (AOKI and FUJIMAKI, 1982).

TABLE 2. Summary of characteristics of ultramafic xenolith.

	Upper mantle origin	Cumulus origin
Major rock type	Lherzolite, Harzburgite, Olivine websterite, Websterite. No transitional between peridotite and pyroxenite.	Olivine clinopyroxenite, Clinopyroxenite, Wehrlite, Dunite. Transitional from peridotite through pyroxenite to mafic rock.
Major constituent	Olivine, Orthopyroxene, Clinopyroxene, Chromian spinel, Garnet.	Clinopyroxene, Olivine, Orthopyroxene, Chromian spinel, Spinel, Kaersutite, Plagioclase.
Texture	Granular, Tabular, Porphyroclastic, Mosaic.	Cumulus, Exsolution.
Composition of constituents		
Olivine	Fo_{94-85}	Fo_{92-65}
Orthopyroxene	En_{93-85}, Ti, Ca, Al poor	En_{91-75}, Ti, Ca, Al, Fe rich
Clinopyroxene	Cr, Mg rich and Ti, Al poor	Ti, Al, Fe rich and Cr poor
Spinel	Cr, Al rich	Fe, Al rich and Cr poor

dispersed mineral grains. In this case it becomes very difficult to distinguish megacryst from xenocryst.

Megacrysts most universally found in basaltic rocks are diopside, olivine, andesine, anorthoclase, kaersutite, orthopyroxene, magnetite, and spinel, while those in kimberlites are diopside, endiopside, ilmenite, garnet, olivine, and orthopyroxene. Minerals assumed to form under the upper mantle condition (above 7–10 kbar) are numerous. Table 3 shows all of them, although more kinds of minerals will surely be found.

TABLE 3. The minerals of upper mantle.

Tectosilicates	
quartz	SiO_2
coesite	SiO_2
andesine-oligoclase	$(Ca, Na)(Al, Si)AlSi_2O_8$
sanidine	$(K, Na)AlSi_3O_8$
orthoclase	$KAlSi_3O_8$
anorthoclase	$(Na, K)AlSi_3O_8$
scapolite	$(Na, Ca)_4[(Al, Si)_4O_8]_3(Cl, CO_3)$
Phyllosilicates	
phlogopite-biotite	$K(Mg, Fe)_3(AlSi_3O_{10})(OH)_2$
antigorite	$Mg_6Si_4O_{10}(OH)_8$
Inosilicates	
orthopyroxene	$(Mg, Fe)SiO_3$
diopside-augite	$Ca(Mg, Fe)Si_2O_6$
endiopside	$Ca_{0.6}Mg_{1.4}Si_2O_6$
soda-bearing tremolite	$Na_{0.8}Ca_{1.6}Mg_5Si_8O_{22}(OH)_2$
pargasite-hornblende-kaersutite	$NaCa_2(Mg, Fe^{+2})_4(Ti, Fe^{+3}, Al)Al_2Si_6O_{22}(OH)_2$
potassic richterite	$KNaCaMg_5Si_8O_{22}(OH)_2$
Nesosilicates	
olivine	$(Mg, Fe)_2SiO_4$
chondrodite	$Mg_5Si_2O_8(OH)_2$
clinohumite	$Mg_9Si_4O_{16}(OH)_2$
pyrope	$Mg_3Al_2Si_3O_{12}$
grossular	$Ca_3Al_2Si_3O_{12}$
uvarovite	$Ca_3Cr_2Si_3O_{12}$
knorringite	$Mg_3Cr_2Si_3O_{12}$
sillimanite	Al_2SiO_5
kyanite	Al_2SiO_5
zircon	$ZrSiO_4$
sphene	$CaTiSiO_5$
Other minerals	
calcite	$CaCO_3$
dolomite	$CaMg(CO)_3$
magnesite	$MgCO_3$
brucite	$Mg(OH)_2$
corundum	Al_2O_3
rutile	TiO_2

(continued)

TABLE 3. (continued)

perovskite	$CaTiO_3$
spinel	$(Mg, Fe)Al_2O_4$
chromian spinel-chromite	$(Mg, Fe)(Cr, Al)_2O_4$
magnetite	Fe_3O_4
ilmenite	$(Mg, Fe)TiO_3$
apatite	$Ca_5(PO_4)_3(F, Cl, OH)$
graphite	C
diamond	C
moissanite	SiC
barite	$BaSO_4$
new potassium-nikel sulphide	$K_{9.3}Cu_{0.7}Fe_{19.3}Ni_{7.2}S_{28.0}$

Most of the minerals in Table 3 are usually found in upper crustal rocks. Except for shock and meteorite minerals, minerals found only in the mantle are simply coesite, knorringite, endiopside, diamond, moissanite, and new potassium-iron-nickel sulphide. Among them, the last new potassium-iron-nickel sulphide has recently been found in a clinopyroxene-ilmenite intergrowth in Frank Smith kimberlites, in South Africa. Its origin is still unknown, but it suggests a transformation of itself into a potassium-bearing mineral in the upper mantle.

The suites and rock types of ultramafic and mafic xenoliths in kimberlites and alkali basalts reflect the characteristics of the lower crust and the upper mantle in the districts where they have been distributed, and megacrysts and cognate xenoliths are also greatly controlled by the chemical composition and temperature and pressure conditions of magmas when crystallizing these minerals. When xenoliths everywhere in the world are investigated by detailed petrographically, garnet peridotite, spinel peridotite, garnet pyroxenite, and eclogite are commonly recognized in kimberlites, while spinel peridotite and pyroxenite are recognized in alkali basalts. And in the latter, garnet-bearing xenoligh is very rarely found. Volumetrically, peridotite is found in a greater amount, and eclogite and pyroxenite are found in less than 10% of peridotite. In the Mid-Atlantic Ridge, an island (St. Paul) composed of peridotite is found, which has appeared directly above the sea from the upper mantle, and large peridotite bodies are distributed in the continental margins and island arcs, which have appeared into the upper crust, accompanied by the orogenic movements from a relatively shallower depth. But they do not contain much pyroxenite or mafic rock. From these facts, the major constituent material of the upper mantle is estimated to be peridotite.

2.3 *Garnet peridotite and eclogite*

Of kimberlites found in South Africa, 95 of xenolith is estimated to be peridotite, and 5 to be eclogite. An exceptional case is found in kimberlites, where 80% is eclogite and 20% is peridotite. In ultramafic rocks harzburgites (±garnet) are found in amazingly large amounts of almost 80%, lherzolites (±garnets) are less than 10% and dunites, pyroxenites, and others are also less than 10%. In most cases, they contain phlogopites in small or minor amounts. Kimberlites found around the Kimberley area of South Africa are especially recognized as unversally containing phlogopites. Table 4 shows the modal composition of peridotites in them.

BOYD (1973) classified garnet peridotites from Lesotho largely into a granular and sheared type according to their texture, structure, bulk composition, and the chemical composition of constituent minerals. The former type shows the coarse and equigranular texture containing phlogopite and chromite, but rarely graphite, as its accessory minerals. On the other hand, the latter megascopically shows the submylonitic texture, and that garnet and diopside porphyroblasts are scattered in the matrix composed of fine grained olivines and orthopyroxenes. The volume percentage of garnet peridotites from Lesotho between the granular and the sheared type is 90:10 to 85:15 approximately, and the former type shows a generally high Mg/Mg + Fe ratio and is poor in Ca; it is mainly composed of harzburgites. The latter is relatively composed of lherzolites. This unique type of garnet peridotite is found not only in Lesotho but also in South Africa, Southwest Africa, the U.S.A., and Siberia.

Eclogites are basically composed of reddish orange garnet and dark green diopside, sometimes containing kyanite, corundum, orthopyroxene, and quartz as their major to minor phase (exceptionally, diamond, graphite, ilmenite, rutile, sanidine, and coestite are contained). But olivines are absent there. According to MATHIAS *et al* (1970), the ratio of volume of eclogite group is 75% of eclogite, 9% of kyanite eclogite, 7% of corundum eclogite, 3% of orthopyroxene eclogite, less than 1% of quartz eclogite, and 6% of others. The modal composition of garnet and clinopyroxene varies largely in the eclogite groups, but, on average, both garnet and clinopyroxene are contained in nearly the same volume. Compared with peridotite, eclogite is very coarse

TABLE 4. Modal composition of peridotite and eclogite xenoliths.

	Kimberley, South Africa			Lesotho	
	Garnet peridotite	Peridotite	Eclogite	Garnet peridotite	Peridotite
Olivine	65(43–86)	69(52–89)		60(45–75)	69(50–88)
Orthopyroxene	27(9–46)	29(11–48)		34(20–50)	30(11–44)
Diopside	2(0–17)	1(0–10)	48(18–72)	1(0–20)	1(0–5)
Garnet	5(0–15)		52(28–82)	5(0–11)	
Phlogopite	1(0–12)	1(0– 4)		<0.3 (0–3)	

grained and, in extreme cases, the individual grain is larger than 10 to 20 cm in diameter.

Diamondiferous eclogite was found by Cohen in South Africa in 1879, and more than 70 specimens have been found in the Roberts Victor and Bellsbank, Kimberley area, Orapa, Botswana, Tanzania, and the U.S.S.R. since then. Accordingly, diamondiferous eclogites are relatively more common than they were formerly considered. Eclogites having a high-pressure phase of silica mineral (coesite), which were first found at Meteor Crater in Arizona, were also found in Roberts Victor kimberlite in 1977. On the other hand, diamond was not found in peridotite xenoliths for a long time, but recently it has been found in Aykhal and Udachnaya (U.S.S.R.), Mothae (Lesotho), Wyoming, and Finsch (South Africa).

According to Sobolev *et al.* (1974) and Reid *et al.* (1976), the volume percentage of garnet and clinopyroxene in diamondiferous eclogites varies to a great degree, being 100–30 (average 60)% to 0–70 (average 40)%. They sometimes contain kyanite and rutile. The chemical variation of garnet and clinopyroxene is also very remarkable, and the Mg–Fe partition coefficient between these two minerals $[K_D = (FeO/MgO)_{gar}/(FeO/MgO)_{cpx}]$ is 2.2–4.0. These diamondiferous eclogites are within the petrographic nature of diamond-free eclogites and seem to offer an important clue to the genetic consideration of eclogites.

Kimberlites in the Kimberley area and Yakutia (U.S.S.R.), contain a mica nodule as their unique xenoliths. This mica nodule is composed of phlogopites of less than 20 cm in size, and of 5 mm euhedral to subhedral form. Usually, more than 95% of phlogopites are contained, and sometimes ilmentites and diopsides or potassic richterites are contained as their major constituent in 50% of the whole. Compared with minerals from peridotite xenoliths, the mica nodule ones are characterized by being lower in Al_2O_3 and Cr_2O_3 and higher in Fe_2O_3, and it is considered that these rocks are a cumulus phase of magma, chemically similar to kimberlite under oxidizing conditions of less than 100 km in depth, if potassic richterite is present (Aoki, 1974; Smith and Dawson, 1976).

2.4 *Spinel peridotite*

Xenolith of alkali basalts is spinel peridotite and the range and average of its modal composition of major constituent minerals is 55–90 (average 66)% of olivine, 5–43 (average 24)% of orthopyroxene, 0–20 (average 8)% of diopside and 0–5 (average 2)% of chromian spinel. Figure 2 shows the ratio of olivine, orthopyroxene, and clinopyroxene (diopside) in 166 spinel peridotites from Kilbourne Hole, New Mexico. As can be clearly seen from this figure, lherzolite appears in amazingly large amounts and harzburgite in very small amounts. In most cases, orthopyroxene is seen in a larger amount than clinopyroxene. The modal composition of spinel peridotite from various

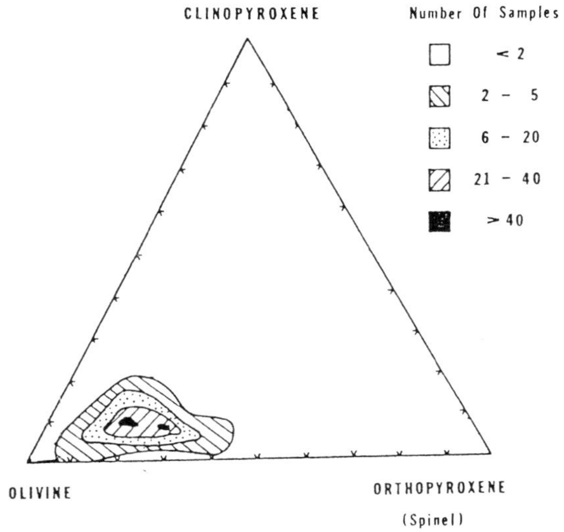

FIG. 2. Countered triangular plot of volume percent olivine-clinopyroxene-orthopyroxene of peridotite xenoliths from Kilbourne Hole, New Mexico (simplified from CARTER, 1970).

places in the world, such as from Itinome-gata, Salt Lake, and San Carlos, tends to show almost the same tendency as that of Kilbourne Hole, and this contrasts to the fact that harzburgite is the main constituent of peridotite xenolith in kimberlites.

The textures, structures and properties of the constituent minerals of spinel peridotites are similar to those of rocks of a high-temperature type and of Alpine-type ultramafic intrusion, and they have usually undergone the complicated and complex metamorphic process containing solid deformations. Generally, they show the granoblastic textures and the fabrics scattered with orthopyroxene porphyroblasts in the matrix of medium to coarse grains, and they are characterized by containing clear green diopsides.

Spinel peridotite xenoliths are found in many localities in the world, but they are rarely found to occur as sole xenoliths. They are usually accompanied by various types of ultramafic and mafic xenoliths and megacrysts. Xenoliths produced at the regions of island arc and continental margin, especially reflect the remarkable igneous activity and orogenic movement at the geologic time and are various in kinds. For example, there are many cumulative ultramafic and mafic rocks, such as xenoliths in alkali basalts in southwestern Japan, which were formed by a crystallization differentiation of magmas, and it is not rare that spinel peridotites are not contained.

2.5 Chemical composition of peridotite and eclogite

In order to make clear the chemical natures of the upper mantle materials, many chemical analyses of peridotite xenoliths have been carried out. The analytical data of peridotites gained from the fully and minutely investigated areas show a large variety of major oxides, not only in those from continental, island arc and oceanic regions but also in those from the same locality, and they also show that the upper mantle is heterogeneous both laterally and vertically. However, the average composition—for example, of spinel peridotites from Itinome-gata (Japan), Hawaii, Dreiser Weiher (Germany), and Central Massief France—is very similar (Table 5). The average composition of spinel peridotites from Western U.S.A., Australia, New Zealand, Antarctica, and England also shows the same tendency, and they have lherzolitic in their composition. An exceptional case is those from Samburu in the Gregory Rift Valley, Kenya, which contain more MgO and less FeO and CaO and have the harzburgitic in their composition. On the other hand, the chemical composition of garnet peridotite and garnet-free peridotite xenoliths in kimberlites from South Africa and Lesotho has a wider range of variation than that of spinel peridotites, and its average composition is rich in MgO and SiO_2 and poor in Al_2O_3, FeO, and CaO, and has harzburgitic. And garnet peridotite xenoliths from Lashaine volcano in the Gregory Rift Valley, Tanzania, have almost the same chemical composition in kimberlites or those of spinel peridotites from Kenya. Figure 3 shows the histograms for major oxides of about 380 spinel peridotites and about 200 garnet peridotites. As clearly seen from this figure, a rather remarkable difference is recognized between the composition of spinel peridotites and that of garnet peridotites.

Eclogites are poor in number and are coarse grained, so that their sample selection is very difficult and their chemical data are not so abundant. Table 6 shows the average chemical composition of eclogites from South Africa and from the U.S.S.R. Eclogites have a rather large chemical variation. Eclogites are composed of garnet and clinopyroxene so that K_2O cannot be contained in them. But, in many cases, eclogites contain K_2O of a negligible amount, owing to biotites produced by metasomatism or contamination. As already pointed out, the chemical compositions of eclogites are basaltic, resembling picritic basalts. Differences cannot be recognized between eclogites from South Africa and those from the U.S.S.R. The composition of diamondiferous eclogites coincides with that of diamond-free eclogites. Kyanite eclogites and grospydites are more abundant in Al_2O_3 and poorer in MgO nd FeO.

2.6 Chemical composition of constituent mineral

The chemical composition of olivines and orthopyroxenes, which are contained in a large amount of upper mantle materials is as follows: that of olivines is approximately expressed in Mg_2SiO_4–Fe_2SiO_4 solid solutions, less than 2% of Ni_2SiO_4, Mn_2SiO_4, and Ca_2SiO_4 being contained as other

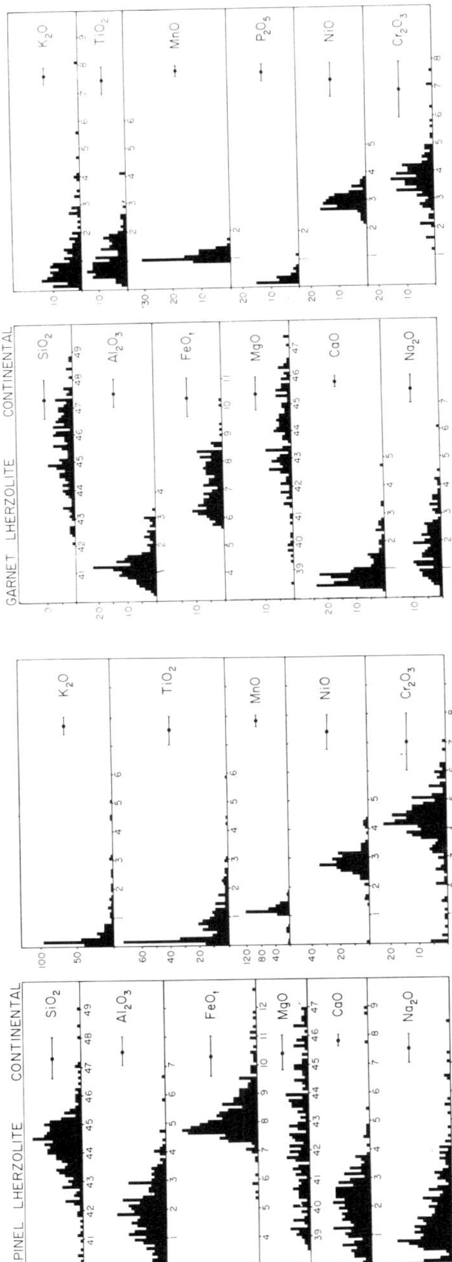

FIG. 3. Histograms for major oxides of continental spinel and garnet peridotite xenoliths (MAALØE and AOKI, 1977).

TABLE 5. Average chemical composition of peridotite xenoliths.

	1	2	3	4	5	6	7	8	9	10	11	12
No. of analyses	21	11	15	130	15	384	30	12	63	13	8	8
SiO_2	44.62	44.70	43.60	44.6	43.67	44.20	46.06	46.25	45.94	45.78	43.97	42.52
TiO_2	0.17	0.13	0.17	0.04	0.06	0.13	0.07	0.16	0.14	0.13	0.07	0.06
Al_2O_3	2.29	2.70	1.63	2.12	0.83	2.05	1.54	0.87	1.16	0.79	1.64	0.43
Cr_2O_3	0.40	0.36	0.30	0.45	0.37	0.44	0.38	0.33	0.35	0.38	0.49	0.28
FeO*	8.80	9.21	9.08	8.13	7.12	8.29	6.41	6.29	6.96	7.01	6.83	7.11
MnO	0.14	0.13	0.14	0.12	0.11	0.13	0.11	0.10	0.11	0.11	0.13	0.12
MgO	40.87	40.35	42.94	41.99	46.92	42.21	43.75	44.03	43.88	44.55	44.73	47.89
NiO				0.27	0.33	0.28	0.29	0.30	0.29	0.29	0.36	0.41
CaO	2.50	1.95	1.92	2.12	0.42	1.92	0.98	1.24	0.92	0.66	1.10	0.58
Na_2O	0.17	0.36	0.18	0.15	0.13	0.27	0.21	0.16	0.18	0.18	0.12	0.10
K_2O	0.03	0.05	0.03	0.03	0.17	0.06	0.22	0.05	0.11	0.09	0.03	0.03
P_2O_5	0.01	0.05	0.00			0.03	0.02	0.04	0.02	0.01	0.06	0.02

$\frac{100\ Mg}{Mg + Fe}$	89.1	88.5	89.2	90.1	92.1	90.1	92.3	92.5	91.7	91.8	92.1	92.3

1. Spinel peridotite, Itinome-gata, Japan (AOKI and SHIBA, 1973).
2. Spinel peridotite, Hawaii (KUNO, 1969).
3. Spinel peridotite, Dreiser Weiher, Germany (KUNO and AOKI, 1970).
4. Spinel peridotite, Central Massief France (HUTCHISON *et al.*, 1970).
5. Spinel peridotite, Gregory Rift Valley, Kenya (AOKI and SUWA, 1977).
6. Spinel peridotite, the approximate composition of the uppermost mantle (MAALØE and AOKI, 1977).
7. Garnet peridotite, South Africa (Aoki, unpubl.).
8. Garnet free peridotite, South Africa (Aoki, unpubl.).
9. Garnet peridotite, Lesotho (Aoki, unpubl.).
10. Garnet free peridotite, Lesotho (Aoki, unpubl.).
11. Garnet peridotite, Lashaine, Tanzania (RHODES and DAWSON, 1975).
12. Garnet free peridotite, Lashaine, Tanzania (RHODES and DAWSON, 1975).

TABLE 6. Average chemical composition of eclogite xenoliths.

	1	2	3	4	5
No. of analyses	15	21	8	5	10
SiO_2	46.71	47.13	46.92	44.06	44.72
TiO_2	0.48	0.61	0.57	0.27	0.07
Al_2O_3	15.85	14.65	14.36	25.14	28.10
Cr_2O_3	0.13	0.06	0.09	0.02	0.02
Fe_2O_3	2.21	4.59	3.48	1.80	1.71
FeO	9.00	7.31	7.83	5.56	2.83
MnO	0.26	0.20	0.21	0.19	0.09
MgO	13.60	13.83	13.84	10.97	7.70
NiO	0.04				
CaO	9.25	9.81	9.69	9.47	12.42
Na_2O	1.82	1.31	2.30	1.85	1.56
K_2O	0.62	0.50	0.69	0.53	0.67
P_2O_5	0.03			0.13	0.09

1. Eclogite, South Africa (Aoki, unpubl.).
2. Eclogite, Yakutia, Siberia (FRANTSESSON, 1970).
3. Diamondiferous eclogite, Yakutia, Siberia (FRANTSESSON, 1970).
4. Kyanite eclogite, Yakutia, Siberia (FRANTSESSON, 1970).
5. Grospydite, Yakutia, Siberia (SOBOLEV *et al.*, 1968).

components, and that of orthopyroxenes in $MgSiO_3$–$FeSiO_3$ solid solutions, 10% (very rarely more than 10%) of $CaSiO_3$, $MgAl_2SiO_6$, and $MgCr_2SiO_6$ contained.

The chemical composition of clinopyroxenes is largely controlled by temperature and pressure, and the chemical environments at the time of their formation. The range of their compositions is also very wide, forming complex solid solutions. $CaSiO_3 + MgSiO_3 + FeSiO_3$ occupy 80% of their components in peridotite, pyroxenite, and megacrystic clinopyroxenes, and they are plotted in diopside, salite, and augite fields in the Di–Hd–Fs–En quadrilateral.

Endiopsides, poor in Ca (less than 30% of the Wo component), are found only in the sheared-type peridotite and in megacrysts. Of these three pyroxene components, the Wo component is largely dependent up on the temperature of their formation, namely, crystallized minerals formed at a high temperature have smaller amounts of Wo component and the Wo components tend to increase with a falling temperature. The other minor components are $NaFeSi_2O_6$, $NaAlSi_2O_6$, $NaCrSi_2O_6$, $CaAl_2SiO_6$, $CaCr_2SiO_6$, $CaTiAl_2O_6$, and so on. Generally, megacrystic and ultramafic and mafic rock clinopyroxenes in alkali basalts are rich in $CaAl_2SiO_6$, sometimes they contain 20% of $CaAl_2SiO_6$. On the other hand, those in kimberlites are rich in $NaAlSi_2O_6$ and $NaCrSi_2O_6$. The relative proportion of (Al + Cr − 2Ti)–(Na + K) of clinopyroxenes are plotted in Fig. 4. As can be clearly seen from the figure, a great difference can be recognized between clinopyroxenes in alkali basalts and those in kimberlites.

Clinopyroxenes in eclogites are mainly expressed in Ca(Mg·Fe)Si_2O_6—$NaAlSi_2O_6$ solid solution, and they do not contain much of the other components. The $NaAlSi_2O_6$-rich omphacite is $(Di + Hd)_{40}Jd_{60}$ and is found in grospydites. Generally, clinopyroxenes do not contain K because of the large ionic radius, but those of diamond inclusions sometimes contain innegligible amounts. The chemical compositions of clinopyroxenes from various occurrences are given in Table 7.

Garnet forms a rather complex solid solution, having $Mg_3Al_2Si_3O_{12}$–$Fe_3Al_2Si_3O_{12}$ series as its main components, with additional $Ca_3Al_2Si_3O_{12}$ and $Ca_3Cr_2Si_3O_{12}$ components. The proportion of the four end members differs by their occurrences and rock types. Representative garnet compositions are given in Table 8. Cr_2O_3 and CaO contents have the largest chemical variation among major oxides in garnet. Among ultramafic rock and megacryst garnets, those of the granular type peridotites have the most Cr_2O_3.

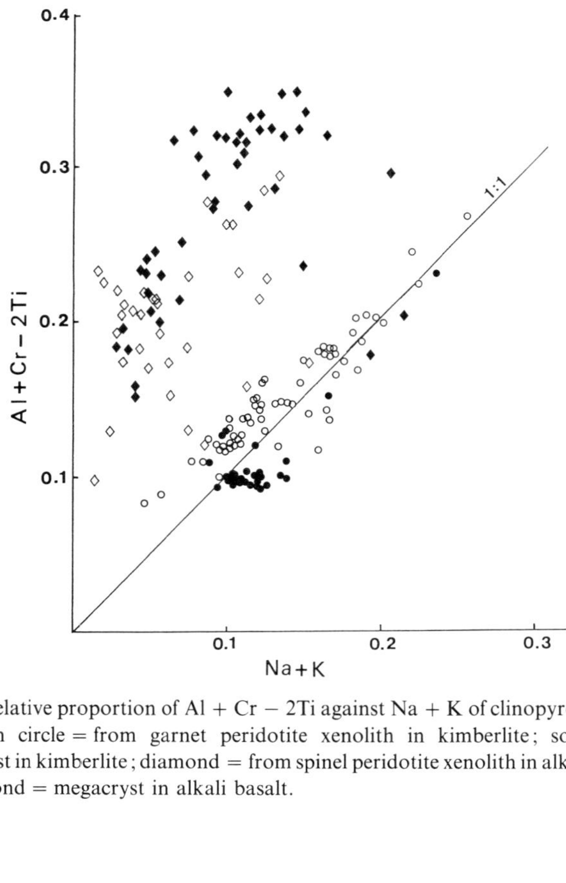

FIG. 4. Relative proportion of Al + Cr − 2Ti against Na + K of clinopyroxenes (O = 6). Open circle = from garnet peridotite xenolith in kimberlite; solid circle = megacryst in kimberlite; diamond = from spinel peridotite xenolith in alkali basalt; solid diamond = megacryst in alkali basalt.

TABLE 7. Chemical analyses of pyroxenes.

	1	2	3	4	5	6	7	8	9	10
SiO_2	55.08	54.22	56.26	48.05	55.39	54.0	52.28	54.61	54.96	50.78
TiO_2	0.02	0.14	0.23	1.76	0.38	0.37	0.29	0.23	0.42	0.95
Al_2O_3	3.45	4.03	1.34	8.02	2.59	9.4	4.25	1.30	8.01	6.17
Cr_2O_3	0.62	0.64	0.21	0.11	0.30	0.01	0.98	0.92	0.06	0.33
Fe_2O_3	0.07	1.15		2.60	0.48	0.05*	1.01	1.14	1.92	2.09
FeO	4.60	4.88	7.04	3.11	4.87	6.2	2.60	3.02	3.13	3.88
MnO	0.10	0.11	0.13	0.10	0.13	0.04	0.08	0.10	0.10	0.08
MgO	33.37	33.60	32.66	13.96	20.17	9.6	18.59	20.88	11.62	14.37
NiO	0.12		0.16		0.08					
CaO	1.71	1.27	1.59	21.15	13.73	15.0	19.26	16.20	14.13	19.68
Na_2O	0.08	0.17	0.34	0.83	1.72	4.5	0.81	1.28	4.78	1.58
K_2O	0.02	0.05		0.04	0.05	0.76	tr	0.12	0.18	tr
H_2O		0.42		0.40			0.45	0.48	0.65	0.35
Total	99.24	100.68	99.96	100.13	99.89	99.9	100.60	100.28	99.96	100.26

*V_2O_3.

1. Enstatite megacryst in Monastery kimberlite, South Africa (Aoki *et al.*, 1980).
2. Enstatite from spinel lherzolite xenolith in basanite, Hualalai volcano, Hawaii (Kuno, 1969).
3. Bronzite from garnet lherzolite (sheared type) xenolith, Thaba Putsoa kimberlite, Lesotho (Nixon and Boyd, 1973).
4. Aluminous augite megacrsyt in basanite, Dreiser Weiher, Germany (Aoki and Kushiro, (1968).
5. Endiopside megacryst in Monastery kimberlite, South Africa (Aoki *et al.*, 1980).
6. Omphacite inclusion in diamond, South Africa (Prinz *et al.*, 1975).
7. Chromian diopside from spinel lherzolite xenolith in basanite, Hualalai volcano, Hawaii (Kuno, 1969).
8. Chromian endiopside from garnet lherzolite (sheared type) xenolith, Thaba Putsoa kimberlite, Lesotho (Nixon *et al.*, 1963).
9. Omphacite from eclogite xenolith, Roberts Victor kimberlite, South Africa (Kushiro and Aoki, 1968).
10. Aluminous augite from garnet websterite xenolith in nephelinite, Salt Lake, Hawaii (Kuno, 1969).

TABLE 8. Chemical analyses of garnets.

	1	2	3	4	5	6	7	8	9	10
SiO_2	41.82	41.25	41.2	38.02	40.4	41.95	42.44	40.92	39.95	41.70
TiO_2	1.18	0.25	0.16	0.40	0.01	0.05	0.52	0.47	0.17	0.26
Al_2O_3	21.30	23.16	10.06	10.49	10.4	18.54	22.24	21.97	21.96	23.47
Cr_2O_3	0.59	0.05	16.78	14.04	17.3	6.76	1.98	0.10	0.04	0.42
Fe_2O_3	0.88	0.16		2.75			1.56	1.10	1.64	1.18
FeO	9.76	9.19	7.73	3.10	5.84	6.07	4.75	15.02	2.84	11.45
MnO	0.25	0.19	0.49	0.37		0.36	0.16	0.45	0.15	0.35
MgO	19.57	13.16	13.80	6.61	24.6	20.80	21.53	15.33	3.14	16.25
NiO	0.02	0.00								
CaO	4.37	12.62	9.16	23.86	0.93	6.04	4.90	5.01	30.13	5.02
Na_2O										0.13
K_2O										0.04
Total	99.74	100.03	99.38	99.64	99.5	100.57	100.08	100.31	100.02	100.27

1. Megacryst in Monastery kimberlite, South Africa (Aoki, unpubl.).
2. Megacryst in Frank Smith kimberlite, South Africa (Aoki, unpubl.).
3. Kao kimberlite mineral concentrates, Lesotho (HORNUNA and NIXON, 1973).
4. Megacryst in Newlands kimberlite, South Africa (CLARKE and CARSWELL, 1977).
5. Diamond inclusion in placer deposit, Brazil (MEYER and SVISERO, 1975).
6. Garnet harzburgite (granular type) xenolith in Mothae kimberlite, Lesotho (NIXON and BOYD, 1973).
7. Garnet lherzolite (sheared type) xenolith in Thaba Putsoa kimberlite, Lesotho (NIXON and BOYD, 1973).
8. Eclogite xenolith in Roberts Victor kimberlite, South Africa (KUSHIRO and AOKI, 1968).
9. Grospydite xenolith in Zagadochnaya kimberlite, Siberia (SOBOLEV *et al.*, 1968).
10. Olivine garnet websterite xenolith in nephelinite, Salt Lake, Hawaii (KUNO, 1969).

In them, Cr_2O_3 is contained as a $Ca_3Cr_2Si_3O_{12}$ component. However, among diamond inclusions and mineral concentrates from the Kao kimberlites, Lesotho, garnets poor in Ca but rich in Cr are rarely found (Nos. 3 and 5 in Table 8). In this case, Cr is considered to form the $Mg_3Cr_2Si_3O_{12}$ component (knorringite) combined with Mg. The Cr content closely affects the hue of garnet. Garnet with fewer Cr offers pale reddish purple (less than 4% of Cr_2O_3), and garnet with more Cr offers bluish-green (15–17%) through violet (10–12%) and reddish purple (4–8%). On the other hand, garnet rich in $Ca_3Cr_2Si_3O_{12}$ components (8–18% of Cr_2O_3) offer green, and these garnets are found in kimberlites from Newlands, South Africa, and Siberia.

3. *Crystallization Temperature and Pressure of Xenolith and Megacryst*

Among xenoliths and megacrysts carried by magmas, those referring to the highest *P–T* condition point at the upper limit of their depth of magma formation. The depth where these magmas were formed can be estimated from various data, combinations of the chemical compositions of minerals, the mineral assemblage and the element partition coefficient among coexisting minerals and from the high temperature and high pressure experimental works.

How deep is the place where those now-obtainable rocks and minerals were formed in the deeper part of the earth? This is one of the most interesting problems in the field of earth sciences. It tries at first to explain the most simple method to assume the depth.

Kimberlites, which are the only mother-rock of diamonds, have so far been considered to be magmas formed at the deepest part. So by investigating diamond itself in magmas, the depth of the formation can be roughly estimated. Diamond is the typical high-pressure mineral, composed of only C, and it is the hardest mineral in the world. It becomes unstable at low pressure and high temperature, transforming itself into graphite (Fig. 5). The temperature rises from the surface towards the deeper part of the earth, but its ratio of rising varies very much in regions. For example, the ratio is rather low in continental regions, and high in island arc and oceanic regions. The diamond-graphite stability curve intersects the continental geotherm at the point 130 km deep under the earth. This fact explains that diamond can only be formed at deeper place than that. But from diamond crystal itself, the lower limit of its formation cannot be obtained.

The stability relation of olivine, pyroxene, garnet, and quartz at the *P–T* condition has recently been obtained and understood in detail. Accordingly, forsterite (Mg_2SiO_4) transforms itself into Mg_2SiO_4 modified spinel at about 130 kbar, and into Mg_2SiO_4 spinel at about 200 kbar, while quartz transforms itself into coesite at about 40 kbar, and into stishovite at about 100 kbar.

When diamond is crystallized, other minerals are also crystallized at the

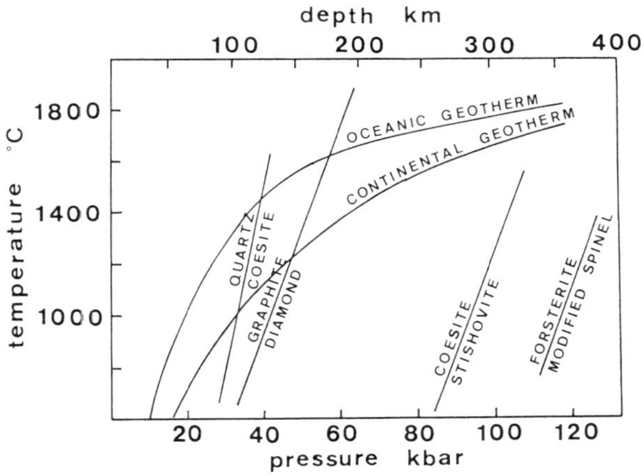

FIG. 5. Pressure-temperature stability fields for quartz-coesite, graphite-diamond, coesite-stishovite and forsterite-modified spinel, and oceanic and continental geotherms.

same time. Therefore, in the course of its growth, diamond sometimes takes in coexisting minerals as inclusions. When these kinds of inclusions are investigated, olivine, pyroxene, and garnet are mainly recognized, but very rarely is coesite found there. However, minerals stable above 100 kbar, such as Mg_2SiO_4 modified spinel and stishovite, are not found. These facts assume that kimberlite magmas were formed at a place shallower than 350 km.

Clinopyroxene-ilmenite and orthopyroxene-ilmenite intergrowths are sometimes found in kimberlites from, U.S.A., South Africa, Lesotho, and Tanzania. RINGWOOD and LOVERING (1970) synthesized garnet having its composition of clinopyroxene-ilmenite intergrowth at 1,000°C and at 105 kbar. They demonstrated that this intergrowth is the break-down product of garnet formed at high pressure, and kimberlite magmas containing this were formed at a place deeper than 300 km. Against RINGWOOD and LOVERING (1970), BOYD and DAWSON (1971), BOYD and NIXON (1973), MITCHELL *et al.* (1973, 1977), and WYATT *et al.* (1975) refuted the results from cooling experiments of melt having a composition of pyroxene-ilmenite intergrowth, the bulk composition of intergrowth, the chemical composition of constituent minerals, the texture and structure of ingergrowth and REE patterns that these intergrowths were formed by eutectoide crystallization from magmas at about 150 km in depth.

The chemical composition of the upper mantle is considered to be lherzolitic. So, the phase relation of lherzolite under an anhydrous condition was determined by GREEN and RINGWOOD (1967a, 1970), and by ITO and KENNEDY (1967). Figure 6 shows the results of Green and Ringwood. The peridotite solidus is of much higher temperature than the geotherm, so that they do not intersect themselves, and the upper mantle shows the solid state so long as it is anhydrous. The upper limit of the pressure of the plagioclase peridotite field (stable at lower pressure condition) is at its highest up to 7–8 kbar (25–30 km). Plagioclase peridotite is found only in the regions where the crust is thin and the geotherm is unusually high, such as in Mid-Oceanic Ridges, Basin, and Range in the U.S.A., and continental margins and island arcs, like in Japan.

Spinel peridotites are found at 25–30 km deep from the surface in both continental and oceanic regions, but the depth where they transform themselves into garnet peridotites varies according to their regions. For example, in the high geotherm region, the depth is 60–70 km deep (the depth is assumed to cover the transition zone of 5–15 km deep (the depth is assumed to cover the transition zone of 5–15 km), while in the low geotherm region, the spinel peridotites are found until a depth of 45 km. Accordingly, spinel peridotites are universally distributed at the top of the mantle in oceanic, island arc and continental regions. But, in stable continents with a thick crust, it is possible that spinel peridotites should be lacking or scanty and that garnet

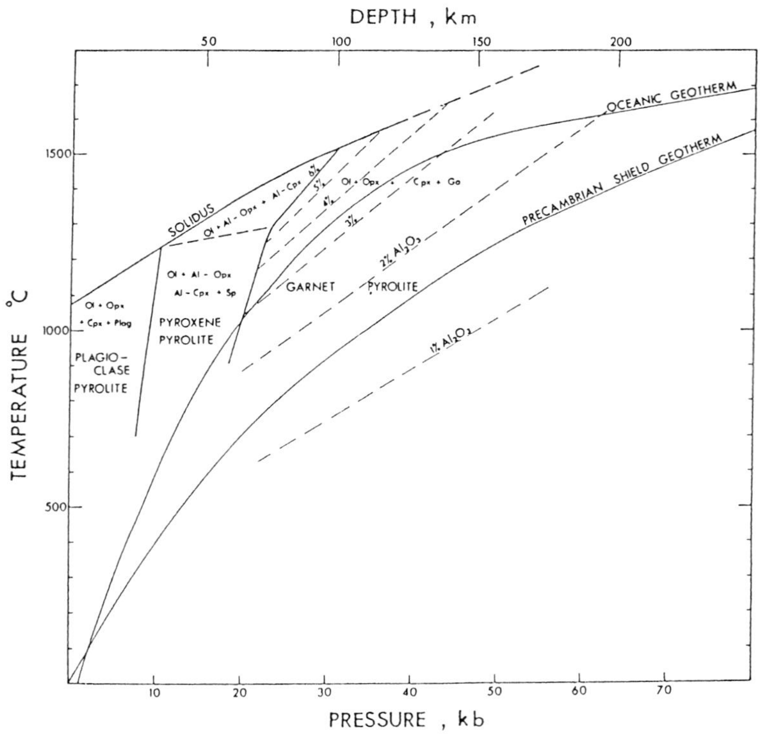

FIG. 6. Stability fields of anhydrous peridotite (pyrolite) (GREEN and RINGWOOD, 1967).

peridotites should be found there instead. The stability fields of garnet peridotites are very wide and are found from 45–75 km deep to about 350 km deep where they transform themselves into modified spinel bearing ultramafic rocks. Quantitatively, most of the upper mantle is composed of garnet peridotite.

BOYD (1973) first proposed the method to estimate the depths of derivation of garnet peridotite xenoliths using the chemical composition of pyroxenes. Since then, various methods (geothermometry and geobarometry) to estimate the equilibration temperatures and pressures of garnet peridotites have been published. These methods are based on the two-pyroxene solvus, Fe^{+2}–Mg^{+2} partitioning between coexisting garnet and clinopyroxene, Fe^{+2}–Mg^{+2} partitioning between coexisting garnet and olivine, and Al^{VI}–Cr partitioning between coexisting pyroxenes. Recently, CARSWELL and GIBB (1980) examined the dependability of these geothermometries in detail, chosing the following five methods of WELLS (1977), MORI and GREEN (1975, 1978), ELIS and GREEN (1979), and O'NEILL and WOOD (1979) as the preferred ones, and calculating the *P*–*T* condition of garnet periodotite xenoliths from Lesotho kimberlites, using these five methods (Fig. 7). As pointed out by BOYD (1973), inflection exists in Lesotho geotherm and sheared-type garnet peridotites are plotted on the side of higher temperatures, and granular-type garnet peridotites on the side of lower temperatures. Further, Lesotho geotherm determined by granular-type ones almost agrees with continental ones, but that gained from sheared-type shows a steep slope. Estimated temperatures and pressures for Lesotho garnet peridotite xenoliths range from 870°–1,450°C and 26–56 kbar respectively. Detailed investigations into diamond inclusions have recently been carried out by TSAI *et al.* (1977), GURNEY *et al.* (1979), and HERVIG *et al.* (1980). According to their results, estimated temperatures and pressures for minerals coexisting in diamonds also range from 960°–1,300°C and 48–59 kbar respectively, and they are within the range of temperatures and pressures of the above-mentioned garnet peridotite xenoliths.

The geothermometries and geobarometries proposed thus far can generally estimate temperatures precisely, while they cannot estimate pressures so precisely. So, an attempt is made here to explain the geobarometry which is quite different from those geobarometries in principle (AOKI *et al.*, in prep.).

AKAOGI and AKIMOTO (1977, 1979) have shown that a continuous increase in the solid solubility of the pyroxene component in garnet is recognized in the $Mg_4Si_4O_{12}$–$Mg_3Al_2Si_3O_{12}$, $Fe_4Si_4O_{12}$–$Fe_3Al_2Si_3O_{12}$ and lherzolite systems at pressures of 40–150 kbar. As very well known, garnet is a typical stoichiometric compound, and its chemical formula can be expressed as $R''_3R'''_2Si_3O_{12}$ where R″ = Ca, Mg, and Fe, and R‴ = Al, Fe‴, and Cr. Its tetrahedral site is occupied only by Si, sometimes being replaced by a trace to

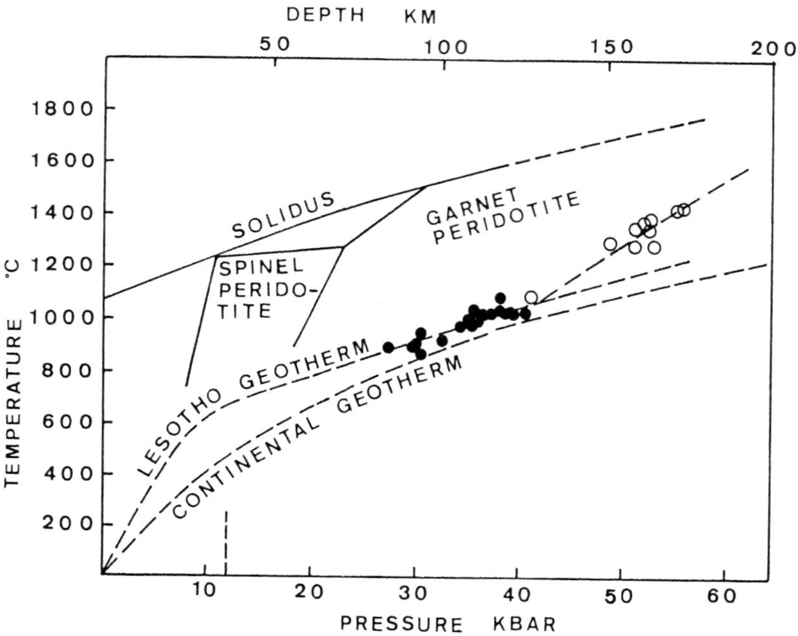

FIG. 7. *P–T* diagram showing the positions for garnet peridotite xenoliths from Lesotho (simplified from CARSWELL and GIBB, 1980). Solid circle = granular type; open circle = sheared type.

negligible amount of Al.

When garnet forms a solid solution with orthopyroxene by the increase of pressures, $2Al \rightleftharpoons Mg \cdot Si$ replacement occurs successively in a octahedral site. Therefore, when garnet contains pyroxene component, its total amounts of Si exceed the value of 3 (when O = 12).

Concerning garnets of various occurrences which are assumed to have been crystallized in the lower crust and the upper mantle; a chemical analyses has been carried out. Their localities are as follows;

(1) kimberlites and their incorporated mantle garnet peridotites and eclogites, and crustal granulites, from South Africa and Lesotho,

(2) kimberlites from the Wyoming-Colorado border,

(3) kimberlites and their incorporated crustal eclogites and granulites from the Colorado Plateau,

(4) crustal eclogites and granulites in latite from Arizona,

(5) garnet pyroxenite xenoliths from Salt Lake, Hawaii.

The histograms for Si in these garnets are shown in Fig. 8. The amounts of Si contained in garnets from all crustal eclogite and granulite and Hawaiian garnet pyroxenite are 3 or slightly lower than 3. This fact shows that these garnets were crystallized at relatively lower pressures. On the other hand, those in garnets which are assumed to have been of the upper mantle origin, are distributed on both sides of the value 3, and they show that, at most, 6% of orthopyroxene components are contained. Garnets in kimberlites contain more orthopyroxene components than those in garnet peridotites and eclogites, and this fact suggests that the former have derived from a deeper

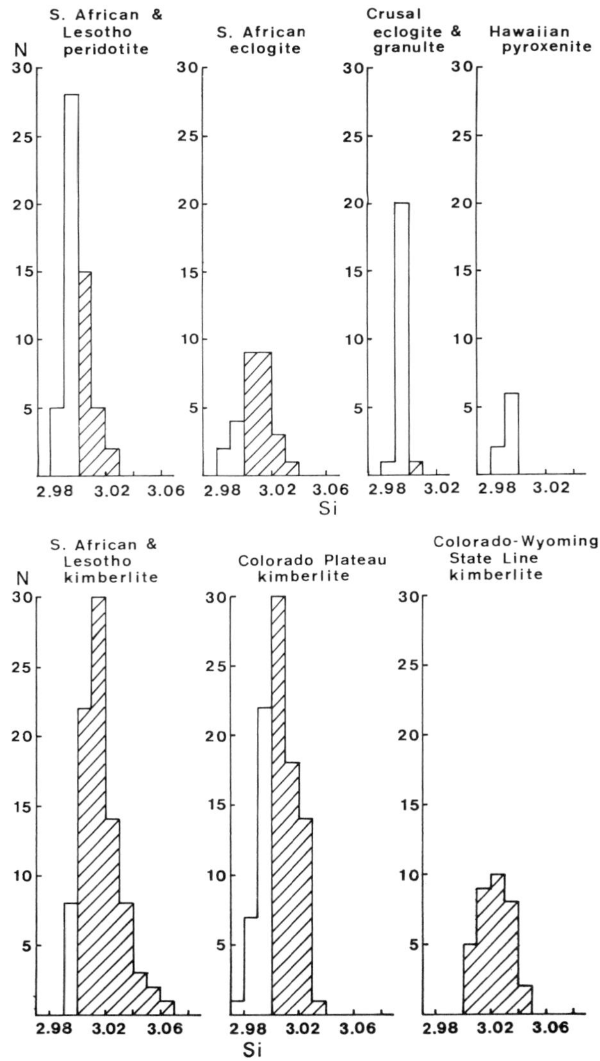

FIG. 8. Histogram for Si in garnet (O = 12) derived from lower crust and upper mantle.

part. Garnets crystallized in lherzolite at 45 kbar and at 1,200°C, respectively, do not contain orthopyroxene components, but those crystallized at 75 kbar and at 1,200°C, contain 10% of orthopyroxene components. Namely, it is shown that all of these garnets have been crystallized at pressures lower than 60 kbar, and this result does not contradict the maximum pressure estimated by CARSWELL and GIBB (1980).

REFERENCES

AKAOGI, M. and AKIMOTO, S. (1977): Pyroxene-garnet solid solution equilibria in the system $Mg_4Si_4O_{12}$–$Mg_3Al_2Si_3O_{12}$ and $Fe_4Si_4O_{12}$–$Fe_3Al_2Si_3O_{12}$ at high pressures and temperatures, *Phys. Earth Planet. Inter.*, **15**, 90–106.

AKAOGI, M. and AKIMOTO, S. (1979): High-pressure phase equilibria in a garnet lherzolite, with special reference to Mg^{2+}–Fe^{2+} partitioning among constituent minerals, *Phys. Earth Planet. Inter.*, **19**, 31–51.

AOKI, K. (1974): Phlogopites and potssic richterites from mica nodules in South African kimberlites, *Contrib. Mineral. Petrl.*, **48**, 1–7.

AOKI, K. and FUJIMAKI, H. (1982): Petrology and geochemistry of calc-alkaline andesite of presumed upper mantle origin from Itinome-gata, Japan, *Am. Mineral.*, **67**, 1–13.

AOKI, K., FUJIMAKI, H. and KITAMURA, M. (1980): Exsolved garnet-bearing pyroxene magacrysts from some South African kimberlites, *Lithos*, **13**, 269–279.

AOKI, K. and KUSHIRO, I. (1968): Some clinopyroxenes from ultramafic inclusions in Dreiser Weiher, Eifel. *Contrib. Mineral. Petrol.*, **18**, 326–337.

AOKI, K. and SHIBA, I. (1973): Pyroxenes from lherzolite inclusions of Itinome-gata, Japan. *Lithos*, **6**, 41–51.

AOKI, K. and SUWA, K. (1977): Major element geochemistry of peridotite nodules from Samburu district, Kenya, *J. Japan. Assoc. Min. Petr. Econ. Geol.*, **72**, 173–180.

BINNS, R. A. (1969): High-pressure magacrysts in basanitic lavas near Armidale, New South Wales, *Am. J. Sci.*, **267-A**, 33–49.

BOYD, F. R. (1973): A pyroxene geotherm, *Geochim. Cosmochim. Acta*, **37**, 2533–2546.

BOYD, F. R. and DAWSON, J. B. (1971): Kimberlite garnets and pyroxene-ilmenite intergrowths, *Carnegie Inst. Wash. Year Book*, **71**, 373–378.

BOYD, F. R. and NIXON, P. H. (1973): Origin of the ilmenite-silicate nodules in kimberlites from Lesotho and South Africa, in *Lesotho Kimberlites*, Ed. P. H. Nixon, Lesotho National Devel. Corp. Maseru, Lesotho, pp. 254–268.

CARSWELL, D. A. and GIBB, F. G. F. (1980): Geothermometry of garnet lherzolite nodules with special reference to those from the kimberlites of northern Lesotho, *Contrib. Mineral. Petrol.*, **74**, 403–416.

CARTER, J. L. (1970): Mineralogy and chemistry of the earth's upper mantle based on the partial fusion-partial crystallization model, *Geol. Soc. Am. Bull.*, **81**, 2021–2034.

CLARKE, D. B. and CARSWELL, D. W. (1977): Green garnets from the Newlands kimberlite, Cape Province, South Africa, *Earth Planet. Sci. Lett.*, **34**, 30–38.

DAWSON, J. B. and HAWTHORNE, J. B. (1973): Magmetic sedimentation and carbonatitic differentiation in kimberlite sills at Benfontein, South Africa, *J. Geol. Soc. Lond.*, **129**, 61–85.

ELLIS, D. .and GREEN, D. H. (1979): An experimental study of the effect of Ca upon garnet-clinopyroxene Fe–Mg exchange equilibria, *Contrib. Mineral. Petrol.*, **71**, 13–22.

FORBES, R. B. and KUNO, H. (1965): Peridotite inclusions and basaltic host rocks, in *Upper Mantle Symposium, New Delhi 1964*, Intern. Union of Geol. Sci. Copenhagen, pp. 161–179.

FRANTSESSON, E. V. (1970): *The Petrology of the Kimberlites* (translated by D. A. Brown), Publication No. 150, Canberra. Dept. of Geology, Australian National University.

GREEN, D. H. and RINGWOOD, A. E. (1967): The stability fields of aluminous pyroxene peridotite and garnet peridotite and their relevance in upper mantle structure, *Earth Planet. Sci. Lett.*, **3**, 151–160.

GREEN, D. H. and RINGWOOD, A. E. (1970): Mineralogy of peridotitic compositions under upper mantle conditions, *Phys. Earth Planet. Inter.* **3**, 359–371.

GURNEY, J. J. and EBRAHIM, S. (1973): Chemical composition of Lesotho kimberlites, in *Lesotho Kimberlites*, Ed. P. H. Nixon, Lesotho National Devel. Corp. Maseru, Lesotho, pp. 280–284.

HERVIG, R. L., SMITH, J. V., STEELE, I. M., GURNEY, J. J., MEYER, H. O. A. and HARRIS, J. W. (1980): Diamonds: Minor elements in silicate inclusions: pressure-temperature implications, *J. Geophys. Res.*, **85**, 6919–6929.

HORNUNG, G. and NIXON, P. H. (1973): Chemical variations in the knorringite-rich garnets, in *Lesotho Kimberlites*, Ed. P. H. Nixon, Lesotho National Devel. Corp. Maseru, Lesotho, pp. 122–127.

HUTCHISON, R., PAUL, D. K. and HARRIS, P. G. (1970): Chemical composition of the upper mantle, *Min. Mag.*, **37**, 726–729.

ILUPIN, I. P. and LUTZ, B. G. (1971): The chemical composition of kimberlite and questions on the origin of kimberlite magma, *Soretskaya Geol.*, **6**, 61–73.

ITO, K. and KENNEDY, G. C. (1967): Melting and phase relations in a natural peridotite to 40 kilobars, *Am. J. Sci.*, **265**, 519–538.

IUGS Subcommission on the systematics of Igneous Rocks (1973): Classification and nomenclature of plutonic rocks: Recommendations. *N. Jahrb. f. Mineral. Monat.*, 149–164.

KUNO, H. (1969): Mafic and ultramafic nodules in basaltic rocks of Hawaii. *Geol. Soc. Am. Mem.* **115,** 189–234.

KUNO, H. and AOKI, K. (1970): Chemistry of ultramafic nodules and their bearing on the origin of basaltic magmas. *Phys. Earth Planet. Interiors,* **3**, 273–301.

KUSHIRO, I. and AOKI, K. (1968): Origin of some eclogite inclusions in kimberlite. *Am. Mineral.*, **53**, 1347–1367.

MAALØE, S. and AOKI, K. (1977): The major element composition of the upper mantle estimated from the composition of lherzolites. *Contrib. Mineral. Petrol.*, **63**, 161–173.

MATHIAS, M., SIBERT, C. and RICKWOOD, P. C. (1970): Some aspects of the mineralogy and petrology of ultramafic xenoliths in kimberlites. *Contrib. Mineral. Petrol.*, **26**, 75–123.

MEYER, H. O. A. and SVISERO, D. P. 1975): Mineral inclusions in Brazilian diamonds. *Phys. Chem. Earth,* **9**, 797–815.

MITCHELL, R. H. (1977): Geochemistry of magnesian ilmenites from kimberlites in South Africaand Lesotho. *Lithos*, **10**, 29–37.

MITCHELL, R. H., CARSWELL, D. A. and BRUNFELT, A. O. (1973): Ilmenite association trace element studies, Part I: Mineralogy and rare earth geochemistry of an ilmenite-clinopyroxene xenolith from the Monastery Mine. in *Lesotho Kimberlites*, Ed. P. H. Nixon, Lesotho National Devel. Corp. Maseru, Lesotho, pp. 224–229.

MORI, T. and GREEN, D. H. (1975): Pyroxenes in the system $Mg_2Si_2O_6$–$CaMgSi_2O_6$ at high pressure. *Earth Planet. Sci. Lett.*, **26**, 277–286.

MORI, T. and GREEN, D. H. (1978): Laboratory duplication of phase equilibria observed in natural garnet lherzolites. *J. Geol.*, **86**, 83–97.

NIXON, P. H. and BOYD, F. R. (1973): Petrogenesis of the granular and sheared ultrabasic nodule suite in kimberlites in *Lesotho kimberlites,* Ed. P. H. Nixon, Lesotho National Devel. Corp. Maseru, Lesotho, pp. 48–56.

NIXON, P. H., VON KNORRING, O. and ROOKE, J. M. (1963): Kimberlites and associated inclusions of Basutoland: a mineralogical and geochemical study. *Am. Mineral.*, **48**, 1090–1132.

O'Neill, H. St. C. and Wood, B. J. (1980): An experimental study of Fe–Mg partitioning between garnet and olivine and its caribration as a geothermometer: corrections, *Contrib. Mineral. Petrol.*, **72**, 337.

Prinz, M., Manson, D. V., Hlava, P. F. and Keil, K. (1975): Inclusions in diamonds, garnet lherzolite and eclogite assemblages, *Phys. Chem. Earth*, **9**, 797–815.

Reid, A. M., Brown, R. W., Dawson, J. B., Whitifield, G. G. and Siebert, J. C. (1976): Garnet and pyroxene compositions in some diamondiferous eclogites, *Contrib. Mineral. Petrol.*, **58**, 203–220.

Rhodes, J. M. and Dawson, J. B. (1975): Major and trace element chemistry of peridotite inclusions from the Lashaine volcano, Tanzania, *Phys. Chem. Earth.* **9**, 545–557.

Ringwood, A. E. and Lovering, J. F. (1970); Singinficance of pyroxene-ilmenite intergrowths among kimberlite xenoliths, *Earth Planet. Sic. Lett.*, **7**, 371–375.

Smith, J. V. and Dawson, J. B. (1976): Chemistry of micas from kimberlites and xenoliths—I. Micaceous kimberlites, *Geochim. Cosmochim. Acta*, **42**, 959–971.

Sobolev, N. V., Botkunov, A. I. and Lavrent'ev, Yu, G. (1974): A diamond-bearing crundum eclogite from the 'Mir' kimberlite pipe, Yakutia, *Geologiya Geofiz, Novosibirsk*, No. 5.

Sobolev, N. V., Kunznetsova, I. K. and Zyuzin, N. I. (1968): The petrology of grospydite xenoliths from the Zagadochnaya kimberlite pipe in Yakutia, *J. Petrol.*, **9**, 253–280.

Tsai, H., Meyer, H. O. A., Moreau, J. and Milledge, H. J. (1979): Mineral inclusions in diamond: Premier, Jergasfontein and Finsch kimberlites, South Africa, and Williamson mine, Tanzaniz, in *Kimberlites, Diatremes and Diamonds: Their Geology, Petrology and Geochemistry*, Eds. F. R. Boyd and H. O. A. Meyer, *Am. Geophys. Union, Washington*, pp. 16–26.

Watson, K. D. (1955): Kimberlite at Bachelor Lake, Quebec, *Am. Mineral.*, **40**, 565–579.

Wells, P. R. A. (1977): Pyroxene thermometry in simple and complex systems, *Contrib. Mineral. Petrol.*, **62**, 129–139.

Wyatt, B., McCallister, R. H., Boyd, F. R. and Ohashi, Y. (1975): An experimentally produced clinopyroxene-ilmenite intergrowth, *Carnegie Inst. Wash. Year Book*, **74**, 536–539.

Materials Science of the Earth's Interior, edited by I. Sunagawa, pp. 445–460.

IGNEOUS MINERAL EQUILIBRIA IN SOME ALPINE-TYPE PERIDOTITES IN JAPAN

Shoji ARAI

Institute of Geoscience, The University of Tsukuba, Ibaraki, Japan

1. *Introduction*

Through the Upper Mantle Project (UMP), basic petrographical information on so-called alpine-type peridotites in Japan has been accumulated (e.g., RESEARCH GROUP OF PERIDOTITE INTRUSION, 1967). The RESEARCH GROUP OF PERIDOTITE INTRUSION (1967) pointed out the fact that one metamorphic (or orogenic) belt is intruded by one characteristic type of peridotite. For example, they insisted that Sangun and Kamuikotan belts (Fig. 1) are characterized by the almost clinopyroxene-free dunite-harzburgite.

FIG. 1. Locality map of metamorphic belts with ultramafic complexes in Japan. K, Kamuikotan belt; H, Hidaka belt; S, Sambagawa belt; S-Y, Sangun-Yamaguchi belt; M, Maizuru belt; Mn-St, Mineoka-Miura-Setogawa zone.

After the UMP, more complicated characters of alpine-type ultramafic intrusions have been revealed by detailed petrographical and field observations and EPMA analyses of constituent minerals (e.g., ARAI, 1975a, b), which corrected the oversimplified conclusions obtained by the UMP. For example, clinopyroxene-free harzburgite, which had been assumed to be characteristic of the Sangun belt, was clarified to be the product of the thermal metamorphism of serpentinite by the granitic intrusion (ARAI, 1975a). ARAI (1975b) found that the primary peridotites of the Sangun-Yamaguchi belt are almost dunite and clinopyroxene-bearing harzburgite with subordinate amounts of lherzolite, wehrlite, clinopyroxenite and websterite.

ISHIWATARI (1978) and ISHIZUKA (1980) reported ophiolitic complexes from the Maizuru belt and the Kamuikotan belt, respectively.

The thermal history and geothermometry of alpine-type ultramafic rocks have been discussed on the basis of EPMA analyses of minerals (e.g., MORI and BANNO, 1973; ARAI, 1975b; UCHIDA and ARAI, 1978). Alpine-type ultramafic complexes, mainly composed of dunite, wehrlite, clinopyroxenite and hornblendite, are relatively well investigated (e.g., NAKAMURA, 1971), and their aspects of igneous petrology are well known. Researches on the igneous petrology of harzburgites and lherzolites, however, have not been sufficiently done (ARAI, 1980), and it is the main object in the following sections to discuss the petrological characters of alpine-type harzburgites and lherzolites.

2. *Harzburgite and Lherzolite*

Petrological natures of harzburgites (or lherzolites) exposed in the orogenic belt as solid intrusions, are of great importance because they can present chemical and physical conditions of the ancient upper mantle (of oceanic or other situations). In Japan, harzburgites or lherzolites are commonly found in Kamuikotan and Hidaka belts (the axial zone of Hokkaido) (KOMATSU, 1975; NIIDA and KATOH, 1978), the Mineoka-Miura-Setogawa zone (UCHIDA and ARAI, 1978; TAKASAWA, 1976) of central Japan and the Sangun-Yamaguchi belt (ARAI, 1980) (Fig. 1). In the Maizuru belt bounding the western margin of the Sangun-Yamaguchi belt, ISHIWATARI (1978) found that the dunite-harzburgite layer forms the lowest member in the Yakuno ophiolitic complex. At the Horokanai area of the Kamuikotan belt, harzburgite together with small amounts of dunite, makes the lowest member of the ophiolitic complex (ISHIZUKA, 1980). Some of harzburgites from the Hidaka belt have been so strongly metamorphosed to form large amounts of tremolite that the modal amounts of primary minerals cannot be estimated (KINUGASA, 1981). Harzburgites from the Mineoka-Miura-Setogawa zone sometimes contain large amounts of hornblende and/or plagioclase (TAKASAWA, 1976; UCHIDA and ARAI, 1978; ARAI and UCHIDA, 1979). Hornblende- or plagioclase-bearing harzburgites are so enriched in Ca and Al that they are "lherzolite" if recrystallized under spinel peridotite facies (Fig. 2). Typical chromian spinel-bearing harzburgites from the Sangun-Yamaguchi belt and from the Kamuikotan and related zone will be mainly discussed.

According to the nomenclature, harzburgite is separated from lherzolite by the boundary of 5 volume % of clinopyroxene. COLEMAN (1977) gave two distinctly separated areas to metamorphic harzburgite and lherzolite in the olivine-orthopyroxene-clinopyroxene modal triangle (Figs. 2 and 3). However, NIIDA and KATOH (1978) reported harzburgite and lherzolite with gradual modal amounts from the axial zone of Hokkaido (Kamuikotan and Hidaka belts). OYAMA (1981) described harzburgite and lherzolite, which partly cover the modal gap of COLEMAN (1977), from the Ochiai-Hokubo complex of the Sangun-Yamaguchi belt (Fig. 3). To be conclusive, harzburgite seems to change gradually to lherzolite without any modal gaps.

2.1 *Mode and chemistry*

Modal amounts of olivine, orthopyroxene and clinopyroxene of some Japanese harzburgites (or lherzolites) are shown in Fig. 2 and 3. Sangun harzburgites are from Tari-Misaka, Ashidachi and Sekinomiya areas (ARAI, 1980) (Fig. 2) and from the Ochiai-Hokubo complex (ARAI, 1975b; OYAMA, 1981) (Fig. 3) of the Sangun-Yamaguchi belt. Kamuikotan harzburgites are taken from the Horokanai, Wassamu, Yamabe, Ashibetsu, Iwanai-dake and Mitsuishi areas of the Kamuikotan and related zone (Figs. 1 and 2).

Olivine and orthopyroxene occupy 90 to 63%, and 35 to 10% in volume,

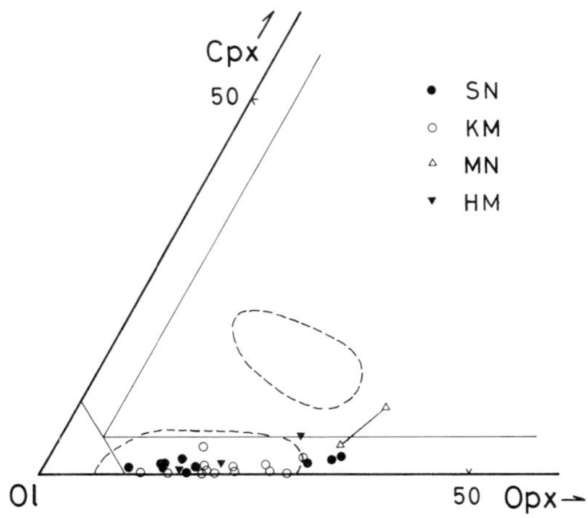

FIG. 2. Modal amounts of some Japanese harzburgites. Ol, olivine; Opx, orthopyroxene; Cpx, clinopyroxene; SN, Sangun-Yamaguchi belt; KM, Kamuikotan belt; MN, Mineoka belt; HM, Horoman complex of the Hidaka belt. The two encircled areas correspond to metamorphic lherzolite and harzburgite of COLEMAN (1977). One lherzolite from MN connected to harzburgite by a light solid line is recalculated at spinel peridotite facies from plagioclase harzburgite.

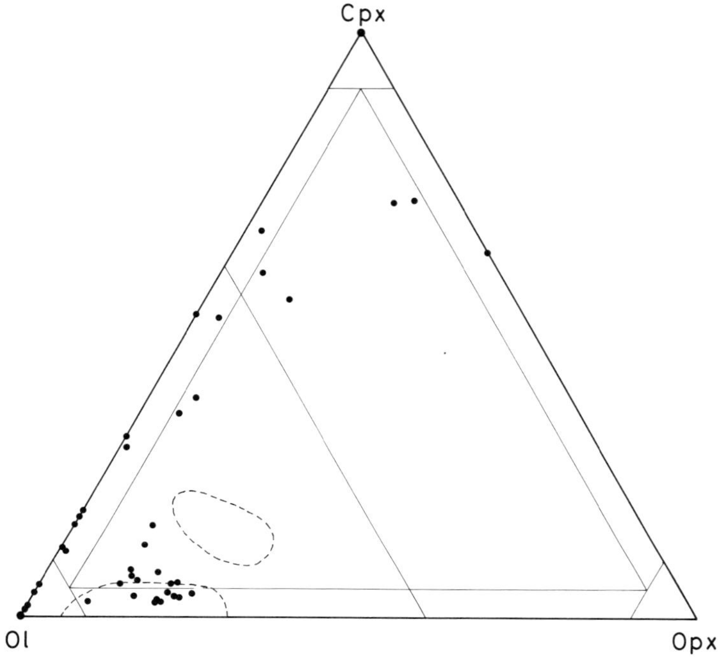

FIG. 3. Modal amounts of ultramafic rocks from the Ochiai-Hokubo complex of the Sangun-Yamaguchi belt (ARAI, 1975b; OYAMA, 1981). The two encircled areas are the same as Fig. 2.

respectively (Fig. 2). Clinopyroxene and chromian spinel are almost always present as minor constituents. Modal amounts of clinopyroxene and chromian spinel are hardly correlated sympathetically with that of orthopyroxene (Figs. 4 and 5).

Harzburgites, which have less than 5 volume % of clinopyroxene, are divided into two types in terms of the presence or absence of modal clinopyroxene at the solidus temperature ("solidus clinopyroxene"). Harzburgite with the solidus clinopyroxene is the same as lherzolite in respect of the phase petrology. Enstatite may contain about 10 mole % of diopside at solidus temperatures (e.g., MORI and GREEN, 1976). Bulk chemical analyses of orthopyroxene in harzburgites of Miura Peninsula and the Horoman complex give 1.0 to 1.2 wt% (Haramura, personal commun.), and 0.8 to 1.0 wt% (NAGASAKI, 1966) of CaO, respectively, which are partly attributed to exsolved lamellae or blebs of diopside within orthopyroxene. If the modal orthopyroxenes of Japanese harzburgites generally contain 0.8 to 1.2 wt% of CaO, the "diopside saturation line" at the solidus temperatures can be drawn as in Fig. 4. Hence, harzburgites plotted far below the region "S" of Fig. 4 may not have modal clinopyroxene at solidus temperatures. It may be possible that

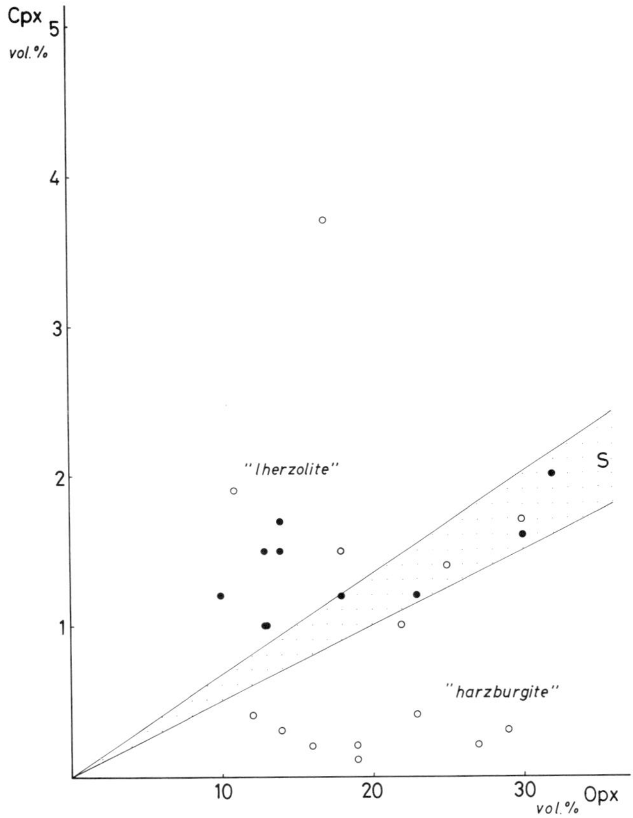

FIG. 4. Relationships between modal amounts of orthopyroxene and clinopyroxene from Sangun and Kamuikotan harzburgites. Dotted region "S" defines the saturation of Ca-rich clinopyroxene in orthopyroxene at the solidus temperature. "Lherzolite" has the solidus clinopyroxene, and "harzburgite" does not. Opx, orthopyroxene; Cpx, clinopyroxene. Open and closed circles, Kamuikotan and Sangun harzburgites, respectively.

Kamuikotan harzburgites usually lose the modal clinopyroxene at the solidus temperature (Fig. 4). On the other hand, Sangun harzburgites may retain the solidus clinopyroxene.

Part of chromian spinel may be subsolidus phase exsolved from pyroxenes judging from the positive correlation between modal amounts of clinopyroxene and chromian spinel (Fig. 6). However, the weak sympathetic relation between modal amounts of orthopyroxene and chromian spinel, and the mode of occurrence of chromian spinel in the thin section may imply that chromian spinel is mainly the solidus phase (Fig. 5). Volumes of solidus clinopyroxene and chromian spinel do not show a distinctive correlation (Fig. 7).

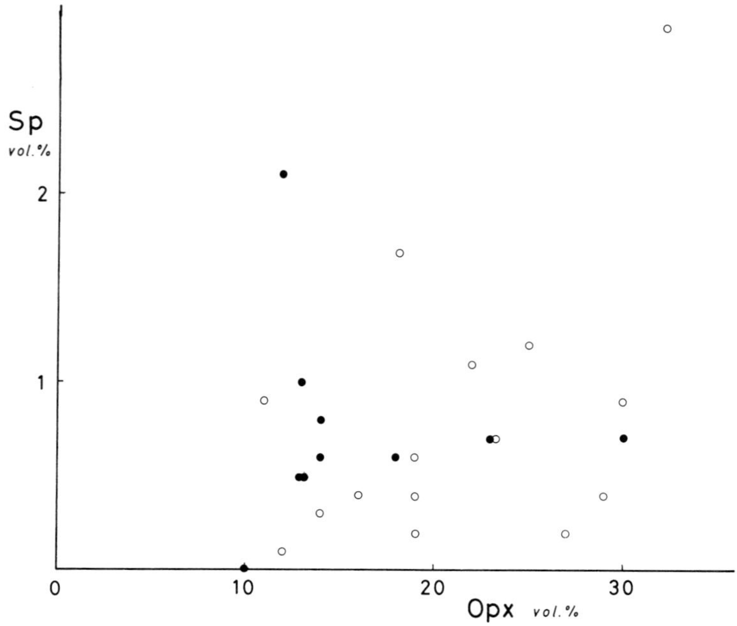

FIG. 5. Relationships between modal amounts of orthopyroxene and chromian spinel in Sangun and Kamuikotan harzburgites. Opx, orthopyroxene. Sp, chromian spinel. Symbols are the same as Fig. 4.

Chemical compositions of constituent minerals of harzburgites have been modified to a degree at subsolidus and metamorphic stages from those of the solidus stage. Olivine effectively has become Mg-rich or Fe-rich by the Mg–Fe redistribution at low temperatures with chromian spinel or clinopyroxene, respectively (ARAI, 1975b). However, olivine sufficiently far from chromian spinel or clinopyroxene, may have preserved the primary Mg/Mg + Fe ratio (or Fo content) of the igneous or solidus stage, which is one of the important petrogenetic indicators. The Cr/Cr + Al ratio of chromian spinel is also more or less changeable at the subsolidus stage through the element redistribution with pyroxenes, though it may be safely assumed that the core of a large chromian spinel grain has preserved the initial Cr/Cr + Al ratio of the solidus stage, which is also an important petrogenetic indicator.

The Fo content of olivine shows a good positive correlation with the Cr/Cr + Al ratio of coexisting chromian spinel in Kamuikotan harzburgites (Fig. 8). On the contrary, the Fo content of olivine and the Cr/Cr + Al ratio of chromian spinel in Sangun harzburgites are relatively constant, as reported by ARAI (1980) (Fig 8). ARAI (1980) found that the Fo content of olivine and the Cr/Cr + Al ratio of chromian spinel in Sangun ultramafic rocks of dunite-harzburgite-chromitite complexes are roughly constant; the former varies

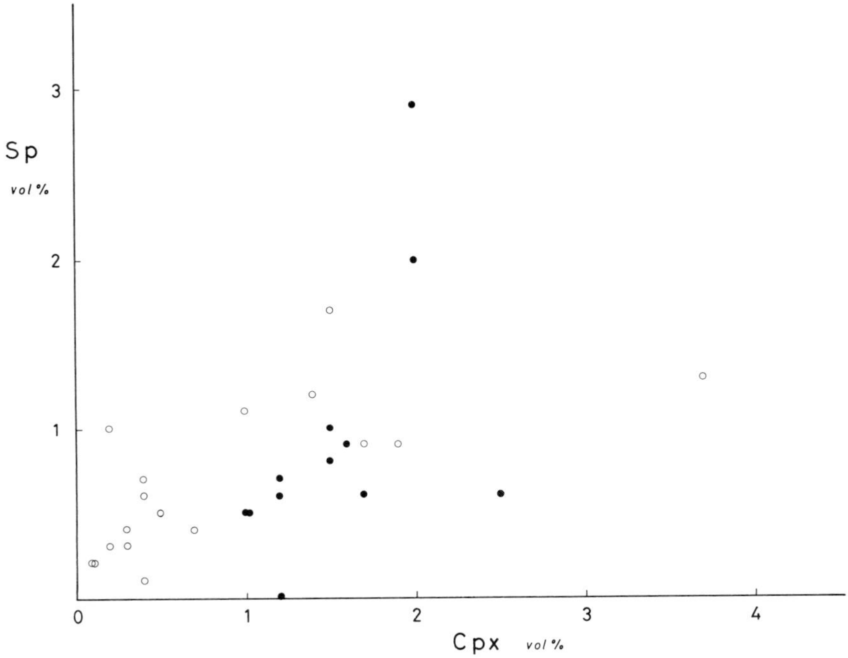

FIG. 6. Relationships between modal amounts of clinopyroxene and chromian spinel in Sangun and Kamuikotan harzburgites. Cpx, clinopyroxene;. Sp, chromian spinel;. Symbols are the same as Fig. 4.

from 88 to 92 and the latter, from 0.4 to 0.6. In Kamuikotan harzburgites, the mode of chromian spinel strongly correlates with the Fo content of olivine and the Cr/Cr + Al ratio of chromian spinel (Figs. 9 and 10). Figure 10 clearly shows a negative correlation of the Cr/Cr + Al ratio of chromian spinel with its volume per cent. In other words, relatively Al-rich spinels with the Cr/Cr + Al ratio less than 0.5, tend to be more than 1% in mode (Fig. 10). On the other hand, chromian spinels with the Cr/Cr + Al atomic ratio more than 0.7, are very small in volume, less than 5 volume % (Fig. 10). The Cr/Cr + Al atomic ratio of chromian spinel in Sangun harzburgites is always around 0.5, although the volume varies from 0.5 to 2.9% (Fig. 10). Figure 9 presents the Fo content in olivine vs. spinel volume % plots in Sangun and Kamuikotan harzburgites. Fo contents of olivine in Kamuikotan harzburgites vary from 91.2 to 92.6, being negatively correlative with the mode of chromian spinel. It is notable that Kamuikotan harzburgites containing primary solidus clinopyroxene, which are, therefore, "lherzolite" in terms of the mineral assemblage, are more enriched in chromian spinel (more than 1.0 volume %), which is Al-rich (Fig. 8), and have more Fe-enriched olivine ($Fo_{91.2}$ to $Fo_{91.5}$) than those free of the primary solidus clinopyroxene.

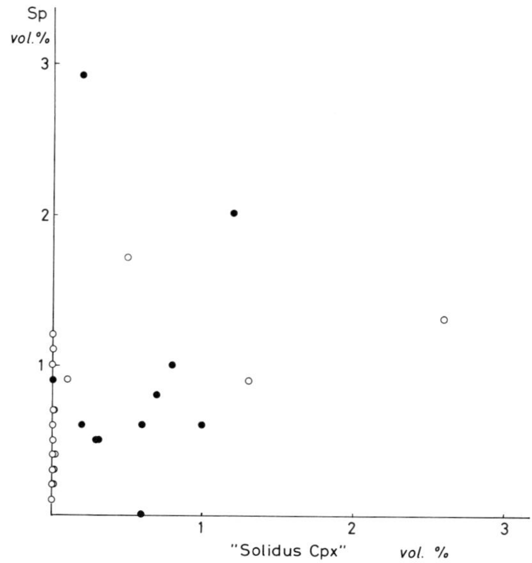

FIG. 7. Relationships between modal amounts of chromian spinel and solidus clinopyroxene in Sangun and Kamuikotan harzburgites. Sp, chromian spinel; Cpx, clinopyroxene. Symbols are the same as Fig. 4.

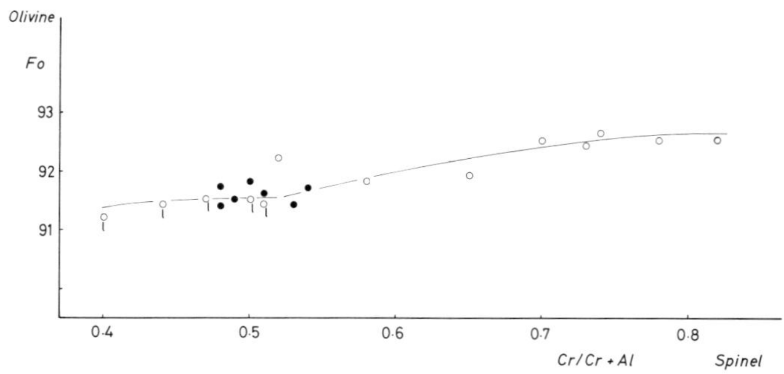

FIG. 8. Relationships between *Fo* content of olivine and Cr/Cr + Al atomic ratio of chromian spinel in Sangun and Kamuikotan harzburgites. Symbols are the same as Fig. 4. *l*, Kamuikotan harzburgite plotted above and around "S" of Fig. 4. Light solid line, possible variation of Fo in olivine with a degree of partial melting of Kamuikotan harzburgites.

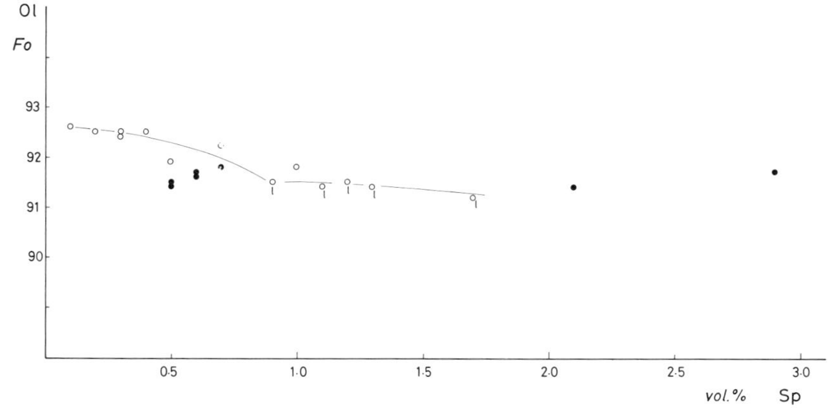

FIG. 9. Relationships between the Fo content of olivine and the volume % of chromian spinel in Sangun and Kamuikotan harzburgites. Symbols are the same as Fig. 4. *l* and light solid line, the same as Fig. 8.

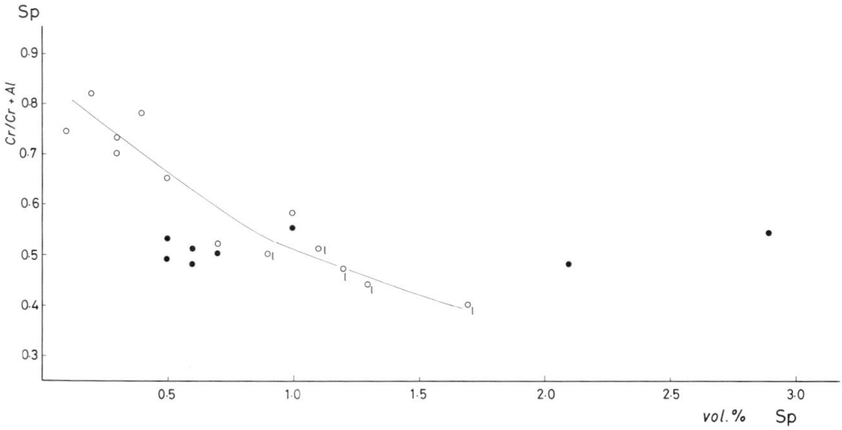

FIG. 10. Relationships between the Cr/Cr + Al atomic ratio and the volume % of chromian spinel in Sangun and Kamuikotan harzburgites. *l*, the same as Fig. 8. Light solid line, possible variation of the Cr/Cr + Al atomic ratio of chromian spinel with a degree of partial melting of Kamuikotan harzburgites. Symbols are the same as Fig. 4.

2.2 Genesis

As the partial melting proceeds, Mg/Mg + Fe and Cr/Cr + Al ratios of constituent minerals of peridotite increase (e.g., MYSEN and KUSHIRO, 1977). Increasing in the Cr/Cr + Al ratio, chromian spinel may continuously decrease in the volume % as the degree of partial melting enhances. Therefore, the mode of chromian spinel can be an indicator of the degree of partial melting, provided that the starting materials are the same.

The relationships among chemical compositions of olivine and chromian spinel and the volume of chromian spinel in Kamuikotan harzburgites indicate that Kamuikotan harzburgites are refractory residua after the partial melting at various degrees. Kamuikotan harzburgites formed by the relatively low degree of partial melting, have the spinel lherzolitic mineral assemblage at the solidus temperature. On the other hand, those formed by the higher degree of partial melting are free from Ca-rich clinopyroxene as the solidus phase and contain Mg-rich olivine and orthopyroxene and Cr-rich chromian spinel (Figs. 8, 9 and 10). In terms of the locality, harzburgites from the Horokanai-Wassamu area are the most refractory, and those from the Mitsuishi area and the Ashibetsu-Yamabe area have the solidus clinopyroxene and are less refractory.

In Sangun harzburgites, the Fo content of olivine and the Cr/Cr + Al ratio of chromian spinel are almost constant, 91.5 and 0.5, respectively, irrespective of the modal amount of chromian spinel. MYSEN and KUSHIRO (1977) showed in melting experiments of peridotites that the Fo content of olivine coexisting with orthopyroxene, Ca-rich clinopyroxene, spinel and liquid, increases far more slowly as the partial melting proceeds than that of olivine coexisting with orthopyroxene, spinel and liquid. The constancy of the Fo content of olivine in Sangun harzburgites, which have the solidus clinopyroxene, is thus in harmony with the results of MYSEN and KUSHIRO (1977). Their experimental results, however, indicate a fairly rapid increase of the Cr/Cr + Al ratio of chromian spinel with an increase of the degree of partial melting, though spinels which appeared in their experiments are highly enriched in Al (Cr/Cr + Al atomic ratio less than 0.05). It is likely to conclude that petrographical and chemical properties of Sangun harzburgites may have been obtained by a nearly equilibrium partial melting (ARAI, 1980). Variation of the modal amounts of chromian spinel in Sangun harzburgites is mainly due to the initial heterogeneity of the primordial lherzolite which had also resulted in the local difference of the degree of partial melting.

3. *Chemical Relationships between Dunite and Harzburgite*

Harzburgite is almost always accompanied by dunite. Sometimes dunite is the only associate of harzburgite, except very small amounts of chromitite, as in the case of the Iwanai-dake complex (BAMBA, 1955; ARAI, 1978) and the Tari-Misaka complex (ARAI, 1980). Dunite and harzburgite occasionally make well-layered alternating beds.

In dunite-harzburgite complexes in the Sangun-Yamaguchi belt, olivine in harzburgite has the relatively constant Fo content, 91 to 92 (ARAI, 1980). The Fo content of olivine in associated dunite varies from 88 to 92 (rarely up to 93) (ARAI, 1980). The Cr/Cr + Al ratio of chromian spinel in harzburgite, however, is almost comparable to that of chromian spinel in dunite. In ultramafic complexes of the Maizuru belt, olivine in harzburgite is also almost unchangeable in composition, Fo_{91} to Fo_{92}, though olivine in dunite is more Fe-enriched, varying in the Fo content from 91 to 88 (HIRANO, 1977; ISHIWATARI, 1978).

In the Ochiai-Hokubo complex (ARAI, 1975b; OYAMA, 1981), harzburgite or lherzolite is accompanied by dunite-wehrlite-clinopyroxenite (Fig. 3). The Ochiai-Hokubo complex is an exceptional one and is different in lithology from ordinary ultramafic complexes in the Sangun-Yamaguchi belt, which are composed of dunite, clinopyroxene-poor harzburgite and a small amount of chromitite (ARAI, 1980). Olivine in lherzolite-harzburgite is relatively constant in chemistry, varying from Fo_{89} to Fo_{91} (ARAI, 1975b). On the other hand, the Fo content of olivine in dunite-wehrlite overlying lherzolite-harzburgite is 91 to 87 with few exceptions (ARAI, 1975b).

In the Iwanai-dake complex of the Kamuikotan belt and in the Kamogawa complexes of the Mineoka belt, olivine in dunite is distinctly more magnesian than that in harzburgite. Dunite and harzburgite form alternating layers, both in Iwanai-dake (KATOH, 1978; MIIDA and KATOH, 1978) and Kamogawa (ARAI and UCHIDA, 1978). In the Iwanai-dake complex, olivine in dunite varies in the Fo content from 91 to 94 (Fig. 11). The Cr/Cr + Al ratio of associated chromian spinel varies in harmony with the Fo content of olivine, from 0.6 to 0.8 (Fig. 11). Modal amounts of chromian spinel are always less than 1%, regardless of the Cr/Cr + Al ratio (Fig. 11). Olivine in harzburgite associated with dunite is usually Fo_{91} to Fo_{92} in composition, being

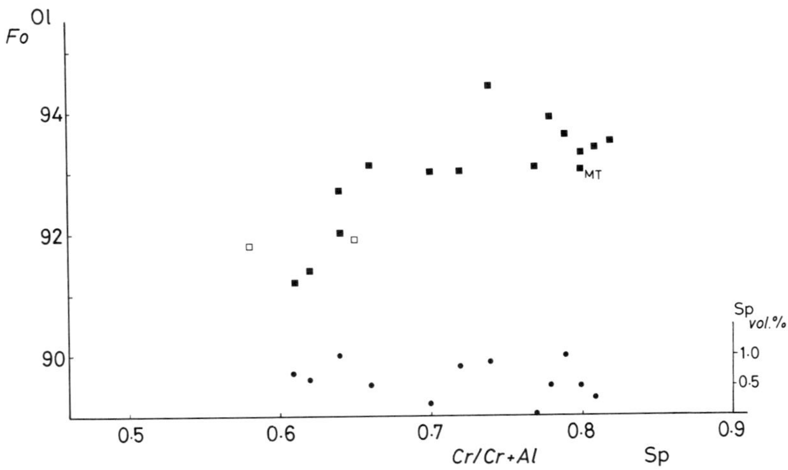

FIG. 11. Relationships between the Fo content of olivine and the Cr/Cr + Al atomic ratio of chromian spinel in Iwanai-dake dunite and harzburgite, and between the volume % and, Cr/Cr + Al atomic ratio of chromian spinel in the Iwanai-dake dunite. Solid square, dunite (MT, dunite from Mitsuishi area); Open square harzburgite. Solid circle, volume % of chromian spinel in dunite.

accompanied by chromian spinel more aluminous than dunite spinel (Fig. 11).

An almost similar chemical relationship between harzburgite and dunite is observed at the Kamogawa complex of the Mineoka belt (Fig. 12). Dunite is quite smaller in volume than harzburgite in the Mineoka belt, though alternating beds of dunite and harzburgite are also reported (Arai and Uchida, 1978). Olivine in dunite changes from 91 to 94 in the Fo content as in the case of the Iwanai-dake dunite (Figs. 11 and 12). The Cr/Cr + Al ratio of chromian spinel increases sympathetically with the Fo content of the associated olivine (Fig. 12). The Fo content of harzburgite olivine is almost identical with that of the least magnesian olivine of dunite (Fig. 12).

In the Wassamu area of the Kamuikotan belt, Fe-rich dunite with olivine of Fo_{83} is found in Mg-rich harzburgite with olivine of Fo_{92} to Fo_{93}. Fe-rich dunite is accompanied by websterite and olivine gabbro which seem to form a differentiated small body with the base of Fe-rich dunite in harzburgite.

In summary, the relationship in chemistry (mainly the Mg/Mg + Fe ratio) of harzburgite and associated dunite is not regular. Olivine in harzburgite is sometimes more magnesian than that in dunite (Ochiai-Hokubo and Wassamu complexes) and is sometimes less magnesian than that in dunite (Iwanai-dake and Kamogawa complexes).

Cryptic layering in terms of Fo and NiO contents of olivine often observed in a dunite layer (e.g., Arai, 1975b) and the sympathetic variation of chromian spinel (Figs. 11 and 12), strongly indicate that the compositional variation in dunite have been obtained by the crystal accumulation during the fractional crystallization of some mafic to ultramafic magmas.

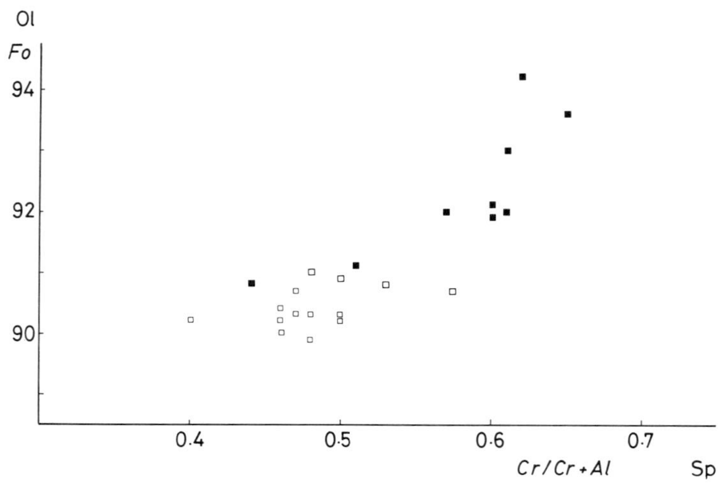

Fig. 12. Relationships between the Fo content of olivine and the Cr/Cr + Al atomic ratio of chromian spinel in dunite and harzburgite from the Mineoka belt. Symbols are the same as Fig. 11.

It is possible that dunite with the Mg/Mg + Fe ratio lower than harzburgite (or lherzolite) (e.g., the Ochiai-Hokubo complex (ARAI, 1975b; OYAMA, 1981), and the Oshima complex (HIRANO, 1977; ISHIWATARI, 1978)) had been formed from some magmas almost equilibrated with the constituent minerals of associated harzburgite (or lherzolite). This process producing dunite and harzburgite (or lherzolite) in the upper mantle can be approximated by the model of partial zone melting (KUSHIRO, 1968) or by the model of the crystal mush intrusion. The crystal mush may have resulted from a partial melting of some primordial lherzolite and had retained crystals with the volume ratio of minerals corresponding to refractory harzburgite or clinopyroxene-poor lherzolite. The residual liquid after the formation of harzburgite or lherzolite by crystal settling, may have precipitated olivine and chromian spinel in consequence of the enlargement of the primary liquidus field of olivine mainly due to the successive release of pressure.

On the other hand, chemical relationships between dunite and harzburgite of Iwanai-dake and Kamogawa complexes connot be explained by the mechanism mentioned above. The liquid forming dunite had not been equilibrated with the harzburgite olivine on account of the low Mg/Mg + Fe ratio of the latter (Figs. 11 and 12). In this case, harzburgite must be interpreted to have been the "wall rock" to the Mg-rich liquid which had precipitated olivine to form dunite more magnesian than harzburgite. Harzburgite alternating with dunite may have been formed by the "disequilibrium" partial zone melting by which the liquid only mechanically dismembered the minerals of harzburgite by stoping to move upwards. Intruded magmas were sometimes relatively Fe-rich to form Fe-rich dunite in Mg-rich harzburgite (e.g. Wassamu complex).

4. *Dunite-Wehrlite-Clinopyroxenite*

Usually, dunite gradually changes to wehrlite, olivine clinopyroxenite and clinopyroxenite with an increase of clinopyronese. Plagioclase sometimes increases in volume together with clinopyroxene. Dunite-wehrlite-clinopyroxenite complexes almost free of harzburgite, are common in the Sambagawa belt (e.g., TAZAKI, 1966; IIZUMI, 1968; RESEARCH GROUP OF PERIDOTITE INTRUSION, 1967; NAKAMURA, 1971; MORI and BANNO, 1973; MAKIMOTO, 1978). Hornblendite and other hornblende-bearing rocks are often associated (e.g., NAKAMURA, 1971).

On the other hand, dunite-wehrlite-clinopyroxenite body is occasionally intimately accompanied by harzburgite-lherzolite. A large body of clinopyroxenite is present in a dunite-harzburgite-lherzolite complex in the Maizuru belt (KUROKAWA, 1972). In the Ochiai-Hokubo complex, dunite, wehrlite and clinopyroxenite are associated with harzburgite-lherzolite, as mentioned above (Fig. 3).

In ophiolitic complexes, olivine-clinopyroxene rocks (dunite-wehrlite-clinopyroxenite) are generally present just on the "petrological Moho" (MALPAS and STRONG, 1975), overlying harzburgite (e.g., JACKSON *et al.*, 1975). ISHIWATARI (1978) reported layered members including dunite, wehrlite and clinopyroxenite lying on the harzburgite member from the Yakuno ophiolitic complex in the Maizuru belt. Ultramafic complexes mainly composed of dunite-wehrlite-clinopyroxenite distributed in the Sambagawa belt may have been formed as ultramafic cumulates on the harzburgite member which had not been exposed.

The Fo content of olivine in dunite associated with wehrlite and clinopyroxenite is variable. Dunites from the Mt. Higashi-Akaishi complex (MORI and BANNO, 1973), the Irisawai complex (MAKIMOTO, 1978) of the Sambagawa belt, and the Ochiai-Hokubo complex have magnesian olivines (Fo_{91} to Fo_{93}). Dunites from the Sambagawa belt of the Kanto Mountains area contain very Fe-rich olivines with the Fo content smaller than 85 (TAZAKI, 1966). Olivine with Fo_{90} to Fo_{85} is the most common in dunite-wehrlite from the Sambagawa belt (MAKIMOTO, 1978). Modal amounts of olivine and clinopyroxene change gradually from dunite to clinopyroxenite (Fig. 3), though olivine-poor wehrlite and olivine clinopyroxenite are much less predominant than dunite, clinopyroxene-poor wehrlite and clinopyroxenite.

Dunite is almost always accompanied by chromitite, of which the volume ratio is quite variable. Dunite-harzburgite complexes (especially the Tari-Misaka complex) in the Sangun-Yamaguchi belt (ARAI, 1980) and the Sarugawa complex (KATOH, 1978), including the Iwanai-dake peridotites of the Kamuikotan belt, possess relatively large amounts of chromitite. The Cr/Cr + Al ratio of chromian spinel in chromitite is almost comparable to that of chromian spinel in dunite. The Cr/Cr + Al atomic ratio of chromian spinel of chromitite in the Sarugawa complex varies from 0.82 to 0.58, which is almost identical to the same value of dunite spinel from Iwanai-dake (Fig. 11). Chromian spinels in Sangun chromitites are also similar to those in dunites in the Cr/Cr + Al ratio (Fig. 13). Modal ratio of chromian spinel in chromitite is quite variable (Fig. 13).

Genesis Ultramafic rocks of the dunite-wehrlite-clinopyroxenite series, and of the dunite-chromitite series, are composed of various amounts of olivine and clinopyroxene, and olivine and chromian spinel, respectively (Figs. 3 and 13). It is natural that they have been formed by the crystal accumulation from some mafic to ultramafic magmas. If they had been produced by crystal settling during the perfect fractional crystallization in the static state, modal amounts would have been expected to be discontinuous (PRESNALL, 1969), which seems to be partly realized. In the Ochiai-Hokubo complex, dunite to clinopyroxene-poor wehrlite and clinopyroxenite clearly predominate over intermediate rocks in volume (OYAMA, 1981). In the Tari-

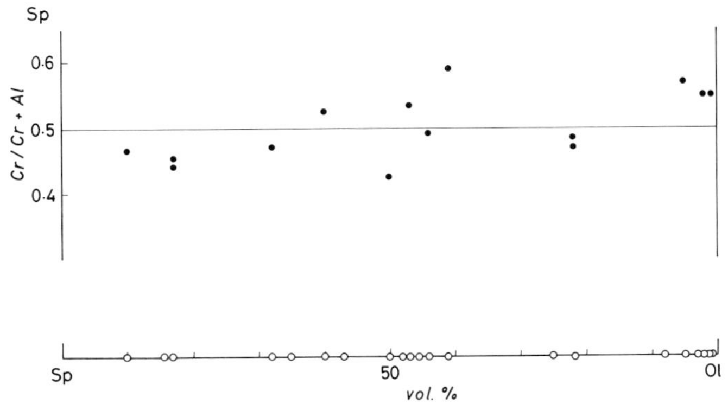

FIG. 13. Relationships between the Cr/Cr + Al atomic ratio of chromian spinel and the volume ratio of olivine and chromian spinel in dunites and chromitites from the Sangun-Yamaguchi belt (ARAI, 1980). Ol, olivine; Sp, chromian spinel.

Misaka complex, the volume of chromian spinel is discontinuous at 10 to 20% in olivine-chromian spinel bimineralic rocks. However, dunite with 1 to 2 volume % of large discrete clinopyroxene (Fig. 3) and chromitite containing 20 to 90 volume % of chromain spinel (Fig. 13), for example, cannot be produced by the cotectic crystal accumulation, at least on the available phase diagrams (e.g., KUSHIRO, 1969). It must be concluded that bimineralic rocks of olivine-clinopyroxene and olivine-chromian spinel have been formed by the imperfect fractional crystallization and/or the blending of settled crystals due to the turbidity current or any other mechanism in the magma chamber.

REFERENCES

ARAI, S. (1975a): Contact metamorphosed dunite-harzburgite complex in the Chugoku district, western Japan, *Contrib. Mineral. Petrol.*, **52**, 1–16.

ARAI, S. (1975b): Petrology of Alpine-type ultramafic complexes in the Sangun zone, western Japan, Unpublished thesis, Univ. of Tokyo.

ARAI, S. (1978): Chromian spinel lamellae in olivine from the Iwanai-dake peridotite mass, Hokkaido, Japan, *Earth Planet. Sci. Lett.*, **39**, 267–273.

ARAI, S. (1980): Dunite-harzbugite-chromitite complexes as refractory residue in the Sangun-Yamaguchi zone, western Japan, *J. Petrol.*, **21**, 141–165.

ARAI, S. and UCHIDA, T. (1978): Highly magnesian dunite from the Mineoka belt, central Japan. *J. Japan. Assoc. Mineral. Petrol. Econ. Geol.*, **73**, 176–179.

ARAI, S. and UCHIDA, T. (1979): Estimation of the equilibrium condition of the ultramafic rocks in the Setogawa Belt, Shizuoka Prefecture, *Geosci. Rep. Shizuoka Univ.*, **4**, 19–24 (in Japanese with English abstract).

BAMBA, T. (1955): Petrological study on the Iwanaidake peridotite mass, *Bull. Geol. Commit. Hokkaido*, **29**, 7–14 (in Japanese).

COLEMAN, R. G. (1977): *Ophiolites*, Springer-Verlag, Berlin.

HIRANO, H. (1977): Genesis of harzburgite- and wehrlite-series ultramafic rocks in the Maizuru belt, *Japan. J. Geol. Soc. Japan*, **83**, 707–718.

IIZUMI, S. (1968): The Ogawara ultrabasic intrusion, Nagano Prefecture in central Japan (Part 1), *Chikyu Kagaku*, **99**, 267–273 (in Japanese).

ISHIWATARI, A. (1978): A preliminary on the Yakuno ophiolite in the Maizuru zone, inner southwest Japan, *Chikyu Kagaku*, **32**, 301–310 (in Japanese with English abstract).

ISHIZUKA, H. (1980): Geology of the Horokanai ophiolite in the Kamuikotan tectonic belt, Hokkaido, Japan, *J. Geol. Soc. Japan*, **86**, 119–134 (in Japanese with English abstract).

JACKSON, E. D., GREEN, H. W. II and MOORES, E. M. (1975): The Vourinos ophiolite, Greece: cyclic units of lineated cumulates overlying harzburgite tectonite, *Geol. Soc. Am. Bull.*, **86**, 390–398.

KATOH, T. (1978): The Saru-gawa ultrabasic massif in Kamuikotan belt, central axial zone of Hokkaido, *Chikyu Kagaku*, **32**, 273–279 (in Japan with English abstract).

KINUGASA, S. (1981): Ultramafic-mafic complexes of the northern part of Hidaka metamorphic belt, Hokkaido, Japan, Unpublished thesis, Shizuoka Univ. (in Japanese with English abstract).

KOMATSU, M. (1975): Recrystallization of the high alumina pyroxene peridotite of the Uenzaru area in Hidaka province, Hokkaido, Japan, *J. Geol. Soc. Japan*, **81**, 11–28.

KUROKAWA, K. (1972): Ultramafic masses in the eastern part of the Maizuru zone and their geological bearings, *Mem. Fac. Sci. Kyoto Univ., Ser. Geol. Mineral.*, **39**, 65–82.

KUSHIRO, I. (1968): Composition of magma formed by partial zone melting of the earth's uppeer mantle, *J. Geophys. Res.*, **73**, 619–634.

KUSHIRO, I. (1969): The system forsterite-diopside-silica with and without water at high pressures, *Am. J. Sci.*, **267A**, 269–294.

MAKIMOTO, H. (1978): Petrology of the Irisawai ultramafic complex in the Oshika district Nagano Prefecture, central Japan—Petrology and chemical character of the Shiokawa peridotite mass—, *J. Geol. Soc. Japan*, **84**, 317–329 (in Japanese with English abstract).

MALPAS, L. and STRONG, D. F. (1975): A comparison of chrome-spinels in ophiolites and mantle diapirs of Newfoundland, *Geochim. Cosmochim. Acta*, **39**, 1045–1060.

MORI, T. and GREEN, D. H. (1976): Subsolidus equilibria between pyroxenes in the $CaO–MgO–SiO_2$ system at high pressures and temperatures, *Am. Mineral.*, **61**, 616–625.

MORI, T. and BANNO, S. (1973): Petrology of peridotite and garnet clinopyroxenite of the Mt. Higashi-Akaizhi mass, central, Shikoku, Japan—subsolidus relation of anhydrous phases, *Contrib. Mineral. Petrol.*, **41**, 301–323.

MYSEN, B. O. and KUSHIRO, I. (1977): Compositional variations of coexisting phases with degree of melting of peridotite in the upper mantle, *Am. Mineral.*, **62**, 843–865.

NAGASAKI, H. (1966): A layered ultrabasic complex at Horoman, Hokkaido, Japan, *J. Fac. Sci., Univ. Tokyo*, **16**, 313–346.

NAKAMURA, Y. (1971): Petrology of the Toba ultrabasic complex, Mie Prefecture, central Japan, *J. Fac. Sci., Univ. Tokyo*, **18**, 1–51.

NIIDA, K. and KATOH, T. (1978): Ultramafic rocks in Hokkaido, *Assoc. Geol. Collab. Japan, Monogr.*, **21**, 61–81.

OYAMA. (1981): Ochiai-Hokubo ultramafic complex in Okayama Prefecture, southwest Japan, Unpublished thesis, Shizuoka Univ. (in Japanese with English abstract).

PRESNALL, D. C. (1969): The geometric analysis of partial fusion, *Am. J. Sci.*, **267**, 1178–1194.

RESEARCH GROUP OF PERIDOTITE INTRUSION (1967): Ultrabasic rocks in Japan, *J. Geol. Soc. Japan*, **73**, 543–553.

TAKASAWA, K. (1976): Anorthite in peridotite from the Setogawa Group, Shizuoka Prefecture, central Japan, *Chikyu Kagaku*, **30**, 163–169 (in Japanese).

TAZAKI, K. (1966): Ultrabasic rocks in the northern Kwanto mountains, central Japan, *Chikyu Kagaku*, **84**, 14–25 (in Japanese).

UCHIDA, T. and ARAI, S. (1978): Petrology of ultramafic rocks from the Boso Peninsula and the Miura Peninsula, *J. Geol. Soc. Japan*, **84**, 561–570.

Chapter 6: Interaction between Solid and Fluid Components

Materials Science of the Earth's Interior, edited by I. Sunagawa, pp. 463–475.

H_2O IN THE EARTH'S INTERIOR

YU HARIYA

Department of Geology and Mineralogy, Hokkaido University, Sapporo, Japan

1. Introduction

In the geological evolution of Earth's Interior, small amounts of water are now widely recognized as playing a highly important role, especially in the generation of magmas and the most satisfactory explanation for the low-velocity and Benioff zones. The content and distribution of water are critical factors in determining mantle and crust properties. Various approaches have been made by different workers to determine the role of water in the upper mantle and lower crust (e.g. WYLLIE, 1971; ALLEN *et al.*, 1972; KUSHIRO, 1972; MYSEN, 1973).

Water in Earth's Interior stabilized in hydrous silicate minerals such as amphibole and mica in the upper mantle and the lower crust; at greater depths, it may exist as an intergranular fluid, probably absorbed on mineral surfaces, or it is dissolved in interstitial silicate magma. Magmas are the parent material of igneous rocks. They may originate through partial or complete fusion of pre-existing rocks. In the gas and fluid phase of magmatic melts, water is the main constituent, followed by CO_2. The composition of the magma generated depends on many variables, including the water content, different H_2O–CO_2 ratio, and the stability of hydrous silicate minerals. Knowledge of the pressure temperature conditions under which hydrous silicate minerals exist is, therefore, fundamental to our understanding of melting and related process in the Earth's Interior. The full importance of volatile constituents in the evolution of igneous rocks can not yet be quantitatively assessed, because neither their total amounts nor their relative proportions in the regions of magma generation are precisely known.

The behavior of volatiles is, to a considerable degree, determined by the capacity of magmas to dissolved them at pressures and temperatures typical of the crust and mantle of the earth. Experimental investigations have created a physico-chemical base, which allows us to understand the influence that temperature and pressure may have upon the solubility of water in magmas of different compositions. Recent studies at high-pressure silicate melts showed significant amounts of water (e.g., BURNHAM and JAHNS, 1962; HAMILTON *et al.*, 1964; MYSEN, 1975, 1976). A water content of more than 5% by weight, which is tolerated by some silicate melts at water pressure of a few kilobars, corresponds to a molar concentration of approximately 30%.

ALLEN *et al.* (1972) reported that we can establish a conceptual model of the mantle in which amphiboles and micas essentially fractionate nearly all of the alkalies and water to depths of 50–75 km. At greater depths, the amphiboles decompose, releasing water, sodium and potassium which will become incorporated into mica. At depths of about 100 to 175 km, the breakdown of the mica will release additional water and potassium. Thus, increase in the amount of melting, and most of the water will then be dissolved in the silicate liquid.

Magmatic water is a solution that has thoroughly exchanged chemically and isotopically at magmatic temperatures with a magma system or with an igneous rock body. Isotope information on hydrogen of minerals and rocks of deep-seated origin is extremely useful in deciphering the origin and nature of water in the Earth's Interior. The ultimate origin of the water—either juvenile, in the sense that it comes from degassing within the lower crust or the upper mantle, recycled meteoric water-is not inherent in this difinition. The concept of juvenile water is a widespread one and has influenced various independent fields in igneous petrology and ore deposits. Juvenile water is commonly defined as water that has never been in contact with surface waters. Testing this concept of juvenile water with D/H in combination with $^{18}O/^{16}O$ messurements is a highly interesting topic.

The isotopes of hydrogen provide a natural tracer for water, and their variations have been useful in elucidating problems in meteorology, oceanography and, in general, with the origins of geologically-important aqueous fluids, such as ore-forming solutions and solution with silicate melts. Hydrogen is nearly omnipresent in the forms H_2O, OH, H_2-gas and CH_4, even at great depths in the Earth's Interiour. Therefore, it is conceivable that hydrogen plays a major role, directly or indirectly, in almost all natural occurring geological processes.

2. *Stability of Hydro-Silicate Minerals*

2.1 *Amphiboles*

Amphibole is a hypersolidus mineral in all compositions, although its pressure-temperature field depends on bulk rock chemistry. Some amphiboles are the most likely minerals as the hydrous phases in the xenoliths derived from the upper mantle and in kimberlite. Recently, K-richterite was found in mica nodules in South Africa kimberlite (ERLANK and FINGER, 1970; AOKI, 1974). KUSHIRO and ERLANK (1970) suggested that K-richterite is stable up to at least 30 kb at 1,000°C, and it may be formed in the upper mantle. HARIYA *et al.* (1974) reported that this amphibole is stable up to 30 kb at 900°C. KUSHIRO and ERLANK (1970) also suggested that the breakdown pressures of K-richterite coexisting with diopside and garnet at 1,000°C, were 24 kb and 20 kb.

AOKI *et al.* (1972) have found tremolite phenocrysts with a high richterite molecule in kimberlite diatrems at Buell Park, Arizona. They suggested that this amphibole is formed as a primary phase from kimberlite magma under upper mantle conditions. HARIYA and TERADA (1973) reported that this amphibole has the highest stability field for pressure among the amphibole group and the pressure limit of complete breakdown of this amphibole is about 34 kb at 900°C.

Amphiboles of richterite series show a wide field of stability as compared with calcium-amphibole. It is evident from these experimental results, that these amphiboles could form at pressures up to about 35 kb or at a depth of about 120 km in the continental area when the water pressure is nearly equal to the total pressure in the upper mantle.

The world-wide occurrence of calcic amphiboles, such as kaersutite or pargasite, as possible mantle-derived minerals in alpine-type ultramafic bodies, in ophiolite complexes, and in ultramafic xenoliths of alkali basalts, suggests that these amphiboles are essential hydrous phases in the upper mantle. Stability limit of pargasite in the system of spinel-bearing lherzolite was determined by KUSHIRO (1970). This experimental result indicates that pargasite could form at a depth of about 90 km in the continental area and 65 km in the oceanic area. Many experimental studies indicate that pargasite can exist at upper mantle temperature and pressure in hydrous peridotite (MYSEN and BOETTCHER, 1975; GREEN, 1973) and in hydrous basaltic magma (ALLEN *et al.*, 1971, 1975, 1978).

Kaersutite should be considered as an important phase which may contain high Ti and some K in the upper mantle, like phlogopite. YAGI *et al.* (1975) reported that kaersutite from mineral breccia of Kakanui, New Zealand and trachybasalts of Iki Island, Japan had fairly wide field of stability up to 23 kb at temperatures less than 700°C. On the basis of the experimental result, they suggested that the kaersutite may have been derived from the upper mantle. MERRILL and WYLLIE (1975) also reported that high-pressure limits of amphibole stability fields with excess water have a negative value of dP/dT, which crosses 25 kb at 1,075°C and 30 kb at 925°C in the kaersutite system.

ALLEN *et al.* (1972) reported that amphiboles in an andesite and in three basalts persist to a maximum temperature of 1,090°C and a maximum pressure of 25 kb. The stability decreases with an increasing silica content of the rocks, and the maximum stability for each composition occurs at f_{O_2} at, or less than those of, the N–NO buffer. The maximum depth to which amphiboles of the types encountered in their experiment can exist is about 75 km.

As shown in Fig. 1, some amphiboles show an extremely wide fields of stability. Assuming that the geotherms in the oceanic and continental areas are represented by those given by CLARK and RINGWOOD (1964), and the water pressure is nearly equal to the total pressure in the upper mantle, it would be predicted that some amphibole could form at pressures up to about 120 km in the continental area, if the water pressure is less than the total pressure amphibole breakdown at low temperatures. However, some amphiboles could persist at great depths, provided high activities of H_2O were maintained locally.

Stability of amphiboles is governed not only by their chemical compositions but also by the presence of volatiles. In particular, F substitution for OH markedly increases the thermal stability of amphibole (HOLLOWAY and FORD, 1975). HOLLOWAY (1973) reported that the stability temperature of

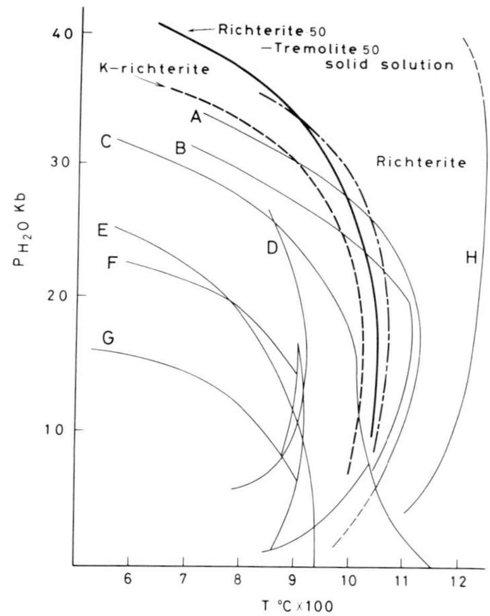

A : Kaersutite (Merrill & Wyllie, 1975)
B : Kaersutite eclogite (Merrill & Wyllie, 1975)
C : Pargasite (Kushiro, 1970)
D : Tschermakite (Oba, 1978)
E : Kaersutite (Yagi et al, 1974)
F : Hornblende (Nishikawa et al, 1971)
G : Hornblende (Lambert & Wyllie, 1968)
H : Fluoro hydroxy pargasite (Holloway & Ford, 1975)

FIG. 1. Stability fields of some amphiboles at high pressures and temperatures.

pargasite increases as $X^{fl}_{H_2O}$ (mixtures of $H_2O + CO_2$) is decreased below one, passes through a temperature maximum and then decreases with a further decrease in $X^{fl}_{H_2O}$.

2.2 *Mica*

Phlogopite, which is mica group, is a possible hydrous mineral in the upper mantle. The possibility of the presence of phlogopite in the Earth's Interior has been suggested recently by several investigators (e.g., KUSHIRO *et al.*, 1967; YODER and KUSHIRO, 1969; WYLLIE, 1970; MODRESKI and BOETTCHER, 1972). Recently, WILSHIRE and TRASK (1971) reported that peridotite inclusions in basanite at Dish Hill and Siberia Craters near Ludlow, California, contain abundant amphiboles and a small amount of phlogopite. Also AOKI (1974) described that phlogopite was commonly present in kimberlite and their cognate or accidental xenoliths in the Kimberly district of South Africa.

A natural phlogopite is stable up to at least 72 kb at about 1,000°C (KUSHIRO *et al.*, 1967). Their experimental results also indicated that a synthetic fluor-phlogopite was stable to at least 450°C higher than a natural hydroxyl-phlogopite at 74 kb. The stability field of phlogopite may change in the presence of other phases. KUSHIRO (1970) indicated that the phlogopite + orthoenstatite assemblage and phlogopite + diopside were stable up to at least 33 kb and about 1,100°C, and at least 32 kb at 1,000°C, respectively. It is suggested that phlogopite may be stable at depths of at least 100 km. MODRESKI and BOETTCHER (1972) reported from their experimental results in the system $K_2O-MgO-CaO-Al_2O_3-SiO_2-H_2O$ that phlogopite will persist to depths of about 75 km with a high geothermal gradient and to about 175 km with a low geothermal gradient.

The effect of additional elements on the stability of phlogopite is important. Natural phlogopite are characterized by wide ranges of TiO_2 (FLOWER, 1969; DAWSON *et al.*, 1970; AOKI, 1974). It is a great matter of regret that the effect of TiO_2 on the stability of phlogopite is unknown. The melting of phlogopite is considered analogous to the partial melting of hydrous parental assemblage in the upper mantle or lower crust.

3. *Behavior of Water in the System of Vapor-Silicate Melts*

3.1 *Solubility of H_2O in silicate melts at high pressure*

The composition of the liquid generated depends on many variables, including the water content and the stability of hydrous minerals. The information about solubility of water in hydrous silicate melts are important for the understanding of deep-seated water and magma generation. Magma generation may depend on the regime of water, rising from the deep mantle in the prefered zones and carried down with the lithosphere in Benioff zones. An understanding of the mechanisms of the solution of volatiles in silicate melts is essential to a full understanding of the role of volatiles in igneous petrogenesis, as the mechanisms of solution largely determine the thermodynamic properties of the solutions and they, in turn, govern the phase equilibrium relations.

The role of volatiles in magma that have appeared since the pioneer efforts of GORANSON (1931, 1932, 1938), emphasis has been placed on the effects of water in lowering melting temperatures of rock-forming minerals and mineral assemblages. It was observed by HAMILTON *et al.* (1964) that a plot of the solubility of water in basalt and andesite to $P_{H_2O} = 6$ kb versus the square root of pressure resulted in a straight line relationship in the pressure region. This straight line relationship between solubility and the square root of pressure might be interpreted to indicate that water entering the solution is fixed in the liquid structure as hydroxyl ions. BURNHAM (1974, 1975) reported that the simple concept of OH^- formation in the melt is not an adequate explanation and he proposed a new model for the H_2O solubility in silicate melts at high pressure. BURNHAM (1974) also reported that for H_2O contents of melts up to 50 mole percent, the solubility of H_2O (mole fraction of H_2O in melt, $X^m_{H_2O}$) was proportional to the square root of the water fugacity, f_{H_2O}. With mole fraction for values of $X^m_{H_2O} > 0.5$, the water solubility was found to vary exponentially with f_{H_2O}. H_2O dissolves in a alminosilicate melts predominantly by a reaction with oxygen ions of the melt, such as $H_2O(v) + O^{2-}(m) = 2OH^-(m)$ (HAMILTON *et al.*, 1964). The drastic effects of H_2O in decreasing the viscosity of these melts (BURNHAM, 1963, 1967.; SHAW, 1963) and in increasing their electrical conductivity (LEBEDEV and KHITAROV, 1964) strongly suggest that this reaction results in depolymerization of melt. Consequently, it appears to be mainly the bridging oxygen ions of the polymeric framework that are involved. Evidence that OH^- is produced in the solution reaction comes directly from spectral studies of quenched glass (ORLOVA, 1964; VELDE and KUSHIRO, 1976) and indirectly from *P-V-T* studies (BURNHAM and DAVIS, 1964). BURNHAM (1974) proposed the following two ideallized solubility mechanism for H_2O, using the system $NaAlSi_3O_8$–H_2O as an example:

For $X^m_{H_2O} < 0.5$

$$NaAlSi_3O_8(m) + H_2O(v) = AlSi_3O_7\,(OH)\,(m) + NaOH\,(m)$$

For $X^m_{H_2O} > 0.5$

$$AlSi_3O_7(OH)\,(m) + nH_2O(v) = AlSi_3O_{7-n}\,(OH)_{2n+1}(m)$$

This model was successfully used to predict the H_2O solubility for a variety of silicate melts to at least 10 kb pressure. It thus appears clear that H_2O at least initially enters the silicate melt by forming hydroxyl groups through an interaction with bridging oxygens in the melt. Additional H_2O in the melt results in further decoupling of the polymers and perhaps also in the entry of molecular H_2O. BURNHAM (1975) also showed that, provided the components of the melt are appropriately chosen, the molar solubility of H_2O does not depend on the bulk composition of the silicate melt, at least within the pressure and temperature range studied by him and his coworkers (10 kb and 1,100°C).

Data of BOETTCHER and WYLLIE (1969), BURNHAM and DAVIS (1971, 1974) and ROSENHAUER and EGGLAR (1975) showed that the solubility in typical melts may increase from about 10 wt% near 10 kb to about 30 wt% at 30 kb. On a basis of wt%, there is a compositional dependence of water solubility, but recalculated to molar solubility using the method described by BURNHAM (1974), this compositional dependence becomes negligible. The temperature dependence of water solubility in silicate melts is also small (KADIK *et al.*, 1971; BURNHAM and DAVIS, 1974) and apparently positive at pressures below 10 kb and temperature below 1,100°C.

The solubilities of water in the some silicate melts in recent preliminary studies (HARIYA *et al.*, 1979; KURODA *et al.*, 1982) are 7 wt% at 1,300°C for quartz, 8.5 wt% at 1,000°C for K-feldspar, 14 wt% at 1,000°C for albite, and 15 wt% at 1,300°C for anorthite in the pressure at 20 kb. These data show the compositional dependence of water solubility in silicate melts.

The break-down of the linkage structure of SiO_4^{4-} due to the dissolution of water into the melt should have definite effect on D/H fractionation between the aqueous fluid and dissolved water. BURNHAM (1979) pointed out that when the water content in an albite melt is more than 6.4%, the mechanism of water dissolution can be considered to be,

$$Si_nO_m + H_2O \rightarrow Si_nO_{m-1}(OH)_2$$

If this mechanism holds for all the types of melts, some fraction of cations, if any, is made free on dissolution of water (e.g., MYSEN and VIRGO, 1980).

3.2 D/H fractionation between water and the silicate melts

Isotopic information on hydrogen of minerals and rocks of deep-seated origin is extremely useful in deciphering the origin and nature of water in the Earth's Interior. Today, there are over 100 laboratories in the world where stable isotope analyses are being made of rocks, minerals and geological fluids. However, experimental determinations of hydrogen isotope fractionations among minerals have only recently been reported (SUZUOKI and EPSTEIN, 1976; O'NEIL and KHARAKA, 1976; SAKAI and TSUTSUMI, 1978; GRAHAM *et al.*, 1980). Mineral studied include muscovite, biotite, amphibole, serpentine, epidote, zoisite, and clay minerals.

As mentioned before, some micas and amphiboles are considered to be the most abundant of the hydrous minerals in the upper mantle, they have been cited as evidence for water in the mantle. An object of experimentally-d etermined D/H fractionation was tremolite$_{50}$-richterite$_{50}$ solid solution and phlogopite, because these minerals show an extremely wide stability fields (HARIYA *et al.*, 1974; KUSHIRO *et al.*, 1967). According to recent preliminary data (Hariya *et al.*, unpublished), δD values of synthetic amphiboles at high pressures were found to be -80 to -90 permil. These data are much lower than that of the starting water (-76 permil), and the equilibrated water enriches deuterium relative to these amphiboles during high pressure experimental runs. Also δD values of synthetic phlogopites were found to be -78 to -90 permil. These data are also much lower than that of starting water. Hydrogen isotope fractionation factors (α) between synthetic amphiboles and phlogopites and water are around 0.9901 and 0.9893, respectively. As a result there is no temperatures and pressures effects on the fractionation factors between both hydrous minerals and water. The range of D/H ratios for synthetic amphiboles and mica are small compare to natural these minerals.

The information about D/H fractionation between silicate melt and water is also important for the understanding of deep-seated water. The only isotopic way of approach to the origin of water in magma or of the juvenile water, can be made based on the information of hydrogen isotopic ratio. Actually, many investigators have tried to make access to this problem through the study of the D/H ratio measurement of hydrous silicate of magmatic origin or mantle origin (e.g., GODFREY, 1962; SHEPPARD and EPSTEIN, 1970; KURODA *et al.*, 1975, 1977; TAYLOR, 1977 MATSUO *et al.*, 1978). In order to estimate the D/H ratio of the water dissolved in magma from the D/H ratio of hydrous minerals crystallized from the magma, it is indispensable to known the D/H fractionation factor between the water dissolved in magma and hydrous minerals. It is, however, almost impossible to get such information by experimental studies, since the separation of quenched glass in which the dissolved water is retained, and authentic crystal is quite difficult.

There may be three types water in the magmas; the dissolved water in a silicate melt, bonded in hydrous minerals, and the aqueous fluid, as shown in Fig. 2 (KURODA *et al.*, 1982). The amount of these kind of water in magma reservoir may be variable, owing to composition and pressure-temperature condition of a magma. SUZUOKI and EPSTEIN (1976) found that the distribution of hydrogen isotopes between hydrous silicates and water depends on the chemical composition of the octahedral cation site in the hydrous silicates, as well as on the temperature of equilibration. This experimental result is only valid for the aqueous fluid-hydrous silicate system. When a magma is undersaturated with water, hydrogen isotopic fractionation takes place in the dissolved water-hydrous silicate system. The energetics of water molecules are different for the aqueous fluid and the dissolved water systems, which implies that the pattern of hydrogen isotopic fractionation is different for aqueous fluid-hydrous silicate system and dissolved water-

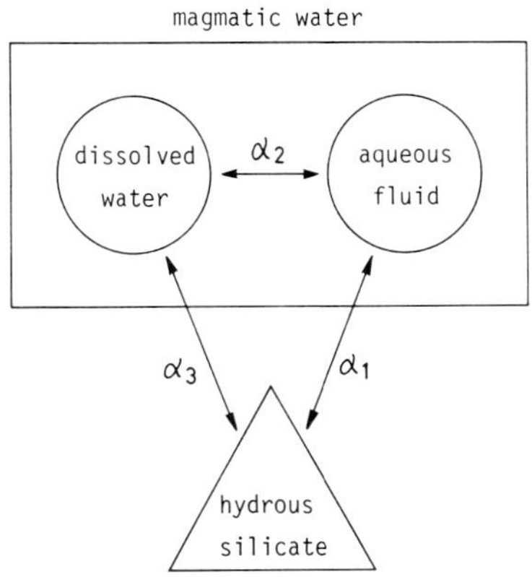

FIG. 2. Fractionation of hydrogen isotopes between various phases in magmatic processes.

hydrous silicate system. SUZUOKI and EPSTEIN (1976) measured α_1 (Fig. 2) for a variety of hydrous silicate-supercritical vapor systems. If the information on α_2, the fractionation factor between dissolved water in silicate melt and supercritical vapor, is available, then we can estimate α_3, the D/H fractionation factor between the water dissolved in magma and hydrous silicate minerals, by combining α_1 and α_2. No data is available for the estimation of the D/H fractionation between aqueous fluid and the dissolved water in silicate melts.

If we define the D/H fractionation factor between the dissolved water and aqueous fluid to be

$$\alpha = (1 + 10^{-3}\delta D_{\text{dissolved water}})/(1 + 10^{-3}\delta D_{\text{aqueous fluid}})$$

In preliminary high-pressures and temperatures experimental data by KURODA *et al.* (1982), a major effect of chemical composition is observed in the fractionation factors between dissolved water in silicate melts and aqueous fluids. The fractionation factors of water in anorthite glass and equilibrated water are close to unity. The dissolved water in quartz melt is depleted in deuterium and enriched in the cases of albite and K-feldspar melts. The isotopic behavior in the plagioclase solid solution system is depend on a change in the stoichiometric formula with CaAl replacing NaSi. Thus, it should be noted that there existed several tens permil difference in deuterium content between albite and anorthite glasses which partained to the same mineral group. Such a wide fractionation might be attributed to the extreme difference in the structural configuration around the hydrogen atom in the hydrous melts of albite and anorthite. Because it is thermodynamically secured, the site preference of D and H is closely related to the energy level of the site where D or H resides. The fractionation of the D content for each silicate mineral melt at a specified pressure-temperature condition may be due to variation of the degree of polymerization of each melt through out high-pressure-temperature runs and quenching process.

Based on the fractionation factors, we may conclude that the D/H fractionation between the dissolved water in silicate melts and aqueous fluid is not significant for granitic intrusions under high-temperature conditions, because granite consists mainly of quartz ($\alpha < 1$), K-feldspar ($\alpha > 1$) and plagioclase (albite $\alpha > 1$ and anorthite $\alpha \sim 1$). It can be said that the fractionation of hydrogen isotopes between hydrous silicate minerals, hydrous silicate melts and aqueous fluid in magma is not so simple, even for the mineral with a relatively simple silicate structure.

MARTIN and DONNAY (1972) submitted that hydroxyl groups, to the extent of less than 1 wt% of water, replace oxygen atoms in the certain crystallographic point-positions on the structures of nominally anhydrous mantle phases such as feldspars, garnets, olivines, spinels, and pyroxenes, or of their high-pressure breakdown products. The small amounts of OH-for-O substitution can reasonably be expected to take place, not only in silicate minerals, especially those with non-bridging oxygen atoms, but also in other oxide structures such as spinels and the modifications of TiO_2 (BERAN and ZEMANN, 1971). Quantitative water determinations made by WILKINS and SABINE (1973) have established the presence of small and variable amounts of OH ions in the structures of various anhydrous silicate minerals. Many detailed studies for deep-seated water, especially reliable water analysis and establish refinements by neutron diffraction of anhydrous silicate minerals, will be need to establish the nature of water in Earth's Interior.

I am grateful to Drs. Kuroda, Matsuo, and Suzuoki, who provided the unpublished data and also for many helpful discussions. The author is responsible, however, for any errors of fact or interpretation.

REFERENCES

ALLEN, J. C. and BOETTCHER, A. L. (1971): The stability of amphiboles in basalts and andesites at high pressures, *Geol. Soc. Am. Abs.*, **3**, 490.

ALLEN, J. C., BOETTCHER, A. L. and MARLAND, G. (1975): Amphiboles in andesite and basalt: I. Stability as a function of P–T–f_{O_2}, *Am. Mineral.*, **60**, 1069–1085.

ALLEN, J. C., BOETTCHER, A. L. (1978): Amphiboles in andesite and basalt: II. Stability as a function of P–T–f_{H_2O}–f_{O_2}, *Am. Mineral.*, **63**, 1074–1087.

ALLEN, J. C., MORDRESKI, P. J., HAYGOOD, C. and BOETTCHER, A. L. (1972): The role of water in the mantle of the earth: The stability of amphibole and mica, *24th IGC*, sec. 2, 231–240.

AOKI, K., FODOR, R. V., KEIL, K. and DOWTH, E. (1972): Tremolite with high richterite-molecule content in kimberlite from Buell Park, Arizona, *Am. Mineral.*, **57**, 1889–1893.

AOKI, K. (1974): Phlogopites and potassic richterites from mica nodules in South African kimberlites, *Contrib. Mineral. Petrol.*, **48**, 1–7.

BERAN, A. and ZEMANN, J. (1971): Messung des ultrarot-pleochroismus von mineralen. XI. Der pleochroixmus der OH-streckfrequenz in rutil, anatas, brookit und cassiterit, *Tschermaks Mineral. Petrogr. Mitt.*, **15**, 71–81.

BOETTCHER, A. L. and WYLLIE, P. J. (1969): Phase relations in the system $NaAlSiO_4$–SiO_2–H_2O to 35 kilobars pressure, *Am. J. Sci.*, **267**, 875–909.

BURNHAM, C. W. and JAHANS, R. H. (1962): A method for determining the solubility of water in silicate melts, *Am. J. Sci.*, **260**, 721–745.

BURNHAM, C. W. (1963): Viscosity of a water-rich pegmatite melt at high pressures (abstr.), *Geol. Soc. Am. Spec. Paper*, **76**, 26.

BURNHAM, C. W. (1967): Hydrothermal fluids at the magmatic stage, in *Geochemistry of Hydrothermal Ore Deposits*, Ed. H. L. Barnes, Holt, Rinehart and Winston, New York, pp. 34–76.

BURNHAM, C. W. and DAVIS, N. F. (1971): The role of H_2O in silicate melts: I. P–V–T relations in the system $NaAlSi_3O_8$–H_2O to 10 kilobars and 1,000°C, *Am. J. Sci.*, **270**, 54–79.

BURNHAM, C. W. (1974): $NaAlSi_3O_8$–H_2O solutions: A thermodynamic model for hydrous magmas, *Bull. Soc. Fr. Mineral. Cryst.*, **97**, 223–230.

BURNHAM, C. W. and DAVIS, N. F. (1974): The role of H_2O in silicate melts: II. Thermodynamic and phase relations in the system $NaAlSi_3O_8$–H_2O to 10 kilobars, 700°C, *Am. J. Sci.*, **274**, 902–940.

BURNHAM, C. W. (1975): Water and magmas; a mixing model, *Geochim. Cosmochim. Acta*, **39**, 1077–1084.

BURNHAM, C. W. (1975): Thermodynamics of melting in experimental silicate-volatile systems, *Fortschr. Mineral.*, **52**, 101–118.

BURNHAM, C. W. (1979): Magmas and hydrothermal fluid, in *Geochemistry of Hydrothermal Ore Deposits*, 2nd ed., Ed. H. L. Barnes, Holt, Rinehart and Winston, New York, pp. 71–135.

CLARK, S. P. and RINGWOOD, A. E. (1964): Density distribution and constitution of the mantle, *Rev. Geophys.*, **2**, 35–88.

DAWSON, J. B., POWELL, D. G. and REID, A. M. (1970): Ultrabasic xenoliths and lava from the Lashaine volcano, Northern Tanzania, *J. Petrol.*, **11**, 519–548.

ERLANK, A. J. and FINGER, L. W. (1970): The occurrence of potassic richterite in a mica nodule from the Wesselton kimberlite, South Africa, *Carnegie Inst. Wash. Year Book*, **68**, 320.

FLOWER, M. F. J. (1959): Phlogopite from Jan Mayen Island (North Atlantic), *Earth Planet. Sci. Lett.*, **6**, 461–466.

GODFREY, J. D. (1962): The deuterium content of hydrous minerals from the East-Central Sierra Nevada and Yosemite National Park, *Geochim. Cosmochim. Acta*, **26**, 1215–1245.

GORANSON, R. W. (1931): The solubility of water of granite, *Am. J. Sci.*, **22**, 481–502.

GORANSON, R. W. (1932): Some notes on the melting of granite, *Am. J. Sci.*, **23**, 227–236.

GORANSON, R. W. (1938): Silicate water systems: Phase equilibria in the $NaAlSi_3O_8$–H_2O and $KAlSi_3O_8$–H_20 systems at high temperatures and pressures, *Am. J. Sci.*, **35 A**, 71–91.

GRAHAM, C. M., SHEPPARD, S. M. F. and HEATON, T. H. E. (1980): Experimental hydrogen isotope studies —I. Systematics of hydrogen isotope fractionation in the systems epidote–H_2O, zoisite–H_2O and AlO(OH)–H_2O, *Geochim. Cosmochim. Acta*, **44**, 353–364.

GREEN, D. H. (1973): Experimental melting studies on a model upper mantle composition at high pressure under water-saturated and water-undersaturated conditions, *Earth Planet. Sci. Lett.*, **19**, 37–53.

HAMILTON, D. L., BURNHAM, C. W. and OSBORN, E. F. (1964): The solubility of water and effects of oxygen fugacity and water content on crystallization of mafic magmas, *J. Petrol.*, **4**, 21–39.

HARIYA, Y. and TERADA, S. (1973): Stability of richterite$_{50}$–tremolite$_{50}$ solid solution at high pressures and possible presence of sodium calcic amphibole in the upper mantle conditions, *Earth Planet. Sci. Lett.*, **18**, 72–76.

HARIYA, Y., OBA, T. and TERADA, S. (1974): Stability relation of some hydro-silicate minerals at high pressure, *Proc. 4th Intern. Conf. on High Press.*, 206–210.

HARIYA, Y., KURODA, Y., MATSUO, S. and SUZUOKI, T. (1979): Deuterium/hydrogen ratios and water content of some silicate melts at high pressure, *High-Press. Sci. Technol.* **1**, 555–560.

HOLLOWAY, J. R. (1973): The system pargasite–H_2O–CO_2: A model for melting of a hydrous mineral with a mixed-volatile fluid —I. Experimental results to 8kb, *Geochim. Cosmochim. Acta*, **37**, 651–666.

HOLLOWAY, J. R. and FORD, C. E. (1975): Fluid-absent melting of the fluorohydroxy amphibole pargasite to 35 kilobars, *Earth Planet. Sci. Lett.*, **25**, 44–48.

KADIK, A. A. and EGGLER, D. H. (1975): Melt vapor relations on the join $NaAlSi_3O_8$–H_2O–CO_2, *Carnegie Inst. Wash. Year Book*, **74**, 479–484.

KURODA, Y., SUZUOKI, T., MATSUO, S. and AOKI, K. (1975): *D/H* ratios of the coexisting phlogopite and richterite from mica nodules and a peridotite in South Africa kimberlites, *Contrib. Mineral. Petrol.*, **52**, 315–318.

KURODA, Y., SUZUOKI, T. and MATSUO, S. (1977): Hydrogen isotope composition of deep-seated water, *Contrib. Mineral. Petrol.*, **60**, 311–315.

KURODA, Y., HARIYA, Y., SUZUOKI, T. and MATSUO, S. (1982): *D/H* fractionation between water and the melts of quartz, K-feldspar, albite and anorthite at high temperature and pressure, *Geochem. J.*, in press.

KURODA, Y., HARIYA, Y., SUZUOKI, T. and MATSUO, S. (1982): *D/H* fractionation between water and the melts of plagioclase and granite at high temperature and pressure, *J. Japan. Assoc. Mineral. Petrol. Econ. Geol.*, in press (in Japanese).

KUSHIRO, I., SYONO, Y. and AKIMOTO, S. (1967): Stability of phlogopite at high pressures and possible presence of phlogopite in the earth's upper mantle, *Earth Planet, Sci. Lett.*, **3**, 197–203.

KUSHIRO, I. (1970): Stability of amphibole and phlogopite in the upper mantle, *Carnegie Inst. Wash. Year Book*, **68**, 245–247.

KUSHIRO, I. and ERLANK, A. J. (1970): Stability of potassic richterite, *Carnegie Inst. Wash. Year Book*, **68**, 231–233.

KUSHIRO, I. (1972): Effect of water on the composition of magmas formed at high pressures, *J. Petrol.*, **13**, 311–334.

LEBEDEV, E. G. and KHITAROV, N. I. (1964): The dependence of electrical conductivity of granite melt and the beginning of granite melting on high pressure of water, *Geokhimiya*, No. 3, 195–201.

MARTIN, R. F. and DONNAY, G. (1972): Hydroxyl in the mantle, *Am. Mineral.*, **57**, 554–570.

MATSUO, S., KURODA, Y., SUZUOKI, T., AOKI. and HARIYA, Y. (1978): Mantle water based on the hydrogen isotopic ratios of hydrous silicate in the mantle, in *4th Inter. Conf. Geochron. Cosmochron. Isotope Geol.*, pp. 278–280.

MERRIL, R. F. and WYLLIE, P. J. (1975): Kaersutite and kaersutite eclogite from Kakanui, New Zealand—Water excess and water deficient melting to 30 kilobars, *Geol. Soc. Am.*, **86**, 555–570.

MODRESKI, P. J. and BOETTCHER, A. L. (1972): The stability of phlogopite and enstatite at high pressures: A model for micas in the interior of the earth, *Am. J. Sci.*, **272**, 852–869.

MYSEN, B. O. (1973): Melting in a hydrous mantle: Phase relations of mantle peridotite with controlled water and oxygen fugacities, *Carnegie Inst. Wash. Year Book*, **72**, 467–478.

MYSEN, B. O. (1975): Solubility of volatiles in silicate melts at high pressure and temperature: The role of carbon dioxide and water in feldspar, pyroxene, and feldspathoid melts. *Carnegie Inst. Wash. Year Book*, **74**, 454–468.

MYSEN, B. O. BOETTCHER, A. L. (1975): Melting in a hydrous mantle: II. Geochemistry of crystals and liquids formed by anatexis of mantle peridotite at high pressures and high temperatures as a function of controlled activities of water, hydrogen, and carbon dioxide,

J. Petrol., **16**, 549–593.

MYSEN, B. O. and BOETTCHER, A. L. (1976): Melting of hydrous mantle: III. Phase relations of garnet websterite + H_2O at high pressures and temperatures, *J. Petrol.*, **17**, 1–14.

MYSEN, B. O. and VIRGO, D. (1980): Solubility mechanisms of water in basalt melt at high pressures and temperatures: NaCaAl–Si_2O_7–H_2O as a model, *Am. Mineral.*, **65**, 1176–1184.

NISHIKAWA, M., KUSHIRO, I. and UEDA, S. (1970): Stability of a natural hornblende at high water pressure: Preliminary experiments, *Japan. J. Geol. Geogra.*, **41**, 41–50.

O'NEIL, J. R. and KHARAKA, Y. K. (1976): Hydrogen and oxygen isotope exchange reactions between clay minerals and water, *Geochim. Cosmochim. Acta*, **40**, 241–246.

ORLOVA, G. P. (1964): The solubility of water in albite melts—Under pressure, *Internat. Geol. Rev.*, **6**, 254–258.

ROSENHAUER, M. and EGGLER, D. E. (1975): Solution of H_2O and CO_2 in diopside melt, *Carnegie Inst. Wash. Year Book*, **74**, 474–479.

SAKAI, H. and TSUTSUMI, M. (1978): *D*/*H* fractionation factors between serpentine and water at 100° to 500°C and 2,000 bar water pressure, and the *D*/*H* ratio of natural serpentines, *Earth Planet, Sci. Lett.*, **40**, 231–242.

SHAW, H. R. (1963): Obsidian–H_2O viscosities at 1,000 and 2,000 bars in the temperature range 700°–900°C, *J. Geophys. Res.*, **68**, 6337–6343.

SHEPPARD, S. M. F. and EPSTEIN, S. (1970): *D*/*H* and $^{18}O/^{16}O$ ratios of minerals of possible mantle or lower crust origin, *Earth Planet. Sci. Lett.*, **9**, 232–239.

SUZUOKI, T. and EPSTEIN, S. (1967): Hydrogen fractionation between OH–bearing minerals and water, *Geochim. Cosmochim. Acta*, **40**, 1229–1240.

TAYLOR, H. P. (1977): Water/rock interactions and origin of H_2O in granitic batholiths, *J. Geol. Soc., Lond.*, **133**, 09–558.

VELDE, B. and KUSHIRO, I. (1976): Infrared spectra of high pressure quenched silicate liquids, *Carnegie Inst. Wash. Year Book*, **75**, 618–619.

WILKINS, R. W. T. and SABINE, W. (1973): Water content of some nominally anhydrous silicates, *Am. Mineral.*, **58**, 508–516.

WILSHIRE, H. G. and TRASK, N. I. (1971): Structural and textural relationships of amphibole and phlogopite in peridotite inclusions, Dish Hill, California, *Am. Mineral.*, **56**, 240–255.

WYLLIE, P. J. (1970): Ultramafic rocks and the upper mantle, Miner. *Soc. Am. Spec. Pap.*, **3**, 3.

WYLLIE, P. J. (1971): Role of water in magma generation and initiation of diapiric uprise in the mantle, *J. Geophys. Res.*, **76**, 1328–1338.

YAGI, K., HARIYA, Y., ONUMA, K. and FUKUSHIMA, N. (1975): Stability relation of kaersutite, *J. Fac. Sci., Hokkaido Univ., Ser IV*, **16**, 331–342.

YODER, H. S. JR. and KUSHIRO, I. (1969): Melting of a hydrous phase: Phlogopite, *Am. J. Sci., Schairer. Vol.* **267 A**, 558–582.

Materials Science of the Earth's Interior, edited by I. Sunagawa, pp. 477–480.

POSSIBLE HYDROUS MAGNESIAN SILICATES IN THE MANTLE TRANSITION ZONE

Syun-iti AKIMOTO and Masaki AKAOGI*

The Institute for Solid State Physics, The University of Tokyo, Roppongi, Minato-ku, Tokyo, Japan

It is generally accepted that small amounts of water in the shallow levels of the upper mantle are stored in hydrous silicate minerals such as amphibole and phlogopite. A few members of humite group minerals, such as clinohumite, $Mg(OH)_2 \cdot 4Mg_2SiO_4$, and chondrodite, $Mg(OH)_2 \cdot 2Mg_2SiO_4$, are also suggested to be stable in the upper mantle at depths down to at most 200 km (YAMAMOTO and AKIMOTO, 1974, 1977). However, very little information has been provided on the stability of hydrous silicate minerals at greater depths corresponding to the mantle transition zone. Very recently, substantial progress in the phase equilibrium study of the system Mg_2SiO_4–MgO–H_2O has been made in our laboratory, particularly in the pressure range above 100 kbar (AKAOGI and AKIMOTO, 1980; AKIMOTO and AKAOGI, 1980). Important results are summarized in this short paper.

The system Mg_2SiO_4–MgO–H_2O was investigated at pressures between 85 and 160 kbar at temperatures between 750 and 1,200°C, using a double-staged, cubic-octahedral-anvil type apparatus. In runs for a gel with Mg/Si ratio of 3 and with 4.0 wt.-percent H_2O, a new dense hydrous magnesian silicate was synthesized at pressures above about 100 kbar in the whole temperature range studied. X-ray diffraction patterns indicated that the new phase is the same as phase B designated by RINGWOOD and MAJOR (1967) based on their preliminary experimental data. The ambiguities of the structure and chemical composition of phase B in their preliminary data are considerably removed through the precession X-ray diffraction photography and the microprobe analysis on a crystal successfully grown to a few tens μm in size. In Table 1, crystallographic parameters of the new dense hydrous magnesian silicate, phase B, are compared with those of hydrous and anhydrous phases appeared in the system Mg_2SiO_4–MgO–H_2O. Although two alternatives, $Mg_{48}Si_{16}O_{84}H_8$ and $Mg_{46}Si_{16}O_{84}H_{12}$, are still open for the

*Present address: Department of Earth Sciences, Kanazawa University, Kanazawa, Japan.

chemical composition of phase B, the density of phase B is the highest in the hydrous silicate phases in the system $MgO-SiO_2-H_2O$.

Stabilities and reaction behaviors of the phases listed in Table 1 are illustrated in Fig. 1 in terms of composition-paragenesis diagrams projected on the $MgSiO_3-MgO-H_2O$ triangular system. Phase B occurs at pressures from about 100 kbar to at least 160 kbar at temperatures of 750–1,200°C, and phase A, $3Mg(OH)_2 \cdot 2Mg_2SiO_4$, also occurs at pressures from about 50 kbar to about 150 kbar at relatively lower temperatures. These results suggest that phase A and phase B could be stable at depths around 400 km in the mantle, particularly in the subducting slab where the temperatures are much lower than the surrounding normal mantle. The experimental data shown in Fig. 1 also indicate that phase B is stable only in rocks richer in MgO than the accepted mantle rocks, since the Mg/Si ratio in phase B is higher than that in forsterite. Even in this case, however, phase B in addition to phase A would be of much interest as a possible hydrous phase in the mantle transition zone. NAKAMURA and KUSHIRO (1974) suggested on the basis of their experimental

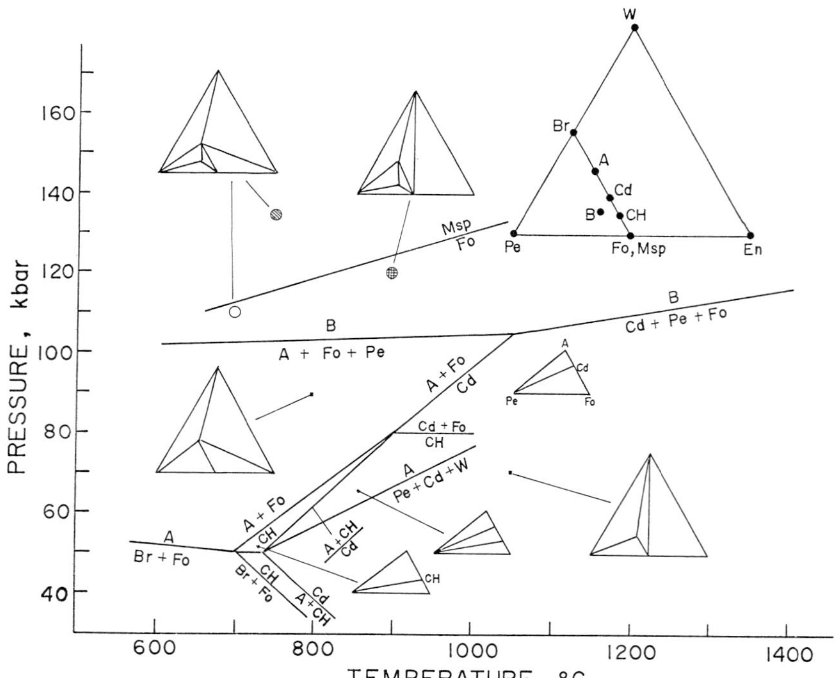

FIG. 1. Composition-paragenesis diagrams projected on the system $MgSiO_3-MgO-H_2O$. Abbreviations: Pe, periclase; Br, brucite; A, B, phases A, B; Cd, chondrodite; CH, clinohumite; Fo, forsterite; Msp, modified spinel; En, enstatite. Forsterite-modified spinel boundary curve is based on the data by KAWADA (1977).

TABLE 1. Crystallographic parameters of hydrous and anhydrous silicate phases in the system Mg_2SiO_4–MgO–H_2O.

Phase	Composition	Space group	Cell parameters				Density (g/cm^3)
			a(Å)	b(Å)	c(Å)	β(°)	
Phase A[1]	$Mg_7Si_2O_{14}H_6$	$P6_3$	7.8603(2)		9.5730(2)		2.96
Phase B[2]	$Mg_{23}Si_8O_{42}H_6$	$P2_1/c$	10.600 (2)	14.098 (2)	10.092 (2)	104.05(2)	3.32*
Phase B[2]	$Mg_{24}Si_8O_{42}H_4$	$P2_1/c$	10.600 (2)	14.098 (2)	10.092 (2)	104.05(2)	3.37*
Chondrodite[3]	$Mg_5Si_2O_{10}H_2$	$P2_1/c$	7.914 (2)	4.752 (1)	10.350 (2)	108.71(5)	3.06
Clinohumite[4]	$Mg_9Si_4O_{18}H_2$	$P2_1/c$	13.695 (1)	4.7474(1)	10.284 (1)	100.64(1)	3.14
Forsterite[5]	α-Mg_2SiO_4	*Pbnm*	4.7553(6)	10.198 (2)	5.9820(7)		3.22
Modified spinel[6]	β-Mg_2SiO_4	*Imma*	5.6983(4)	11.4380(7)	8.2566(8)		3.47
Spinel[7]	γ-Mg_2SiO_4	*Fd3m*	8.076 (1)				3.55

References: 1, Horiuchi *et al.* (1979); 2, Akimoto and Akaogi (1980); 3, Yamamoto and Akimoto (1974); 4, Yamamoto and Akimoto (1977); 5, Akimoto *et al.* (1976); 6, Horiuchi and Sawamoto (1981); 7, Ito *et al.* (1974).

*Observed value of the density: 3.32 ± 0.04 g/cm^3.

data on the system Mg_2SiO_4–SiO_2–H_2O that water passing upward through the mantle peridotite will carry large amounts of SiO_2 in solution, leaving the mantle as a whole relatively depleted in SiO_2. This effect substantially increases the stability fields of magnesian silicates with higher Mg/Si ratios than 2, such as phase A and phase B, in the mantle. The natural occurrence of clinohumite with Mg/Si ratio of 2.25 and chondrodite with Mg/Si ratio of 2.5 in kimberlites derived from the upper mantle supports the present idea (McGetchin *et al.*, 1970; Aoki *et al.*, 1976).

REFERENCES

Akaogi, M. and Akimoto, S., (1980): High pressure stability of a dense hydrous magnesian silicate $Mg_{23}Si_8O_{42}H_6$ and some geophysical implications, *J. Geophys. Res.*, **85**, 6944–6948.

Akimoto, S. and Akaogi, M. (1980): The system Mg_2SiO_4–MgO–H_2O at high pressures and temperatures—possible hydrous magnesian silicates in the mantle transition zone, *Phys. Earth Planet. Inter.*, **23**, 268–275.

Akimoto, S., Matsui, Y. and Syono, Y. (1976): High-pressure crystal chemistry of orthosilicates and formation of the mantle transition zone, in *The Physics and Chemistry of Minerals and Rocks*, Ed. R. G. J. Strens, Wiley, London, pp. 327–363.

Aoki, K., Fujino, K. and Akaogi, M. (1976): Titanochondrodite and titanoclinohumite derived from the upper mantle in the Buell Park kimberlite, Arizona, USA, *Contrib. Mineral. Petrol.*, **56**, 243–253.

Horiuchi, H. and Sawamoto, H. (1981): β–Mg_2SiO_4: Single crystal X-ray diffraction study, *Am. Mineral.*, **66**, 568–575.

Horiuchi, H., Morimoto, N., Yamamoto, K. and Akimoto, S. (1979): Crystal structure of $2Mg_2SiO_4 \cdot 3Mg(OH)_2$, a new high-pressure structure type, *Am. Mineral.*, **64**, 593–598.

Ito, E., Matsui, Y., Suito, K. and Kawai, N. (1974): Synthesis of γ–Mg_2SiO_4. *Phys. Earth Planet. Inter.*, **8**, 342–344.

Kawada, K. (1977): The system Mg_2SiO_4–Fe_2SiO_4 at high pressures and temperatures and the earth's interior, Ph. D. Thesis, Univ. of Tokyo, 187pp.

McGetchin, T. R., Silver, L. T. and Chodos, A. A. (1970): Titanoclinohumite: a possible mineralogical site for water in the upper mantle, *J. Geophys. Res.*, **75**, 255–259.

Nakamura, Y. and Kushiro, I. (1974): Composition of the gas phase in Mg_2SiO_4–SiO_2–H_2O at 15 kilobars, *Carnegie Inst. Wash. Year Book*, **73**, 255–258.

Ringwood, A. E. and Major, A. (1967): High-pressure reconnaissance investigation in the system Mg_2SiO_4–MgO–H_2O, *Earth Planet. Sci. Lett.*, **2**, 130–133.

Yamamoto, K. and Akimoto, S. (1974): High-pressure and high-temperature investigations in the system MgO–SiO_2–H_2O, *J. Solid State Chem.*, **9**, 187–195.

Yamamoto, K. and Akimoto, S. (1977): The system MgO–SiO_2–H_2O at high pressures and temperatures—stability field for hydroxyl-chondrodite, hydroxyl-clinohumite and 10Å-phase, *Am. J. Sci.*, **277**, 288–312.

Materials Science of the Earth's Interior, edited by I. Sunagawa, pp. 481–492.

PROBLEMS ON WATER IN GRANITE INTRUSIONS IN VIEW OF HYDROGEN ISOTOPE STUDY

Yoshimasu KURODA[1], Tetsuro SUZUOKI[2], and Sadao MATSUO[3]

[1]*Department of Geology, Shinshu University, Asahi, Matsumoto, Japan*
[2]*Marine Department, Japan Meteorological Agency, Otemachi, Chiyoda-ku, Tokyo, Japan*
[3]*Department of Chemistry, Tokyo Institute of Technology, O-okayama, Meguro-ku, Tokyo, Japan*

1. Introduction

As we reported several times previously (KURODA *et al.*, 1974, 1975a, 1977a, 1978a, 1979, 1980a, etc.), granitic rocks can be divided into two types, equilibrium type and disequilibrium type, based on the pattern of coexisting biotite and hornblende plotted on $\delta D - X_{Fe}$ diagram*. A typical example of the equilibrium type is shown in Fig. 1. This is for the Hitokabe mass, which is one of the Cretaceous granitic masses in the Kitakami mountainous district, Japan (KURODA *et al.*, 1974). A typical example of the disequilibrium type is seen in Fig. 2. This is for the rapakivi granites in Finland (KURODA *et al.*, 1978b).

SUZUOKI and EPSTEIN (1976) experimentally determined the D/H fractionation factor between water and hydrous silicates, such as biotite and hornblende, under hydrothermal conditions, 400°C–800°C and 1 kbar. The result is expressed by the following equation**,

$$10^3 \ln\alpha_{\text{mineral-water}} = -22.4(10^6 T^{-2}) + 28.2 + (2X_{Al} - 4X_{Mg} - 68X_{Fe}).$$

This equation means that the fractionation factor depends not only on equilibrium temperature but also on the chemistry of minerals, especially

*$\delta D = [(D/H)_{sample} - (D/H)_{SMOW}]/(D/H)_{SMOW} \times 1{,}000$, where SMOW denotes standard mean oceanic water. $X_{Fe} = Fe/(Fe + Mg)$ in octahedral cations.

**GRAHAM and SHEPPARD (1978) described that D/H fractionation between hornblende and water was essentially independent of temperature over a wide range (700°–350°C). However, their description is based on only single experiment. Experimental work must be done more precisely and carefully on amphiboles and micas of various compositions, but here we consider the Suzuoki-Epstein's equation is applicable for all kinds of mica and amphibole. At least it is valid for mica and hornblende of usual compositions used in this paper.

Fe/Mg in octahedral cation sites. The chemistry-dependent term can be approximated to $X_{Fe} = Fe/(Fe + Mg)$, so that the above equation turns out to be $\delta D_{bi} - \delta D_{ho} \simeq 64\{(X_{Fe})_{ho} - (X_{Fe})_{bi}\}$ for equilibrium pairs of biotite and hornblende. Therefore, the tie-line of coexisting biotite and hornblende on the $\delta D - X_{Fe}$ diagram has the inclination of 6.4‰ for 0.1 difference in X_{Fe}, if isotopic equilibration has been attained. Provided that the temperature of isotopic equilibration, X_{Fe} and the δD values of biotite and hornblende are settled, the δD value of the ambient water (free phase of water) existing during the equilibration can be estimated. On the other hand, the equilibration temperature can be estimated when X_{Fe} and δD values of the minerals, and the δD value of the ambient water are given.

The tie-lines of coexisting biotite and hornblende from the natural equilibrium type rocks clearly show the inclination of 6.4‰ for 0.1 difference in X_{Fe} on $\delta D - X_{Fe}$ diagram (Fig. 1). This implies that in the magma the coexisting biotite and hornblende have occurred, attaining equilibrium relationship with respect to the partitioning of hydrogen isotopes between two minerals. On the other hand, the tie-lines of the natural disequilibrium type rocks are nearly vertical to the abscissa on the $\delta D - X_{Fe}$ diagram (Fig. 2). Both biotite and hornblende can be considered not to be in the isotope exchange equilibrium in the disequilibrium type rocks, even though one of them could be in equilibrium with water in magma, from which biotite and hornblende are crystallized.

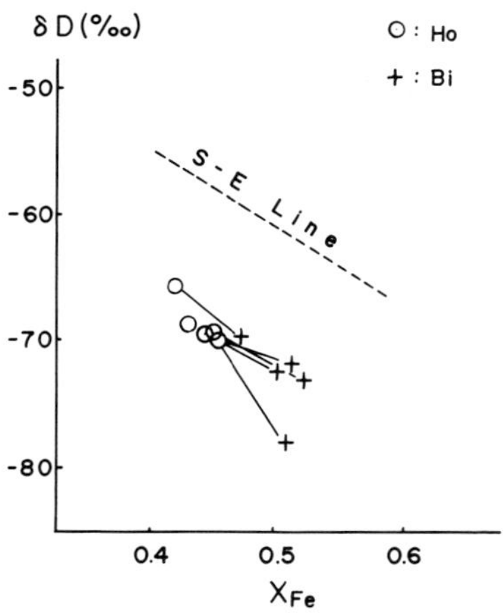

FIG. 1. $\delta D - X_{Fe}$ diagram for the Hitokabe mass from the Kitakami mountainous district, Northeast Japan. S-E line (broken line) in all figures (Figs. 1–8) gives the inclination of the tie-line under the hydrogen isotope exchange equilibrium. The portion of the tie-line varies when the temperature of equilibration and the δD value of water in equilibrium with the minerals are different. (Analytical data have been presented in KURODA *et al.*, 1974.)

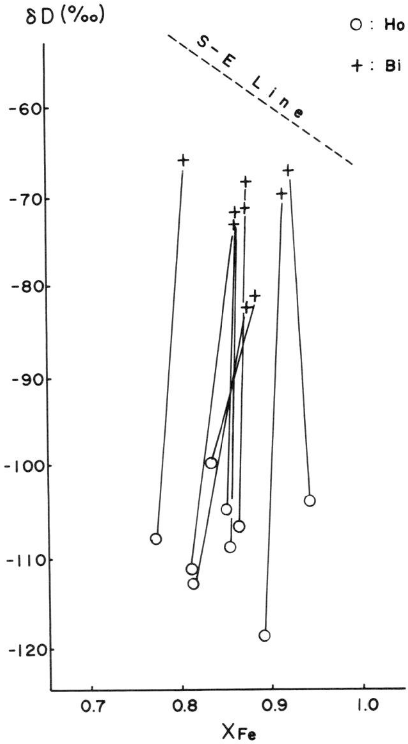

FIG. 2. δD–X_{Fe} diagram for rapakivi granites in Finland. (Analytical data have been presented in KURODA *et al.*, 1978b.)

2. *Equilibrium-Type Rocks*

In Northeast Japan, there are many Cretaceous granitic masses. They belong to the typical equilibrium type as far as we have investigated and the δD value of the ambient water for all of them can be estimated to be about -30‰, if the equilibrium temperature is presumed to be 800°C for the highest and 500°C for the lowest. All the data are shown in Fig. 3.

In Southwest Japan, there are also many Cretaceous to Paleogene granitic masses. Geologically they can be divided into two groups, i.e., the Ryoke plutonic rocks, which are closely associated with the Ryoke metamorphic rocks, and the Sanyo-San'in plutonic rocks, which are not associated with regional metamorphic rocks. Although there are only a few analytical results on the Sanyo-San'in plutonic rocks, they may belong to the equilibrium type and the δD value of the presumed ambient water, $-60 \sim -80$‰, may be different from that for the Cretaceous granites of Northeast Japan (KURODA *et al.*, 1977b). In the Ryoke belt there are many granitic

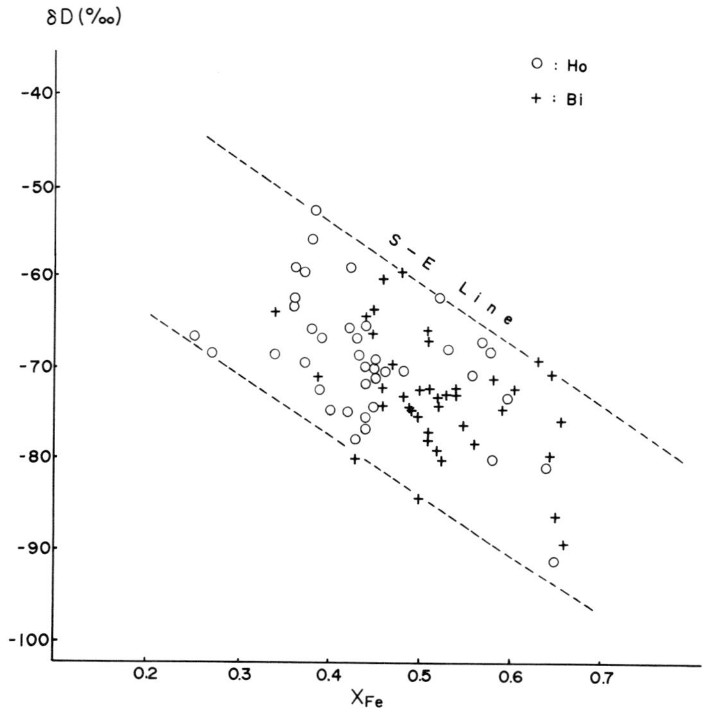

FIG. 3. δD–X_{Fe} diagram for the Cretaceous granitic rocks from Northeast Japan. Tie-lines are not shown, but all are paired hornblende and biotite. (Analytical data have been presented in KURODA *et al.*, 1974, 1975a, 1976, 1980a, 1981.)

masses, of which the petrological characters are different from each other and the intrusive sequence is complicated. Some of them probably belong to the equilibrium type, but others are composed of the distinct disequilibrium type rocks (KURODA *et al.*, 1982).

X_{Fe} values of coexisting biotite and hornblende from the typical equilibrium type rocks are usually around 0.5, the difference between them is about 0.1, and the value of hornblende is smaller than that of biotite. This pattern on the δD–X_{Fe} diagram is called the normal subtype of the equilibrium type.

There is another subtype of the equilibrium type. The granitic rocks from Mt. Evans in the Rocky Mountains, U.S., which is of Precambrian age (about 1,700 m.y.) (PETERMAN and HEDGE, 1968), show the inclination of tie-lines of coexisting biotite and hornblende in good accordance with Suzuoki-Epstein's experimental result, but X_{Fe} of biotite is smaller than that of hornblende (Fig. 4). The difference of the X_{Fe} values between them is same (0.1) as that of the normal equilibrium-type rocks. This is called the reverse subtype of the

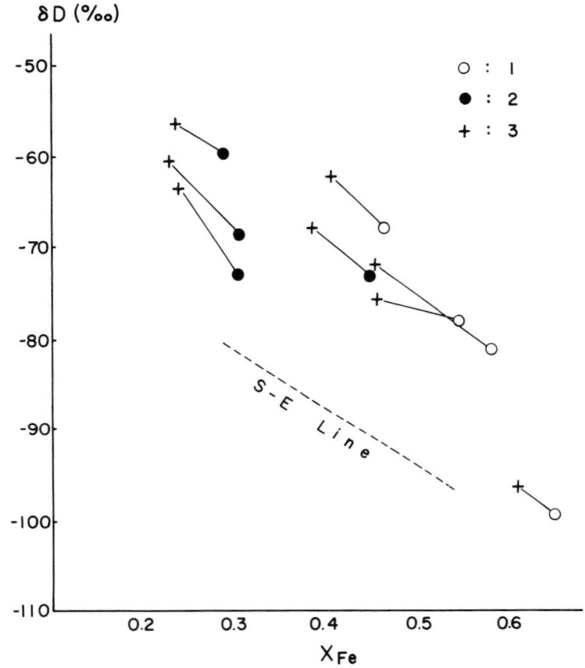

FIG. 4. δD–X_{Fe} diagram for reverse equilibrium type rocks. (1): hornblende from Mt. Evans pluton, (2): hornblende from Antarctica, (3): biotite. (Data are not published, but the chemistry of those of Antarctica is presented in KANISAWA *et al.*, 1979.)

equilibrium type. In some basic gneissic-migmatitic rocks from Antarctica, rocks of this subtype are also found (Fig. 4). In Japan, the rocks of this subtype are scarcely found in some masses of the Ryoke belt.

Besides the typical equilibrium-type rocks, sometimes we can find modified equilibrium-type rocks, which are always close to the normal equilibrium type. In the modified equilibrium-type rocks δD value of biotite is usually lower than the value expected from the value of hornblende; that is, the tie-line is steeper than the normal inclination. This may suggest that the equilibrium temperature for biotite might be lower than that of hornblende.

We have only a small number of data on metamorphic rocks, but coexisting biotite and hornblende from metamorphic rocks, at least derived from sedimentary formations, show the equilibrium situation (Fig. 5). The relationship for muscovite and biotite is also close to the equilibrium pattern (Fig. 5). In this context, the data of coexisting biotite and muscovite in the granitic and metamorphic rocks described by SHIEH and TAYLOR (1969), may also show the equilibrium situation, namely deuterium is always more

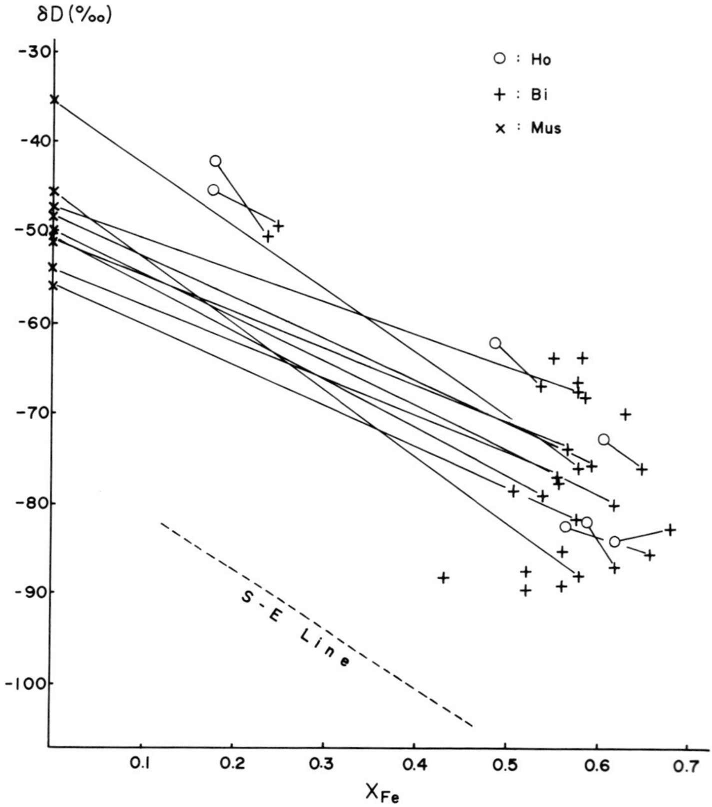

FIG. 5. δD–X_{Fe} diagram for metamorphic rocks from the Ryoke belt. (Analytical data have been presented in KURODA *et al.*, 1982.)

enriched in muscovite than in biotite, though it cannot be plotted on the diagram because of the absence of the chemical composition of biotites.

3. *Disequilibrium-Type Rocks*

In the disequilibrium-type rocks tie-lines of coexisting biotite and hornblende are nearly vertical to the abscissa on the δD–X_{Fe} diagram (Fig. 2). This means that the difference of X_{Fe} values between biotite and hornblende is very small. Usually, the δD value of biotite is higher than that of hornblende. This is called the normal disequilibrium type. The opposite relationship, e.g., lower δD value of biotite and higher of hornblende, is sometimes seen and this is called the reverse disequilibrium type.

Of the normal disequilibrium-type granitic rocks, X_{Fe} values are usually higher than that of the equilibrium-type granitic rocks.

As seen in Fig. 2, the Finnish rapakivi granites are typical normal disequilibrium-type rocks. It is interesting to note that the rocks from Pikes Peak batholith (1,040 m.y.) in the Rocky Mountains, of which petrological characteristics are the same as rapakivi, according to BARKER *et al.* (1975), also belong to the normal disequilibrium type (Fig. 6). The X_{Fe} values are also high, as like as Finnish rapakivi, but the difference of the δD values between biotite and hornblende is smaller. In Japan, we found the normal disequilibrium-type rocks in some masses only in the Ryoke belt, such as the Inagawa, Ikuta, etc., which do not show the rapakivi texture.

Reverse disequilibrium-type rocks are found in the Ishikawa and Miyamoto masses from the central Abukuma, (KURODA *et al.*, 1980a) (Fig. 7). However, it is not easy to distinguish this subtype of rocks from some modified normal equilibrium type. We can just distinguish them after careful investigation of the geological situations.

Other than granitic rocks, there are rocks, of which coexisting mica and amphibole show the normal disequilibrium situation on the δD$-X_{Fe}$ diagram. Phlogopite and richterite pairs in the mica nodules from South African kimberlites are the typical examples (KURODA *et al.*, 1975b). In addition, coexisting mica and amphibole in the alnoites from Sweden, some charnockitic rocks from Antarctica, and some anorthositic rocks from Africa also belong to this type (Fig. 8).

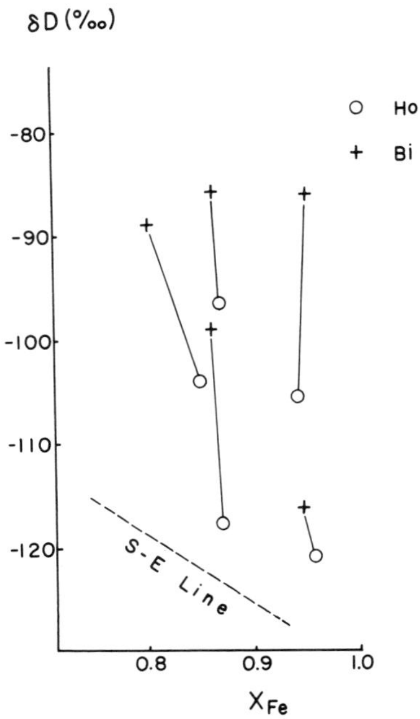

FIG. 6. δD$-X_{Fe}$ diagram for the Pikes Peak batholith. (Data are not yet published.)

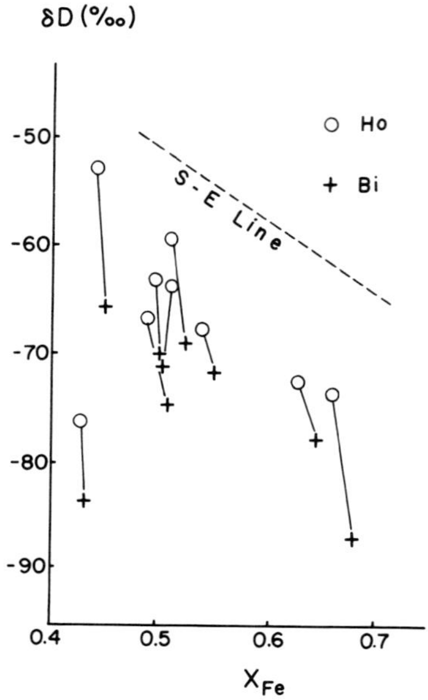

FIG. 7. δD–X_{Fe} diagram for the Ishikawa and Miyamoto masses from the Abukuma plateau, Japan. (Analytical data have been presented in KURODA *et al.*, 1980a.)

4. *Discussion*

We do not have many examples for discussion of the origin of the equilibrium and disequilibrium situations. The δ^{18}O data for each mineral should also be useful for the discussion on the problem, but we have no data about it. However, we think we can discuss the relationship between water and magma only through the D/H data on hydrous minerals at the present time. The voluminous works of Taylor and his coworkers (TAYLOR, 1974, 1978; MAGARITZ and TAYLOR, 1976; SHEPPARD and TAYLOR, 1974, etc.) on oxygen and hydrogen isotopes of granitic rocks and associated formations are very useful for the discussion of the relationship between magma and water, but our works show a slightly different way of approaching the problem. We believe these works must be combined with each other.

Based on the results of Suzuoki-Epstein's experimental work, we can conclude whether or not a pair of hydrous silicates is in isotopic exchange equilibrium. Since their experiments have been conducted under hydrothermal condition with large excess of water, we can estimate the δD value of water coexisting with hydrous silicates from δD values of the equilibrated hydrous silicates pairs when the free phase of water is present.

A geological condition which allows the presence of free phase of water in a magma may not be possible in the mantle. The presence of free water in a

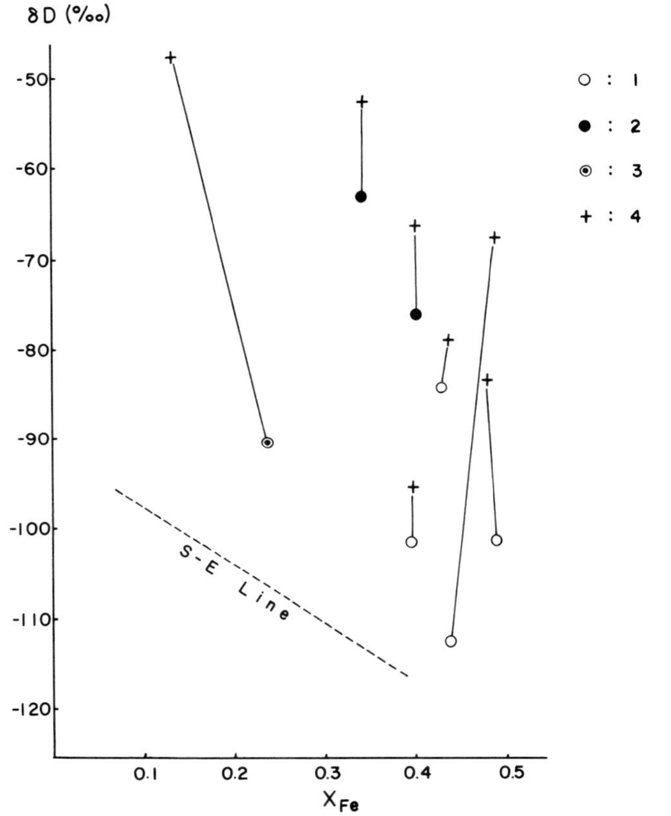

FIG. 8. $\delta D - X_{Fe}$ diagram of amphibole-mica pairs in the charnockites from Antarctica (1), anorthosites in Africa (2), and alnoite from Sweden (3). The symbol for micas is commonly denoted by the cross (4). (Analytical results have not yet been published.)

magma may be realized in the upper crust. In the former case, we cannot estimate the δD value of the water in magma even the data on δD and X_{Fe} of the equilibrium pairs of hydrous silicates are obtained, because there is no information on the hydrogen isotope fractionation factor between hydrous silicate and the dissolved water (structural water).

Since there is no experimental work producing a biotite-hornblende pair, the plot of which on the $\delta D - X_{Fe}$ diagram shows the disequilibrium type, we cannot identify the origin of the disequilibrium-type rocks. However, we tentatively consider that they must be formed under water deficient conditions. Because mica-amphibole pairs from the mica nodules in the kimberlites show the disequilibrium situation (KURODA *et al.*, 1975b), and the water content of hornblende from the disequilibrium-type rocks is generally

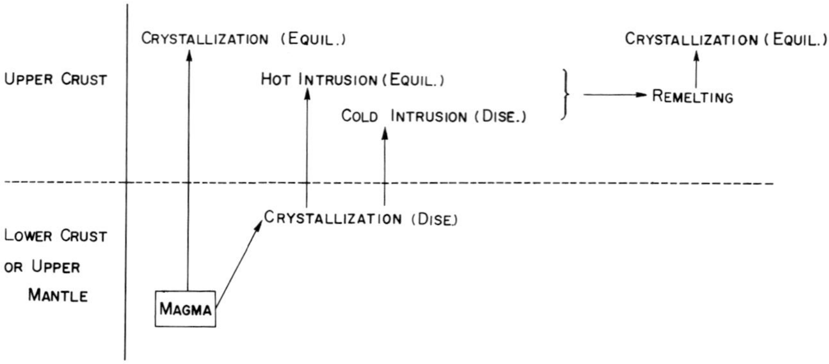

FIG. 9. Schematic diagram for geologic settings of equilibrium- and disequilibrium-type rocks.

lower than that from the equilibrium-type rocks (KURODA *et al.*, 1980b). Under such condition, if biotite or hornblende crystallizes earlier than the other, the available water dissolved in magma should be reduced significantly to give a different δD value for the remaining water. This means that two minerals are formed using the water with a different δD, and then the δD values of the two minerals are different from each other. Such a condition must be realistic in the mantle or in the lower crust. Therefore, the schematic relationship between the equilibrium type and disequilibrium type is shown in Fig. 9.

A small difference of X_{Fe} values between coexisting biotite and hornblende in the disequilibrium type rocks may indicate that the rocks were formed under high-temperature conditions. A larger difference of them in the equilibrium type rocks might be due to the formation or equilibration under lower temperature than in the disequilibrium-type rocks.

In relation to the equilibrium- and disequilibrium-type rocks we must mention the important classification of granitic rocks (or bodies) proposed by CHAPPELL and WHITE (1974) and ISHIHARA (1977) from geological and mineralogical characteristics. CHAPPELL and WHITE classified them to S-type and I-type, and ISHIHARA to ilmenite-series and magnetite series. The relationship between the two categories has not yet been clear, but it seems there may be some correlations. We have not yet investigated the relationship between our classification and the others, but it might be mentioned that the equilibrium-type rocks would be seen in the magnetite-series rather than in the ilmenite-series.

We are very much obliged to Drs. S. Kanisawa, T. Yamada, K. Suwa, and many students of the Department of Geology, Shinshu University, who contributed the unpublished data used in this paper.

REFERENCES

BARKER, F., WONES, D. R., SHARP, W. N. and DESBOROUGH, G. A. (1975): The Pikes Peak batholith, Colorado Front Range, and a model for the origin of the gabbro-anorthosite-syenite-potassic granite suite, *Precamb. Res.*, **2**, 97–160.

CHAPPELL, B. W. and WHITE, A. J. R. (1974): Two contrasting granite types, *Pacific Geology*, **8**, 173–174.

GRAHAM, C. M. and SHEPPARD, S. M. F. (1978): Hydrogen isotope fractionation between aluminous hornblende and water, *Progress in Experim. Petrol., Natural Environ. Res. Contrib., Pub. Ser., D*, **11**, 152–153.

ISHIHARA, S. (1977): The magnetite-series and ilmenite-series granitic rocks, *Mining Geology*, **27**, 293–306.

KANISAWA, S., ONUKI H. and YAMAI, K. (1979): Chemical characteristics of biotite and hornblende from metamorphic rocks around Lützow-Holmbukta, East Antarctica, *Proc. 1st Sympo. Antarc. Geosci., Mem. Nation. Inst., Polar Res.*, sp. No. 14, 153–163.

KURODA, Y., SUZUOKI, T., MATSUO, S. and KANISAWA, S. (1974): D/H fractionation of coexisting biotite and hornblende in some granitic rock masses, *J. Japan. Assoc. Min. Pet. Econ. Geol.*, **69**, 95–102.

KURODA, Y., SUZUOKI, T., MATSUO, S. and KANISAWA, S. (1975a): D/H fractionation of coexisting biotite and hornblende in some granitic masses—A supplement, *J. Japan. Assoc. Min. Pet. Econ. Geol.*, **70**, 352–362.

KURODA, Y., SUZUOKI, T., MATSUO, S. and AOKI, K. (1975b): D/H ratios of the coexisting phlogopite and richterite from mica nodules and a peridotite in South African kimberlites, *Contrib. Mineral. Petrol.*, **52**, 315–318.

KURODA, Y., SUZUOKI, T. and MATSUO, S. (1977a): Equilibrium and disequilibrium type of the Japanese granitic rocks derived from hydrogen isotope study. Plutonism in relation to volcanism and metamorphism, *7th Meeting CPPP (IGCP), Japan*, 184–193.

KURODA, Y., SUZUOKI, T., MURAKAMI, N., KANISAWA, S. and KINUGAWA, T. (1977b): D/H ratios of biotites and hornblendes from some granitic rocks in the Chugoku district, Southwest Japan, *J. Geol. Soc. Japan.*, **83**, 719–724.

KURODA, Y., SUZUOKI, T. and MATSUO, S. (1978a): Hydrogen isotope fractionation in granitic magma, in *Abst. Papers, Intern. Geodyn. Conf., West. Pacif. and Magma Genesis, Tokyo*, pp. 282–283.

KURODA, Y., YAMADA, T., FUJIMOTO, K., SUZUOKI, T. and MATSUO, S. (1978b): Hydrogen isotope study on biotite and hornblende from Finnish granitic rocks, *Geochem. J.*, **12**, 259–263.

KURODA, Y., YAMADA, T., SUZUOKI, T. and MATSUO, S. (1979): Granitic magma and water. *The Earth Monthly*, **1**, 433–441 (in Japanese).

KURODA, Y., ARAI, T., MARUYAMA, T., TANAKA, H., SUZUOKI, T. and MATSUO, S. (1980a): Hydrogen isotope ratios of biotite and hornblende from some granitic masses in Abukuma plateau, Japan., *J. Japan. Assoc. Min. Pet. Econ. Geol.*, Special Issue, **2**, 235–242.

KURODA, Y., SUZUOKI, T. and MATSUO, S. (1980b): Water content, X_{Fe} and D/H of amphibole, *Sci. Rep. Dept. Geol., Kyushu Univ.*, **13**, 191–200 (in Japanese with English abstract).

KURODA, Y., SHIRAI, H., SUZUOKI, T. and MATSUO, S. (1981): Origin of water participated in the formation of the cupriferous iron-sulfide deposit, Hitachi Mine, Northeast Japan, based on hydrogen isotope study, *J. Japan. Assoc. Min. Pet. Econ. Geol.*, **76**, 155–162.

KURODA, Y., YAMADA, T., FUJIMOTO, K., O'HASHI, H., TAKAHASHI, K., OKANO, O., MOCHIZUKI, Y., YAMANA, S. and OHTOMO, S. (1982): A proposal for classification of the granitic rocks from the Ryoke belt in the Chubu and Kinki districts in view of hydrogen isotope study, *J. Geol. Soc. Japan*, **88**, 541–554 (in Japanese with English abstract).

MAGARITZ, M. and TAYLOR, H. P., JR. (1976): $^{18}O/^{16}O$ and D/H studies along a 500km traverse across the Coast Range batholith and its country rocks, central British Columbia, *Can. J. Earth Sci.*, **13**, 1514–1536.

PETERMAN, Z. E. and HEDGE, C. E. (1968): Chronolgy of Precambrian events in the Front Range, Colorado, *Can. J. Earth Sci.*, **5**, 749–756.

SHEPPARD, S. M. F. and TAYLOR, H. P., JR. (1974): Hydrogen and oxygen isotope evidence for the origin of water in the Boulder batholith and Butte ore deposits, Montana, *Econ. Geol.*, **69**, 926–926.

SHIEH, Y. N. and TAYLOR, H. P., JR. (1969): Oxygen and hydrogen isotope studies of contact metamorphism in the Santa Rosa Range, Nevada and other areas, *Contrib. Mineral. Petrol.*, **20**, 306–356.

SUZUOKI, T. and EPSTEIN, S. (1976): Hydrogen fractionation between OH–bearing silicate minerals and water, *Geochim. Cosmochim. Acta*, **40**, 1229–1240.

TAYLOR, H. P., JR (1974): The application of oxygen and hydrogen isotope studies to problems of hydrothermal alteration and ore deposition, *Econ. Geol.*, **69**, 843–883.

TAYLOR, H. P., JR. (1978): Oxygen and hydrogen isotope studies of plutonic granitic rocks, *Earth Planet. Sci. Lett.*, **38**, 177–210.

Materials Science of the Earth's Interior, edited by I. Sunagawa, pp. 493–513.

HYDROTHERMAL SOLUTION-MINERAL EQUILIBRIA AS A CLUE FOR STUDIES OF GEOCHEMICAL TRANSPORT PHENOMENA IN THE EARTH'S CRUST

J. Toshimichi IIYAMA

Geological Institute, Faculty of Science, The University of Tokyo, Hongo, Bunkyo-ku, Tokyo, Japan

1. *Introduction*

In spite of the general recognition of the metasomatic material transfer in the Earth's crust, the analysis of this process occurring in nature is still very qualitative and sometimes even speculative. Theories on this phenomena were first developed by KORZHINSKII (1936, 1950, 1959, 1970), RAMBERG (1952, for example) and THOMPSON (1955, 1959). They are reexplained by several authors (HOFMANN, 1972; FISHER, 1970, 1973; FISHER and ELLIOT, 1974; FLETCHER and HOFMANN, 1974; and FONTEILLES, 1978, for example) in a more comprehensive way than the previous authors did. The theoretical background is now sufficient for the application of field observations.

The practice of this study is, however, hindered because of the scarcity of information on aqueous solution (vapour or fluid state)-mineral equilibria indispensible for quantitative analyses.

It is the aim of this paper to explain as simply as possible the basic concepts of a geochemical material transfer and to describe the experimental procedure of this kind of research. The paper will try to show how to apply the obtained results for the practical analyses of a geochemical material transfer occurring in nature.

2. *Geochemical Material Transport in the Earth's Crust*

Except for the material addition from the deep by magma intrusions, most of the movement of chemical elements in the Earth's crust is controlled by aqueous solutions in a vapour or liquid state.

Weathering and hydrothermal alteration of rocks, formation of veins of diverse minerals, and ore deposition are examples of such movement accompanied by the displacement of the solutions themselves.

The metamorphic recrystallization of rocks is considered, in most cases, to be proceeded by the diffusion of elements in a fine networks of aqueous solutions existing along the boundaries of the constituent minerals. The diffusion of elements across the crystal-crystal boundary surfaces without water seems ineffective for the recrystallization because of the low temperature conditions of the metamorphism in general.

Aqueous solutions in rock systems react on contact with minerals and change their composition. Some constituents of the mineral will enter into the solution, while others in the solution will be fixed in the mineral. The reaction continues until the establishment of an equilibrium between the solid (mineral) and fluid (aqueous solution) phases. The net effect of the reaction can sometimes change the mineral species existing in the rock. In other cases, minerals will change their composition without modifying the species. Thus, the rocks gradually change their composition according to the mobility of elements in the solution and to the mobility of the solution itself in the rock.

In general, the mobility of elements in a fine network of intergranular "pore solution" in rock systems is not great. The degree of interconnection of channels of the solution is also very restricted. The resulting modification of the rock composition by the diffusion of elements in a static aqueous solution is very limited. It is for this reason that the most of metamorphism appears as isochemical without a perceptible modification of rock compositions.

Contrary to the case of the diffusion alone, the modification of rock compositions by aqueous solutions permeating in the rocks can be very great.

Under the effect of geothermal water, a cubic kilometer or more of lava can be transformed into a series of rocks containing different kinds of secondary minerals such as chlorite, sericite, montmorillonite, zeolite, albite, adularia, quartz and kaolinite in extreme cases.

Another eloquent example of the material transfer by moving solutions is the skarnization. A considerable amount of limestone can be transformed into a series of rocks containing calc silicates and ferromagnesian minerals such as olivines, clinopyroxenes, epidotes, garnets, and amphiboles accompanied by minerals reflecting special elements contained in the solution (tourmaline, fluorite, for example). The modification of rock composition is not limited to limestone but it can also attain to neighbouring igneous rocks. The igneous rock is often calcified, thus transforming sodic plagioclases to calcic ones and forming minerals such as grandite garnets and clinopyroxes. The skarnization is often succeeded by the deposition of ores such as W, Mo, Cu, Zn, pb, Ag or Fe as sulfides, oxides and silicates.

3. *How and to What Extent the Mineral-Aqueous Solution Reaction Takes Place*

A rapid glance on the material transfer in the Earth's crust shows how important is the information about the composition of an aqueous solution in equilibrium with a given mineral or assemblage of minerals.

As theories on the phase equilibria in an open system constitute the basic concept throughout this paper, they will first be summarized in the forthcoming paragraphs.

Let us now examine a little more in detail the process by taking as an example the classical equilibrium of calcite, quartz, wollastonite and CO_2 (DANIELSSON, 1950; MIYASHIRO, 1960). In difference to the previous authors, the present discussion takes the coexisting aqueous fluid phase into consideration.

The first simplest case will be of a closed system at a given pressure P and temperature T conditions (no movement of the fluid phase). The phases envisaged belong to a 4-component (i.e., CaO, SiO_2, H_2O and CO_2) system. Quartz, calcite and wollastonite can be in equilibrium with a supercritical fluid (one fluid phase only) in a certain P, T field. At a specific pressure and temperature, the variance (degree of freedom) of the system being consumed, the system is perfectly determined and particularly the composition of the fluid. The equilibrium is expressed by the following reaction:

$$\underset{\text{calcite}}{CaCO_3} + \underset{\text{quartz}}{SiO_2} \rightleftharpoons \underset{\text{wollastonite}}{CaSiO_3} + \underset{\text{fluid}}{CO_2} \qquad (1)$$

The activity $a(CO_2)$ of carbon dioxide in the fluid phase is determined by a relation:

$$RT\ln a(CO_2) = -\Delta G^{\circ}_{(1)P,T} \qquad (2)$$

where $\Delta G^{\circ}_{(1)P,T}$ designates the standard free energy change accompanied by the reaction (1). R is the gas constant. Temperature T is expressed in °K.

The reaction will be blocked as soon as the activity $a(CO_2)$ attain to a value satisfying the relation (2).

Let us now try to open the system to one of the four constituents. If the carbon dioxide could escape from the system, the activity $a(CO_2)$ of the fluid will be lowered. The reaction (1) blocked in the previous example, will take place again in the direction of the wollastonite formation to establish the equilibrium.

If the activity $a(CO_2)$ of the fluid increases, wollastonite will be decomposed into calcite and quartz to reestablish the equilibrium. The three mineral assemblage, therefore has, a capacity to buffer (FONTEILLES, 1979) the activity $a(CO_2)$ of the fluid. The activities of other constituents are also buffered as long as the solid phases are composed of the three minerals. After Korzhinskii's definition, the component CO_2 of the system are mobile, but not perfectly mobile, because its chemical potential, hence activity, are determined by the interior of the system.

The continuous escape of CO_2 from the system will make either quartz or calcite disappear. In this state, the system looses its capacity to buffer $a(CO_2)$ of the fluid. The influx of a fluid having low $a(CO_2)$ does not affect the system of calcite (or quartz) + wollastonite + fluid. In the same way, the continuous influx of a fluid with very high $a(CO_2)$ will finally make wollastonite disapper and the system (calcite + quartz + fluid) will become insensible to this elevated value of $a(CO_2)$. The carbon dioxide in the last two examples is perfectly mobile in Korzhinskii's sense, because its chemical potential is determined entirely from the exterior.

Since the three minerals, wollastonite, quartz and calcite, in the above examples do not form solid solutions in the system, the maximum possible number φ of solid phases is three. When one constituent, CO_2 in the last two example cases, became perfectly mobile, the maximum number of solid phases was reduced by one.

If two of the three phases could form solid solutions, and if the system were composed of these two phases, the situation would be different. The number of solid phases are two, but there is no perfectly-mobile component. The influx of a fluid containing substitutable constituents for the solid-solution minerals will modify the compositions of these two minerals and, consequently, that of the fluid. The constituents in question are not perfectly mobile, because their chemical potential is still determined by the interior of the system. The situation is schematically represented by Fig. 1.

For this reason, the Korzhinskii's mineralogical phase rule indicating the number φ of mineral phases appearing in a n component system with an m perfectly mobile component at given P and T conditions, is an inequality $\varphi \leqslant n - m$.

4. *Buffered Fluid Composition and Mineral Assemblage*

Let us now consider wollastonite + quartz assemblage suffering the influx of a fluid containing both magnesium and calcium, but without CO_2. The chemical potential of Si in the fluid is fixed by the presence of quartz. In this case, the system at given P and T conditions will not become perfectly determined as long as another phase, diopside for example, does not appear in the system. The reaction corresponding to the diopside formation is:

$$\underset{\text{woll.}}{2CaSiO_3} + \underset{\text{fluid}}{Mg^{++}} \rightleftharpoons \underset{\text{diopside}}{CaMgSi} + \underset{\text{fluid}}{Ca^{++}} \qquad (3)$$

and the equilibrium is determined by a relation

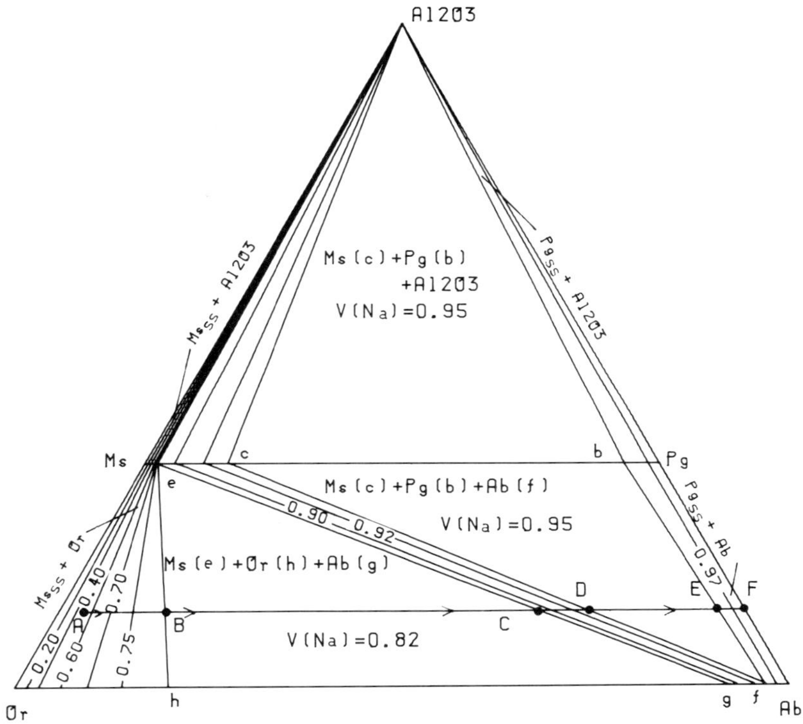

FIG. 1. Phase assemblages and corresponding fluid composition $v(\text{Na})_{aq} = m(\text{Na})/(m(\text{Na}) + m(\text{K}))$ in the system Al_2O_3–$KAlSi_3O_8$–$NaAlSi_3O_8$–(Na, K)Cl–H_2O at 500°C, 1 kb (IIYAMA, 1964)
Composition $V(\text{Na})_{aq}$ of fluids in equilibrium with solid phase assemblages are indicated by arabic numerals. The infiltration of a sodic fluid into a rock of the composition A (muscovite ss + orthoclase ss) will change the rocks along the line A–B–C–D–E–F. During the period in which the rock composition varyies from A to B, the rock is composed of two mineral phases, orthoclase ss + muscovite ss. Each phases including fluid change their compositions. Between B and C, the rock is composed of three minerals of fixed composition muscovite e, orthoclase ss, h, and albite ss, g and fluid having $V(\text{Na})_{aq} = 0.82$. Along the segment C–D, two minerals (albite ss + muscovite ss) and fluids of varying composition coexist. Along the segment D–E, the system is again composed of three solid phases and a fluid of fixed composition (muscovite ss c + albite ss, f + paragonite ss, b + fluid with $V(\text{Na})_{aq} = 0.95$. After the rock attains the composition E, its composition and fluid again changes composition to attain the final composition F. During this variation from A to F, Na and K are mobile but not perfectly mobile, in spite of the variation of the number of mineral phases.

$$RT\ln\frac{a(\mathrm{Ca})}{a(\mathrm{Mg})} = -\Delta G^{\circ}_{(3)\mathrm{P,T}} \tag{4}$$

Thus, the fluid composition blocking the reaction (3) can be theoretically formulated, but we do not know what is the value of $[a(\mathrm{Ca})/a(\mathrm{Mg})]$ or the proportion $[m(\mathrm{Ca})/m(\mathrm{Mg})]$ of molality m (ion g/(kg of solution)) of Ca and Mg in the fluid. Is the fluid very rich in Ca or in Mg? The lack of knowledge on thermodynamical properties of supercritical aqueous solutions does not permit us to estimate this composition. It is necessary to determine it experimentally.

It should be remarked that the influx of a fluid more enriched in Mg than the value defined by the relation (4), can transform diopside into less calcic minerals such as amphibole, orthopyroxene, or talc and forsterite (or anthophyllite).

$$\underset{\text{diop.}}{4\mathrm{CaMgSi_2O_6}} + \underset{\text{fluid}}{\mathrm{Mg^{++}}} + \mathrm{H_2O} \rightleftharpoons \underset{\text{tremolite}}{\mathrm{Ca_2Mg_5Si_8O_{22}(OH)_2}} + \underset{\text{fluid}}{2\mathrm{Ca^{++}}} \tag{5}$$

$$\underset{\text{diop.}}{\mathrm{CaMgSi_2O_6}} + \underset{\text{fluid}}{\mathrm{Mg^{++}}} \rightleftharpoons \underset{\text{opx.}}{2\mathrm{MgSiO_3}} + \underset{\text{fluid}}{\mathrm{Ca^{++}}} \tag{6}$$

$$\underset{\text{diop.}}{5\mathrm{CaMgSi_2O_6}} + \underset{\text{fluid}}{5\mathrm{Mg^{++}}} + \mathrm{H_2O} \rightleftharpoons \underset{\text{talc}}{2\mathrm{Mg_3Si_4O_{10}(OH)_2}} + \underset{\text{forsterite}}{2\mathrm{Mg_2SiO_4}} + \underset{\text{fluid}}{5\mathrm{Ca^{++}}} \tag{7}$$

Limiting conditions for these reactions are formulated by the following equations:

For reaction (5)

$$RT\ln\frac{[a(\mathrm{Ca})]^2}{a(\mathrm{Mg})\cdot a(\mathrm{H_2O})} \rightleftharpoons -\Delta G^{\circ}_{(5)\mathrm{P,T}} \tag{8}$$

For reaction (6)

$$RT\ln\frac{a(\mathrm{Ca})}{a(Mg)} \rightleftharpoons -\Delta G^{\circ}_{(6)\mathrm{P,T}} \tag{9}$$

and for reaction (7)

$$RT\ln\frac{a(\mathrm{Ca})}{a(\mathrm{Mg})} \rightleftharpoons -\frac{1}{5}G^{\circ}_{(7)\mathrm{P,T}}. \tag{10}$$

All of these conditions should be experimentally determined to establish the relation between the composition of fluid and corresponding mineral assemblages.

The next section deals with the experimental procedure to determine these relations and some results.

5. *Experimental Determination of Fluid Compositions Corresponding to Particular Mineral Assemblages.*

The experiments were performed according to the ordinary hydrothermal techniques operated by using autoclaves of the Morey type and, in some cases, internally heated high pressure vessels.

The procedures consist of synthetizing, in sealed gold tubes, the minerals and mineral assemblages of interest under the presence of aqueous solutions of Mg and Ca and to analyze, after the synthesis, compositions of solution and solid phases. The Si content of the fluid was supposed to be in an equilibrium value for each mineral assemblage inspected. Aqueous solutions of $MgCl_2$ and $CaCl_2$ were used to facilitate the introduction of Mg^{++} and Ca^{++} in the system. The use of chlorides renders experiments realistic for the study of the material transfer in the Earth's crust, because most of the phenomena are supposed to occur with chloride solutions (STOLLERY *et al.*, 1971; HOLLAND, 1972, for examples).

The introduction of chlorides into the system, hence an increase of one independent constituent, necessitates the imposition of a condition to define the state of equilibrium. The necessary restriction was introduced by always using aqueous solutions having molalities $m(\mathrm{Mg} + \mathrm{Ca}) = 1$ ion g/(kg of solution).

The starting solid material were gels having compositions corresponding to wollastonite, diopside and enstatite. They were prepared after the method described by HAMILTON and HENDERSON (1968).

The experimental P and T conditions were 1 kb, 600°C; 3 kb, 700°C; 3 kb, 800°C and 3 kb, 900°C. (IIYAMA *et al.*, 1980; UCHIDA and IIYAMA, 1980). The choice of 3 kb for experiments at high temperatures was to facilitate the introduction of the necessary amount (100 μl) of solution in gold tubings of a limited internal volume (450 μl).

The system was supposed to attain an equilibrium after 7 days of treatment at given P, T conditions. The system was then quenched by stopping the heating. The run product, solid and liquid, were transferred from the gold tubing into a plastic vial of 25 ml by injecting 5 ml of distilled water into the tube whose one end was cut and opened.

The solid product was separated from the liquid by filtration on a "milipore" disk. The liquid solution diluted by a washing solution (distilled water) of the solid phase was analyzed by the atomic absorption technique. After dissolution of an aliquote of the solid product by HCl + HF solution, the analysis of the solid was also performed by the atomic absorption method.

Figures 2 to 5 represent the results of these experiments.

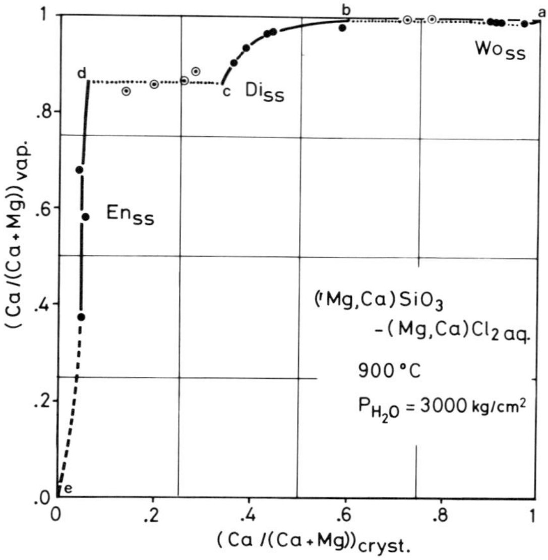

FIG. 2. Isotherm of the system $CaSiO_3$–$MgSiO_3$–(Ca, Mg)Cl_2 – H_2O at 900°C, 3 kb (IIYAMA *et al.*, 1980)

Figure 2 represents the results obtained at 900°C and 3 kb. The composition of the solid phase expressed by the proportion of the number of atoms $X(\mathrm{Ca}) = \mathrm{Ca}/(\mathrm{Ca} + \mathrm{Mg})$ of Ca and Mg in the solid was taken in abscissa and those of fluid expressed by a proportion $V(\mathrm{Ca})_{\mathrm{aq}} = m(\mathrm{Ca})/(m(\mathrm{Ca}) + m(\mathrm{Mg}))$ of molality m ion g/(kg of solution) of Ca and Mg in the fluid was taken in coordinate. The point a corresponds to the equilibrium of the coexistence, wollastonite + fluid. An addition of an extremely small amount of Mg introduced in the fluid immediately makes a small amount of diopside of the composition b of Fig. 2 appear in the solid. Composition $V(\mathrm{Ca})_{\mathrm{aq}}$ of the fluid corresponding to the assemblage wollastonite + diopside + fluid is very difficult to determine because of the very high content of Ca and very low content of Mg. It is at least higher than 0.995.

The further increase of Mg/Ca in the starting solution of the run provokes an increase of the amount of diopside and a decrease of wollastonite in the solid product. The composition of the fluid is buffered by this two-mineral assemblage and remains constant as long as the two crystalline species coexist in the solid.

When all of wollastonite is consumed for diopside formation, the fluid phase can change its composition from the point b to c of Fig. 2. The solid phase is composed of diopside alone with a slight variation of the Ca/Mg ratio. The presence at this temperature of sub-calcic diopside in a compositional range as wide as is shown in Fig. 2 could be interpreted by a metastable formation of the solid solutions. The most important aspect remarked on here is in the fact that diopside, $CaMgSi_2O_6$, is stable only with a highly calcic fluid having $V(Ca)_{aq} = 0.995$. Sub-calcic diopside solid solutions exist in these experiments with fluids having $0.995 \geqslant V(Ca)_{aq} \geqslant 0.875$. The dependency of the compositions of these solid solutions to fluid compositions is represented by the segment b-c of Fig. 2. The two phase assemblage subcalcic diopside + enstatite solid solution containing about 5 mole % of diopside "molecule" buffers the fluid to the value $V(Ca)_{aq} = 0.875$. A further decrease of Ca in the fluid renders the solid phase to be composed of enstatite solid solutions of variable composition (segment d-e of Fig. 2). The experiments performed at lower temperatures (Figs. 3, 4, and 5) show similar relations as Figure 2. The difference remarked upon is that the solid products corresponding to high magnesian compositions (enstatite solid solutions in Fig. 2) are an assemblage of talc and orthorhombic amphibole of undetermined species (a kind of anthophyllite variety). The domain of subcalcic diopside are narrowed considerably at low temperatures. It is also remarked upon that the composition of fluids buffered by diopside solid solutions + Mg rich and Ca poor phases are displaced to a less calcic side than at 900°C. Table 1 summarizes the observed buffered fluid compositions and corresponding mineral assemblages.

An inspection of Table 1 shows that wollastonite is only stable with fluids extremely enriched in Ca. Diopside can stand alone with a less calcic fluid.

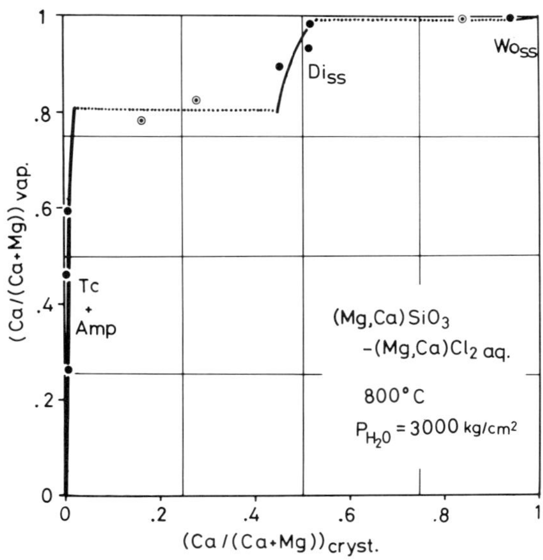

FIG. 3. Isotherm of the system $CaSiO_3$–$MgSiO_3$–(Ca, Mg)Cl_2–H_2O at 800°C 3 kb (IIYAMA *et al.*, 1980)

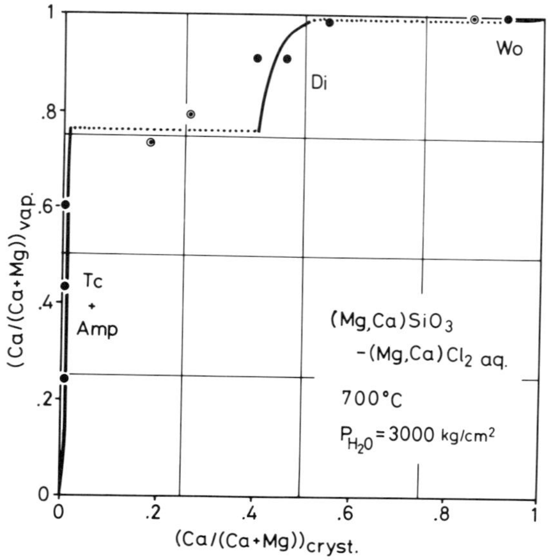

FIG. 4. Isotherm of the system $CaSiO_3$–$MgSiO_3$–(Ca, Mg)Cl_2–H_2O at 700°C 3 kb. (IIYAMA *et al.*, 1980).

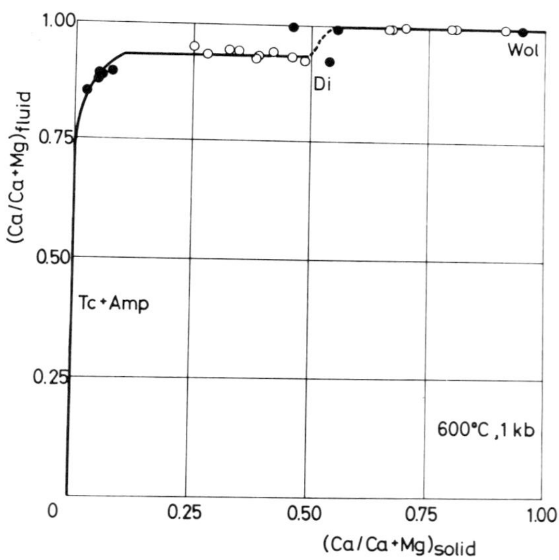

FIG. 5. Isotherm of the system $CaSiO_3$–$MgSiO_3$–(Ca, Mg)Cl_2–H_2O, at 600°C, 1 kb (IIYAMA *et al.*, 1980).

TABLE 1. Mineral assemblages and fluid compositions in the systems $CaSiO_3$–$MgSiO_3$–(Ca, Mg)Cl_2–H_2O and CaSiO–$FeSiO_3$–(Ca, Fe)Cl_2–H_2O.

P (kb)	T(°C)	Mineral assemblage	Fluid composition
3	900	Wo + Di ss	$V(Ca)_{aq} \geq 0.995$
3	800	"	$V(Ca)_{aq} \geq 0.995$
3	700	"	$V(Ca)_{aq} \geq 0.990$
1	600	"	$V(Ca)_{aq} \geq 0.999$
1	600	Wo + Hd ss	$V(Ca)_{aq} \geq 0.999$
3	900	Di ss	$0.870 \leq V(Ca)_{aq} \leq 0.995$
3	800	"	$0.800 \leq V(Ca)_{aq} \leq 0.995$
3	700	"	$0.760 \leq V(Ca)_{aq} \leq 0.995$
1	600	"	$0.930 \leq V(Ca)_{aq} \leq 0.999$
1	600	Hd ss	$0.610 \leq V(Ca)_{aq} \leq 0.999$
3	900	Di ss + En ss	$V(Ca)_{aq} = 0.870$
3	800	Di ss + Talc + amphibole	$V(Ca)_{aq} = 0.800$
3	700	"	$V(Ca)_{aq} = 0.760$
1	600	"	$V(Ca)_{aq} = 0.930$
1	600	Hd ss + Fa ss + Q	$V(Ca)_{aq} = 0.610$
3	900	En ss	$0.00 \leq V(Ca)_{aq} \leq 0.870$
3	800	Talc + amphibole ss	$0.00 \leq V(Ca)_{aq} \leq 0.800$
3	700	"	$0.00 \leq V(Ca)_{aq} \leq 0.760$
1	600	"	$0.00 \leq V(Ca)_{aq} \leq 0.930$
1	600	Fa + Q	$0.00 \leq V(Ca)_{aq} \leq 0.610$

Wo: Wollastonite; Di: diopside; Hd: hedenbergite; En: enstatite Fa: fayalite; Q: quartz.

Even though the mineral necessitates a fluid having a $V(Ca)_{aq}$ value higher than about 0.800 at low temperatures and more at high temperatures. The lowering of the Ca content in the fluid permeating through diopsidic rock transforms the mineral into less calcic species, such as amphiboles, talc or orthopyroxenes. The amphibolization of clinopyroxene skarn often observed in fields, can be explained in some cases by a change in temperature condition, but it can also be interpreted by the lowering of the Ca content in the penetrating fluid.

Experiments with Fe–Ca solutions were also performed to compare the results with those performed with Mg–Ca solutions. Metallic iron powder was mixed with silica gel (for ferrosilite composition) or with $CaSi_2O_5$ gel (for hedenbergite) and used as starting materials. Use of metallic iron in starting materials and the introduction of a very small amount (5 mg for 50 mg of solid material and 100 μl of solution) of oxalic acid permitted the maintenance of the iron species in the system in a divalent state, thus avoiding the formation of magnetite or hematite in the solid product. It was, however, necessary not to put too great amount of oxalic acid in the starting preparation; if not carbonates will appear in the product.

The obtained results at 600°C, 1 kb (Fig. 6) are similar to the case with Ca–Mg aqueous solutions. Wollastonite, in spite of the observed presence of ferrobustamite and its presumed solid solutions with wollastonite, can stand only in fluids extremely enriched in Ca. Hedenbergite solid solutions can also stand in fluids of a limited Ca/Fe range. Only the fayalite + quartz assemblage observed in the present study is stable for fluids of a wide compositional variation range ($0.60 \geqslant Ca/(Ca + Fe) \geqslant 0.0$). The results are summarized in Table 1 together with the case of Ca–Mg aqueous solutions.

6. *Solid Solutions and Fluid Compositions*

Studies of the preceding section was focussed on the fluid compositions buffered by assemblages of minerals. Let us now examine the cases where a solid solution mineral $(A, B)R$ is in equilibrium with fluids containing constituents A and B. A and B here represent exchangeable constituents of the solid solutions and R is silicate radical in consideration.

Numerous studies of this kind were performed with tecto- and phyllo-s ilicates with alkali and alkali- earths. (Wyart and Sabatier, 1956a, b, 1962, 1965, 1974; Sabatier, 1971; Orville, 1962, 1963, 1972; Debron, 1965; Iiyama, 1964, 1965, 1968, 1970, 1972, 1974; Lagache, 1968, 1969, 1971,

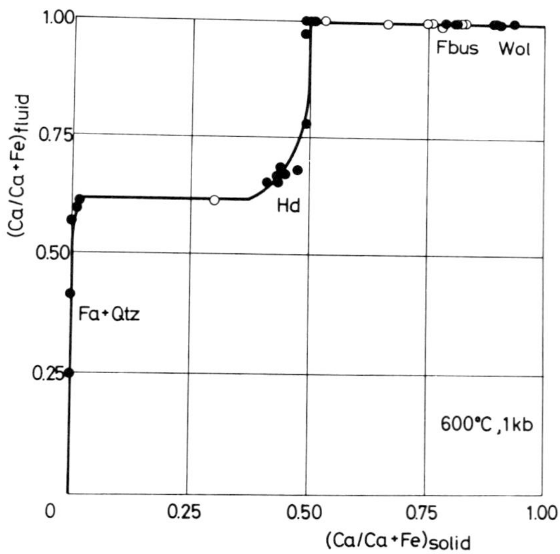

Fig. 6. Isotherm of the system $CaSiO_3$–$FeSiO_3$–(Ca, Fe)Cl_2–H_2O at 600°, 1 kb (Iiyama *et al.*, 1980).

1974a, b, 1972, 1980; GHELIS and LAGACHE, 1972; LAGACHE and SABATIER, 1973; VOLFINGER, 1969, 1970, 1974, 1975a, b; ROUX, 1972, 1979; PASCAL and ROUX, in press, for example).

Experiments with systems containing Mg and Fe were performed with olivines (SCHULIEN *et al.*, 1970), orthopyroxenes (MEDARIS, 1969) and with biotite (SCHULIEN, 1980).

The present section deals with the experimental results hitherto obtained with clinopyroxens (UCHIDA and IIYAMA, 1980) and with chlorite (TAMURA and IIYAMA, 1980).

Diopside submitted in contact with an aqueous fluid containing Fe will exchange its Mg with Fe without modifying its Ca/(Mg + Fe) atomic ratio if the chemical potential μ (Ca) in the fluid is high enough to stabilize the clinopyroxene. The reaction can be formulated by:

$$CaMgSi_2O_6 + Fe^{++} \rightleftharpoons CaFeSi_2O_6 + Mg^{++}. \tag{11}$$

diopside "molecule" in CPX.	fluid	hedenbergite "molecule" in CPX	fluid

The equilibrium condition of this reaction is:

$$RT\ln\frac{a(\text{Hd})a(\text{Mg})}{a(\text{Di})a(\text{Fe})} = -\Delta G^{\circ}_{(11)\text{P,T}} \tag{12}$$

where a(Hd) and a(Di) designate, respectively, the activities of hedenbergite and diopside "molecules" in clinopyroxene solid solutions and a(Mg) and a(Fe) are activities of Mg and Fe in the fluid phase. Equation (12) shows that the composition of clinopyroxene expressed, for example, by atomic fraction $X(\text{Mg}) = \text{Mg}/(\text{Mg} + \text{Fe})$ is a function of the composition of the fluid expressed by a proportion $V(\text{Mg})_{\text{aq}} = m(\text{Mg})/(m(Mg) + m(\text{Fe}))$ of molality m of Mg and Fe in the solution. The necessary amount of Ca to stabilize clinopyroxene was determined by consulting the experimental results of the preceding section.

The experiments were proceeded in the same way as was explained in the last section by using a mixture of metallic iron and a gel of $CaSi_2O_5$ for hedenbergite composition and a gel of $CaMgSi_2O_6$ for diopside composition. Aqueous solutions containing, at the same time, $CaCl_2$ and $MgCl_2$ or $CaCl_2$ and $FeCl_2$ were used as starting solutions. The total concentration of chlorides were always maintained at 1 mole/(kg of solution). The value 0.800 of $m(\text{Ca})/(m(\text{Ca}) + m(\text{Mg}) + m(\text{Fe}))$ ratio of the solution was chosen to stabilize the clinopyroxenes. Because of the fact that the chemical potential μ(Ca) of the fluid was not buffered by a coexisting mineral, the 1:1 proportion of Ca/(Mg + Fe) of the solid product of each run was not ascertained. This point was examined by analyzing the solid run product and only experimental results with composition $X(\text{Ca}) = \text{Ca}/(\text{Ca} + \text{Mg} + \text{Fe})$ falling in a range $0.65 \geqslant X(\text{Ca}) \geqslant 0.40$ were taken into consideration. A small amount of oxalic acid was also used to maintain the iron species in the system in a divalent state. A temperature of 600°C and a pressure of 1 kb were chosen for the condition of the experiments.

Figure 7 represents the results of the study. The isotherm constructed by taking $X(\text{Mg})$ of the clinopyroxene obtained in abscissa and $V(\text{Mg})_{aq}$ in ordinate shows a strong tendency of iron to be concentrated in the fluid phase. For a salite of $Di_{75}Hd_{25}$ corresponds a fluid already enriched in iron as high as $V(\text{Fe})_{aq} = 1 - V(\text{Mg})_{aq} = 0.87$. This tendency of iron to remain preferentially in the aqueous fluid phase was also observed in the case of olivine solid solutions and biotite solid solutions (SCHULIEN *et al.*, 1970; SCHULIEN, 1980).

The fact that diopside or forsterite containing small amounts of iron are in equilibrium with fluids already enriched in Fe, can explain the formation of hematite-bearing marble in zonal sequence of some types of skarn (observed at Costabone in Pyrenée, France):

|diopside|diopside, forsterite|forsterite|hematite, calcite|

Mg and Si in the incoming fluid are consumed by the formation of diopside and olivine solid solutions. The fluid is enriched in Fe which precipitates magnetite or hematite in this zonal sequence.

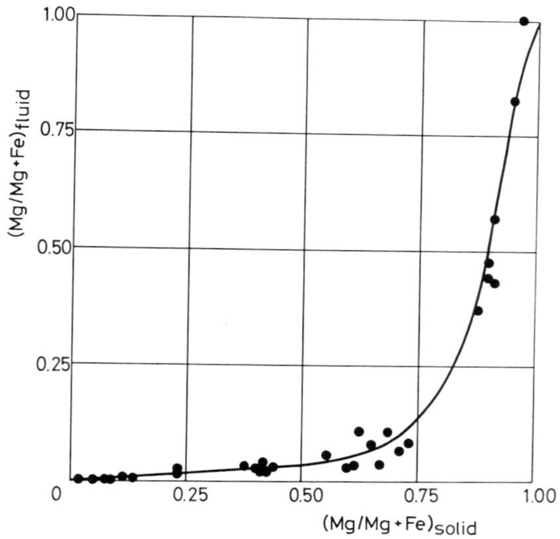

FIG. 7. Isotherm of the system $CaMgSi_2O_6$–$CaFeSi_2O_6$–(Mg, Fe, Ca)Cl_2–H_2O at 600°C, 1 kb. $V(\text{Ca})_{aq}$ of the fluid was controlled to keep clinopyroxene solid solutions in their theoretical composition in regards to the Ca/(Mg + Fe) ratio (UCHIDA and IIYAMA, 1980).

The same kind of studies with clinochlore-daphinite series, (Mg, Fe)$_5$Al$_2$Si$_3$O$_{10}$(OH)$_8$ at 1 kb, 400°C (Fig. 8) shows also the tendency of iron to remain in the fluid, but not so strongly as in the cases of olivines, biotites or clinopyroxenes. For a chlorite with $X(\mathrm{Mg}) = 0.70$ corresponds a fluid with $V(\mathrm{Mg})_{aq} = 0.50$, instead of $V(\mathrm{Mg})_{aq} = 0.13$ in the case of clinopyroxenes. This contrasted behaviour of iron and magnesium may arise from the difference in structures of minerals, especially from that of the surroundings of Mg or Fe sites. It cannot be excluded, however, that the difference observed is caused by the variation with temperature of Mg and Fe species present in the fluid phase. The comparison of isotherms obtained with olivine solid solutions (SCHULIEN *et al.*, 1970) at 500°C and 400°C under the pressure of 1 kb suggests this eventuality. In effect, the standard free energy change $\Delta G^\circ_{(11)}$ of the reaction (11) for example, can be further developed as:

$$\Delta G^\circ_{(11)P,T} = (\mu^\circ(\mathrm{Mg}) - \mu^\circ(\mathrm{Fe})) - (\mu^\circ(Di) - \mu^\circ(Hd)) \tag{13}$$

where $\mu^\circ(i)$ designates the free energy at a given P, T condition of constituent i in the standard state. The variation with temperature of the term $\mu^\circ(\mathrm{Di}) - \mu^\circ(\mathrm{Hd})$ is not so great in the temperature intervals of 100 ~ 200°C except in the cases where some structural transformation occurs in the interval. On the contrary, the behaviour of Mg and Fe in the aqueous fluid can vary considerably if the state of complex ions is changed in certain conditions of pressure and temperature. Accumulation of careful experimental data on aqueous (Mg, Fe) solution-ferromagnesian mineral equilibria will provide this kind of informations in future. The actual state of our knowledge cannot, unfortunately, permit us to clarify this question.

The shape of isotherms discussed in the preceding paragraphs depends, in

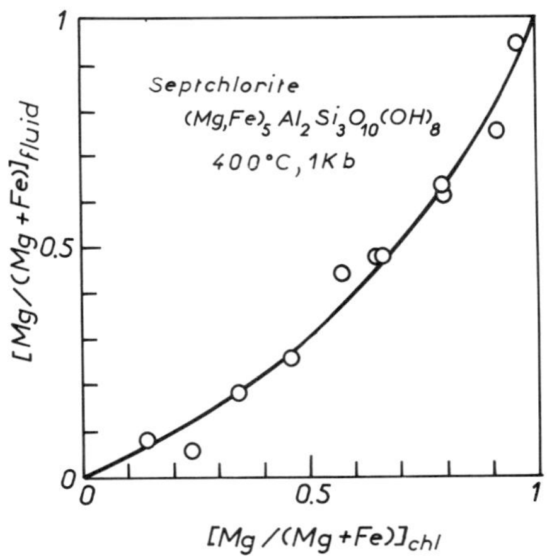

FIG. 8. Isotherm of the system Mg$_5$Al$_2$Si$_3$O$_{10}$(OH)$_8$–Fe$_5$Al$_2$Si$_3$O$_{10}$(OH)$_8$–(Mg, Fe)Cl$_2$–H$_2$O at 400°C, 1 kb (TAMURA and IIYAMA, 1981)

a great part, on the value of $\Delta G^{\circ}_{(i)P,T}$ of the reaction considered, but it also depends on the thermodynamic properties of the mixing of solid solutions. A strong deviation from the ideality of the solid solution (existence of a strong excess-free energy of mixing) will provoke, an immiscibility gap in the solid solutions series as in the case of alkali feldspar or paragonite-muscovite system.

As was predicted by MATSUI and BANNO (1967), most of the ferromagnesian solid solution silicates presented here show very little excess-free energy of mixing. An estimation of the excess-free energy G^{ex} of mixing of the diopside-hedenbergite series, performed after a method first proposed by DELBOVE (1970), and by using the isotherm of Fig. 7, indicates that G^{ex} of the series attains a maximum at $Hd_{30}Di_{70}$, nevertheless, the value is only 0.3 kcal/mole, which is quite small compared with the case of alkali feldspar solid solutions (G^{ex} maximum = 1.0 kcal/mole).

7. *Aqueous Fluid-Mineral Equilibria and Compositional Profile of Infiltration Metasomatic Column*

Fluid-mineral equilibria discussed above can be used, not only to analyze the material transfer occurring with quasi-static movement of aqueous fluid, but also to estimate the compositional profile of a metasomatic column formed by the infiltration of aqueous fluid in general.

The theory of this process was first developed by KORZHINSKII (1951) and reexplained by HOFMANN (1972) by using theories developed in chromatography.

Let us now examine here the process by taking as an example the isotherm obtained with diopside-hedenbergite series (Fig. 7). As a model, we shall take a cylindrical rock column (with direction z of the axis) composed of clinopyroxene (Fig. 9). Its cross-section is 1 cm^2. A fluid containing Fe, Mg, Ca and Si infilltrates this column along the direction z. The chemical potentials μ(Ca) and μ(Si) of Ca and Si are high enough to keep clinopyroxene stable. The compositions of the rock column and the fluid are changing during the passage of the aqueous phase, according to the Mg–Fe exchange reaction.

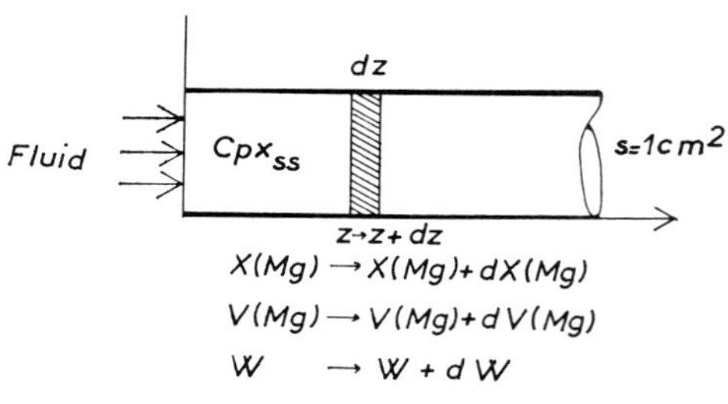

FIG. 9. Infiltration of a pluid into a rock column. Schematic representation.

Let us now focus our attention on a thin rock plate of a thickness dz cm perpendicular to the axis of the cylinder and situated at a distance z from the inflow side of the column. At a given time t, the plate had already suffered the infiltration of W kg of fluid and got a composition $X(\text{Mg}) = \text{Mg}/(\text{Mg} + \text{Fe})$. Further infiltration of dW kg of the fluid through this plate modifies the composition of this plate to $X(\text{Mg}) + \mathrm{d}X(\text{Mg})$, and at the same time, the fluid changes its composition from $V(\text{Mg}) = (m(\text{Mg})/(m(\text{Mg}) + m(\text{Fe}))$ to $V(\text{Mg}) + \mathrm{d}V(\text{Mg})$.

As the reaction proceeds by ion exchange, the total amount $V_{\text{tot}} = \text{Mg} + \text{Fe}$ (ion grammes) contained in dW kg of the fluid and that (M ion grammes) of Mg + Fe contained in the rock plate, is unchanged during the passage. This condition is formulated by:
Amount dM_1 of Mg entered from the fluid to the rock plate

$$\mathrm{d}M_1 = \mathrm{d}z \cdot M \cdot \mathrm{d}X(\text{Mg}).$$

Amount dM_2 of Mg passed from the rock column to the fluid

$$\mathrm{d}M_2 = \mathrm{d}V(\text{Mg}) \cdot \mathrm{d}W \cdot V_{\text{tot}}$$

Because $\mathrm{d}M_1 = -\,\mathrm{d}M_2$, we get:

$$\frac{\mathrm{d}z}{\mathrm{d}w} = -\frac{V_{\text{tot}}}{M} \cdot \frac{\mathrm{d}V(\text{Mg})}{\mathrm{d}X(\text{Mg})} \tag{14}$$

This relation shows that the advance dz of a section of a rock column having the composition $X(\text{Mg})$ with an inflow of dW kg of the fluid, is proportional to the slope $\mathrm{d}V(\text{Mg})/\mathrm{d}X(\text{Mg})$ of the isotherm of the exchange reaction at a point corresponding to the composition $X(\text{Mg})$ of the solid phase.

In the case where the rock column was initially composed of diopside $CaMgSi_2O_6$ and the fluid did not contain Mg, the advance of an Fe-rich

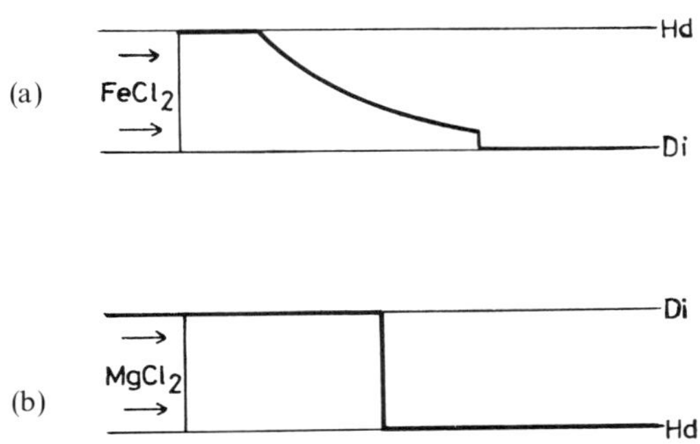

FIG. 10. Compositional profile of infiltration metasomatism (a) Profile produced by the infiltration of a fluid containing Fe into a diopside rock. (b) Profile produced by the infiltration of a fluid containing Mg into a hedenbergite rock.

section of the column will be smaller than an Fe poor section, because the isotherm of Fig. 7 is a strongly concaved curve. The resulting compositional profile of the column after sufficient time of infiltration, will show a gradual variation with z (Fig. 10a). On the contrary, if a rock column composed of hedenbergite suffer the infiltration of magnesian fluid, the advance of the Fe-rich section will be faster than that of the Mg rich section. Therefore, a sharp front of diopside separating the diopsidic part on the inflowing side and the hedenbergitic part on the outflowing side (Fig. 10b), will appear in the column.

The two extreme cases discussed here are only an example and many other shapes of compositional profiles can be expected, according to the shapes of the isotherms of equilibria concerned. These aspects are already well discussed by HOFMANN (1974) and by FLETCHER and HOFMANN (1972). They are not repeated here.

8. *Coclusions*

Two fundamental principles controlling the geochemical material transfer in the Earth's crust are explained by taking equilibria in the system wollastonite-clinopyroxene-orthopyroxene-(Mg, Fe)Cl_2–H_2O which were studied recently in the research group of the present author.

The fixtation of elements into solid phases from the fluid or liberation of elements into the fluid from the solid, always occur when the composition of the fluid is buffered by solid phases. In another words, the geochemical material transfer does occur during the period in which the element concerned is still mobile, but is not perfectly mobile in the system. A component is perfectly mobile if its chemical potential in incoming and outgoing flux are equal in a given open system. The perfect mobility of elements must be judged from the view point of the classical phase rule formulated by Gibbs and must take into consideration the principle of Korzhinskii's mineralogical phase rule.

The compositional profile of an infiltration metasomatic column is governed by the shape of the isotherm of mineral-aqueous fluid equilibrium of the system concerned. The appearance of a compositional front, or its absence, in the column depends on the composition of the infiltrating fluid and the composition of the solid phase in one part, and on the shape of the isotherm in the other part.

The shape of the isotherm of the mineral-aqueous fluid equilibria depends on the thermodynamical properties of minerals, including the mixing properties of solid solutions, and on those of the elements in the aqueous solution (fluid) of the system concerned. These last aspects were not developed in detail here because of the limitation on the length of this paper. It should be remarked, however, that these aspects can be explored by analyzing the experimental data to establish the isotherms of the mineral-aqueous fluid equilibria of the systems concerned. Examples of these analyses are now frequently appearing in mineralogical and geochemical literatures.

REFERENCES

DANIELSSON, A. (1950): Das Calcit-Wollasonitgleichgewicht, *Geoch. Cosmochim. Acta*, **1**, 55–69.

DEBRON, G. (1965): Contribution à l'étude des réactions d'échange des ions alcalins et alcalinoterreux dans les feldspathoides, *Bull. Soc. Franç. Miner. Crist.*, **88**, 69–96.

DELBOVE, F. (1971): Equilibre d'échange d'ions entre feldspaths alcalins et halogénures sodi-potassiques fondus. Application au calcul des propriétés thermodynamiques de la série des feldspaths alcalins, *Bull. Soc. Franç. Miner. Crist.*, **94**, 456–466.

FISHER, G. W. (1970): The application of ionic equilibria to métamorphic différentiation: an example, *Contrib. Mineral. Petrol.*, **29**, 91–103.

FISHER, G. W. (1973): Nonequilibrium thermodynamics as a model for diffusion-controlled metamorphic process, *Am. J. Sci.*, **273**, 897–924.

FISHER, G. W. and ELLIOT, D. (1974): Criteria for quasi-steady diffusion and local equilibrium in metamorphism, in *Geochemical Transport and Kinetics*, Eds. Hofmann, Giletti, Yoder, and Yund, Carnegie Institution of Washington Publication 634, Washington D. C., U.S.A., 231–241.

FLETCHER, R. C. and HOFMANN, A. W. (1974): Simple models of diffusion and combined diffusion-infiltration metasomatism, in *Geochemical Transport and Kinetics*, Eds. Hofmann, Giletti, Yoder and Yund, Carnegie Institution of Washington Publication 634, Washington, D.C., U.S.A., pp. 243–259.

FONTEILLES, M. (1978): Les mécanismes de la métasomatose, *Bull. Minéral.*, **101**, 166–194.

GHELIS, M. and LAGACHE, M. (1972): Etude de l'équilibre entre les solutions hydrothermales et les feldspaths de la série $KAlSi_3O_8$–$RbAlSi_3O_8$ à 600°C sous une préssion de 1,000 bars, *Bull. Soc. Franç. Miner. Crist.*, **94**, 157–158.

HAMILTON, D. and HENDERSON, M. (1968): The preparation of silicate compositions by a gelling method, *Mineral. Mag.*, **36**, 832–838.

HOFMANN, A. (1972): Chromatographic theory of infiltration metasomatism and its application to feldspars, *Am. J. Sci.*, **272**, 69–90.

HOLLAND, H. D. (1972): Granites, solutions, and base metal deposits, *Econ. Geol.*, **67**, 281–301.

IIYAMA, J. T. (1964): Etude de réaction d'échange d'ions Na–K dans la serie muscovite-paragonite, *Bull. Soc. Franç. Miner. Crist.*, **87**, 532–541.

IIYAMA, J. T. (1965): Influence des anions sur les équilibres d'échange d'ions Na–K dans les feldspaths alcalins à 600°C sous une pression de 1,000 bars, *Bull. Soc. Franç. Miner. Crist.*, **88**, 618–622.

IIYAMA, J. T. (1968): Etude expérimentale de la distribution d'éléments en traces entre deux feldspaths-feldspath potassique et plagioclases-coexistants. 1. Distribution de Rb, Cs, Sr et Ba à 600°C, *Bull. Soc. Franç. Miner. Crist.*, **91**, 130–140.

IIYAMA, J. T. (1970): Influence de la préssion sur la composition de la solution hydrothermale sodi-potassique de divers sels en équilibre avec feldspaths perthitiques à 600°C. Observation expérimentale, *C. R. Acad. Sci. Paris, Ser. D.*, **271**, 1925–1927.

IIYAMA, J. T. (1972): Fixations des éléments alcalinoterreux Ba, Sr et Ca dans les feldspaths—Etudes expérmentales—, in *Proc. 24th Intern. Geological Congress, Montreal, Section 10, Geochemistry*, pp. 122–130.

IIYAMA, J. T. (1974): Substitution, déformation locale de la maille et équilibre de distribution des éléments en traces entre silicates et solution hydrohermale, *Bull. Soc. Franç. Miner. Crist.*, **97**, 143–151.

IIYAMA, J. T., KAWAMURA, K. and UCHIDA, E. (1980): Composition of aqueous fluid during skarnization. Collected abstracts of annual meetings of 3 Japanese societies of Mineralogy, in *Mineralogists, Petrologists and Economic Geologists; and Mining Geology*, p. 99.

KORZHINSKII, D. S. (1936): Mobility and inertness of components in metasomatosis, *Bull. Akad. Nauk. SSSR. Ser. Geol.*, No. 1, 35–60.

KORZHINSKII, D. S. (1950): Phase rule and geochemical mobility of elements, in *Rep. 18th Sess., International Geological Congress, G. B., 1948, Pt. 2*, pp. 50–65.

KORZHINSKII, D. S. (1959): *Physicochemical Basis of the Analysis of the Paragenesis of Minerals*, Consultants Bureau, New York, U.S.A.

KORZHINSKII, D. S. (1970): *Theory of Metasomatic Zoning*, Translated by J. AGRELL, Clarendon Press, Oxford.

LAGACHE, M. (1968): Etude expérimentale de la répartition des éléments traces entre la leucite, l'orthose et des solutions hydrothermales. Le rubidium à 600°C, *C. R. Acad. Sci. Paris*, **267**, 141–144.

LAGACHE, M. (1969): Etude expérimentale de la répartition des elements traces entre la leucite, l'orthose et des solutions hydrothermales à 600°C, *C. R. Acad. Sci. Paris*, **268**, 1241–1243.

LAGACHE, M. (1971): Etude experimentale de la répartition de cesium entre les feldspaths sodi-potassiques et des solutions hydrothermales à 700°C 1kbar., *C. R. Acad. Sci. Paris*, **272**, 1328–1330.

LAGACHE, M. (1974a): Calcul des propriétés thermodynamiques des solutions ternaires de leucite (K, Rb, Cs) par le modèle de Köhler, *C. R. Acad. Sci. Paris*, **278**, 2497–2500.

LAGACHE, M. (1974b): Utilisation du modèle de F. Köhler pour le calcul de la distribution d'un éléments en traces entre une solution solide binaire et une solution hydrothermale. Application à quelques tecto-silicates, *Bull. Soc. Franç. Miner. Crist.*, **97**, 117–121.

LAGACHE, M. (1980): Etude expérimentale des systemes albite–orthose–NaCl–KCl–H_2O, albite–orthose–anorthite–NaCl–KCl–H_2O, à 500°C et pression variables (0,5 à 2 kbars), in *Mineralisations lies aux Granitoïdes*, Mémoire de B. R. G. M., No. 99, pp. 121–127.

LAGACHE, M. and SABATIER, G. (1973): Distribution des éléments Na, K, Rb et Cs à l'état de trace entre feldspaths alcalins et solutions hydrothermales à 650°C 1 kbar: données experimentales et interpretation thermodynamique, *Geochim. Cosmochim. Acta*, **37**, 2617–2640.

MATSUI, Y. and BANNO, S. (1968): Distribution of elements between crystalline phases, *Kagaku no Ryoiki*, **22**, 56–174; 256–265 (in Japanese).

MEDARIS, L. G. (1969): Partitioning of Mg^{+2} and Fe^{+2} between coexisting synthetic olivine and orthopyroxene, *Am. J. Sci.*, **267**, 945–968.

MIYASHIRO, A. (1960): Thermodynamics of reactions of rock forming minerals with silica. I–VI, *Japan, J. Geol.*, **31**, 71–78, 79–84, 107–111, 113–120, 241–246, 247–252.

ORIVILLE, P. M. (1962): Alkali metasomatism and feldspars, *Norsk. Geol. Tidsskr.*, **42**, 283–316.

ORVILLE, P. M. (1963): Alkali ion exchange between vapor and feldspar phases, *Am. J. Sci.*, **261**, 201–237.

ORVILLE, P. M. (1972): Plagioclase cation exchange equilibria with aqueous chloride solution: results at 700°C and 2,000 bars in the presence of quartz, *Am. J. Sci.*, **272**, 234–272.

PASCAL, M. L. and J. ROUX (1982): Propriétes thermodynamiques des solutions (Na, K) Cl–H_2O entre 400 et 800°C, 1–2 Kb: une revue des équilibres d'échange avec les silicates sodi-potassiques, *Geoch. Cosmochim. Acta*, in press.

RAMBERG, H. (1952): *The Origin of Metamorphic and Metasomatic Rocks*, The University of Chicago Press, Chicago.

ROUX, J. (1974): Etude des solutions solides des néphélines (Na, K)$AlSiO_4$ et (Na, Rb)$AlSiO_4$, *Geoch. Cosmochim. Acta*, **38**, 1213–1244.

ROUX, J. (1979): Etude physicochimiques des feldsparhoïdes et application aux problèmes pétrographiques, Thèse. Université Paris sud.

SABATIER, G. (1971): Théorie de la distribution d'un élément entre solutin hydrothermale et solution solide. Application à la distribution du cesium dans la serie des feldspaths alcalins, *Bull. Soc. Franç. Miner. Crist.*, **94**, 451–455.

SCHLIEN, S. (1980): Mg–Fe partitioning between biotite and a supercritical chloride solution, *Contrib. Mineral. Petrol.*, **74**, 85–93.

SCHLIEN, S., FRIEDRICHSEN, H. und HELLNER, E. (1970): Das Mischkristallverhalten des Olivins 450°C und 650°C bei 1kb Druck, *N. Jb. Miner. Mh.*, **4**, 141–147.

STOLLEY, G., BORSIK, M. and HOLLAND, H. D. (1971): Chlorine in intrusives: A possible prospecting tool, *Econ. Geol.*, **66**, 361–367.

TAMURA, M. and IIYAMA, J. T. (1980): Mg–Fe partitioning between chlorite and hydrothermal solutions (abstract in Japanese), *Mining Geol.*, **31**, p. 43.

THOMPSON, J. B., JR. (1955): The thermodynamic basis for the mineral facies concept, *Am. J. Sci.*, **253**, 65–103.

THOMPSON, J. B., JR. (1959): Local equilibrium in metasomatic processes, in *Researches in Geochemistry*, Ed. P. H. Aberson, John Willey, New York, pp. 452–457.

UCHIDA, E. and IIYAMA, J. T. (1980): Echanges ioniques Mg–Fe entre la serie diopside-hedenbergite et une solution hydrothermale, in *Résume de 26 Congres Geologique International Paris*, vol. 1, p. 97.

VOLFINGER, M. (1969): Partage de Rb et Cs entre sanidine, muscovite, et solution hydrothermale à 600°C et 1,000 bars *C. R. Acad. Sci. Paris*, **269**, 1–3.

VOLFINGER, M. (1970): Partage de Na et Li entre sanidine, muscovite et solution hydrothermale à 600°C et 1,000 bars, *C. R. Acad. Sci. Paris*, **271**, 1345–1347.

VOLFINGER, M. (1974): Effect de la composition des micas trioctaédriques sur les distributions de Rb et Cs a l'état de traces, *Earth Planet. Sci. Lett.*, **24**, 99–304.

VOLFINGER, M. (1975a): Effet de la composition des micas, dioctédriques sur les distributins des éléments alcalins en traces, *Bull. Soc. Franç. Miner. Crist.*, **98**, 49–52.

VOLFINGER, M. (1975b): Effet da la température sur les distributions de Na, Rb et Cs entre la sanidine, la muscovite, la phlogopite et une solution hydrothermale sous une pression de 1 kbar, *Geoch. Cosmochim. Acta.*, **40**, 267–282.

WYART, J. and SABATIER, G. (1956a): Mobilité des ions alcalins et alcalinoterreux dans les feldspaths, *Bull. Soc. Franç. Miner. Crist.*, **80**, 574–581.

WYART, J. and SABATIER, G. (1956b): Transformations mutuelle des feldspaths alcalins. Reproduction du microcline et de l'albite, *Bull. Soc. Franç. Miner. Crist.*, **80**, 574–581.

WYART, J. and SABATIER, G. (1962): L'équilibre des feldspaths et des feldspathoïdes en présence de solutions sodi-potassiques, *Norsk. Geol. Tidsskr.*, **42**, 317–329.

WYART, J. and SABATIER, G. (1965): Réactions des feldspaths alcalins avec des solutions hydrothermales de $CaCl_2$, *C. R. Acad. Sci. Paris*, **260**, 1681–1685.

WYART, J. and SABATIER, G. (1974): Can Schwantke's compound be formed by reaction of alkalic feldspars with hydrothermal calcic solution? A comment on recent paper by P. M. Orville: Plagioclase cation exchange equilibria with aqueous chloride solutions at 700°C and 2,000 bars, *Am. J. Sci.*, **274**, 633–637.

Materials Science of the Earth's Interior, edited by I. Sunagawa, pp. 515–544.

HYDROTHERMAL FLUIDS INFERRED FROM FLUID INCLUSIONS

Sukune TAKENOUCHI

Department of Mineral Development Engineering, The University of Tokyo, Tokyo, Japan

1. Introduction

When a crystal grows in a medium, it traps the medium in the defects which were formed along the growth surface of crystal or among dendritic crystals. Furthermore, if the crystal is fractured during or after crystallization, the medium invades the fractures and is then trapped when the fractures are heal. Those vacuoles containing the media which existed during or after the crystal growth, are generally found in minerals grown from silicate melts or hydrothermal solutions, and are called glass or fluid inclusions.

In general, they are too small in size to be observed with the naked eye, but, in some cases, they are as large as the order of centimeters. They randomly distribute in crystal along growth planes or curved and branched planes, and in clouds. These fluid inclusions are various in the mode of occurrence, condition of formation, chemical composition and so on, but offer valuable information on the environment of crystal growth, if they are carefully studied. In order to evaluate the fluid inclusion data, some assumptions should be satisfied (ROEDDER, 1979).

Fluid inclusions are generally classified by the constituent phases at room temperatures and by the genetic relation to the host crystal, as shown in Tables 1 and 2, respectively. Classification and categories for identification of fluid inclusions are described in detail elsewhere (ROEDDER, 1979). For the historical or general description of the fluid inclusion studies, readers should refer to the reviews written by SMITH (1953), ROEDDER (1967, 1976, and 1979), YERMAKOV (1965), and TAKENOUCHI (1978).

Fluid inclusions provide us with important information on the temperature, pressure, density, chemical composition, and origin of water of the fluids, from which the host minerals crystallized. As the information is valuable not only for the consideration of the genesis of ore deposits, but also for the indicator of mineral exploration, fluid inclusion studies have been especially

TABLE 1. Classification of fluid inclusions by the constituent phases at room temperatures.

1.	Gaseous inclusion
2.	Liquid inclusion
3.	Polyphase fluid inclusion
	3-a High-salinity fluid inclusion
4.	CO_2-bearing fluid inclusion
5.	H_2S-bearing fluid inclusion
6.	Hydrocarbon-bearing fluid inclusion

TABLE 2. Classification of fluid inclusions by the genetic relation to the host crystal.

1.	Primary inclusion
2.	Secondary inclusion
3.	Pseudosecondary inclusion

carried out in the field of ore deposits geology since the middle of the nineteenth century.

Vein-type deposits are the most suitable for fluid inclusion studies because the optical properties and grain size of such constituent minerals as quartz, calcite, fluorite, sphalerite, etc., are appropriate for the observation and the paragenetic sequence is generally distinct concerning the structure of deposits. Valuable data of fluid inclusions of this type of deposits have been accumulated throughout the world. Recently, fluid inclusion data of porphyry copper deposits have been accumulated, especially, from the southwestern United States. In strata-bound deposits, particularly, the data of the Mississippi Valley-type lead-zinc deposits are important for the identification of the type (ROEDDER, 1967).

Data on chemical and stable isotopic compositions of fluid inclusions are also important for the consideration of hydrothermal fluids in the earth's crust as well as the temperature and salinity, but as the analyses of fluid inclusions are not so easy as in the determination of temperature and salinity, data are still scarce.

Figures 1 and 2 show some of the ranges of homogenization temperatures of fluid inclusions from deposits of various types in Japan. Temperature and salinity data are also reported from some vein-type deposits such as the Higashiyama (YOSHIDA, 1975), Ani (NAMBU *et al.*, 1978), Kaneuchi (NAMBU *et al.*, 1979), and Ohe (SATO *et al.*, 1980), as well as form southern Hokkaido (OKAMURA, 1980) and some Neogene deposits (WATANABE, 1979). ENJOJI and TAKENOUCHI (1976) summarized the data of temperature, salinity, and composition of fluid inclusions from various vein-type deposits in Japan. Analyses of fluid inclusions were carried out by ENJOJI (1972).

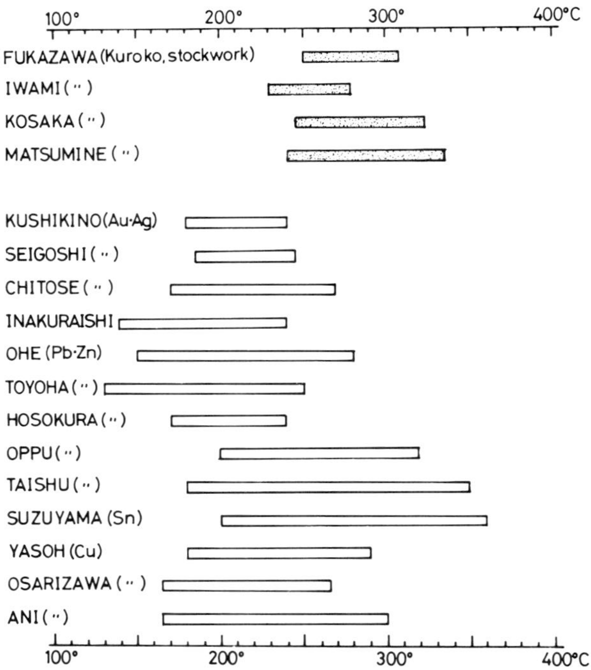

FIG. 1. Range of homogenization temperature of fluid inclusions from some Tertiary vein-type deposits and stockwork orebodies of Kuroko deposits.

HATTORI and SAKAI (1979, 1980) discussed the origin of ore-forming fluids based on the data of a stable isotope composition of fluid inclusions, and concluded that the ore-forming fluids of gold-silver veins were generated in the water-rock interaction of meteoric water having a low rock/water ratio. Meanwhile, those of copper-lead-zinc veins were the products of a high rock/water ratio interaction.

2. *Vein and Skarn Deposits*

Fluid inclusion data have been reported from many vein-type deposits, but the fluid inclusion studies on the igneous rocks, which are genetically related to the mineralization, are also useful for the consideration of the generation of ore-forming fluids.

High-salinity polyphase inclusions are observable in quartz phenocrysts of porphyries and granitic rocks in the mineralized areas. The mode of occurrence of polyphase inclusions in igneous rocks was reported from the Suzuyama tin deposit, Kagoshima, and Kamioka lead-zinc deposit, Gifu (TAKENOUCHI and IMAI, 1975), Akagane copper deposit, Iwate (TAKENOUCHI,

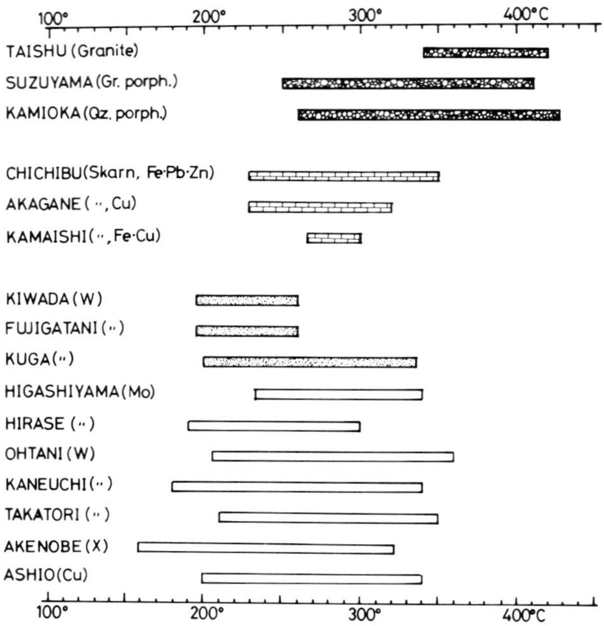

FIG. 2. Range of homogenization temperature of fluid inclusions from igneous rocks, skarn deposits, and late Cretaceous vein-type deposits.

1975), Ani copper deposit, Akita (NAMBU *et al.*, 1978), and from the Tsumo copper deposit, Shimane (WATANABE *et al.*, 1981).

The coexistence of polyphase and gaseous inclusions, that is, the evidence of boiling, was often observed but not necessarily in all cases. In general, as they are very small in size, it is hard to perform a number of temperature determinations on them. The obtained data, however, are valuable for the consideration of hydrothermal activity in the rocks genetically related to the mineralization.

Suzuyama In the Suzuyama tin deposit, high-salinity polyphase inclusions were found in quartz phenocrysts of granite porphyry which exposes in the center of a mineralized district in a small area of 200 × 300 m. Cassiterite-bearing quartz veins are found to the east-southeast of the stock and cassiterite-bearing sulfide veins to the west-northwest. Polyphase inclusions coexist with gaseous inclusions indicating the boiling phenomena and show the homogenization temperature in a range between 310° and 500°C, with the maximum frequency at about 370°C. Homogenization temperatures of liquid inclusions found in quartz phenocrysts, are in a range of 230°–350°C with a frequency peak at 280°C (Fig. 3). On the other hand, those of vein quartz range from 200° to 350°C, with two frequency peaks at 230° and 330°C,

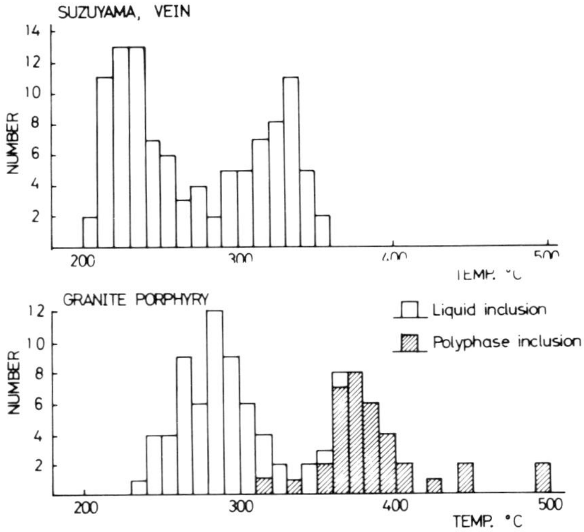

FIG. 3. Homogenization temperature of fluid inclusions in granite porphyry and veins of the Zuzuyama deposit, Kagoshima.

indicating a temperature difference of several tens of degrees lower than those of porphyry. No fluid inclusion having halite was recognized in vein quartz.

It is inferred that the polyphase inclusions in the porphyry represent the original ore-forming fluids which were released from the igneous rock, and that the high-temperature liquid inclusions in the veins indicate the temperature condition at the site of ore deposition. Liquid inclusions in the porphyry and those of a lower temperature in the veins, would be hydrothermal solutions of later stages, probably of meteoric origin.

Taishu The similar relation between the ore deposit and genetically related igneous rock is observed at the Taishu lead-zinc deposit, Nagasaki (IMAI *et al.*, 1971). It is inferred geologically that the mineralization was related to the intrusion of Miocene granitic rock which is situated to the south of the veins. In this deposit, polyphase inclusions are observable in lower parts of the southernmost veins, which are close to the granitic rock. A great number of polyphase inclusions are found in quartz of the granitic rock, especially in the northern part of the body. The homogenization temperature of polyphase inclusions in the igneous rock ranges from 340° to 420°C, and salinity from 32 to 57 wt.% NaCl equevalent concentration. The highest temperature and salinity of the southernmost vein is 370° and 32 wt.%, respectively. Both the temperature and salinity in the ore deposit decrease generally from about 370°C and 32 wt.% at lower levels of the southernmost vein to about 230°C and less than 10 wt.% at upper levels of the northernmost vein (Fig. 4). This indicates that when the ore-forming fluids migrated along fissures leaving the igneous rock, the fluids were mixed with meteoric water and the temperature and salinity of the fluids decreased as they moved on further from the source.

Akagane Mineralization of the Akagane copper deposit is presumably related to the intrusion of Cretaceous granodiorite. Most ore bodies are of a skarn-type deposit formed in Paleozoic limestone and Mesozoic gabbro, but a part of the Jishakuyama ore body is a disseminated vein-type formed in silicified quartz porphyry. The Koganetsubo breccia pipe, which bears native gold and scheelite, is formed in the gabbroic rock and consists of a quartz matrix and fragments of gabbro.

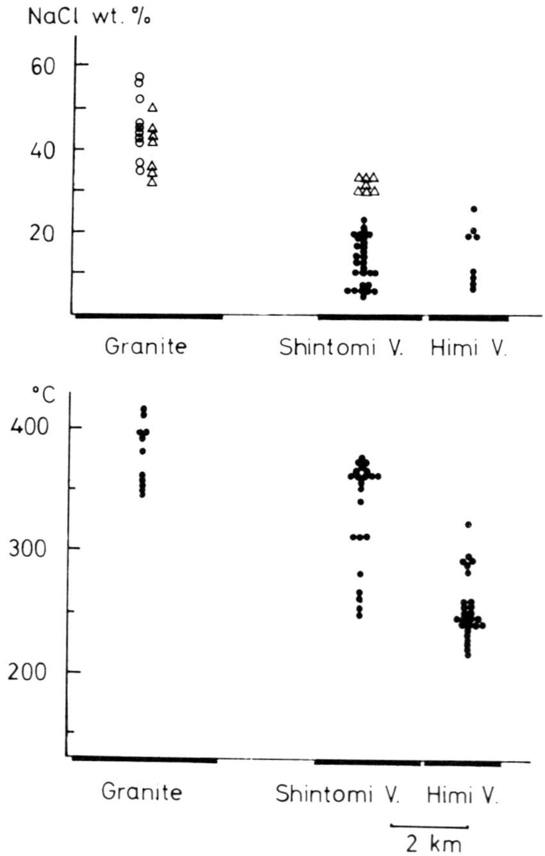

FIG. 4. Homogenization temperature and salinity of liquid and polyphase inclusions in quartz from veins and granite of the Taishu deposit, Nagasaki. The open triangle shows the salinity of polyphase inclusion determined by the heating-stage, and the open circule shows the one determined by the volumetric calculation. The black circle represents the salinity of liquid inclusion determined by the freezing-stage, and homogenization temperature in the lower figure.

CO_2-rich fluid inclusions are commonly observed, in addition to liquid inclusions, in quartz and scheelite from the Koganetsubo breccia pipe. Partial homogenization of the CO_2 phases in primary inclusions occurs to the CO_2 liquid phase, indicating high CO_2 contents, but CO_2-rich secondary inclusions show the partial homogenization to the CO_2 gas phase indicating lower CO_2 contents.

It is inferred from the behavior of the CO_2 phases that the fluid density of the CO_2-rich secondary inclusion is lower than that of the primary inclusion. Liquid inclusions are apparently of secondary origin and all have high degrees of filling. From the microscopic observation of fluid inclusions, it is presumed that the CO_2 concentration of ore-forming fluids varied remarkably during the formation of the Koganetsubo breccia pipe.

Three types of inclusions, that is, liquid, polyphase, and CO_2-rich inclusions, are recognized in quartz from granodiorite. Liquid inclusions are by far the most abundant in both northern and southern granodiorite stocks. In the southern stock, CO_2-rich inclusions are abundant but polyphase inclusions are rare. The mode of occurrence of the CO_2-rich inclusions is similar to that of the Koganetsubo orebody, which is situated about 500 m to the south of the southern stock.

On the other hand, in the northern stock, around which several skarn-type ore bodies are found, CO_2-rich inclusions are less abundant.

The volume of halite in polyphase inclusions from the Akagane area is generally small, and the salinity is assumed to be about 30–35 wt.% from the volume ratio. As the size of these inclusions is small, temperature and salinity were not determined. Ranges of homogenization temperature and salinity, however, were reported from some skarn-type ore bodies as 230°–320°C and 13–21 wt.%, respectively (MURAMATSU and NAMBU, 1976).

The copper mineralization of the Akagane area would have been brought about by the ore-forming fluids released from the granodiorite stocks, but the study of CO_2-rich inclusions suggests an intimate genetic relation of the Koganetsubo tungsten mineralization with the southern stock and CO_2-rich fluids.

Skarn deposits ENJOJI (1977) summarized the temperature and salinity data of fluid inclusions from some skarn deposits in Japan. Both the temperature and salinity are generally higher than those of ore-forming fluids of Neogene mineralization. The data on skarn minerals, however, are very limited owing to the hardness of microscopic observation of inclusion mainly because of the small size of inclusion and poor optical condition of the host minerals.

3. *Kuroko Deposits*

Geologic features of the Kuroko deposits in Japan are described in detail elsewhere (for example, ISHIHARA, 1974). The genesis of Kuroko deposits is related to the submarine volcanism of the Miocene age. The deposits generally comprise two types of orebody, that is, bedded-type and stockwork-type. The bedded-type orebodies lie on the flanks of rhyolite domes or on acidic pyroclastic beds, and involve Kuroko (black ore), Oko (yellow ore) and/or Keiko (siliceous ore), successively from the upper level to lower, with gypsum bodies at the lowest horizon or peripheries of deposits.

The stockwork-type orebodies, which generally consists of ores with pyrite and chalcopyrite but, in some cases, with sphalerite, extend downward in altered rhyolite. The siliceous stockwork orebodies are generally continuous with the siliceous-bedded orebodies which lie laterally on the stockwork. In some deposits, the ore reserve of a stockwork orebody is more than that of the bedded orebody. It is considered that the fluid inclusions of these stockwork orebodies would give us significant information on the physicochemical condition of ore-forming fluids before their outflow to the sea floor.

The temperature and salinity of fluid inclusions from bedded orebodies were summarized by TOKUNAGA and HONMA (1974). Fluid inclusion studies of stockwork orebodies were carried out by MARUTANI and TAKENOUCHI (1978) at the Kosaka mine, Akita, and by YOSHIDA (1979) at the Iwami mine, Shimane.

In the Uwamuki No. 4 Orebody of the Kosaka mine, the range of homogenization temperature of fluid inclusion in quartz from the sphalerite-rich bedded-type siliceous ore was 260°–290°C. On the other hand, that of the chalcopyrite-rich siliceous ore was 280°–310°C. In the stockwork-type orebody, however, no remarkable difference of temperature was recognized between the upper and lower levels. The temperature range was approximately 275°–315°C, with the maximum frequency at about 300°C. But the temperature obtained from drilling cores at a depth 60 m lower than the lowest level, showed a range of 300°–325°C, indicating a slight increase of temperature at depth (MARUTANI and TAKENOUCHI, 1978). The difference of homogenization temperature between the bedded and stockwork siliceous ores probably depends on the environment of deposition of the two types. A similar result was obtained for the Uchinotai No. 7 Orebody.

In the Matsumine deposit, four rhyolite domes are found underneath the bedded-type orebodies. Stockwork orebodies consisting of pyrite-rich siliceous ores were found in the rhyolite domes, but they are much smaller than the above-situated bedded-type orebodies. The range of the homogenization temperature of fluid inclusion was approximately 275°–335°C with the peak of maximum frequency changing from about 310°C at the −225 ml to 325°C at −449 ml, for the deepest sample. The temperature gradient was much less than that of the above-mentioned Uwamuki No. 4 and Uchinotai No. 7 stockwork orebodies (Fig. 5).

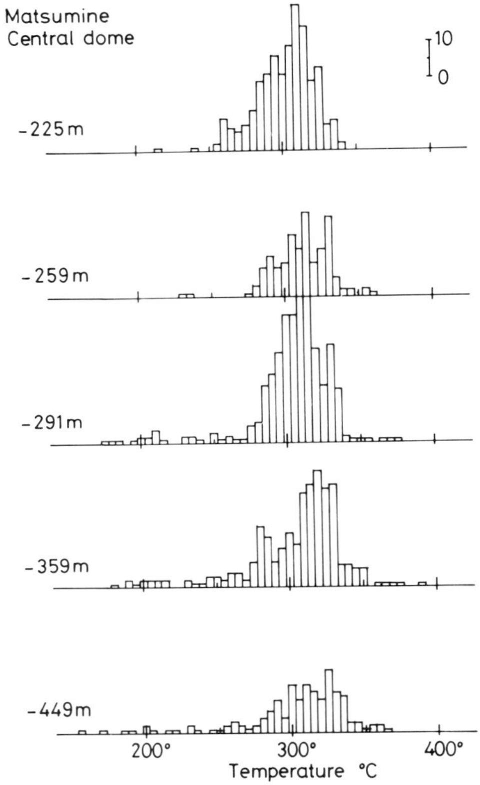

FIG. 5. Homogenization temperature of liquid inclusions from quartz of stockwork in the central rhyolite dome of the Matsumine Kuroko deposit, Akita.

A fairly constant thermal condition in the stockwork orebodies would have been caused by a high heat flow brought up by the ore-forming fluids which ascended through the brecciated parts of rhyolite domes in which the stockwork orebodies were formed.

No evidence of boiling of fluids was observed microscopically on the mode of occurrence of fluid inclusions from both of the bedded and stockwork-type siliceous orebodies. Therefore, it is inferred that the pressure of fluids was higher than the vapor pressure of the system.

As it is considered that the stockwork siliceous orebodies were the vents of hydrothermal fluids, the hydrostatic pressure of sea water must have been greater than that of hydrothermal fluids in order to suppress the boiling of fluids. If it is assumed that the temperature increase of sea water by the outflow of hydrothermal fluids from the sea floor was limited to the neighborhood of the vents, the estimated depth of the sea floor where Kuroko deposits were formed, should be greater than 850 m in the Hokuroku area (Fig. 6).

The salinity of fluid inclusion ranges from 2.5 to 5.5 wt.%, having a tendency to increase slightly at depth (Figs. 7, 8, and 9). YOSHIDA (1979) carried out a fluid inclusion study of the stockwork orebodies of the Iwami mine, and concluded that they were formed at temperatures around 270°C and at the depth of sea not shallower than 540 m.

Temperature data of stockwork orebodies indicate that the temperature gradient was generally gentle, but become steeper in upper parts, especially near the sea floor. It is pressumed that the temperature gradient would have been controlled by the flow rate of ascending fluids and by the permeability of wall-rocks which controlled the incursion of sea water through the wall-rock.

As the brecciated structures formed in rhyolite domes would have served as channels for the ascending fluids, vertically extended high-temperature zones must have been formed around the channels (CATHLES, 1978). The gentle increase of temperature with depth would suggest a deeper origin of ore-forming fluids.

At the Kosaka and Fukazawa mines, Akita, several breccia dikes are recognized. The breccia dikes consist of many angular or subangular breccias of the volcanic and pyroclastic wall rocks as well as the basement rocks such as phyllite, quartzite, quartz porphyry, and quartz diorite, and of an ash-like matrix. In general, those igneous rocks are highly argillized. In quartz phenocrysts or grains of the porphyry and quartz-diorite, polyphase inclusions are commonly observed though they are very small in size. They

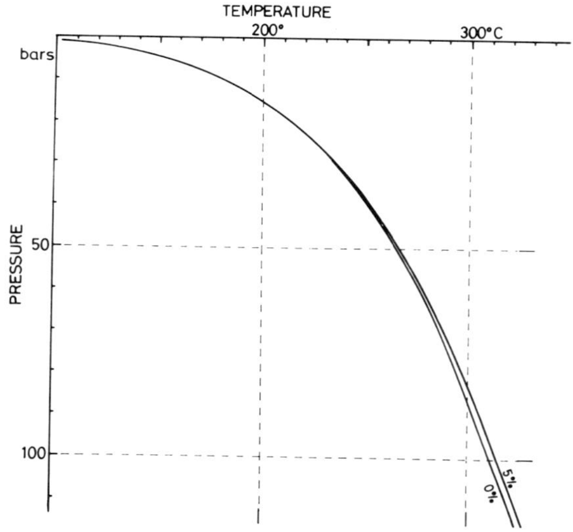

FIG. 6. Vapor pressure curve of pure water and 5 wt.% NaCl aqueous solution (HAAS, 1970).

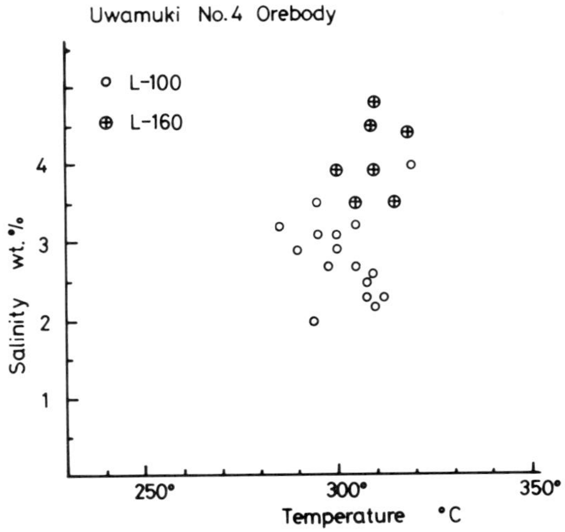

FIG. 7. Salinity-temperature diagram of fluid inclusions from the Uwamuki No. 4 stockwork orebody of the Kosaka mine, Akita.

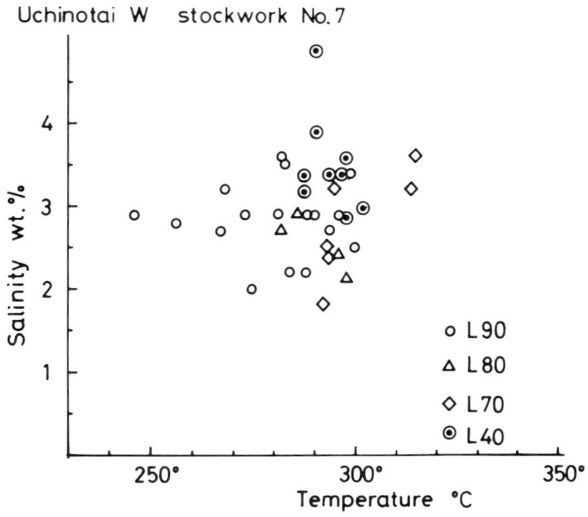

FIG. 8. Salinity-temperature diagram of fluid inclusions from the Uchinotai West No. 7 stockwork orebody of the Kosaka mine, Akita.

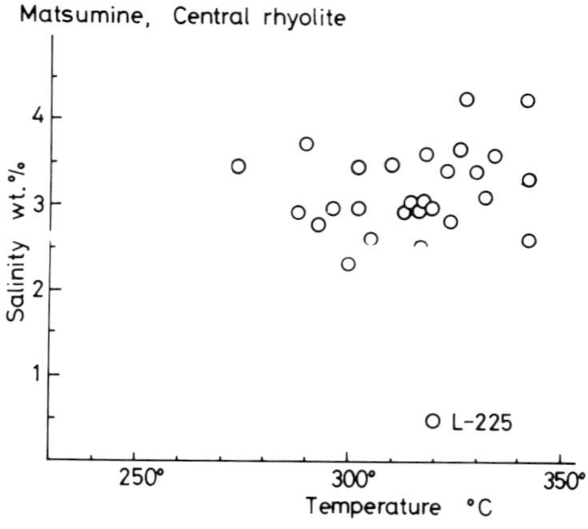

FIG. 9. Salinity-temperature diagram of fluid inclusions from the stockwork orebody in the central rhyolite dome of the Matsumine mine, Akita.

comprise halite, and sylvite or opaque minerals in some cases. The homogenization temperatures of polyphase inclusion varied from 250° to 450°C and the homogenization was caused by the disappearance of halite. Presumed salinity is 35–50 wt.%.

It is evident that the intrusion of the breccia dikes is later than the mineralization, but there is a possibility that the breccias of porphyry and quartz diorite would have been brought up from the deeper origin and that they would have a genetic relation with the rhyolitic volcanism which has brought the Kuroko mineralization. If it is the case, the polyphase inclusions contained in porphyry or quartz diorite may represent the original ore-forming fluids of Kuroko deposits. At present, however, there is no data that positively prove the genetic relation of polyphase inclusions. Further geologic and fluid inclusion data will be needed for elucidation.

4. *Porphyry Type Deposits*

Since LINDGREN (1905) reported the existence of halite-bearing fluid inclusions in the Clifton-Morenci district, Arizona, there had been no elaborated work on fluid inclusions in porphyry-type deposits until ROEDDER (1971) studied them at Bingham, Utah, Climax, Colorado, and Butte, Montana. He identified five types of fluid inclusions (1) halite-bearing hypersaline inclusion, (2) gaseous inclusion, (3) liquid inclusion, (4) CO_2-bearing inclusion, and (5) H_2S-bearing inclusion. Fluid inclusions from Bingham were also studied by MOORE and NASH (1974). Temperature and salinity of fluid inclusions from the Copper Canyon, Nevada, were studied by NASH and THEODORE (1971) and BATCHELDER (1977).

HALL *et al.* (1974) added the data of stable isotopic composition of water to Roedder's, and concluded that the Climax deposit was formed by fluids of three stages which were related, respectively, to the successive intrusion of different stocks. According to them, the salinity of ore-forming fluids decreased and the isotopic composition of water changed to a lighter value when the mineralization proceeded, suggesting an increasing participation of meteoric water at later stages of mineralization.

NASH (1976) summarized the data of fluid inclusions of 37 porphyry copper deposits in the United States, and synthesized a general conception of ore-forming fluids which formed porphyry copper deposits (Fig. 10). Generally, three types of fluid inclusion, are found in this type of deposit.

(1) Liquid inclusions of high densities and various homogenization temperatures.

(2) Gaseous inclusions having low and various degrees of filling.

(3) Polyphase fluid inclusions of high salinities and high homogenization temperatures (Fig. 11).

In general, both fluids of high-salinity and intermediate-salinity participated in the formation of deposits and the boiling of high-salinity fluids is characteristic to this type of deposit. Nash emphasized the application of the

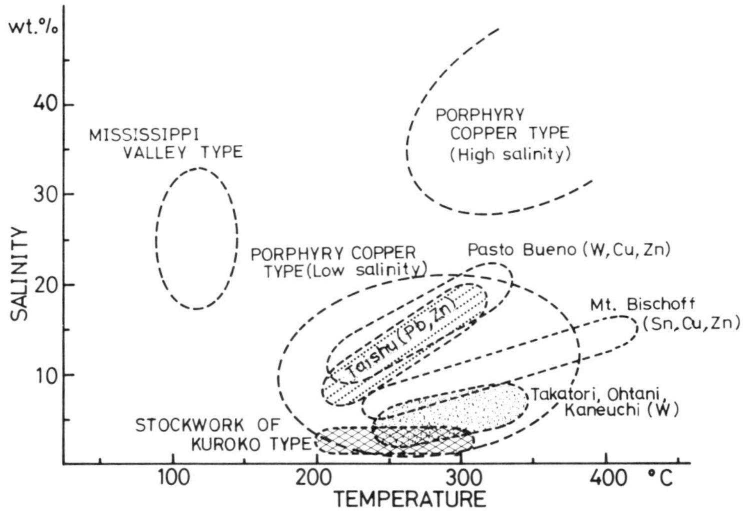

FIG. 10. Salinity-temperature ranges for high-salinity polyphase inclusions and low-salinity liquid inclusions from porphyry copper deposits compared with those of some other types. Porphyry copper deposits (after NASH, 1976), Mississippi Valley type deposits (ROEDDER, 1967), Pasto Buena, Peru (LANDIS and RYE, 1974), and Mt. Bischoff, Tasmania (GROVES and SOLOMON, 1969). The Takatori, Ibaragi, Ohtani, Kyoto, and Kaneuchi, Kyoto, are of hypothermal tungsten deposits (TAKENOUCHI and IMAI, 1971).

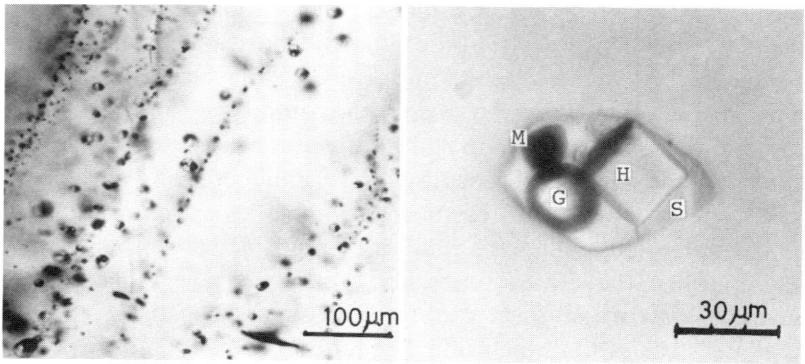

FIG. 11. Photographs showing the distribution of polyphase inclusions (a) and a large polyphase inclusion (b) in quartz from the Santo Tomas II porphyry copper deposit, Philippines. G: bubble, H: halite, S: sylvite, M: hematite.

local and temporal distribution of fluid inclusions to the exploration of porphyry copper deposits in relation to the mineralization and wall-rock alteration. He also put a stress upon the detailed description of the following items at the studies: (1) type of fluid inclusion, (2) mode of occurrence of each type of inclusion, (3) species and size of daughter minerals in inclusion, (4) volume ratio of each phase constituting inclusion, and (5) total number of inclusions in a volume of host mineral.

General models of mineralization and wall-rock alteration of porphyry copper deposits were given by LOWELL and GUILBERT (1970), but recent studies revealed that the zoning of wall-rock alteration is not the result of a single sequence of hydrothermal activity but of more than two systems. A pair of potassic and propylitic alteration is mainly the result of magmatic water and the overlap of phyllic and argillic alteration brought by circulated meteoric water, makes the alteration pattern more complex (SHEPPARD and GUSTAFSON, 1976; CHIVAS and WILKINS, 1977).

Distribution of polyphase, gaseous, and liquid inclusions in and around porphyry copper deposits is characteristic (Fig. 12). In general, high-salinity polyphase inclusions occur in the core part intimately associated with gaseous inclusions, whereas liquid inclusions of intermediate- or low-salinity predominate outside the ore shell.

Stable isotope studies of water in minerals and fluid inclusions of porphyry copper deposits revealed that the core part contained fluids rich in heavy water (magmatic water). Meanwhile the outside of the ore shell contained much light water (meteoric water) (SHEPPARD and TAYLER, 1974; HALL *et al.*, 1974; SHEPPARD and GUSTAFSON, 1976; BATCHELDER, 1977).

More than 23 occurrences of porphyry copper mineralization have been found in the Philippines. They are genetically classified into two groups. Deposits found in the Cebu and Southwestern Negros belong to the one group

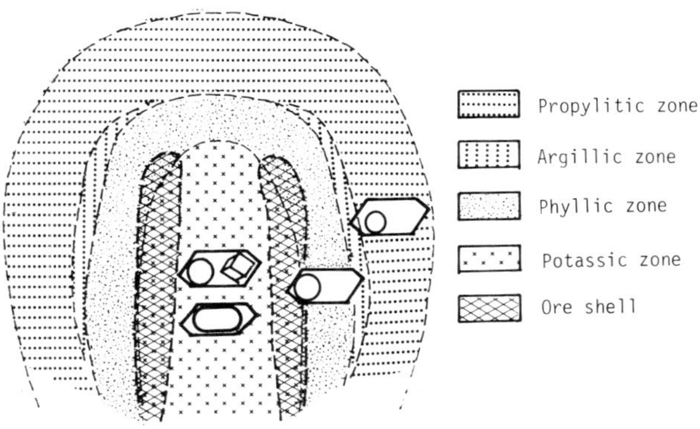

FIG. 12. Lowell-Guilbert model (LOWELL and GUILBERT, 1970) and the distribution of fluid inclusions in porphyry copper deposits.

which was formed in Paleocene, and those found in the Luzon and Marinduque belong to the other of Late Miocene (MOTEGI, 1975, 1977).

In general, either biotitization or sericitization predominates in the wall-rock alteration of the ore zone and silicification accompanies sulfide mineralization. Zoning of wall-rock alteration is irregular or indistinct.

Fluid inclusion studies were carried out on some of these deposits by BALCE (1978) and TAKENOUCHI (1980, 1981). Balce's work includes the studies of the Santo Tomas II, Santo Niño, and Kennon which are of Miocene age, and Takenouchi's work covers some of the Paleocene deposits, as well as the Miocene.

Fluid inclusion studies of Philippine porphyry copper deposits revealed that the variety of the mode of occurrence of fluid inclusions depends on the geologic features of deposits. GUSTAFSON (1978) discussed the space-time relationship of igneous intrusion and mineralization of porphyry copper deposits, and concluded that depth and size of intrusions, volatile evolution of magma, geologic structure and ground water incursion could be the factors controlling the pattern and range of temperature and pressure conditions. A similar result to Gustafson's was obtained from fluid inclusion studies of Philippine porphyry copper deposits.

Polyphase and gaseous inclusions are abundant but liquid inclusions are few in a vertically extended pipe-shaped orebody such as the Santo Tomas II orebody of the Philex mine.

Upon heating, sylvite in polyphase inclusion dissolved at a temperature between 100° and 200°C, and halite between 450° and 550°C. The temperature data show that the NaCl concentration of fluids was in a range between 30 and 51 wt.%, especially clustering between 43 and 51 wt.%, and that the KCl concentration varied from 13 to 27 wt.%, but mostly from 13 to 21 wt.%.

Homogenization of inclusion was achieved by the disappearance of bubble at a temperature range between 500° and 600°C, but in some inclusions, bubble was still observed, even at a temperature higher than 600°C which was the upper limit of operation of the used heating-stage. Homogenization temperature of secondary liquid inclusion concentrated in a range between 180° and 280°C, and the salinity was between 1 and 3 wt.%.

The mode of occurrence of fluid inclusion indicates that high temperature and high pressure fluids of boiling condition would have mainly contributed to the formation of the deposit, and that the incursion of low-salinity meteoric water was relatively less, even in late stages (Fig. 13). This type of hydrothermal activity resulted in the formation of pervassive biotite alteration of the ore zone.

On the other hand, liquid inclusions are abundant in a laterally extended deposit such as the Tapian orebody of the Marcopper mine, but polyphase inclusions are scarce. Liquid inclusions in deposits of this type generally show a wide range of homogenization temperature. This fact suggests that the liquid inclusions were formed at various stages of meteoric water incursion, which caused extensive sericitization of the wall-rocks (TAKENOUCHI, 1981). At present, however, it is not clear whether the participation of meteoric water happened in a single hydrothermal activity forming the zonation of biotitization and sericitization as a result of mixing with magmatic water, or it happened in a sequence of stages, in which biotitization was first formed by magmatic water and then sericitization was formed overlapping the former alteration products. More data on the occurrence of fluid inclusions and the chemical and isotopic composition of fluids will be necessary to arrive at any conclusion, but the present result indicates that the features of ore-forming fluids of porphyry copper deposits are in close relation with the fracture pattern, wall-rock alteration, and evolution of hydrothermal systems.

The geology of the Mamut porphyry copper deposit, Sabah, Malaysia, was reported by KOSAKA and WAKITA (1975, 1978) and a fluid inclusion study was performed by NAGANO *et al.* (1977). The deposit occurs in the top of

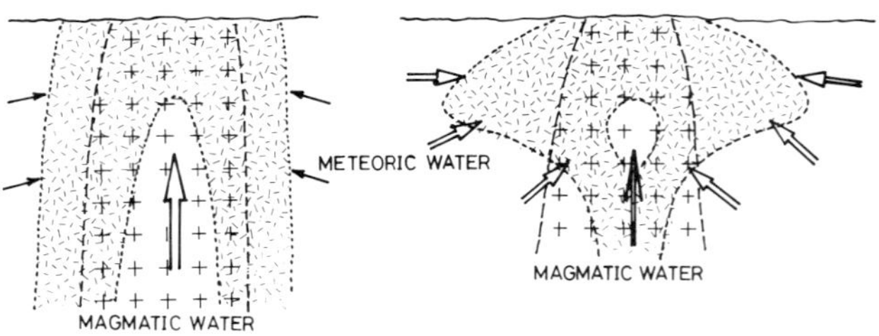

FIG. 13. Schematic section showing two types of participation of magmatic and meteoric waters at the formation of porphyry copper deposits. Stippled area represents orebody. Left: pipe-shaped orebody. Right: mushroom-shaped orebody.

adamellite porphyry as well as in siltstone and serpentinite, and has the core part and ore shell. The alteration zoning is slightly modified from the typical one, and comprises a core of fresh hornblende, and zones of actinolite-tremolite, biotite, and chlorite from the inner zone to the outer. The biotite zone roughly corresponds to the ore zone.

Fluid inclusion data indicate that high-salinity polyphase inclusions predominate in the orebody and that boiling occurred in the ore-forming fluids. The homogenization temperature of polyphase inclusion ranges between 300° and 480°C, and salinity was in a range between 35 and 55 wt.%.

Evolution of high-salinity hydrothermal solutions in the earth crust was discussed, based on the fluid inclusion studies of porphyry copper deposits. CLOKE and KESLER (1979) studied the trend of NaCl and KCl concentration in hydrothermal solutions during the formation of porphyry copper deposits. BODNER and BEANE (1980) studied the time-temperature-salinity relationships of fluid inclusions in the upper level cover overlying a deeply buried porphyry copper deposit at Red Mountain, Arizona. They obtained the result that early fluids of moderate salinity and high temperature had sources in the nearby wall rocks and later fluids of high salinity and moderate temperature originated in intrusives at depth.

5. *CO_2-Rich Fluids*

5.1 *Phase relation at low temperatures*

In general, CO_2 is the most abundant gas component in fluid inclusions. Most fluid inclusions contain CO_2 more or less, but some inclusions are so rich in CO_2 that CO_2 liquid or CO_2 gas hydrate is observable under the microscope, if the temperature is low enough to form those phases. Figure 14 shows some types of CO_2-rich fluid inclusions at room temperature.

As the critical temperature of CO_2 is 31.3°C, the CO_2 fluid in fluid inclusion is in the supercritical state under a strong ilumination, and the meniscus between the CO_2 gas and liquid phase is not identified (Fig. 14 Ay, By). In such a case, a drop of acetone on the polished thin section will help to identify the two CO_2 phases as the latent heat of vaporization of acetone lowers the temperature to as low as about 15°C. When acetone dries and the temperature raises gradually, the two CO_2 phases homogenize to a single phase. This is called the partial homogenization of the CO_2 phases.

The behavior of the CO_2 phases is interpreted on the H_2O–CO_2 phase diagram at low temperatures, and indicates an approximate CO_2 concentration in fluid inclusions. Figure 15 shows the H_2O–CO_2 phase diagram and the phase change in a CO_2-rich fluid inclusion. The HGL_1 and GL_1L_2 univariant curves almost overlap with the vapor pressure curve of pure CO_2 on the scale of this diagram, because the H_2O solubility in CO_2 liquid is very low.

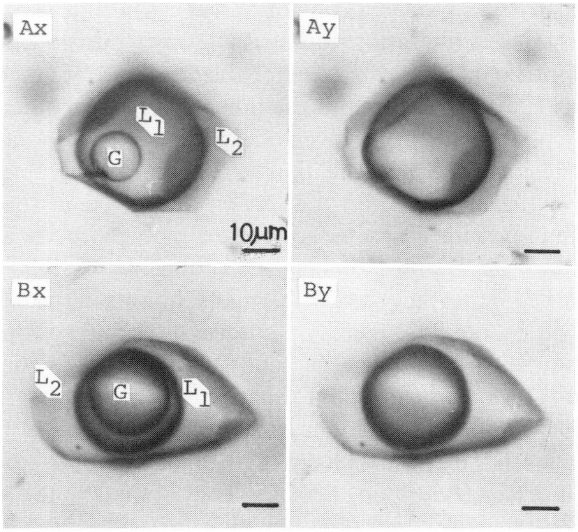

FIG. 14. Photographs showing the behavior of CO_2 phases in CO_2-rich fluid inclusions. Ax: $G + L_1 + L_2$, Ay: $L_1 + L_2$, Bx: $G + L_1 + L_2$, By: $G + L_2$. A and B correspond to those in Fig. 15. G: gas phase, L_1: CO_2-rich liquid phase, L_2: H_2O-rich liquid phase.

At a temperature range between 10° and 31°C, two types of partial homogenization are observed, as shown by A and B arrow on the diagram. At the pressure and temperature condition along the GL_1L_2 curve, gas, CO_2 liquid and aqueous solution coexist. But, when the gas phase homogenizes to the CO_2 liquid phase as the temperature increases, (the partial homogenization shown by arrow A), the inclusion consists of a CO_2 liquid (L_1) and aqueous solution (L_2). After the partial homogenization, the pressure of inclusion increases rapidly with temperature, because of the low compressibility of the two liquid phases. In this case, the inside pressure of inclusion reaches as high as several tens of bars, even at the room temperature.

In the case of the partial homogenization shown by B arrow, the CO_2 liquid phase homogenizes to a gas phase, and the inclusion consists of a gas and aqueous solution. In this case, the inside pressure of the inclusion would not increase remarkably with temperature, because the gas bubble buffers the increase of pressure.

Therefore, the partial homogenization at a temperature along the GL_1L_2 curve indicates two different CO_2 concentrations, and the former case (A) represents a higher CO_2 concentration than that of the latter case (B). In the former case, the lower the temperature of the partial homogenization, the higher the CO_2 concentration, and in the latter case, *vice versa*. In the neighborhood of the critical end point, the appearance or disappearance of

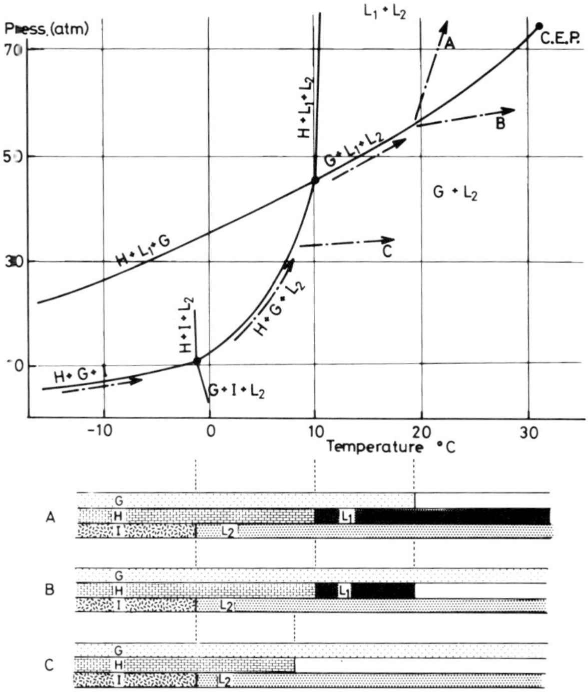

FIG. 15. Phase diagram for the H_2O–CO_2 system at low temperatures and phase change in CO_2-rich fluid inclusion. I : ice, H : CO_2 gas hydrate, G : gas phase, L_1 : CO_2-rich liquid phase, L_2 : H_2O-rich liquid phase, C.E.P. : critical end point.

the CO_2 gas phase (or CO_2 liquid phase) is so sensitive to the temperature change that even the operation of the iris diaphragm of iluminator can control this phenomena.

If a CO_2-rich fluid inclusion is frozen, it contains gas, ice, and CO_2 gas hydrate, and the pressure and temperature condition is shown by a point along the HGI univariant curve. At the invariant point (−1,5°C and 10.3 atm) where ice begins to melt, the gas bubble, CO_2 gas hydrate, ice, and aqueous solution coexist, and the temperature of the system is kept constant until the ice melts completely. At a temperature higher than the invariant point, the pressure of inclusion increases along the HGL_2 curve up to the second invariant point at 10.0°C and 44.4 atm.

If CO_2 content in the inclusion is enough to keep CO_2 gas hydrate up to 10°C, the gas hydrate decomposes to H_2O and CO_2 at the second invariant point and CO_2 liquid is formed. If the CO_2 content is not enough to keep the above-mentioned condition, the gas hydrate decomposes completely at a temperature lower than 10°C, and H_2O and CO_2 released from the hydrate go into an aqueous solution and gas bubble, as shown by C arrow. In this case, the CO_2 concentration of the inclusion is much less than the former two cases, but it is still possible to microscopically check the existence of CO_2 in the inclusion.

The existence of CO_2-rich inclusions is reported from tin or tungsten deposits as well as from some base metal deposits (SMITH and LITTLE, 1959; TAKENOUCHI, 1975), and also from olivine nodules in basalts (ROEDDER, 1965) and granitic rocks (ROEDDER and COOMBS, 1967).

CO_2-rich fluids are not uncommon in high-grade metamorphic rocks (YERMAKOV, 1965). Recently, KREULEN (1980) studied CO_2-rich fluid inclusions in metamorphic rocks of Nexos, Greece, and concluded that CO_2 with $\delta^{13}C$ values of -1 to -5 ‰ came from a deep-seated source and CO_2 with high $\delta^{13}C$ values ($+1$ to $+5$ ‰) was formed in situ by decarbonation of dolomite.

5.2 Gas hydrate

Gas hydrate is a kind of clathrate consisting of water and gas molecules. Two types of structure, that is, single hydrate and double hydrate, are reported for gas hydrate (CLAUSSEN, 1951a, b; STACKELBERG *et al.*, 1951, 1954a, b).

Single gas hydrate has a body-centered cubic lattice containing 46 tetrahedrally coordinated water molecules and eight cavities, in which gas molecules of a small diameter such as CO_2, H_2S, CH_4, Br_2 or Cl_2, are trapped. Therefore, the composition is expressed as 8 (Gas) · $46H_2O$.

As shown in Fig. 16, the body-centered cubic lattice is composed of nine pentagonal dodecahedra, each of which consists of 20 water molecules and occupies the center and each corner of the lattice. The pentagonal dodecahedron at each corner is connected with those of the nearest corners through the additional two water molecules, forming two tetrakaidecahedra on each face of the lattice. Thus, the unit cell has a lattice constant of 12.0 Å, and contains two pentagonal dodecahedra(1) which trap gas molecules smaller than 5.1 Å in diameter, and six tetrakaidecahedra(2) which trap gas molecules smaller than 5.8 Å in diameter. The trapped gas molecules stabilize the lattice made of water molecules by filling cavities, and the water molecules forming the cavities do not interact with the trapped gas molecules.

The second type, double gas hydrate, has a face-centered cubic lattice and consists of 136 water molecules involving 16 pentagonal dodecahedra, which trap smaller gas molecules, and eight hexakaidecahedra(3) which trap larger gas molecules such as CCl_4 or C_6H_6. The lattice constant of the cell is 17.4 Å.

(1)Pentagonal dodecahedron consists of 12 pentagonal faces.
(2)Tetrakaidecahedron consists of 12 pentagonal and 2 hexagonal faces.
(3)Hexakaidecahedron consists of 12 pentagonal and 4 hexagonal faces.

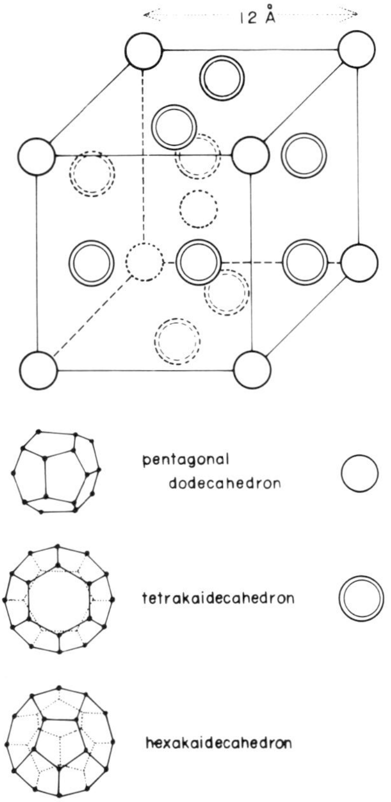

FIG. 16. Structure of single gas hydrate. Black circle represents water molecule.

COLLINS (1979) discussed the formation of CO_2 hydrate in a CO_2-rich fluid inclusion, and suggested the use of the decomposition temperature of the CO_2 hydrate coexisting with CO_2 liquid and gas phases, for the estimation of the salinity of the fluid inclusion. As the formation of CO_2 hydrate takes out some molecules of H_2O from the chloride aqueous solution, the residual aqueous solution becomes more saline than the original one after the clathration of CO_2 hydrate and, as the result, the melting temperature of ice will give a false salinity of the fluid inclusion.

Therefore, the formation of CO_2 hydrate in the fluid inclusion invalidates the estimation of salinity by means of the depression of the melting temperature of ice, but, as shown in Fig. 17, the depression of temperature of the invariant point, at which CO_2 hydrate, CO_2 liquid, CO_2 gas, and aqueous solution coexist, gives information of the salinity of inclusion, instead of the melting temperature of ice.

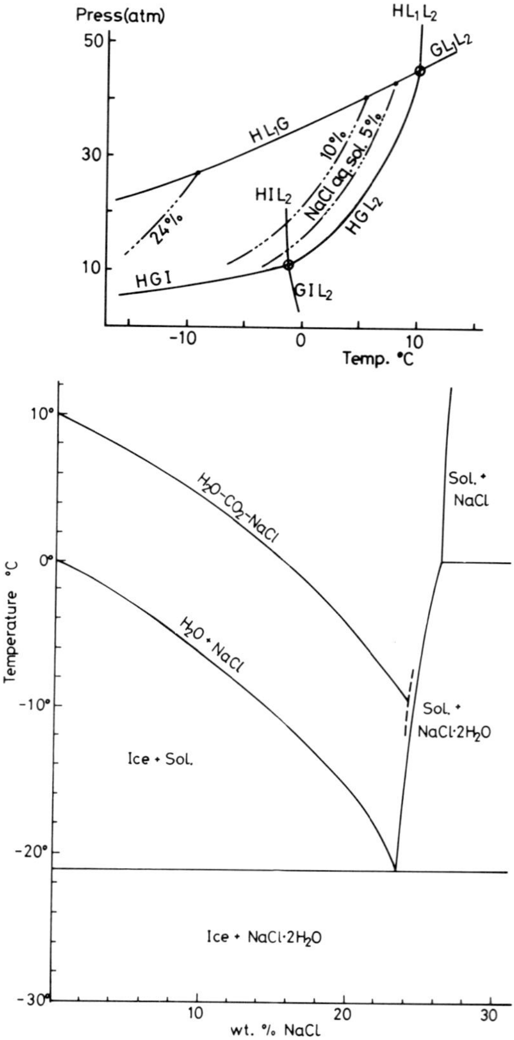

FIG. 17. Effect of NaCl on the invariant point of the H_2O–CO_2 system. A: *P-T* diagram, B: *T-X* diagram (after COLLINS, 1980).

6. *H_2O–NaCl System*

The compositions of fluid inclusions which were analyzed by various methods, indicate that a large part of hydrothermal solutions in the earth's crust are chloride aqueous solutions (ROEDDER, 1972). It can be reasonably said that the physicochemical properties of chloride hydrothermal solutions are represented with those of the H_2O–NaCl binary system, although they may contain some other compounds. Therefore, the data on the pressure (P)-specific volume(v)-temperature(T), and pressure-composition (X)-temperature relation of the H_2O–NaCl system are the basic knowledge on the hydrothermal solutions in the earth's crust.

P-v-T relation LEMMLEIN and KLEVTSOV (1961) studied the *P-v-T* relation of a NaCl aqueous solution at various concentrations (Fig. 18).

The diagrams show the following characteristics.

1. Filling temperature increases with the increase of NaCl concentration if the degree of filling is the same.

2. The slope of iso-specific volume line of NaCl aqueous solution is slightly gentler than that of pure water.

3. The critical temperature and pressure increase with the increase of NaCl concentration.

P-X-T relation SOURIRAJAN and KENNEDY (1962) studied the *P-X-T* relation of the H_2O–NaCl system.

Figure 19 shows the three-dimensional diagram of the system with isotherms, isobars, and isochors of the gas and liquid phase. The left surface of the complex curved surface in Fig. 19 represents the composition of gas phase and the right one, which is separated by the critical curve, represents that of the liquid phase.

Figure 20 shows the *P-X* diagram with isotherms as a parameter, and it is more convenient for reading the compositions of the coexisted gas and liquid phase. A and B lines on the diagram represent the saturated composition of the gas and liquid phase which coexists with NaCl crystal, respectively. On the diagram, the critical points of the system at various NaCl concentrations are clearly read on the critical curve.

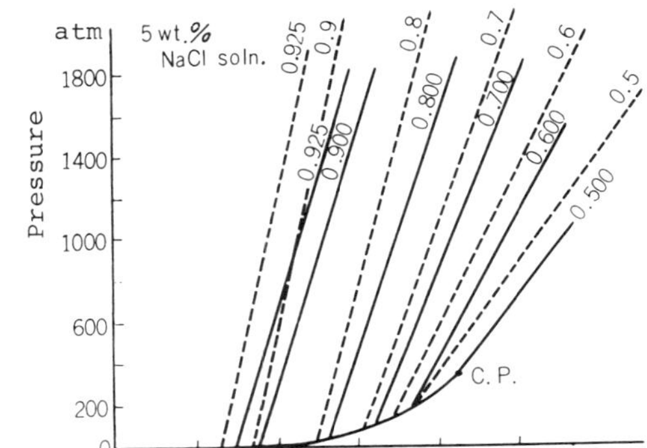

FIG. 18. Comparison of the *P-v-T* relation of 5 wt.% NaCl aqueous solution with that of pure water. Solid line: 5 wt.% aqueous solution, Broken line: pure water. Figure of parameter represents the degree of filling (after LEMMELIN and KLEVTSOV, 1961).

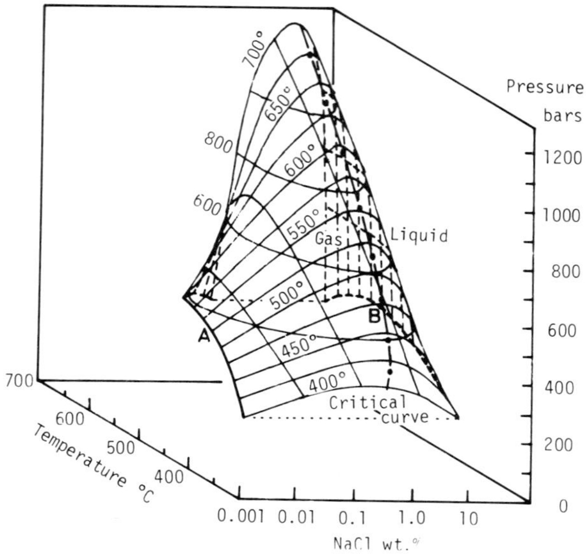

FIG. 19. *P-X-T* diagram of the H_2O–NaCl system (after SOURIRAJAN and KENNEDY, 1962).

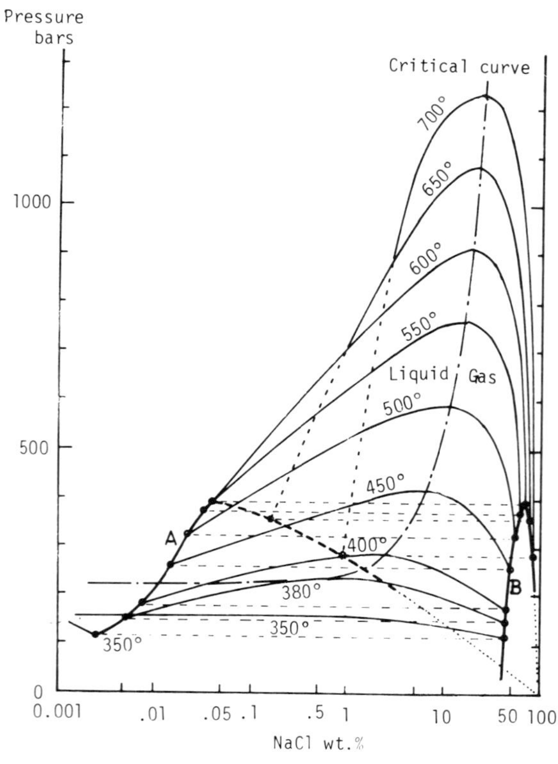

FIG. 20. *P-X* diagram of the H_2O–NaCl system with isotherms (after SOURIRAJAN and KENNEDY, 1962).

7. *H_2O–CO_2 System*

Some fluid inclusions are essentially rich in CO_2. The existence of CO_2 in fluid inclusions is optically recognizable as the liquid CO_2 phase or CO_2 gas hydrate, if the CO_2 concentration is fairly high. Therefore, a part of fluids in the earth's crust would be represented with the H_2O–CO_2 system.

The *P-v-T* relation of the H_2O–CO_2 system at the two-phase region, that is, at the temperature and pressure condition in which the CO_2-rich gas phase and H_2O-rich liquid phase coexist, is not available because of the technical difficulties of the experiments.

P-X-T relation TAKENOUCHI and KENNEDY (1964) studied the *P-X-T* relation of the H_2O–CO_2 system.

Figure 21 shows the three-dimensional diagram of the system. On this diagram, the critical curve separates the liquid phase region which is shown on the right-hand side of the curved surface, and the gas phase on the left-hand side.

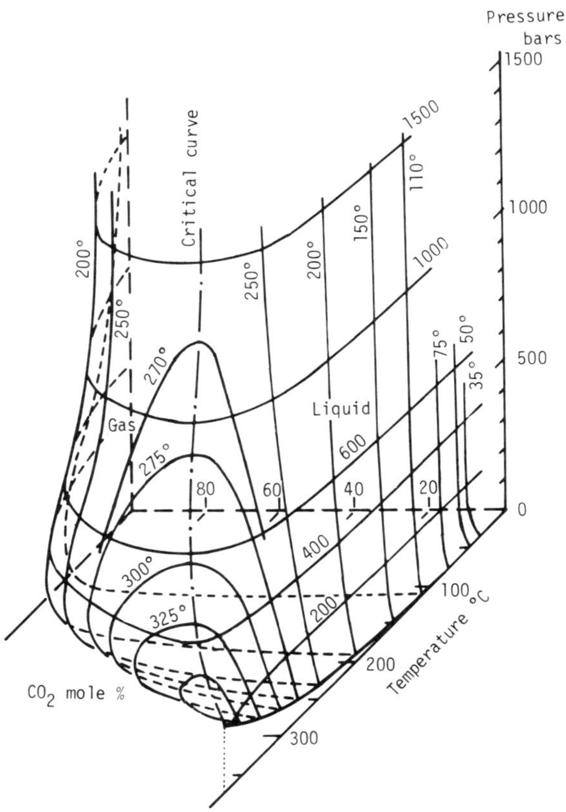

FIG. 21. *P-X-T* diagram of the H_2O–CO_2 system (after TAKENOUCHI and KENNEDY, 1964).

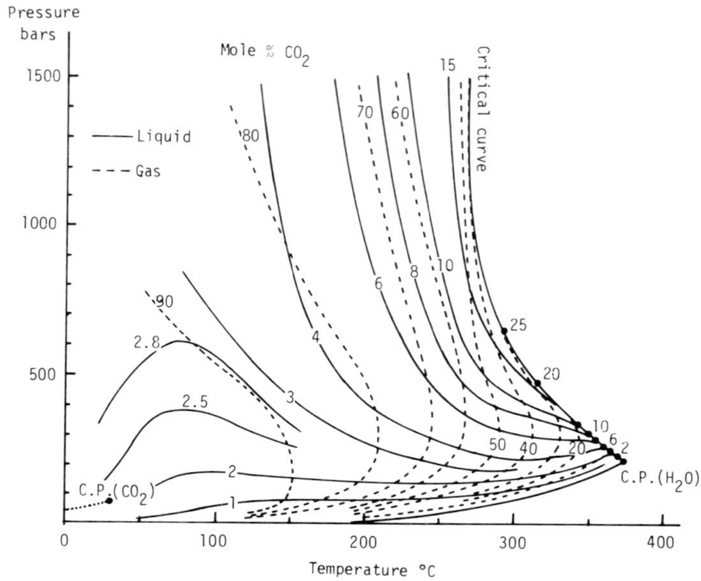

FIG. 22. *P-T* diagram of the H_2O-CO_2 system with isopleths (after TAKENOUCHI and KENNEDY, 1964).

The *P-T* diagram is shown in Fig. 22. From the diagram, it is known that the critical temperature decreases with the increase of the CO_2 concentration to as low as 265°C at about 30 mole%CO_2, and that the CO_2 concentration of CO_2-rich aqueous solutions would scarcely exceed 20 mole%CO_2 in the ordinary hydrothermal condition.

8. $H_2O-NaCl-CO_2$ System

Although two binary systems $H_2O-NaCl$ and H_2O-CO_2 may represent simplified solutions in the earth's crust, the natural solutions are more complex in the composition, and ternary or quaternary systems would be necessary for more elaborated discussion. NaCl or CO_2 effects on the critical temperature and pressure of water. NaCl increases the critical temperature but CO_2 decreases it to as low as 260°C. Accordingly, the combined effect of NaCl and CO_2 on the critical temperature and pressure should be considered in the consideration of natural hydrothermal solutions.

No systematically compiled diagram of the $H_2O-NaCl-CO_2$ ternary system is available but, as Fig. 23 shows, the data of CO_2 solubility in NaCl aqueous solutions indicate that CO_2 solubility decreases remarkably with the increase of the NaCl concentration (TAKENOUCHI and KENNEDY, 1965). Therefore, it can be said that homogeneous CO_2-rich hypersaline fluids exist

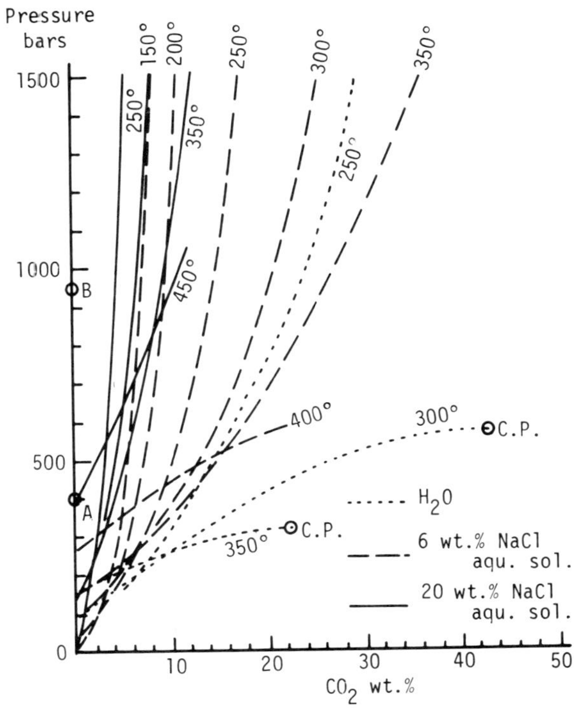

FIG. 23. Solubility of CO_2 in NaCl aqueous solutions (after TAKENOUCHI and KENNEDY, 1965).

only at very high pressures and temperatures. As a matter of fact, the coexistence of halite and CO_2 liquid in a fluid inclusion is very rare.

REFERENCES

BALCE, G. R. (1978): Geology and ore genesis of the prophyry copper deposits in Baguio district, Luzon Island, Philippines, *J. Geol. Soc. Philip.*, **33**, 1–43.

BATCHELDER, J. (1977): Light stable isotope and fluid inclusion study of the porphyry cooper deposit at Copper Canyon, Nevada, *Econ. Geol.*, **72**, 60–70.

BODNER, R. J., BEANE, R. E. (1980): Temporal and spatial variations in hydrothermal fluid characteristic during vein filling in preore cover overlying deeply buried porphyry copper-type minerlization at Red Mountain, Arizona, *Econ. Geol.*, **75**, 876–893.

CATHLES, L. M. (1978): Hydrothermal constraints on the formation of Kuroko deposits, *Mining Geol.*, **28**, 257–265.

CHIVAS, A. R. and WILKINS, R. W. T. (1977): Fluid inclusion studies in relation to hydrothermal alteration and mineralization at the Koloula porphyry copper prospect, Guadalcanal, *Econ. Geol.*, **72**, 153–169.

CLAUSSEN, W. F. (1951a): Suggested structures of water in inert gas hydrates, *J. Chem. Phys.*, **19**, 259–260.

CLAUSSEN, W. F. (1951b): A second water structure for inert gas hydrates, *J. Chem. Phys.*, **19**, 1425–1426.

CLOKE, P. L. and KESLER, S. E. (1979): The halite trend in hydrothermal solutions, *Econ. Geol.*, **74**, 1823–1831.

COLLINS, P. L. F. (1979): Gas hydrates in CO_2-bearing fluid inclusions and the use of freezing data for estimation of salinity, *Econ. Geol.*, **74**, 1435–1444.

ENJOJI, M. (1972): Studies on fiuid inclusions as the media of the ore formation, *Sci. Rept. Tokyo Kyoiku Daigaku, Sec. C*, **11**, 79–126.

ENJOJI, M. (1978): Fluid inclusions in minerals in some contact metasomatic deposits, in *Prof. T. Miyazawa Memorial Volume*, B49–B64 (in Japanese).

ENJOJI, M. and TAKENOUCHI, S. (1976): Present and future researches of fluid inclusions from vein-type deposits, *Mining Geol.*, Special Issue No. 7, 85–100 (in Japanese).

GROVES, D. I. and SOLOMON, M. (1969): Fluid inclusion studies at Mt. Bischoff, Tasmania, *Trans. Inst. Mining Metall.*, **78**, B1–B11.

GUSTAFSON, L. B. (1978): Some major factors of porphyry copper genesis, *Econ. Geol.*, **73**, 600–607.

HAAS, J. L., JR. (1970): An equation for the density of vapor-saturated $NaCl-H_2O$ solutions from 75° to 325°C, *Am. J. Sci.*, **269**, 489–493.

HALL, W. E., FRIEDMAN, I. and NASH, J. T. (1974): Fluid inclusion and light stable isotope study of the Climax molybdenum deposits, Colorado, *Econ. Geol.*, **69**, 884–901.

HATTORI, K. and SAKAI, H. (1979): D/H ratios, origins, and evolution of the ore-forming fluids for the Neogene veins and Kuroko deposits of Japan, *Econ. Geol.*, **74**, 535–555.

HATTORI, K. and SAKAI, H. (1980): Implications of D/H and $^{18}O/^{16}O$ ratios of ore fluids for the Neogene vein-type and Kuroko mineralization of Japan, in *I.A.G.O.D. 5th Symp. 1978, Snowbird, Proc. II*, pp. 101–108.

IMAI, H., TAKENOUCHI, S. and KIHARA, T. (1971): Fluid inclusion study at the Taishu mine, Japan, as related to geologic structure, *Mining Geol.*, Special Issue No. 3, 321–326.

ISHIHARA, S. (Ed.) (1974): Geology of Kuroko Deposits, *Mining Geol.*, Special Issue No. 6.

KOSAKA, H. and WAKITA, K. (1975): Geology and mineralization of the Mamut mine, Sabah, Malaysia, *Mining Geol.*, **25**, 303–320 (in Japanese).

KOSAKA, H. and WAKITA, K. (1978): Some geologic features of the Mamut porphyry cooper deposit, Sabah, Malaysia, *Econ. Geol.*, **73**, 618–627.

KREULEN, R. (1980): CO_2–rich fluids during regional metamorphism on Naxos (Greece). Carbon isotopes and fluid inclusions, *Am. J. Sci.*, **280**, 745–771.

LANDIS, G. P. and RYE, R. O. (1974): Geologic, fluid inclusion, and stable isotope studies of the Pasto Buena tungsten-base metal ore deposit, northern Peru, *Econ. Geol.*, **69**, 1025–1058.

LEMMLEIN, G. G. and KLEVTSOV, P. V. (1961): Relations among the principal thermodynamic parameters in a part of the system H_2O–NaCl, *Geokhimiya*, 148–158 (in Russian).

LINDGREN, W. (1905): The copper deposits of the Clifton-Morenci district of Arizona, *U.S. Geol. Survey*, Prof. Paper 43, 213–218.

LOWELL, J. D. and GUILBERT, J. M. (1970): Lateral and vertical alteration—mineralization zoning in porphyry ore deposits, *Econ. Geol.*, **65**, 373–408.

MARUTANI, M. and TAKENOUCHI, S. (1978): Fluid inclusion study of stockwork siliceous orebodies of Kuroko deposits at the Kosaka mine, Akita, Japan, *Mining Geol.*, **28**, 349–360.

MOORE, W. J. and NASH, J. T. (1974): Alteration and fluid inclusion studies of the porphyry copper ore body at Bingham, Utah, *Econ. Geol.*, **69**, 631–645.

MOTEGI, M. (1975): Mineralization of the Philippines—A geohistrical review—Contributions to the geology and paleontology of Southeast Asia, in *Geology and Paleontology of Southeast Asia*, Eds., T. Kobayashi and R. Toriyama, Vol. 15, Tokyo, Univ. of Tokyo Press., pp. 393–417.

MOTEGI, M. (1977): Porphyry copper deposits in the Philippines. Their tectonic setting and present status of development, *Mining Geol.*, **27**, 221–230 (in Japanese).

MURAMATSU, Y. and NAMBU, M. (1976): Studies on the Mineralization of the pyrometasomatic copper and iron ore deposits of the Akagane mine, Iwate Prefecture, Japan (1), *J. Japan. Assoc. Petrol. Mineral. Econ. Geol.*, **71**, 264–272 (in Japanese).

NAGANO, K., TAKENOUCHI, S., Imai, H. and SHOJI, T. (1977): Fluid inclusion study of the Mamut Porphyry copper deposit, Sabah, Malaysia, *Mining Geol.*, **27**, 201–212.

NAMBU, M., HARA, K. and HAYAKAWA, N. (1978): Fluid inclusion study of Inari copper vein, Ani mine, Akita Prefecture, *Mining Geol.*, **28**, 25–34 (in Japanese).

NAMBU, M., HAYAKAWA, N. and OKA, H. (1979): Fluid inclusion study of the Kaneuchi tungsten deposit, Kyoto Prefecture, *Mining Geol.*, **29**, 43–47 (in Japanese).

NASH, J. T. and THEODORE, T. G. (1971): Ore fluid in the porphyry copper deposit at Copper Canyon, Nevada, *Econ. Geol.*, **66**, 385–399.

NASH, J. T. (1976): Fluid inclusion petrology—Data from porphyry copper deposits and applications to exploration, *U.S. Geol. Survery, Prof. Paper 907–D.*

OKAMURA, S. (1980): Studies on the fluid inclusions in minerals from epithermal manganese-lead-zinc vein-type deposits in the southwestern part of Hokkaido, Japan, *Bull. Tokyo Gakugei Univ., Ser. IV*, **32**, 173–194.

ROEDDER, E.: Liquid CO_2 inclusions in olivine-bearing nodules and phenocrysts from basalts, *Am. Mineral.*, **50**, 1746–1782.

ROEDDER, E. and COOMBS, D. S. (1967): Immiscibility in granitic melts indicated by fluid inclusions in ejected granitic blocks from Ascention Island, *J. Petrol.*, **8**, 417–451.

ROEDDER, E. (1967): Environment of deposition of stratiform (Mississippi Valley-type) ore deposits, from studies of fluid inclusions, *Econ. Geol.*, Monograph 3, 349–363.

ROEDDER, E. (1971): Fluid inclusion studies on the porphyry-type deposits at Bingham, Utah, Butte, Montana, and Climax, Colorado, *Econ. Geol.*, **66**, 98–120.

ROEDDER, E. (1972): Composition of Fluid Inclusions, in *data of Geochemistry* (6th ed.), Ed. M. Fleisher, *U.S. Geol. Survey, Prof. Paper*, 440–JJ, 1972.

ROEDDER, E. (1976): Fluid-inclusion evidence on the genesis of ores in sedimentary and volcanic rocks, in *Handbook of Strata-Bound and Stratiform Ore Deposits, Vol. 2, Geochemical Studies*, Ed. K. H. Wolf, Elsevier, Amsterdam, pp. 67–110.

ROEDDER, E. (1979): Fluid inclusions as samples of ore fluids, in *Geochemistry of Hydrothermal Ore Deposits*, Ed. H. L. Barnes, 2nd. ed., John Wiley & Sons, New York, pp. 684–737.

SATO, J., MAEDA, H., KINRYU, Y. and ONO, S. (1980): Mineral paragenesis and fluid inclusion data of the Ohe polymetallic vein-type deposits, Hokkaido, Japan, *Mining Geol.*, **30**, 277–288.

SHEPPARD, S. M. F. and TAYLOR, H. P. JR. (1974): Hydrogen and oxygen isotope evidence for the origin of water in the Boulder Batholith and the Butte ore deposits, Montana, *Econ. Geol*, **69**, 926–946.

SHEPPARD, S. M. F. and GUSTAFSON, L. B. (1976): Oxygen and hydrogen isotopes in the porphyry copper deposit at El Salvador, Chile, *Econ. Geol.*, **71**, 1549–1559.

SMITH, F. G. (1953): *Historical Development of Inclusion Thermometry*. Univ. of Toronto Press, Toronto.

SMITH, F. G. and LITTLE, W. M. (1959): Filling temperatures of H_2O–CO_2 fluid inclusions and their significance in geothermometry, *Can. Mineral.*, **6**, 380–388.

SOURIRAJAN, S. and KENNEDY, G. C. (1962): The system H_2O–NaCl at elevated temperatures and pressures, *Am. J. Sci.*, **260**, 115–141.

STACKLBERG, M. V. and MÜLLER, H. R. (1951): On the structure of gas hydrate, *J. Chem. Phys.*, **19**, 1319–1320.

STACKELBERG, M. V. and MÜLLER, H. R. (1954): Feste Gashydrate (II), Struktur und Raumschemie. *Zeit, Elektrochem.*, **58**, 25–39.

STACKELBERG, M. V. and JAHNS, W. (1954): Feste gashydrate (VI), die Gitteraufweitungsarbeit, *Zeit. Elektrochem.*, **58**, 162–164.

TAKENOUCHI, S. (1971): Study of CO_2–bearing fluid inclusions by means of the freezing stage microscope, *Mining geol.*, **21**, 286–300 (in Japanese).

TAKENOUCHI, S. (1975): Fluid inclusion study of the Jishakuyama and Koganetsubo ore deposits and acidic igneous rocks at the Akagane mine, Iwate Prefecture, *Mining Geol.*, **25**, 247–259 (in Japanese).

TAKENOUCHI, S. (1978): Fluid inclusion studies. In: *Geological Studies of the Mineral Deposits in Japan and East Asia*, Ed. H. Imai, Tokyo, Univ. of Tokyo Press, pp. 312–334.

TAKENOUCHI, S. (1978): Fluid inclusions and ore-forming fluids of porphyry copper deposits, *Mining Geol.*, **28**, 117–130 (in Japanese).

TAKENOUCHI, S. (1980): Preliminary studies on fluid inclusions of the Santo Tomas II deposit (Philex) and Tapian (Marcopper) porphyry copper deposits in the Philippines, *Mining Geol.*, Special Issue No. 8, 141–150.

TAKENOUCHI, S. (1981): Fluid inclusion studies of several Philippine porphyry copper deposits, in *Geol. Survey of Japan, Rept. No. 261*, pp. 149–167.

TAKENOUCHI, S. and IMAI, H. (1971): Fluid inclusion study of some tungsten-quartz veins in Japan, *Mining Geol.*, Special Issue No. 3, 345–350.

TAKENOUCHI, S. and IMAI, H. (1975): Glass and fluid inclusions in acidic igneous rocks from some mining areas in Japan, *Econ. Geol.*, **70**, 750–769.

TAKENOUCHI, S. and KENNEDY, G. C. (1964): The binary system H_2O–CO_2 at high temperatures and pressures, *Am. J. Sci.*, **362**, 1055–1074.

TAKENOUCHI, S. and KENNEDY, G. C. (1965): The solubility of carbon dioxide in NaCl solutions at high temperatures and pressures, *Am. J. Sci.*, **263**, 445–454.

TOKUNAGA, M. and HONMA, H. (1974): Fluid inclusions in the minerals from some Kuroko deposits, in *Geology of Kuroko Deposits*, Ed. S. Ishihara, *Mining Geol.*, Special Issue No. 6, 385–388.

WATANABE, M. (1979): Fluid inclusions in some Neogene ore deposits in the Green Tuff Region, Japan, *Mining Geol.*, **29**, 307–321.

WATANABE, M., SHIMADA, M. and YOSHIDA, T. (1981): Fluid inclusion study on the granitic rocks and possibly related ore deposits in the Tsumo mining district, Southwest Japan, *Mining Geol.*, Special Issue No. 9, 145–162 (in Japanese).

YERMAKOV, N. P. (1965): Research on the Nature of Mineral-Forming Soluitons, in *International Series, Monograph in Earth Sci.*, 22, Ed. D. E. Ingerson, New York, Pergamon Press.

YOSHIDA, T. (1975): Fluid inclusions of the molybdenum deposits at the Higashiyama mine, Shimane Prefecture, Japan, *Mining Geol.*, **25**, 285–292.

YOSHIDA, T. (1979): Fluid inclusion study and ore forming process of the Iwami deposit, Shimane Prefecture, Japan, *Mining Geol.*, **29**, 21–31 (in Japanese).

Materials Science of the Earth's Interior, edited by I. Sunagawa, pp. 545–583.

HYDROTHERMAL SYNTHESIS AND PHASE RELATIONS OF THE POLYMETALLIC SULFIDE SYSTEM, ESPECIALLY ON THE Cu–Fe–Bi–S SYSTEM

Asahiko SUGAKI, Arashi KITAKAZE, and Kenichiro HAYASHI

Institute of Mineralogy, Petrology, and Economic Geology, Faculty of Science, Tohoku University, Sendai, Japan

1. *Introduction*

Metallic sulfide and sulfosalt minerals are very important as ore minerals and, in many cases, they occur from a hydrothermal solution. It is a very interesting geological problem in making clear the forming process and condition of them in nature. Experimental studies of hydrothermal synthesis and phase equilibrium of sulfide minerals are significant in order to know their forming conditions, and have been carried out over the past 20 years by several authors.

BRYSTOV and KUZ'MINA (1959) produced galena and sphalerite from a powder of simple sulfide in an aqueous solution of lithium and sodium chlorides at 350° and 450°C. KUZ'MINA (1961) and IKORNIKOVA (1962) found that an aqueous solution of ammonium chloride was the best solvent for the synthesis of galena and sphalerite. LAUDISE *et al.* (1965) succeeded in synthesizing large crystals of sphalerite using a potassium hydroxide aqueous solution at 350°C. The synthetic method employed by them was that of thermal gradient transportation using a difference of solubility due to temperature. BARNARD and CHRISTOPHER (1966) first synthesized the ternary sulfide minerals, chalcopyrite $CuFeS_2$, in a sodium chloride aqueous solution *in situ*. CHERNYSHEV *et al.* (1968) performed the hydrothermal synthesis of sphalerite, pyrrhotite, and pyrite of the system Fe–Zn–S in an ammonium chloride solution at temperatures from 250° to 500°C and obtained the solvus of the sphalerite solid solution for pyrrhotite and pyrite. SCOTT and BARNES (1971) carried out a work on the synthesis of sphalerite solid solution coexisting with pyrrhotite and pyrite in ammonium iodite and chloride aqueous solutions by the thermal gradient transporting method and made clear the phase relations of the sphalerite solid solution, troilite, pyrrhotite, and pyrite in the system Fe–Zn–S and the relationship between the sphalerite composition, temperature, and sulfur fugacity. Godovikov and Ptitsin (1969) and Ptitsin and Godovikov (1971) synthesized the phases in the Cu_2S–Bi_2S_3 join by using the hydrothermal *in-situ* crystallizing method with an ammonium chloride solution at temperatures from 300° to 500°C, and from 250° to 300°C, respectively. Sugaki and Kitakaze (1972) synthesized alabandite solid solution associated with pyrrhotite and pyrite in an ammonium chloride solution and determined their phase relationship. Sugaki *et al.* (1975) succeeded in the hydrothermal synthesis of the three component phases of the Cu–Fe–S system, such as bornite, chalcopyrite, intermediate solid solution, and nukundamite etc. by the thermal gradient method, and obtained an isothermal phase diagram of the system. Sugaki *et al.* (1980) also synthesized wittichenite, emplectite, cuprobismutite, and phases $CuBi_3S_5$, and $Cu_8Bi_8S_{19}$, etc. belonging to the system Cu–Bi–S, by using the hydrothermal transporting method, and made clear the phase relations in the system. Recently, Sugaki *et al.* (1981b) studied on phase equilibrium in the quaternary system Cu–Fe–Bi–S in the same way and partly determined the complicated phase relations.

Although studies of the phase equilibrium of most of the sulfide system by the dry method, using an evacuated glass tube, have been carried out, as summarized by Barton and Skinner (1967) and Craig and Scott (1974), experimental works on hydrothermal synthesis and the phase relations of the sulfide system are not mentioned above. That is because only such systems as Fe–Zn–S, Fe–Mn–S, Cu–Fe–S, Cu–Bi–S, and Cu–Fe–Bi–S have been studied hydrothermally. However, hydrothermal synthesis is significant in getting near to natural conditions and has the advantage that its reaction is faster than the dry method and, as a result, equilibrium is obtained relatively easily. Accordingly, the phase relation at low temperatures, which cannot be solved by the dry method, may be made clear by this hydrothermal method. Thus, for example, from the polymetallic sulfide system studied under hydrothermal conditions, the phase relations in the Cu–Fe–Bi–S system are described in this paper. The sulfide phases in this study are shown in the Cu–Fe–Bi–S tetrahedron of Fig. 1. Their crystal data, such as chemical formula, crystal system, space group, and cell parameters are also given in Table 1.

2. *Experimental Procedures*

2.1 *Hydrothermal synthesis*

The thermal gradient transporting method is generally employed as a way of hydrothermally synthesizing sulfide minerals. In this method, sulfides as a nutrient are immediately dissolved at the hot end of a reaction tube, transported by convection across a small temperature gradient in an aqueous

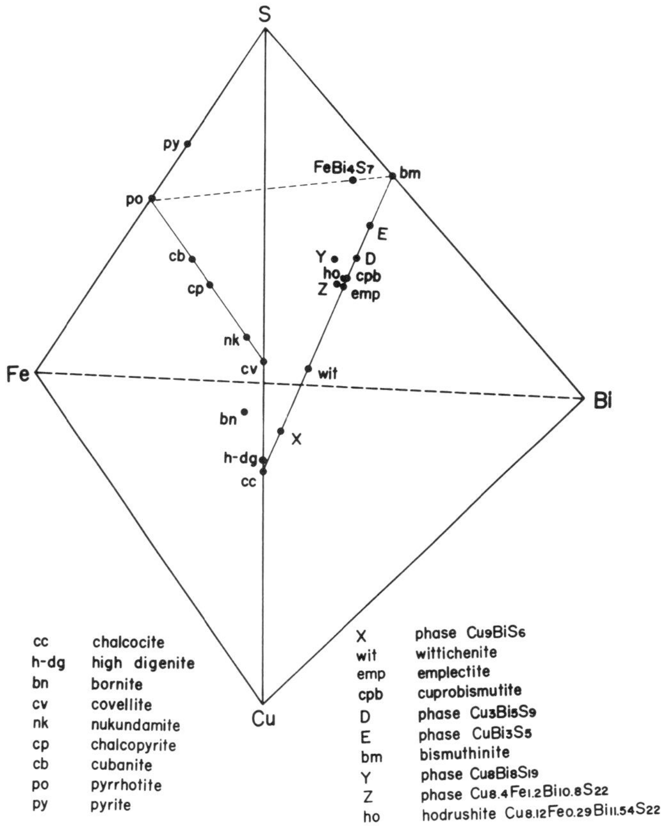

FIG. 1. Minerals and synthetic phases in the Cu–Fe–Bi–S system.

solution, and then reprecipitated as coexisting crystals at the cool end of the tube.

As a nutrient material, a mixture of powder sulfides was used which had been synthesized in advance by a reaction between pure metals (over 99.99% in purity) and 99.99% pure sulfur in an evacuated silica glass tube. Of the dry method using an evacuated glass tube, the papers of KULLERUD (1953, 1971), SUGAKI and SHIMA (1965a, b), and CRAIG and SCOTT (1974), furnish us with much information. Natural pure sulfides were sometimes used as a nutrient.

As a transporting solution, several kinds of aqueous solutions, such as potassium and sodium chloride or hydroxide, ammonium iodite and chloride, and lithium chloride were usually employed in varing concentrations. Among them, an ammonium chloride solution was the most adequate solvent, as reported by KUZ'MINA (1961), IKORNIKOVA (1962), CHERNYSHEV *et al.* (1968), SCOTT and BARNES (1971), and SUGAKI *et al.* (1975, 1976a). Most experiments

TABLE 1. The synthetic phases appeared in the experiments of the Cu–Fe–Bi–S system and their chemical compositions and crystallographic data.

Phases	Abb.	Chemical formula	Crystal system	Space group	Cell parameters a(A)	b(A)	c(A)	β
Chalcocite	cc	Cu_2S	mon	$P2_1/c$	15.246	11.884	13.494	116.35°
Phase X	X	Cu_9BiS_6	hex	$P6_322$	20.42	-	6.67	-
Wittichenite	wit	Cu_3BiS_3	Orth	$P2_12_12_1$	7.694	10.393	6.710	-
Emplectite	emp	$CuBiS_2$	Orth	Pnam	6.144	14.525	3.925	-
Cuprobismutite	cpb	$Cu_{10}Bi_{12}S_{23}$	mon	C2/m	17.513	3.909	15.255	101.09°
Phase D	D	$Cu_3Bi_5S_9$	mon	C2/m	13.08	4.00	14.70	99.4°
Phase E	E	$CuBi_3S_7$	mon	C2/m	13.221	4.023	14.077	115.46°
Phase Y	Y	$Cu_8Bi_8S_{19}$	orth	Pbnb	11.591	32.05	3.953	-
Bismuthinite	bm	Bi_2S_3	Orth	Pbnm	11.146	11.298	3.980	-
	-	$FeBi_4S_7$	mon	C2/m	12.72	3.96	11.80	104.5°
Phase Z	Z	$Cu_{8.4}Fe_{1.2}Bi_{10.8}S_{22}$	mon	C2/m	17.483	3.902	12.869	108.11°
Hodorushite	ho	$Cu_{8.1}Fe_{0.3}Bi_{11.5}S_{22}$	mon	A2/m	27.21	3.93	17.58	92.9°
Covellite	cv	CuS	hex	$P6_3mmc$	3.794	-	16.351	-
Digenite (high form)	h-dg	$Cu_{7-x}S_4(Cu_{6.6}S_4)$	cub	Fm3m	(5.51	-	-	-)
Bornite (high form)	bn	Cu_5FeS_4	cub	Fd3m	5.50	-	-	-)
Nukundamite	nk	$Cu_{3.39}Fe_{0.61}S_4$	hex	$P\bar{3}m1$	3.7830	-	11.1950	-
Chalcopyrite	cp	$CuFeS_2$	tet	$I\bar{4}2d$	5.278	-	10.430	-
Intermediate solid solution	iss	$CuFe_2S_3$	cub	$F\bar{4}3m$	(5.272	-	-	-)
Pyrrhotite	po	$Fe_{1-x}S$ ($Fe_{0.9}S$)	hex	P6/mmc	(3.45	-	5.75	-)
Pyrite	py	FeS_2	cub	Pa3	5.410	-	-	-)

on synthesis in the system Fe–Mn–S, Cu–Fe–S, and Cu–Fe–Bi–S by Sugaki *et al.* (1972, 1975, 1981b) were thus performed using a 5 m ammonium chloride solution as the transporting solvent. Scott (1975) discussed the necessary factors for the aqueous solution as the recrystallizing agent, and mentioned that an ammonium halide solution is the most effective and stable solvent for metal sulfide transportation.

As a reaction vessel, a gold tube of 4 to 5 mm in inside diameter and about 50 mm in length was commonly used. The reaction tube must not react to either the transporting solution nor to the nutrient sulfides. Gold is stable enough in a 5 m ammonium chloride solution to 500°C, in 5 m ammonium iodite to 400°C, in 6 m sodium chloride to at least 500°C, in 4.5 m potassium chloride to at least 700°C, in 2 to 15 m sodium hydroxide to 700°C, and in 11.5 m potassium hydroxide solution to at least 500°C (Scott, 1974). Also, gold is not attacked in the general sulfide system, but it becomes unstable in the system containing large amounts of such metals as mercury, bismuth, and arsenic. In such cases, a sealed silica tube is used, but this permits only a limited range of temperatures, up to 500°C.

Several ten milligrams, 10 to 50 mg in common, of nutrient and 0.2 to 0.3 ml of solution were put into the gold tube and the tube was sealed by welding. The sealed gold tube was placed into a test tube-type pressure vessel and heated in a vertical electric furnace designed so as to provide a linear thermal gradient by independently-controlled separate upper and lower heating elements. The temperature during the runs was measured by chromel-almel thermocouples inserted in two thermocouple walls at the top (or middle) and bottom portions of the outside wall of the pressure vessel. However, the inside temperature of the vessel was about 20° to 30°C lower than that of the outside at about 300°C. So the temperature crystallizing sulfide phases in the gold tube need to be directly measured in the pressure vessel. SUGAKI *et al.* (1980) measured temperature reprecipitated sulfide crystals by a thermocouple inserted into the inside of the vessel so as to get close to the top of the gold tube. When the thermal gradient is steep, the amount of sulfide crystallized was large, but metastable phases occurred. Therefore, a gentle thermal gradient was necessary in order to obtain an equilibrium condition, and the temperature difference between the top and bottom of the gold tube, 5 cm long, was kept between 5° to 20°C; that is, a thermal gradient of 1° to 4°C/cm. Pressure is measured by a Bourdon's tube gauge. At the termination of an experiment, the furnace was pulled down and the pressure vessel was cooled rapidly to room temperature in an air blast.

2.2 *Identification of phases*

Synthesized products were usually found in aggregate of several kinds of sulfides grown near the top of the gold tube. They were identified by means of a stereomicroscope, ore microscope, Gandolfi and Guinier cameras, and an electron probe microanalyser.

Because of the well-developed idiomorphic crystal forms, some synthesized phases could be quickly identified only by observation under a binocular microscope. Bornite commonly appeared as cubic, polyhedral crystals up to 300 μm in size, chalcopyrite as tetrahedral form, and pyrrhotite as hexagonal plates or prismatic crystals. Nukundamite occurred as beautiful, thin hexagonal platy forms up to 0.3 mm in diameter. However, it was difficult to distinguish intermediate solid solution from chalcopyrite by crystal form alone. Also, wittichenite appeared as rhombic, polyhedral forms of pale lead-gray in color, emplectite as thin platy or long prismatic crystals showing a lead-gray color, and phase $Cu_{8.4}Fe_{1.2}Bi_{10.8}S_{22}$ (Z) showed as short prismatic crystals of steel gray. Emplectite, cuprobismutite, phase $CuBi_3S_5$ (E), and phase $Cu_8Bi_8S_{19}$ (Y) were hard to distinguish from each other by their crystal shapes because they show forms similar to long prismatic or acicular forms. Metallic bismuth occurred as melted globules of about 200 μm in diameter. The scanning electron micrographs of the sulfide minerals crystallized in these synthetic experiments are shown in Figs. 2 and 3.

The ore microscope was very effective in identifying the mineral phases. A small amount of the product was mounted in a polyester resin, polished carefully, and observed under a reflecting microscope. Most hydrothermally

FIG. 2 A–D. Scanning electron micrographs of sulfide crystals synthesized hydrothermally. A, Pyrite and intermediate solid solution at 400°C; B, Pyrrhotite and intermediate solid solution at 300°C; C, Bornite at 420°C; D, Nukundamite at 420°C.

synthesized phases have optical properties similar to those of the equivalent natural minerals. The reflecting color of bornite solid solution varies from pinkish brown in copper-poor members to bluish gray in copper-rich ones, and also becomes a lighter brown with a bluish tint with increasing bismuth contents. Intermediate solid solution and chalcopyrite look very similar, but the former shows rather pinkish and less yellowish tints. Also, an iron-rich intermediate solid solution becomes lighter and less pinkish in color and is difficult to distinguish from chalcopyrite. Meanwhile, the copper-rich intermediate solid solution has usually fine lamellae of bornite or chalcopyrite

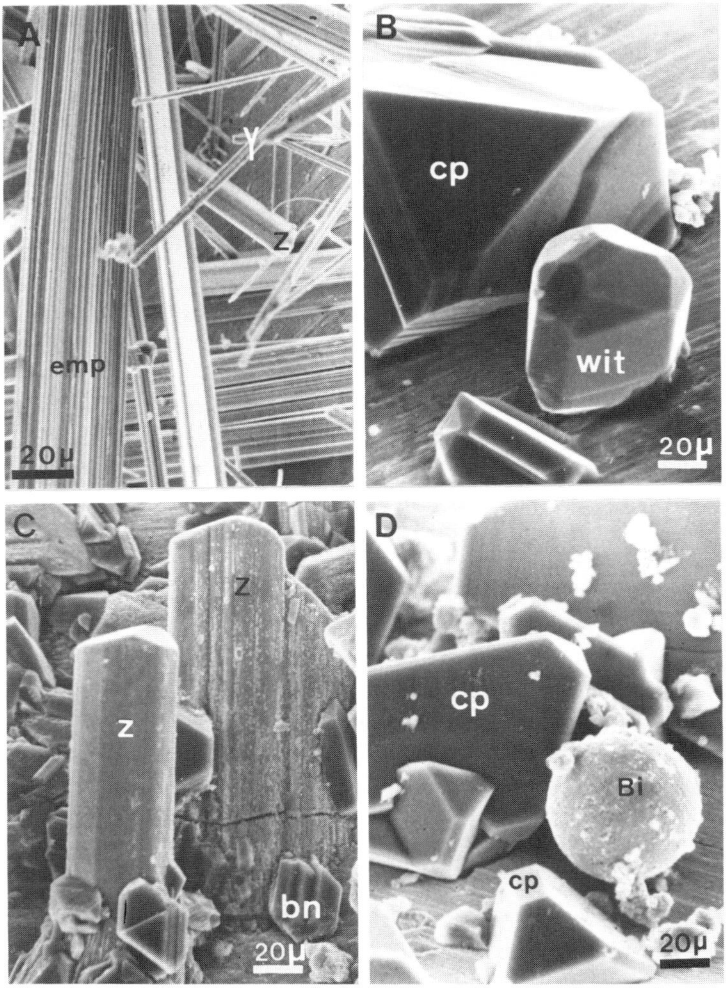

FIG. 3 A–D. Scanning electron micrographs of sulfide and sulfosalt crystals synthesized hydrothermally. A, Emplectite (tabular), phase Z (long prismatic) and phase Y (acicular) at 300°C; B, Wittichenite and chalcopyrite at 300°C; C, phase Z and bornite at 420°C; D, Chalcopyrite and bismuth liquid at 300°C.

as exsolution products segregated during cooling. Pyrrhotite is lighter and pinkish in color in comparison with iron-rich intermediate solid solution.

On the other hand, synthetic phases belonging to copper-bismuth sulfides have nearly the same optical properties, but some phases can be distinguished from others by their reflecting color and anisotropy. Wittichenite is darker gray color than emplectite, cuprobismutite, and phase $CuBi_3S_5$ (E). Also it shows anisotropy, but is very weak in comparison with emplectite. Emplectite and cuprobismutite are gray with a creamy tint in the reflecting color and have a strong anisotropism changing its interference colors from gray to reddish-brown. But it is difficult to distinguish both phases. Phase $CuBi_3S_5$ has almost a similar color to emplectite, but it shows very strong anisotropism, changing colors from bluish purple to pale brown. Phase $Cu_8Bi_8S_{19}$ (Y) is nearly similar to cuprobismutite, but is slightly brownish and shows a distinct bireflectance and strong anisotropism, the same as the phase $CuBi_3S_5$. The four-component phase $Cu_{8.4}Fe_{1.2}Bi_{10.8}S_{22}$ (Z) has the same optical properties as phase $CuBi_3S_5$ and cannot be distinguished from phase $CuBi_3S_5$ under a microscope. Bismuthinite shows a light color of grayish white to white, with a stronger bireflectance and anisotropism than those of the copper-bismuth sulfide minerals described above.

An X-ray diffractometer was not practical because of the small quantity of synthesized products. Thus, a Gandorfi camera was used to identify the phases, but the precise data of powder diffractions for the phases were measured by a Guinier camera. A precession camera and automatic four circle diffractometer for a single crystal were occasionally employed to determine the type of superstructure, crystal system, space group, and cell constants, etc. for synthesized phases.

An electron probe microanalyser played a most important role in the phase study. Because many phases in the system Cu–Fe–Bi–S have some range of solid solutions at elevated temperatures, it is necessary to determine the chemical composition of each phase in order to construct the phase diagram. Thus, all of the synthesized phases were quantitatively analysed by the electron probe microanalyser. The analyses were carried out by a Shimazu-ARL EMX-II instrument using 20 kV accelerating voltage, 0.02 μA specimen current on chalcopyrite, 3 μm spot size, and curved crystals LiF for Cu–Kα, Fe–Kα, and Bi–Lα radiations, and ADP for S–Kα. As standard samples, synthesized chalcocite and bismuthinite, and natural chalcopyrite were used for Cu, Bi, Fe, and S, respectively. The measured X-ray intensities, mean values of five times about the counting data for 10 sec on each mineral grain and standard, were corrected for background and dead time effects. Chemical compositions were calculated by the BENCE and ALBEE (1968) procedure using the alfa-factor for the Cu–Fe–Bi sulfide system (SUGAKI *et al.*, 1974, 1976b). Mineral grains having exsolved phases were first heated in an evacuated glass tube at the same temperature as those synthesized, and then quenched in ice water. In some specimens, such as intermediate solid solution and bornite solid solution, complete homogenization was not achieved in this way, but even so, the exsolution texture became sufficiently fine to allow analysis of the bulk composition of the phase by an electron probe microanalyser.

3. *Experimental Results and Phase Relation*

3.1 *The system Cu–Fe–S*

Minerals which belong to the system Cu–Fe–S are the most important among ore minerals because they are persistent minerals occurring in many kinds of ore deposits. Accordingly, phase relations in the Cu–Fe–S system are of great geological interest, and have been studied for more than 30 years since the first study was made by MERWIN and LOMBARD (1937) by such authors as SCHLEGEL and SCHÜLLER (1952), ROSEBOOM and KULLERUD (1958), BRETT (1963), YUND and KULLERUD (1966), KULLERUD *et al.* (1969), MUKAIYAMA and IZAWA (1970), CABRI (1973), and BARTON (1973). They carried out a large number of systematic experiments on this system and determined the phase relations to be between 700° and 200°C. Bornite, pyrrhotite, and intermediate solid solutions occur in the system at high temperatures. The latter forms an extensive field in the central portion of the system. These solid solutions associate with pyrite and chalcopyrite. However, as the temperature falls, various phase-changing reactions take place and the phase relations become very complicated, but have not been elucidated in detail. Especially after the new phases, talnakhite, mooihoekite, and haycockite, had been found by CABRI and his co-workers (1967, 1971, 1972), the phase relations at low temperatures were seen to be much more complicated. Many problems remain unsolved. There is also the problem of sluggish reaction rates at low temperatures in addition to the complexity of the phase relations in the system. All the experiments done by the authors mentioned above were performed by the dry method using the evacuated silica glass tube. SUGAKI *et al.* (1975) tried hydrothermal synthesis of copper and iron sulfides using the thermal gradient method and obtained isothermal phase diagrams of the system Cu–Fe–S at 300° and 350°C.

3.1.1 *The isothermal section at 400°C*

In order to determine the 400°C isothermal phase relation of the Cu–Fe–S system as a continuous work of SUGAKI *et al.* (1975), hydrothermal experiments were performed by the present authors. Crystallized mineral phases were determined from crystal shapes, optical properties, and X-ray powder data etc., and microprobe analyses were performed on each of the phases synthesized. Figure 4 is a part of the phase diagram of the system Cu–Fe–S at 400°C obtained from the experimental data. Seven stable phases are shown, but do not include the three end member elements. They are bornite solid solution, chalcopyrite solid solution, intermediate solid solution, pyrrhotite solid solution, covellite, nukundamite, and pyrite. Among them, the intermediate and bornite solid solutions have extensive composition fields, while chalcopyrite and pyrrhotite solid solutions have limited ones. The following ten univariant phase assemblages are as stable at 400°C as those at 350°C in the experiments carried out by SUGAKI *et al.* (1975): covellite +

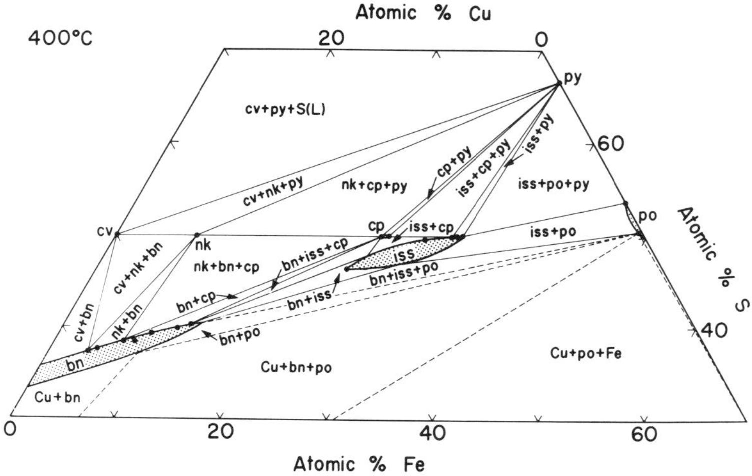

FIG. 4. Phase relations in the central portion of the Cu–Fe–S system at 400°C under hydrothermal conditions. Solid circles show the analytical compositions of the synthetic phases. Abbreviations in this figure are shown in Table 1 and Fig. 1.

pyrite + sulfur (L), covellite + nukundamite + pyrite, nukundamite + chalcopyrite + pyrite, intermediate solid solution + chalcopyrite + pyrite, intermediate solid solution + pyrrhotite + pyrite, covellite + bornite solid solution + nukundamite, bornite solid solution + nukundamite + chalcopyrite, bornite solid solution + chalcopyrite + intermediate solid solution, bornite solid solution + intermediate solid solution + pyrrhotite, bornite solid solution + pyrrhotite + copper. The last two assemblages, which have low sulfur fugacities, have not been observed in the present study.

The chalcopyrite solid solution field at 400°C is not so wide as YUND and KULLERUD (1966) and BARTON and SKINNER (1967) have suggested, and has a limited range from nearly stoichiometric $CuFeS_2$ to $Cu_{0.98}Fe_{1.02}S_{2.00}$. The range corresponds to 1.00 to 0.96 in the Cu/Fe atomic ratio along a line of metal to a sulfur ratio of approximately 1.0. Compositions of chalcopyrite coexisting with bornite and/or nukundamite concentrate into stoichiometric $CuFeS_2$ and the chalcopyrite in the univariant assemblage with the intermediate solid solution and pyrite becomes most iron-rich. The intermediate solid solution, as stated above, has a considerably wide field extending from near the stoichiometric cubanite composition toward the copper-rich and sulfur-deficient mooihoekite composition at 400°C, as is clearly seen in Fig. 4. The field includes the high temperature cubic phase of cubanite as well as the composition of mooihoekite and haycockite described by CABRI and HALL (1972). Although the iron-rich and copper-rich ends of the solid solution field have not been determined precisely, the most copper-rich phase of the intermediate solid solution synthesized at 400°C was the composition of Cu 40.5, Fe 27.9, and S 31.7 in a weight percent which gives the formula of $Cu_{1.80}Fe_{1.41}S_{2.79}$ with Cu/Fe atomic ratio of 1.28. The intermediate solid solution coexists stably with pyrite, pyrrhotite, chalcopyrite, and bornite solid solution. The composition of the intermediate solid solution in equilibrium with pyrite and pyrrhotite is Cu 24.1, Fe 40.1, and S 35.2 in a weight percent corresponding to $Cu_{1.04}Fe_{1.96}S_{3.00}$, which is very close to stoichiometric cubanite ($CuFe_2S_3$). Also, the composition of the intermediate solid solution in the univariant assemblage with chalcopyrite and pyrite is Cu 25.5, Fe 38.2, and S 34.7 in weight percent ($Cu_{1.11}Fe_{1.90}S_{3.00}$).

The boundary of the sulfur-rich part of the bornite solid solution field at 400°C was determined by means of microprobe analysis on the bornite coexisting with other sulfide phases. The boundary coincides approximately with the $Cu_{1.8}S$–Cu_5FeS_4 join, but there are no data for the metal-rich parts of the composition field because synthesis of the bornite solid solution with low sulfur fugacity assemblages is difficult. The bornite solid solution field extends far beyond its stoichiometric composition toward a more iron-rich side. The maximum iron content of the solid solution was 13.4 in weight percent (3.97 in Cu/Fe atomic ratio, $Cu_{4.76}Fe_{1.20}S_{4.04}$). The bornite solid solution associates with covellite, nukundamite, chalcopyrite, intermediate solid solution (and pyrrhotite). Bornite with a composition of more than 13.4 weight percent iron (less than 3.97 in the Cu/Fe atomic ratio) was in an equilibrium with only the intermediate solid solution. Also, bornite assembled with only chalcopyrite has a considerable composition range from 13.4 ($Cu_{4.76}Fe_{1.20}S_{4.04}$) to 7.1 ($Cu_{5.52}Fe_{0.65}S_{3.83}$) in weight percent of iron (3.97 to 8.49 in Cu/Fe atomic ratio), which includes the stoichiometric composition of bornite. The composition of bornite in the univariant assemblage of bornite, chalcopyrite, and intermediate solid solution is Cu 60.7, Fe 13.4, and S 26.0 in weight percent ($Cu_{4.76}Fe_{1.20}S_{4.04}$). Bornite associated with chalcopyrite and nukundamite has a composition of Cu 68.5, Fe 7.1, and S 24.0 in weight percent (8.49 in Cu/Fe atomic ratio, $Cu_{5.52}Fe_{0.65}S_{3.83}$). The bornite composition in equilibrium with covellite and nukundamite was given as Cu 72.2, Fe 3.4, and S 22.9 in weight percent (18.6 in the Cu/Fe atomic ratio, $Cu_{5.95}Fe_{0.32}S_{3.73}$) nearly close to the digenite composition.

Nukundamite always occurs in a crystal of thin hexagonal plates, as shown in Fig. 2-D. The chemical composition of nukundamite synthesized at 400°C is Cu 56.5, Fe 9.4, and S 34.0 in weight percent and is almost equivalent to $Cu_{3.38}Fe_{0.62}S_{4.00}$ ($Cu_{5.5}FeS_{6.5}$). These values are in good accordance with those of idaite (nukundamite) produced by the dry method using the evacuated glass tube (YUND, 1963) and synthesized hydrothermally at 350°C by SUGAKI *et al.* (1975). The crystal structure data of nukundamite synthesized hydrothermally were also given by SUGAKI *et al.* (1981a). Nukundamite can coexist stably with pyrite, covellite, bornite, and chalcopyrite. Because the tie line between nukundamite and chalcopyrite occurs in a stable assemblage at

400°C, the same as the isothermal section at 350° and 300°C reported by SUGAKI *et al.* (1975), the assemblage of bornite and pyrite previously accepted by YUND and KULLERUD (1966), was not found in this study.

3.1.2 The 300°C isothermal section

In order to construct an isothermal phase diagram of the system Cu–Fe–S at 300°C, hydrothermal experimental runs were carried out by SUGAKI *et al.* (1975). The work has been continued by the present authors using the same method because the experimental runs were insufficient. The phase diagram in the central portion of the system Cu–Fe–S, which is the result of data compiled from both experiments, is shown in Fig. 5. Except for a small reduction of the solid solution fields and the tie line change from the intermediate solid solution-pyrite join to the chalcopyrite-pyrrhotite join, the phase relations of the system are essentially the same as those at 400°C, as mentioned above. The stable phases in the system at 300°C are chalcopyrite, bornite solid solution, intermediate solid solution, pyrrhotite, nukundamite, covellite, and pyrite excluding digenite (high form) solid solution and end-member elements.

Solid solution ranges of chalcopyrite at 300°C are about the same or slightly increased in comparison with those at 400°C. The most iron-rich chalcopyrite, which coexists with pyrrhotite and intermediate solid solution, has a composition of $Cu_{0.97}Fe_{1.04}S_{1.99}$. The chalcopyrite solid solution extends from 1.0 to approximately 0.93 in the Cu/Fe atomic ratio along the line of the metal/sulfur atomic ratio 1.0. The intermediate solid solution, still having a remarkable area in the system at 300°C, extends between the Cu/Fe

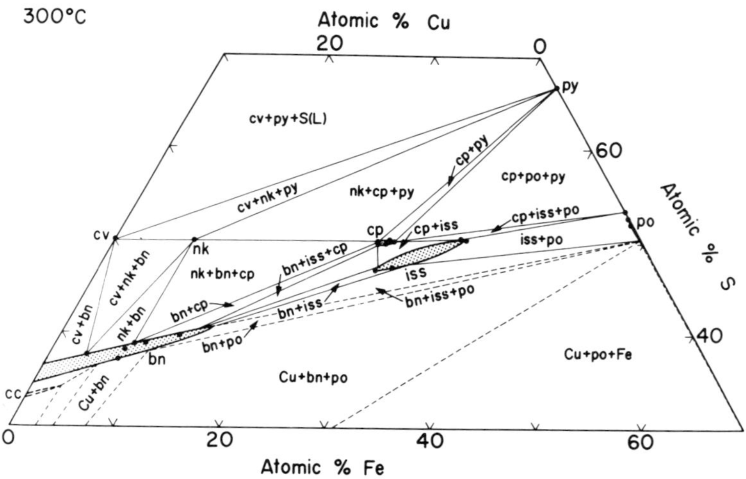

FIG. 5. Phase relations in the central portion of the Cu–Fe–S system at 300°C under hydrothermal conditions.

atomic ratio of 0.51 and 1.04, and has a composition of $Cu_{1.01}Fe_{1.97}S_{3.02}$ in equilibrium with hexagonal pyrrhotite and chalcopyrite, while the phase $Cu_{1.63}Fe_{1.57}S_{2.81}$ coexists with bornite and chalcopyrite. The solid solution field includes the compositions of mooihoekite $Cu_9Fe_9S_{16}$ and haycockite $Cu_4Fe_5S_8$. The composition of the intermediate solid solution, coexisting with chalcopyrite and pyrrhotite, is Cu 23.5, Fe 40.1, and S 35.4 in weight percent (Cu/Fe atomic ratio of 0.51, $Cu_{1.01}Fe_{1.97}S_{3.02}$), which is very close to stoichiometric cubanite $CuFe_2S_3$. The tie line between the intermediate solid solution and pyrite, existing stably at 350° and 400°C, is replaced by the tie line between chalcopyrite and pyrrhotite at 300°C. The tie line change was determined at 328° $\pm$ 5°C by YUND and KULLERUD (1966). Also, SUGAKI *et al.* (1982) recently found that the tie line was not changed at 330°C but changed to the chalcopyrite-pyrrhotite join at 320°C under hydrothermal conditions at pressures of 1,000 to 2,000 kg/cm^2.

Although the bornite solid solution becomes less extensive at 300°C than at 400°C, the field extends still far beyond the stoichiometric composition of bornite toward the iron-rich composition. The maximum iron content of the bornite solid solution is approximated by the composition of bornite coexisting with intermediate solid solution, that is, Cu 57.2, Fe 15.7, and S 26.2 in weight percent (Cu/Fe atomic ratio of 3.19, $Cu_{4.50}Fe_{1.41}S_{4.09}$). Meanwhile, bornite in equilibrium with chalcopyrite has a wide compositional range from about 3.5 to 7.2 in the Cu/Fe atomic ratio at 300°C. The composition of bornite coexisting with chalcopyrite and nukundamite is Cu 69.2, Fe 7.6, and S 24.3 in weight percent (Cu/Fe atomic ratio of 7.96, $Cu_{5.49}Fe_{0.69}S_{3.82}$). Nukundamite occurs in equilibrium associations with covellite, bornite, chalcopyrite, and pyrite. The assemblage of nukundamite and chalcopyrite is also stable at 300°C, the same as at 350° and 400°C. The mean values of the microprobe analysis for nukundamite synthesized at 300°C are Cu 57.1 $\pm$ 0.2, Fe 9.0 $\pm$ 0.4, and S 33.9 $\pm$ 0.2 in weight percent (Cu/Fe atomic ratio of 5.5 $\pm$ 0.2, a metal sulfur ratio of approximately 1.0). Nukundamite has no field of solid solution at 300°C, and its composition is very close to $Cu_{3.38}Fe_{0.62}S_{4.00}$ or $Cu_{5.5}FeS_{6.5}$, as proposed by SUGAKI *et al.* (1981a), YUND (1963), and YUND and KULLERUD (1966). All pyrrhotite synthesized at 300°C were found to be of a hexagonal type as a result of re-examination by this study and SUGAKI *et al.* (1977), although SUGAKI *et al.* (1975) reported that pyrrhotite has two structure types at 300°C, hexagonal and monoclinic. Pyrrhotite assembles with pyrite, chalcopyrite, intermediate solid solution, and bornite solid solution. But, the last assemblage could not be seen in these experimental runs.

3.2 *The system Cu–Bi–S*

In the system Cu–Bi–S, several common copper bismuth sulfosalt minerals, such as wittichenite, emplectite, and cuprobismutite, etc. were known widely. Synthetic experiments on these minerals have been carried out by several authors (GAUDIN and DICKE, 1939; NUFFIELD, 1947, 1952; GODOVIKOV and PTITSIN, 1969; CHEN and CHANG, 1974). However, the phase equilibrium of the system has not been performed by so many authors. The binary Cu_2S–Bi_2S_3 join was studied by BUHLMANN (1971), SUGAKI and SHIMA (1972), FEDOROVA (1972), and SUGAKI *et al.* (1980), and the ternary Cu–Bi–S system was investigated by SUGAKI and SHIMA (1971). However, these studies were performed almost exclusively by the dry method using an evacuated glass tube. One exception is the work by SUGAKI *et al.* (1980) who employed the hydrothermal transport method for the synthesis of copper bismuth sulfosalts. The results of the experiments on phase relations in the binary Cu_2S–Bi_2S_3 and the ternary Cu–Bi–S system are mentioned below.

3.2.1 The binary Cu_2S–Bi_2S_3

The phase diagram of the binary Cu_2S–Bi_2S_3 experimental data, summarized by SUGAKI and SHIMA (1972) and SUGAKI *et al.* (1980), is shown in Fig. 6. As the crystalline phases, chalcocite (Cu_2S), phase Cu_9BiS_6 (X), wittichenite (Cu_3BiS_3), emplectite ($CuBiS_2$), cuprobismutite ($Cu_{10}Bi_{12}S_{23}$), phase $Cu_3Bi_5S_9$ (D), phase $CuBi_3S_5$ (E), and bismuthinite (Bi_2S_3) are found in the binary of Fig. 6. The phases Cu_9BiS_6, $Cu_3Bi_5S_9$, and $CuBi_3S_5$ have not been found in nature. Phase $CuBi_3S_5$ is stable below $649 \pm 5°C$ and incongruently melts at this temperature to Bi_2S_3 and liquid. The phase $Cu_3Bi_5S_9$ is stable in the temperature range between $442° \pm 5°C$ and $620° \pm 5°C$, at which it incongruently melts to phase $CuBi_3S_5$ and liquid. It dissolves a considerable amount of the Cu_2S molecule in solid solution, and its maximum limit is 56.5 mole percent Bi_2S_3 at 523°C. Synthetic wittichenite is identical with the natural one in every respect. Though it has the composition of stoichiometric Cu_3BiS_3 at room temperature, it forms a solid solution in a

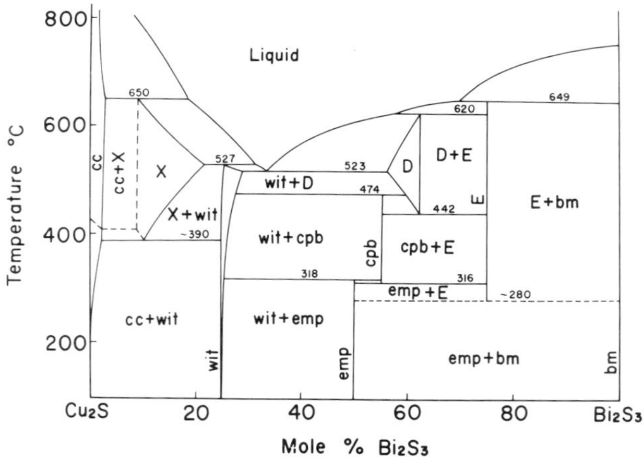

FIG. 6. Phase diagram of the Cu_2S–Bi_2S_3 system after SUGAKI and SHIMA (1972) and SUGAKI *et al.* (1980).

very limited range, shifting toward Cu_2S poor composition at high temperature (Fig. 6). The solid solution of wittichenite extends from 26.5 to 28.5 mole percent Bi_2S_3 at 500°C. The solid solution of nearly 27.5 mole percent Bi_2S_3 incongruently melts at 527 ± 5°C. The eutectic point between wittichenite solid solution and the solid solution of phase $CuBi_3S_5$ is 527 ± 5°C at about 35 mole percent Bi_2S_3. The phase Cu_9BiS_6 is stable between about 650° and 390°C, but a lower stability limit has not been determined. It dissolves a considerable amount of Bi_2S_3 molecules and the solid solution field extends to 13 mole percent Bi_2S_3 at 450°C, 22 mole percent Bi_2S_3 in maximum at 527°C. The phase relation between emplectite and cuprobismutite was one of the important problems in this join. Cuprobismutite, found by Hillebrand in 1884, was first given a chemical composition of 3 $(Cu, Ag)_2S \cdot 4Bi_2S_3$. DANA (1892) and SCHNEIDERHÖHN and RAMDOHR (1931) described this mineral, but SHORT (1931) and PALACHE (1940) denied its existence. However, NUFFIELD (1952), by re-examination on the type specimens from the Missouri mine, identified cuprobismutite as a varied species with a chemical composition of $CuBiS_2$ and suggested a dimorphous relationship with emplectite. BUHLMANN (1971) stated that the low temperature polymorph of emplectite (β-$CuBiS_2$) inverts to the high form of cuprobismutite (α-$CuBiS_2$) at 290°C, while SUGAKI and SHIMA (1972) reported that the composition of cuprobismutite is not $CuBiS_2$ but $Cu_{24}Bi_{26}S_{51}$ and the inversion temperature between both phases might be about 360°C from the DTA experiment. In order to make more clear the phase relation between them, SUGAKI *et al.* (1980) have carried out detailed experiments under the dry and hydrothermal conditions. Figure 7 shows a central portion of the binary Cu_2S–Bi_2S_3 at a low temperature obtained by both the dry and hydrothermal methods, and compiled from the experimental data of SUGAKI *et al.* (1980) and from recent work by the present authors. From the results of these experiments, emplectite and cuprobismutite are not in dimorph, but are two independent minerals with different compositions. Emplectite has a stoichiometric composition of $CuBiS_2$, but cuprobismutite has a little bismuth-rich composition of $Cu_9Bi_{11}S_{21}$, described by TAYLOR *et al.* (1973), as being $Cu_{10}Bi_{12}S_{23}$, as is the composition of natural cuprobismutite. Emplectite is only stable below 318° ± 3°C, while cuprobismutite is never synthesized below 316° ± 3°C. There is a narrow field of coexistence of both minerals at temperatures from 316° to 318°C in a composition range from 50 to 54.5 mole percent Bi_2S_3. The phase $CuBi_3S_5$ is stable above approximately 280°C and does not appear below this temperature, but the assemblage of emplectite and bismuthinite forms as those found in natural ores. Thus, the phase $CuBi_3S_5$ could not be expected to be found in nature, though SUGAKI and SHIMA (1972) mentioned that it could possibility be found so in the future, while wittichenite could not associate with bismuthinite in an equilibrium state. During all the experiments on the

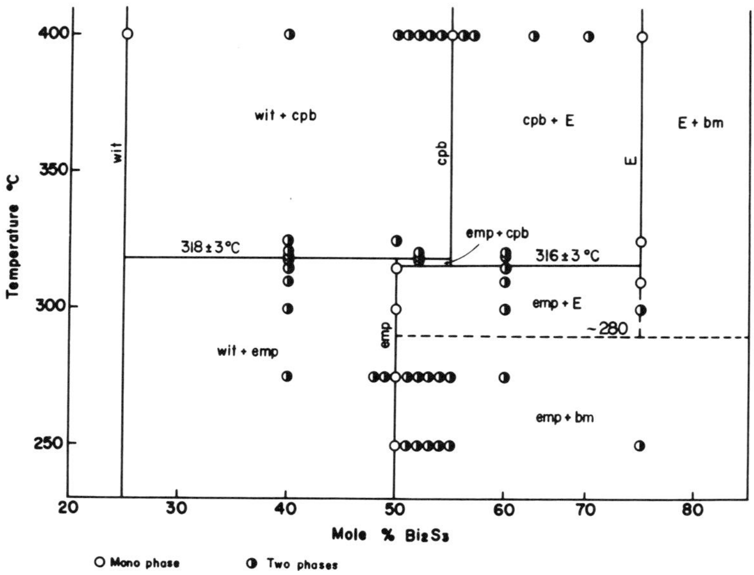

FIG. 7. The phase relations of a central part in the join Cu_2S–Bi_2S_3 of Fig. 6.

synthesis and phase equilibrium of the binary Cu_2S–Bi_2S_3, klaprothite and dognacskite have not appeared. Both minerals, having $Cu_6Bi_4S_9$ and $Cu_2Bi_4S_7$ in composition, could not be expected to be found hereafter.

3.2.1 The ternary Cu–Bi–S

The phase relations of the ternary system Cu–Bi–S at 300°, 400°, and 500°C were determined by using the method of the evacuated glass tube (SUGAKI and SHIMA, 1971). The isothermal section of the phase diagram in the system at 300°C is shown in Fig. 8. In this figure, high digenite solid solution, covellite, wittichenite, emplectite, phase $CuBi_3S_5$ (E), bismuthinite, and phase $Cu_8Bi_8S_{19}$ (Y) are shown as stably crystalline phases, and 12 univariant assemblages are found as follows: sulfur (L) + covellite + phase $Cu_8Bi_8S_{19}$, sulfur (L) + phase $Cu_8Bi_8S_{19}$ + bismuthinite, phase $Cu_8Bi_8S_{19}$ + bismuthinite + phase $CuBi_3S_5$, phase $Cu_8Bi_8S_{19}$ + phase $CuBi_3S_5$ + emplectite, phase $Cu_8Bi_8S_{19}$ + emplectite + wittichenite, covellite + phase $Cu_8Bi_8S_{19}$ + wittichenite, high digenite + covellite + wittichenite, bismuthinite + phase $CuBi_3S_5$ + bismuth (L), phase $CuBi_3S_5$ + emplectite + bismuth (L), wittichenite + emplectite + bismuth (L), wittichenite + high digenite + bismuth (L), and high digenite + copper + bismuth (L). The phase $Cu_8Bi_8S_{19}$, which was found by SUGAKI and SHIMA (1971), GODOVIKOV *et al.* (1971), and SUGAKI *et al.* (1972a), assembles with many phases, such as covellite, wittichenite, emplectite, phase $CuBi_3S_5$, bismuthinite, and sulfur at 300°C, but cannot coexist with high digenite, copper, and bismuth. Bismuthinite, em-

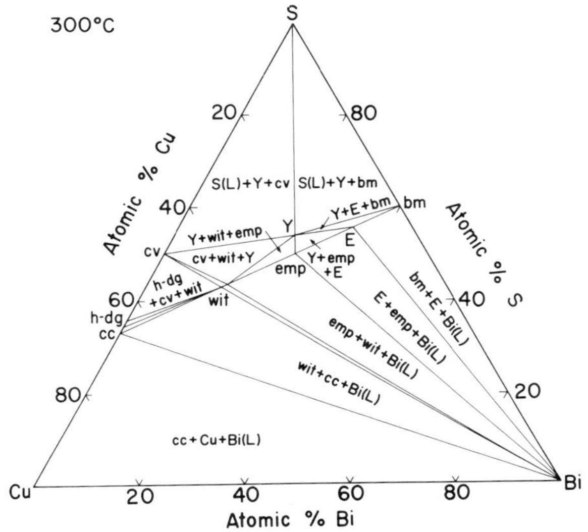

FIG. 8. Phase relations of the Cu–Bi–S system at 300°C obtained under dry conditions by SUGAKI and SHIMA (1971).

plectite, wittichenite, and high digenite can coexist with bismuth at 300°C, as is commonly seen in bismuth bearing ores.

At 400°C, emplectite does not form, but cuprobismutite appears. The phase relations in the isothermal section at 400°C are similar to those at 300°C, being exclusive of the appearance of phase Cu_9BiS_6 and cuprobismutite. There are 15 univariant assemblages. These are as follows: phase $Cu_8Bi_8S_{19}$ + bismuthinite + sulfur (L), phase $Cu_8Bi_8S_{19}$ + covellite + sulfur (L), phase $Cu_8Bi_8S_{19}$ + phase $CuBi_3S_5$ + bismuthinite, phase $Cu_8Bi_8S_{19}$ + cuprobismutite + phase $CuBi_3S_5$, wittichenite + phase $Cu_8Bi_8S_{19}$ + cuprobismutite, covellite + wittichenite + phase $Cu_8Bi_8S_{19}$, covellite + phase Cu_9BiS_6 + wittichenite, covellite + high digenite + phase Cu_9BiS_6, (chalcocite + high digenite + phase Cu_9BiS_6), bismuthinite + phase $CuBi_3S_5$ + bismuth (L), phase $CuBi_3S_5$ + cuprobismutite + bismuth (L), cuprobismutite + wittichenite + bismuth (L), wittichenite + phase Cu_9BiS_6 + bismuth (L), phase Cu_9BiS_6 + chalcocite + bismuth (L), and chalcocite + copper + bismuth (L).

Wittichenite, cuprobismutite and phase Cu_9BiS_6 form a very limited range of the solid solution at 400°C, while phases $Cu_8Bi_8S_{19}$ and $CuBi_3S_5$ have a stoichiometric composition at this temperature. Bismuth liquid dissolves some of the copper and sulfur.

3.3 *The system Fe–Bi–S*

Phase relations in the Fe–Bi–S system have been investigated by SUGAKI *et al.* (1970, 1972a), CRAIG *et al.* (1971), and SHIMA *et al.* (1977). Synthetic phase $FeBi_4S_7$ was found in the join $FeS-Bi_2S_3$ of the system, as shown in Fig. 9. It is stable in the temperature range between 623° ± 3°C and 726° ± 3°C, which is the congruent melting point and is decomposed to the assemblage of pyrrhotite + bismuthinite + bismuth-rich liquid at temperatures below 623°C. Therefore, phase $FeBi_4S_7$ cannot be expected to be found in nature. The isothermal sections of the system Fe–Bi–S at temperatures below 623°C, are very simple because there is no three-component phase in the ternary system. Figure 10 is the isothermal section at 300°C, and four univariant assemblages, pyrite + bismuthinite + sulfur (L), pyrite + bismuthinite + pyrrhotite, pyrrhotite + bismuthinite + bismuth (L), and iron + pyrrhotite + bismuth (L), exist in the system. According to the data of experiments by CRAIG *et al.* (1971), it is inferred that the tie line of pyrrhotite and bismuthinite is replaced by the tie line of pyrite and bismuth below about 235°C.

3.4 *The system Cu–Fe–Bi–S*

In order to determine the isothermal phase relation in the quaternary system Cu–Fe–Bi–S, the experiments of hydrothermal syntheses, as mentioned above, were carried out by SUGAKI *et al.* (1981b) at 420° and 300°C. The phases that appeared in the experiment were bornite, chalcopyrite, intermediate solid solution, pyrrhotite, and nukundamite belonging to the ternary Cu–Fe–S, phase Cu_9BiS_6, wittichenite, emplectite, cuprobismutite, phase $CuBi_3S_5$, phase $Cu_8Bi_8S_{19}$, bismuthinite and bismuth liquid in the ternary Cu–Bi–S, and phase $Cu_{8.4}Fe_{1.2}Bi_{10.8}S_{22}$ (Z) in the quaternary Cu–Fe–Bi–S system. Phases $Cu_3Bi_5S_9$ (D) and $FeBi_4S_7$ did not appear

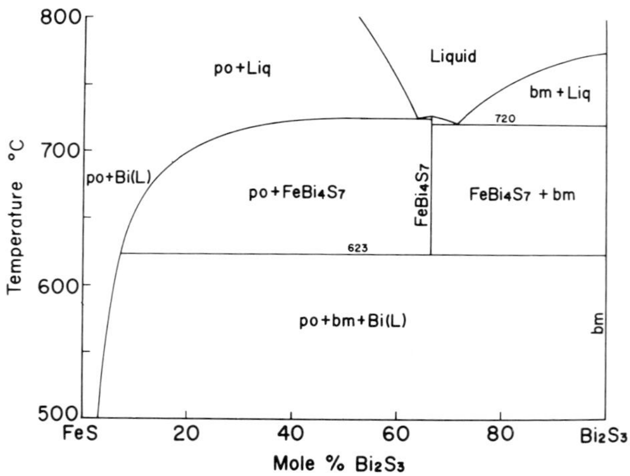

FIG. 9. Phase diagram of pseudo-binary system $FeS-Bi_2S_3$ under dry conditions after SHIMA *et al.* (1977).

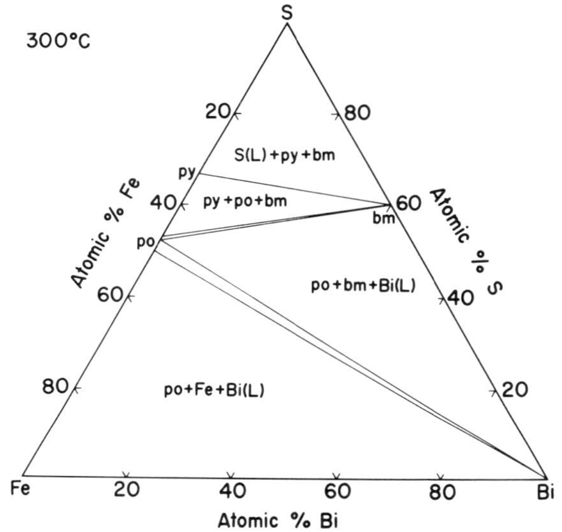

FIG. 10. Phase relations of the Fe–Bi–S system under dry conditions obtained by SUGAKI *et al.* (1970).

because they are unstable at 420°C or below. The experiments of SUGAKI *et al.* (1981b) were mainly performed under comparatively high sulfur fugacity because most of the bulk composition of nutrient materials were sulfur-rich and iron-poor. Thus, the appearance of bismuth and pyrrhotite was rare. The present study continues the work of SUGAKI *et al.* (1981b) and focusses on the phase relation of comparatively low sulfur fugacity to obtain the tie line with bismuth (liquid). A gold tube was generally used as a sample container in this experiment, but a silica tube had to be employed in the case of the experiments containing a fair amount of bismuth metals, because bismuth reacts on the gold tube. In this paper, the phase relations obtained by both experimental runs are described together.

3.4.1 The phase relations at 420°C

The results of the experimental runs at 420°C, together with the bulk composition and the phase assemblages of nutrient materials, the experimental conditions, and the synthesized phase assemblages, are given in Table 2. Synthesized products were usually composed of three or four solid phases, and occasionally of two phases, which seemed to be in equilibrium with each other. The phase assemblages at 420°C, seen in the table, were controlled principally by a composition of nutrient and sulfur fugacity within the gold or silica tubes. Sulfur fugacity was directly fixed by the bulk composition of nutrients. The principle mineral phases in the system, such as bornite, chalcopyrite, intermediate solid solution, pyrrhotite, nukundamite, wittichenite, cuprobismutite, phases Cu_9BiS_6, $CuBi_3S_5$, $Cu_8Bi_8S_{19}$, and

TABLE 2. Hydrothermal experimental results at 420°C in the Cu–Fe–Bi–S system.

Run No.	Nutrient materials: Composition (Atomic %) Cu	Fe	Bi	S	Reactants	Pressure (kg/cm^2)	Heating period (days)	Synthesized phases
HCFB 027	39.33	4.91	11.48	44.28	bn + wit + Z	960	5	bn + cpb
HCFB 028	32.64	14.28	6.12	46.96	bn + cp + Z	960	5	bn + cp + Z
HCFB 031	15.08	0.74	30.20	53.98	D + E + Z	1000	5	cpb + E + Z
HCFB 032	20.14	1.01	26.92	51.92	wit + D + E	1000	5	cpb + E
HCFB 037	24.99	12.50	12.50	50.01	bn + cp + Z	1020	6	cpb + E + Z
HCFB 038	24.98	4.22	20.80	50.00	bn + wit + Z	1020	6	bn + cpb + Z
HCFB 043	22.40	1.39	25.19	51.02	wit + D + Z	980	6	bn + cpb
HCFB 044	16.46	6.32	23.87	53.40	cp + E + Z	980	6	bn + cp + nk + Z
HCFB 045	10.13	4.45	30.37	55.06	po + E	1000	12	cp + bm + E + Y
HCFB 046	44.74	2.63	10.53	42.10	bn + wit	1000	12	bn + wit
HCFB 047	40.59	3.13	12.50	43.78	bn + wit	1000	12	bn + wit + cpb
HCFB 048	9.08	9.21	27.21	54.51	po + E	1000	12	cp + E + Z
HCFB 052	18.71	28.24	6.24	46.81	po + wit	990	11	cp + E
HCFB 053	19.86	4.91	23.82	51.42	po + wit + D	960	11	bn + Z
HCFB 077*	12.59	1.39	55.99	30.03	cpb + Z + Bi	400	7	wit + cpb + Bi
HCFB 078*	7.68	1.41	60.77	30.13	iss + bm + E + Z	400	7	E
HCFB 079*	26.59	1.41	42.00	29.99	bn + wit + Bi	400	5	wit + Bi
HCFB 080*	17.47	1.48	51.01	30.05	bn + cpb + Bi	400	5	wit + Bi
HCFB 089*	17.48	17.53	34.97	30.02	bn + iss + Bi	400	4	bn + Bi
HCFB 090*	14.99	17.50	37.48	30.02	iss + Bi	400	4	bn + iss + Bi
HCFB 091*	15.99	14.98	38.99	30.04	bn + iss + Bi	400	4	bn + Z + Bi
HCFB 092*	12.49	17.52	39.97	30.02	iss + Bi	400	4	iss + po + Bi
HCFB 093*	24.07	7.01	38.48	30.07	bn + Z + Bi	400	11	bn + wit + Bi
HCFB 094*	21.00	6.98	42.00	30.02	bn + Z + Bi	400	11	bn + wit + Bi
HCFB 095*	20.99	10.52	38.49	30.00	bn + iss + Z + Bi	400	11	bn + Bi
HCFB 096*	17.48	10.54	41.98	30.01	bn + iss + Z + Bi	400	11	bn + wit + Bi
HCFB 107*	5.00	22.48	42.48	30.04	iss + po + bm + Bi	400	8	iss + po + Bi
HCFB 108*	33.57	1.39	34.96	30.08	bn + wit + Bi	400	13	wit + X + Bi
HCFB 109*	6.99	7.00	55.95	30.06	iss + Z + Bi	400	13	po + E + Bi
HCFB 110*	22.47	14.02	33.47	30.04	bn + iss + Bi	400	13	bn + Bi
HCFB 123*	10.50	10.51	49.00	29.99	iss + bm + E + Bi	400	9	bn + Z + Bi
HCFB 124*	14.00	7.02	49.00	29.99	bn + iss + Z + Bi	400	9	wit + Bi
HCFB 125*	30.09	1.34	38.49	30.09	bn + wit + cpb + Bi	400	9	wit + X + Bi
HCFB 126*	44.96	1.40	23.60	30.04	bn + X + Bi	400	9	wit + X + Bi
HCFB 127*	10.00	22.50	37.51	29.99	iss + po + Bi	400	9	iss + po + Bi
HCFB 128*	16.90	13.54	40.57	28.97	bn + iss + Z + Bi	400	9	bn + Z + Bi
HCFB 130*	10.14	13.51	47.32	29.03	iss + Z + Bi	400	7	bn + Z + Bi
HCFB 131*	6.76	13.52	50.69	29.03	iss + bm + Bi	400	7	iss + po + Bi
HCFB 132*	10.49	1.41	58.05	30.05	cpb + Z + Bi	400	6	cpb + E + Bi
HCFB 133*	19.99	14.01	35.98	30.02	bn + iss + Bi	400	6	bn + Z + Bi
HCBS 043	19.05	0.00	28.57	52.38	cpb + E	350	5	cpb + wit + Y
HCBS 077	15.50	0.00	30.00	54.50	cpb + E + Y	350	5	bm + E + Y

Solvent : 5 m NH_4Cl aqueous solution, * Silica tube was used as reaction vessel

$Cu_{8.4}Fe_{1.2}Bi_{10.8}S_{22}$, bismuthinite, and bismuth (liquid), were produced, but pyrite, covellite, and sulfur (liquid) did not appear in these runs, because the experiments under such high sulfur fugacity, as forming them, had not yet been performed. Also, emplectite was not found at this temperature. The phase relations in the quaternary system are shown in Fig. 11, but they are too complicated to be of any use in solving the phase diagram. Thus, the stable phase assemblages at 420°C are shown as the exploded diagram as Fig. 12. In the figure, heavy solid and dotted lines show the tie lines confirmed in the runs. Meanwhile, the tie lines expected are shown by light solid and dotted lines. As

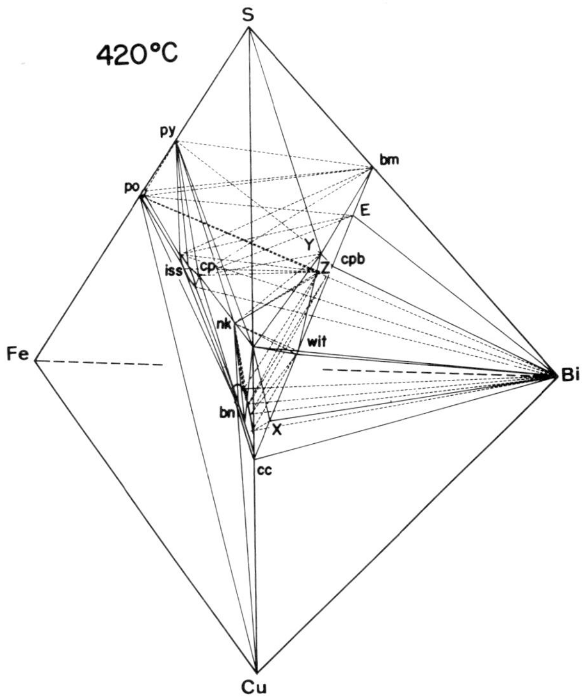

FIG. 11. Schematic phase diagram in the quaternary Cu–Fe–Bi–S system at 420°C under hydrothermal conditions.

seen in Fig. 12, the following univariant assemblages were recognized: bornite + phase Cu_9BiS_6 + wittichenite + bismuth (L), bornite + wittichenite + cuprobismutite + bismuth (L), bornite + cuprobismutite + phase $Cu_{8.4}Fe_{1.2}Bi_{10.8}S_{22}$ + bismuth (L), bornite + phase $Cu_{8.4}Fe_{1.2}Bi_{10.8}S_{22}$ + intermediate solid solution + bismuth (L), phase $Cu_{8.4}Fe_{1.2}Bi_{10.8}S_{22}$ + intermediate solid solution + pyrrhotite + bismuth (L), phase $Cu_{8.4}Fe_{1.2}Bi_{10.8}S_{22}$ + pyrrhotite + phase $CuBi_3S_5$ + bismuth (L), pyrrhotite + phase $CuBi_3S_5$ + bismuthinite + bismuth (L), cuprobismutite + phase $Cu_{8.4}Fe_{1.2}Bi_{10.8}S_{22}$ + phase $CuBi_3S_5$ + bismuth (L), (bornite + intermediate solid solution + pyrrhotite + bismuth (L))*, (bornite + pyrrhotite + copper + bismuth (L)), (pyrrhotite + copper + iron + bismuth (L)), (bornite + phase Cu_9BiS_6 + wittichenite + covellite), (bornite + wittichenite + covellite + nukundamite), (wittichenite + covellite + nukundamite + phase $Cu_8Bi_8S_{19}$), (wittichenite + cuprobismutite + phase $Cu_8Bi_8S_{19}$ + nukundamite), (bornite + wittichenite + cuprobismutite + nukundamite), bornite +

*The parentheses mean that the phase assemblage is inferred from experimental data.

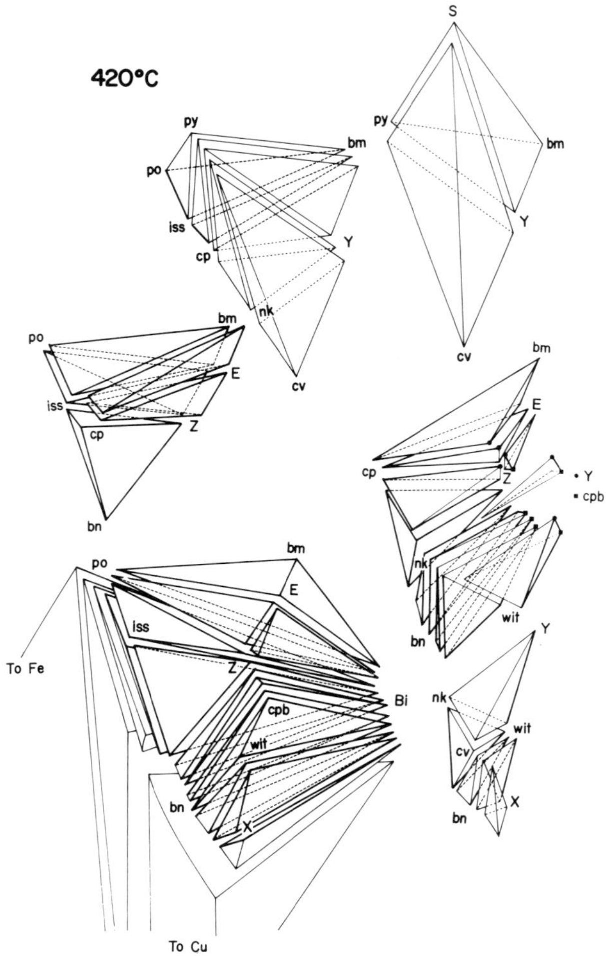

FIG. 12. Exploded phase diagram of phase relations in the system Cu–Fe–Bi–S at 420°C. Heavy solid and dot lines show the tie lines appeared in the hydrothermal experiments.

cuprobismutite + phase $Cu_{8.4}Fe_{1.2}Bi_{10.8}S_{22}$ + nukundamite, bornite + phase $Cu_{8.4}Fe_{1.2}Bi_{10.8}S_{22}$ + chalcopyrite + nukundamite, chalcopyrite + phase $Cu_{8.4}Fe_{1.2}Bi_{10.8}S_{22}$ + phase $Cu_8Bi_8S_{19}$ + nukundamite, cuprobismutite + phase $Cu_{8.4}Fe_{1.2}Bi_{10.8}S_{22}$ + phase $Cu_8Bi_8S_{19}$ + nukundamite, cuprobismutite + phase $Cu_{8.4}Fe_{1.2}Bi_{10.8}S_{22}$ + phase $Cu_8Bi_8S_{19}$ + phase $CuBi_3S_5$, chalcopyrite + phase $Cu_{8.4}Fe_{1.2}Bi_{10.8}S_{22}$ + phase $CuBi_3S_5$ + phase $Cu_8Bi_8S_{19}$, chalcopyrite + phase $CuBi_3S_5$ + bismuthinite + phase $Cu_8Bi_8S_{19}$, bornite + chalcopyrite + intermediate solid solution +

phase $Cu_{8.4}Fe_{1.2}Bi_{10.8}S_{22}$, chalcopyrite + intermediate solid solution + phase $Cu_{8.4}Fe_{1.2}Bi_{10.8}S_{22}$ + phase $CuBi_3S_5$, chalcopyrite + intermediate solid solution + bismuthinite + phase $CuBi_3S_5$, pyrrhotite + intermediate solid solution + bismuthinite + phase $CuBi_3S_5$, pyrrhotite + intermediate solid solution + phase $CuBi_3S_5$ + phase $Cu_{8.4}Fe_{1.2}Bi_{10.8}S_{22}$, (covellite + nukundamite + phase $Cu_8Bi_8S_{19}$ + pyrite), (chalcopyrite + nukundamite + phase $Cu_8Bi_8S_{19}$ + pyrite), (chalcopyrite + phase $Cu_8Bi_8S_{19}$ + bismuthinite + pyrite), (chalcopyrite + intermediate solid solution + bismuthinite + pyrite), (pyrrhotite + intermediate solid solution + bismuthinite + pyrite), (covellite + phase $Cu_8Bi_8S_{19}$ + pyrite + sulfur (L)), and (pyrite + bismuthinite + phase $Cu_8Bi_8S_{19}$ + sulfur (L)).

Phase $Cu_{8.4}Fe_{1.2}Bi_{10.8}S_{22}$ (Z), which is the only four-component phase in the quaternary system, occurs as a monoclinic polyhedral form, as shown in Fig. 3, though its appearance in nature is yet unknown. Its optical properties and X-ray powder data, lattice constants and DTA data were described in detail by SUGAKI *et al.* (1972, 1976a, 1981b). The crystal data of this phase is also summarized, together with other crystalline phases synthesized, in Table 1.

Among these phases, bornite, intermediate solid solution, wittichenite and cuprobismutite form some limited ranges of the solid solution. Thus, microprobe analyses were carried out on them and the coexisting phases with them. Results of the analyses are given in Table 3.

In order to make the phase relations clearly understandable, they can be presented in the triangle Cu–Fe–Bi plane by a projection from the sulfur summit to the metal basal plane in the tetrahedron of the quaternary system, as shown in Fig. 13. Figure 13-A shows the phase assemblages, which are produced under conditions of such comparatively higher sulfur fugacity as coexisting with nukundamite and phase $Cu_8Bi_8S_{19}$. On the other hand, Fig. 13-B represents the phase relations assembled with bismuth liquid, indicating a low sulfur fugacity condition. Bornite has a considerably wide space of solid solution and assembles stably with many phases, such as wittichenite, cuprobismutite, phase $Cu_{8.4}Fe_{1.2}Bi_{10.8}S_{22}$, chalcopyrite, intermediate solid solution, nukundamite and bismuth liquid. It does not associate with phase $CuBi_3S_5$, bismuthinite (and pyrite). While chalcopyrite coexists with phase $Cu_{8.4}Fe_{1.2}Bi_{10.8}S_{22}$, phase $CuBi_3S_5$, and bismuthinite, as well as the Cu–Fe–S minerals, the assemblage of chalcopyrite and wittichenite or cuprobismutite are not found. Wittichenite associates with cuprobismutite, phase $Cu_8Bi_8S_{19}$, bornite solid solution, nukundamite and bismuth liquid, but not with an assemblage of wittichenite with chalcopyrite, intermediate solid solution, phase $Cu_{8.4}Fe_{1.2}Bi_{10.8}S_{22}$, phase $CuBi_3S_5$, and bismuthinite. Cuprobismutite has the tie lines with bornite solid solution, nukundamite, wittichenite, phase $CuBi_3S_5$, phase $Cu_8Bi_8S_{19}$, phase $Cu_{8.4}Fe_{1.2}Bi_{10.8}S_{22}$, and bismuth liquid. Phase $Cu_{8.4}Fe_{1.2}Bi_{10.8}S_{22}$ (Z) appears in so stable a

TABLE 3. Analytical data for the phases synthesized by hydrothermal experimental runs at 420°C.

	Phase	Weight percent Cu	Fe	Bi	S	Total	Atomic percent Cu	Fe	Bi	S	Phase assemblage
HCFB 027	bn	53.1	8.5	13.9	23.6	99.1	46.7	8.5	3.7	41.1	bn + cpb
HCFB 028	cp	34.2	30.0	0.0	34.8	99.0	24.9	24.9	0.0	50.2	bn + cp + Z
HCFB 038	bn	52.2	12.1	10.5	25.2	100.0	43.6	11.4	2.7	42.3	bn + cpb + Z
	Z	15.4	1.4	63.1	19.9	99.8	20.4	2.1	25.4	52.2	
HCFB 043	cpb	15.5	0.3	64.0	19.3	99.1	21.1	0.5	26.5	51.9	bn + cpb
HCFB 044	Z	14.6	1.5	64.6	19.9	100.6	19.4	2.3	26.1	52.3	bn + cp + rk + Z
HCFB 045	cp	34.2	30.5	0.0	34.6	99.3	24.9	25.2	0.0	49.9	cp + bm + E + Y
	bm	0.9	0.4	79.6	19.0	99.9	1.5	0.8	38.2	59.6	
HCFB 046	bn	63.0	6.0	7.5	23.7	100.2	53.0	5.7	1.9	39.5	bn + wit
	wit	36.5	0.0	44.9	19.3	100.7	41.3	0.0	15.4	43.3	
HCFB 047	bn	54.5	7.0	13.8	23.8	99.1	47.9	7.0	3.7	41.4	bn + cpb + wit
	cpb	15.5	0.5	64.7	19.0	99.7	21.1	0.8	26.8	51.3	
	wit	37.1	0.1	42.4	19.3	98.9	42.1	0.1	14.6	43.3	
HCFB 048	cp	34.0	30.4	0.0	34.8	99.2	24.7	25.1	0.0	50.2	cp + Z + E
	E	7.2	0.2	73.6	19.2	100.2	10.6	0.3	33.0	56.1	
HCFB 053	bn	48.1	13.7	11.3	26.4	99.5	40.3	13.1	2.9	43.8	bn + Z
	Z	15.1	1.4	63.7	19.5	99.7	20.2	2.1	25.9	51.8	
HCFB 077	wit	36.4	0.1	44.4	18.9	99.8	41.6	0.1	15.4	42.8	wit + cpb + Bi
	cpb	15.4	0.1	66.0	19.0	100.6	21.1	0.2	27.4	51.3	
HCFB 078	E	8.4	0.0	74.2	18.9	101.5	12.3	0.0	33.0	54.7	E
HCFB 079	wit	36.0	0.3	44.5	19.5	100.3	40.7	0.4	15.3	43.7	wit + Bi
HCFB 089	bn	43.9	16.3	12.9	27.3	100.4	36.4	15.4	3.3	44.9	bn + Bi
HCFB 090	bn	37.0	17.6	17.2	27.2	99.6	31.6	17.1	4.5	46.9	bn + iss + Bi
	iss	31.6	32.9	1.9	34.5	100.9	22.9	27.2	0.4	49.6	
HCFB 091	bn	39.4	17.2	15.2	27.4	99.2	33.4	16.6	3.9	46.0	bn + Bi
HCFB 092	iss	27.8	37.4	0.4	34.8	100.4	19.9	30.5	0.1	49.5	iss + po + Bi
	po	0.7	62.1	0.0	37.2	100.0	0.5	48.7	0.0	50.8	
HCFB 093	bn	52.3	13.3	9.2	25.2	100.1	43.5	12.6	2.3	41.6	bn + wit + Bi
	wit	36.4	0.1	45.2	19.2	100.9	41.2	0.1	15.6	43.1	
HCFB 094	bn	52.3	13.8	10.1	24.6	100.8	43.6	13.1	2.6	40.7	bn + wit + Bi
	wit	35.7	0.3	44.8	18.9	99.7	40.9	0.4	15.6	43.0	
HCFB 095	bn	49.4	14.2	10.9	26.5	101.0	40.7	13.3	2.7	43.3	bn + Bi
HCFB 096	bn	50.0	14.0	12.0	25.2	101.2	41.8	13.3	3.1	41.8	bn + wit + Bi
HCFB 107	iss	23.3	40.6	0.2	35.4	99.5	16.7	33.1	0.0	50.2	iss + po + Bi
	po	0.7	61.3	0.3	38.0	100.3	0.5	47.8	0.1	51.6	
HCFB 108	wit	39.7	0.1	42.4	18.4	100.6	44.5	0.1	14.5	40.9	wit + X + Bi
	X	59.7	0.1	21.2	19.0	100.0	57.4	0.1	6.2	36.2	
HCFB 109	E	8.7	0.3	72.3	18.8	100.1	12.7	0.5	32.2	54.6	po + E
HCFB 123	bn	43.4	16.2	15.1	26.5	101.2	36.5	15.5	3.9	44.2	bn + Z + Bi
	Z	13.1	2.3	66.0	19.5	100.9	17.6	3.5	27.0	51.9	
HCFB 125	wit	38.2	0.1	41.8	19.2	99.3	42.9	0.1	14.3	42.7	wit + X + Bi
HCFB 126	X	57.8	0.2	22.3	19.9	100.2	55.4	0.2	6.5	37.8	bn + X + Bi
HCFB 128	bn	48.1	15.1	12.5	25.7	101.4	40.1	14.3	3.2	42.4	bn + Z + Bi
	Z	14.2	2.3	63.0	19.6	99.1	19.0	3.5	25.6	51.9	
HCFB 132	E	8.7	0.2	72.3	18.8	100.0	12.8	0.3	32.2	54.7	cpb + E + Bi
	cpb	15.1	0.1	65.9	19.0	100.1	20.7	0.2	27.5	51.7	

form as to coexist with a lot of phases, such as bornite solid solution, intermediate solid solution, chalcopyrite, nukundamite, cuprobismutite, phase $CuBi_3S_5$, phase $Cu_8Bi_8S_{19}$, and bismuth liquid. Bismuthinite assembles with chalcopyrite, intermediate solid solution, pyrrhotite, phase $CuBi_3S_5$, phase $Cu_8Bi_8S_{19}$, bismuth liquid, and pyrite.

The bornite solid solution at 420°C contains considerable quantities of iron and bismuth. A composition of bornite, which nearly corresponds to the maximum contents of iron and bismuth, is Cu 37.0, Fe 17.6, Bi 17.2, and S 27.2 in weight percent, as is presented in the chemical formula $Cu_{3.16}Fe_{1.71}Bi_{0.45}S_{4.69}$ as the total atomic number ten. It associates with

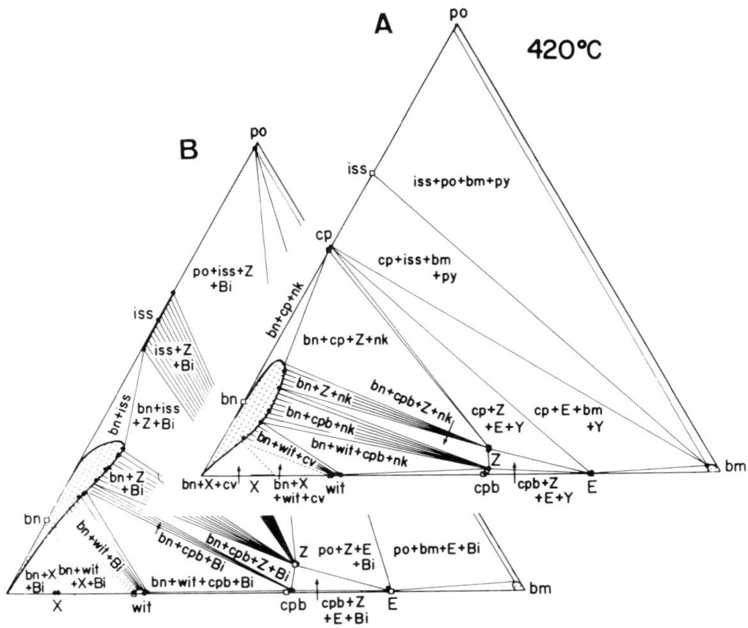

FIG. 13. Phase relations of relatively sulfur-rich (A) and metal rich (B) parts of the Cu–Fe–Bi–S system at 420°C obtained by projection from sulfur corner to the basal Cu–Fe–Bi plane in the tetrahedron of the quaternary Cu–Fe–Bi–S.

intermediate solid solution and bismuth liquid, as given in Table 3 (Run No. 090), very close to the point of univariant assemblage with intermediate solid solution and phase $Cu_{8.4}Fe_{1.2}Bi_{10.8}S_{22}$, approached at a univariant point of the association with chalcopyrite, phase $Cu_{8.4}Fe_{1.2}Bi_{10.8}S_{22}$, and nukundamite, and has Cu 48.1, Fe 13.7, Bi 11.3, and S 26.4 in a weight percent corresponding to $Cu_{4.03}Fe_{1.31}Bi_{0.29}S_{4.38}$ (Run No. 053 in Table 3). Also bornite, which assembles with wittichenite and bismuth liquid, got very near to a univariant point of the assemblage with wittichenite, cuprobismutite, and bismuth liquid and contains Cu 52.3, Fe 13.8, Bi 10.1, and S 24.6 in weight percent ($Cu_{4.36}Fe_{1.31}Bi_{0.26}S_{4.07}$). Meanwhile, the composition of bornite with wittichenite and cuprobismutite, get very close to the univariant point of the association of bornite + wittichenite + cuprobismutite + nukundamite and is Cu 54.5, Fe 7.0, Bi 13.8, and S 23.8 in weight percent ($Cu_{4.79}Fe_{0.70}Bi_{0.37}S_{4.14}$). The bornite solid solution compositions in equilibrium with bismuth liquid along the boundary at 420°C are, in order of bismuth-richness, as follows: $Cu_{3.16}Fe_{1.71}Bi_{0.45}S_{4.69}$ (Run No. 090) $Cu_{3.64}Fe_{1.54}Bi_{0.33}S_{4.49}$ (Run No. 089), $Cu_{3.79}Fe_{1.50}Bi_{0.35}S_{4.36}$ (Run No. 130), $Cu_{4.01}Fe_{1.43}Bi_{0.32}S_{4.24}$ (Run No. 128), $Cu_{4.07}Fe_{1.33}Bi_{0.27}S_{4.33}$ (Run No. 095), $Cu_{4.36}Fe_{1.31}Bi_{0.26}$ $S_{4.07}$ (Run No. 094), $Cu_{4.35}Fe_{1.26}$

$Bi_{0.23}S_{4.16}$ (Run No. 093), and $Cu_{4.57}Fe_{1.21}Bi_{0.22}S_{4.00}$ (Run No. 133) etc. They occur under the condition of relatively low sulfur fugacity, so as to appear as bismuth (liquid). On the other hand, the bornite solid solution with nukundamite, which forms under comparatively high sulfur fugacity, is bordered with the compositions as follows: $Cu_{4.03}Fe_{1.31}Bi_{0.29}S_{4.38}$ (Run No. 053), $Cu_{4.36}Fe_{1.14}Bi_{0.27}S_{4.23}$ (Run No. 038), $Cu_{4.67}Fe_{0.85}Bi_{0.37}S_{4.11}$ (Run No. 027), $Cu_{4.79}Fe_{0.70}Bi_{0.37}S_{4.14}$ (Run No. 047), $Cu_{5.30}Fe_{0.57}Bi_{0.19}S_{3.95}$ (Run No. 046) etc.

As shown in Fig. 14, bornite associated with bismuth (liquid), as to be stable under low sulfur fugacity, has more bismuth and iron than one assembled to nukundamite, which occurs in relatively high sulfur fugacity.

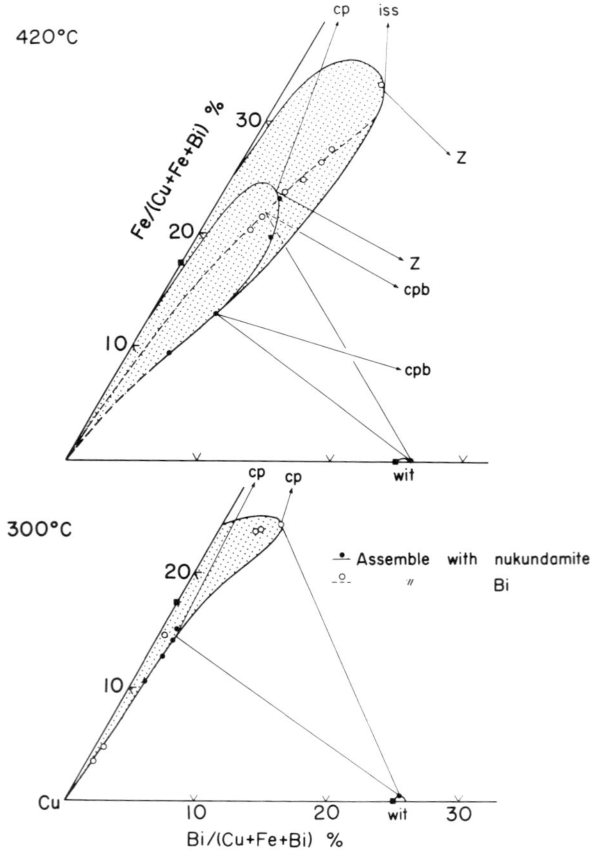

FIG. 14. The area of bornite solid solution at 420° and 300°C. Open circle show the composition of bornite associated with bismuth liquid, while solid one, bornite with nukundamite.

3.4.2 The phase relations at 300°C

The results of the experiments at 300°C are given in Table 4. The phases produced at 300°C are almost the same as 420°C except that emplectite appears instead of cuprobismutite. Phase relations in the quaternary system at 300°C are shown in Fig. 15, but this figure is complicated. Thus, in order to solve the phase relations the figure was simplified, as Fig. 16. This figure shows the phase relations of the quaternary system at 300°C presented by projection from the sulfur corner to the basal Cu–Fe–Bi plane of the tetrahedron Cu–Fe–Bi–S. The stable phase assemblages under conditions of comparatively high sulfur fugacity, so as to coexist with nukundamite and phase $Cu_8Bi_8S_{19}$, are given in Fig. 16-A, while the phase relations under low sulfur fugacity with bismuth (liquid) are shown in Fig. 16-B. As found in Fig. 16, the association of chalcopyrite and wittichenite appears instead of the assemblages of bornite + cuprobismutite and bornite + phase $Cu_{8.4}Fe_{1.2}Bi_{10.8}S_{22}$ at 420°C. The univariant assemblages of bornite + chalcopyrite + wittichenite + nukundamite or bismuth liquid, chalcopyrite + wittichenite + phase $Cu_{8.4}Fe_{1.2}Bi_{10.8}S_{22}$ + nukundamite or bismuth liquid, and wittichenite + emplectite + phase $Cu_{8.4}Fe_{1.2}Bi_{10.8}S_{22}$ + phase $Cu_8Bi_8S_{19}$ or bismuth liquid become stable. In addition to them, the following univariant assemblages in the quaternary system at 300°C are found in Fig. 17 shown as a decomposed diagram of the quaternary tetrahedron in Fig. 15. Chalcopyrite + phase $Cu_{8.4}Fe_{1.2}Bi_{10.8}S_{22}$ + phase $CuBi_3S_5$ + bismuth (L), chalcopyrite + phase $CuBi_3S_5$ + bismuthinite + bismuth (L), chalcopyrite + intermediate solid solution + bismuthinite + bismuth (L), pyrrhotite + intermediate solid solution + bismuthinite + bismuth (L), emplectite + phase $Cu_{8.4}Fe_{1.2}Bi_{10.8}S_{22}$ + phase $CuBi_3S_5$ + bismuth (L), (bornite + chalcopyrite + intermediate solid solution + bismuth (L)), (bornite + intermediate solid solution + pyrrhotite + bismuth (L)), (bornite + pyrrhotite + copper + bismuth (L)), (pyrrhotite + copper + iron + bismuth (L)), (bornite + wittichenite + nukundamite + covellite), (covellite + wittichenite + nukundamite + phase $Cu_8Bi_8S_{19}$), wittichenite + nukundamite + phase $Cu_{8.4}Fe_{1.2}Bi_{10.8}S_{22}$ + phase $Cu_8Bi_8S_{19}$, chalcopyrite + nukundamite + phase $Cu_{8.4}Fe_{1.2}Bi_{10.8}S_{22}$ + phase $Cu_8Bi_8S_{19}$, chalcopyrite + phase $CuBi_3S_5$ + phase $Cu_{8.4}Fe_{1.2}Bi_{10.8}S_{22}$ + phase $Cu_8Bi_8S_{19}$, chalcopyrite + bismuthinite + phase $CuBi_3S_5$ + phase $Cu_8Bu_8S_{19}$, chalcopyrite + intermediate solid solution + pyrrhotite + bismuthinite, emplectite + phase $Cu_{8.4}Fe_{1.2}Bi_{10.8}S_{22}$ + phase $CuBi_3S_5$ + phase $Cu_8Bi_8S_{19}$, (covellite + nukundamite + phase $Cu_8Bi_8S_{19}$ + pyrite), (chalcopyrite + nukundamite + phase $Cu_8Bi_8S_{19}$ + pyrite), (chalcopyrite + bismuthinite + phase $Cu_8Bi_8S_{19}$ + pyrite), (chalcopyrite + pyrrhotite + bismuthinite + pyrite), (covellite + phase $Cu_8Bi_8S_{19}$ + pyrite + sulfur (L)), and (bismuthinite + phase $Cu_8Bi_8S_{19}$ + pyrite + sulfur (L)).

Emplectite, which first appears at 300°C, can coexist with wittichenite,

TABLE 4. Hydrothermal experimental results at 300°C in the Cu–Fe–Bi–S system.

Run No.	Nutrient materials					Pressure	Heating period	Synthesized phases
	Composition (Atomic %)				Reactants			
	Cu	Fe	Bi	S		(kg/cm²)	(days)	
HCFB 002	39.33	4.91	11.48	44.48	bn + wit + Z	395	8	bn + cp + wit
HCFB 004	32.64	14.28	6.12	46.96	bn + cp + Bi	412	11	bn + cp + wit
HCFB 005	40.59	3.13	12.50	43.78	bn + wit + Bi	412	11	bn + cp + wit
HCFB 006	30.42	17.39	4.35	47.84	bn + cp + Z	412	11	cp + wit + nk + Z
HCFB 007	36.35	9.10	9.90	45.47	bn + Z	396	10	cp + wit + nk
HCFB 008	34.61	11.54	7.69	46.16	bn + cp + Z	396	10	cp + wit + nk + Z
HCFB 009	44.74	2.63	10.53	42.10	bn + wit	400	11	bn + cp + wit
HCFB 010	46.33	4.88	7.31	41.47	bn + wit	400	11	bn + wit
HCFB 011	47.74	6.80	4.55	40.91	bn	410	13	bn + cp + wit
HCFB o12	48.94	8.51	2.14	40.42	bn	410	13	bn + cp + Bi
HCFB 014	24.99	12.50	12.50	50.01	bn + cp + Z	420	10	cp + wit + Z
HCFB 015	15.05	0.77	30.20	53.98	D + E + Z	385	13	emp + E + Z + Y
HCFB 016	20.14	1.01	26.92	51.92	wit + D + E	385	13	emp + E + Z
HCFB 018	24.98	4.22	20.80	50.00	bn + wit + Z	390	13	cp + Z + Y
HCFB 019	19.12	2.87	25.69	52.32	cp + Z	380	9	cp + wit + Z
HCFB 020	16.46	6.32	23.83	53.40	cp + E + Z	380	9	cp + Z + Y
HCFB 021	45.62	4.81	8.64	40.92	bn + wit + Bi	395	9	bn + cp
HCFB 023	24.99	6.45	18.56	50.00	bn + Z	395	8	cp + Z + Y
HCFB 024	24.59	6.32	19.81	49.28	bn + Z + Bi	395	8	bn + cp + wit + nk
HCFB 025	16.51	30.82	5.50	47.17	wit + po	410	14	iss + Bi
HCFB 026	23.03	23.20	7.68	46.09	wit + po	410	14	cp + Z
HCFB 033	40.21	10.98	4.87	43.94	bn + cp + wit	400	25	cp + wit
HCFB 044	16.51	30.82	5.50	47.17	wit + po	400	25	iss + bm + Bi
HCFB 035	20.80	25.55	6.94	46.72	wit + po	410	25	iss + po + bm + Bi
HCFB 036	27.24	18.29	9.08	45.39	wit + po	410	25	cp + wit + Bi
HCFB 041	8.05	13.79	24.12	54.04	po + E	400	21	iss + po + bm + Bi
HCFB 042	5.33	26.03	15.97	52.67	po + E	400	21	po + bm
HCFB 070 *	7.68	1.41	60.77	30.13	iss + E + Bi	300	7	bm + E + Bi
HCFB 073 *	17.47	1.48	51.01	30.05	bn + cpb + Bi	300	10	bn + cp + Bi
HCFB 074 *	24.49	3.49	41.99	30.03	bn + cpb + Bi	300	10	cp + wit + Bi
HCFB 083 *	10.50	3.49	56.02	29.99	iss + Z + Bi	300	21	bn + cp + Bi
HCFB 084 *	17.48	17.53	34.97	30.02	bn + iss + Bi	300	21	cp + Z + Bi
HCFB 085 *	14.99	17.50	37.48	30.02	iss + Bi	300	21	bn + cp + Bi
HCFB 086 *	12.49	17.52	39.97	30.02	iss + Bi	300	21	iss + bm + Bi
HCFB 087 *	22.47	14.02	33.47	30.04	bn + iss + Bi	300	19	bn + cp + Bi
HCFB 088 *	15.99	14.98	38.99	30.04	bn + iss + Z + Bi	300	19	cp + wit + Bi
HCFB 097 *	44.96	1.40	23.60	30.04	bn + X + Bi	300	18	bn + wit + Bi
HCFB 098 *	20.99	10.52	38.49	30.00	bn + iss + Z + Bi	300	18	cp + wit + Bi
HCFB 099 *	10.00	22.50	37.51	29.99	iss + po + Bi	300	18	iss + Bi
HCFB 100 *	6.99	7.00	55.95	30.06	iss + Z + Bi	300	18	cp + bm + Bi
HCFB 101 *	33.57	1.39	34.96	30.08	bn + wit + Bi	300	18	bn + wit + Bi
HCFB 102 *	5.00	22.48	42.48	30.04	iss + po + bm + Bi	300	18	iss + po + bm + Bi
HCFB 103 *	24.47	7.01	38.45	30.07	bn + Z + Bi	300	18	bn + cp + wit + Bi
HCFB 104 *	21.00	6.98	42.00	30.02	bn + Z + Bi	300	18	cp + Bi
HCFB 105 *	17.48	10.54	41.98	30.01	bn + iss + Z + Bi	300	18	cp + wit + Bi
HCFB 114 *	10.50	10.51	49.00	29.99	iss + bm + E + Bi	300	18	cp + bm + Bi
HCFB 115 *	14.00	7.02	49.00	29.99	bn + iss + Z + Bi	300	18	cp + bm + Bi
HCFB 117 *	30.09	1.34	38.49	30.08	bn + wit + cpb + Bi	300	18	cp + wit + Bi
HCFB 118 *	19.99	14.01	35.98	30.02	bn + iss + Bi	300	18	bn + cp + wit + Bi
HCFB 119 *	10.14	13.51	47.32	29.03	iss + Z + Bi	300	18	cp + bm + Bi
HCFB 121 *	13.51	13.53	43.90	29.05	bn + iss + Z + Bi	300	18	cp + Bi
HCFB 122 *	16.90	13.54	40.57	28.97	bn + iss + Z + Bi	300	18	cp + wit + Bi
HCBF 003	19.70	3.00	25.50	51.80	cp + Z	300	10	cp + Z
HCBS 069	28.20	0.00	23.08	48.72	wit + cpb	300	13	emp + wit
HCBS 084	18.82	0.00	28.23	52.94	cpb + E	280	21	wit + emp + Y
HCBS 095	15.50	0.00	30.00	54.50	cpb + E + Y	350	15	bm + E + Y

Solvent : 5 m NH_4Cl aqueous solution, * Silica tube was used as reaction vessel

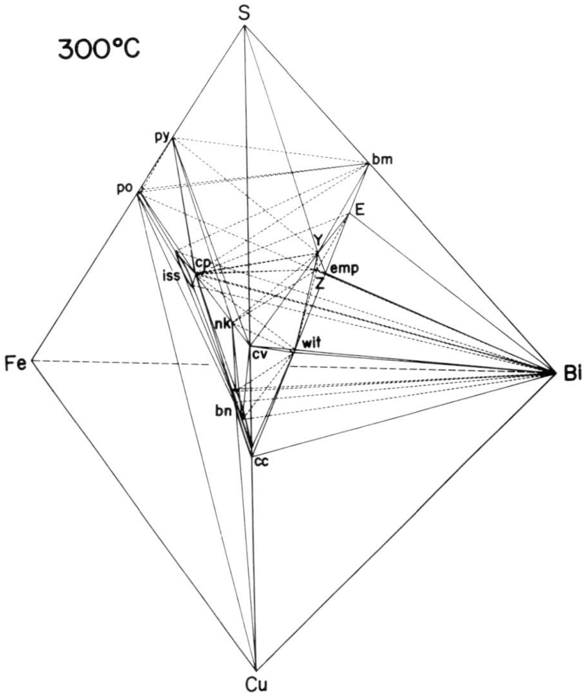

FIG. 15. Schematic diagram in the quaternary Cu–Fe–Bi–S system at 300°C under hydrothermal conditions.

phase $CuBi_3S_5$, phase $Cu_8Bi_8S_{19}$, phase $Cu_{8.4}Fe_{1.2}Bi_{10.8}S_{22}$, nukundamite, and bismuth liquid, but it cannot associate with bornite, chalcopyrite, intermediate solid solution, pyrrhotite, and bismuthinite, etc., although it occurs in intimate association with bornite, chalcopyrite, and bismuthinite in natural copper-bismuth ores.

Analytical data for the solid phases synthesized in each run by microprobe analyser are given in Table 5. As seen in the table, bornite still has a passable space of the solid solution although it becomes smaller in comparison with that at 420°C. Bornite, which occurs just as a univariant assemblage with chalcopyrite, wittichenite, and bismuth liquid, has the composition of Cu 47.8, Fe 14.4, Bi 11.4, and S 25.9 in weight percent corresponding to the formula $Cu_{4.02}Fe_{1.37}Bi_{0.29}S_{4.32}$ (Run No. 103). This composition is nearly the maximum content of iron and bismuth at 300°C, while bornite with chalcopyrite and wittichenite, close to a univariant point of the assemblage bornite + chalcopyrite + wittichenite + nukundamite, contains Cu 63.6, Fe 9.2, Bi 3.4, and S 24.1 in weight percent corresponding to $Cu_{5.18}Fe_{0.85}Bi_{0.08}S_{3.89}$ (Run No. 011). The same as at

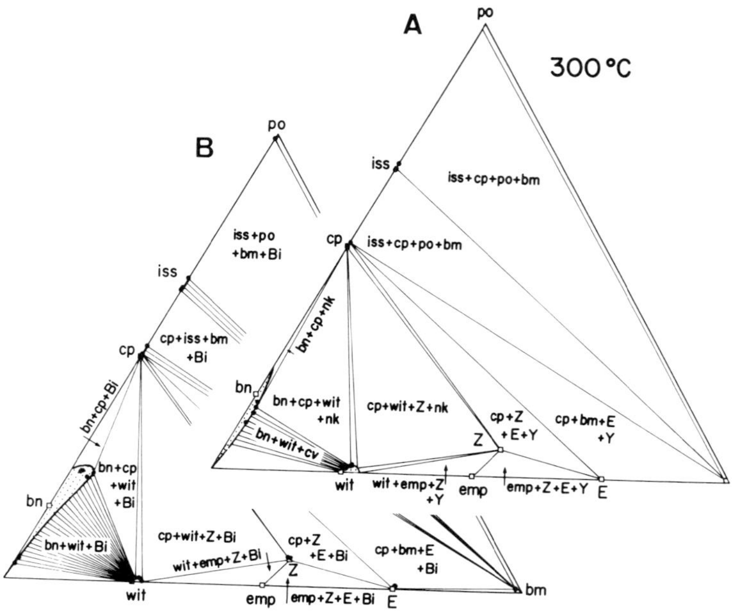

FIG. 16. Phase relations of relatively sulfur-rich (A) and metal rich (B) parts of the Cu–Fe–Bi–S system at 300°C shown by projection from sulfur corner to the basal metal plane in the Cu–Fe–Bi–S tetrahedron.

420°C, bornite with bismuth (liquid) also contains more iron and bismuth than bornite with nukundamite, as shown in Fig. 14. Wittichenite has a very limited range of solid solution at 300°C from Cu_3BiS_3 to $Cu_{2.83}Bi_{1.10}S_{3.07}$ in the binary Cu_2S–Bi_2S_3, as reported by SUGAKI and SHIMA (1971, 1972b), and includes 0.6 to 1.0 weight percent iron (0.8 to 1.3 atomic percent iron) in the case associated with chalcopyrite. Emplectite and phase $CuBi_3S_5$ almost contain no iron. Intermediate solid solution has very small amounts of bismuth, 0.9 weight percent or less, and chalcopyrite also contains only 0.2 weight percent or less.

4. *Discussion*

4.1 *Bornite solid solution*

As mentioned above, bornite synthesized hydrothermally in the Cu–Fe–Bi–S system has a relatively extensive field of solid solution, containing a maximum of 17.2 weight percent bismuth at 420°C. Similarly, NANRI *et al.* (1978) carried out experiments on the system Cu–Fe–Bi–S using the dry method, and reported that bornite solid solution appeared in a large area, having a maximum of 18.2 weight percent bismuth at 500°C. Bornite

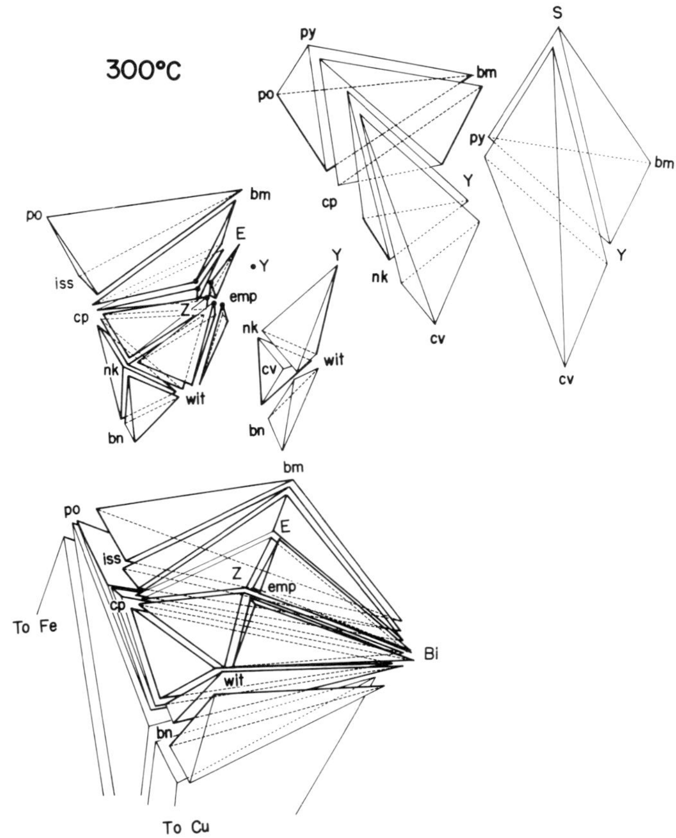

FIG. 17. Exploded phase diagrams showing stable phase assemblages in the Cu–Fe–Bi–S system at 300°C.

forms a fairly wider solid solution under low sulfur fugacity condition, as coexisting with bismuth (liquid), than its field under a high sulfur fugacity accompanied by unkundamite at constant temperatures, as shown in Fig. 14. The bornite solid solution in the quaternary system becomes more iron-and sulfur-rich besides bismuth than it does in the system Cu–Fe–S, as shown in Fig. 18 which shows the bornite solid solution by projection on the Cu–Fe–S plane from the bismuth corner of the tetrahedron, and contains 14.4 and 17.6 weight percent iron, and 25.9 and 27.2 weight percent sulfur in maximum at 300° and 420°C, respectively. Figure 18 shows a field of bornite solid solution at 420°C in comparison with that of bornite at 400°C in the system Cu–Fe–S. Bornite, which is mostly iron-and sulfur-rich, is produced in the case coexisting with phase $Cu_{8.4}Fe_{1.2}Bi_{10.8}S_{22}$ (Z), and the iron and sulfur contents in the bornite solid solution decrease in order of phase assemblage

TABLE 5. Analytical data for the phases synthesized by experimental runs at 300°C.

Run No.	Phase	Weight percent Cu	Fe	Bi	S	Total	Atomic percent Cu	Fe	Bi	S	Phase assemblage
HCFB 002	cp	34.7	30.1	0.0	34.7	99.5	25.2	24.9	0.0	49.9	bn + cp + wit
HCFB 004	cp	34.8	30.3	0.0	34.8	99.0	25.2	24.9	0.0	49.9	cp + wit + bn
HCFB 005	cp	34.3	30.1	0.0	34.8	99.2	25.0	24.9	0.0	50.1	bn + cp + wit
HCFB 006	cp	34.6	30.1	0.3	34.9	99.9	25.1	24.8	0.1	50.0	cp + wit + nk + Z
	wit	36.9	0.6	43.3	19.4	100.2	41.3	0.8	14.8	43.1	
HCFB 007	cp	34.6	30.1	0.0	34.9	99.6	25.1	24.8	0.0	50.1	cp + wit + nk
	wit	36.9	1.0	43.0	19.5	100.4	41.1	1.3	14.6	43.1	
HCFB 008	cp	34.6	30.2	0.1	34.7	99.6	25.1	24.9	0.0	50.0	cp + wit + nk + Z
	wit	36.5	0.7	44.1	19.5	100.8	40.8	0.9	15.0	43.3	
	nk	55.9	10.0	0.0	33.8	99.7	41.6	8.5	0.0	49.9	
HCFB 009	bn	66.7	6.9	2.3	23.1	99.0	55.1	6.5	0.6	37.8	bn + cp + wit
HCFB 010	bn	64.3	8.2	3.2	23.8	99.5	52.8	7.7	0.8	38.7	bn + wit
	wit	37.5	0.4	43.2	19.5	100.6	41.8	0.5	14.6	43.1	
HCFB 011	cp	34.2	30.4	0.1	34.7	99.4	24.9	25.1	0.0	50.0	bn + cp + wit
	bn	63.6	9.2	3.4	24.1	100.3	51.8	8.5	0.8	38.9	
HCFB 012	cp	34.4	30.1	0.2	34.7	99.4	25.0	24.9	0.0	50.0	bn + cp + Bi
	bn	63.4	9.8	2.5	24.5	100.2	51.2	9.0	0.6	39.2	
HCFB 025	iss	22.7	41.2	0.4	35.2	99.5	16.3	33.6	0.1	50.0	iss + Bi
HCFB 026	cp	33.6	30.6	0.0	35.0	99.2	24.4	25.3	0.0	50.3	cp + Z
HCFB 033	cp	34.6	30.5	0.1	35.0	100.2	24.9	25.0	0.0	50.0	cp + wit
HCFB 034	iss	22.3	41.9	0.1	35.1	99.4	16.0	34.2	0.0	49.8	iss + bm + Bi
HCFB 035	iss	22.8	40.9	0.2	35.2	99.1	16.4	33.4	0.1	50.1	iss + po + bm + Bi
HCFB 036	cp	34.3	30.4	0.1	35.0	99.8	24.8	25.0	0.0	50.2	cp + wit + Bi
HCFB 041	iss	22.4	42.1	0.3	35.7	100.5	15.9	33.9	0.1	50.1	iss + po + bm + Bi
	bm	0.0	0.1	82.1	19.1	101.3	0.0	0.1	39.7	60.2	
HCFB 042	po	0.2	60.7	0.1	39.4	100.4	0.2	46.8	0.0	53.0	po + bm
HCFB 070	E	7.2	0.3	73.1	18.6	99.2	10.8	0.5	33.4	55.3	bm + E + Bi
HCFB 073	bn	51.9	15.0	7.1	26.5	100.4	42.0	13.8	1.8	42.5	bn + cp + Bi
	cp	34.7	30.3	0.0	35.1	100.1	25.0	24.9	0.0	50.1	
HCFB 074	wit	36.5	0.1	43.2	19.2	99.0	41.6	0.1	15.0	43.3	cp + wit + Bi
HCFB 083	cp	34.7	31.0	0.1	35.1	100.9	24.9	25.3	0.0	49.8	bn + cp + Bi
	bn	52.5	14.8	6.4	26.0	99.7	42.8	13.7	1.6	42.0	
HCFB 084	cp	34.8	30.9	0.0	35.1	100.8	24.9	25.2	0.0	49.9	cp + Z + Bi
	Z	15.2	2.1	64.0	19.6	100.9	20.0	3.1	25.6	51.2	
HCFB 086	iss	24.5	39.6	0.2	35.4	99.7	17.5	32.2	0.0	50.2	iss + bm + Bi
	bm	0.1	0.1	81.1	18.8	100.1	0.2	0.2	39.7	60.0	
HCFB 087	cp	34.6	30.5	0.2	34.9	100.2	25.0	25.1	0.0	49.9	bn + cp + Bi
HCFB 088	cp	34.5	30.6	0.2	34.8	100.1	24.9	25.2	0.0	49.9	cp + wit + Bi
HCFB 097	bn	75.9	2.3	1.2	20.9	100.3	63.1	2.2	0.3	34.4	bn + wit + Bi
	wit	39.0	0.1	42.5	19.4	101.0	43.1	0.1	14.3	42.5	
HCFB 098	wit	37.9	0.2	43.2	19.0	100.3	42.6	0.3	14.8	42.4	cp + wit + Bi
	cp	34.5	30.4	0.2	34.7	99.8	25.0	25.1	0.0	49.9	
HCFB 099	iss	24.2	40.2	0.1	35.5	100.0	17.2	32.6	0.0	50.1	iss + Bi
HCFB 100	bm	1.4	0.5	79.4	18.5	99.8	2.2	0.9	38.5	58.4	cp + bm + Bi
HCFB 101	bn	74.5	3.3	1.6	21.6	101.0	61.3	3.1	0.4	35.2	bn + wit + Bi
HCFB 102	iss	22.4	41.7	0.9	35.5	100.5	16.0	33.8	0.2	50.1	iss + po + bm + Bi
	po	0.2	61.3	0.3	38.6	100.4	0.1	47.6	0.1	52.2	
	bm	0.1	0.1	80.9	18.5	99.6	0.2	0.2	40.0	59.6	
HCFB 103	bn	47.8	14.4	11.4	25.9	99.5	40.2	13.7	2.9	43.2	bn + cp + wit + Bi
	wit	38.6	0.4	42.5	19.2	100.7	42.9	0.5	14.4	42.3	
HCFB 119	cp	33.2	32.0	0.2	35.1	100.4	23.8	26.1	0.0	49.9	cp + bm + Bi
	bm	0.1	0.1	81.6	18.6	100.4	0.2	0.2	40.1	59.6	
HCFB 003	Z	14.9	1.8	63.3	19.9	99.9	19.7	2.7	25.4	52.1	cp + Z
HCBS 069	wit	36.1	0.0	44.3	19.7	100.1	40.7	0.0	15.2	44.1	emp + wit
	emp	18.8	0.0	61.9	19.0	99.7	25.0	0.0	25.0	50.0	
HCBS 095	E	7.7	0.0	73.5	19.2	100.4	11.3	0.0	32.8	55.9	bm + E + Y

with cuprobismutite and wittichenite, as seen in the figure. The upper curve bordering the bornite solid solution represents the solvus of the solid solution associating with the phases having a comparatively high sulfur fugacity, such as nukundamite and phase $Cu_8Bi_8S_{19}$. The low curve corresponds to the solvus associating with bismuth liquid, which is stable under low sulfur fugacity. Although it can contain more bismuth, iron, and sulfur than its

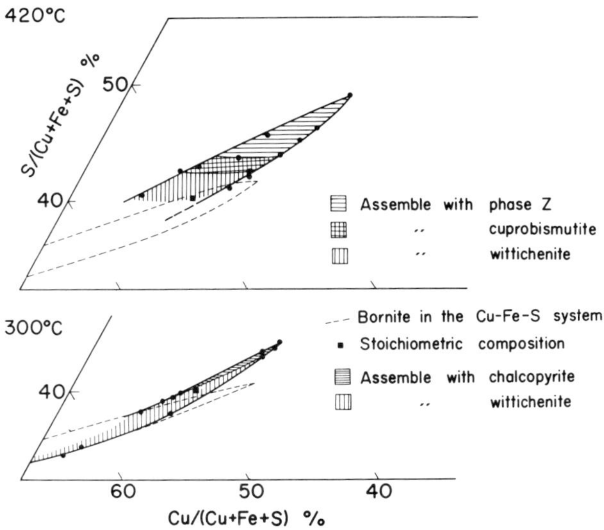

FIG. 18. The regions of bornite solid solution at 420° and 300°C obtained by projection from bismuth corner to Cu–Fe–S plane in the tetrahedron. The areas of bornite solid solution under hydrothermal conditions in the Cu–Fe–S system are also shown together.

stoichiometric composition at high temperatures, bornite in natural ores has a stoichiometric composition with no bismuth, as reported by SUGAKI *et al.* (1973). Thus, it is a possibility that bornite in bismuth-bearing copper ores, formed at higher temperature than about 300°C, exsolves bismuth minerals such as wittichenite during cooling. In fact, wittichenite (klaprothite) occasionally occurs as exsolution lamella in natural bornite, as described by WATANABE (1937, 1951) and KRIEGER (1940). In order to experimentally confirm the exsolution of bismuth minerals from bornite solid solution, bornite with Cu 54.5, Fe 7.0, Bi 13.8, and S 23.8 in weight percent ($Cu_{4.79}Fe_{0.70}Bi_{0.37}S_{4.14}$) which was synthesized hydrothermally at 420°C, was annealed at 300°C for 12 hours under a pyrrhotite (48.0 atomic percent iron) buffer. Consequently it is found that wittichenite is exsoluted as lamella in bornite, as shown in Fig. 19-A. In this figure (Fig. 19-B) exsolution lamella of wittichenite in bornite from Sannotake-Yokozuru mine, Fukuoka prefecture, Japan, is shown in comparison with the experimental product above.

4.2 *Chalcopyrite and wittichenite assemblage*

Chalcopyrite and bornite occur in intimate association with wittichenite in bismuth-bearing copper ores. In the experiments at 420°C, although wittichenite coexists with bornite solid solution, it is not found in the assem-

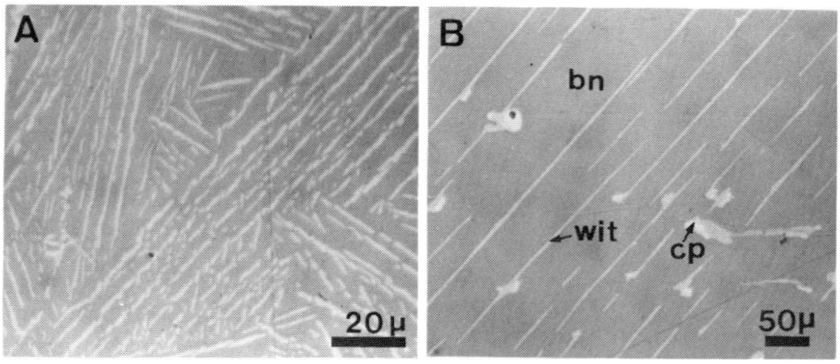

FIG. 19 A and B. Photomicrographs of exsolution lamellae of wittichenite in bornite. A, Wittichenite lamellae appeared in bornite in which produced at 420°C after annealed at 300°C; B, Wittichenite lamellae in bornite from the Sannotake-Yokozuru mine.

blage with chalcopyrite because the assemblages of bornite + cuprobismutite + nukundamite (or bismuth liquid) and bornite + phase $Cu_{8.4}Fe_{1.2}Bi_{10.8}S_{22}$ + nukundamite (or bismuth liquid) form in an equilibrium state. However, wittichenite appears in equilibrium with chalcopyrite in the experimental runs at 300°C. SUGAKI *et al.* (1972b) carried out the study on the phase relation in the pseudo-binary $CuFeS_2$–Cu_3BiS_3 using an evacuated glass tube method and found that the assemblage of chalcopyrite and wittichenite is stable below 395°C. Their result does not conflict with that of the experiments by the present authors. From these data, the chalcopyrite and wittichenite assemblage found in natural ores is interpreted to have been formed at temperatures below 395°C.

4.3 *Emplectite, chalcopyrite and bornite assemblage*

Although emplectite often occurs in intimate association with chalcopyrite and bornite in bismuth and copper ores, the assemblages of emplectite, chalcopyrite and/or bornite are not found in the experimental runs at 300°C because the assemblages of wittichenite, phase $Cu_{8.4}Fe_{1.2}Bi_{10.8}S_{22}$, and phase $Cu_8Bi_8S_{19}$ or bismuth liquid are stable, as shown in Fig. 15. It is not certain wherther the emplectite, chalcopyrite and/or bornite assemblages form at lower temperatures than 300°C or not, but it is possible that they appear as a stable association below 300°C. Also, in natural ores, emplectite commonly coexists with wittichenite, bismuthinite, and native bismuth. Among them, the assemblage of emplectite, wittichenite, and bismuth (liquid) occurs in the experiments at 300°C, but the association of emplectite and bismuthinite is not found. SUGAKI *et al.* (1980) reported from their experimental data that emplectite forms in association with bismuthinite at temperatures below approximately 280°C because of the decomposition of phase $CuBi_3S_5$ at that temperature, as shown in Fig. 7. Therefore, the emplectite and bismuthinite assemblage found in ores seems to have been formed at lower temperatures than about 280°C. SUGAKI *et al.* (1980) also carried out the study on the relation between emplectite and cuprobismutite in the join Cu_2S–Bi_2S_3 as already mentioned, and found that cuprobismutite is more bismuth-rich than emplectite, which has stoichiometric composition $CuBiS_2$ and becomes unstable as to decomposed into emplectite and phase $CuBi_3S_5$ at 318°C or below. So cuprobismutite should not occur in nature, but it sometimes appears in ores. For this reason, it is thought that minor elements such as iron, lead, and silver contained in cuprobismutite may act as a stabilizer to keep cuprobismutite at room temperature.

4.4 *Phase assemblages with bismuth*

As bismuth minerals in natural ores, native bismuth is common as well as bismuthinite. In the present experiments of the system Cu–Fe–Bi–S at 300° and 420°C, bismuth (liquid) appears in company with all solid phases, except phase $Cu_8Bi_8S_{19}$, pyrite, covellite, and nukundamite, under the condition of relatively low sulfur fugacity, as seen in Figs. 13-B and 16-B. The tie lines from most of the sulfide phases in the quaternary space are concentrated upon a bismuth corner of the tetrahedron Cu–Fe–Bi–S. It means that almost all phases in the system can coexist as stable assemblages with bismuth under low sulfur fugacity. SUGAKI *et al.* (1972b) and NANRI *et al.* (1978) investigated phase relations in the Cu–Fe–Bi–S system under low sulfur fugacity by the dry method using an evacuated glass tube and obtained the results that bismuth (liquid) appears as a stable phase in all of the phase assemblages at 400°C. These results are in good agreement with those obtained under hydrothermal conditions by the present study. On the other hand, in the case of relatively-high sulfur fugacity, the tie lines are connected to phase $Cu_8Bi_8S_{19}$ (Y) and nukundamite instead of bismuth, as shown in Figs. 13-A and 16-A, and bismuth is not found.

4.5 *Relation between phase $Cu_{8.4}Fe_{1.2}Bi_{10.8}S_{22}$ (Z) and hodrushite.*

Phase Z is the only four-element compound in the Cu–Fe–Bi–S system synthesized by both the dry and hydrothermal methods. It occurs in a stable state at temperatures below 600°C. KODĚRA *et al.* (1970) found a new sulfosalt mineral, hodrushite $Cu_{8.12}Fe_{0.29}Bi_{11.54}S_{22}$, from Banska Hodoruša, Czechoslovakia. Hodorushite has a similar composition to that of phase Z. Its symmetry is monoclinic, space group $A2/m$, lattice constants $a = 27.21$ Å, $b = 3.39$ Å, $c = 17.58$ Å, and $\beta = 92.9°$. However, phase Z is monoclinic, $C2/m$, $a = 17.483$ Å, $b = 3.902$ Å, $c = 12.869$ Å, and $\beta = 108.11°$. Therefore, it is thought that phase Z is a different compound from hodorushite and has not yet been found in nature. Phase Z is stable at low temperatures and does not change, even after being kept in an evacuated glass tube for about 10 years at room temperature. Therefore, it is possible that phase Z will be found as a new mineral in nature.

A part of the expense for this study was defrayed by Grant in Aid for Scientific Research from the Ministry of Education of Japan. The writers offer their sincere thanks to the Ministry of Education.

REFERENCES

BARNARD, W. M. and CHRISTOPHER, P. A. (1966): Hydrothermal synthesis of chalcopyrite, *Econ. Geol.*, **61**, 897–902.

BARTON, P. B. (1973): Solid solutions in the system Cu–Fe–S, Part I: The Cu–S and CuFe–S join, *Econ. Geol.*, **68**, 455–465.

BARTON, P. B. and SKINNER, B. J. (1967): Sulfide mineral stabilities, in *Geochemistry of Hydothermal Ore Deposits*, Ed., H. L. Barnes, Holt. Reinhart and Winston, New York, pp. 236–333.

BENCE, A. E. and ALBEE, A. L. (1968): Empirical correction factors for the electron micro-analysis of silicate and oxides, *J. Geol.*, **76**, 382–403.

BRETT, P. R. (1963): The Cu–Fe–S system, *Carnegie Inst. Wash. Year Book*, **62**, 1963–1966.

BRYSTOV, L. V. and KUZ'MINA, I. P. (1959): *Growth of crystals*, Vol. 3, Consultants Bureau, New York, p. 357.

BUHLMANN, E. (1971): Untersuchungen im System Bi_2S_3–Cu_2S und geologische Schlussfologerungen, *Neu. Jahrb. Miner., Monat.*, **3**, 137–141.

CABRI, L. J. (1967): A new copper-iron sulfide, *Econ. Geol.*, **62**, 910–925.

CABRI, L. J. (1973): New data on phase relations in the Cu–Fe–S system, *Econ. Geol.*, **68**, 443–454.

CABRI, L. J. and HALL, S. R. (1971): Phase relations in the chalcopyrite region of the Cu–Fe–S system (abstr.), *Can. Mineral.*, **11**, 569.

CABRI, L. J. and HALL, S. R. (1972): Mooihoekite and haycockite, two new copper-iron sulfides, and their relationship to chalcopyrite and talnakhite, *Am. Mineral.*, **57**, 689–708.

CABRI, L. J. and HARRIS, D. C. (1971): New compositional data on talnakhite, *Econ. Geol.*, **66**, 673–675.

CHEN, T. T. and CHANG, L. L. Y. (1974): Investigations in the system Ag_2S–Cu_2S–Bi_2S_3 and Ag_2S–Cu_2S–Sb_2S_3, *Can. Mineral.*, **12**, 404–410.

CHERNYSHEV, L. V., ANFILOGOV, V. N., PASTUSHKOVA, T. M. and SUTURINA, T. A. (1968): A study of the system Fe–Zn–S under hydrothermal conditions, *Geochem. Intern.*, **5**, 196–209.

CRAIG, J. R. and SCOTT, S. D. (1974): Sulfide phase equilibria, in *Sulfide Mineralogy, Min. Soc. Am. Short Course Notes*, **1**, CS1–110.

CRAIG, J. R., BARTON, P. B. and SEPNUK, B. H. (1971): Experimental investigation in the Bi–Fe–S system (abstr.), *Geol. Soc. Am. Abstr. Prog.*, **3**, 305.

DANA, E. S. (1892): *Dana's System of Mineralogy*, John Wiley and Sons, New York, pp. 110–111.

FEDROVA, ZH. N. (1972): Investigation of phase digram of the Cu_2S–Bi_2S_3 system, in *Exp. Invest. in Mineralogy, 1970–1971*, Acad. Sci. USSR., Siber. Div., Inst. Geol. Geophy., Nobosibirsk, pp. 24–30 (in Russian).

GAUDIN, A. M. and DICKE, G. (1939): The pyrosynthesis, microscope study and irridescent filming of sulfide compounds of copper with arsenic, antimony and bismuth, *Econ. Geol.*, **34**, 214–230.

GODOVIKOV, A. A. and PTITSIN, A. B. (1969): Synthesis of copper-bismuth sulfosalts in hydrothermal conditions, in *Exp. Invest. in Mineralogy 1968–1969*, Acad. Sci. USSR., Siber. Div., Inst. Geol. Geophys., Novosibirsk, pp. 26–41 (in Russian).

GODOVIKOV, A. A., FEDOROVA, ZH. N., PAVLJUCHENKO, V. S. and PTITSIN, A. B. (1971): New synthetic bismuth sulfosalt of copper–$Cu^{+1}Cu_2^{+2}Bi_3S_7$, in *Exp. Invest. in Mineralogy, 1969–1970*, Acad. Sci. USSR., Siber. Div., Inst. Geol. Geophy., Novosibirsk, pp. 3–9 (in Russian).

HILLEBRAND, W. F. (1884): Löllingite and other minerals, *Am. J. Sci.*, **27**, 355.

IKORNIKOVA, N. YU. (1962): The role of chlorides in the hydrothermal transportation of metals during the mineralization process (abstr.), *Econ. Geol.*, **58**, 1011.

KODERA, M., KUPCIK, V. and MAKOVICKY, E. (1970): Hodrushite—a new sulfosalt, *Mineral. Mag.*, **37**, 641–648.

KRIEGER, P. (1940): Bornite-klaprotholite relations at Concepcion del Oro, Mexico, *Econ. Geol.*, **35**, 687–697.

KULLERUD, G. (1953): The FeS–ZnS system, a geological thermometer, *Norsk Geol. Tiddsskr.*, **32**, 61–147.

KULLERUD, G. (1971): Experimental techniques in dry sulfide research, in *Research Techniques for High Pressure and High Temperature*, Ed. G. C. Ulmer, Springer, New York, pp. 288–315.

KULLERUD, G., YUND, R. A. and MOH, G. H. (1969): Phase relations in the Cu–Fe–S, Cu–Ni–S and Fe–Ni–S system, *Econ. Geol., Monogr. Ser.*, **4**, (magmatic ore deposits), 323–343.

KUZ'MINA, I. P. (1961): Experimental study of PbS and ZnS formation from aqueous solution of chlorides (abstr.), *Econ. Geol.*, **57**, 990–991.

LAUDISE, R. A., KOLB, E. D. and DENEUFVILLE, J. P. (1965): Hydrothermal solubility and growth of sphalerite, *Am. Mineral.*, **50**, 382–391.

MERWIN, H. E. and LOMBARD, R. H. (1937): The system Cu–Fe–S, *Econ. Geol.*, **32**, 203–284.

MUKAIYAMA, H. and IZAWA, E. (1970): Phase relations in the Cu–Fe–S system, the copper-deficient part, in *Volcanism and Ore Genesis*, Ed. T. Tatsumi, Univ. Tokyo Press, Tokyo, pp. 339–355.

NANRI, H., SUGAKI, A., SHIMA, H. and KITAKAZE, A. (1978): Phase relations of Cu–Fe–Bi–S system (V)—bornite solid solution field at 400°C and 500°C (abstr.), *J. Japan. Assoc. Mineral. Petrol, Econ. Geol.*, **73**, 83 (in Japanese).

NUFFIELD, E. W. (1947): Studies of mineral sulphosalts: XI—wittichenite (klaprothite), *Econ. Geol.*, **42**, 147–160.

NUFFIELD, E. W. (1952): Studies of mineral sulphosalts: XVI—cuprobismutite, *Am. Mineral.*, **37**, 447–452.

PALACHE, C. (1940): Cuprobismutite—a mixture, *Am. Mineral.*, **25**, 611–613.

PTITSIN, A. B. and GODOVIKOV, A. A. (1971): Hydrothermal synthesis of bismuth sulfosalts of copper in acid solutions and oxidative medium, in *Exp. Invest. in Mineralogy 1969–1970*, Acad. Sci. USSR., Siber. Div., Inst. Geol. Geophys., Novosibirsk, pp. 52–60 (in Russian).

ROSEBOOM, E. H. and KULLERUD, G. (1958): The solidus in the system Cu–Fe–S between 400°C and 800°C, *Carnegie Inst. Washington Year Book*, **57**, 222–227.

SCHLEGEL, H. and SCHÜLLER, A. (1952): Die Schmelz- und Kristallisations-gleischgewichte im System Cu–Fe–S und ihre Bedeutung für Kupfergewinnug, *Freiberger Forschung, sec. B*, 1–32.

SCHNEIDERHÖHN, H. and RAMDOHR, P. (1931): *Lehrbuch der Erzmikroskopie* Bd. 2, Berlin, pp. 378–379.

SCOTT, S. D. (1974): Experimental methods in sulfide synthesis, in *Sulfide Mineralogy, Min. Soc. Am. Short Course Notes*, **1**, S1–S38.

SCOTT, S. D. (1975): Hydrothermal synthesis of refractory sulfide minerals, *Fortschr. Mineral.*, **52**, 185–195.

SCOTT, S. D. and BARNES, H. L. (1971): Sphalerite geothermometry and geobarometry, *Econ. Geol.*, **66**, 653–669.

SHIMA, H., SUGAKI, A. and KITAKAZE, A. (1977): Phase equilibrium study in the Fe–Bi–S system—synthetic phase $FeBi_4S_7$, *Abstr. Ann. Meet. Mineral. Soc. Japan*, p. 128 (in Japanese).

SHORT, M. N. (1931): Microscopic determination of the ore minerals, *U.S. Geol. Survery Bull.*, **914**, U.S. Geol Survery, Washington.

SUGAKI, A. and KITAKAZE, A. (1972): Chemical composition of synthetic alabandite solid solution and its phase relations in the system Mn–Fe–S, in *Proc. 6'th Intern. Conf. X-ray Optics and microanalysis, Osaka*, pp. 755–761.

SUGAKI, A. and SHIMA, H. (1965a): Synthetic sulfide minerals (I), *Mem. Fac. Eng. Yamaguchi Univ.*, **15**, 15–31.

SUGAKI, A. and SHIMA, H. (1965b): Synthetic sulfide minerals (II), *Mem. Fac. Eng. Yamaguchi Univ.*, **15**, 33–47.

SUGAKI, A. and SHIMA, H. (1971): The phase equilibrium study of the Cu–Bi–S system, in *Proc. IMA–IAGOD Meeting '70, IMA Vol. Spec. Pap. No. 1, Tokyo.*, pp. 270–271.

SUGAKI, A. and SHIMA, H. (1972): Phase relations of the Cu_2S–Bi_2S_3 system, *Tech. Rept. Yamaguchi Univ.*, **1**, 45–70.

SUGAKI, A., SHIMA, H. and NOMIYAMA, S. (1970): Phase equilibrium study in the Fe–Bi–S system (abstr.), *J. Mineral. Soc. Japan.*, **10**, 110–111 (in Japanese).

SUGAKI, A., SHIMA, H. and KITAKAZE, A. (1972a): Synthetic sulfide minerals (IV), *Tech. Rept. Yamaguchi Univ.*, **1**, 71–85.

SUGAKI, A., KITAKAZE, A. and YAMAMOTO, T. (1972b): Phase equilibrium study of the Cu–Fe–Bi–S system (II)—wittichenite-chalcopyrite (abstr.), *J. Japan. Assoc. Mineral. Petrol. Econ. Geol.*, **65**, 105 (in Japanese).

SUGAKI, A., SHIMA, H. and KITAKAZE, A. (1973): Chemical composition and phase relations of naturally occurring Cu–Fe sulfides (abstr.), *Mining Geol. Japan.*, **23**, 61 (in Japanese).

SUGAKI, A., SHIMA, H. and KITAKAZE, A. (1974): Fundamental study on the quantitative analysis of sulfide minerals by electron probe microanalyser (V), *Mem. Fac. Eng. Yamaguchi Univ.*, **24**, 225–230 (in Japanese).

SUGAKI, A., SHIMA, H., KIKAKAZE, A. and HARADA, H. (1975): Isothermal phase relations in the system Cu–Fe–S under hydrothermal conditions at 350°C and 300°C, *Econ. Geol.*, **70**, 806–823.

SUGAKI, A., SHIMA, H. and KITAKAZE, A. (1976a): Hydrothermal synthesis of sulfide minerals with complex compositions, *Sci. Rept. Tohoku Univ., Ser. 3*, **13**, 115–129.

SUGAKI, A., SHIMA, H. and KITAKAZE, A. (1976b): Application of Bence and Albee method to analyses of sulfide minerals; Cu–Fe–Pb–Bi–S system, *J. Mineral. Soc. Japan*, **12**, Spec. Issue, 85–92 (in Japanese).

SUGAKI, A., SHIMA, H., KITAKAZE, A. and FUKUOKA, M. (1977): Hydrothermal synthesis of pyrrhotite and their phase relations at low temperature, *Sci. Rept. Tohoku Univ., Ser. 3*, **13**, 165–182.

SUGAKI, A., SHIMA, H. and KITAKAZE, A. (1980): The phase equilibrium of the system Cu–Bi–S below 400°C, especially the relation between emplectite and cuprobismutite, in *Proc. 9'th IMA Meeting 1978*, sulphosalts, plutinum minerals and ore microscopy volume, pp. 100–109.

SUGAKI, A., SHIMA. H., KITAKAZE, A. and MIZOTA, T. (1981a): Hydrothermal synthesis of nukundamite and its crystal structure, *Am. Mineral.*, **66**, 398–402.

SUGAKI, A., KITAKAZE, A. and HAYASHI, K. (1981b): Synthesis of minerals in the Cu–Fe–Bi–S system under hydrothermal condition and their phase relation, *Bull. Mineral.* **104**, 484–495.

SUGAKI, A., KITAKAZE, A. and UENO, T. (1982): Hydrothermal synthesis of minerals in the system Cu–Fe–S and their phase equilibrium at 400°C and 500°C, *J. Japan. Assoc. Mineral. Petrol. Econ. Geol., Special Issue*, **3**, 257–269 (in Japanese).

TAYLOR, C. M., RADTKE, A. S. and CHRIST, C. L. (1973): New data on cuprobismutite, *J. Res. U.S. Geol. Survey*, **1**, 99–103.

YUND, R. A. (1963): Crystal data for synthetic $Cu_{5.5x}Fe_xS_{6.5x}$ (idaite), *Am. Mineral.*, **48**, 672–676.

YUND, R. A. and KULLERUD, G. (1966): Thermal stability of assemblages in the Cu–Fe–S system, *J. Petrol.*, **7**, 454–488.

WATANABE, M. (1937): Geology and ore deposits of the Obari and Hongo mines, Yamagata Prefecture, Japan; especially mineral paragenesis of copper and copper-bismuth sulfides, *J. Japan. Assoc. Mineral. Petrol. Econ. Geol.*, **18**, 211–269; **19**, 27–32, 70–88 (in Japanese).

WATANABE, M. (1951): Lattice and graphic intergrowth of some copper minerals from the Obari mine, Japan, *Sci. Rept. Tohoku Univ., Ser. 3*, **4**, 1–10.

Chapter 7: Technical Developments

Materials Science of the Earth's Interior, edited by I. Sunagawa, pp. 587–603.

RECENT PROGRESS IN THE GENERATION OF STATIC HIGH PRESSURE BY MEANS OF THE MULTIPLE ANVIL

Mineo KUMAZAWA[1] and Shoichi ENDO[2]

[1]*Department of Earth Sciences, Faculty of Science, Nagoya University, Chikusa, Nagoya, Japan*
[2]*High Pressure Research Laboratory, Faculty of Engineering Science, Osaka University, Toyonaka, Japan*

1. *Introduction*

The generation of a higher pressure in a larger volume is a continuing requirement, not only in earth science, but also in many other disciplines. A characteristic feature in Japan in the past few decades is an extensive development of multiple anvil-type apparatus which is driven by means of a large press. Despite recent progress in the development of the handy and versatile diamond-anvil cell, the multiple anvil principle, which enables us to compress a larger volume, possesses an indispensable utility which is complementary to that of the diamond cell. In this paper, we review the present state of the multiple anvil technique in Japan and its related subjects. There are five important Ms that play the essential roles in the high pressure experimentation: Man, Mechanism, Mechanics, Material, and Measurement. The last four are discussed in the first four sections, and a new high temperature technique useful for geologic application, is presented in the last section.

An approach to the multiple anvil has been pioneered in Japan by the late Professor Naoto Kawai, and developmental work on multiple anvil systems has been mostly promoted by his distinguished contributions. Unfortunately, however, he died on 3 July 1979. We have lost the most important M as a leading scholar in this discipline. We would like to express our gratitude to Professor Kawai for his leading and pioneering contributions during his academic life, and we deeply mourn his untimely death.

2. *Mechanism*

Distinctive features of multiple anvil-type apparatus are (1) the compression of a large volume, (2) a geometry that is flexible enough to be modified for a cascading compression to higher pressures, and (3) the favorable configuration of anvils for reducing the stress deviator by lateral and massive supports.

Depending on the difference in the mechanisms of volume reduction of a pressure cell, multiple anvils are classified into two groups: the ordinary multiple anvil with compressible gaskets (abbreviated to MA) and the multiple anvil sliding system without compressible gaskets (abbreviated to MASS). In addition, there are a large number of possible mechanisms that are supposed to be used in pressure generation. For practical uses, however, we have some geometrical constraints, originated from the stable and convenient driving system, which provide the synchronized displacement of anvils with a necessary force. The multiple anvil devices currently in common use, have simple outside configuration in adapting to any convenient driving system.

2.1 Driving system

The early multiple anvil apparatus (e.g., HALL, 1958) employs several hydraulic rams, linked by a steel framework, to drive many anvils. However, the adjustment of alignment and the synchronized advancement of each ram were difficult for stable operation. Nowadays, this problem has been removed by confining the multiple anvil device in an appropriate low pressure cell, which is usually a couple of buckets similar to the NBS type (LLOYD *et al.*, 1959), set in a uniaxial press (Fig. 1). The driving system of this type has been extensively improved and refined in Japan for accurate alignment, excellent

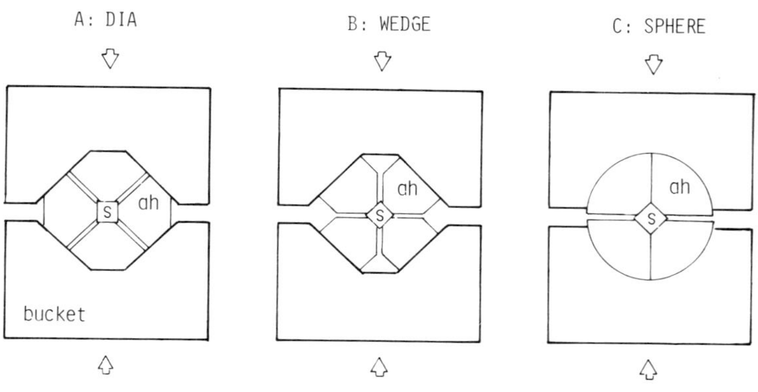

FIG. 1. A schematic illustration of driving systems of the multiple anvil by means of a couple of buckets in a uniaxial press. A: DIA apparatus, B: WEDGE apparatus, C: split sphere apparatus. ah: anvil head, s: space to be compressed. Anvil heads slide on the inside surface of the buckets and their alignment is kept accurately in A and B, while they are not in C. The stress concentration in the buckets is smaller for C, and the smaller buckets are used to sustain the same force than for A and B.

reproducibility, and easy and stable operation. The cubic anvil apparatuses developed along these lines are DIA apparatus (OSUGI *et al.*, 1964) and WEDGE apparatus (WAKATSUKI and ICHINOSE, 1982). These apparatuses with six anvil heads of cemented tungsten carbide WC, are designed for a single-stage compression. Pressures of up to 10 GPa are attained very easily on a routine basis by means of a uniaxial press of several hundred tons allowance.

2.2 *Cascading compression*

In order to reach high pressure without cutting down the volume of a cell, a cascading compression is employed. A smaller system is compressed in another larger multiple anvil system. Both the DIA and WEDGE apparatus mentioned above work as a very useful and appropriate compressor of the first stage in a cascading system for the second stage compressor with a cube in external configuration. Other systems are also developed along these lines of approach to allow us to compress the second-stage anvil assemblies of a trigonal bipyramid (KUMAZAWA *et al.*, 1972), of a sphere and of a tetragonal bipyramid (KAWAI *et al.*, 1973). In the driving system for a cascading compression, the anvil heads of the first stage, usually made of hardened steel, are very large (40 to 220 mm in edge length) and a compression of up to 1 to 2 GPa is made by a very large press (10^3-10^4 tons or more). The multiple anvil for the second or the higher-stage should have a simple geometry in the external configuration in order to fit the inside configuration of the lower-stage compressor, as is exemplified in Fig. 2 (OGAWA *et al.*, 1982).

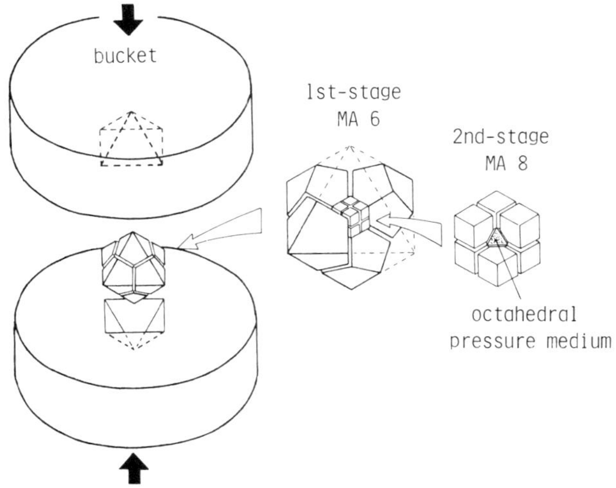

FIG. 2. An example of multiple stage apparatus. A pressure medium of an octahedron is compressed in an MA8 (cube) anvil assembly, which is compressed in an MA6 (tetragonal bipyramid), which is further driven in a couple of buckets set in a uniaxial press. See also Figs. 1 and 3.

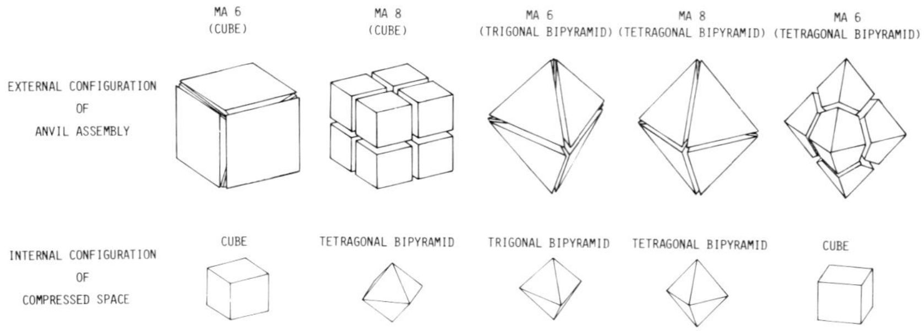

FIG. 3. Multiple anvils of different geometry. External and internal configurations are shown together with the basic arrangement of the anvils.

2.3 *Multiple anvils (MA) with compressible gaskets*

Some of the useful geometries of MA are shown in Fig. 3. A cube, hexahedron (trigonal bipyramid), or octahedron (tetragonal bipyramid) as an external configuration of the anvil assembly, is cut into pieces. The inside corner of each piece is truncated to produce a space for a pressure cell or for another pressure-generating system of the higher stage.

A basic principle of MA is schematically shown in Fig. 4a. The volume surrounded by a group of anvils is reduced with the thickness reduction of gaskets between the anvils. The gaskets possess three roles in pressure generation: (1) the sealing of the material in the pressure cell, (2) giving the inward displacement of the anvils by thickness reduction through volume reduction and extrusion, and (3) keeping the appropriate pressure gradient at the flanks of anvils to prevent their failure. The appropriate choice and setting of the gaskets are the most important specifications for the desired pressure generation in this type of multiple anvil apparatus. The most serious problem in the current gasket system is that the gaskets are too soft to keep the desired setting at the initial stage of compression, while they are too hard to flow or to contract in giving rise to the effective thickness reduction for further compression, particularly in the pressure range above 10 GPa. These problems are resolved by the careful designing of composite gaskets to satisfy the necessary mechanical characteristics (YONEDA *et al.*, 1982).

When the volume of a pressure cell is allowed to be small ($\sim 1\ mm^3$), the pressure gradient in the anvils and the necessary advancement of the anvils are provided by the elastic deformation of the anvils themselves, using the elastic design proposed by KUMAZAWA (1973). ITO and MATSUI (1977) and TOGAYA *et al.* (Personal Communication, 1979) were successful with their elastic design and were able to reach pressures far above 20 GPa.

The most important MA commonly used is a so called 6–8 type device (KAWAI and ENDO, 1970); a two stage cascading system of an octahedral anvil

(Fig. 2) compressed in a cubic anvil. The significant advantage of this system is the use of cube-shaped anvils, which are very simple in geometry, for fabrication and are suitable for lateral support. With this device, a number of important contributions have been achieved in Japan; olivine-spinel transformation in natural forsterite (ITO *et al.*, 1971), synthesis and phase relations of post spinel phases in silicate (ITO and MATSUI, 1977, 1978), synthesis of orthorhombic SnO_2 (SUITO *et al.*, 1975), synthesis of large, single crystal ferromagnesian spinels (HORIUCHI and SAWAMOTO, 1981), melting of fayalite up to 20 GPa (OHTANI, 1979), phase relation in ice (MISHIMA and ENDO, 1980), detection of metallic transitions in a number of insulators (KAWAI *et al.*, 1975a, b; ONODERA *et al.*, 1974), and many other works, which cannot be cited here.

2.4 *Multiple anvil sliding system (MASS)*

The geometry of MASS and its operational mechanisms are shown by schematic illustrations in two dimensions in Fig. 4(b) and (c). The actual devices in three dimensions are the complicated combination of the two mechanisms; b and c in Fig. 4 so as to be geometrically compatible (KUMAZAWA, 1971; KUMAZAWA *et al.*, 1972). When an anvil assembly is compressed, the volume of the cell decreases to vanish in association with the thickness reduction of the compressible pads and the relative sliding between anvils. There is neither a gap nor a compressible gasket which can be exerted to very high pressures between the neighboring anvils. Instead, compressible pads are set at the base of several anvils. The compression of the pads gives rise to the synchronized displacement of anvils, and the compressibility of the pads controls the compression characteristics of the whole system. When the pads are very soft, the large fraction of the applied force is used for the direct compression of the material in the cell, while the force is used mainly for the lateral support of anvils on their flanks when the pads are incompressible. In

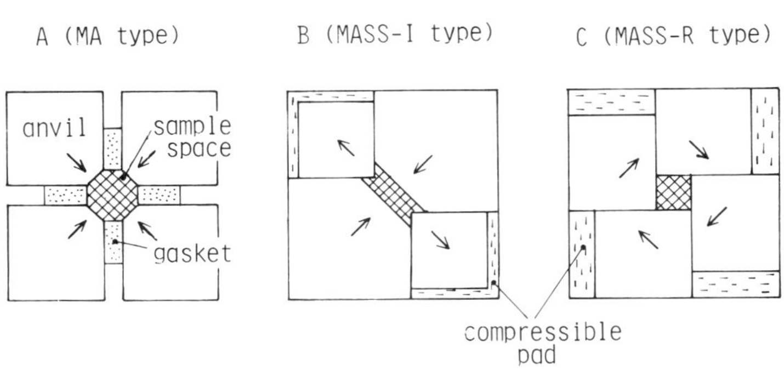

FIG. 4. A schemmatic illustration showing the mechanism of pressure generation in an ordinary multiple anvil with gaskets (A) and in two basic types of multiple anvil sliding systems (B and C). Arrow shows the displacement of anvils.

order to attain the appropriate distribution of the applied force to the pressure generation in the cell and to the lateral support, we have to have an optimum compressibility of the pads for a state of "compressibility matching" between the pressure medium and the compressible pads (KUMAZAWA *et al.*, 1972).

There are many different types of MASS mechanisms which have so far been tested (MASAKI *et al.*, 1975), and the types called 3I8 (MASAKI *et al.*, 1975, SAWAMOTO, 1977) and OB8 (YAMAMOTO *et al.* 1982; SAWAMOTO *et al.*, 1982) appear to be the most useful in practical uses. In a 3I8, the outside configuration of the anvil assembly is a cube and is driven by means of a cubic anvil. In an OB8, on the other hand, it is a trigonal bipyramid and is driven by a couple of PY3 type guide blocks (KUMAZAWA *et al.*, 1972). In the both apparatuses, pressures far above the GaP point (22 GPa) are easily achieved, though the mechanical behavior of the compressible pads and friction between the anvils has been difficult to control accurately, and they have not been ideal yet for general use on a routine basis. Recent technical studies on the OB8 equipped with an X-ray diffraction system for in-situ pressure measurements (SAWAMOTO *et al.*, 1982; and YAMAMOTO *et al.*, 1982) have made it more stable.

3. *Mechanics*

The effective generation of high pressure by using a material of finite strength requires consideration of the mechanics of a pressure-generating system. The basic principle allowing one to generate pressures higher than the strength of the material for anvils is a lateral support, by which the stress deviator in the anvils is reduced to below the material strength by laterally applying an additional stress to the direction of the maximum compression (RUOFF, 1972; KUMAZAWA, 1973).

For simplicity of explanation, the high pressure-generating device is regarded as a spherically-symmetric system with radius R_0 set in an zero pressure environment. Pressure P or the radial component of stress, and the lateral pressure P_L are defined as functions of position r in the spherical system. The force balance equation in a spherical coordinates is expressed by

$$\frac{\mathrm{d}P(r)}{\mathrm{d}r} = -2\frac{[P(r) - P_L(r)]}{r}. \tag{1}$$

The distribution of $P(r)$ and $P_L(r)$ are solved with a boundary condition $P(r) = 0$ at $r = R_0$, provided that an additional constraint on P_L is given. $P(r) - P_L(r)$ is the stress difference in the system, and can never exceed the material strength, or

$$P(r) - P_L(r) \leqq \tau \tag{2}$$

where τ, defined as the maximum stress difference sustainable by the anvils, is given by a function of r for an actual system with a specific design. The aim of the mechanics is to find a condition to optimize the distribution of $P(r)$ and $P_L(r)$ in the system by controlling the operational condition within the limitations of available material and of possible geometry. When τ is regarded as a representative constant specifying the material strength, and an idealized system with an equality in (2) is considered, we have from (1) and (2)

$$P(r) = 2\tau \ln(R_0/r)$$

and

$$P_L(r) = 2\tau \ln(R_0/r) - \tau.$$

The total force needed to drive this system is given by

$$F = 4\pi r^2 P(r)]_{R_0/r=\sqrt{e}}.$$

Then we have a couple of constraints on the generated pressure P in a volume $v(=4\pi r^3/3)$ by means of apparatus of size $V_0(=4\pi R_0^3/3)$ in volume, material strength τ, and force F (press allowance);

$$v = V_0 \exp(-3P/2\tau) \tag{3a}$$

$$F = (36\pi)^{1/3} e^{-1} \tau V_0^{2/3}. \tag{3b}$$

Equation (3a) shows how the material strength is important in generating high pressure in a given volume. The necessity of a very large system and a large force for the compression of a large volume is also indicated, particularly when the material strength is comparable or less than the pressure to be generated. As a consequence of this consideration, the construction of a very large press ($\sim 10^4$ ton allowance) has been carried out at Osaka and Nagoya Universities. A large force is mostly devoted to the lateral support of anvils, and also to compress a large volume. The use of a large force by a large press possesses an additional utility in the technical developments of high pressure generation. This is the syntheses of the cemented or sintered diamond or some other hard materials for anvils, as will be exemplified in the next section.

The derivation of (3a, b) is too crude to be applied to the actual system, since it is intended to show only the basic idea of the mechanics of a pressure-generating system in general. On the basis of more realistic situations and assumptions, the relation of the generated pressure to the compressed volume for given material strength and the tonnage of a uniaxial press is evaluated, and shown in Fig. 5. The performance of the actual experiments to date is also shown in the same figure. The pressure-volume attained so far would be further improved through technical refinements. The volume in Fig. 5 is the total volume including the pressure-transmitting medium, and the actual sample volume is much smaller, particularly for high temperature work. If, for example, we are to synthesize a silicate perovskite sample of 1 mm in size, we have to reserve a volume of 10 mm across for the pressure cell. Then a press tonnage of $\sim$10,000 tons is necessary, if the cemented WC with a strength of 5 GPa is used for the anvils. The necessary force is, however, reduced to

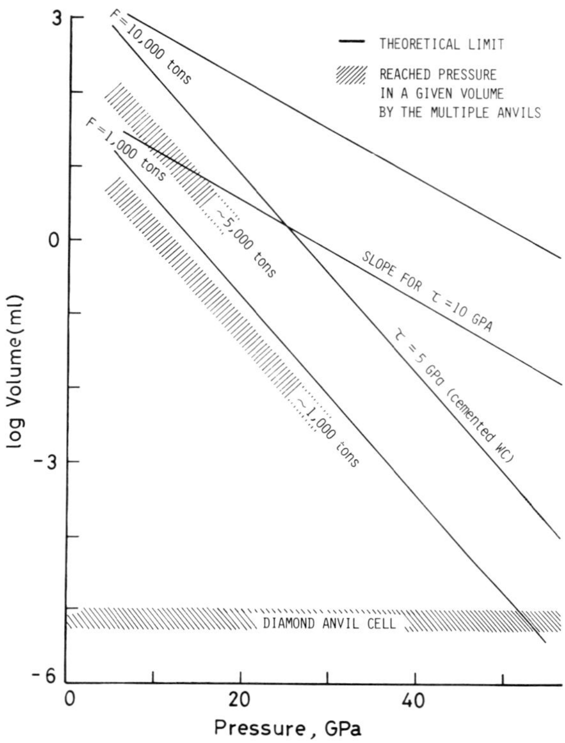

FIG. 5. Relation between the pressure generated and the volume compressed. The current performance and theoretical limit are shown.

~1,000 tons when the material strength is increased to 10 GPa. Suppose that a force of 10,000 tons is used to compress a volume of 1 mm^3 by means of anvils with a strength of 5 GPa. The attainable pressure is ~50 GPa, while it could be extended to ~80 GPa with ~1,000 tons provided that the material strength is doubled. Such an argument clearly shows the importance of developing tough material for anvils.

4. *Material-Synthesis and Application for Anvils*

The synthesis of sintered diamond as one of the most promising hard materials is made by means of a large press and the obtained polycrystalline diamond is used to generate high pressures. Such an approach was first made by BUNDY (1975).

Cemented WC, having been used usually as an anvil material, has the maximum hardness of about 93 in H_RC. Pressures of ~30 GPa have been attained by using this material for multiple anvil apparatus, in which the neighboring anvils. The generation of pressures higher than 30 GPa is, however, difficult in practice because of the plastic deformation of the anvil.

Some of the ceramic materials are harder than cemented WC. The anvils made of sintered alumina Al_2O_3 in a cube were tested as second-stage anvils of the 6–8 type multiple anvil apparatus. They were easily broken into pieces and behaved as if they were pressure-transmitting medium. Toughness, in addition to hardness, is indispensable for anvil material.

A diamond cell with a couple of single-crystalline diamonds has been recorded with a maximum pressure of 172 GPa (Mao and Bell, 1978). Its transparency for visible and invisible light makes various physical measurements possible. However, the dimensions of available single-crystal diamonds are limited. Recently polycrystalline diamond aggregate is in the limelight because of its industrial uses as cutting-tools and drawing dies. The material available on the market is not adequate for anvils in size, quality, and shape. The polycrystalline aggregates of 9 mm are synthesized from diamond particles with cobalt as a binder at about 6.5 GPa and 1,500°C in multiple anvil apparatus driven by means of a 15 kton press (Mishima and Endo, 1981). The sufficiently isotropic and homogeneous aggregates obtained, are partly cemented by cobalt and are sintered to a considerable degree, as inferred from their hardness, so that they are not so inferior to a single crystal diamond. In the case of sintered diamond aggregate, pressures of 100 GPa or higher can be generated by the strong lateral support, when we consider (1) the significant increase in the maximum pressure attainable with cemented WC anvils by the lateral support to the anvil flanks, and (2) the extraordinary hardness of diamond compared with WC. Item (1) is not applied to a single-crystal diamond. It has an intrinsic cleavage, along which cracks easily develop unless appropriate lateral support is provided. Polycrystalline diamond does not have such a structural weakness.
strong lateral support is given to the flanks of anvils by the gasket between the

The most useful application of sintered diamond appears to be the replacement of cemented WC for second-stage cube anvils of the 6–8 type apparatus. As a first step, a sintered diamond was tried as the last-stage anvil of the triple-stage apparatus. The configuration of the cell is shown in Fig. 6. A couple of anvils made of sintered diamond were set in an opposed way in an octahedral pyrophyllite medium of the 6–8 type multiple anvil apparatus. When the pyrophyllite medium was compressed, the sintered diamond anvils were kept in a high-pressure environment and were supported laterally. The compressive strength increased appreciably and the pointed end of each anvil withstood ultrahigh pressure. This triple-stage system is denoted as "6–8–2" type (Endo and Ito, 1982).

A comparison of the pressure generation in the 6–8–2 system with the third stage anvils of sintered diamond and of cemented WC is shown in Fig. 7. A metallic transition in GaP is realized with a substantially small force by the

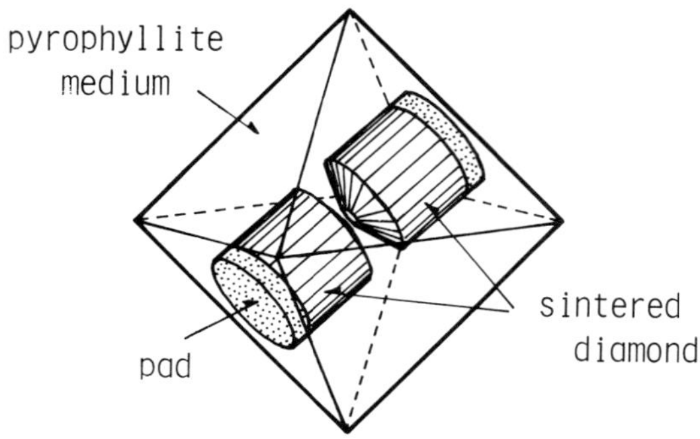

FIG. 6. Arrangement of a couple of sintered diamond anvils in an octahedral pressure medium in a 6–8 type apparatus.

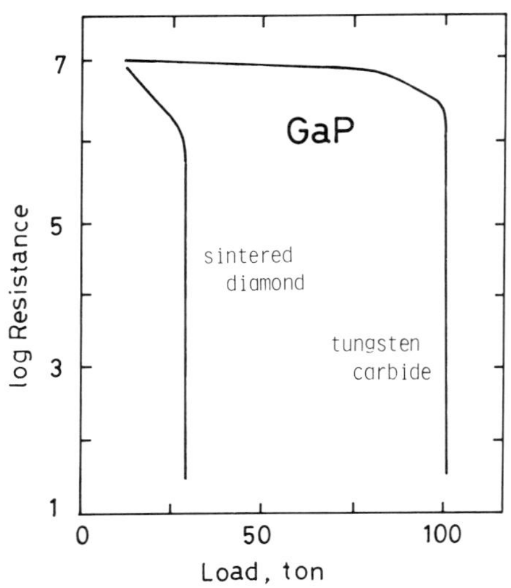

FIG. 7. Comparison of pressure-generating efficiencies in the 6–8–2 type apparatus with tungsten carbide and sintered diamond anvils. The press load neccessary to reach a GaP point is significantly lowered by use of sintered diamond anvils.

use of sintered diamond in comparison with the use of cemented WC.

The sample space of the present 6–8–2 system is essentially of two dimensions and is small in volume. Therefore, it does not always allow us to make various kinds of physical measurements. However, almost all the experiments made so far with ordinary opposed anvils up to a few 10 GPa, can be made at ultrahigh pressures of the order of 100 GPa with further technical improvements. Two examples having been put into practice are presented for demonstration.

Electric resistance measurement: A central part of the cell for the resistance measurement of insulators and semiconductors in the 6–8–2 system is shown in Fig. 8. The result on synthesized high-purity hematite α-Fe_2O_3 is shown in Fig. 9 (ENDO and ITO, 1980). The pressure value of 50 GPa, shown at the top of the figure, is determined on the basis of the α-ε transition in an Fe-V alloy (DUNN and BUNDY, 1979), detected electrically at the same time in the cell shown in Fig. 8. The present data suggest that α-Fe_2O_3 possesses two phase transitions accompanied by a resistivity decrease.

Mineral synthesis: The 6–8–2 system with a couple of sintered diamond anvils for the last stage, enables us to make mineral synthesis at ultrahigh pressures (AKAHAMA *et al.*, Personal Communication, 1981). A sample filled in the hole of a pyropyllite disk between the sintered diamond anvils is electrically heated with tantalum foil heaters placed at the upper and lower sides of the disk. Temperature is determined by a thermocouple inserted between the foils. The sample, exposed to high pressure and temperature, is recovered and examined by X-ray diffraction. In the test run, the synthesis of stishovite from silica gel is confirmed, suggesting the possibility of this type of apparatus for mineral synthesis and related subjects at high temperatures.

5. *Measurement*

The determination of pressure in multiple anvil apparatus is either made by detecting the pressure-induced phase transition as a fixed point or by an *in-situ* X-ray diffraction of standard materials, such as NaCl. Though a continuous determination of pressure by X-ray diffraction is advantageous, it

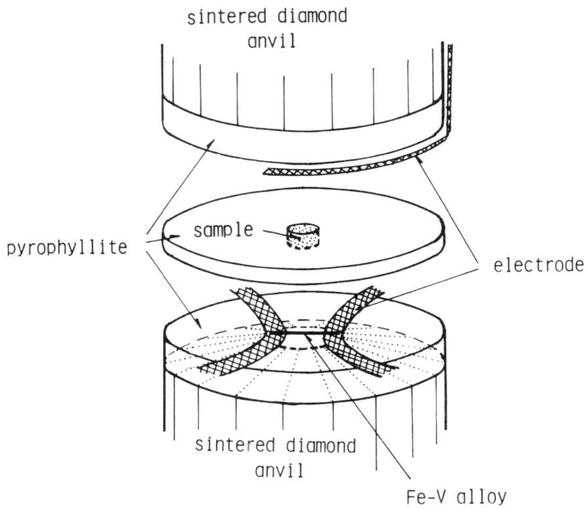

FIG. 8. Cell assembly in a 6–8–2 type apparatus for resistance measurement of the sample and of the pressure standard material.

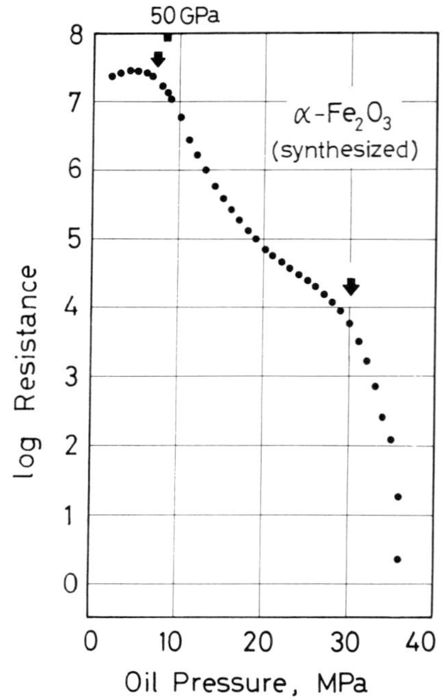

FIG. 9. The variation of electric resistance in synthetic α-Fe_2O_3 compressed in a 6–8–2 type apparatus. The pressure value indicated at the top of the figure is given by the phase transition in Fe-V alloy.

usually takes a long time to acquire the necessary data. In addition, it is not always applied to apparatus of multiple-stage compression. Therefore, both the detection of the phase transition in the calibrating material and the X-ray diffraction are complementary in the determination of the pressure generated in the multiple anvil. Recent progress in this subject is presented below.

5.1 X-ray diffraction

The X-ray diffraction technique of compressed material in a high-pressure cell possesses two important utilities: the determination of pressure by means of an NaCl scale and the study of the sample at high pressures. A small version of DIA apparatus specially designed for X-ray work (DIA-6, which is commercially available) enables us to make measurements of up to 1,000°C and 10 GPa on a routine basis (INOUE and ASADA, 1973). When the cell volume is reduced, measurements up to 20 GPa at room temperature are made very stably (YAGI and AKIMOTO, 1976).

Two new systems with a 360° aperture for incident and diffracted beams of X-ray have been developed. They are an octahedral anvil with a tetragonal bipyramid in the external configuration, driven in an octahedral anvil

(OHTANI *et al*, 1979), and MASS-OB8 driven in a couple of buckets of trigonal bipyramid at the inside configuration (SAWAMOTO *et al*., 1982; YAMAMOTO *et al*., 1982). These apparatuses enable us easily to generate and to determine the pressures above the GaP point.

5.2 Detection of the phase transition by a resistance wire probe

The detection of the phase transition in pressure calibrants such as Bi, Ba, and ZnS is usually carried out by means of a resistance measurement on the material embedded in the pressure cell. However, the change of electric resistivity associated with the transition is not always sharp or large enough to be detected clearly. For example, an expected B1-B2 transition in NaCl (30 GPa) would be difficult to detect by this method, because the value of the resistivity is considerably high and its change would not be very large. Therefore, a new and simple technique of detecting the volume change of phase transition has been developed by means of a resistance wire probe (YONEDA and ENDO, 1980).

The cell used for this method is shown in Fig. 10(a). A manganin wire is wound around a cylindrical sample embedded in a pressure cell. When the

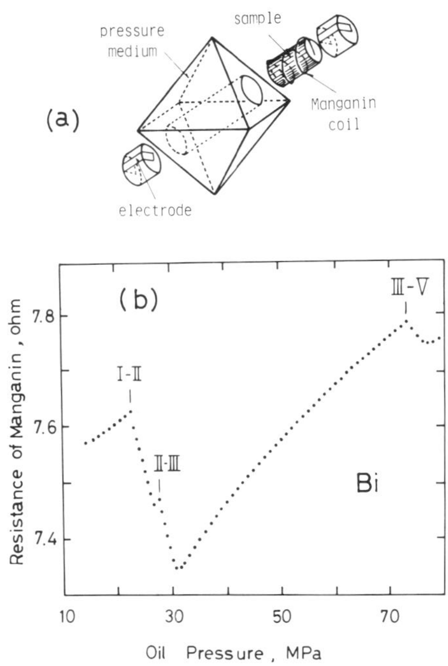

FIG. 10. A resistance wire probe for the detection of volume change associated with phase transition. (a) a configuration of a cell assembly in an octahedral pressure medium, and (b) an experimental result showing the three-phase transitions in Bi. The step heights are well correlated to the volume changes in the transitions.

sample of volume V shrinks at transition, the coil deforms and its resistance R decreases, mainly due to a decrease in the total length of the wire and the increase of its cross-sectional area. The volume change of the sample ΔV and the resistance change of the detecting coil ΔR are related by the following expression

$$\Delta R/R = K \cdot \Delta V/V. \qquad (K \fallingdotseq 2/3)$$

As an example, the result obtained for bismuth metal is shown in Fig. 10(b). Once a calibration is made for the value of K, the volume change associated with phase transition is determined accurately. It is noted that the present technique is sensitive enough to show the presence of a phase transition in barium metal of around 8 GPa, which has been questioned for a long time.

The fixed point at high pressures above 22 GPa is rare, and this situation has made pressure evaluation in the multiple anvil difficult. We have, however, a possibility of finding out useful fixed points in dielectrics and in insulators, which may have some phase transitions without a significant change in the electric resistivity but with a volume change detectable by the present method.

6. *Very High Temperature Technique*

Another important specification of a high-pressure device useful for geological applications, is the stable attainment and accurate measurement of very high temperatures at high pressures. Such a technique enables us to experimentally study the melting and chemical fractionation processes in silicate systems in the deep interior of the earth.

The high temperature technique customarily employs a carbon resistor and a Pt/PtRh thermocouple. However, when the temperature reaches ~2,000°C above 10 GPa, the carbon heater does not work because of a quick conversion into diamond and the thermocouple suffers from deterioration because of melting. An alternative heating element may be refractory metals. However, it also suffers from damage due to local melting, since the thickness of the metal heater (usually a tube) required for the appropriate resistivity, is very thin (~0.03 mm or less) and a large electric power should be fed into such a small heating element.

Recent progress in the high-temperature technique is the attainment and measurement of temperatures up to 2,500°C at 20 GPa by means of a semiconducting heater and a W/WRe thermocouple (Ohtani, 1979, and Ohtani *et al.*, 1981). The best material for a heater is found to be semi-sintered lanthanum chromite $LaCrO_3$, which is a semiconductor with a cubic perovskite structure and does not possess any phase changes, at least up to 20 GPa and 2,500°C. It is machinable to shape into a desired configuration by laboratory lace, saw and file, and it has an appropriate electric property for use in an ordinary multiple-anvil device. The material provided by the Kyoto

Ceramic Co. possesses a specific resistivity of ~0.1 ohm · cm at 1,000°C and an activation energy of 0.12 eV below 2,200°C and 0.5 eV above this temperature. Lanthanum chromite is reactive with oxides and silicates so that an appropriate sealing for the sample should be provided. Chemically carbon is an excellent sealing material for ordinary use, though it is not so physically, particularly when an Fe-containing sample (e.g., Fe_2SiO_4) is contained at a very high temperature and pressure. It turns into a sintered diamond and recovery of the sample after a run in a diamond container is difficult.

The most appropriate thermocouple for high-temperature use is W/WRe. The emf of W/W25%Re and W/W26%Re thermocouples up to 2,205°C in hydrogen have been reported (LACMAN and MCGURTY, 1962). OHTANI (1979) has reported preliminary data on the simultaneous mesurement of temperatures up to 1,800°C at 7 GPa by two thermocouples, W/25%Re and Pt/Pt13%Rh. However, the accurate emf and pressure effect on the W/WRe thermocouple are not yet determined and they are left for future study.

7. *Summary and Prospect*

The technical progress in multiple anvil-type apparatus for pressure generation above 10 GPa in a large volume is reviewed with a special emphasis on recent Japanese contributions.

The multiple anvil of the first generation was developed by Hall. It was technically improved in its driving system by employing a couple of sophisticated buckets. The recent cubic press developed in Japan (DIA and WEDGE apparatus) is very stable and easy to use in making routine experiments in a large volume (a few cm^3) up to 10 GPa.

The multiple anvil of the second generation may be the cascading system, going up to 20 GPa in 1 cubic centimeter. The most stable and established system is the so-called 6–8 type, a cascading cubic-octahedral anvil driven by a couple of buckets in a uniaxial press. The most important experimental works above 10 GPa in Japan have mainly been carried out in this system.

A number of new pressure-generating systems aiming at the multiple anvil of the third generation are being extensively studied. They are (1) further improvement of a 6–8 type device for the compression of a very large volume to the higher pressure above several tens of GPa by means of a very large force, (2) the development of several different types of multiple anvils and multiple cascading systems, (3) the development of a multiple anvil sliding system with better operational stability, (4) the synthesis of tough material such as sintered diamond of a large size for anvil use (5) high temperature technique for the melting experiments of silicate systems above 10 GPa, and (6) many other subjects, which are supposed to extend the technical possibilities.

Although we do not yet know what type of system would be the multiple anvil of the third generation, further important progress is expected to be made in near future to facilitate the experimental works that otherwise could not be carried out by other devices.

REFERENCES

BUNDY, F. P. (1975): Ultrahigh pressure apparatus using cemented tungsten carbide pistons with sintered diamond tips, *Rev. Sci. Instrum.*, **46**, 1318–1324.

DUNN, K. J. and BUNDY, F. P. (1979): Materials and techniques for Pressure Calibration by Resistance-Jump Transitions up to 500 Kilobars, in *High-Pressure Science and Technology*, Vol. **1**, Eds.K. T. Timmerhaus and M. S. Barber, Plenum Press, New York, pp. 773–778.

ENDO, S. and ITO, K. (1980): Electrical resistance of αFe_2O_3 under ultrahigh static pressure, *Solid State Commun.* **36**, 189–190.

ENDO, S. and ITO, K. (1982): Triple-stage high pressure appararus with sintered diamond anvils, in *High-Pressure Research in Geophysics,*, Eds. S. Akimoto and M. H. Manghnani, Center for Academic Publication Japan/D. Reidal, Tokyo/Dordrecht, pp. 3–12.

HALL, H. T. (1958): Some high-pressure, high temperature apparatus design consideration: equipment for use of 100,000 atmospheres and 3,000°C, *Rev. Sci. Instrum.*, **29**, 267–275.

HORIUCHI, H. and SAWAMOTO, H. (1981): β–Mg_2SiO_4: single-crystal X-ray diffraction study, *Am. Mineral.*, **66**, 568–575.

INOUE, K. and ASADA, T. (1973): Cubic anvil X-ray diffraction press up to 100 kb and 1,000°C, *Jap. J. Appl. Phys.*, **12**, 1786–1973.

ITO, E. and MATSUI, Y. (1977): Silicate ilmenites and the post-spinel transformations, in *High Pressure Research: Applications in Geophysics*, Eds. M. H. Manghnani and S. Akimoto, Academic Press, New York, pp. 193–208.

ITO, E. and MATSUI, Y. (1978): Synthesis and crystal-chemical characterization of $MgSiO_3$ perovskite, *Earth Planet Sci. Lett.*, **38**, 443–450.

ITO, K., ENDO, S. and KAWAI, N. (1971): Olivine-spinel transformation in a natural forsterite, *Phys. Earth Planet. Inter. 4*, 425–428.

KAWAI, N. and ENDO, S. (1970): The generation of ultrahigh hydrostatic pressures by a spilit sphere apparatus, *Rev. Sci. Instrum.*, **41**, 1178–1181.

KAWAI, N., MISHIMA, O., TOGAYA, M., and LE NEINDRE, B. (1975): H_2O rendered metallic under high pressure, *Proc. Japan Acad.*, **51**, 627–629.

KAWAI, N., TOGAYA, M. and ONODERA, A. (1973): New device for pressure vessels, *Proc. Japan Acad.*, **49**, 623–626.

KAWAI, N., TOGAYA, M., and MISHIMA, O. (1975): A study of the metallic hydrogen, *Proc. Japan Acad.*, **51**, 630–633.

KUMAZAWA, M. (1971): Multiple-anvil sliding system—a new mechanism of producing very high pressure in a large volume, *High Temp. High Press.*, **3**, 243–260.

KUMAZAWA, M. (1973): Theory of generation of very high static pressure by an external force, *High Temp. High Press*, **5**, 599–610.

KUMAZAWA, M., MASAKI, K., SAWAMOTO, H. and KATO, M. (1972): Guide block and compressible pads for the practical operation of multiple-anvil sliding system for the production of high pressure, *High Temp. High Press.* **4**, 293–310.

LACHMAN, J. C. and MCGURTY, J. A (1962): The use of refractory metals for ultra high temperature thermocouples, *Temperature-Its Measurement and Control in Science and Industry*, **3**, 117–182.

LLOYD, E. C., HUTTON, U. O., and JOHNSON, D. P. (1959): Compact multi-anvil wedge-type high pressure apparatus, *J. Res. NBS*, **63C**, 59–64.

MAO, H. K. and BELL, P. M. (1978): High-pressure physics: sustained static generation of 1.36 to

1.72 megabars, *Science*, **200**, 1145–1147.

MASAKI, K., SAWAMOTO, H. and KUMAZAWA, M. (1975): High pressure generation by MASS apparatus, *Rev. Phys. Chem., Special issure*, 832–839.

MASAKI, K., SAWAMOTO, H., OHTANI, E., KUMAZAWA, M., MACHIDA, M., MIZUKASA, S. and NAKAYAWA, N. (1975): High pressure generation by MASS–318–90 type apparatus, *Rev. Sci., Instrum.*, **46**, 84–88.

MISHIMA, O. and ENDO, S. (1980): Phase relations of ice under pressure, *J. Chem. Phys.*, **73**, 2454–2456.

MISHIMA, O. and ENDO, S. (1981): Sintering of polycrystalline diamond cube for high pressure anvil, to be published in *Jpn. J. Appl. Phys.*, **20**, 2197–2200.

OGAWA, Y., NUNOGAKI, K., ENDO, S., KIRITANI, M. and FUJITA, F. E. (1982): Crystallization of amorphous Fe-B alloys under pressure, in *Proc. 4th Int. Conf. on Lapidly Quenched Metals, Sendai, Vol. 1*, pp. 675–678.

OHTANI, A., ONODERA, A., KAWAI, N. (1979): Pressure apparatus of split-octahedron type for X-ray diffraction studies, *Rev. Sci. Instrum.*, **50**, 308–315.

OHTANI, E. (1979): Melting relation of Fe_2SiO_4 up to about 200 kbar, *J. Phys. Earth*, **27**, 189–208.

OHTANI, E. and KUMAZAWA, M. (1981): Melting of forsterite Mg_2SiO_4 up to 15 GPa, *Phy. Earth Planet Interior.*, **27**, 32–38.

ONODERA, A., KAWAI, N., ISHIZAKI, K. and SPAIN, I. L. (1974): Semiconductor-to-metal transition in GaP under high pressure, *Solid State Commun.*, **14**, 803–806.

OSUGI, J., SHIMIZU, K., INOUE, K. and YASUNAMI, K. (1964): A compact cubic anvil high pressure apparatus, *Rev. Phys. Chem., Japan*, **34**, 1–6.

RUOFF, A. L. (1972): Penultimate static pressure containment consideration and possible applications to metallic hydrogen preparation, *Advances Cryogenic Eng.*, **18**, 435–440.

SAWAMOTO, H. (1977): Orthorhombic perovskite (Mg, Fe)SiO_3 and constitution of the lower mantle, in *High Pressure Research: Applications in Geophysics*, Eds. M. H. Manghnani, and S. Akimoto, Academic Press, New York, pp. 219–244.

SAWAMOTO, H., HASEGAWA, A. and YAMAMOTO, S. (1982): Generation of 240 kbar in MASS–OB8 apparatus and its measurement by X-ray diffraction of NaCl, (in prepartion).

SUITO, K., KAWAI, N. and MASUDA, Y. (1975): High pressure synthesis of orthorhombic SnO_2, *Mat. Res. Bull.*, **10**, 677–680.

WAKATSUKI, M. and ICHINOSE, K. (1982): A wedge-type cubic anvil high-pressure apparatus and its application to material synthesis research, in *High-Pressure Research in Geophysics*, Eds. S Akimoto and M. H. Manghnani, Center for Academic Publication Japan/D. Reidel, Tokyo/Dordrecht, pp. 13–26.

YAGI, T. and AKIMOTO, S. (1976): Pressure fixed points between 100 and 200 kb hased on the compression of NaCl, *J. Appli. Phys.*, **47**, 3350–3354.

YAMAMOTO, S., SAWAMOTO, H., HASEGAWA, A. and KUMAZAWA, M. (1982): Generation of high pressures to 10 GPa by means of MASS–OB8 apparatus equipped with X-ray diffraction system, *High Temp. High Press.*, **14**, (in Press).

YONEDA, A. and ENDO, S. (1980): Phase transition in barium and bismuth under high pressure, *J. Appl. Phys.*, **51**, 3216–3222.

YONEDA, A., YAMAMOTO, S., KATO, M., SAWAMOTO, H. and KUMAZAWA, M. (1982): Conditional intensification of pressure in MA8 by means of composite gaskets, (in preparation).

Materials Science of the Earth's Interior, edited by I. Sunagawa, pp. 605–619.

TECHNICAL ASPECT OF SHOCK COMPRESSION EXPERIMENTS USING THE GUN METHOD

Tsuneaki GOTO and Yasuhiko SYONO

The Research Institute for Iron, Steel and Other Metals, Tohoku University, Sendai, Japan

1. *Generation of Dynamic High Pressure*

Dynamic high pressures are practically generated either by transmitting a detonation wave of explosives or by the impact of a high velocity flyer plate. Hypervelocity acceleration of the flyer plate is achieved with compressed gas, or burnt gas of explosives or propellants, although other means, such as lasers or electromagnetic propulsion, are now coming in for practical use. The explosive method has been widely used for shock compression experiments at an early stage (GOTO *et al.*, 1975). After two-stage light gas gun techniques were developed at NASA's Ames Research Center to launch small models of reentry vehicles into various planetary atmospheres, the gun method began to be used for acceleration of flyer plates up to 8 km/sec to generate plane shock waves. The method extended the range of shock pressures and also made shock-wave experiments possible in the ordinary laboratory. Figure 1 shows the pressure-particle velocity curves for various substances to estimate the shock pressures produced in them by the impact of a flyer plate. The intersection point of positive sloping curves (of samples) with negative sloping curves (of flyer plates of tungsten and copper), centered at various impact velocities, gives the shock state induced upon impact. Pressures in excess of 200 GPa in silicate minerals and about 400 GPa in core materials of the earth, are now available through this method.

For the acceleration of a projectile by the gun method, use of smokeless powder is more practical than compressed gas. The simplest is a single-stage propellant gun which propels the flyer-plate-bearing projectile with the burnt gas of smokeless powder. Figure 2 is an illustrative view of a 25 mm bore propellant gun installed at RIISOM, Tohoku University (SYONO and GOTO, 1982a). The powder chamber is designed for a maximum powder charge of 120 g, which can accelerate a 20 gram projectile to 2.4 km/s. The length of the launch tube is 2 m, which corresponds to 80 calibers. The inside of the launch

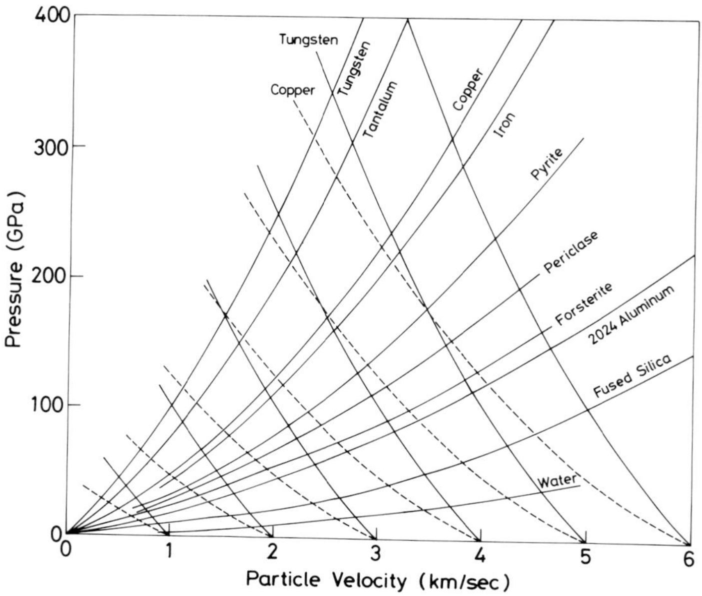

FIG. 1. Shock characteristic curves for various materials. Negative-sloping curves centered at various impact velocities indicate the reflected-shock characteristics of flyer plates. The shock state induced in the sample by impact of the flyer plate is given by the intersection of the shock characteristic curve of the sample and the reflected-shock characteristic curve of the flyer plate.

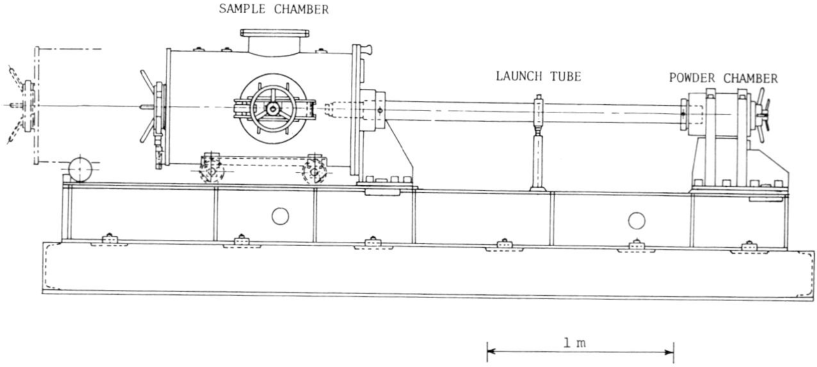

FIG. 2. Construction of the single-stage propellant gun.

tube and the specimen chamber is to be evacuated to a 1 torr range for minimal shot noise. The momentum of the accelerated projectile is finally caught by damping materials inserted in the catcher tail. The specimen chamber is equipped with two optical windows for streak photography, a terminal plate for electrical measurement, outlets for burnt gas and for evacuation, and an inlet for compressed air. The relation of projectile velocity versus propellant mass is shown in Fig. 3. The reproducibility of the gun performance is proved to be excellent.

The maximum velocity obtainable with the single-stage propellant gun is limited to about a 2.5 km/s range, because the burnt gas behind the projectile is accelerated to the same velocity as the projectile, and the increase of the propellant mass does not effectively contribute to the acceleration of the projectile. In order to get rid of this limitation, the idea of using a lighter gas as an acceleration medium was proposed. However, it is not practical to use compressed gas, as it requires a large high pressure reservoir. A two-stage light gas gun was designed on the basis of an accelerated reservoir concept, in which high-pressure light gas is instantaneously generated by compression with a

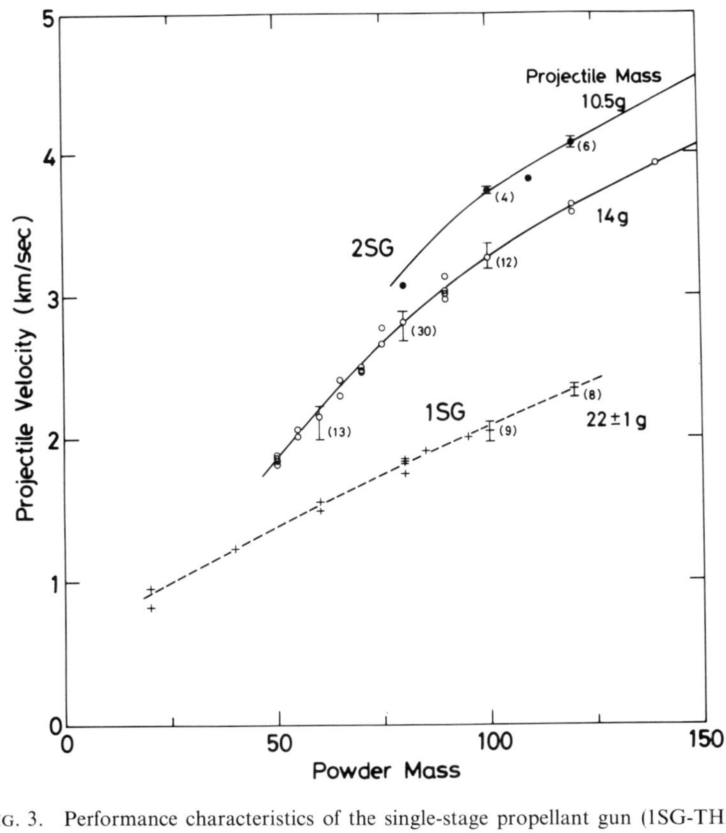

FIG. 3. Performance characteristics of the single-stage propellant gun (1SG-TH2) and two-stage light gas gun (2SG-TH1). Mean values of projectile velocities are indicated with velocity uncertainties for short numbers (shown in brackets) greater than 3.

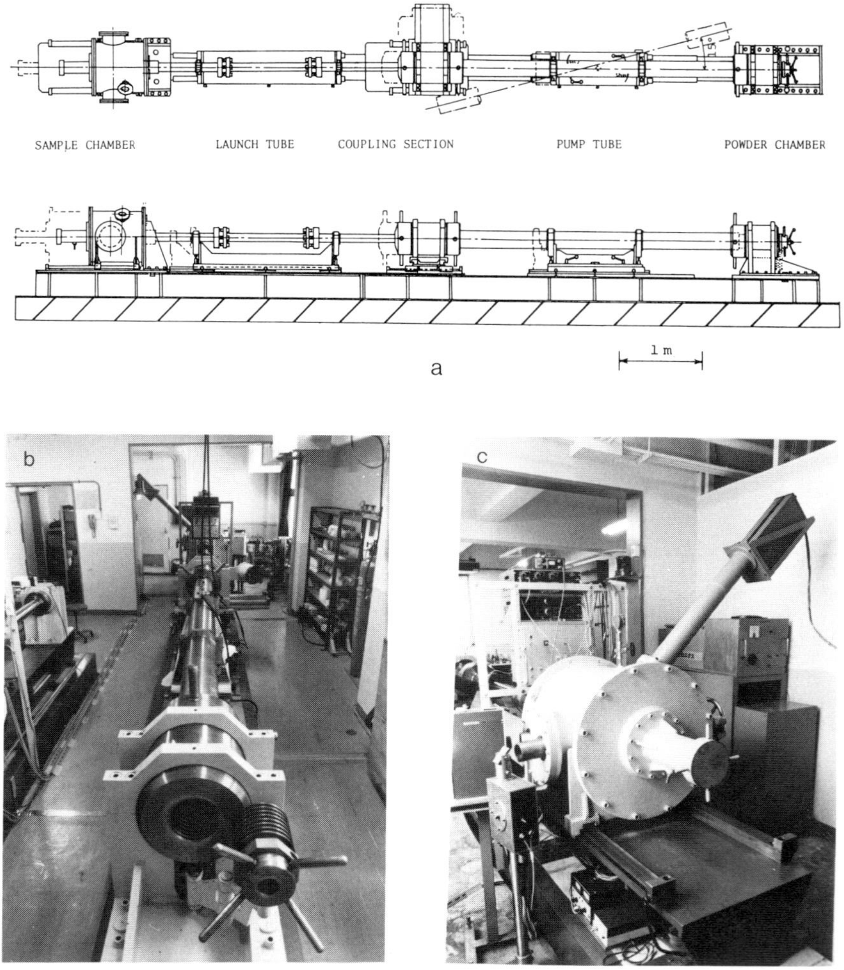

FIG. 4. (a) Construction of the two-stage light gas gun. (b) Overall view of the two-stage light gas gun. (c) Sample chamber viewed from the optical window side.

propellant-driven plastic piston. The design of the two-stage light gas gun (2SG-TH1) installed at RIISOM, Tohoku University, is shown in Fig. 4 (SYONO and GOTO, 1980, 1982a). It consists of a pump tube (60 mm in inner diameter and 3.6 m in length) and a launch tube (20 mm in inner diameter and 3.4 m in length), which are coupled by a tapered high-pressure coupling.

The principle of operation of the two-stage light gas gun is schematically illustrated in Fig. 5. In the first stage, a plastic piston is driven by the burnt gas of the propellant into the coupling, thereby compressing the precompressed light gas to several bars in the pump tube section. The piston is stopped by the

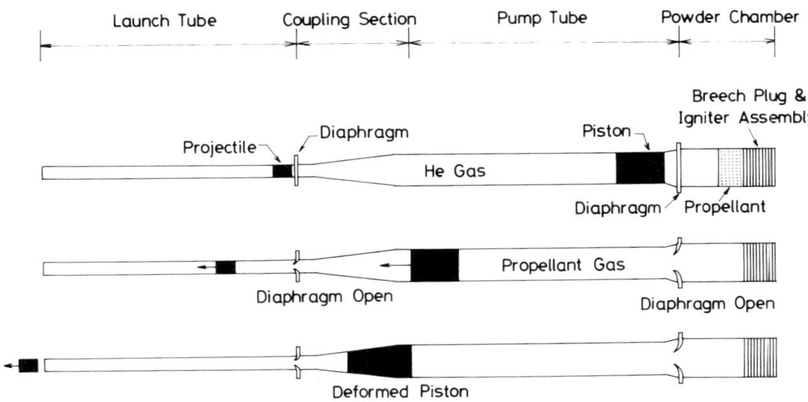

FIG. 5. An illustration of the operating sequence of the two-stage light gas gun (from top to bottom).

tapered section. The instantaneously-generated high-pressure light gas bursts a diaphragm and accelerates the small projectile in the evacuated launch tube to hypervelocity. The projectile velocity is mainly determined by the mass of the propellant and of the projectile. The velocity characteristics of the two-stage light gas gun (2SG-TH1) are also summarized in Fig. 3. The apparatus covers the range of projectile velocity for 1.5–4.5 km/s, using helium as the light gas. Details of the construction and operation of the apparatus have been described elsewhere (SYONO and GOTO, 1980).

2. *Reduction of Shock Data*

As was shown in (SYONO, 1982), the shock compressed state can be determined by applying conservation relations of mass, momentum and energy across the shock front, if the shock and particle velocities were measured. Generally, the particle velocity cannot be directly determined and is usually estimated from the measured free-surface velocities using the free-surface approximation, $u_{fs} \sim 2u_p$. This approximation is known to hold to within a precision of a few percent.

In the case of the gun method, the velocity of the flyer (projectile), u_{imp}, is accurately measured and the shocked state of the test specimen placed on the standard material (target) can be determined from the impedance match solution. Let us consider the case as illustrated in Fig. 6. When a plane shock wave $\mathscr{S}_T$, travelling in the target arrives at the interface with the test specimen, a reflected wave $\mathscr{S}'_T$ and transmitted shock wave $\mathscr{S}_s$ are produced in the target and sample, respectively. The requirement for a continuity relation across the interface forces a condition where pressures and particle velocities are common on both sides of the boundary. The situation can best be illustrated in

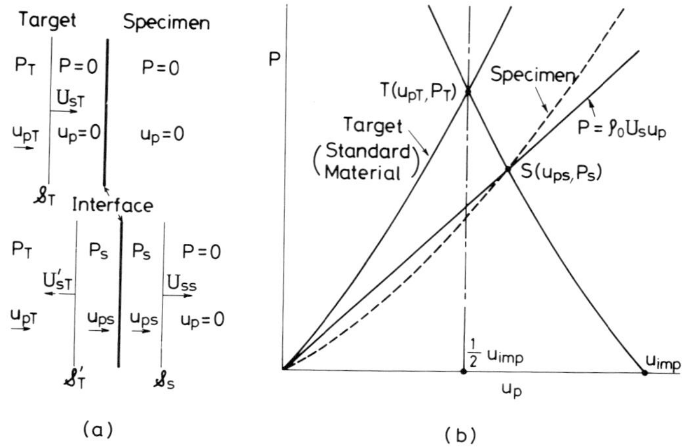

FIG. 6. (a) Shock state produced in the specimen by a transmitted shock wave through the interface between the target and specimen. (b) Pressure-particle velocity plane representation of the impedance match method.

the pressure-particle velocity plane (Fig. 6). In the case of a symmetric impact (i.e., target and flyer plates of the same material), the shock state realized in the target is at the point T with $u_p = \frac{1}{2}u_{imp}$. The state achieved in the test specimen is represented by the intersection S of the reflected Hugoniot of the standard material (the mirror image of the Hugoniot with respect to $u_p = \frac{1}{2}u_{imp}$ in the $P - u_p$ plane) and the line $P = \rho_0 U_s u_p$, where ρ_0 and U_s are the initial density and measured shock velocity of the test specimen. The multiple, $\rho_0 U_s$, which gives the slope of the line, is called the shock impedance. The Hugoniot of the test specimen is thus determined as a locus of several experimental points, as represented by a broken line in the figure.

If a diagnostic plate whose shock characteristics are accurately known, is placed upon the test specimen and one of the shock parameters of it (for example, shock velocity) is measured, the secondarily-produced state in the specimen, due to the reflection of the shock wave at the interface with the diagnostic plate, can be determined again by making use of the impedance match method (Fig. 7). The possible new state achieved by the reflected wave from the initially shocked state S is represented by the intersection of the characteristic curve of the reflected shock state or release adiabat originating from S, and the Hugoniot of the diagnostic plate. If the shock impedance of the diagnostic plate is higher than that of the test specimen, the reflected shock wave intensifies the pressure from P_s to P_{Dh}. If the shock impedance of the diagnostic plate is lower than that of the test specimen, the reflected wave is a rarefaction which reduces the pressure from P_s to P_{Dl} in the test specimen along the release adiabat. The technique can be used to determine how the shocked state S is released (release adiabat measurements).

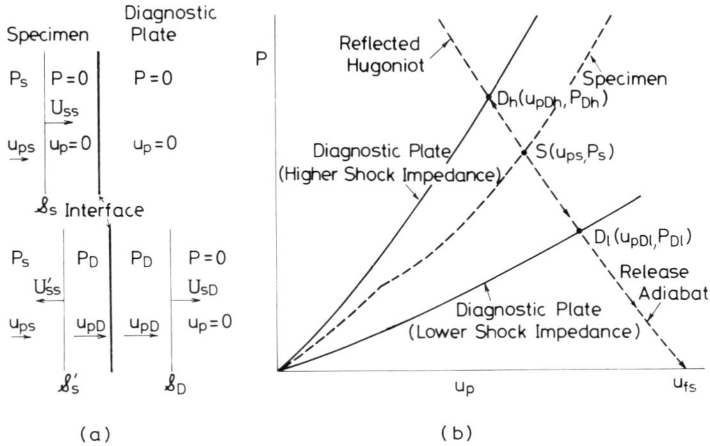

FIG. 7. (a) The reflected wave state achieved in the specimen due to the reflection of plane wave from the interface between the specimen and diagnostic plate. (b) The pressure-particle velocity plane representation of the reflected wave state in the specimen.

3. *Measurements of Shock Parameters*

3.1 *Measurement system*

The instrumental system for shock experiments using the gun method is schematically shown in Fig. 8 (GOTO and SYONO, 1980). The projectile velocity is determined by measuring electric signals generated when the projectile, in which a small magnetized magnet is embedded, passes through a pair of coils placed at a distance of 130–140 mm. The attitude of the projectile in flight is monitored by taking a flash X-ray shadowgraph with an X-ray source of 150 kV and 30 ns exposure. Both the transient recorder for projectile velocity measurement and the flash X-ray source, together with an intense xenon flash light source for streak measurements, are triggered by breaking a wire with the edge of the projectile.

In order to observe shock phenomena, high speed response of the measurement system is required, as most of the shock phenomena are completed within a microsecond time scale. Both electrical and optical measurement techniques have been applied for observation of shock phenomena. Although the electrical measurement system has an advantage that equipment with high frequency response, such as an oscilloscope or transient recorder, are easily available, the optical method using high speed photography provides more direct and abundant information of the various shock phenomena.

Standard equipment for optical observation is a high-speed streak camera with a writing speed higher than several mm/μs on the recording film.

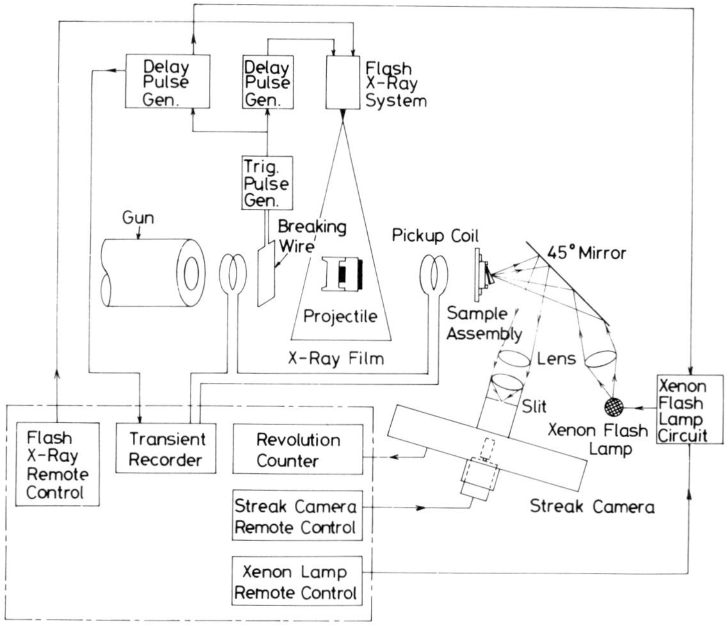

FIG. 8. A block diagram of the instrumentation for optical measurements under shock compression by means of streak photography.

Both rotating mirror type and electronic image converter cameras have been developed. The image converter camera has various advantages, such as high writing speed, high sensitivity, etc., but is very expensive. In our laboratory, a continuous writing streak camera of the rotating mirror-type has been developed for recording non-synchronizable shock phenomena produced by the impact of the gun method (GOTO and SYONO, 1980). Figure 9 shows a schematic illustration of the streak camera. An image of the object is focused onto a slit plane by an objective lens L_1. The slit image is then focused onto a 35 mm recording film placed along the semi-circle wall of the housing by a relay lens, *via* a 45° relay mirror and turbine-driven rectangular rotating mirror. The length of the writing arm is 536.5 mm, which results in a maximum writing speed of 10 mm/μs on the film with a rotation frequency of 90,000 rpm. Utmost care was paid to get a good dynamic balance of the rotating mirror system at the maximum revolution speed. When a slit width of 23 μm was used, time resolution on the film was estimated to be 3 ns. The f number of optical system is 34. As an intense light source is necessary to photographically record the time variation of high-speed phenomena, a pulsed xenon flash lamp was also manufactured. Electric energy up to 2 kJ stored in a 125 μF capacitor, is to be discharged for 50 μs duration.

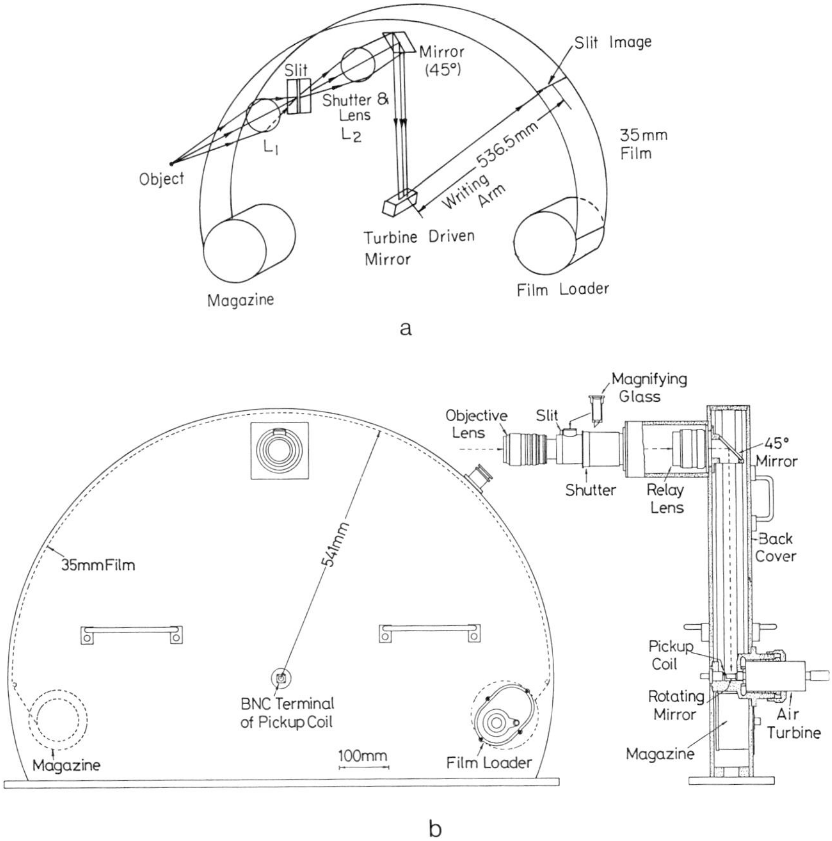

FIG. 9. Optical system (a) and construction (b) of the rotating mirror streak camera.

3.2 *Specimen assembly*

A typical specimen assembly for the measurement of shock parameters adopting the inclined mirror technique is shown in Fig. 10. A specimen of the standard size, 10 × 14 mm in lateral dimension and 2–3 mm in thickness, is polished on both parallel surfaces and mounted on the target made of standard materials whose shock characteristics are accurately known. Small pieces of aluminized mirror are placed on both the specimen and target surfaces. An inclined mirror is fixed on the specimen surface with a small angle α. The specimen assembly is illuminated by the intense xenon light source and the reflected light is adjusted for introduction to the streak camera.

An example of the inclined mirror record is reproduced in Fig. 10. Upon arrival of the shock waves at the mirror, its reflectivity suddenly drops, resulting in a sharp extinction on the streak record. Shock velocity can be

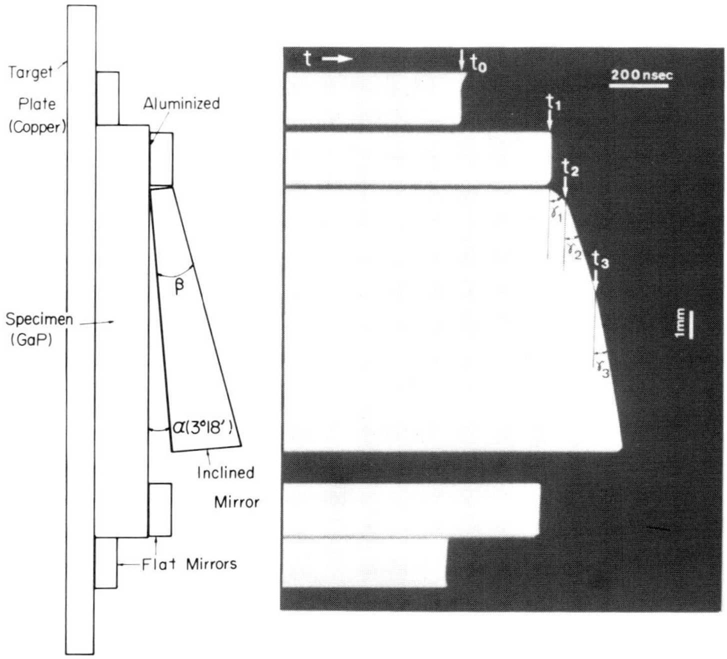

FIG. 10. Sample assembly and streak photograph of the inclined mirror technique. The time of incidence of the shock wave into the specimen, and times of arrival of three successive shock waves (elastic precursor, plastic I and II waves) on the free surface of the specimen are indicated by t_0, t_1, t_2 and t_3, respectively.

determined from the thickness of the specimen and elapse time of the shock wave in it. An inclined mirror image recorded in the streak record monitors the free surface motion of the specimen. The free surface velocity is determined as

$$u_{fs} = \frac{W \tan \alpha}{M \tan \gamma}$$

where γ, M, W are the angle of the slanted image of the inclined mirror in the streak record, magnification and writing speed of the streak camera, respectively. As the inclined mirror technique provides continuous information of the free surface motion, it is capable of detecting the structure of a multiple shock wave. If a double shock wave is produced in the specimen by a phase transition (see SYONO, 1982), free surface of the specimen is discontinuously accelerated at the arrival of successive shock waves on the free surface, as schematically shown in Fig. 11. The shock and free surface velocities of the first and second waves are given by the equations

$$U_1 = d/(t_1 - t_0)$$
$$u_{f1} = W \tan \alpha / M \tan \gamma_1$$
$$U_2 = [d + u_{f1}(t_2 - t_1)]/(t_2 - t_0)$$
$$u_{f2} = W \tan \alpha / M \tan \gamma_2,$$

where d is the thickness of the specimen, and t_0, t_1 and t_2 are the incident time of shock wave into the specimen and arrival times of the first and second shock waves on the free surface of the specimen, respectively. On the basis of the free surface approximation, shock pressures and densities of the specimen corresponding to the first and second shock waves are given by the following equations

$$P_1 = \rho_0 U_1 u_1$$
$$\rho_1 = \rho_0 U_1/(U_1 - u_1)$$
$$P_2 = P_1 + \rho_1(U_2 - u_1)(u_2 - u_1)$$
$$\rho_2 = \rho_1(U_2 - u_1)/(U_2 - u_2),$$

where $u_1 = u_{f1}/2$ and $u_2 = u_{f2}/2$.
The first shock state (ρ_1, P_1) is the phase transition point. Successive arrivals of shock waves are clearly recorded on the streak photograph in Fig. 10, indicating the occurrence of the phase transition.

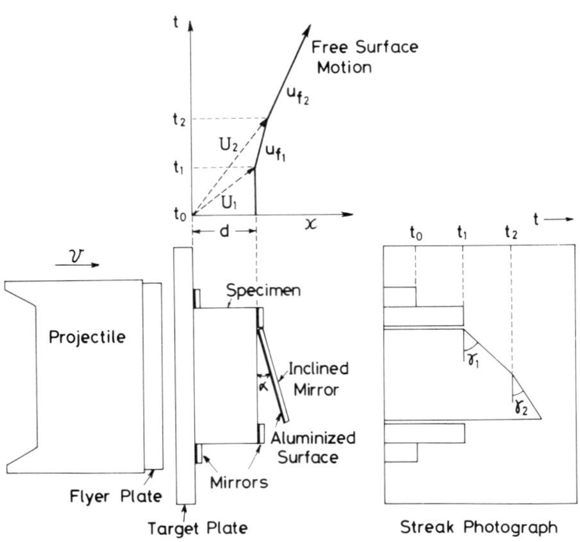

FIG. 11. An illustration of the inclined mirror technique for detection of a double shock wave.

3.3 *Spectroscopy under shock compression*

Recently, a spectrographic system for measurements of optical absorption spectra under shock loading was developed and applied for the measurement of pressure effect on the crystal field spectra of transition metal ions in oxides (GOTO *et al.*, 1977, 1980). Information obtained provides knowledge of changes in the electronic states in transition metal ions at high pressures, which have significant meanings for the study of thermal properties in the earth's mantle.

A time-resolved spectrograph designed with the streak camera and xenon flash lamp is shown in Fig. 12. The light from the xenon flash lamp enters the sample by a lens L_1 via a 45° mirror. The transparent sample is polished on both surfaces, one of which is aluminized so as to form a back surface mirror. An extra layer of a single crystal sapphire of 1 mm thick is inserted between the sample and target plate in order to smooth out the shockfront during the passage. The light reflected from the mirror coating on the sample is focused by a lens L_2 on the entrance slit of a spectroscope (Nikon G250 modified with a grating of 300 lines/mm blazed at 500 nm or 300 nm). The image of the dispersed spectrum is recorded by the streak camera. Before the shockfront enters the sample, the absorption is due to the unshocked sample. When the shock wave propagates through the sample, the spectrum of the shocked

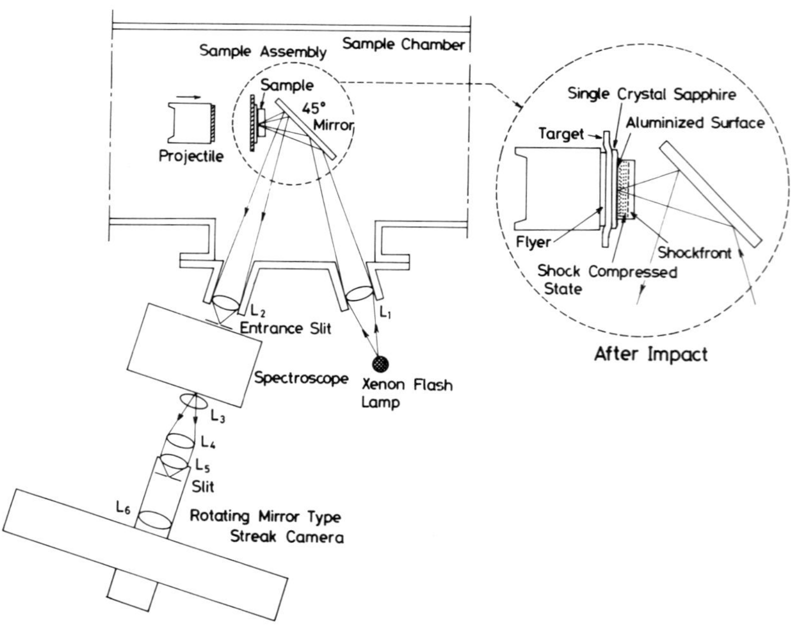

FIG. 12. Optical system and sample assembly for measurements of the optical absorption spectra of solids under shock compression.

region is observed through the unshocked region. Immediately after the shockfront arrives at the free surface of the sample, the light intensity of the spectrum sharply decreases, due to a pulverization of the sample by rarefaction waves originating from the free surface. The spectrographic system is installed at the two-stage light gas gun and can observe absorption spectra of transparent material in the Megabar range (GOTO and SYONO, 1980).

4. *Recovery Technique*

Shock loaded materials are generally broken to pieces by a distension wave in the shock release process. In order to retain shock-loaded substances for the study of residual shock effects, a special precaution is required. For minimizing the destructive effects, the momentum trap concept has been successfully applied. The specimen container is surrounded for protection with a spalling ring and backing plate, as shown in Fig. 13, which is designed to take away the greater part of the large momentum when the distension wave interacts with the interface.

Recovery experiments using the gun method have considerable difficulty in comparison with the explosive method (SYONO *et al.*, 1977), because the shock stress is inevitably concentrated on a relatively small area, thus resulting in a crater formation. Presently, a stainless steel container has been found most successful for recovery of brittle substances such as oxides and silicates. Complete recovery of the specimen is possible up to a projectile velocity of about 2 km/s, which corresponds to 50 GPa. If a high shock impedance metal, such as tantalum, is used for container material, a higher pressure can be achieved by a relatively low velocity impact. A limited amount of shocked materials can be recovered from higher pressures approaching 1 Mbar (100 GPa). Specimens of 10 mg order, which are plenty amount for X-ray analysis and TEM observation, are obtained as embedded in the rear plug made of stainless steel, while other parts of the assembly are completely destroyed.

Shock pressures in the recovery experiments can be approximately

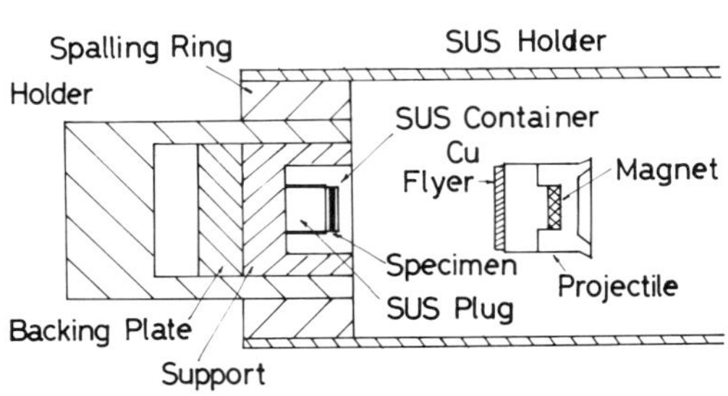

FIG. 13. Configuration of shock recovery experiments.

estimated from the impedance match solution. The thickness ratio of the specimen plus the front cover of the container and flyer plate should be chosen so that the rarefaction wave originating from the back surface of the flyer plate should not overtake the compression wave on the back of the specimen. The maximum thickness ratio allowed for this requirement is given by

$$R = \frac{\rho_{0f}[(\rho_0 U_s)^{-1} - (\rho c)^{-1}]_f}{\rho_{0t}[(\rho_0 U_s)^{-1} - (\rho c)^{-1}]_t}$$

where U_s and c are shock and sound velocities and subscripts f and t indicate the states in the flyer and target, respectively. In the case of utilizing multiple shock reflection to establish equilibrium between the specimen and specimen container, the thickness of the specimen should be thin enough, compared with the specimen container. The practical design of the recovery assembly can

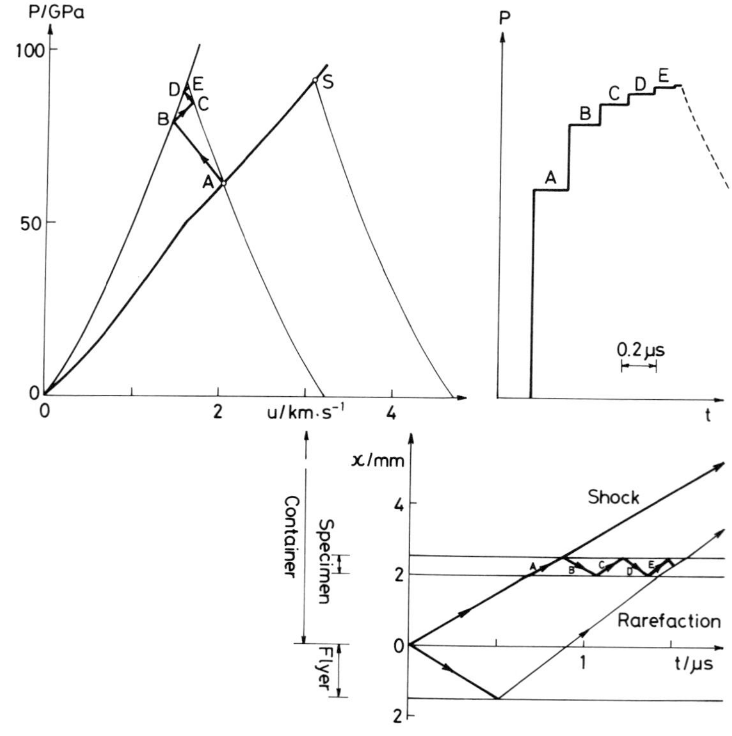

FIG. 14. Shock state of the specimen attained by multiple shock reflection in the shock recovery experiments. Thin and thick solid lines indicate shock characteristics of the container as well as flyer and the specimen respectively (upper left figure). The trajectory A → B → C → D → E → shows how the pressure in the specimen increases, due to multiple shock reflection between the specimen and container, as illustrated in the upper right figure. The point S corresponds to the state attained by a single shock in Hugoniot measurements.

be made graphically, as shown in Fig. 14, using available shock data.

It is important to note that the final shock state achieved by multiple shock reflection considerably differs from the shock state attained by a single shock as in the case of Hugoniot measurements (see Fig. 14). This is particularly important when the shock residual effects are compared with the results of Hugoniot measurements. The temperature achieved in the multiply-shocked state is remarkably lower than that attained in the singly-shocked state, even if the achieved pressure is the same. In the case of a reconstructive type phase transformation, the reaction rate is primarily determined by temperature rather than by pressure itself (SYONO and GOTO, 1982b).

REFERENCES

GOTO, T., SYONO, Y., NAKAI, J. and NAKAGAWA, Y. (1975): Shock compression experiments in solids using high explosives, *Sci. Rep. Res. Inst, Tohoku Univ.*, **A25**, pp. 186–199.

GOTO, T., AHRENS, T. J. and ROSSMAN, G. R. (1977): Absorption spectra of Cr^{3+} in Al_2O_3 under shock compression, *Phys. Chem. Miner.*, **4**, 253–263.

GOTO, T., AHRENS, T. J., ROSSMAN, G. R. and SYONO, Y. (1980): Absorption specturm of shock-compressed Fe^{2+}-bearing MgO and the radiative conductivity of the lower mantle, *Phys. Earth. Planet. Inter.*, **22**, 277–288.

GOTO, T. and SYONO, Y. (1980): Optical measurement techniques under shock compression using rotating mirror type streak camera, *Sci. Rep. Res. Inst. Tohoku Univ.*, **A29**, 32–49.

SYONO, Y. (1982): Shock-induced phase transition in oxides and silicates, this volume, pp. 395–415.

SYONO, Y., GOTO, T., NAKAGAWA, Y. and KITAMURA, M. (1977): Formation of diaplectic glass in anorthite by shock-loading experiment, in *High Pressure Research: Applications in Geophysics*, Eds. M. H. Manghnani and S. Akimoto, Academic Press, New York, pp. 477–489.

SYONO, Y. and GOTO, T. (1980): A two-stage light gas gun for shock wave research, *Sci. Rep. Res. Inst. Tohoku Univ.*, **A29**, 17–31.

SYONO, Y. and GOTO, T. (1982a): Shock wave facilities for high-pressure experiments at Tohoku University, in *AIP Conf. Proc.*, No. 78, pp. 701–705.

SYONO, Y. and GOTO, T. (1982b): Behavior of single-crystal forsterite under dynamic compression, in *High Pressure Research in Geophysics*, Eds. S. Akimoto and M. H. Manghnani, Center for Academic Publication Japan/D. Reidel, Tokyo/Dordrecht, pp. 563–578.

Materials Science of the Earth's Interior, edited by I. Sunagawa, pp. 621–631.

TECHNIQUE AND GAS-FLAME HEATER FOR SINGLE-CRYSTAL X-RAY DIFFRACTION STUDIES AT HIGH TEMPERATURES UP TO AROUND 2,500°C

Takamitsu YAMANAKA

Mineralogical Institute, Faculty of Science, The University of Tokyo, Hongo, Tokyo, Japan

1. Introduction

Many types of heating devices for high temperature X-ray studies have been developed, since crystal structure analyses as a function of temperature attracted mineralogists and crystallographers. Furnaces of the most common type utilize the electric resistance heaters as those described by FOIT and PEACOR (1967), QUARENI (1969), LYNCH and MOROSIN (1971), SMYTH (1972) and BROWN *et al.* (1973). A number of high temperature X-ray diffraction studies have been carried out by making use of these heaters or their modifications (CAMERON *et al.* 1973, HAZEN, 1976; YAMANAKA, 1976; TAKEDA and MOROSIN, 1975).

The various heaters reported to date could generate temperatures up to 1,200°C at the most. However, many silicate and oxide crystals, significant for the earth sciences or material sciences generally, have melting points higher than 1,200°C. It may be of interest to study, by X-ray means, various physical phenomena appearing at temperatures higher than 1,000°C, such as transition, intracrystalline cation exchange, decomposition, melting etc.

In connection with these problems, only a few experiments have been conducted to date. GUBSER *et al.* (1963) developed an acetylene furnace for their study of anorthite by the precession photographs in the temperature range from 300 to 1,400°C. SMYTH (1974) used a microtorch for the study of enstatite transformation up to 1,400°C. NUKUI *et al.* (1972) and MIYATA *et al.* (1979) constructed a gas-flame heating device with a mixture of propane and oxygen gases which could provide temperatures of up to 2,300°C. They installed the device on a four-circle diffractometer to study the crystal structure of α-Al_2O_3 at 1,900°C.

Recently, a YAG laser (LIU and RINGWOOD, 1975) and a CO_2-gas laser

(Ohsumi, 1981) were applied as heat sources for high-temperature studies. In combination with the diamond high-pressure cell, the fomer is applicable for study under high pressure. For the purpose of the X-ray diffraction study at high temperatures under room or high pressure, newly designed temperature regulation systems and optical systems of the laser heaters are detailed elesewhere in this volume by Ohsumi *et al.* (1982). Further an infrared image furnace using a halogen lamp may be another candidate (Morimoto and Kitamura, 1981).

This article describes a gas-flame heating apparatus, which has been reported elsewhere (Yamanaka *et al.*, 1981), together with an improved model of a furnace and discusses, in general, high-temperature techniques, including temperature control, temperature measurement and control of the atmosphere surrounding samples.

Note that the above-mentioned heating systems, including the gas-flame heater, are difficult to provide temperatures lower than about 800°C. Accordingly, high temperature studies at temperatures lower than about 800°C have to be supplemented by conventional electric resistance heaters.

2. *Gas-Flame Heater for the Single Crystal X-Ray Diffraction Study*

Gas furnaces have, in general, the advantage in providing heating temperatures much higher than electric resistance furnaces, but they have the disadvantage of having difficulties in their temperature control, particularly at temperatures lower than about 1,500°C. In order to overcome these difficulties, a new type of gas flow controller has been designed, which enables a gas-flame heater to be an effective apparatus for the single crystal X-ray diffraction study. Details of the flow-control system, such as a bellows pressure valve, flow controller, flow gauge, pressure gauge, etc., are presented in Yamanaka *et al.* (1981). The gas-flame heating system designed by them consists of a microburner and gas flow controller. This proved capable of controlling the heating temperature and supplying a steady heat.

The temperature control of the heater is carried out by changing the ratio of the flow rates of propane, oxygen and nitrogen gases; the third gas plays a role in increasing the blower pressure of the flame and reducing fluctuation of the sample temperature. Even at a fixed flow rate of propane and oxygen, the sample temperature still varies with the blower pressure at the outlet of the microburner. An addition of nitrogen gas to the above two gases can continuously change the pressure. As the pressure increases, the sample temperature rises and the isothermal zone of the flame is enlarged, due to a smaller temperature fluctuation. This event was more noticeable, especially at low temperatures around, 1,000°C. The heater can thus stably provide a temperature ranging from about 1,000 to 2,000°C within a deviation of ±20°C. In this system, it is an important factor to accurately control the flow rates of the respective gases. The oxygen, propane and nitrogen gases, whose flow rates are precisely regulated, are led into a mixing chamber having a turbulant stream stirrer. The homogeneous mixed gases are taken out of the gas-flow controller. A schematical illustration of this heating system applying to a four-circle X-ray diffractometer is presented in Fig. 1.

The gas mixture is led to a microburner sat on a burner stand, which permits horizontal and vertical shifts of the microburner. Finally, the gas mixture is ejected from a nozzle attached to the end of the microburner. After many tests, the most effective nozzle for the X-ray diffraction study was found to have the following specification: five ports are in the nozzle head; the diameter of each port is 0.3 mm; they are arranged at regular intervals on a circle, 3 mm in diameter; These ejection ports are tapered in order to converge the gas-flame at a point in the distance of 10 to 15 mm apart from the nozzle head, as presented in Fig. 2. Two types of nozzle heads are prepared for the different uses, X-ray precession photography (type B in Fig. 2) and four-circle diffractometer (type A in Fig. 2). They are different in the direction of the flame ejection.

The temperature stability is graphically presented in Fig. 3 as a function of the flow rate of nitrogen gas under the condition where the optimum mixing ratio of propane and oxygen to obtain a high calorific value is fixed at 1:3.

3. *Joint Assembly of Electric Resistance Heater and Gas-Flame Heater*

Heating by the above-mentioned gas-flame heating system can be applied to measuring X-ray diffraction intensities by the four-circle diffractometer. Where, however, the measurement is carried on, a relative direction between the flame and the φ axis varies, because the microburner sits on the stand at the

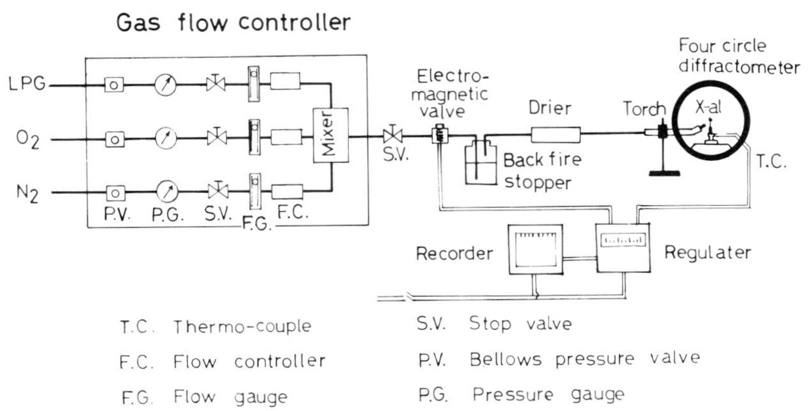

FIG. 1. Schematical illustration of the gas-flame heating system and design of the gas-flow controller of propane, oxygen and nitrogen gases.

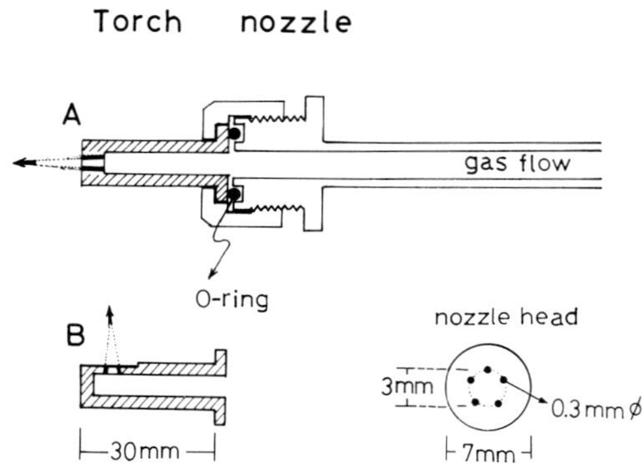

FIG. 2. Specification of the two types of nozzles: type A for a four-circle diffractometer and type B for a precession camera.

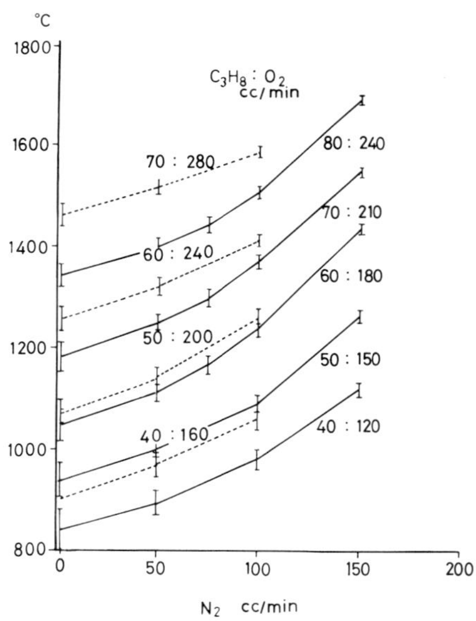

FIG. 3. A test of temperature fluctuation in terms of flow rate of nitrogen gas at the constant mixing ratio of propane and oxygen gases.

fixed position. To keep the specimen temperature as stable as possible, it is desirable to retain the same relative direction throughout the collection of a set of diffraction intensities. Therefore, in the improved furnace, the microburner of a gas-flame heater has been set on the χ-circle of a diffractometer.

Since, as previously described, the gas-flame heater can not produce temperatures below 1,000°C, it is preferred for the gas-flame heating system to assemble an electric heater and the microburner into a joint system, which is set on the χ-circle of a four-circle diffractometer.

A newly constructed apparatus is described below. A hole having a diameter of 3 mm is formed in the center portion of a brass rod, 20 mm long and 20 mm in diameter, supporting the electric heater and a nozzle set at the outer end of the hole. And a mixed gas inlet pipe is connected to the other end on the hole by means of an O ring. On the other, the electric heater suspended by the brass rod is provided with two heat generators, which are constructed by coiling a Pt (Rh 20%) wire, 0.6 mm in diameter, extended about 20 mm and spaced in relation to each other at a distance of 7 mm. A thermocouple and gas-flame nozzle are placed between the two heat generators. The assembly of the electric heater and microburner is provided with a mechanism capable of vertical and horizontal shift in order to enable the assembly to always occupy an optimum position with respect to the sample (Fig. 4).

This assembly has made it possible to carry out X-ray structure analyses at any temperatures ranging from room temperature to 2,000°C. The electric resistance heater is applied to the experiments at temperatures lower than 1,000°C and the gas-flame heater is used for the experiments conducted at temperatures higher than 1,000°C. Even in the latter experiment, application of the electric heater as a supplementary heater providing the heating temperature around 800°C, gives the merit of reducing a temperature gradient and broadening an isothermal region.

By using the joint assembly, measurement of the X-ray diffraction intensity can be made throughout the same sample setting from room temperature to elevated temperatures of up to 2,300°C.

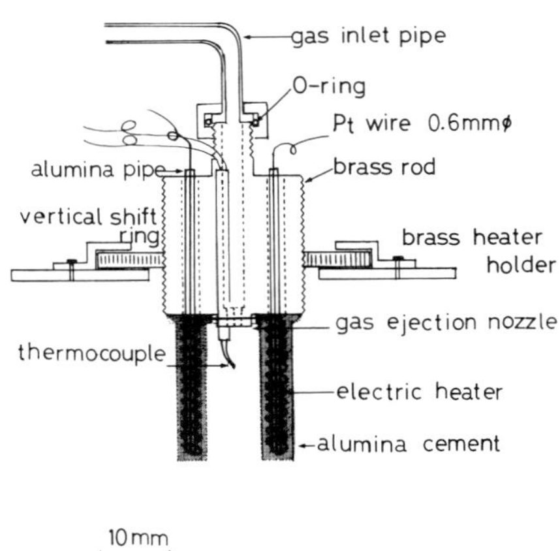

FIG. 4. Design of the joint assembly of electric resistance heater and gas-flame heater. The assembly is attached to the χ-circle of the four-circle diffractometer so as to conduct the constant heat supply to the sample, regardless of the axial driving of the diffractometer.

4. System of the Temperature Control and Measurement

In the conventional high temperature X-ray diffraction studies, temperature control was carried out by a thermocouple attached to an electric heater. Where an electric heater is fitted to a four-circle diffractometer, the electric heater, during measurement of the diffraction intensities, is rotated with the rotations around three axes, 2θ, ω and χ. Particularly with the χ rotation, thermal convection current varies with the rotated position of the electric heater, thereby giving rise to a large difference between the temperature of the sample and that of the electric heater.

To solve this problem it is preferred to control the sample temperature by a thermocouple placed in a sample holding rod. A sample is fixed immediately on the thermocouple by alumina cement, thereby allowing the thermocouple to be moved with the rotation of a sample.

Since the apparatus of the present experiment is intended to be capable of carrying out structure analyses at high temperatures up to 2,000°C, it is necessary to select a proper thermocouple to meet this requirement. A thermocouple of Pt-13%Rh is suitable for the experiment at lower temperatures than 1,400°C. In the case of the experiment at higher temperatures than 1,400°C, a thermocouple is prepared from Ir-40%Rh. These thermocouples do not give rise to difficulties, even in an oxidizing atmosphere. However, the latter thermocouple lacks flexibility and is brittle, and must be handled with great care.

The thickness of a single line of the thermocouple used in the experiment is set at 0.1 mm in diameter in consideration of the size of samples provided. Where determination of the sample temperature is made under a small electric heater used for the high temperature X-ray studies, which has a very small heat capacity than the ordinary electric furnaces, difficulties are presented in carrying out accurate temperature determination due to a heat release from the thermocouple itself. On this matter, the above-mentioned thermocouple is composed of sufficiently thin lines and then it is little affected by such a problem. From this point of view, therefore, it is preferable to apply an electric heater having as broad an isothermal region as possible.

The thermocouple is supported by a two-holed alumina rod which is attached to the top of the goniometer head. The two lines of the thermocouple are respectively connected to the corresponding ring terminals fixed to the goniometer head to prevent the lead lines of the thermocouple being tangled upon rotation of the goniometer head around the ϕ rotation axis. Thermoelectric current is conducted to a temperature indicator and control device from the ring terminals through the copper brushes as shown in Fig. 5. One of the ends of each copper brush is fixed on the χ circle of the diffractometer. In this way the goniometer assembly including the thermocouple is free to rotate around any of the four rotation axes.

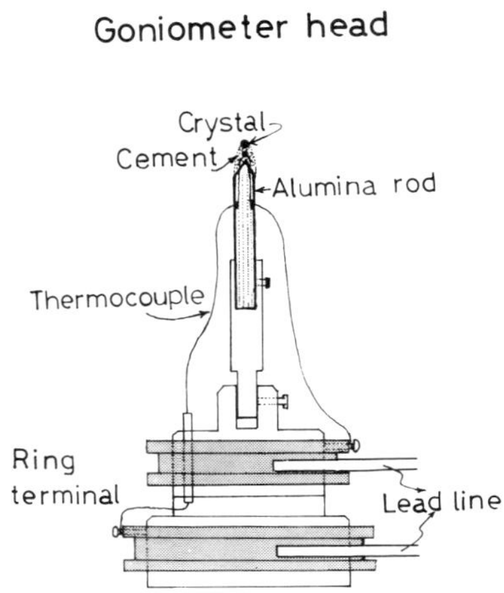

FIG. 5. Assembly of goniometer head, indicating the system of the temperature detection. The thermocouple is connected to the ring terminal, to which the copper brushes are attached, these copper brushes are fixed on the χ-circle.

Measurement of the sample temperature by heat radiation is carried out besides the measurement by the thermocouple. The former measurement is a noncontact process with respect to a heated substance and is considered a very useful means for remote sensing. The instruments for this purpose include an infrared heat radiation thermometer, dichromatic heat radiation thermometer and luminosity thermometer. The former two are largely affected by the rate of heat radiation from the heated sample. Therefore, compensation has to be made for the rate of heat radiation from various samples used in the high temperature X-ray studies. It is troublesome to accurately measure the sample temperatures when we deal with different kinds of materials, because the rate of heat radiation from these samples is generally unknown.

On the other hand, the luminosity thermometer is used with reference to the standard luminosity and is, therefore, practically applicable for the high temperature X-ray studies.

Since very fine samples, mostly less than 0.4 mm in length, are used in the studies, it is necessary to magnify the sample by a convex lens and shut off externally radiated energy. For our experiments, we use a micropyrometer manufactured by Pyro-Werk Ltd., West Germany. The lens used is of the 60 times magnification type. Measurement is made at a distance ranging from 90 to 100 mm. This micropyrometer is capable of measuring sample temperatures ranging from 600° to 3,000°C.

Calibration of the present temperature-measuring system was carried out with reference to the known melting points of the standard materials of Au, Pd and Pt, these are 1,063°, 1,552° and 1,769°C, respectively. The calibration was further conducted by repeating the melting and solidification of the materials.

The calibration was also undertaken with reference to the known melting points of other samples Fe_2O_3, CuO, Co_2SiO_4, $CaSiO_3$. After these calibration, the temperature-measuring system including thermocouple was compensated to conduct the precise measurement.

Since, during the collection of the X-ray diffraction intensities, the temperature measurement is not carried on by using the luminosity thermometer, the thermocouple plays a role of an indicator of the sample temperature, though it is not capable of a precise temperature measurement, especially at temperatures higher than 1,450°C, as previously described.

5. *Control of the Atmosphere Surrounding Samples*

In the case of the X-ray structure analyses, especially at high temperature, the atmosphere surrounding the sample is a large factor for measurement of the diffraction intensities. It is greatly important to refrain the sample from redox reaction with the atmosphere, because quite a few minerals or crystals contain transition elements as major or minor components. Since, further, most of them suffer from the sublimation, even at temperatures much lower than their melting points, partial pressures of the constituent elements in the crystals, if possible, had better be controlled. In this connection, it is necessary for oxides or silicates to control the partial pressure of oxygen, Po_2 in the atmosphere. Device of the Po_2 controller is required as an indispensable apparatus for the high temperature studies, because those sublimation or redox reactions are easily activated, especially at high temperatures.

In order to control the atmosphere surrounding the small sample in the open space, gas with the Po_2 set by the control system is supplied to the sample through an ajutage or ejection nozzle. The following control systems are possible to be applied for the present purpose; (1) changing the mixing ratio of inert gas and oxygen gas, (2) using the dissociation of oxygen from oxide gas as a function of temperature, (3) application of the redox reaction of the mixture of oxide gases, such as CO_2 and H_2, CO and CO_2 or CO and H_2.

In the first system, a desired Po_2 is given simply by the mixing ratio, but when the value of Po_2 is lowered to some extent, it is almost impossible to produce the atmosphere set at the precise value. The second system also has a problem in setting the Po_2 at an arbitrary value at various temperatures. For two decades, several computations of the fugacity of oxygen, fo_2, produced by gaseous mixtures at various temperatures within the C–H–O system have been reported by Muan and Osborn (1965), Deins, *et al.* (1974). They published the diagrams or tables of fo_2 vs temperature for mixtures of $CO-CO_2$, H_2-H_2O, H_2-CO_2. By making use of their data, an arbitrary Po_2 value in the atmosphere around the sample is given in terms of the mixing ratio of those pair gases, if only to precisely control their respective flow rates.

In the present high temperature studies, a gaseous mixture of CO_2 and H_2

is employed with changing their respective flow rates from 3 to 300 ml/min. CO_2 and H_2 gases, whose pressures are preset at 2 kg/cm^2, by a pressure reduction valve, are led to the flow controller. Further, those pressures are accurately set at 1 kg/cm^2 by a bellows pressure valve. Then the flow rate comprises a needle valve, which allows for a free control. The flow rates are indicated on a high-precision flow gauge. These gases are brought to the gas-ejection nozzle through a gas-mixing chamber.

This control system requires, to a certain extent, the isothermal region around the sample because temperature had an effect on the redox reaction of the gases, which defined the value of Po_2. In this sense, the supplemental heater played a role as an effective apparatus.

Although, in the present high temperature X-ray diffraction studies, this Po_2 control system could not satisfactorily surpress the sublimation of the sample, yet it seemed to be effective to the problem. On this matter, in order to solve the problem, a new device or an improvement of the controller of the atmosphere, in the future, will enable us to succeed the high temperature study up to the temperature of the melting point of the samples. It is reserved for further study.

6. *Experimental Performance*

Some of the experiments carried out using the present high temperature system include:

(1) second order transition in $MgAl_2O_4$ and observation of the anharmonic thermal vibration of the atoms (YAMANAKA *et al.*, 1979),

(2) crystal structure of protoenstatite as a function of temperature (MURAKAMI *et al.*, 1980),

(3) effects of temperature on the several olivine structures including Mg_2GeO_4 and Ca_2GeO_4 (YAMANAKA *et al.*, 1980), Al_2BeO_4 (HAGA and TAKÉUCHI, 1980) and $LiScSiO_4$ (SATO, 1980).

These studies have proved the usefulness of our gas-flame heater, which produced much higher temperatures than those of conventional heaters. The results of some of these studies, in particular those of forsterite and spinel, are presented in this volume by TAKÉUCHI *et al.* (1982).

REFERENCES

BROWN, G. E., SUENO, S. and PREWITT, C. T. (1970): A new single-crystal heater for the precession camera and four-circle diffractometer, *Am. Mineral.*, **58**, 698–704.

CAMERON, M., SUENO, S., PREWITT, C. T. and PAPIKE, J. J. (1973): High temperature crystal chemistry of acmite, diopside, hedenbergite, Jadeite, spodumane and ureyite *Am. Mineral.*, **58**, 594–618.

DEINES, P., NAFZIGER, R. H., ULMER, G. C. and WOEMRMANN, E. (1974): Temperature-oxygen fugacity tables for selected gas mixtures in the system C–H–O at one atmosphere total pressure, *Bull. Earth and Mineral Sciences Experiment Station 88*, Pennsylvania State Univ.

FOIT, F. F. JR. and PEACOR, D. R. (1967): A high temperature furnace for a single crystal X-ray diffractometer, *Rev. Sci. Instrum.*, **44**, 183–185.

GUBSER, R. A., HOFFMAN, W. and NISSEN, H. U. (1963): Röntgenaufnahmen mit der Buergerschen Präzessionskamera bei Temperaturen zwischen 1,000°C und 2,000°C, *Z. Kristallogr.*, **119**, 264–272.

HAGA, N. and TAKÉUCHI, Y. (1980): Crystal structure of chrysoberyl, Al_2BeO_4, at high temperature, *Mineral. Soc. Japan 1980, Annual Meeting Abst.*, **A–8**.

HAZEN, R. M. (1976): Effects of temperature and pressure on the crystal structure of forsterite, *Am. Mineral.*, **61**, 1280–1293.

LAGER, G. A. and MEAGHER, E. P. (1978): High-temperature structure study of six olivines, *Am. Mineral.*, **63**, 365–377.

LIU, L. G. and RINGWOOD, A. E. (1975): Synthesis of a perovskite-type polymorph of $CaSiO_3$, *Earth. Planet. Sci. Lett.*, **28**, 209–211.

LYNCH, R. W. and MOROSIN, B. (1971): A hemispherical furnace for high-temperature single-crystal X-ray diffraction studies, *J. Appl. Crystallogr.*, **4**, 352–356.

MIYATA, T., ISHIZAWA, N., MINATO, I. and IWAI, S. (1979): Gas-flame heating equipment providing temperatures up to 2,600 K for the four-circle diffractometer, *J. Appl. Crystallogr.*, **12**, 303–305.

MORIMOTO, N. and KITAMURA, M. (1981): Infrared reflection heater for the X-ray diffraction study, *KEK. Internal. 81–4. (National Laboratory for High Energy Physics, Japan)*, 71–74.

MUAN, A. and OSBORN, E. F. (1965): *Phase Equilibria Among Oxides in Steel Making*, Addison-Wesley, Reading, Mass.

MURAKAMI, T. (1980): Crystal-chemical characteristics of protoenstatite and its related minerals at high temperature, D. Sc. Thesis, Faculty of Science, Univ. of Tokyo.

NUKUI, A., IWAI, S. and TAGAI, H. (1972): Gas flame heating equipment providing up to 2300°C for an X-ray diffractometer, *Rev. Sci. Instrum.*, **43**, 1299–1301.

OHSUMI, K. (1981): Laser heating system for high temeparture single crystal X-ray diffraction study, *KEK. Internal. 81–4 (National Labolatory for High Energy Physics, Japan)*, 60–63.

OHSUMI, K., SAWADA, T., TAKÉUCHI, Y. and SADANAGA, R. (1982): Laser-heating device for single-crystal diffractometry and its application to the structural study of high cristobalite, this volume, pp. 635–645.

QUARENI, S. (1969): A high-temperature apparatus for Weissenberg and precession diffractometers, *Z. Kristallogr*, **128**, 294–299.

SATO. G. (1980): Effects of temperature or pressure on the crystal structure of $LiScSiO_4$, D. Sci. Thesis, Faculty of Science, Univ. of Tokyo.

SMYTH, J. R. (1972): A simple heating stage for single-crystal diffraction studies up to 1,000°C, *Am. mineral*, **42**, 39–55.

SMYTH, J. R. (1974): Experimental study on the polymorphism of enstatite, *Am. Mineral.*, **59**, 345–352.

TAKEDA, H. and MOROSIN, B. (1975): Comparison of observed and predicted structureal parameters of mica at high temperature, *Acta. Cryst.*, **B, 31**, 2444–2452.

TAKÉUCHI, Y., YAMANAKA, T., HAGA, N. and HIRANO, M. (1982): High-temperature crystallography of olivines and spinels, this volume, pp. 191–231.

YAGI, T. and AKIMOTO, S. (1982): In-situ observation of crystal structure and phase transformation under pressure, this volume, pp. 233–247.

YAMANAKA, T. and SDANAGA, R. (1976): High temperature X-ray study of the Jahn-Teller effect on the tetragonal-cuic transition in Mn_3O_4, *Crystallogr. Soc. Japan 1976, Annual Meeting Abst, PA–1.*

YAMANAKA, T., TAKÉUCHI, Y. and SADANAGA, R. (1979): The crystal structure of $MgAl_2O_4$ spinel at elevated temperatures up to 1,800°C, *Crystallogr. Soc. Japan. 1979, Annual Meeting Abst.*, **3B–7**.

YAMANAKA, T., HIRANO, M., TAKÉUCHI, Y. and SADANAGA, R. (1980): Comparison of thermal variation in the olivine structures of Me_2GeO_4 and Me_2SiO_4, *Mineral. Soc Japan 1981, Annual Meeting Abst*, **A–10**.

YAMANAKA, T., TAKÉUCHI, Y. and SADANAGA, R. (1981): Gas-flame, high-temperature apparatus for single-crystal X-ray diffraction studies, *Z. Kristalogr.*, **154**, 147–153.

Materials Science of the Earth's Interior, edited by I. Sunagawa, pp. 633–643.

LASER-HEATING DEVICE FOR SINGLE-CRYSTAL DIFFRACTOMETRY AND ITS APPLICATION TO THE STRUCTURAL STUDY OF HIGH CRISTOBALITE

Kazumasa OHSUMI, Toshiyuki SAWADA*, Yoshio TAKÉUCHI, and Ryoichi SADANAGA

Mineralogical Institute, Faculty of Science, University of Tokyo, Hongo, Tokyo, Japan

1. *Introduction*

In order to carry out single-crystal diffraction studies at very high temperatures, like those above 1,000°C, it is of essential importance to develop devices to heat crystal specimens. In addition to the gas-flame heater as described elsewhere (pp. 623–633), a laser beam is thought to be a useful candidate as a heat source. A laser beam, in fact, has various advantages in generating high temperatures and in controlling temperatures. The present paper describes our first model of a heating system utilizing a CO_2 gas laser for single-crystal X-ray studies and demonstrates its application to the structural study of high cristobalite.

The high-temperature form of cristobalite was first studied by WYCKOFF (1925) using a powder specimen at both temperatures of 290° and 430°C. The space group and cell dimensions were determined to be $Fd3m$ and 7.12 $\pm$ 0.01Å, respectively. The structure was described as a build up of ABC stacking sheets consisting of SiO_4 tetrahedra, each having an Si–O bond length of 1.541Å and Si–O–Si angle of 180°. BARTH (1932) reinvestigated the structure from powder diffraction data obtained at 500°C and reported that the space group was $P2_13$ rather than $Fd3m$, the mean Si–O bond length being 1.63Å, and the Si–O–Si angle less than 180°. NIEUWENKAMP (1937) studied the data from a rotation camera at 270°C and verified WYCKOFF's determination of the space group and the position of silicon atoms, concluding that oxygen atoms were distributed randomly on a circumference of a radius of

*Present address: Electron Device Engineering Laboratory, Toshiba Corporation, Saiwai-ku, Kawasaki, Japan

about a 0.4Å bisecting Si–Si line. In connection with the transition in cristobalite, PEACOR (1973) analyzed the structure in detail by a single-crystal X-ray method at temperatures between 221° and 310°C. The resulting structures were almost the same as those obtained by NIEUVENKAMP (1937) and the oxygen atoms were statistically distributed in six sites, each having an occupancy factor of 1/6 on the circumference of radius 0.45Å. LEADBETTER *et al.* (1973) and WRIGHT and LEADBETTER (1975) studied the powder data obtained at temperatures between 300° and 1,200°C and determined the distribution of the oxygen atoms on the circumference. Detailed interpretation of the statistical distribution of oxygen atoms, however, has still been reserved for further study.

2. *Laser-Heating Device*

The laser-heating system (Fig. 1) consists of three parts; the power supply, optical system, and temperature detector and controller. The CO_2 gas laser ($\lambda = 10.6\ \mu m$) is utilized as a power supply to directly heat up a specimen on the goniometer head. The laser beam from its oscillator is led to a gold-coated stainless mirror set up just above the X-ray tube housing and is reflected towards a specimen at the center of the χ circle of a four-circle diffractometer in such a manner that the beam path is in the plane, as defined by the X-ray beam and the ω-axis of the diffractometer. For the case of the precession camera, the laser beam is so aligned that it vertically hits the crystal

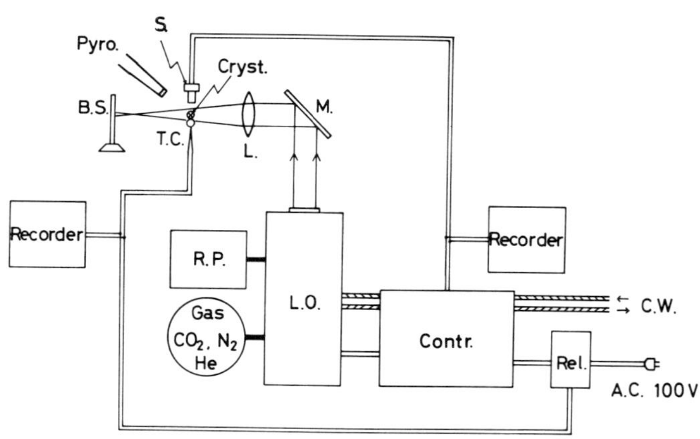

FIG. 1. A schematic diagram showing the whole system for heating specimens by means of a focused CO_2 gas laser. The output power of the laser beam is controlled to make the specimen temperature stable by feedback of electric current of silicon photodiode. S.; sensor, Pyro.; pyrometer, B.S.; beam stopper, T.C.; thermocouple, L.; Ge-lens, M.; gold-coated stainless mirror, R.P.; rotary pump, L.O.; laster oscillator, Rel.; reliant switch, C.W.; cooling water.

specimen through the Ge-lens. These two settings of the optical system impose no limitations on the movement of the X-ray diffraction apparatus.

Any desired temperature of the specimen may be obtained by adjusting the position of the Ge-lens with respect to the specimen to control the diameter (1.5 ~ 2.5 mm) of the focused laser beam and the amount of electric current applied to the oscillator. These adjustments are manually operated, before data collection, according to the size, shape or reactivity to the laser beam of the specimen. The temperature of the specimen can be stabilized in the following way. The output power of the laser oscillator is easily and rapidly controlled by the amount of electric current applied to the oscillator. Therefore, the controller of the laser oscillator is so designed that the electric current applied to it is automatically changed in accordance with the fluctuation of electric current generated by a sensor. The sensor used in our device is composed of a silicon photodiode, a filter to cut off visible light, and a slit to receive infrared radiation only from the specimen. It is small and light enough to be installed in the χ circle of a four-circle diffractometer or on the plate holding the goniometer head of the precession camera. The location of the sensor is such that it is on the ϕ-axis and at a distance of about 1.5 mm from the crystal specimen, as shown in Fig. 2. In this way, the position of the sensor relative to the crystal is kept constant during the intensity data collection. This situation may be rephrased so that the cross-section of the crystal viewed from the sensor is invariable regardless of the rotation of the crystal for intensity measurement. Since the temperature of the crystal corresponds to the amount of infrared radiations received by the sensor, the power of laser can be regulated by automatically controlling the electric

FIG. 2. Photograph showing the lay out of the Ge-lens (L) and sensor (S) with respect to a crystal specimen (C) on a goniometer head mounted on the χ circle of a four-circle diffractometer.

current applied to the oscillator in such a manner that the amount of infrared radiation received by the sensor is kept constant, thus the crystal temperature is stabilized.

3. *Temperature Measurement*

The specimen temperature was measured by a pyrometer with a telescope ($\times 60$). Although the emissivity of a material is usually unknown, its value at the melting point can be calculated from a measured melting-temperature T', once the true melting-temperature T is known (in general, $T' < T$). Then the emissivity thus obtained for the melting point, may be used to correct the measured temperatures of the material, at least in a high temperature region in which the temperature dependence of emissivity would be negligible. In the present case of high cristobalite, the sample temperatures, for example, of 835° and 1,250°C measured by the pyrometer, were corrected to be 1,000° and 1,600°C, respectively. The fluctuation of temperature during data collection in the present study is estimated to be less than ± 20°C by conversion from that of the electric current gained by the sensor and that of the thermoelectromotive force of a thermocouple holding the specimen with cement. Possible heterogeneity in temperature within a specimen was examined using high cristobalite at temperatures around 1,180°C. The result showed that the maximum difference was within 10°C, thus being compatible with the fluctuation during the data collection. Therefore, the effect caused by the directed application of a laser beam was found to be negligible.

The following advantages of the new laser-heating device have been confirmed. First, the device can be installed in the X-ray diffraction equipment without giving it any mechanical hindrance; second, the laser with a power of as small as 20W can heat specimens over 2,000°C in temperature; third, the stability of the specimen's temperatures can be automatically accomplished by the feedback method; fourth, a heated specimen on a goniometer head can always be seen by a telescope, and so on.

For further improvement of the system, the pyrometer applicable to a small size of the spot on a specimen, as well as generating electric current or voltage in accordance with temperature, and the method to bind a specimen on thin rod without cement, are required to enable the specimen temperature to be more stable within 10°C during data collection.

4. *Experimental*

The sample of cristobalite from Ito, Shizuoka Prefecture, Japan was used for the present study (The University Museum, The University of Tokyo, PM40447). A description of the material from this locality was given by KUNO (1933). They are irregular in shape and have maximum dimensions up to about 0.5 mm. Most of them are twinned in the manner described by DALLASE (1965); upon heating at temperatures above the transition point they turned to single crystals (Fig. 3). Precession photographs at 1,170°C (Fig. 3) showed no reflections which violated the diffraction symmetry and extinction rule characteristic of the space group $Fd3m$, thus giving diffraction patterns of a high cristobalite structure. Sets of graphite-monochromatized Mo$K\alpha$ diffraction intensities were collected at 1,000° and 1,140°C using a four-circle diffractometer (Syntex $P2_1$) and the laser-heating device mentioned previously. The collections of intensity data were also made at 1,540, 1,580 and 1,600°C. Between the first two and last three cases, however, it was necessary to change the crystal specimen during data collection because of a rapid sublimation of the crystal at temperatures above around 1,400°C. For the collections of intensity data, the ω-2θ scan technique was employed and 55 ~ 85 symmetrically independent reflections were obtained up to $2\theta_{max}$ = 65 ~ 80°. After Lorentz and polarization corrections, the intensities greater than $2\sigma(I)$ were selected for the structure refinement.

5. *Refinement of the Structure*

The unit-cell dimensions were obtained by the least-squares procedure applied to sin2θ values of 9 ~ 15 reflections measured on the four-circle diffractometer and listed in Table 1 and shown in Fig. 4. As observed in the table and the figure, they retain nearly the same in the temperature range examined; they are even smaller than the value at 220°C (PEACOR, 1973). The least-squares program LINUS (COPPENS and HAMILTON, 1970) was used to refine the structures at each temperature. The structure at 1,600°C was first refined using 38 reflections selected from the data at 1,600°C and scattering

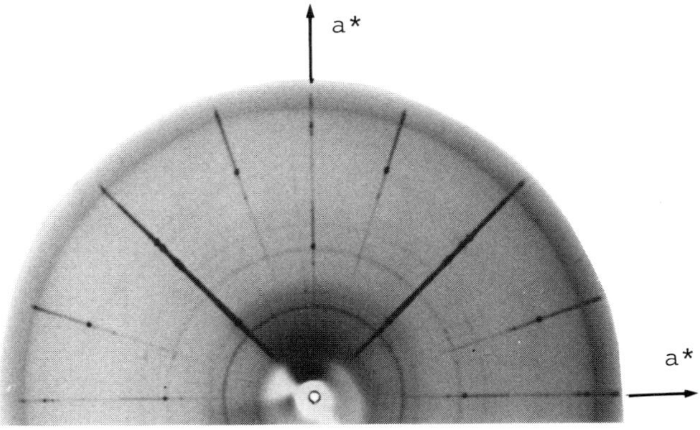

FIG. 3. *hk0* precession photograph of high cristobalite at 1,170°C (Mo$K\alpha$, unfiltered).

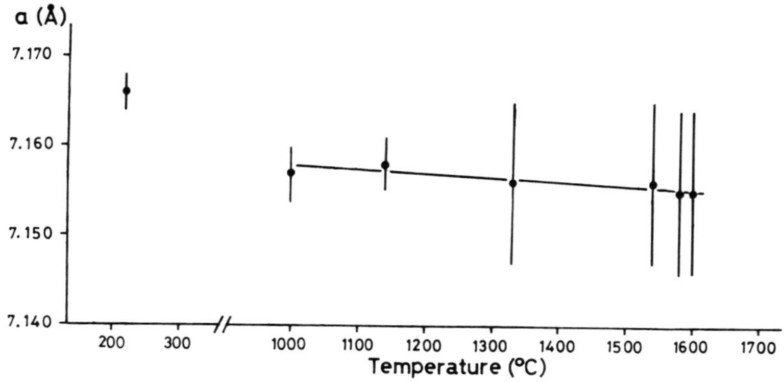

FIG. 4. A plot of the cubic cell edge of high cristobalite versus temperatures. The entry for 210°C was taken from PEACOR (1973).

TABLE 1. Experimental data and atomic parameters of high cristobalite at five temperatures.

T(°C)	Sample No.	Number of refl. ($2\theta_{max}$)	a(Å)	O* y	O* B(Å^2)	Si* B(Å^2)	R
1,000	1	46(80°)	7.157(3)	0.0443(14)	3.9(2)	3.83(6)	0.040
1,140	1	47(80°)	7.158(3)	0.0439(13)	4.1(2)	3.99(6)	0.041
1,540	2	39(65°)	7.156(9)	0.0438(16)	4.9(2)	4.72(7)	0.040
1,580	2	43(65°)	7.155(9)	0.0441(15)	4.8(2)	4.59(7)	0.045
1,600	2	38(65°)	7.155(9)	0.0440(14)	4.9(2)	4.60(7)	0.034

Anisotropic temperature factor coefficients** in the form of $\exp[-(h^2\beta_{11} + k^2\beta_{22} + 1^2\beta_{33} + 2hk\beta_{12} + 2hl\beta_{13} + 2kl\beta_{23})]$, for the oxygen atom

T(°C)	β_{11}	β_{22}	β_{33}	β_{12}	β_{13}	β_{23}	R
1,000	0.019(2)	0.020(3)	$(=\beta_{22})$	−0.001(2)	$(=\beta_{12})$	−0.003(1)	0.037
1,600	0.026(2)	0.025(3)	$(=\beta_{22})$	−0.000(1)	$(=\beta_{12})$	−0.006(1)	0.027

*These atoms are at special positions; Si: $x = y = z = 1/8$, O; $x = 0$, $z = -y$.
**Other atomic parameters were fixed.

factors for half-ionized silicon and oxygen atoms. Several cycles of refinement with the unit weight gave $R = 0.035$ based on starting atomic parameters at 310°C provided by PEACOR (1973). At this stage of refinement, there remained ambiguity in the determination of oxygen positions. Regarding these, NIEUWENKAMP (1937) had concluded that each oxygen atom was radially distributed about the Si–Si separation. Later, PEACOR (1973) tried to clarify the nature of the radial distribution of the oxygen atom at temperatures between 221° and 310°C by the single-crystal X-ray method, and revealed that three different locations of oxygen atoms, 96h, 96g and 192i yielded the same result of refinement. While, LEADBETTER *et al.* (1973) independently studied the problem at temperatures between 300° and 1,200°C by powder diffraction, reporting that oxygen atoms were located at equipoint 96h.

In the present study, four cases of oxygen positions were examined on a circumference representing radial distribution. These cases were 96h, 96g, 96h + 96g and 192i, and the results of refinement are listed in Table 2. Although the differences between the results of these cases appear to be insignificant, the results of the last two cases would nearly correspond to that of the case in which a radial distribution of the oxygen atom about the Si–Si separation is assumed. On the other hand, the oxygen atom in the last case moved towards the equipoint 96h from its initial position, which was closer to the equipoint 96g rather than 96h. This trend of oxygen-atom movement through refinement was different from that obtained by PEACOR (1973).

An averaged picture of high cristobalite structure may then be drawn from these results as follows. Silicon atoms are located at the equipoint 8a with rather high temperature factors and the oxygen atoms distributed along the circumference of radius 0.45Å with a continuous but slight change in their weight. The mode of statistic distribution of the oxygen atoms on the circumference at 1,600°C might be different from that studied with the data obtained at temperatures between 221° and 310°C by PEACOR (1973). Although the temperature dependence of the mode of statistical distribution is uncertain, the structure now considered has a salient feature that the Si–O bond lengths and Si–O–Si angles are determined only by the radius of the circumference, on which the oxygen atoms are distributed, independently from the specific oxygen positions on the circumference. Consequently, for the subsequent refinements of the structures at temperatures other than 1,600°C,

TABLE 2. Refined atomic parameters and the values of R for four possible positions of the oxygen atoms at 1,600°C.

		O				Si	R
	Position	x	y	z	$B(Å^2)$	$B(Å^2)$	
1	96h	0	0.0440(14)	$(=-y)$	4.9(2)	4.60(7)	0.034
2	96g	$(=y)$	0.0267(45)	−0.0484(25)	5.0(2)	4.59(7)	0.035
3*	96h + 96g	0	0.0439(21)	$(=-y)$	4.9(4)	4.60(6)	0.035
4**	192i	0.0015(787)	0.0438(19)	−0.0440(104)	4.9(2)	4.60(6)	0.034

*The 96h and 96g positions were respectively expressed by $(O, x, -x)$ and $(x/\sqrt{3}, x/\sqrt{3}, -2x/\sqrt{3})$ so that they are located on the same circumference.

**Since correlations between x, y and z for the oxygen atom were very strong, refinement was carried out step by step.

the oxygen atoms were assumed to be located at 96h, which, as mentioned above, may well represent the oxygen positions at 1,600°C. The results are listed in Tables 1 and 3. The refined values of bond lengths and angles are nearly the same as those obtained at temperatures between 221° and 310°C by PEACOR (1973) and WRIGHT and LEADBETTER (1975) as shown in Fig. 5. The isotropic temperature factors of silicon atom is somewhat smaller than those extrapolated from the values reported by WRIGHT and LEADBETTER (1975), based on powder data up to 1,200°C.

TABLE 3. Bond lengths (Å) and angles (°) of high cristobalite.

T(°C)	1,000	1,140	1,540	1,580	1,600
Si–O	1.613(4)	1.612(4)	1.612(5)	1.612(5)	1.612(4)
∠Si–O–Si	147.7(5)	148.0(5)	148.1(6)	147.9(7)	147.9(6)

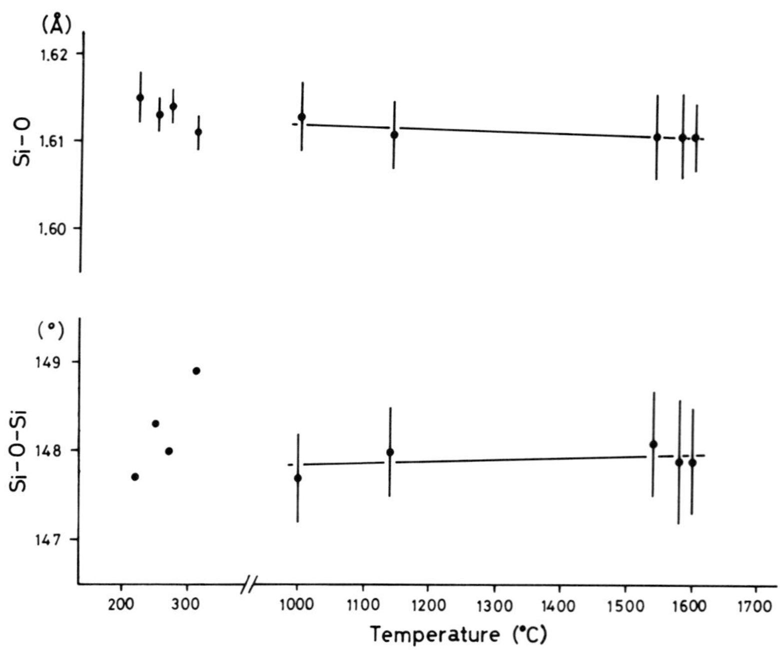

FIG. 5. Variations of the Si–O bond lengths and Si–O–Si bond angles of high cristobalite with temperatures. Those for the temperature 210°∼310°C were taken from PEACOR (1973).

6. *Discussion*

With respect to a given silicon atom, each bridge oxygen atom is statistically located, as a close approximation, at six positions on a circumference (Fig. 6) bisecting a line defined by the silicon atom and that of the adjacent tetrahedron. As shown in Fig. 6, the statistical distribution of the oxygen atom implies a statistical orientation of the silicate tetrahedron in six different directions. The tetrahedron at 1,600°C is slightly distorted, four of the six tetrahedral edges having a length of 2.607Å and two 2.682Å; the angles subtended to the former edges have a value of 107.9°, while those subtended to the latter, 112.6°. These values are almost the same to those for the structures between 221° and 310°C (PEACOR, 1973). An average of the above edge lengths and that of the tetrahedral angles are also compatible respectively with the corresponding values (2.63Å and 109.5°) of the hexagonal modification of tridymite at 460°C (KIHARA, 1978, 1980). Our study has brought out that the shape of the silicate tetrahedra retains nearly the same over the examined temperature range from 1,000° up to 1,600°C.

Thus, the structure of high cristobalite may be looked upon as composed of 'domains' (or twin individuals) primarily in a set of six different orientations as mentioned by WRIGHT and LEADBETTER (1975).Such a composite feature of the structure would very likely be generated upon transition, at approximately 220°C (PEACOR, 1973), from the low-temperature form of the structure. An inspection of the framework of high cristobalite shows that the continuation of the framework across the domain boundaries can be achieved by tilting but with insignificant distortion of the tetrahedra. At high temperatures, the domain boundaries would probably be liable to move in a way which is similar to the process of the α-β transition in quartz, as proposed by BUERGER (1971).

Since high-temperature crystallographic studies of silicates like olivines

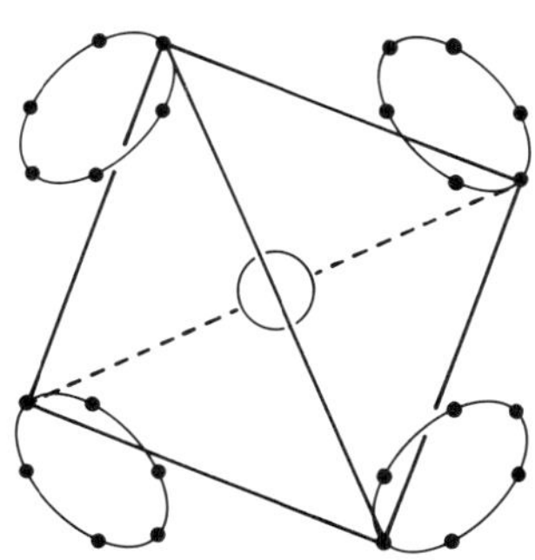

FIG. 6. Showing a tetrahedron formed by a set of four oxygen positions each of which being taken from six on each circumference, large open circle and small solid circles representing silicon and oxygen atoms, respectively.

(see TAKÉUCHI *et al.*, 1982) have revealed that the Si–O bond lengths tend to retain nearly the same all through the temperature range from room condition up to melting point. The tetrahedral-angle distortion becomes even smaller with rising temperatures in the case of forsterite. These facts may permit the following geometrical interpretation of the high-cristobalite structure. Based on the results of the present study and those of high- and low-cristobalite (PEACOR, 1973), we assume that (1) the Si atoms are at the 8a position of *Fd*3*m*, (2) the radius of the oxygen circumference is 0.45Å (Si – O = 1.612Å), (3) selection of a set of four oxygen positions to form a tetrahedron about the Si atom is made free from the symmetry restriction of *Fd*3*m*, (4) the edge lengths of the tetrahedron may vary in the range from 2.6Å to 2.7Å, which means that the tetrahedra are basically regular with tolerance of edge length.

Now consider what oxygen position on a circumference can form a tetrahedron, about an Si atom, together with other three oxygen positions, each locating on one of the remaining three circumferences about the Si atom. Then, the 24 positions locating with an equal interval on the circumference were examined one by one with the result that a greatest number of tetrahedra (satisfying the above assumptions) can be formed when oxygen atoms are located at the positions corresponding to the 96h equipoints, thus supporting the results obtained by X-ray structure analysis. The numbers of tetrahedra which can be formed by other sets of positions are significantly smaller than that of the above case (Fig. 7). Those oxygen positions other than 96h would

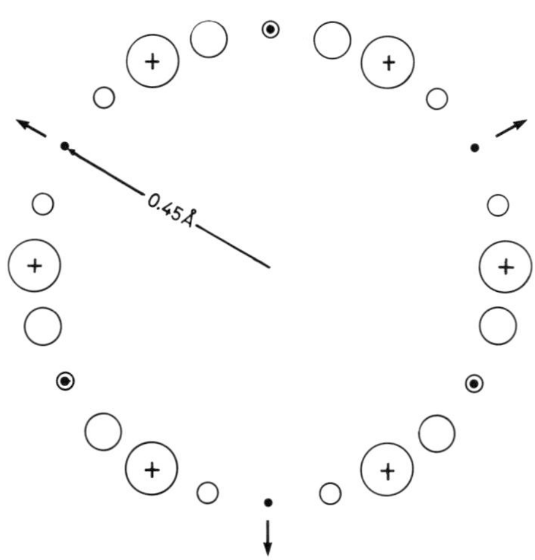

FIG. 7. Showing equally spaced 24 oxygen positions on a cicumference of radius 0.45Å as shown in Fig. 6; crosses and dots indicate the 96h and 96g equipoint positions, respectively. Arrows represent the direction to other three oxygen circumferences. The size of the circle at each position is proportional to the relative number of tetrahedra which can be formed (making allowance of the range, 2.60 ~ 2.70Å, for the edge length) by choosing it as one of their corners.

possibly represent those of tetrahedra in the domain boundaries.

A new model of the laser-heating system is now under construction with the hope that further details of the structure of high cristobalite at temperatures close to the melting point will be brought out with the use of the new heating system.

We wish to thank Dr. Y. Nakada, Department of Astronomy, University of Tokyo, for making available his CO_2 gas laser for our preliminary test and Dr. T. Murakami for his assistance during the early stage of our study. Computations were performed on HITAC8700/8800 at the Computer Center of the University of Tokyo.

REFERENCES

Barth, T. F. W. (1932): The cristobalite structures, *I. High cristobalite, Am. J. Sci.*, **23**, 350–356.

Buerger, M. J. (1971): Crystal-structure aspects of phase transformations, Ed. S. Block, *Mechanism of Phase Transitions*, Trans. Am. Cryst. Assoc., pp. 1–24.

Coppens, P. and Hamilton, W. C. (1970): Anisotropic extinction corrections in the Zachariazen approximation, *Acta. Cryst, A*, **26**, 71–83

Dallase, W. A. (1965): Reinvestigation of the structure of low cristobalite, *Zeit. Krist.*, **121**, 369–377.

Kihara, K. (1978): Thermal change in unit-cell dimensions, and a hexagonal structure of tridymite, *Zeit. Krist.*, **148**, 237–253.

Kihara, K. (1980): On the split-atom model for hexagonal trigymite, *Zeit. Krist.*, **152**, 95–101.

Kuno, H. (1933): On silica minerals occuring in the groundmass of common Japanese volcanic rocks, *Bull. Earthq. Res., Inst. Tokyo. Imp. Univ.*, **11**, 382–390.

Leadbetter, A. J., Smith, T. W. and Wright, A. F. (1973): Structures of high cristobalite, *Nature. Phy. Sci.*, **244**, 125–126.

Nieuwenkamp, W. (1937): Über die Struktur von Hoch-Cristobalit, *Zeit. Krist.*, **96**, 454–458.

Peacor, D. R. (1973): High temperature single-crystal study of the cristobalite inversion, *Zeit. Krist.*, **138**, 274–298.

Takéuchi, Y., Yamanaka, T., Haga, N. and Hirano, M. (1982): High-temperature crystallography of olivines and spinels, this volume, pp. 189–230.

Wright, A. F. and Leadbetter, A. J. (1975): The structures of the β-cristobalite phases of SiO_2 and $AlPO_4$, *Phil. Mag.*, **31**, 1391–1401.

Wyckoff, R. W. G. (1925): The Crystal structure of the high temperature form of cristobalite (SiO_4), *Am. J. Sci.*, **9**, 448–459.

MATERIALS INDEX

SUBJECT INDEX